## About the Cover

The design shown on the cover of this book was created by using double-exposure photography and two cameras. A camera with a 150-mm lens was used to photograph, from 30 feet, a large black panel on which Christmas tree lights had been arranged. The shot was purposely out of focus so that the colored lights would appear larger and spherical. The film magazine was consequently put into another camera with a 40-mm lens, positioned on the floor, directed straight up. Above, a small penlight was suspended on a string from the ceiling. During the second exposure of 30 seconds, this light source was allowed to swing freely while various color filters were placed over the camera lens — this explains the change of color in the light tracery.

## About the Title Page

The illustration on the title page is a graph drawn by a computer-directed plotter. The plot, entitled "Infinity," was written for Hewlett-Packard by a high school student. The equations used were:

$$R = \tfrac{4}{3} \sin \theta' + 1$$
$$\theta = \tfrac{4}{5} \sin \theta' + \cos \theta'$$

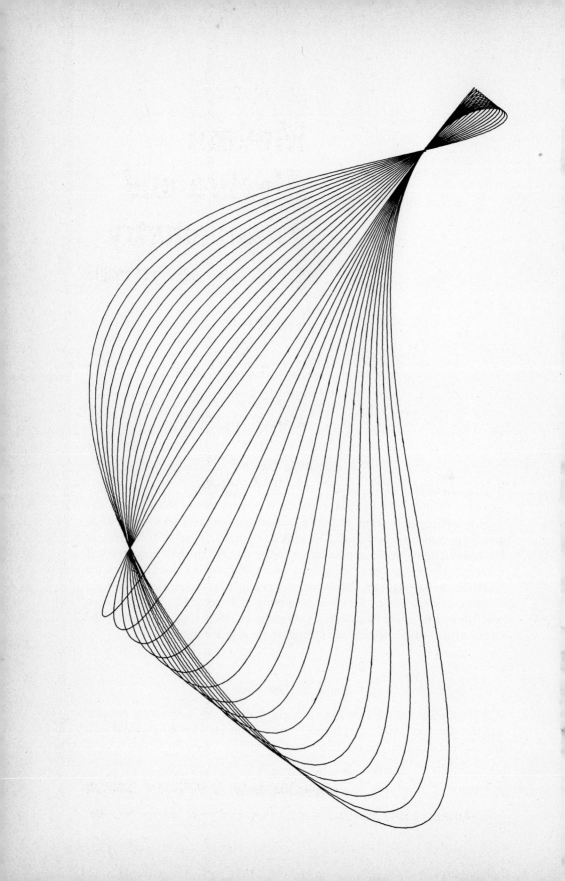

BOOK 2

# Modern Algebra and Trigonometry

## STRUCTURE AND METHOD

**New Edition**

Robert H. Sorgenfrey
William Wooton
Mary P. Dolciani

*Editorial Adviser*
*Albert E. Meder, Jr.*

**HOUGHTON MIFFLIN COMPANY · BOSTON**

Atlanta · Dallas · Geneva, Illinois · Hopewell, New Jersey · Palo Alto

## ABOUT THE AUTHORS

ROBERT H. SORGENFREY, Professor of Mathematics, University of California, Los Angeles. Dr. Sorgenfrey does research work in topology and has won the Distinguished Teaching Award at U.C.L.A. He has been a team member of the National Council of Teachers of Mathematics (NCTM) summer writing projects.

WILLIAM WOOTON, Mathematics Consultant, Vista, California, Unified School District. Mr. Wooton has been a teacher at the high school and college levels. He has also been a member of the SMSG writing team and a team member of the National Council of Teachers of Mathematics summer writing projects.

MARY P. DOLCIANI, Professor and Chairman, Department of Mathematics, Hunter College of the City University of New York. Dr. Dolciani has been a member of the School Mathematics Study Group (SMSG) and a director and teacher in numerous National Science Foundation and New York State Education Department institutes for mathematics teachers.

### EDITORIAL ADVISER

ALBERT E. MEDER, JR., Dean and Vice Provost and Professor of Mathematics, Emeritus, Rutgers University, The State University of New Jersey. Dr. Meder was Executive Director of the Commission on Mathematics of the College Entrance Examination Board, and has been an advisory member of SMSG.

Library of Congress Catalog Card No. 72–553

ISBN 0–395–14399–3

# 1 Real Numbers 1

# 2 Applications of Real-Number Properties 39

# 3 Systems of Linear Open Sentences 83

# 7 Irrational and Complex Numbers 261

# 8 Quadratic Equations and Functions 309

# 9 Quadratic Relations and Systems 351

# 13  Circular Functions and Their Inverses   513

# 14  Sequences, Series, and Binomial Expansions   543

# 15  Permutations, Combinations, and Probability   573

# 16   Matrices and Determinants     603

| | | PAGE | | | PAGE |
|---|---|---|---|---|---|
| $a^n$ | the $n$th power of $a$ | 148 | ! | factorial | 565 |
| $A^T$ | $A$ transpose | 604 | $\sum$ | sigma (summation sign) | 550 |
| $\lvert a \rvert$ | absolute value | 39 | $>$ | is greater than | 12 |
| $-a$ | additive inverse of $a$, or | | $<$ | is less than | 12 |
| | the negative of $a$ | 20 | $\geq$ | is greater than or equal to | 13 |
| $\overset{\frown}{TP}$ | arc $TP$ | 513 | $\leq$ | is less than or equal to | 13 |
| $\mathcal{C}$ | the set of complex numbers | 291 | $\cap$ | intersection | 65 |
| $\cos^{-1}$ | arccosine | 528 | $A^{-1}$ | inverse matrix | 613 |
| $_nC_r$ | combinations of $n$ elements | | $_nP_r$ | permutations of $n$ things | |
| | taken $r$ at a time | 572 | | taken $r$ at a time | 577 |
| $\ldots$ | continue unendingly | 6 | $\mathcal{R}$ | the set of real numbers | 2 |
| $^\circ$ | degree | 428 | $1^R$ | 1 radian | 514 |
| det | determinant function | 611 | $\{\ \}$ | set | 5 |
| $'$ | minute | 428 | $\subset$ | subset | 5 |
| $''$ | second | 428 | $\therefore$ | therefore | 24 |
| $\in$ | is an element of | 5 | $\cup$ | union | 66 |
| $\notin$ | is not an element of | 5 | $\overrightarrow{AB}$ | the vector $AB$ | 457 |
| $\emptyset$ | empty set | 6 | $x_1$ | $x$ sub one | 93 |
| $=$ | equals $or$ is equal to | 2 | | | |
| $\neq$ | does not equal | 2 | Greek letters | | |
| $\doteq$ | equals approximately | 244 | $\alpha, \beta$ | alpha, beta | |
| $f(x)$ | $f$ of $x$ $or$ the value of $f$ at $x$ | 13 | $\theta, \pi, \varphi$ | theta, pi, phi | |
| $f^{-1}$ | inverse function of $f$ | 398 | | | |

## Acknowledgment

ACKNOWLEDGMENT is given to Mr. Albert Posamentier for his assistance in preparing exercises for Chapter 5.

## Photograph Credits

COVER photograph by Jack O'Mahony; TITLE PAGE photograph is by courtesy of Hewlett-Packard Company.

PHOTOGRAPHS are by the courtesy of: Ed Cooper Photo, p. xii; Historical Pictures Service — Chicago, p. 35; National Oceanic and Atmospheric Administration, p. 36; Stock, Boston, p. 38; Patricia Hollander Gross, p. 82; Bryn Mawr College Library, p. 126; Harold E. Edgerton, p. 130; Stock, Boston, p. 168; Eric M. Sanford, p. 172; Stock, Boston, p. 208; Historical Pictures Service — Chicago, p. 258; American Cancer Society, p. 259; David S. Haas, p. 260; Culver Pictures, Inc., p. 307; David S. Haas, p. 308; Charles D. Druss, p. 350; The Bettman Archive, Inc., p. 387; Hal H. Harrison, p. 388; Charles D. Druss, p. 426; ARP Instruments, Inc., p. 470; ARP Instruments, Inc., p. 512; Patricia Hollander Gross, p. 542; ARP Instruments, Inc., p. 571; Charles D. Druss, p. 572; Derrick Te Paske, p. 602

# 1

# Real Numbers

## 1–1 Numbers and Operations

You began your mathematical studies when you learned the **counting numbers**:

$$1, 2, 3, 4, \text{ and so on.}$$

These numbers are also called **natural numbers** or **positive integers**. The counting numbers, their negatives, and zero make up the **integers**:

$$0, 1, -1, 2, -2, 3, -3, \text{ and so on.}$$

When you divide an integer by a nonzero integer, the quotient is a **rational number**;

$$\frac{2}{3}, \frac{-8}{5}, 7 \text{ or } \frac{7}{1}, \text{ and } 2.6 \text{ or } \frac{26}{10}$$

are some examples of rational numbers. Do you see that every integer is a rational number?

The numbers mentioned above are called **real numbers**. There are, however, real numbers which are not rational; $\sqrt{2}$ and $\pi$ are examples. In most of this book we shall work with real numbers.

1

Real numbers can be combined using the operations of arithmetic: addition (+), multiplication (×), subtraction (−), and division (÷).

Symbols which name numbers are called **numerals** or **numerical expressions**. Do you see that the numerals 6 ÷ 2, 7 − 4, 1 + 2, and 3 all name the same number? The number named by a numerical expression is called the **value** of the expression. To indicate that two expressions have the same value, you use the symbol =, read "equals" or "is equal to"; for example,

$$\frac{2 + 6}{2} = 4.$$

Any statement of equality is called an **equation**.

The following properties result from our interpretation of equality. We ordinarily use them without specific mention, but we list them here for occasional reference.

### Properties of Equality

If $a$, $b$, $c \in \mathcal{R}$, then:

| | |
|---|---|
| **Reflexive Property.** | $a = a$ |
| **Symmetric Property.** | If $a = b$, then $b = a$. |
| **Transitive Property.** | If $a = b$ and $b = c$, then $a = c$. |
| **Substitution Property.** | Changing the numeral by which a number is named in an expression does not change the value of the expression. |

To indicate that two expressions, such as 2 + 3 and 2 × 3, have different values, you write

$$2 + 3 \neq 2 \times 3.$$

(≠ is read "**does not equal**" or "**is not equal to**.") In general, a slant bar through a symbol means "not."

In the expression 4 × (5 − 2) the parentheses, ( ), indicate that you are to do the subtraction *first*, then the multiplication:

$$4 \times (5 - 2) = 4 \times 3 = 12.$$

Note that (4 × 5) − 2 = 20 − 2 = 18, so that 4 × (5 − 2) ≠ (4 × 5) − 2. Other *grouping symbols*, like brackets, [ ], and braces, { }, are sometimes needed. To remove such symbols, start with the innermost ones.

**Example.** Write each numerical expression in simple form by removing the grouping symbols:

a. $2 \times [5 - (3 - 1)]$     b. $2 \times [(5 - 3) - 1]$

*Solution:*    a. $2 \times [5 - (3 - 1)] = 2 \times [5 - 2] = 2 \times 3 = 6$
b. $2 \times [(5 - 3) - 1] = 2 \times [2 - 1] = 2 \times 1 = 2$

A quotient like $(3 + 7) \div 2$ can be written in the form of a fraction:

$$(3 + 7) \div 2 = \frac{3 + 7}{2}.$$

Notice that the fraction line acts as a grouping symbol. Thus, $\frac{3 + 7}{2} = \frac{10}{2} = 5$.

*ORAL EXERCISES*

Tell whether or not each of the following is: a natural number, an integer, a rational number, a real number.

**Sample.** $\frac{3}{5}$     *Solution:* $\frac{3}{5}$ is not a natural number or an integer, but $\frac{3}{5}$ is a rational number and a real number.

**1.** 10          **3.** 0          **5.** $-\frac{3}{4}$          **7.** $-2.5$          **9.** $\sqrt{7}$

**2.** $-2$          **4.** $\pi$          **6.** 7.2          **8.** $\frac{35}{5}$          **10.** $\sqrt{4}$

Find the numerical expression which does *not* have the same value as the other expressions in the list.

**11.** $2 \times 2, 36 - 31, 16 \div 4, 3 + 1$     **14.** $9 \times 0, 9 - 0, \frac{0}{9}, 0$

**12.** $\frac{35}{7}, 3 + 2, 3 \times 2, 41 - 36$     **15.** $\frac{1}{2} + \frac{3}{2}, 3 - 1, \frac{3}{4}, \frac{1}{2} \times 4$

**13.** $7, 4 + 3, 12 - 5, 49 \div 8$     **16.** $\frac{3}{4} - \frac{1}{4}, \frac{7}{14}, \frac{1}{2} \times 6, \frac{1}{4} + \frac{1}{4}$

What number is named by each of the following numerical expressions?

**17.** $3 \times (7 - 2)$          **20.** $\frac{3 + 9}{4}$          **22.** $(3 - 2) \times (35 \div 5)$

**18.** $(4 - 2) \times 5$                              **23.** $(1 + 2) \times (2 \times 2)$

**19.** $(6 - 2) \div 2$          **21.** $\frac{21 - 9}{3}$

## WRITTEN EXERCISES

Write each numerical expression in simple form by removing the grouping symbols.

A

**1.** $6 + [(3 + 4) + 2]$     **3.** $5 - [(3 - 1) + 1]$     **5.** $\dfrac{8 + 14}{11}$

**2.** $[(8 + 3) + 2] + 1$     **4.** $[7 - (3 + 2)] - 1$     **6.** $\dfrac{7 + 18}{4 + 1}$

**7.** $[25 + (18 + 25)] + 12$     **14.** $0[3 + 2(1 + 5)]$

**8.** $18 + [(14 + 75) + 25]$     **15.** $2[6 - 3(2 - 1)]$

**9.** $5 + \dfrac{18 + 27}{6 + 3}$     **16.** $6 + 2[1 + 2(3 - 1)]$

**10.** $\dfrac{23 + 7}{7 + 3} + \dfrac{8 + 42}{10 + 15}$     **17.** $7 + 3[(2 - 1) + 3]$

**11.** $2[3 + (1 + 4)]$     **18.** $8 + 2[\{3 + (-3)\} + 1]$

**12.** $4[(6 + 1) + 3]$     **19.** $2(4 - 3) + 3(1 + 2)$

**13.** $5[6 + (-6)]$     **20.** $2[3(1 + 2) - 2(1 + 1)]$

Copy each sentence, making it true by replacing each question mark with a numeral or with the sign $=$ or $\neq$.

**21.** $12 + 6 = \underline{\ ?\ } + 12$     **25.** $8 \div \underline{\ ?\ } = 1$

**22.** $4 \times 15 = 15 \times \underline{\ ?\ }$     **26.** $12 + \underline{\ ?\ } = 12$

**23.** $\underline{\ ?\ } \times 25 = 7 - 7$     **27.** $41 \times \underline{\ ?\ } = 41$

**24.** $6 + 6 = 6 \times \underline{\ ?\ }$     **28.** $4 \times \underline{\ ?\ } = 0$

**29.** $15 \times 2 \underline{\ ?\ } 15 + 2$     **32.** $\dfrac{16 + 2}{3} \underline{\ ?\ } 6 + 6$

**30.** $4 + 4 \underline{\ ?\ } 4 \times 2$     **33.** $\dfrac{9 \times 4}{6} \underline{\ ?\ } 4 \times 6$

**31.** $1 + 1 \underline{\ ?\ } 1 \times 1$     **34.** $\dfrac{4 \times 16}{8} \underline{\ ?\ } \dfrac{16}{2}$

Simplify each expression by removing the grouping symbols.

B

**35.** $3 \times [4 + 2(10 - \{3 \times 2\})]$     **37.** $2 + 7[5 - (10 - \{12 \div 2\})]$

**36.** $2 \times [3(5 + \frac{20}{4}\{3 - 1\})]$     **38.** $18 - 4[\{8 \div (2 \times 2)\} + \frac{1}{2}(2)]$

## 1-2 Sets and Set Membership

The idea of a set, or collection, of objects is so basic that mathematicians do not try to define it. For example, your algebra class is a set of people, a dictionary contains a listing of a set of words, and a line is a set of points.

The objects which make up a set are called its members, or elements, and are said to *belong to* the set. The symbol $\in$ is used to mean "is an element of" and $\notin$ is used to mean "is not an element of." Thus if $J$ is the set of integers,

$$5 \in J \text{ and } \tfrac{2}{3} \notin J.$$

You can sometimes describe or specify a set by listing all of its elements and enclosing the list, or roster, in braces, $\{\ \}$. For example, if $S$ is the set of even integers between 1 and 11, you can write

$$S = \{2, 4, 6, 8, 10\},$$

read "$S$ is the set 2, 4, 6, 8, 10." Another way of describing a set is to give a *rule* or *condition* which enables one to decide whether or not a given object belongs to the set. Thus $\{a, e, i, o, u, y\}$ can be described by

$$\{\text{vowels in the English alphabet}\},$$

read "the set of all vowels in the English alphabet."

**Example.**　**a.** Describe $\{\text{days of the week}\}$ by roster.

　　　　　　**b.** Describe $\{0, 1, 2, 3, 4, 5, 6, 7, 8, 9\}$ by a rule.

*Solution:*　**a.** $\{$Sunday, Monday, Tuesday, Wednesday, Thursday, Friday, Saturday$\}$

　　　　　**b.** $\{$digits used in the base-ten system$\}$

Two sets are equal if they have the same members. For example,

$$\{2, 4, 6\} = \{4, 6, 2\};$$

but

$$\{2, 4, 6\} \neq \{2, 4, 5\},$$

and

$$\{2, 4, 6\} \neq \{2, 4, 6, 8\}.$$

Notice that every element of the set $A = \{1, 3, 5\}$ is also a member of the set $B = \{1, 3, 5, 7\}$; in such a situation we call $A$ a subset of $B$ and write $A \subset B$. Thus $\{\text{vowels}\} \subset \{\text{letters}\}$, and $\{\text{dogs}\} \subset \{\text{animals}\}$.

Diagrams like the one in Figure 1–1 can be used to illustrate subset relationships. One of the subsets of any set $S$ is $S$ itself, that is, $S \subset S$; every other subset of $S$ is a **proper subset**.

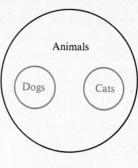

The set which has no members at all is called the **empty** or **null set** and is denoted by $\emptyset$. By agreement, there is only one empty set; thus

$$\{\text{integers between } \tfrac{1}{3} \text{ and } \tfrac{2}{3}\} = \emptyset,$$

and

$$\{\text{purple cows}\} = \emptyset.$$

**FIGURE 1–1**

The empty set is considered to be a proper subset of every set except itself.

In this book we shall always use the symbols $\Re$, $N$, $W$, $J$, and $Q$ to name the sets described below:

$$\Re = \{\text{real numbers}\}$$
$$N = \{\text{natural numbers}\} = \{1, 2, 3, \ldots\}$$
$$W = \{\text{whole numbers}\} = \{0, 1, 2, 3, \ldots\}$$
$$J = \{\text{integers}\} = \{\ldots, -2, -1, 0, 1, 2, 3, \ldots\}$$
$$Q = \{\text{rational numbers}\} = \{\text{quotients of integers}\}$$

Each of the rosters above is only a partial listing. The three dots, ... (often read "and so on") indicate that the listing never ends, that is, that the sets are **infinite**. Any set that is not infinite is called a **finite** set. Do you see that

$$N \subset W \subset J \subset Q \subset \Re?$$

*ORAL EXERCISES*

Describe each set by roster.

**Sample.** {the letters in the word "decent"}

*Solution:* {d, e, c, n, t}. (Note that you do not list the same letter twice.)

**1.** {the numerals appearing on the face of a clock}

**2.** {the months in a year}

**3.** {the authors of this textbook}

**4.** {the teacher of your mathematics class}

**5.** {the letters in the word "field"}

6. {the letters in the word "honest"}

7. {the letters in the word "revenge"}

8. {the letters in the word "savage"}

Tell whether or not each statement is true. Give a reason for your answer.

9. $\{3\} \in \{3, 6\}$

10. $6 \subset \{3, 6\}$

11. $7 \in \{4 + 1, 4 + 2, 4 + 3\}$

12. $\{15\} \subset \{2 + 12, 3 \cdot 5, 15 \div 3\}$

13. $\{5, 7\} = \{7, 5\}$

14. $\{3, 5\} \neq \{2 + 1, 3 + 3\}$

15. $7 \notin \{3, 6, 7, 8\}$

16. $\{3\} \not\subset \{3, 6, 7, 8\}$

## WRITTEN EXERCISES

Describe each set by roster. If the set described is ∅, state this.

**A** 1. {the letters in the word "algebra"}

2. {the letters in the word "attention"}

3. {the first three Presidents of the United States}

4. {the capital cities of Texas, California, and New York}

5. {the countries bordering on Canada}

6. {all horses that are cows}

7. {the letters after $z$ in the English alphabet}

8. {the letters between $s$ and $v$ in the English alphabet}

Describe each set by a rule.

9. $\{2, 4, 6\}$

10. $\{1, 3, 5\}$

11. $\{1, 2, 3, 4, \ldots\}$

12. $\{\ldots, -2, -1, 0, 1, 2, 3, \ldots\}$

13. $\emptyset$

14. $\{0\}$

15. {Ottawa; Washington, D.C.}

16. {Superior, Michigan, Huron, Erie, Ontario}

Copy each sentence, making it true by replacing the question mark with a numeral or with one of the symbols = or ≠.

**17.** {1, 2, 3} __?__ {3, 1, 2}

**18.** {r, e, d} __?__ {r, c, d, e}

**19.** {0, 1, 2} __?__ {1, 2}

**20.** {Counting numbers that are negative} __?__ {counting numbers that are not integers}

**B** **21.** {0} ⊂ {5, 6, __?__ }　　　**23.** {0} __?__ ∅

**22.** 7 ∈ {3, 4, 5, __?__ }　　　**24.** {1} __?__ {2}

**25.** {4 + 1, 4 + 2} __?__ {5, 7}

**26.** {7 × 3, 4 + 0} __?__ {25 − 21, 3 × 7}

Let $U = \{x, y, z\}$. List all subsets of $U$ that have:

**27.** Exactly 1 element　　　**29.** No elements

**28.** Exactly 2 elements　　　**30.** Three elements

**C** **31.** Copy and complete the following table comparing the number of elements of a set $U$ and the number of subsets of $U$.

| $U$ | Number of Elements of $U$ | Number of Subsets of $U$ |
|---|---|---|
| {a} | 1 | $2 = 2^1$ |
| {a, b} | 2 | $4 = 2^2$ |
| {a, b, c} | 3 | __?__ |
| {a, b, c, d} | 4 | __?__ |

**32.** If $U$ is a set with $n$ elements, use the results of Exercise 31 to find a formula which expresses the number of subsets of $U$ in terms of $n$.

## 1–3　Variables and Open Sentences

Of the six sentences displayed below, the two on line (1) are *true*, while those on line (2) are *false*.

|  Sentences about Cities of the World |  | Sentences about Real Numbers |
|---|---|---|
| (1) Paris is in France. | ⟵True⟶ | $8 - 2 = 6$ |
| (2) Tokyo is in England. | ⟵False⟶ | $7 + 2 = 5$ |
| (3) It is in Canada. | ⟵Open⟶ | $x + 3 = 8$ |

However, the sentence "It is in Canada" is true when you replace *It* by *Montreal*, for example, but the sentence is false if you use *Chicago* as a replacement for *It*. Similarly, the sentence "$x + 3 = 8$" is true when $x$ is replaced by 5 but is false if $x$ is replaced by 2, 9, or any other numeral not naming 5.

In the sentences on line (3) on page 8, the pronoun *It* and the letter $x$ play similar roles. Each symbol is a **variable**; a variable may represent any element of a specified set. Thus you may replace *It* by the name of any city in the world, and $x$ by any real number.* The set whose elements may serve as replacements for the variable is called the **domain** or **replacement set** of the variable. To indicate that the domain of a variable $x$ is $D$, you may write $x \in D$. The members of the domain of a variable are called the **values** of the variable. A variable having just one value is a **constant**.

A sentence containing a variable is called an **open sentence**, for example,

$$x + 3 = 8, \quad x \in \Re.$$

An open sentence serves as a pattern for the various *statements,* true or false, which you obtain by substituting for the variable the various members of its domain. The subset of the domain of the variable for which the sentence is true is called the **truth set** or **solution set** of the sentence *over that domain.* Each member of the solution set is said to **satisfy** the open sentence and to be a **root** or **solution** of it.

Here are the solution sets of some open sentences over $N$, the set of natural numbers.

| Open Sentence | Solution Set |
|---|---|
| $x + 3 = 8$ | $\{5\}$ |
| $x + 2 = 2 + x$ | $N$ |
| $z + 2 = z + 1$ | $\emptyset$ |
| $4 - n \in N$ | $\{1, 2, 3\}$ |

You will often find it convenient to use *set-builder notation* in describing sets. For example, the first solution set, $\{5\}$, displayed above can be written

$$\{x : x + 3 = 8, x \in N\}.$$

This is read "the set of all $x$ such that $x + 3 = 8$ and $x$ is a member of $N$." In general, the $\{x : \quad\}$ part of set-builder notation is read "the set of all $x$ such that."

---

*For brevity we use the phrase "any real number" in place of "any numeral naming a real number."

In most problems the domain of the variable has been specified in advance; the part of set-builder notation corresponding to "$x \in N$" at the bottom of page 9 can then be omitted.

**Example.** Give a roster of each of the following subsets of $J$, the set of integers.
  a. $\{x: x + 5 = 5\}$
  b. $\{x: x \text{ is between } -1\frac{1}{2} \text{ and } 2\frac{1}{2}\}$
  c. $\{n: n \text{ is between } -1\frac{1}{2} \text{ and } 2\frac{1}{2}\}$

*Solution:*    a. $\{0\}$        b. $\{-1, 0, 1, 2\}$        c. $\{-1, 0, 1, 2\}$

Parts **b** and **c** of the Example illustrate that the particular symbol used for a variable is unimportant.

## ORAL EXERCISES

Tell whether each of the following sentences is a true statement, a false statement, or an open sentence.

1. February is a month of the year.

2. She is an opera singer.

3. The moon is made of blue cheese.

4. He was a member of Congress.

5. $\{1, 2\} \subset \{1, 4, 2\}$        8. $p \in \{1, 2, 3, 4, 5\}$

6. $\dfrac{6 - 2}{2} = \dfrac{6 - 3}{3}$        9. $1 \subset \{0, 1\}$

7. $x - 3 = 7$        10. $z + 1 = 2z - 3$

Recall that $N = \{$natural numbers$\}$, $J = \{$integers$\}$, and $\Re = \{$real numbers$\}$. Determine by inspection the solution set over (a) $N$, (b) $J$, and (c) $\Re$ of each sentence.

11. $x + 3 = 3$        16. $7 \cdot y = 0$

12. $n - 2 = 5$        17. $z + 3 = z - 1$

13. $3 \cdot n = 1$        18. $k - k = 0$

14. $\dfrac{t}{3} \in N$        19. $x + x = 2$

15. $3 + r = 2$        20. $x + x = 3$

Specify each set by roster. Assume that the domain of each variable is *N*.

**21.** $\{x: x + 1 = 5\}$

**22.** $\{y: 2 \cdot y = 10\}$

**23.** $\{n: 3 - n \in N\}$

**24.** $\{n: n - 3 \notin N\}$

**25.** $\{z: z \text{ is between } 0 \text{ and } 4\frac{1}{2}\}$

**26.** $\{n: n \text{ is odd and between 2 and 10}\}$

**27.** $\{t: t \text{ is even and less than 10}\}$

**28.** $\{x: x \text{ is odd and } x + 2 = 4\}$

**29.** $\{y: y + 1 = 2 \text{ or } y + 2 = 4\}$

**30.** $\{z: 3 \cdot z = 2\}$

## *WRITTEN EXERCISES*

Specify the solution set of each sentence over the given domain.

**A**

**1.** $2 \cdot y + 1 = 7; \{1, 2, 3, 4\}$

**2.** $n + 3 = 3 + n; \{1, 2, 3, 4\}$

**3.** $n \in N; \{\frac{1}{2}, \frac{2}{2}, \frac{3}{2}, \frac{4}{2}\}$

**4.** $t + 2 = t; \{0, 1, 2, 3\}$

**5.** $r - r = r; J$

**6.** $t + t = t; N$

**7.** $x - 5 = 0; N$

**8.** $x + 0 = x; J$

Specify each set using set-builder notation. There may be several ways to do this.

**Sample.** $\{3, 4, 5\}$

*Solution:* Two possibilities are:
$\{x: x \text{ is an integer between 2 and 6}\}$, or
$\{y: y \text{ is a natural number between 2 and 6}\}$.
There are other answers also.

**9.** $\{2, 3, 4\}$

**10.** $\{2, 4, 6\}$

**11.** $\{1, 3, 5\}$

**12.** $\{3, 6, 9\}$

**13.** $\{10, 11, 12, \ldots\}$

**14.** $\{10, 12, 14, \ldots\}$

**15.** $\{11, 13, 15, \ldots\}$

**16.** $\emptyset$

Specify each set by roster. Use three dots, . . . , to describe infinite sets.

**17.** $\{t: t + 3 = 5, t \in N\}$

**18.** $\{x: x - 5 = 4, x \in J\}$

**19.** $\{y: 2y = 5, y \in Q\}$

**20.** $\{z: 3z = 6, z \in Q\}$

**21.** $\{n: n \text{ is a multiple of } 5, n \in N\}$

**22.** $\{t: t \text{ is a multiple of } 7, t \in N\}$

**23.** $\{x: 3x \text{ is even}, x \in N\}$

**24.** $\{y: \frac{y}{3} \text{ is odd}, y \in N\}$

Determine by inspection the solution set over ℜ of each equation.

    **Sample.**  $3 \cdot x + 2 = 17$

   *Solution:*  Because the sum of $3 \cdot x$ and 2 is 17, $3 \cdot x$ must represent 15. But, since $3 \times 5 = 15$, $x$ must represent 5. Therefore, $\{5\}$ is the solution set.

**B**  **25.** $2 \cdot x + 1 = 11$           **29.** $21 - 3 \cdot y = 15$

    **26.** $4 \cdot r - 2 = 18$         **30.** $15 + 3 \cdot y = 18$

    **27.** $7 \cdot s - 4 = 25$         **31.** $24 + 5 \cdot y = 34$

    **28.** $5 \cdot t + 7 = 42$         **32.** $8 - 2 \cdot y = 0$

## PROPERTIES OF REAL NUMBERS

### 1–4  Order Relations

When you discuss real numbers, you often find it helpful to picture them as points on a *number line*, as indicated in Figure 1–2. The point which corresponds to any real number $a$ is called the **graph of a** or the **point a**; the number corresponding to a point on the line is the **coordinate** of the point. The point $O$ is called the **origin**.

FIGURE 1–2

    The geometric fact that the point 3 lies to the *left* of the point 5 corresponds to the algebraic fact that 3 is *less* than 5, in symbols, $3 < 5$. Similarly, the fact that 4 is *greater* than 1, $4 > 1$, is reflected in the fact that the point 4 lies to the *right* of the point 1. The relationships

$$< \text{ ("is less than")}$$

and

$$> \text{ ("is greater than")}$$

are called **order relations**. To avoid confusing the **inequality symbols** $<$ and $>$, you can think of them as arrowheads pointing toward the smaller number. For example,

$$-3 < 1, 1 > -3, 0 < 3, \tfrac{2}{3} > -\tfrac{3}{2}, \text{ and } -2 < 0.$$

You usually write a compound sentence like

$$x < 5 \; or \; x = 5$$

in the form

$$x \leq 5,$$

read "$x$ is less than or equal to 5." Similarly,

$$b \geq a$$

stands for

$$b > a \; or \; b = a.$$

Do you see that the statements $-3 \leq 2$, $2 \leq 2$, $2 \geq 2$, and $0 \geq -3$ are all true, and that $2 > 2$ and $2 \leq -3$ are false?

A sentence of the type

$$a < x \; and \; x < b$$

is usually written in the combined form

$$a < x < b,$$

read "$a$ is less than $x$, which is less than $b$," or "$x$ is between $a$ and $b$."

You can use a number line to picture any subset $S$ of $\Re$ as a set of points, called the **graph** of $S$. Graphs of sets of the type $\{x: a < x < b\}$ or, sometimes, the sets themselves, are called **intervals**. Two other kinds of intervals occur in the **b** and **c** parts of the Example below.

**Example.** Graph (that is, draw the graph of) the following subsets of $\Re$:

    **a.** $\{x: x < 2\}$
    **b.** $\{x: -3 \leq x \leq 2\}$
    **c.** $\{t: -3 < t \leq 2\}$

*Solution:*

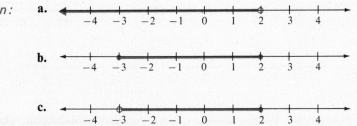

*Note:* We use solid dots and heavy portions of the line to indicate points of the graph and hollow dots or thin portions of the line to indicate points not belonging to the graph.

With respect to the relation $<$, $\mathcal{R}$ is a **linearly ordered set**, that is, one which has the following properties:

---

**Comparison or Trichotomy Property.** Let $a \in \mathcal{R}$ and $b \in \mathcal{R}$. Then exactly one of the following statements is true:

$$a < b \qquad a = b \qquad b < a$$

---

**Transitive Property.** If $a$, $b$, and $c$ are elements of $\mathcal{R}$, and $a < b$ and $b < c$, then $a < c$.

---

*ORAL EXERCISES*

Tell whether each of the following statements is true or false.

**1.** $1 \geq -3$

**2.** $-4 > -2$

**3.** $0 \geq -1$

**4.** $-7 < -2$

**5.** $5 \geq 5$

**6.** $5 \leq 5$

Describe the interval whose graph is shown.

**Sample.**

*Solution:* $\{x: -2 < x \leq 2\}$

**7.**

**8.**

**9.**

**10.**

**11.**

**12.**

**13.**

**14.**

Specify each set by roster. Assume that the domain of each variable is *W*, the set of whole numbers.

**Sample.** $\{x: 1 < x \leq 5\}$     *Solution:* $\{2, 3, 4, 5\}$

**15.** $\{t: t < 3\}$            **18.** $\{y: 6 < y < 7\}$

**16.** $\{n: 0 \leq n < 5\}$     **19.** $\{z: 4 < z \leq 10\}$

**17.** $\{x: 1 \leq x \leq 2\}$     **20.** $\{r: r \leq 1\}$

## WRITTEN EXERCISES

Graph the specified subset of ℛ.

[A]

**1.** $\{x: -1 < x < 1\}$       **6.** $\{x: x > 3\}$

**2.** $\{y: -1 < y \leq 1\}$      **7.** $\{y: 0 \leq y \leq 1\}$

**3.** $\{z: -1 \leq z < 1\}$      **8.** $\{z: -1 \leq z \leq 0\}$

**4.** $\{t: -1 \leq t \leq 1\}$      **9.** $\{r: r = 4\}$

**5.** $\{z: z \leq 0\}$          **10.** $\{t: t = -3\}$

Specify each set of whole numbers by roster.

**11.** $\{z: 4 < z \leq 8\}$      **14.** $\{v: 5 \leq v < 6\}$

**12.** $\{t: t < 6\}$          **15.** $\{k: 5 < k < 6\}$

**13.** $\{n: 6 < n \leq 7\}$     **16.** $\{s: 9 \leq s \leq 12\}$

Express each relation using symbols.

**17.** *n* is less than 5 or *n* is equal to 5.

**18.** *t* is greater than or equal to 6.

**19.** *z* is between 8 and 10.

**20.** *r* is less than 7 and *r* is greater than 0.

**21.** $-5$ is less than *s* which is less than 0.

**22.** 8 is greater than *m* which is greater than $-1$.

**23.** *x* is greater than or equal to 4 and *x* is less than 10.

**24.** *z* is less than $-2$ and *z* is greater than or equal to $-4$.

## 1–5  Sums and Products

We choose addition and multiplication as the *basic* operations of algebra and define subtraction and division in terms of them (see Sections 2–2 and 2–3). When you add two real numbers $a$ and $b$, you obtain their **sum**, $a + b$, and $a$ and $b$ are called **terms** or **addends** in the sum.

The result of multiplying two numbers $a$ and $b$ is their **product,**

$$a \cdot b,$$

and $a$ and $b$ are called **factors** of the product. Other notations for the product are

$$ab, a \times b, a(b), \text{ and } (a)(b).$$

You must not, of course, write the product of 8 and 5 as 85 [which means $(8 \cdot 10) + 5$]; use some other notation such as $8 \cdot 5$ or $(8)(5)$.

In every branch of mathematics you accept as true without proof certain statements called **axioms** or **postulates**. We take as axioms the "self-evident" facts that whenever you add or multiply two real numbers, the result in each case is a unique (one and only one) real number.

---

**Axioms of Closure.** Let $a \in \Re$ and $b \in \Re$. Then $a + b$ and $a \cdot b$ are unique elements of $\Re$.

---

The term *closure* is used because any set of numbers is said to be **closed** under an operation performed on pairs of its elements (including pairs formed using the same element) if each result of the operation is a member of the set.

The examples $2 + 5 = 5 + 2$ and $3 \cdot 7 = 7 \cdot 3$ illustrate the *commutative laws* for addition and multiplication.

---

**Commutative Axioms.** Let $a \in \Re$ and $b \in \Re$. Then

$$a + b = b + a$$

and

$$ab = ba.$$

---

Addition and multiplication are called binary operations because each is applied to two numbers at a time. Thus if you are given *three* numbers, say 8, 5, and 14, you can first add 8 to 5 and then add the result to 14:

$$(8 + 5) + 14 = 13 + 14 = 27.$$

Or you can first add 5 to 14 and then add 8 to the sum:

$$8 + (5 + 14) = 8 + 19 = 27.$$

The general fact that these two methods always lead to the same result is the *associative law* for addition. This law and the corresponding one for multiplication are axioms for the real numbers.

---

**Associative Axioms.** Let $a$, $b$, and $c \in \mathfrak{R}$. Then

$$(a + b) + c = a + (b + c)$$

and

$$(ab)c = a(bc).$$

---

Because of the associative laws we can dispense with many grouping symbols by defining $a + b + c$ to be the sum $(a + b) + c$, $abc$ to be $(ab)c$, $a + b + c + d$ to be $(a + b + c) + d = ([a + b] + c) + d$, and so on.

Since addition is commutative as well as associative, you may rearrange the terms in a sum, group them together, and add in any order. For example,

$$62 + 2\tfrac{1}{2} + 5 + 3\tfrac{1}{2} + 38 = (62 + 38) + (2\tfrac{1}{2} + 3\tfrac{1}{2}) + 5$$
$$= 100 + 6 + 5$$
$$= 111.$$

Similarly, factors in products can be rearranged and regrouped; for example,

$$\tfrac{1}{2}c(6d) = (\tfrac{1}{2} \times 6)(cd) = 3cd.$$

When you rewrite expressions for sums and products such as

$$62 + 2\tfrac{1}{2} + 5 + 3\tfrac{1}{2} + 38 = 111$$

and

$$\tfrac{1}{2}c(6d) = 3cd,$$

you are **simplifying** the original expressions.

## *ORAL EXERCISES*

State whether or not the given set is closed under the given operation. Give a reason for your answer.

> **Sample.** $\{0, 1\}$; addition
>
> *Solution:* The set is not closed under addition because $1 + 1 = 2$, and $2 \notin \{0, 1\}$.

**1.** $\{0, 1\}$; multiplication

**2.** $\{1, 2\}$; multiplication

**3.** {odd natural numbers}; addition

**4.** {even natural numbers}; addition

**5.** {odd natural numbers}; multiplication

**6.** {even natural numbers}; multiplication

**7.** $N$; addition

**8.** $N$; multiplication

In Exercises 9–16, tell which axiom justifies the given statement. Assume that the domain of each variable is $\mathcal{R}$.

> **Sample.** $(x + y) + 3 = x + (y + 3)$
>
> *Solution:* Associative Axiom of Addition

**9.** $rs = sr$

**10.** $(xy)z = x(yz)$

**11.** $t + 3 \in \mathcal{R}$

**12.** $4n \in \mathcal{R}$

**13.** $(x + y) + 2 = 2 + (x + y)$

**14.** $(4r)s = 4(rs)$

**15.** $rs + t = t + rs$

**16.** $5 \cdot (x + y) = (x + y) \cdot 5$

Simplify.

**17.** $9\frac{1}{4} + 12 + 3$

**18.** $\frac{1}{3} + 12 + 4\frac{2}{3}$

**19.** $21 + \frac{2}{5} + 6\frac{3}{5} + 79 + 3$

**20.** $6\frac{1}{2} + 54 + 2 + 3\frac{1}{2} + 46$

**21.** $\frac{1}{17}(3 \cdot 51) \times \frac{1}{27}$

**22.** $144(\frac{1}{15}) \times 75(\frac{1}{12})$

**23.** $\frac{1}{5}m(10n)$

**24.** $\frac{1}{4}e \times (24f)$

## WRITTEN EXERCISES

In Exercises 1–10, state the axiom justifying each statement. Assume that the domain of each variable is $\Re$.

   **Sample.**   $3(x + 2) = (x + 2) \cdot 3$

   *Solution:*  Commutative Axiom of Multiplication

**A**

1. $3 \cdot y \in \Re$
2. $4 + t \in \Re$
3. $3 + (x + 2) = (3 + x) + 2$
4. $(7y)z = 7(yz)$
5. $x + (2 + z) = (2 + z) + x$
6. $x + (2 + z) = (x + 2) + z$
7. $(3 + y) + (z + 6) = [(3 + y) + z] + 6$
8. $(4t)(3z) = (4t)(z \cdot 3)$
9. $(t + 1) + (z + 3) = (z + 3) + (t + 1)$
10. $(3t)(5r) = 3[t(5r)]$

Simplify.

11. $3\frac{1}{3} + 18 + 2\frac{2}{3} + 31$
12. $8\frac{1}{2} + 3\frac{3}{5} + 4\frac{1}{2} + 1\frac{2}{5}$
13. $(3z)(4k)$
14. $(2y)(15w)$

15. $4\frac{3}{7} + (2y) \cdot 3 + 1\frac{1}{7} + 2\frac{3}{7}$
16. $(4t)(7u) + 4\frac{3}{8} + 1 + 2\frac{5}{8}$
17. $\frac{1}{13}(3\frac{1}{9} + 5\frac{2}{9} + 3\frac{4}{9} + 1\frac{2}{9})$
18. $\frac{1}{12}(8\frac{3}{13} + 1\frac{2}{13} + 1\frac{7}{13} + 1\frac{1}{13})$

State whether or not the given set is closed with respect to **(a)** addition and **(b)** multiplication. If the set is not closed, give an example which shows this.

19. $\{0\}$
20. $\{1\}$

21. $\{3, 6, 9, \ldots\}$
22. $\{5, 10, 15, \ldots\}$

**B**

23. $\{1, \frac{1}{2}, \frac{1}{3}, \frac{1}{4}, \ldots\}$
24. $\{1, \frac{1}{2}, \frac{1}{4}, \frac{1}{8}, \ldots\}$

25. $\{1, \frac{1}{3}, \frac{1}{9}, \frac{1}{27}, \ldots\}$
26. $\{1, 4, 9, 16, \ldots\}$

In each exercise an operation ∘ is defined over *N*. In each case **(a)** find
3 ∘ 2; **(b)** determine whether or not *N* is closed under ∘; **(c)** state whether ∘
is (1) commutative; (2) associative.

**27.** $x \circ y = x + (y + 1)$             **29.** $x \circ y = x + (x + y)$

**28.** $x \circ y = (x + y) + x$             **30.** $x \circ y = x - y$

## 1–6   Identities and Inverses

The statements $0 + 3 = 3, 3 + 0 = 3, -2 + 0 = -2$, and $0 + 0 = 0$
illustrate the fact that when 0 is added to any number, the sum is
*identical* with the given number. For this reason we call 0 the **additive
identity**. The role of 1 as the **multiplicative identity** is illustrated by the
statement $1 \cdot 5 = 5 \cdot 1 = 5$.

---

**Axioms of Zero and One.** There are two elements, 0 and 1, of ℛ
such that for every $a \in$ ℛ:

$$0 + a = a \qquad a + 0 = a$$

$$1 \cdot a = a \qquad a \cdot 1 = a$$

---

If the sum of two numbers is 0, then the numbers are **negatives** of or
**additive inverses** of one another. Since $3 + (-3) = 0$, not only is $-3$
the additive inverse of 3, but also 3 is the additive inverse of $-3$.

---

**Axiom of Negatives.** For each $a \in$ ℛ there is a unique member $-a$
of ℛ such that

$$a + (-a) = 0 \quad \text{and} \quad -a + a = 0.$$

---

The symbol $-a$ should be read "the negative of $a$" or "the additive
inverse of $a$"; it may represent a negative number, zero, or a positive
number. For example:

$-2$ is the negative of 2 because $-2 + 2 = 0$.
0 is the negative of 0 because $0 + 0 = 0$.
3 is the negative of $-3$ because $3 + (-3) = 0$.

Thus $-(-3) = 3$. Do you see that $-(-a) = a$?

Two numbers whose product is 1 are called **reciprocals** or **multiplicative inverses** of each other. For example, $\frac{1}{6}$ is the reciprocal of 6, and 6 is the reciprocal of $\frac{1}{6}$. Another pair of reciprocals is 2 and 0.5. The multiplicative identity 1 is its own reciprocal. Every real number except 0 has a reciprocal.

---

**Axiom of Reciprocals.** For each nonzero element $a$ of ℛ there is a unique member $\frac{1}{a}$ of ℛ such that

$$a \cdot \frac{1}{a} = 1 \quad \text{and} \quad \frac{1}{a} \cdot a = 1.$$

---

In algebra we agree that *whenever there is a choice between performing an addition or a multiplication,* the multiplication is to be done first. In the calculation labeled **a** below, the addition within the grouping symbols is done first, in **b** the multiplications are done first.

**a.** $3 \cdot (2 + 5) = 3 \cdot 7 = 21$     **b.** $3 \cdot 2 + 3 \cdot 5 = 6 + 15 = 21$

The fact that the results are the same, $3 \cdot (2 + 5) = 3 \cdot 2 + 3 \cdot 5$, illustrates the distributive law: $a \cdot (b + c) = a \cdot b + a \cdot c$.

---

**Distributive Axiom.** Let $a$, $b$, and $c \in$ ℛ. Then

$$a(b + c) = ab + ac \quad \text{and} \quad (b + c)a = ba + ca.$$

---

Notice how the distributive law is used in the following calculations:

$$(\tfrac{1}{4} + \tfrac{1}{3}) \cdot 36 = \tfrac{1}{4} \cdot 36 + \tfrac{1}{3} \cdot 36 = 9 + 12 = 21,$$
$$8 \cdot 13 + 8 \cdot 7 = 8(13 + 7) = 8 \cdot 20 = 160.$$

## ORAL EXERCISES

State the **(a)** negative and **(b)** reciprocal of the given real number. If no reciprocal exists, so state.

| | | |
|---|---|---|
| **1.** 4 | **5.** $\frac{1}{2}$ | **9.** 0 |
| **2.** $-3$ | **6.** $-\frac{2}{3}$ | **10.** $-2.2$ |
| **3.** $-2$ | **7.** $-\frac{6}{5}$ | **11.** $-(-3)$ |
| **4.** 1 | **8.** 1.5 | **12.** $-(-\frac{7}{4})$ |

Each of the statements in Exercises 13–20 is justified by one of the real-number axioms. In each case assume that the domain of any variable is $\Re$ and state the appropriate axiom.

**13.** $7 + (-7) = 0$     **17.** $7 \cdot (\frac{1}{7}) = 1$

**14.** $x + 0 = x$     **18.** $(-3) \cdot 1 = -3$

**15.** $12(\frac{1}{3} + \frac{1}{4}) = 4 + 3$     **19.** $-3 + 3 = 0$

**16.** $(7 \cdot 8) \cdot 1 = 7 \cdot 8$     **20.** $4(3 + a) = 12 + 4a$

Give the solution set of each equation. Assume that the domain of each variable is $\Re$.

**21.** $-x = 3$     **25.** $n + 0 = 2$

**22.** $-n = -4$     **26.** $-3 + 0 = z$

**23.** $\dfrac{1}{t} = 3$     **27.** $-2 + t = 0$

**24.** $\dfrac{1}{z} = \dfrac{2}{3}$     **28.** $r + 3 = 0$

## WRITTEN EXERCISES

Each of the statements in Exercises 1–10 is justified by one of the real-number axioms. In each case, assume that the domain of any variable or variables involved is $\Re$ and state the appropriate axiom.

**A**   **1.** $5 + (-5) = 0$     **5.** $3 + (\frac{1}{2} \cdot 2) = 3 + 1$

   **2.** $7 \cdot 1 = 7$     **6.** $6 \cdot [7 + (-7)] = 6 \cdot 0$

   **3.** $5 \cdot \frac{1}{5} = 1$     **7.** $\dfrac{1}{-6}(-6) + 3 = 1 + 3$

   **4.** $-3 \cdot \dfrac{1}{-3} = 1$     **8.** $-3 \cdot \dfrac{1}{-3} = \dfrac{1}{-3} \cdot -3$

   **9.** $4 + (-4) = (-4) + 4$

   **10.** $3 + [(-3) + 2] = [3 + (-3)] + 2$

List the members of the solution set of each sentence over the given domain.

**Sample.**  $-z < 0, z \in J$     *Solution*:  $\{1, 2, 3, \ldots\}$

**11.** $k + (-4) = 0, k \in \mathcal{R}$

**12.** $n + 3 = 0, n \in \mathcal{R}$

**13.** $r + 0 = -8, r \in \mathcal{R}$

**14.** $\frac{1}{3} \cdot s = 1, s \in \mathcal{R}$

**15.** $\frac{1}{6} \cdot 6 + r = 0, r \in \mathcal{R}$

**16.** $\frac{1}{2} \cdot 2 + 0 = n, n \in \mathcal{R}$

**17.** $r + (-1) < 1, r \in W$

**18.** $-(-m) < 2, m \in N$

**19.** $-n > 0, n \in J$

**20.** $-4 < -n < -2, n \in J$

**B** **21.** $-(t + 3) = -7, t \in \mathcal{R}$

**22.** $-(5 - t) = -3, t \in \mathcal{R}$

**23.** $-z = z, z \in \mathcal{R}$

**24.** $-[-n + (-1)] = 0, n \in J$

**25.** $\frac{1}{5}(z + 1) = 1, z \in \mathcal{R}$

**26.** $\frac{1}{8}(x + 2) = 1, z \in \mathcal{R}$

**27.** $-3 \cdot \dfrac{1}{t} = 1, t \in \mathcal{R}$

**28.** $-3 \cdot \dfrac{1}{-n} = 1, n \in \mathcal{R}$

**29.** $-z \cdot \dfrac{1}{-z} = 1, z \in \mathcal{R}$

**30.** $5 \cdot \dfrac{1}{z} - 3 \cdot \dfrac{1}{3} = 0, z \in \mathcal{R}$

## 1–7  Theorems about Real Numbers

In this section you will see that many of the familiar properties of real numbers can be proved from the axioms and definitions already stated.  At this stage, however, it is more important that you understand what these properties mean than be able to prove them.  In giving the proofs we shall need to use some facts about equality, for example:

> Let $x$, $y$, and $z$ denote real numbers. Then, if $x = y$:
>
> $x + z = y + z$     (Addition Property of Equality)
> $xz = yz$     (Multiplication Property of Equality)

In properties like these the symbol $=$ is used to indicate that the two expressions have the *same value*.

A statement to be proved is called a **theorem** or **proposition**. (In the theorems of this section, *a*, *b*, *c*, and *d* denote real numbers.)

---

**THEOREM.** **Cancellation Law for Addition.**

If $a + c = b + c$, then $a = b$.

---

A *proof* of this theorem consists in reasoning from the *hypothesis*, $a + c = b + c$, to the *conclusion*, $a = b$. It is arranged below as a sequence of statements, each with its justification.

*Proof*

1. $a + c = b + c$      **1.** Hypothesis
2. $-c$ is a real number.      **2.** Axiom of Negatives
3. $(a + c) + (-c) = (b + c) + (-c)$      **3.** Addition Property of Equality
4. $a + [c + (-c)] = b + [c + (-c)]$      **4.** Associative Axiom
5. $a + 0 = b + 0$      **5.** Axiom of Negatives
6. $\therefore a = b$      **6.** Axiom of Zero

A theorem which follows easily from another is called a **corollary**. You can use the commutative law for addition to prove the following corollary to the cancellation law for addition.

---

**COROLLARY.** If $c + a = c + b$, then $a = b$.

---

There are similar results for multiplication:

---

**THEOREM.** **Cancellation Law for Multiplication.**

If $ac = bc$ and $c \neq 0$, then $a = b$.

**COROLLARY.** If $ca = cb$ and $c \neq 0$, then $a = b$.

---

Proofs of these are left as exercises (see Exercise 5, page 27 and Exercise 10, page 28).

A theorem introduced mainly to help in proving another theorem is called a **lemma**.

**LEMMA.**   If $d + d = d$, then $d = 0$.

*Proof*

1. $d + d = d$
2. $d = 0 + d$
3. $d + d = 0 + d$
4. $\therefore d = 0$

1. Hypothesis
2. Axiom of Zero
3. Transitive Property of Equality
4. Cancellation Law for Addition

**THEOREM.   Zero-factor Property.**   $a \cdot 0 = 0$ and $0 \cdot a = 0$

*Proof*

1. $0 + 0 = 0$
2. $a \cdot (0 + 0) = a \cdot 0$
3. $a \cdot 0 + a \cdot 0 = a \cdot 0$
4. $\therefore a \cdot 0 = 0$
5. $\therefore 0 \cdot a = 0$

1. Axiom of Zero
2. Multiplication Property of Equality
3. Distributive Axiom
4. Lemma (using $a \cdot 0$ in place of $d$)
5. Commutative Axiom

**COROLLARY.**   0 has no reciprocal.

*Proof:*   If $r$ were the reciprocal of 0, then $0 \cdot r = 1$, which contradicts the theorem just proved.

**THEOREM.   Multiplication Property of** $-1$.   $(-1)a = -a$.

*Proof*

1. $(-1)a + a = (-1)a + 1 \cdot a$
2. $(-1)a + a = [(-1) + 1]a$
3. $(-1)a + a = 0 \cdot a$
4. $(-1)a + a = 0$

1. Addition Property of Equality, Axiom of One
2. Distributive Axiom
3. Axiom of Negatives
4. Zero-factor Property

Since the sum of $(-1)a$ and $a$ is 0, $(-1)a$ must be the additive inverse of $a$, that is, $(-1)a = -a$.

For example, $(-1)6 = -6$, $(-1)(-3) = -(-3) = 3$, and $(-1)(-1) = -(-1) = 1$.

Notice that in proving the last theorem and the corollary preceding it, words as well as symbols were used. This is entirely permissible.

*ORAL EXERCISES*

In each of the following statements identify the *hypothesis* and the *conclusion.*

**1.** If it rains, then I will take my umbrella.

**2.** I will take my umbrella if it rains.

**3.** If I can finish my homework in time and John doesn't have to baby-sit, we are going to go to the movies tonight.

**4.** If we do not make the installment payment by next Friday, the stereo set will be repossessed and we will lose our down payment.

**5.** If $x$, $y$, and $z$ denote real numbers and $x = y$, then $xz = yz$.

**6.** If $a \in \Re$, then $a \cdot 0 = 0$ and $0 \cdot a = 0$.

What number is named by each of the following numerical expressions?

**7.** $(-1)5$

**8.** $(-1)(-5)$

**9.** $-(-5)$

**10.** $-[3(-\frac{1}{3})]$

**11.** $(-1)[(0)(-2)]$

**12.** $-[(1)(-1)]$

In Exercises 13–20, name the axiom or theorem which justifies the given statement.

**13.** $0 \cdot x = 0$

**14.** If $x + 3 = 7 + 3$, then $x = 7$.

**15.** $-a = (-1)a$

**16.** If $x = y$, then $3x = 3y$.

**17.** $(-1)(-3) = -(-3)$

**18.** If $a = b$, then $a + 7 = b + 7$.

**19.** If $3x = 3 \cdot 7$, then $x = 7$.

**20.** $0 \cdot (z + 1) = 0$

*WRITTEN EXERCISES*

Supply the missing reasons in each proof. Assume that the domain of each variable is $\Re$, and that no denominator is 0.

**B** **1.** Prove: $(b + c) + (-c) = b$

1. $(b + c) + (-c) = b + [c + (-c)]$     1. __?__
2. $(b + c) + (-c) = b + 0$     2. __?__
3. $(b + c) + (-c) = b$     3. __?__

**2.** Prove: If $a + b = 0$, then $a = -b$. (**Uniqueness of Additive Inverse**)

1. $a + b = 0$      1. Hypothesis
2. $(a + b) + (-b) = 0 + (-b)$      2. _?_
3. $a + [b + (-b)] = 0 + (-b)$      3. _?_
4. $a + 0 = 0 + (-b)$      4. _?_
5. $a = -b$      5. _?_

**3.** Prove: If $ab = 1$, then $a = \dfrac{1}{b}$. (**Uniqueness of Multiplicative Inverse**)

1. $ab = 1$      1. Hypothesis
2. $a \neq 0, b \neq 0$      2. Zero-factor Property
3. $(ab)\dfrac{1}{b} = 1 \cdot \dfrac{1}{b}$      3. _?_
4. $a\left[b \cdot \dfrac{1}{b}\right] = 1 \cdot \dfrac{1}{b}$      4. _?_
5. $a \cdot 1 = 1 \cdot \dfrac{1}{b}$      5. _?_
6. $a = \dfrac{1}{b}$      6. _?_

**4.** Prove: If $a \neq 0$, then $\dfrac{1}{\frac{1}{a}} = a$.

1. $a \cdot \dfrac{1}{a} = 1$      1. _?_
2. $a = \dfrac{1}{\frac{1}{a}}$      2. Uniqueness of Multiplicative Inverse
3. $\dfrac{1}{\frac{1}{a}} = a$      3. _?_

**5.** Prove: If $ac = bc$ and $c \neq 0$, then $a = b$.      (**Cancellation Law for Multiplication**)

1. $ac = bc$      1. _?_
2. $[ac]\left(\dfrac{1}{c}\right) = [bc]\left(\dfrac{1}{c}\right)$      2. Multiplication Property of Equality
3. $a\left[c\left(\dfrac{1}{c}\right)\right] = b\left[c\left(\dfrac{1}{c}\right)\right]$      3. _?_
4. $a \cdot 1 = b \cdot 1$      4. _?_
5. $a = b$      5. _?_

**6.** Prove: $-(a + b) = (-a) + (-b)$. (**Property of the Negative of a Sum**)

1. $-(a + b) = (-1)(a + b)$      1. _?_
2. $-(a + b) = (-1)a + (-1)b$      2. _?_
3. $-(a + b) = (-a) + (-b)$      3. _?_

**7.** Prove: If $a \neq 0$ and $b \neq 0$, then $\dfrac{1}{ab} = \dfrac{1}{a} \cdot \dfrac{1}{b}$. **(Property of the Reciprocal of a Product)**

1. $\left(\dfrac{1}{a} \cdot \dfrac{1}{b}\right) ab = \left(\dfrac{1}{a} \cdot \dfrac{1}{b}\right) ba$      1. ___?___

2. $\left(\dfrac{1}{a} \cdot \dfrac{1}{b}\right) ab = \dfrac{1}{a}\left[\dfrac{1}{b}(ba)\right]$      2. ___?___

3. $\left(\dfrac{1}{a} \cdot \dfrac{1}{b}\right) ab = \dfrac{1}{a}\left[\left(\dfrac{1}{b} \cdot b\right)a\right]$      3. ___?___

4. $\left(\dfrac{1}{a} \cdot \dfrac{1}{b}\right) ab = \dfrac{1}{a}(1 \cdot a)$      4. ___?___

5. $\left(\dfrac{1}{a} \cdot \dfrac{1}{b}\right) ab = \dfrac{1}{a} \cdot a$      5. ___?___

6. $\left(\dfrac{1}{a} \cdot \dfrac{1}{b}\right) ab = 1$      6. ___?___

7. $\dfrac{1}{a} \cdot \dfrac{1}{b} = \dfrac{1}{ab}$      7. Uniqueness of Multiplicative Inverse

8. $\dfrac{1}{ab} = \dfrac{1}{a} \cdot \dfrac{1}{b}$      8. ___?___

**8.** Prove: $(a + b) + [(-b) + c] = a + c$
1. $(a + b) + [(-b) + c] = [(a + b) + (-b)] + c$      1. ___?___
2. $(a + b) + [(-b) + c] = [a + (b + (-b))] + c$      2. ___?___
3. $(a + b) + [(-b) + c] = [a + 0] + c$      3. ___?___
4. $(a + b) + [(-b) + c] = a + c$      4. ___?___

Prove each of the following statements.

**C**    **9.** If $c + a = c + b$, then $a = b$.

**10.** If $ca = cb$, and $c \neq 0$, then $a = b$.

**11.** If $a = b$, then $-a = -b$.

**12.** If $a + b = 0$, then $b = -a$.

**13.** $a(b + c) = ba + ca$

**14.** If $a + (-b) = c$, then $b + c = a$.

**15.** If $a \cdot \dfrac{1}{b} = c$, then $bc = a$.

**16.** If $ax = a$, then $x = 1$. **(Uniqueness of Multiplicative Identity)**

**17.** If $a + x = a$, then $x = 0$. **(Uniqueness of Additive Identity)**

**18.** If $a = c$ and $a + b = c + d$, then $b = d$.

---

# Chapter Summary

---

1. We make certain assumptions about $\Re$, the set of real numbers, which are called **axioms** or **postulates**. From these you can derive other properties, called **theorems**. Every step in the proof of a theorem can be justified by an axiom, a definition, a given fact, or a theorem previously proved.

2. The fundamental operations in $\Re$ are addition and multiplication. In the following, $a$, $b$, and $c$ denote any real numbers.

## AXIOMS OF ADDITION AND MULTIPLICATION

| | **For Addition** | **For Multiplication** |
|---|---|---|
| *Closure Axioms* | $a + b$ represents a unique real number. | $ab$ represents a unique real number. |
| *Commutative Axioms* | $a + b = b + a$ | $ab = ba$ |
| *Associative Axioms* | $(a + b) + c$ $= a + (b + c)$ | $(ab)c = a(bc)$ |
| *Identity Elements* | 0 is the unique element such that $0 + a = a$ and $a + 0 = a$. | 1 is the unique element such that $1 \cdot a = a$ and $a \cdot 1 = a$. |
| *Inverse Elements* | There is a unique real number $-a$, the additive inverse of $a$, such that $a + (-a) = 0$ and $-a + a = 0$. | Provided $a$ is not 0, there is a unique real number $\frac{1}{a}$, the multiplicative inverse of $a$, such that $a \cdot \frac{1}{a} = 1$ and $\frac{1}{a} \cdot a = 1$. |
| *Distributive Axiom* | $a(b + c) = ab + ac$ $(b + c)a = ba + ca$ | |

### Properties of Equality

| | |
|---|---|
| *Reflexive Property* | $a = a$ |
| *Symmetric Property* | If $a = b$, then $b = a$. |
| *Transitive Property* | If $a = b$ and $b = c$, then $a = c$. |
| *Substitution Property* | Changing the numeral by which a number is named in an expression does not change the value of the expression. |

**3.** There is an order relation in $\mathcal{R}$.

### Axioms of Order

*Axiom of Comparison*  One and only one of the following statements is true:

$$a < b,\, a = b,\, a > b.$$

*Transitive Property*  **1.** If $a < b$ and $b < c$, then $a < c$.
**2.** If $a > b$ and $b > c$, then $a > c$.

(The remaining axioms characterizing the real number system will be stated in Chapters 2 and 7.)

**4.** Some important theorems about real numbers are:

*Property of Negative of a Sum*  $-(a + b) = (-a) + (-b)$

*Reciprocal of a Product*  $\dfrac{1}{ab} = \dfrac{1}{a} \cdot \dfrac{1}{b}$

*Cancellation Laws*  If $a + c = b + c$, then $a = b$.
If $ac = bc$, and $c \neq 0$, then $a = b$.

*Zero-factor Property*  $a \cdot 0 = 0$ and $0 \cdot a = 0$

*Multiplication Property of* $-1$  $(-1)a = -a$

**5.** A **variable** is a symbol representing any element of a specified set, called the **domain** or **replacement set** of the variable. A sentence containing a variable is called an **open sentence**. A value of the variable for which the sentence becomes a true statement is a **solution** or **root** of the sentence, and the subset of the domain of the variable consisting of all solutions of the sentence is the **solution set** of the sentence over that domain.

### Vocabulary and Spelling

Review the meaning of each term by reference to the page listed.

counting numbers (*p. 1*)
natural numbers (*p. 1*)
positive integers (*p. 1*)
integers (*p. 1*)
rational numbers (*p. 1*)
real numbers (*p. 1*)
numerals (*p. 2*)
numerical expression (*p. 2*)
value of an expression (*p. 2*)

equation (*p. 2*)
grouping symbols (*p. 2*)
set (*p. 5*)
members (elements) of a set (*p. 5*)
roster (*p. 5*)
equal sets (*p. 5*)
subset (*p. 5*)
proper subset (*p. 6*)
empty (null) set (*p. 6*)

infinite set (*p. 6*)
finite set (*p. 6*)
variable (*p. 9*)
domain (replacement set) of a
   variable (*p. 9*)
constant (*p. 9*)
open sentence (*p. 9*)
solution set (*p. 9*)
root (solution) (*p. 9*)
set-builder notation (*p. 9*)
number line (*p. 12*)
graph of a number (*p. 12*)
coordinate of a point (*p. 12*)
interval (*p. 13*)
linearly ordered set (*p. 14*)
sum (*p. 16*)

term (addend) (*p. 16*)
product (*p. 16*)
factor (*p. 16*)
axiom (postulate) (*p. 16*)
closed (*p. 16*)
simplifying an expression (*p. 17*)
negative of a number (*p. 20*)
additive inverse (*p. 20*)
reciprocal (*p. 21*)
multiplicative inverse (*p. 21*)
theorem (*p. 24*)
proof of a theorem (*p. 24*)
hypothesis (*p. 24*)
conclusion (*p. 24*)
corollary (*p. 24*)
lemma (*p. 24*)

## *Chapter Test*

**1–1**   **1.** Identify which of the following are integers:

$$-3, -\tfrac{3}{5}, 0, \sqrt{1}, \sqrt{2}, \sqrt{4}$$

**Write in simple form.**

**2.** $3\left[2 + \dfrac{8+1}{3}\right]$          **3.** $3[2(5+1) + 3(2+1)]$

**1–2**   **4.** Specify {the letters in the word "tense"} by roster.

      **5.** Specify {6, 7, 8} by rule.

      **6.** Specify all subsets of {1, 2} by roster, and identify each proper
subset.

**1–3**   **7.** Find the solution set of $2z - 1 = 7$ over $\Re$.

      **8.** Specify $\{z: z \text{ is between 5 and 9}, z \in N\}$ by roster.

**1–4**   **Graph the specified subset of $\Re$.**

      **9.** $\{n: -1 \leq n < 3\}$        **10.** $\{r: r \geq 2\}$

**1–5**   State the axiom or property justifying each assertion.

**11.** $3 + (x + 7) = (x + 7) + 3$

**12.** $4 \cdot \pi \in \Re$

**13.** $(x + y) + (w + z) = [(x + y) + w] + z$

**14.** $7 \cdot (3 + x) = (3 + x) \cdot 7$

**1–6**   **15.** $5(z + 2) = 5z + 10$     **16.** $3(-8 + 8) = 3 \cdot 0$

List the members of the solution set of each sentence over $\Re$.

**17.** $\frac{1}{3}(x - 1) = 1$     **18.** $3 \cdot t - 1 = 0$

**1–7**   Justify each step in the proof by citing the appropriate property.

**19.** Prove: If $a \neq 0$, and $a(b + c) = 0$, then $b = -c$.

1. $a \neq 0, a(b + c) = 0$     1. __?__

2. $\frac{1}{a}[a(b + c)] = \frac{1}{a} \cdot 0$     2. __?__

3. $\left(\frac{1}{a} \cdot a\right)(b + c) = \frac{1}{a} \cdot 0$     3. __?__

4. $1 \cdot (b + c) = \frac{1}{a} \cdot 0$     4. __?__

5. $b + c = \frac{1}{a} \cdot 0$     5. __?__

6. $b + c = 0$     6. __?__

7. $b = -c$     7. Uniqueness of Additive Inverse.

---

## *Chapter Review*

**1–1**   **Numbers and Operations**     *Pages 1–4*

**1.** $\{1, 2, 3, \ldots\}$ is called the set of counting numbers or the set of __?__ numbers.

**2.** $\{0, 1, -1, 2, -2, 3, -3, \ldots\}$ is called the set of __?__.

**3.** Symbols which name numbers are called __?__, or __?__ expressions.

**4.** The number named by a numerical expression is called the __?__ of the expression.

**5.** Symbols like parentheses, brackets, and braces are called __?__ symbols.

**6.** Write $5[(8 - 3) + 6]$ in simple form.

**7.** Write $3[2(6 - 1) + 2(5 - 2)]$ in simple form.

**8.** $8 - 8 = 23 \times \underline{\phantom{?}}$.

## 1-2 Sets and Set Membership      *Pages 5–8*

**9.** The objects which make up a set are called its $\underline{\phantom{?}}$ or $\underline{\phantom{?}}$.

**10.** If two sets have exactly the same members they are called $\underline{\phantom{?}}$ sets.

**11.** If every member of a set $A$ is also a member of set $B$, then $A$ is called a $\underline{\phantom{?}}$ of $B$.

**12.** If every member of a set $A$ is also a member of set $B$, and if $A \neq B$, then $A$ is called a $\underline{\phantom{?}}$ $\underline{\phantom{?}}$ of $B$.

**13.** Specify {the letters in the word "error"} by roster.

**14.** Specify $\{d, e, f\}$ by rule.

**15.** The empty or null set is denoted by the symbol $\underline{\phantom{?}}$.

**16.** $\{1, 2, 3, \ldots\}$ is (a finite/an infinite) set.

## 1-3 Variables and Open Sentences      *Pages 8–12*

**17.** A sentence containing a variable is called an $\underline{\phantom{?}}$ $\underline{\phantom{?}}$.

**18.** A variable having just one value is a $\underline{\phantom{?}}$.

**19.** The solution set of an open sentence must be a subset of the $\underline{\phantom{?}}$ of the variable.

**20.** $\{z: z + 2 = 12, z \in \mathcal{R}\}$ is an example of $\underline{\phantom{?}}$ $\underline{\phantom{?}}$ notation.

**21.** Specify $\{n: n + 2 = 5, n \in N\}$ by roster.

**22.** Specify the solution set of $3y + 1 = 7$ over $\{1, 2, 3, 4\}$.

## 1-4 Order Relations      *Pages 12–15*

**23.** On a number line, the point representing a number is called the $\underline{\phantom{?}}$ of the number, and the number is called the $\underline{\phantom{?}}$ of the point.

**24.** If $a < d$ and $d < m$, then $a \underline{\phantom{?}} m$.

**25.** The compound sentence $x > 2$ and $x < 4$ can be written in the form $\underline{\phantom{?}} < \underline{\phantom{?}} < \underline{\phantom{?}}$.

**26.** If the number $y$ is not less than 2 and does not exceed 5, then $2 \underline{\phantom{?}} y \underline{\phantom{?}} 5$.

Graph the specified subset of $\mathcal{R}$.

**27.** $\{z: z \leq -1\}$        **28.** $\{t: 2 \leq t < 7\}$.

### 1–5  Sums and Products    *Pages 16–20*

**29.** The basic operations of algebra are __?__ and __?__ .

**30.** A statement accepted as true without proof is called an __?__ or a __?__ .

In each case state the property justifying the assertion.

**31.** $3 \cdot 5 \in \Re$

**33.** $3 + (5 + 2) = (3 + 5) + 2$

**32.** $t + 3 \cdot 5 = t + 5 \cdot 3$

**34.** $2(5 + 6) = 2(6 + 5)$

### 1–6  Identities and Inverses    *Pages 20–23*

**35.** In the set of real numbers, 1 is the __?__ __?__ for multiplication, and 0 is the __?__ __?__ for addition.

**36.** The __?__ __?__ of 1 is $-1$, and the __?__ __?__ of 3 is $\frac{1}{3}$.

**37.** If $a + 7 = 0$, then $a =$ __?__ , and if $7 \cdot a = 1$, then $a =$ __?__ .

**38.** The negative of $-\frac{3}{4}$ is __?__ , and the reciprocal of $-3$ is __?__ .

State the axiom or property justifying each assertion.

**39.** $3(x + 2) = 3 \cdot x + 6$

**41.** $1 \cdot (x + 5) = x + 5$

**40.** $\frac{1}{5} \cdot 5 + 5 = 1 + 5$

**42.** $\frac{1}{7} + (-\frac{1}{7}) = 0$

List the members of the solution set of each sentence over $\Re$.

**43.** $x \cdot \frac{1}{3} = 1$

**45.** $3 \cdot (z + 2) = 0$

**44.** $-y + 5 = 0$

**46.** $n + 0 = -4$

### 1–7  Theorems about Real Numbers    *Pages 23–28*

**47.** A statement to be proved is called a __?__ .

**48.** "For all $a, b, c \in \Re$, if $a + c = b + c$, then $a = b$." is called the __?__ law for addition.

**49.** The zero-factor property states "For all $a \in \Re, a \cdot$ __?__ $=$ __?__ ."

**50.** The assertion that $-1 \cdot 3 = -3$ is justified by the __?__ __?__ of $-1$.

**51.** A theorem which follows easily from another is called a __?__ .

**52.** A theorem whose main purpose is in proving another theorem is called a __?__ .

# Gottfried Wilhelm Leibniz

Gottfried Wilhelm Leibniz (pronounced Libe-nitz) showed genius in many fields: law, religion, diplomacy, history, philosophy, mathematics. His brilliant mind could not resist venturing into all these areas. Although he was trained in law and made his living as a diplomat and historian, his lasting contributions to mathematics mark him as a first-rate mathematician.

Leibniz was born in Leipzig, Germany, in 1646, the son of a philosophy professor whose family had served the government of Saxony for three generations. His earliest years were spent in an atmosphere of scholarship weighted with politics. Studying philosophy and the classics led him as a young man to a study of logic. He attempted to reform classical logic, and his work foreshadowed that of George Boole, who invented symbolic logic (Boolean algebra) about 1850. Had Leibniz pursued his idea of reducing logic to symbols, the history of modern mathematics might have been greatly altered. However, his dream of a universal symbolic language and system of reasoning, which was considered absurd in his time, has been largely realized in modern abstract algebra and symbolic logic.

It was not until the age of twenty-six that Leibniz became interested in mathematics and physics. He is credited with discovering, independently of Newton, the fundamental theorem uniting differential and integral calculus. He also did significant work in combinatorial analysis, devised a computer which performed all arithmetic operations, and stressed the importance of the binary system of numeration which is used extensively in modern electronic computers.

Leibniz died in Hanover in 1716.

# Mathematics and Meteorology

To strangers at a party the weather is a topic of conversation; to skiers anxiously awaiting the first snows of the winter, it is a source of pleasure or dismay; but to astronauts preparing to lift off on a journey to the moon, or to farmers waiting to plant their first crops in spring, or to the passengers of a small boat in the path of a hurricane, the weather is a much more serious matter, and in cases such as these accurate weather forecasting is vital.

Weather forecasting is just one area of the complex science of *meteorology*, the study of the atmospheres of the Earth and other planets. Actually, only about a third of the meteorologists in the United States are directly concerned with predicting the weather. Other meteorologists are involved in observing the weather, in analyzing weather maps, in preparing summaries of weather data, in teaching,

in research, in developing improved weather instruments, and in consulting with other individuals whose problems involve the atmosphere in some way. But for practically everyone who works in the field of meteorology, some knowledge of mathematics, including statistics, is essential.

This mathematics may be relatively simple. For example, weather observers use graphs, tables, and specialized slide rules in routine calculations to determine the upper-air wind, temperature, pressure, and humidity from data transmitted by instruments carried by balloons into the atmosphere.

On the other hand, some of the mathematics involved in meteorology is extremely complex, and the research meteorologists who use it are essentially mathematical physicists. In fact, the development of modern meteorology is closely associated with the application of mathematical physics to atmospheric phenomena such as cyclones, thunderstorms, and clouds, and to atmospheric processes such as condensation, convection, and other air motions.

Like the other earth sciences, meteorology deals mainly with the observation and interpretation of natural phenomena on a comparatively large scale. It is difficult, if not actually impossible, to reproduce many phenomena of the atmosphere completely in the laboratory. Therefore, meteorologists have relied heavily on statistical relationships between weather variables to explain changes in weather and climate.

Some research meteorologists construct *mathematical models* of the atmosphere based on physical laws. These models usually involve systems of *differential equations* that must be solved simultaneously. Sometimes the differential equations can be reduced to systems of algebraic equations that can be solved by the methods you will learn in Chapter 3. Other equations are solved by the methods of numerical analysis with the aid of large computers. Currently, at the National Meteorological Center in Suitland, Maryland, observations from all over the world are put into such a mathematical model of the atmosphere to produce forecasts of the air temperature, pressure, and motion at various levels in the atmosphere. The predicted maps serve as a basis for forecasting the weather and the *air pollution potential*—the conditions that favor the concentration of pollutants near the ground and near their source.

One of the more dramatic recent applications of mathematics to meteorology involved the development of a way of observing the vertical temperature distribution of the atmosphere by means of observations from *weather satellites*. This method of *remote sensing* of the atmosphere uses observations of the thermal (heat) radiation from the atmosphere at several wave lengths and involves the solution of the "equation of radiative (radiation) transfer." Scientists found that it was possible to reduce the very complicated radiative transfer equation to a set of algebraic equations applicable to each of the observed wave lengths. This set of equations could then be solved by matrix methods (presented in Chapter 16) to yield the temperature from the ground upward into the atmosphere.

In ways like this, mathematically inclined scientists have helped develop new ways of observing, understanding, and predicting the changes of the atmosphere that we collectively call *weather*.

# 2

# Applications of Real-Number Properties

## APPLICATIONS TO ARITHMETIC

### 2–1 Absolute Value

Positive and negative numbers can be defined in terms of the order relations $<$ and $>$:

$$a \text{ is positive if and only if } a > 0.$$
$$a \text{ is negative if and only if } a < 0.$$

Zero is neither positive nor negative.

You should realize that a symbol like $-a$ need not name a negative number. If $a = -5$, for example, then $-a = -(-5) = 5$.

It can be shown (Oral Exercise 17, page 63) that if $a$ is a nonzero real number, then either $a$ *is positive* or $-a$ *is positive*, but not both. For every nonzero real number $a$, we define its absolute value, $|a|$, to be the positive number of the pair $a$ and $-a$ (see Figure 2–1).

FIGURE 2–1

For example, $|2| = 2, |-5| = 5$, and $|-\frac{1}{3}| = \frac{1}{3}$. The absolute value of 0 is defined to be 0.

Scientists call $|a|$ the **magnitude** of $a$. On the number line it measures the distance between the graph of $a$ and the origin.

$|a| = 2$ is equivalent to "$a$ is 2 units from 0"; that is, $a = 2$ or $a = -2$.

$|a| < 2$ is equivalent to "$a$ is within 2 units of 0"; that is, $a > -2$ and $a < 2$, or $-2 < a < 2$.

$|a| > 2$ is equivalent to "$a$ is more than 2 units from 0"; that is, $a < -2$ or $a > 2$.

> **In general, for any positive number $p$:**
>
> **1.** $|a| = p$ **is equivalent to** $a = -p$ **or** $a = p$.
>
> **2.** $|a| < p$ **is equivalent to** $-p < a < p$.
>
> **3.** $|a| > p$ **is equivalent to** $a < -p$ **or** $a > p$.

A second way of depicting real numbers is by means of *arrows*, rather than points, on the number line. You can represent the number 2, for example, as an arrow of length 2 pointing to the right. There are many such arrows, of course (see Figure 2–2), and each of them represents 2. You represent a negative number, say $-3$, by any arrow of length 3

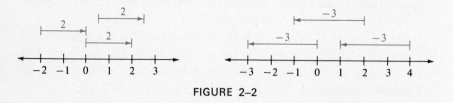

FIGURE 2–2

pointing to the left. The nonzero real number $a$ is represented by any arrow of length $|a|$ pointing to the right if $a > 0$, or to the left if $a < 0$.

You can construct the sum of two numbers, say 7 and $-3$, geometrically as follows: Draw the arrow representing 7 which starts at the origin. Then draw the arrow representing $-3$ which starts at the head of the first arrow. The arrow from the origin to the head of the second arrow (shown dotted in Figure 2–3) represents the sum $7 + (-3)$.

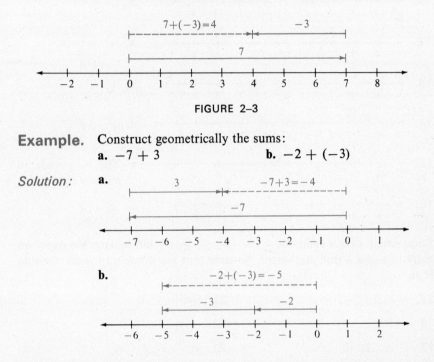

FIGURE 2–3

**Example.** Construct geometrically the sums:

    **a.** $-7 + 3$          **b.** $-2 + (-3)$

*Solution:* **a.**

**b.**

Notice in part **b** of the Example above that to avoid confusion we wrote $-2 + (-3)$ instead of $-2 + -3$.

**ORAL EXERCISES**

Simplify each expression.

    **Sample.**   **a.** $|-3|$     **b.** $-|7-4|$     **c.** $|7| - |4|$

    *Solution:*  **a.** 3       **b.** $-3$       **c.** 3

**1.** $|7|$     **3.** $-|3|$     **5.** $|8 + 3|$     **7.** $|6| + |-5|$     **9.** $2|-1|$

**2.** $|-6|$    **4.** $-|0|$     **6.** $|7 - 5|$     **8.** $|7| - |-3|$    **10.** $3|-4|$

Each graph in Exercises 11–16 is specified by one of the open sentences in the column at the right. Match each graph with the corresponding open sentence. (Assume that all variables represent real numbers.)

**11.**

$$\xleftarrow{\hspace{0.3cm}}\overset{\underset{-4\ \ -3\ \ -2\ \ -1\ \ \ 0\ \ \ 1\ \ \ 2\ \ \ 3\ \ \ 4}{|\ \ |\ \ |\ \ |\ \ |\ \ |\ \ |\ \ |\ \ |}}{\hspace{0.3cm}}\xrightarrow{\hspace{0.3cm}}$$

**a.** $|z| = 0$

**12.**

$$\xleftarrow{\hspace{0.3cm}}\overset{\underset{-4\ \ -3\ \ -2\ \ -1\ \ \ 0\ \ \ 1\ \ \ 2\ \ \ 3\ \ \ 4}{|\ \ |\ \ |\ \ |\ \ |\ \ |\ \ |\ \ |\ \ |}}{\hspace{0.3cm}}\xrightarrow{\hspace{0.3cm}}$$

**b.** $|y| = |-3|$

**13.**

$$\xleftarrow{\hspace{0.3cm}}\overset{\underset{-4\ \ -3\ \ -2\ \ -1\ \ \ 0\ \ \ 1\ \ \ 2\ \ \ 3\ \ \ 4}{|\ \ |\ \ |\ \ |\ \ |\ \ |\ \ |\ \ |\ \ |}}{\hspace{0.3cm}}\xrightarrow{\hspace{0.3cm}}$$

**c.** $|x| = 1$

**14.**

$$\xleftarrow{\hspace{0.3cm}}\overset{\underset{-4\ \ -3\ \ -2\ \ -1\ \ \ 0\ \ \ 1\ \ \ 2\ \ \ 3\ \ \ 4}{|\ \ |\ \ |\ \ |\ \ |\ \ |\ \ |\ \ |\ \ |}}{\hspace{0.3cm}}\xrightarrow{\hspace{0.3cm}}$$

**d.** $|a| < 3$

**15.**

$$\xleftarrow{\hspace{0.3cm}}\overset{\underset{-4\ \ -3\ \ -2\ \ -1\ \ \ 0\ \ \ 1\ \ \ 2\ \ \ 3\ \ \ 4}{|\ \ |\ \ |\ \ |\ \ |\ \ |\ \ |\ \ |\ \ |}}{\hspace{0.3cm}}\xrightarrow{\hspace{0.3cm}}$$

**e.** The magnitude of $n$ is less than 2.

**16.**

$$\xleftarrow{\hspace{0.3cm}}\overset{\underset{-4\ \ -3\ \ -2\ \ -1\ \ \ 0\ \ \ 1\ \ \ 2\ \ \ 3\ \ \ 4}{|\ \ |\ \ |\ \ |\ \ |\ \ |\ \ |\ \ |\ \ |}}{\hspace{0.3cm}}\xrightarrow{\hspace{0.3cm}}$$

**f.** $|a| > 3$

State which of the symbols, $<, =, >, \leq, \geq$, should replace the question mark to make a true statement. Assume that the domain of each variable is $\mathfrak{R}$.

**17.** $-2 \underline{\ \ ?\ \ } |-2|$

**18.** $|-3| \underline{\ \ ?\ \ } -|-3|$

**19.** $|-a| \underline{\ \ ?\ \ } |a|$

**20.** $-|b| \underline{\ \ ?\ \ } |-b|$

**21.** $|-x| \underline{\ \ ?\ \ } 0$

**22.** $-|y| \underline{\ \ ?\ \ } 0$

**23.** $|a| + 2 \underline{\ \ ?\ \ } |a|$

**24.** $2|a| \underline{\ \ ?\ \ } |a|$

State the solution set of each equation over $\mathfrak{R}$.

   **Sample.** $|y| = 5$   *Solution:* $\{5, -5\}$

**25.** $|n| = 0$   **26.** $|t| = -5$   **27.** $|x + 1| = 3$   **28.** $|x - 1| = 5$

*WRITTEN EXERCISES*

Rewrite each of the following without using absolute-value notation.

   **Sample.** $|p| > 5$   *Solution:* $p > 5$ or $p < -5$

A   **1.** $|a| < 4$       **2.** $|t| = 2$       **3.** $|n| > \frac{3}{2}$       **4.** $|x| = 0$

Rewrite each of the following using absolute-value notation.

**Sample.** $-5 < z < 5$   *Solution:* $|z| < 5$

**5.** $x = -\frac{5}{6}$ or $x = \frac{5}{6}$

**8.** $z$ is less than 1 and greater than $-1$.

**6.** $-2 \le t \le 2$

**9.** $y$ is less than $-3$ or greater than 3.

**7.** $v$ is 5 units from 0.

**10.** $a$ is more than 7 units from 0.

Specify the solution set of each sentence over $\mathcal{R}$.

**11.** $|z| = 7$

**15.** $-|k| = 3$

**19.** $|r + 3| = 3$

**12.** $|m| = 10$

**16.** $|-t| = 4$

**20.** $|k + 2| = 6$

**13.** $|t| = |-1|$

**17.** $|x| + |-3| = 4$

**21.** $-|x| + 8 = 6$

**14.** $-|y| = -9$

**18.** $|z| - |2| = 5$

**22.** $4 - |p| = 0$

Use arrows to construct each sum geometrically.

**23.** $2 + 5$

**29.** $-4 + (2 + 1)$

**24.** $-3 + (-1)$

**30.** $7 + (-2) + (-3)$

**25.** $-4 + 3$

**31.** $-1 + (-3) + (2 + 2)$

**26.** $6 + (-7)$

**32.** $3 + (-4) + (-4) + 3$

**27.** $(1 + 2) + 3$

**33.** $3 + (-4) + -7 + 5$

**28.** $(-2) + (-5) + (-1)$

**34.** $(-1) + 2 + (-2) + 1$

Graph each nonempty solution set of the following sentences over $\mathcal{R}$. If the solution set is $\emptyset$, state this.

**Sample.** $|z| < 3$

*Solution:* Each point on the graph must lie closer to the origin than the graphs of $-3$ and 3. The endpoints are not included in the graph.

**B** **35.** $|x| < 1$

**37.** $|t| > 1$

**39.** $|y| \le 4$

**36.** $|z| < |-2|$

**38.** $|n| > 3$

**40.** $|q| \ge 2$

**C** **41.** $|x| \ge x$   **42.** $|r| < r$   **43.** $|x| < -x$   **44.** $|z| \ge -z$

## 2–2 Addition and Subtraction

The multiplication property of $-1$ (page 25) and the distributive axiom (page 21) are used to justify these steps:

$$-(a + b) = (-1)(a + b) = (-1)a + (-1)b = -a + (-b)$$

(see Exercise 6, page 27). This result may be restated as follows:

---

**Property of the Negative of a Sum.** The negative of the sum of two numbers *a* and *b* is the sum of their negatives:

$$-(a + b) = -a + (-b).$$

---

You can use this property, the real-number axioms, and familiar addition facts to find sums of the type constructed geometrically in Section 2–1. For example:

$$
\begin{aligned}
-7 + 3 &= -(4 + 3) + 3 && \text{Addition fact} \\
&= [-4 + (-3)] + 3 && \text{Property of the Negative of a Sum} \\
&= -4 + [(-3) + 3] && \text{Associative Axiom} \\
&= -4 + 0 && \text{Axiom of Negatives} \\
&= -4 && \text{Axiom of Zero}
\end{aligned}
$$

In practice you do not go through all the steps illustrated above but use the rules displayed on page 45.

The operation of subtraction is defined in terms of addition and additive inverses:

$$\text{If } a \text{ and } b \in \mathcal{R}, \text{ then } a - b = a + (-b).$$

For example,

$$7 - 3 = 7 + (-3),$$
$$(-2) - 5 = (-2) + (-5),$$

and

$$(-3) - (-8) = (-3) + [-(-8)] = -3 + 8.$$

Since you can convert subtraction into addition, the following rules enable you to add or subtract nonzero real numbers by finding sums and differences of *positive* numbers.

1. If $a > 0$ and $b > 0$, then $a + b = |a| + |b|$.

   Example. $7 + 4 = 11$

2. If $a < 0$ and $b < 0$, then $a + b = -(|a| + |b|)$.

   Example. $-7 + (-4) = -(7 + 4) = -11$

3. If $a > 0$ and $b < 0$, and $|a| \geq |b|$, then $a + b = |a| - |b|$.

   Example. $7 + (-4) = 7 - 4 = 3$

4. If $a > 0$ and $b < 0$, and $|a| < |b|$, then $a + b = -(|b| - |a|)$.

   Example. $4 + (-7) = -(7 - 4) = -3$

You can say that the *sign* of a positive number is *plus* ($+$) and of a negative number is *minus* ($-$). Thus, two numbers of the *same sign* are either both positive or both negative; if two numbers are of *opposite sign*, then one of them is positive and the other one is negative. The rules displayed above can now be restated as follows:

To add two numbers of the same sign, add their absolute values and prefix their common sign. (*Rules 1 and 2*)

To add two numbers of opposite sign, subtract the lesser absolute value from the greater absolute value and prefix the sign of the number having the greater absolute value. (*Rules 3 and 4*)

It is not difficult to see that the set $\mathcal{R}$ of real numbers is closed under subtraction (Exercise 38, page 48). However, the examples $7 - 3 \neq 3 - 7$ and $(7 - 3) - 2 \neq 7 - (3 - 2)$ show that subtraction is neither commutative nor associative. We must therefore be sure what is meant by an expression such as $7 - 3 - 2$. By an extension of the definition of subtraction you have $7 - 3 - 2 = 7 + (-3) + (-2)$, and you may now use the associative axiom for *addition* to see that

$$7 - 3 - 2 = 7 + (-3) + (-2) = [7 + (-3)] + (-2) = (7 - 3) - 2$$

or

$$7 - 3 - 2 = 7 + (-3) + (-2) = 7 + [(-3) + (-2)]$$
$$= 7 + [-(3 + 2)] = 7 - (3 + 2).$$

That is, either you may group *from left to right*, or you may group together all the positive terms and all the negative terms.

**Example.**     Evaluate $13 - 7 - 2 + 4$.

*First Solution:*     $13 - 7 - 2 + 4 = [(13 - 7) - 2] + 4$
$$= [6 - 2] + 4$$
$$= 4 + 4 = 8$$

*Second Solution:*   $13 - 7 - 2 + 4 = (13 + 4) - (7 + 2)$
$$= 17 - 9 = 8$$

*ORAL EXERCISES*

In Exercises 1–10, **(a)** add the given numbers; **(b)** subtract the number listed below from the one listed above.

| | | | | |
|---|---|---|---|---|
| **1.** $\begin{array}{r} 5 \\ \underline{7} \end{array}$ | **3.** $\begin{array}{r} 2 \\ \underline{-5} \end{array}$ | **5.** $\begin{array}{r} -5 \\ \underline{-5} \end{array}$ | **7.** $\begin{array}{r} -11 \\ \underline{-12} \end{array}$ | **9.** $\begin{array}{r} -8 \\ \underline{-11} \end{array}$ |
| **2.** $\begin{array}{r} -3 \\ \underline{-4} \end{array}$ | **4.** $\begin{array}{r} -7 \\ \underline{3} \end{array}$ | **6.** $\begin{array}{r} 6 \\ \underline{-1} \end{array}$ | **8.** $\begin{array}{r} -7 \\ \underline{1} \end{array}$ | **10.** $\begin{array}{r} 3 \\ \underline{-10} \end{array}$ |

Simplify each expression.

**11.** $8 + 7$

**12.** $4 - 6$

**13.** $-3 + 7$

**14.** $-8 - 5$

**15.** $3 - (-7)$

**16.** $0 - (-2)$

**17.** $4 + (-12)$

**18.** $-8 + (-3)$

**19.** $-(3 - 5)$

**20.** $25 - 25$

**21.** $7 - (-7)$

**22.** $8 - (3 + 2)$

**23.** $(8 - 3) + 2$

**24.** $5 - (-2 + 3)$

**25.** $7 - 4 - 2$

**26.** $10 - 5 + 3 - 2$

**27.** $-8 + 10 - 9 + 6$

**28.** $6 - (5 - 3) + (9 - 3)$

*WRITTEN EXERCISES*

Simplify each expression.

|A|

**1.** $18 + 73$

**2.** $-15 + (-38)$

**3.** $-112 - (-82)$

**4.** $-53 - 76$

**5.** $3.8 + 9.1$

**6.** $-8.7 + 1.6$

**7.** $-21.3 - (-8.2)$

**8.** $-0 + (-18.7)$

**9.** $21.3 + (8.1 - 2.7)$

**10.** $(9.6 - 3.8) - 21.7$

**11.** $13.2 - (8.1 - 2.2)$

**12.** $12 - (-7.1 + 2.3)$

**13.** $3 + (-4) - (-7) + 8$

**14.** $-12 + 8 - 11 + (-15)$

**15.** $-17 - (-3) + 7 - (9)$

**16.** $-22 + (-3) - (-12) + 13$

**17.** $6 - 8 + 18 + 5 - 27$

**18.** $-7 + 3 - 11 - 12 + 6$

**19.** $18 + 6 - 8 - 7 - 1$

**20.** $-27 - 6 + 32 + 11 - 10$

Find the indicated sums.

| 21. | 22. | 23. | 24. |
|---|---|---|---|
| 18 | $-37$ | 81.3 | 21.6 |
| $-27$ | $-16$ | $-21.6$ | $-18.7$ |
| 45 | 112 | $-17.8$ | $-1.8$ |
| $-71$ | $-52$ | 2.4 | $-2.3$ |

**25.** Find the change in altitude from 112 feet below sea level to 213 feet above sea level.

**26.** If a submarine cruising at a depth of 150 feet dives an additional 75 feet and then climbs 23 feet, what is its depth then?

**27.** At 7:30:00 A.M., the countdown on a rocket launch stood at $-57$ seconds. If there was a 26-second delay in resuming the count, at what time did the rocket take off?

**28.** If the voltage across an electric circuit changes from $+230$ volts to $-175$ volts, what is the net change in voltage?

Which of the following sets are closed under subtraction?

**29.** {integers}

**30.** {natural numbers}

**31.** {even integers}

**32.** {odd integers}

Supply the missing reasons in each proof. Assume that the domain of each variable is $\Re$.

**B** **33.** Show that $-2 - 3 = -5$.

    1. $-2 - 3 = -2 + (-3)$          1. __?__

    2.         $= -(2 + 3)$          2. __?__

    3.         $= -5$          3. __?__

**34.** Show that $-3 - (-5) = 2$.

    1. $-3 - (-5) = -3 + [-(-5)]$          1. __?__

    2.         $= -3 + 5$          2. __?__

    3.         $= -3 + (3 + 2)$          3. __?__

    4.         $= (-3 + 3) + 2$          4. __?__

    5.         $= 0 + 2$          5. __?__

    6.         $= 2$          6. __?__

**35.** Prove: $-(a + b) = -a - b$

| | | |
|---|---|---|
| 1. | $-(a + b) = -a + (-b)$ | 1. _?_ |
| 2. | $= -a - b$ | 2. _?_ |

**36.** Prove: $-(a - b) = b - a$. **(Property of the Negative of a Difference)**

| | | |
|---|---|---|
| 1. | $-(a - b) = -[a + (-b)]$ | 1. _?_ |
| 2. | $= -a + [-(-b)]$ | 2. _?_ |
| 3. | $= -a + b$ | 3. _?_ |
| 4. | $= b + (-a)$ | 4. _?_ |
| 5. | $= b - a$ | 5. _?_ |

**37.** Prove: $(a - b) - c = a - (b + c)$

| | | |
|---|---|---|
| 1. | $(a - b) - c = [a + (-b)] + (-c)$ | 1. _?_ |
| 2. | $= a + [(-b) + (-c)]$ | 2. _?_ |
| 3. | $= a + [-(b + c)]$ | 3. _?_ |
| 4. | $= a - (b + c)$ | 4. _?_ |

**38.** Prove that the set $\Re$ of real numbers is closed under subtraction.

*Proof:* Let $a$ and $b$ be any two real numbers. Then

| | |
|---|---|
| 1. $a - b = a + (-b)$ | 1. _?_ |
| 2. $-b$ is a unique real number. | 2. _?_ |
| 3. Thus $a + (-b)$ is a unique real number. | 3. _?_ |

Since the difference of any two real numbers is itself a real number, $\Re$ is closed under subtraction.

[C] **39.** Prove: If $a = b$, then $a - c = b - c$.

**40.** Prove: If $a - c = b - c$, then $a = b$.

Prove that Rules 1–4 on page 45 are valid.

**Sample.**   (*Rule 4*) If $a > 0$ and $b < 0$, and $|a| < |b|$, then

$$a + b = -(|b| - |a|).$$

*Solution:*

| | |
|---|---|
| 1. $a > 0$ and $b < 0$ | 1. Hypothesis |
| 2. $-b > 0$ | 2. If $b$ is a nonzero real number, then either $b$ or else $-b$ is positive (page 39). |
| 3. $|a| = a$ and $-|b| = b$ | 3. Definition of absolute value |
| 4. Then $-(|b| - |a|) = |a| - |b|$ | 4. Property of the Negative of a Difference (Exercise 36) |
| 5.    $= |a| + (-|b|)$ | 5. Definition of Subtraction |
| 6.    $= a + b$ | 6. Substitution |
| 7. Thus $a + b = -(|b| - |a|)$. | 7. Symmetric Property |

**41.** (*Rule 1*) If $a > 0$ and $b > 0$, then $a + b = |a| + |b|$.

**42.** (*Rule 2*) If $a < 0$ and $b < 0$, then $a + b = -(|a| + |b|)$.

**43.** (*Rule 3*) If $a > 0$ and $b < 0$, and $|a| > |b|$, then $a + b = |a| - |b|$.

**44.** Show that for all real numbers $a$ and $b$, $|a - b| = |b - a|$.

## 2–3  Multiplication and Division

Notice how the multiplication property of $-1$ (page 25) is used below.

$$3 \cdot 5 = 15$$
$$-3 \cdot 5 = [(-1)3] \cdot 5 = (-1)(3 \cdot 5) = (-1)15 = -15$$
$$3 \cdot (-5) = 3 \cdot [(-1)5] = (-1)(3 \cdot 5) = (-1)15 = -15$$
$$-3 \cdot (-5) = [(-1)3][(-1)5] = [(-1)(-1)] \cdot (3 \cdot 5) = 1 \cdot 15 = 15$$

In general, for any real numbers $a$ and $b$,

$$-a \cdot b = [(-1)a]b = (-1)(ab) = -(ab) = -ab$$
$$a \cdot (-b) = a \cdot [(-1)b] = (-1)(ab) = -(ab) = -ab$$
$$(-a)(-b) = [(-1)a][(-1)b] = [(-1)(-1)](ab) = 1 \cdot ab = ab$$

We can show that *the product of two numbers of the same sign is positive.* For example, if $x$ and $y$ are *both negative numbers*, then by the definition of absolute value, $x = -|x|$ and $y = -|y|$. Hence $xy = (-|x|)(-|y|) = |x| \, |y|$ and the product of the positive numbers $|x|$ and $|y|$ is *positive* (Oral Exercise 16, page 63). It can be shown similarly that *the product of two numbers of opposite sign is negative.* We can extend these results to any number of factors to give the following rules:

---

1. The absolute value of a product of two or more numbers is the product of their absolute values.

2. A product of nonzero numbers is positive if the number of negative factors is even. A product of nonzero numbers is negative if the number of negative factors is odd.

---

**Example 1.**  Evaluate **a.** $-1 \cdot 3 \cdot (-\frac{1}{3})$  **b.** $(-2) \cdot 3(-2) \cdot (-5)$

*Solution:*  **a.** $(-1) \cdot 3 \cdot (-\frac{1}{3}) = +(1 \cdot 3 \cdot \frac{1}{3}) = 1$

**b.** $(-2) \cdot 3 \cdot (-2) \cdot (-5) = -(2 \cdot 3 \cdot 2 \cdot 5) = -60$

Division is defined in terms of multiplication and multiplicative inverses. For example, $12 \div 3 = 12 \times \frac{1}{3}$, $-6 \div 2 = -6 \cdot \frac{1}{2}$, and $4 \div \frac{1}{2} = 4 \times \dfrac{1}{\frac{1}{2}} = 4 \times 2$. In general:

**If $a$ and $b \in \mathcal{R}$ and $b \neq 0$, then $a \div b = a \cdot \dfrac{1}{b}$.**

Thus, to obtain the *quotient*, you multiply the *dividend* by the reciprocal of the *divisor*. We often use the symbol $\dfrac{a}{b}$ for $a \div b$ and call it the *fraction* with *numerator a* and *denominator b*.

Since 0 has no reciprocal (page 21), *division by 0 is not defined*, that is, a fraction can *not* have denominator 0. Simple examples show that division is neither commutative nor associative. The notation $a \div b \div c$ is seldom used, but when it is, its meaning is given by

$$(a \div b) \div c = a \cdot \frac{1}{b} \cdot \frac{1}{c} \qquad (b \neq 0, c \neq 0).$$

A nonzero number $b$ and its reciprocal $\dfrac{1}{b}$ have the same sign (Exercise 15, page 63) and therefore, since $\dfrac{a}{b} = a \cdot \dfrac{1}{b}$, we have the rule:

---

The quotient $\dfrac{a}{b}$ is positive if *a* and *b* have the same sign; it is negative if *a* and *b* have the opposite sign.

---

For example, $\dfrac{-2 \cdot 3 \cdot (-5)}{4 \cdot (-7)}$ is negative because the numerator is positive and the denominator negative.

A fraction which is a quotient of integers is in lowest terms if the numerator and denominator have in common no integral* factor other than 1 or $-1$. You can reduce a fraction to lowest terms by using the fact that if $b \neq 0$ and $c \neq 0$, then $\dfrac{ac}{bc} = \dfrac{a}{b}$ (see Section 6–1).

For example, $\dfrac{-6}{15} = -\dfrac{2 \cdot 3}{5 \cdot 3} = -\dfrac{2}{5}$ and $\dfrac{-21}{-14} = \dfrac{3 \cdot 7}{2 \cdot 7} = \dfrac{3}{2}$.

---

*Integral* is the adjective form of *integer*.

**Example 2.** Reduce to a fraction in lowest terms:

$$\frac{5[2 + 3(6 - 4)] + 2}{3(6 \div 2 - 1) - 2(8 + 9)}$$

*Solution:*

$$\frac{5[2 + 3(6 - 4)] + 2}{3(6 \div 2 - 1) - 2(8 + 9)} = \frac{5[2 + 3 \cdot 2] + 2}{3(3 - 1) - 2 \cdot 17}$$

$$= \frac{5[2 + 6] + 2}{3 \cdot 2 - 34} = \frac{5 \cdot 8 + 2}{6 - 34}$$

$$= \frac{40 + 2}{-28} = -\frac{42}{28}$$

$$= -\frac{21 \cdot 2}{14 \cdot 2} = -\frac{3 \cdot 7 \cdot 2}{2 \cdot 7 \cdot 2}$$

$$= -\frac{3}{2}$$

In working Example 2, we followed these rules concerning the order in which operations are to be performed.

---

1. Perform the operations within each set of grouping symbols, beginning with the innermost ones.

2. Whenever there is a choice between performing an addition or subtraction on the one hand or a multiplication or division on the other, perform the multiplication or division first.

---

*ORAL EXERCISES*

Simplify each expression. (Reduce fractions to lowest terms.)

**1.** $(-5)(-8)$

**2.** $6(-2)$

**3.** $-4(7)$

**4.** $-3(0)$

**5.** $(\frac{1}{3})(-6)$

**6.** $(-8)(-\frac{1}{2})$

**7.** $(-1)(2)(-3)$

**8.** $(-5)(-1)(-1)$

**9.** $18 \div (-9)$

**10.** $-27 \div 3$

**11.** $-24 \div (-6)$

**12.** $(-36) \div (-6)$

**13.** $0 \div 12$

**14.** $0 \div (-3)$

**15.** $(12 + 3) \div (-5)$

**16.** $8 \div (6 - 8)$

Without doing the computation, state whether each expression represents a positive number, a negative number, or zero.

**17.** $(-25.1)(-3)(-17.2)$

**20.** $\dfrac{(-42)(-18)(-1.3)}{(-7.1)(-8.6)}$

**18.** $(-18)(-3)(15)(0)$

**21.** $\dfrac{(1.3)(-2.1)(0)(-7.6)}{(-14)(1.3)}$

**19.** $\dfrac{(3.2)(-7.1)(-11)}{(-17)(-3.8)}$

**22.** $\dfrac{(127)(18 - 18)(4.7)}{(-3)(6 + 7)(-1)}$

## WRITTEN EXERCISES

Simplify each expression.

**1.** $(-7)(10)(-3)$

**2.** $(-2)(-6)(-3)(0)$

**3.** $(-6)(\frac{1}{5})(-25)$

**4.** $(18)(-14)(\frac{1}{7})$

**5.** $18 \cdot 7 + 18 \cdot 3$

**6.** $27 \cdot 4 + 27 \cdot 6$

**7.** $12 - 3(8 + 12)$

**8.** $2(18 - 3) - 29$

**9.** $16 - 2 \cdot 5 + 3$

**10.** $12 \cdot 3 - 2 \cdot 11 + 6$

**11.** $0 \div 7 \div 14$

**12.** $-15 \cdot 0 \div 6$

**13.** $\dfrac{18 - 3}{7 - 4}$

**14.** $\dfrac{27 + 13}{12 - 4}$

**15.** $\dfrac{6[8 + 2(3 - 1)]}{3[6 + 8]}$

**16.** $\dfrac{7[3 - (8 + 1)]}{3[9 - 10]}$

**17.** $36 - [(6 \div 2) \cdot 4] - 24$

**18.** $48 + [(18 \div 6) \cdot 2] - 55$

**19.** $[3 \cdot 6 - 12 \div 2] \div 6 + (-2)$

**20.** $6 \cdot (-8) \div [(-3)(2) + 24 \div 2]$

Which of the following sets are closed under division?

**21.** {natural numbers}

**22.** {positive rational numbers}

**23.** $\{1, -1\}$

**24.** $\{2, 1, \frac{1}{2}\}$

**25.** Give examples to show that division of real numbers is (a) not commutative, (b) not associative.

Supply the missing reasons in each proof. Assume that the domain of each variable is $\Re$.

**B**  **26.** Prove: If division by 0 is *excluded*, then the set $\Re$ is closed under division.

*Proof:* Let $a$ and $b$ be any two real numbers with $b \neq 0$.

1. $\dfrac{1}{b}$ is a unique real number.        1. __?__

2. $a \div b = a \cdot \dfrac{1}{b}$             2. __?__

3. Therefore $a \div b$ is a unique real number.     3. __?__

*If division by zero is excluded,* then the quotient of any two real numbers is itself a real number and hence $\Re$ is closed under division.

**27.** Prove: If $a \neq 0, \dfrac{a}{a} = 1$.

1. $\dfrac{a}{a} = a\left(\dfrac{1}{a}\right)$         1. __?__

2. $\quad = 1$            2. __?__

**28.** Prove: If $b \neq 0$, then $\dfrac{0}{b} = 0$.

1. $\dfrac{0}{b} = 0 \cdot \dfrac{1}{b}$         1. __?__

2. $\quad = 0$            2. __?__

**29. a.** Prove: $\dfrac{1}{-1} = -1$.

1. $(-1)(-1) = 1 \cdot 1$       1. __?__
2. $(-1)(-1) = 1$         2. __?__
3. $-1 = \dfrac{1}{-1}$         3. __?__

**b.** Prove: $\dfrac{1}{-a} = -\dfrac{1}{a}$.

1. $\dfrac{1}{-a} = \dfrac{1}{-1 \cdot a}$       1. __?__

2. $\quad = \dfrac{1}{-1} \cdot \dfrac{1}{a}$      2. Property of the Reciprocal of a Product (Exercise 7, page 28)

3. $\quad = -1 \cdot \dfrac{1}{a}$        3. __?__

4. $\quad = -\dfrac{1}{a}$          4. __?__

**30.** Use the fact that the product of two positive numbers is positive to prove that the product of two numbers of opposite sign is negative.

*Proof:* Let $a$ and $b$ be two real numbers with $a > 0$ and $b < 0$.

1. $|a| = a$ and $-|b| = b$       1. __?__
2. $ab = |a|(-|b|)$       2. __?__
3. $= -|a|\,|b|$       3. __?__
4. $|a|$ and $|b|$ are positive.       4. __?__
5. Therefore $|a|\,|b|$ is positive.       5. Given facts
6. Hence $-|a|\,|b|$ is negative.       6. __?__

C **31.** Prove that $a \div 1 = a$.

**32.** Prove that if $a \neq 0$, then $\dfrac{-a}{a} = -1$.

**33.** Prove that $a(b - c) = ab - ac$. (Distributive Law for Subtraction)

**34.** Prove that $(a + b) \div c = \dfrac{a}{c} + \dfrac{b}{c}$.

**35.** Prove that $(a - b) \div c = \dfrac{a}{c} - \dfrac{b}{c}$.

**36.** Is division of real numbers distributive with respect to addition? Why or why not?

**37.** Find the mistake in the following proof that $1 = 0$.

1. $c = 0$       1. Assumption
2. $\dfrac{1}{c} \cdot c = \dfrac{1}{c} \cdot 0$       2. Substitution
3. $\dfrac{1}{c} \cdot c = 1$       3. Axiom of Reciprocals
4. $\dfrac{1}{c} \cdot 0 = 0$       4. Zero-factor Property
5. $1 = 0$       5. Substitution

## OPEN SENTENCES IN ONE VARIABLE

## 2–4 Equivalent Equations

Symbols like $6$, $x$, $2(x + 1) - 7$, $\dfrac{x + y}{x - y}$, and $ax + b$ are called **algebraic expressions**; they are formed from numerals and variables using the operations of addition, multiplication, subtraction, and division.

Because variables represent real numbers, you can use the properties of real numbers to simplify algebraic expressions. Notice how the distributive and associative axioms are used below:

$$2(x + a) - 7a = (2x + 2a) - 7a = 2x + (2a - 7a)$$
$$= 2x + (2 - 7)a = 2x - 5a.$$

These steps guarantee that in the equation

$$2(x + a) - 7a = 2x - 5a,$$

the *left member*, $2(x + a) - 7a$, and the *right member*, $2x - 5a$, have the same value for every numerical replacement of the variables. We say that the two expressions are **equivalent** and call the equation an **identity**.

Consider the following two sequences of equations:

| | |
|---|---|
| $3x + 14 = 26$ | $x = 4$ |
| $3x + 14 - 14 = 26 - 14$ | $3x = 3 \cdot 4$ |
| $3x = 12$ | $3x = 12$ |
| $\frac{1}{3} \cdot 3x = \frac{1}{3} \cdot 12$ | $3x + 14 = 12 + 14$ |
| $x = 4$ | $3x + 14 = 26$ |

Do you see that if the first statement in either sequence is true for some value of $x$, then the last statement in the sequence is also true for that value of $x$? This means that $3x + 14 = 26$ and $x = 4$ have the *same solution set*, namely $\{4\}$; that is, they are **equivalent open sentences**.

Each of the statements "If $x = 4$, then $3x + 14 = 26$" and "If $3x + 14 = 26$, then $x = 4$" is the **converse** of the other because the hypothesis of each statement is the conclusion of the other. Alternative ways of expressing these statements are shown below.

If $x = 4$, then $3x + 14 = 26$      If $3x + 14 = 26$, then $x = 4$

*or*                                   *or*

$3x + 14 = 26$ if $x = 4$.      $3x + 14 = 26$ only if $x = 4$.

Thus the sentence

$$3x + 14 = 26 \text{ if and only if } x = 4$$

combines a statement and its converse. Do you see that this "if and only if" statement is a way of saying that $3x + 14 = 26$ and $x = 4$ are equivalent equations?

Often you can solve an equation by transforming it into an equivalent one whose solution set can be found by inspection. The following transformations on page 56 are useful.

---

### Transformations Which Produce Equivalent Equations

**1.** Adding to or subtracting from each member of an equation the same number or the same expression in any variables appearing in the equation.

**2.** Multiplying or dividing each member by the same *nonzero* number.

**3.** Replacing either member by an expression equivalent to it.

---

In Example 1 below we solve an equation and then check our work against error by showing that both members of the equation reduce to the same number when the root replaces the variable, that is, when the root is substituted for the variable.

**Example 1.** Solve $4(2z - 1) + 2 = 5z - 8$.

*Solution:*

**1.** Copy the equation.  $\qquad 4(2z - 1) + 2 = 5z - 8$

**2.** Use the distributive law and simplify the left member.
$$8z - 4 + 2 = 5z - 8$$
$$8z - 2 = 5z - 8$$

**3.** Add 2 to each member and simplify both members.
$$8z - 2 + 2 = 5z - 8 + 2$$
$$8z = 5z - 6$$

**4.** Subtract $5z$ from both members so that the variable will appear only in the left member.
$$8z - 5z = 5z - 6 - 5z$$
$$3z = -6$$

**5.** Divide each member by 3.
$$\frac{3z}{3} = \frac{-6}{3}$$
$$z = -2 \quad \{-2\}$$

*Check:*
$$4(2z - 1) + 2 = 4[2(-2) - 1] + 2$$
$$= 4(-4 - 1) + 2 = 4(-5) + 2$$
$$= -20 + 2 = -18$$
$$5z - 8 = 5(-2) - 8 = -10 - 8 = -18$$

When you are asked to solve an equation for a certain variable, you regard all other variables in the equation as constants.

**Example 2.** If $a \neq 0$, solve for $x$: $ax + b = 0$.

*Solution:*

$$ax + b = 0$$
$$ax + b - b = 0 - b$$
$$ax = -b$$
$$\frac{1}{a} \cdot ax = \frac{1}{a}(-b)$$
$$x = -\frac{b}{a}$$

## ORAL EXERCISES

State the transformation used to transform the first equation into the second equation.

**Sample.** $3x + 5 + x = 8$; $4x + 5 = 8$

*Solution:* Substitution of $4x$ for $3x + x$ in the left-hand member.

**1.** $3x + 2 = 8$; $3x = 6$

**2.** $4z = 12$; $z = 3$

**3.** $3x + 8 = x + 1$; $2x + 8 = 1$

**4.** $6y + 3 = y + 7y$; $6y + 3 = 8y$

**5.** $\frac{1}{5}(t + 3) = 2$; $t + 3 = 10$

**6.** $7x + 3 = 8x - 2$; $3 = x - 2$

**7.** $3(n - 2) = 2(n + 6)$; $3n - 6 = 2n + 12$

**8.** $\frac{4v}{7} = v + 2$; $4v = 7v + 14$

**9.** $7k + 8 + 3k = k + 2$; $10k + 8 = k + 2$

**10.** $7t + 8 = 2t - 5$; $5t + 8 = -5$

Solve each equation for the specified variable.

**11.** $t - 3s = 7$; for $t$

**12.** $u + 4v = -7$; for $u$

**13.** $2n + a = b$; for $n$

**14.** $3k + 3 = t$; for $k$

**15.** $3z + n = 5 + 2z$; for $z$

**16.** $4m + 2n = 3 + 3m$; for $m$

**17.** $8 - (-3) = 11$, because __?__ + __?__ = __?__.

**18.** $18 \div (-6) = -3$, because __?__ $\cdot$ __?__ = __?__.

*WRITTEN EXERCISES*

Solve each equation. Check your answer.

**A**

**1.** $5y - 4 = 31$

**2.** $7z + 8 = 1$

**3.** $3x + x - 2 = 18$

**4.** $7y - 3 - 3y = 45$

**5.** $7 - 2k = k + 22$

**6.** $18 - 4t = 3 - 7t$

**7.** $8(t - 3) = 6t + 8$

**8.** $3(z + 5) = 11 - z$

**9.** $4(x - 3) = 5(x + 7)$

**10.** $-9(x + 3) = 8(3 - x)$

**11.** $7(2a + 3) - 2(a + 1) = -5$

**12.** $-3(2 - 2b) + 5(b - 4) = 7$

**13.** $\dfrac{2t}{3} = 12$

**14.** $\dfrac{6z}{5} = 7$

**15.** $\dfrac{2k}{5} = -3.2$

**16.** $\dfrac{3x}{4.1} = 6$

**17.** $\dfrac{n}{7} - \dfrac{2n}{7} = 5$

**18.** $\dfrac{3n}{5} + \dfrac{n}{5} = 8$

**19.** $\dfrac{5}{t} - 3 - \dfrac{2}{t} = 0$

**20.** $\dfrac{6}{r} + 5 = \dfrac{3}{r} + 11$

In each of the following formulas, substitute values for the variable as indicated, and then solve the resulting equation for the remaining variable.

**Sample.** $S = 2r(r + h)$  when $S = 3200$ and $r = 20$.

*Solution:*  $S = 2r(r + h)$
$3200 = 2(20)(20 + h)$
$3200 = 800 + 40h$
$2400 = 40h$
$h = 60; \quad \{60\}$

**21.** $v = k + gt$ when $v = 70$, $k = 6$, and $g = 32$.

**22.** $S = 3\pi d + 7\pi D$ when $S = 18\pi$ and $d = \frac{4}{3}$.

**23.** $A = \dfrac{h}{2}(b + c)$ when $A = 120$, $h = 5$, and $c = 5$.

**24.** $A = P(1 + rt)$ when $A = 2400$, $P = 1600$, and $r = 0.05$.

**25.** $C = \dfrac{5(F - 32)}{9}$, when $C = 100$.

**26.** $A = 2\pi rh + 2\pi r^2$, when $A = 138\pi$ and $r = 3$.

Specify each of the following subsets of $\mathfrak{R}$ by roster.

**27.** $\{x\colon 3x - (2 + x) = 3(2x - 1)\}$

**28.** $\{y\colon 3(7 - 2y) = 30 - 7(y + 1)\}$

**29.** $\{z\colon 5z - (3 + z) = 2(2z + 5)\}$

**30.** $\{b\colon 6b - 2(2b + 5) = 6(5 + b)\}$

**B** **31.** $\{t\colon -3[t - (2t + 3) - 2t] = -9\}$

**32.** $\{r\colon -2[r - (r - 1)] = -3(r + 1)\}$

**33.** $\{n\colon 16 + 2n + n^2 = 28 - n(4 - n)\}$

**34.** $\{x\colon x(4 - x) = 2 + x(1 - x)\}$

**35.** $\{k\colon 2k(k - 3) = 50 + 2k^2 - k\}$

**36.** $\{s\colon s(3 - 4s) = -4s^2 - s + 2\}$

Solve each equation for the specified variable.

**C** **37.** $z(r - s) - z(r + s) - 8s^2 = 0$; for $z$

**38.** $xy - xz - x(y + z) + z^2 = 0$; for $x$

**39.** $v(g + h) + r = 0$; for $h$

**40.** $k^2(m - p) = 2k^2p$; for $p$

**41.** $A = \frac{1}{2}h(a + b)$; for $b$

**42.** $P = 3d\pi + 5D\pi$; for $d$

## 2–5  Equivalent Inequalities

Figure 2–4 illustrates the fact that adding the same number to both members of an inequality does not change its sense, or direction.

$$-3 < 2 \qquad\qquad -3 < 2$$
$$-3 + 4 < 2 + 4 \qquad -3 + (-2) < 2 + (-2)$$
$$1 < 6 \qquad\qquad -5 < 0$$

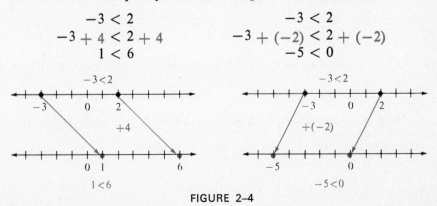

FIGURE 2–4

The effect of multiplying both members of an inequality by the same number is illustrated in Figure 2–5.

$$-3 < 2 \qquad\qquad -3 < 2$$
$$-3 \cdot 2 < 2 \cdot 2 \qquad\qquad -3(-2) > 2(-2)$$
$$-6 < 4 \qquad\qquad 6 > -4$$

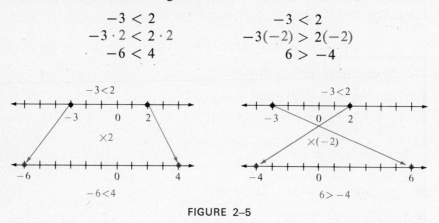

FIGURE 2–5

Notice that when both members of an inequality are multiplied by a *negative* number, the inequality sign must be *reversed*.

The examples above suggest the following properties, which we take as axioms for the set of real numbers.

---

**Addition and Multiplication Axioms of Order.** Let $a$, $b$, and $c$ denote real numbers.

I.  If $a < b$, then $a + c < b + c$.
    If $a > b$, then $a + c > b + c$.

II. If $a < b$ and $c > 0$, then $ac < bc$.
    If $a > b$ and $c > 0$, then $ac > bc$.

III. If $a < b$ and $c < 0$, then $ac > bc$.
     If $a > b$ and $c < 0$, then $ac < bc$.

---

Notice that dividing an inequality by $q$ is the same as multiplying by $\frac{1}{q}$. If $q$ is a negative number, so is $\frac{1}{q}$. Therefore, when you divide both members of an inequality by a negative number, you must reverse the inequality sign. For example, you have $8 > 6$, but $\frac{8}{-2} < \frac{6}{-2}$, that is, $-4 < -3$.

The properties of order and equality guarantee that the following transformations of a given inequality always produce an **equivalent inequality**, that is, one with the same solution set.

---

**Transformations Which Produce Equivalent Inequalities**

**1.** Adding to or subtracting from each member of an inequality the same number or the same expression in the variable appearing in the inequality.

**2.** Multiplying or dividing each member by the same positive number.

**3.** Multiplying or dividing each member by the same negative number and reversing the inequality sign.

**4.** Substituting for either member an expression equivalent to it.

---

**Example.** Solve each inequality and graph the solution set.

$$\textbf{a. } 5x + 2 \le x + 10 \qquad \textbf{b. } 3(3 - u) - 10 < 2$$

*Solution:*

**a.**

$$5x + 2 \le x + 10$$

$$5x + 2 - 2 \le x + 10 - 2$$

$$5x \le x + 8$$

$$5x - x \le x + 8 - x$$

$$4x \le 8$$

$$\frac{4x}{4} \le \frac{8}{4}$$

$$x \le 2$$

$$\{x : x \le 2\}$$

**b.** $3(3 - u) - 10 < 2$

$$9 - 3u - 10 < 2$$

$$-3u - 1 < 2$$

$$-3u - 1 + 1 < 2 + 1$$

$$-3u < 3$$

$$\frac{-3u}{-3} > \frac{3}{-3}$$

$$u > -1$$

$$\{u : u > -1\}$$

## ORAL EXERCISES

State the transformation used to transform the first inequality into the second inequality.

**Sample.**   $2x < -8;\ x < -4$

*Solution:*   Each member is divided by the positive number 2.

**1.** $x - 7 \geq 3;\ x \geq 10$

**2.** $n + 6 < 8;\ n < 2$

**3.** $2n + 3 + n \leq 4;\ 3n + 3 \leq 4$

**4.** $\dfrac{t}{3} > 5;\ t > 15$

**5.** $\dfrac{r}{-7} < 6;\ r > -42$

**6.** $-3y > 12;\ y < -4$

**7.** $4z - 3 \geq 3z;\ z \geq 3$

**8.** $2t + 1 < t + 2;\ t < 1$

**9.** $3 - 2z > 6;\ -3 + 2z < -6$

**10.** $3 - 2z > 6;\ -2z > 3$

Supply the missing steps in each proof.  In each case assume that the domain of each variable is $\mathcal{R}$.

**11.** Prove: If $a < b$, then $a - c < b - c$.

   1. $a < b$               1. Hypothesis

   2. $a + (-c) < b + (-c)$    2. ___?___

   3. $a - c < b - c$       3. ___?___

**12.** Prove: If $a > b$, then $-a < -b$.

   1. $a > b$               1. Hypothesis

   2. $(-1)a < (-1)b$      2. ___?___

   3. $-a < -b$         3. ___?___

**13.** Prove: If $a \neq 0$, then $a \cdot a > 0$.

   1. $a \neq 0$              1. Hypothesis

   2. Either $a > 0$ or $a < 0$.   2. ___?___

   3. Suppose $a > 0$, then    3. ___?___
      $a \cdot a > a \cdot 0$.

   4. $a \cdot a > 0$          4. ___?___

   5. Suppose $a < 0$, then    5. ___?___
      $a \cdot a > a \cdot 0$.

   6. $a \cdot a > 0$          6. ___?___

**14.** Prove: $1 > 0$

   1. $1 \cdot 1 > 0$         1. ___?___

   2. $1 \cdot 1 = 1$        2. ___?___

   3. Thus $1 > 0$       3. ___?___

**15.** Prove: A nonzero number $b$ and its reciprocal $\dfrac{1}{b}$ have the same sign.

1. $b \cdot \dfrac{1}{b} = 1$

2. Since $1 > 0$, 1 is positive.

3. Therefore the number of negative factors in the product $b \cdot \dfrac{1}{b}$ must be even and hence $b$ and $\dfrac{1}{b}$ must have the same sign.

1. __?__

2. Exercise 14, and definition of *positive*, page 39.

3. __?__

**16.** Prove: The product of any two positive numbers $a$ and $c$ is positive.

1. $a > 0$ and $c > 0$

2. $ac > 0 \cdot c$

3. $ac > 0$

4. $ac$ is positive.

1. Definition of *positive*, page 39.

2. __?__

3. __?__

4. __?__

**17.** Prove: If $a \neq 0$, then either $a$ is positive or $-a$ is positive, but not both. By the Comparison Property (page 14) we have *exactly one* of the following cases:

*Case 1.* $a > 0$. Then $a$ is positive by __?__.

*Case 2.* $a = 0$. This case is ruled out by __?__.

*Case 3.* $a < 0$

1. $(-1)a > (-1) \cdot 0$

2. $-a > (-1) \cdot 0$

3. $-a > 0$

4. $-a$ is positive.

1. __?__

2. __?__

3. __?__

4. __?__

Give the solution set over $\Re$ of each inequality.

**Sample.**  $4z > 20$

*Solution:*  $\{z : z > 5\}$

**18.** $-3n > 9$

**19.** $r + 7 \leq 11$

**20.** $\dfrac{k}{6} > -4$

**21.** $\frac{1}{3}y < 7$

**22.** $2y + 1 \leq 5$

**23.** $3 - 3x > 0$

*WRITTEN EXERCISES*

Solve each inequality over $\Re$ and graph its solution set.

**A**

**1.** $z - 3 > 2z + 1$

**2.** $2x - 4 < x - 3$

**3.** $4n - 7 \leq 9$

**4.** $5k + 8 \geq 38$

**5.** $6 - 3z < 2(z + 5)$

**6.** $4 - 5b \geq 3(4 - 2b)$

**7.** $3(5 - r) \leq 3r + 7$

**8.** $-2(r + 6) > 4r - 18$

**9.** $3r - (2 + r) \leq r - (2 + 3r)$

**10.** $-5s + (3 - 2s) \geq 2s + 2(4 - 3s)$

**11.** $12m - 2(3 + 5m) > 7m + 9$

**12.** $23x + 5(2 - 4x) < 3x + 6$

**13.** $3t + 2(2 + 3t) \leq t - 2(t + 7)$

**14.** $4k - 3(k + 5) > 2k + 3(5 - 2k)$

**B**

**15.** $2[3s - 2(s + 1)] > 2s - 3[s + (2 + s)]$

**16.** $-3[x + 4(2 - x)] \leq 2[x - 3(x + 1)] + 8x$

**17.** $x(3 - x) + 2x(x - 1) \leq x^2 - 3x + 12$

**18.** $z(5 - 2z) - 3z(z + 2) > z(3 - 5z) + 6$

**19.** $3y[2 - 2(3 - y)] + 4 \geq 6y(y - 4) - 12$

**20.** $-2p[3p + 4(1 - p)] - 3 < p(3 + 2p) + 21$

Prove each theorem. Assume that the replacement set for each variable is $\Re$.

**21.** If $a < b$ and $c < 0$, then $\dfrac{a}{c} > \dfrac{b}{c}$.

**22.** If $a > b$, then $a - c > b - c$.

**23.** If $a < b$, then $a - b < 0$.

**24.** If $a - b < 0$, then $a < b$.

**25.** If $a - b > 0$, then $a > b$.

**26.** If $a > b$, then $a - b > 0$.

**C**

**27.** If $a < 0$, then $\dfrac{1}{a} < 0$.

**28.** If $a > b$ and $c > d$, then $a + c > b + d$.
(*Hint:* Show that $a + c > b + c$, and $b + c > b + d$.)

**29.** If $a > b > 0$, and $c > d > 0$, then $ac > bd$.

**30.** If $0 < a < b$, then $a^2 < b^2$.

**31.** If $a < b < 0$, then $a^2 > b^2$.

**32.** If $-1 < a < 0$, then $\dfrac{1}{a} < -1$.

## 2–6  Compound Open Sentences

In Section 2–5 you solved inequalities like $x - 3 < 2$, $x - 3 > -1$, and $3x - 1 < x$. When you join two simple open sentences like these by one of the connective words "and" or "or," you obtain a compound sentence. Recall from page 13 that an "*and*-type" compound sentence can sometimes be written as a continued inequality; for example,

$$x - 3 < 2 \quad \text{and} \quad x - 3 > -1$$

can be written as

$$-1 < x - 3 < 2.$$

**Example 1.**   Solve, and graph the solution set.

**a.** $-1 < x - 3 < 2$

**b.** $x - 3 > -1$ or $3x - 1 < x$

*Solution:*   **a.** $-1 < x - 3 < 2$ is equivalent to

$$-1 < x - 3 \qquad \text{and} \qquad x - 3 < 2$$
$$-1 + 3 < x - 3 + 3 \quad \text{and} \quad x - 3 + 3 < 2 + 3$$
$$2 < x \qquad \qquad \text{and} \qquad \qquad x < 5$$

The solution set is $\{x: 2 < x < 5\}$.

**b.**   $\quad x - 3 > -1 \qquad \text{or} \qquad 3x - 1 < x$

$$x - 3 + 3 > -1 + 3 \text{ or } 3x - 1 + 1 - x < x + 1 - x$$
$$x > 2 \qquad \text{or} \qquad 2x < 1, \text{ or } x < \tfrac{1}{2}$$

The solution set is $\{x: x > 2 \text{ or } x < \tfrac{1}{2}\}$.

In solving an "*and*-type" compound sentence, you want all numbers which belong to the solution sets of *both* of the simple sentences. On the other hand, the solution set of an "*or*-type" sentence consists of the numbers belonging to *at least one* of these sets. You can use the language of sets to describe these situations. The **intersection** of two sets $A$ and $B$ (in symbols, $A \cap B$) is the set of all elements belonging to *both* of the given sets, while the **union** of $A$ and $B$ (in symbols, $A \cup B$) is made up of elements belonging to *at least one* of $A$ and $B$. Figure 2–6 applies these ideas to Example 1.

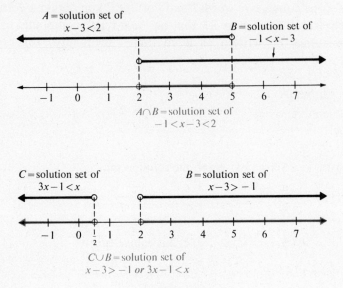

**FIGURE 2–6**

Inequalities involving absolute values, such as $|x - 3| < 2$, can be solved with the help of statements **1**, **2**, and **3** on page 40. Statement **2** is:

$$|a| < p \quad \text{is equivalent to} \quad -p < a < p.$$

By comparison you have:

$$|x - 3| < 2 \quad \text{is equivalent to} \quad -2 < x - 3 < 2.$$

**Example 2.** Solve over $\mathfrak{R}$:

        **a.** $|x - 3| < 2$        **b.** $|x - 3| \geq 2$

*Solution:*     **a.** $|x - 3| < 2$

                $-2 < x - 3 < 2$

                $-2 < x - 3 \quad \text{and} \quad x - 3 < 2$

                $1 < x \qquad \text{and} \qquad x < 5$

          $\therefore$ the solution set is $\{x: 1 < x < 5\}$.

    **b.** By statements **1** and **3**, page 40, $|x - 3| \geq 2$ is equivalent to:

          $x - 3 \leq -2 \quad \text{or} \quad x - 3 \geq 2$

             $x \leq 1 \qquad \text{or} \qquad x \geq 5$

         $\therefore$ the solution set is $\{x: x \leq 1 \text{ or } x \geq 5\}$.

You can shorten the work in part **a** of Example 2 as follows:

$$-2 < x - 3 < 2$$
$$-2 + 3 < x - 3 + 3 < 2 + 3$$
$$1 < x < 5$$

## ORAL EXERCISES

Restate each sentence as an "*or*-type" or as an "*and*-type" compound sentence.

**Sample.** $-3 < 3x < 2$     *Solution:* $-3 < 3x$ and $3x < 2$.

**1.** $-3 < y < 5$     **3.** $|p - 1| > 3$     **5.** $|3x| \leq 4$

**2.** $-6 < 3v + 1 \leq 8$     **4.** $|2y - 2| < 4$     **6.** $|z - 5| \geq 6$

Restate each compound sentence using absolute value notation.

**Sample.** $-3 < x$ and $x < 3$     *Solution:* $|x| < 3$

**7.** $y > 2$ or $y < -2$     **10.** $3z > 5$ or $3z < -5$

**8.** $z + 1 < 5$ and $z + 1 > -5$     **11.** $x - 7 < 8$ and $x - 7 > -8$

**9.** $x + 2 > 3$ or $x + 2 < -3$     **12.** $x - 1 > 0$ or $x - 1 < 0$

## WRITTEN EXERCISES

Graph the solution set of each sentence over $\mathfrak{R}$.

[A]

**1.** $|y| < 3$     **7.** $x - 2 < -1$ or $x + 1 \geq 3$

**2.** $|z| > 1$     **8.** $x + 6 \leq 2$ or $x + 5 > 7$

**3.** $|y - 1| \geq 3$     **9.** $|2x - 3| \leq 11$

**4.** $|z + 3| \leq 5$     **10.** $|3y + 5| \leq 12$

**5.** $-2 < x + 1 < 5$     **11.** $|2 - 3x| > 4$

**6.** $2 \leq x - 2 \leq 7$     **12.** $|1 - 4x| > 7$

Specify by roster each of the following subsets of $J$, the set of integers.

B
13. $\{z: -4 < 3z - 1 \le 8\}$

14. $\{h: -2 \le 4 - 2h < 0\}$

15. $\{n: 8 \ge 4 - 2n \ge -5\}$

16. $\{t: -1 \ge 1 + 2t \ge -7\}$

17. $\{r: 2r + 1 \ge 3 \text{ and } r - 4 < -1\}$

18. $\{x: 5x - 2 \le 8 \text{ and } 2x - 1 > -5\}$

C
19. Prove that if $x, b, c \in \mathcal{R}$, and $|x + b| < c$, then $-c - b < x < c - b$.

20. Prove that if $x, b, c \in \mathcal{R}$, and $|x - b| \le c$, then $-c + b \le x \le c + b$.

## PRACTICAL APPLICATIONS

## 2–7  Problem Solving

By using expressions and sentences involving variables, you can often find a mathematical description, or **model** of a practical problem so that you can solve it by algebraic methods. Notice how, in the examples below, English phrases and sentences about numbers are translated into mathematical language.

**Example 1.**  The sum of the perimeters of a square and an equilateral triangle is 20 feet, and the edges of the triangle are each two feet longer than those of the square. How long are the edges of the square?

*Solution:*

1. The first step in the solution is to introduce a variable to use in representing the numbers described in the problem. (A sketch may be helpful in this step and the next.)

Let $x$ = the length in feet of each edge of the square.

Then $x + 2$ = the length in feet of each edge of the triangle.

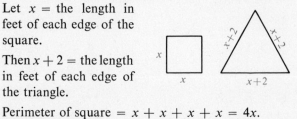

Perimeter of square $= x + x + x + x = 4x$.

Perimeter of triangle $= (x + 2) + (x + 2) + (x + 2)$
$$= 3(x + 2).$$

2. The second step is to write an open sentence showing the relationship given in the problem.

| Perimeter of the square | plus | perimeter of the triangle | is | 20. |
|:---:|:---:|:---:|:---:|:---:|
| ↓ | ↓ | ↓ | ↓ | ↓ |
| $4x$ | $+$ | $3(x + 2)$ | $=$ | $20$ |

3. The third step is to solve the open sentence.

$$4x + 3x + 6 = 20$$
$$7x + 6 = 20$$
$$7x = 14, \quad x = 2$$

∴ each edge of the square is 2 feet long.

*Check:*   4. The last step is to check your result with the requirements stated in the original problem.

If each side of the square is 2 feet long, its perimeter is $4 \cdot 2$, or 8, feet. Each side of the triangle is $2 + 2$, or 4, feet long; its perimeter therefore is $3 \cdot 4$, or 12, feet. Since $8 + 12 = 20$, the conditions of the problem are met.

The four steps used in Example 1 provide a general outline for dealing with many "word problems."

**Example 2.**   A student will receive a "B" grade in a certain course if he has an average of 75 on three examinations. He has grades of 65 and 82 on the first two examinations. At least how high a grade must he make on the third to receive a "B"?

*Solution:*   1. Let $g$ = grade student must make on third examination. Then $\frac{1}{3}(65 + 82 + g)$ = average on the three examinations.

2.
| The average | must be at least | 75. |
|:---:|:---:|:---:|
| | | ↓ |
| $\frac{1}{3}(65 + 82 + g)$ | $\geq$ | $75$ |

3.
$$3 \cdot \tfrac{1}{3}(65 + 82 + g) \geq 3 \cdot 75$$
$$65 + 82 + g \geq 225$$
$$g \geq 225 - 147$$
$$g \geq 78$$

∴ his grade on the third examination must be at least 78.

*Check:*  **4.** If $g \geq 78$, then

$$65 + 82 + g \geq 65 + 82 + 78 = 225.$$

$\therefore$ average $= \frac{1}{3}(65 + 82 + g) \geq \frac{1}{3} \cdot 225 = 75.$
If $g < 78$, then

$$65 + 82 + g < 65 + 82 + 78 = 225.$$

$\therefore$ average $= \frac{1}{3}(65 + 82 + g) < \frac{1}{3} \cdot 225 = 75.$
Hence the conditions are met if and only if $g \geq 78$.

Charts are often helpful in organizing the data in a problem.

**Example 3.**  Originally there were five times as many men as women at a meeting. Later, after three more men and 12 more women arrived, there were twice as many men as women. How many men and how many women were present originally?

*Solution:*  **1.** Let $w$ be the number of women present originally. Then

|  | Women | Men |
|---|---|---|
| **Number present originally** | $w$ | $5w$ |
| **Number present later** | $w + 12$ | $5w + 3$ |

**2.** Later, the number of men was twice the number of women.

$$5w + 3 \quad = \quad 2 \times \quad (w + 12)$$

**3.** $5w + 3 = 2w + 24$
$3w = 21$
$w = 7 \qquad 5w = 35$

$\therefore$ 7 women and 35 men were present originally.

*Check:*  **4.** Is 35 five times 7? Yes.
Is $35 + 3$ twice $7 + 12$? Yes, since $38 = 2 \times 19$.

Formulas taken from geometry and the sciences are often used in solving problems.

**Example 4.**  At noon a plane leaves Central City Airport and flies due east at 350 miles per hour (mph). At 1 P.M. a second plane leaves Anthony Airport, 250 miles due west of Central City Airport and flies east at 650 mph. At what time will the second plane overtake the first?

*Solution:*

1. Let *t* be the number of hours after noon when the second plane overtakes the first.

We use the formula

$$\text{distance} = \text{rate} \times \text{time}$$

and arrange some of the given facts in a chart.

|  | Rate (mph) | Time in flight (hrs) | Distance flown (miles) |
|---|---|---|---|
| **First plane** | 350 | $t$ | $350t$ |
| **Second plane** | 650 | $t - 1$ | $650(t - 1)$ |

The diagrams below may be helpful. (Distances are in miles.)

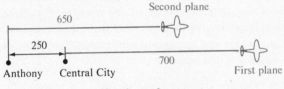

Situation at 2 P.M.

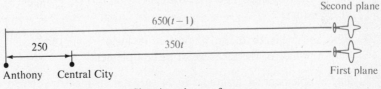

Situation *t* hours after noon

2. The second plane will overtake the first when the distances of the two planes from Anthony Airport are the same. Therefore the requirement is:

| Distance of the first plane from Anthony Airport. | is the same as | the distance of the second plane from Anthony Airport. |
|---|---|---|
| $350t + 250$ | $=$ | $650(t - 1)$ |

Steps **3** and **4** are left to you. The result will be $t = 3$, that is, the second plane will overtake the first at 3 P.M.

## PROBLEMS

Table I in the appendix gives several useful facts from geometry. Use it as necessary to complete the problems in these exercises. Recall that two angles are **complementary** if the sum of their degree measures is 90, **supplementary** if this sum is 180. The sum of the degree measures of three angles of a triangle is 180. An **isosceles** triangle is a triangle in which two angles have the same degree measure.

**A**
1. Find three consecutive odd integers whose sum is 57.

2. Find four consecutive even integers whose sum is 68.

3. Find the dimensions of a rectangle whose length is 3 feet greater than twice its width if its perimeter is 54 feet.

4. The length of the base of an isosceles triangle is 4 inches less than the length of one of the two congruent sides of the triangle. Find the length of each side of the triangle if its perimeter is 32 inches.

5. The length of the base of an isosceles triangle is 2 inches less than the length of each of the congruent sides of the triangle. If each side of a square is equal in length to the base of the triangle, and if the sum of the perimeter of the triangle and the square is 60 inches, find the dimensions of both the triangle and the square.

6. The base of an isosceles triangle rests on top of a square. If the perimeter of the triangle is 24 inches, and the perimeter of the five-sided figure is 36 inches, what is the perimeter of the square?

7. Find the measure of an angle whose measure is 18° less than one-half of the measure of its complement.

8. The supplement of an angle is 24° greater in measure than twice the complement of the angle. Find the measure of the angle.

9. One angle of a triangle measures 11° less than twice the measure of a second angle. This second angle measures 23° less than the third angle. Find the measure of each angle in the triangle.

10. The sum of the measures of two congruent angles in an isosceles triangle exceeds the measure of the third angle by 32°. Find the measure of each angle in the triangle.

11. A tape-cassette contains 13 musical compositions. Of these 13 compositions, 8 of them last for the same length of time and each of the other 5 lasts 1 minute longer than any one of the 8. There are 15 seconds between compositions. How many minutes does each composition last if the entire tape takes 73 minutes to play?

12. In a run-off election between two candidates, a total of 25,200 votes were cast. If 300 voters had switched from the winner to the loser, the loser would have won by 200 votes. How many votes did each candidate actually receive.

13. The budget of the Central City Chapter of the League of Women Voters was increased by $350. If the dues are raised from $5 to $6, how many new members must be acquired to meet the budget if there were 90 members in the chapter to begin with?

14. A smog control device on an automobile reduces the emission of an air pollutant by 64 ppm (parts per million) per hour. After installation the car takes 12 hours to emit the same amount of pollutant it formerly emitted in 4 hours. How many ppm did it originally emit per hour?

15. Mrs. Chambers has $12,000 in the bank with 5% interest paid annually. If the bank drops its rate to 4% how much additional money must Mrs. Chambers deposit to maintain the same return each year?

16. Mrs. Katz invested $2000, part at 6% and the remainder at 8%. Find the amount invested at each rate, if her yearly return from the two investments is $132.

17. One gravel truck has a capacity 4 tons greater than that of a smaller truck. If the smaller truck makes 5 trips and the larger truck makes 7 trips, together they can transport 112 tons of gravel. Find the capacity of each truck.

18. A 10-ton truck and a 12-ton truck deliver sand to a concrete mixing plant. If the smaller truck makes 3 trips more than the larger truck, how many trips does each truck make to deliver 140 tons of sand?

19. Mr. Call and his wife are moving to a distant city. They leave their old home at the same time traveling in two cars. Since Mr. Call is pulling a trailer, he can only travel one-half as fast as his wife. If they are 96 miles apart after 3 hours on the road, how fast is each driving?

20. A freight train traveling at 40 miles per hour leaves Anson City at 9:00 A.M. At 11:00 A.M. a passenger train leaves Anson City on a parallel track and travels the same direction as the freight train at 80 miles per hour. At what time does the passenger train pass the freight train?

21. Mary has grades of 98, 76, 86, and 92 on four tests. Within what range must her grade on a fifth test fall if she wishes an average between 80 and 90 on all five tests?

22. Jane can sail upstream in a river at an average rate of 4 miles per hour, and downstream at an average rate of 6 miles per hour. If she starts at 10:00 A.M., within what period of time must she turn around if she is to return to her point of departure between 6:00 P.M. and 8.00 P.M.?

**B** **23.** In processing a certain metal, an acid bath used in the treatment must be kept between 40% and 50% acid. How much of a 10% solution of acid must be added to 200 gallons of a 60% acid solution to reduce the bath to its proper range?

**24.** A large rectangular piece of fabric is washed to pre-shrink it, and is woven so that it shrinks in one direction only, along its length. The fabric was cut so that originally it was 4 times as long as it was wide. If the fabric shrinks 20 inches in length and decreases in area by 2400 square inches, what were its original demensions.

**25.** In serving dinners on an airplane, it takes twice as long for a stewardess to serve a meal to a passenger in the first-class section as it does to serve a meal to a passenger in the coach section. On a two-hour flight, what is the greatest amount of time a stewardess can average serving a coach passenger on a flight carrying 15 first class passengers and 30 coach passengers?

**26.** The cooling system of a stationary engine contains 16 quarts of coolant which is 35% antifreeze. How much of this coolant must be drained off and replaced with pure antifreeze if the resulting 16 quarts of liquid is to be between 50% and 60% antifreeze?

---

## Chapter Summary

---

**1.** The **absolute value** of a nonzero real number $a$ is the positive number of the pair $a$ and $-a$. The absolute value of 0 is 0. On the number line, $|a|$ measures the distance between the graph of $a$ and the origin.

**2.** Real numbers can be depicted by **arrows** on the number line (page 40), and sums of numbers can be constructed geometrically using arrows.

**3.** **Subtraction** of real numbers is defined in terms of addition and additive inverses:

$$a - b = a + (-b)$$

**Division** of real numbers is defined in terms of multiplication and multiplicative inverses:

$$a \div b = a \cdot \frac{1}{b} \quad (b \neq 0)$$

Division by 0 is not defined.

**4.** The **Property of the Negative of a Sum**,

$$-(a + b) = -a + (-b),$$

together with the real-number axioms and familiar number facts, is used in finding sums and differences of real numbers. You have the following rules for adding real numbers:
1. To add two numbers of the same sign, add their absolute values and prefix their common sign.
2. To add two numbers of opposite sign, subtract the lesser absolute value from the greater, and prefix the sign of the number having the greater absolute value.

**5.** The following rules are used in determining products and quotients of real numbers:
1. The absolute value of a product of two or more numbers is the product of their absolute values.
2. A product of nonzero numbers is positive or negative according as the number of negative factors is even or odd.
3. The quotient $\frac{a}{b}$ is positive or negative according as $a$ and $b$ have the same or opposite sign.

**6.** The following rules concern the order in which operations are to be performed.
1. Perform the operations within each set of grouping symbols, beginning with the innermost ones.
2. Whenever there is a choice between performing an addition or subtraction on the one hand or a multiplication or division on the other, perform the multiplication or division first.

**7.** **Algebraic expressions** are **equivalent** if they have the same value for every numerical replacement of the variables. Open sentences are **equivalent** if they have the same solution set. If the members of an equation are equivalent expressions, then the equation is an **identity**. **To solve an equation**, transform it into an equivalent equation whose solution set can be determined by inspection. The following transformations produce equivalent equations:
1. Adding to or subtracting from each member of an equation the same number or the same expression in any variables appearing in the equation.
2. Multiplying or dividing each member by the same *nonzero* number.
3. Replacing either member by an expression equivalent to it. To check that you have solved an equation correctly, substitute each root for the variable in the original equation and show that both members of the equation reduce to the same number.

8. The **Addition and Multiplication Axioms of Order** and the properties of equality guarantee that the following transformations used in solving inequalities produce equivalent inequalities:
   1. Adding to or subtracting from each member of an inequality the same number or the same expression in the variable appearing in the inequality.
   2. Multiplying or dividing each member by the same positive number.
   3. Multiplying or dividing each member by the same negative number and reversing the inequality sign.
   4. Substituting for either member an expression equivalent to it.

9. The solution set of an "*and*-type" compound sentence is the **intersection** of the solution sets of the two simple sentences. The solution set of an "*or*-type" compound sentence is the **union** of the solution sets of the simple sentences.

10. Some practical problems can be restated in mathematical language and then solved by algebraic methods. Four steps to use in solving such problems are as follows.
    1. Introduce a variable to use in representing the numbers described in the problem.
    2. Write an open sentence showing the given relationship.
    3. Solve the open sentence.
    4. Check the results with the requirements stated in the problem.

### Vocabulary and Spelling

Review each term by reference to the page listed.

positive number (*p. 39*)
negative number (*p. 39*)
absolute value (*p. 39*)
magnitude (*p. 40*)
arrow (*p. 40*)
sign of a number (*p. 45*)
algebraic expression (*p. 54*)
equivalent expressions (*p. 55*)
identity (*p. 55*)
equivalent open sentences (*p. 55*)
converse (*p. 55*)

if and only if (*p. 55*)
transformations of equations (*p. 56*)
sense, or direction, of inequality (*p. 59*)
equivalent inequalities (*p. 61*)
transformations of inequalities (*p.61*)
compound open sentence (*p. 65*)
intersection of sets (*p. 65*)
union of sets (*p. 66*)
mathematical model (*p. 68*)

## Chapter Test

**2–1**   **1.** Which of the following are true statements for all real numbers $a$?
   **a.** $|-a| = |a|$   **b.** $|a| > 0$   **c.** $-|a| \leq |a|$

   **2.** Graph the solution set over $\Re$ of the inequality $|x| \geq 2$.

   **3.** Construct the sum $-3 + 5$ geometrically, using arrows on the number line.

**2–2**   Find the indicated sums.

   **4.** $10 - (3 - 5)$                **5.** $-7 + 5 - 30 + 15 - 3$

   Supply the missing reasons in the proof.

   **6. a.** $-5 + 2 = -(3 + 2) + 2$           **a.** Addition fact
       **b.**          $= [-3 + (-2)] + 2$     **b.** _?_
       **c.**          $= -3 + [(-2) + 2]$     **c.** _?_
       **d.**          $= -3 + 0$              **d.** _?_
       **e.**          $= -3$                  **e.** _?_

**2–3**   Simplify each expression.

   **7.** $(-5)(2)(-\frac{1}{2})(-3)$        **8.** $\dfrac{(72)(-2) + 2}{(-9)(-10 + 6)}$

**2–4**   **9.** Solve for $x$: $3(x + 5) = 9x + 24$

   Solve each inequality over $\Re$ and graph the solution set.

**2–5**   **10.** $2(2 - 2z) - 7 \geq 6z + 2$

**2–6**   **11.** $|3x + 2| < 11$              **12.** $|1 - 2y| \geq 5$

**2–7**   **13.** Betty made a rectangular pen for her dog using 28 feet of fencing. If the width of the pen is 2 feet more than one-half the length, what are the dimensions of the pen?

   **14.** On the morning of an important business conference, Mr. Adams dashed out of the house onto the bus that stopped outside, accidentally leaving his briefcase behind. His wife discovered the briefcase 5 minutes later and set out in the family car to overtake the bus. If the bus traveled at 20 mph and Mrs. Adams, traveling the same route, drove at 40 mph, how long did it take her to catch up to the bus?

## Chapter Review

### 2–1 Absolute Value    Pages 39–43

1. The absolute value of every nonzero number $a$ is the __?__ number of the pair $a$ and $-a$. Thus if $a > 0$, $|a| = $ __?__, and if $a < 0$, $|a| = $ __?__. The absolute value of 0 is __?__.

2. The solution set over $\Re$ of the equation $|a| = 3$ is __?__.

3. On the number line the real number $a$ is represented by any arrow of length __?__, pointing to the __?__ if $a > 0$ or to the __?__ if $a < 0$.

Which of the following sentences are true for all real numbers?

4. If $a < 0$, then $-a > 0$.

5. $a + |a| \geq 0$.

6. $|a| < 5$ is equivalent to "$a > 5$ or $a < -5$."

Graph each nonempty solution set over $\Re$. If the solution set is $\emptyset$, state this.

7. $|z| \leq 3$      8. $|-y| = |-3|$      9. $|a| < a$

### 2–2 Addition and Subtraction    Pages 44–48

10. The negative of the sum of real numbers $a$ and $b$ is the sum of the __?__ of $a$ and $b$.

11. The difference $a - b$ is equal to the sum of $a$ and the __?__ __?__ of $b$.

12. Is subtraction of real numbers (a) commutative, (b) associative?

Simplify each expression.

13. $-(-10) + (-3)$

14. $12 - (3 - 5)$

15. $6 - 27 + 4 - 3 - (-2)$

16. $-8.8 + 17.5 + (-5.2) - 3([-1.1] - 0.2)$

17. State the property justifying each step. Assume that $a \in \Re$ and $b \in \Re$.

1. $-(a - b) = -[a + (-b)]$     1. __?__
2. $\qquad = -a + [-(-b)]$     2. __?__
3. $\qquad = -a + b$     3. __?__

**2–3  Multiplication and Division**    *Pages 49–54*

**18.** A product of nonzero numbers is positive if the number of negative factors is __?__. A product of nonzero numbers is negative if the number of negative factors is __?__.

**19.** The product of the absolute values of two numbers is equal to the absolute value of the __?__ of the two numbers.

**20.** Dividing a real number $a$ by a nonzero real number $b$ is equivalent to __?__ $a$ by the __?__ __?__ of $b$.

Simplify each expression.

**21.** $-36 \div 9$      **22.** $0 \div (-2)$      **23.** $\dfrac{3[5 + (6 \div 3 - 1)]}{2[4 - (1 + 3 \cdot 4)]}$

**2–4  Equivalent Equations**    *Pages 54–59*

**24.** Equations are equivalent if they have the same __?__ __?__.

**25.** What is the converse of the statement "If $a$ and $b$ are positive, then $ab$ is positive"? Is the converse true?

**26.** Solve $3(a + 5) = 7 - a$.

**27.** If $S = 2\pi rh$, find $r$ when $h = 2$ and $S = 24$. Use $\pi = \frac{22}{7}$.

**2–5  Equivalent Inequalities**    *Pages 59–64*

**28.** If $a < b$ and $c > 0$, then $a + c$ __?__ $b + c$ and $ac$ __?__ $bc$. If $a < b$ and $c < 0$, then $a + c$ __?__ $b + c$ and $ac$ __?__ $bc$.

**29.** Solve over $\Re$ and graph the solution set: $2(2 - t) + 16 < 8t + 10$.

**2–6  Compound Open Sentences**    *Pages 65–68*

**30.** The __?__ of sets $A$ and $B$ is the set of all elements belonging to at least one of $A$ and $B$; the __?__ of sets $A$ and $B$ is set of all elements belonging to both $A$ and $B$.

Graph the solution set over $\Re$ of each inequality.

**31.** $-11 < 2x - 5 \leq 5$      **32.** $|9 - z| > 3$

**2–7  Problem Solving**    *Pages 68–74*

**33.** At the beginning of the school year there were 6 more boys than girls in the mathematics class. At midyear, after 2 new girls joined the class and 1 boy left, the number of girls was 6 more than two-fifths the number of boys. How many boys and girls were in the class originally?

# Explorations with a Computer

(To be used if you have access to an electronic computer that will accept BASIC.)

In the next chapter, you will be working with open sentences in two variables which can be expressed in the form:

$$AX + BY = C$$

If you put the following program into your computer, and input the values 1, 1, 6 for A, B, C, you should get the print-out shown at the top of page 81. (Note that this program is for $A \neq 0$ and $B \neq 0$.)

```
10    PRINT "WHAT VALUES DO YOU WANT FOR A(<>0), B(<>0), C";
15    INPUT A,B,C
20    PRINT
25    LET X1=C/A
30    LET Y1=C/B
35    PRINT A;"X +";B;"Y =";C,"Y-INTERCEPT =";Y1
40    PRINT
45    LET E=INT(ABS(X1))+5
50    IF E>15 THEN 170
55    LET D=INT(ABS(Y1))+5
60    LET M=2*E
65    LET N=2*M
70    PRINT TAB(M);"Y"
75    FOR Y=D TO -D STEP -1
80    LET X=(C-B*Y)/A
85    LET X2=2*X+M
90    IF X2>M THEN 145
95    IF X2=M THEN 130
100   IF X2<0 THEN 160
105   IF Y=0 THEN 120
110   PRINT TAB(X2);"*"; TAB(M);"!"; TAB(N+3);"(";X;",";Y;")"
115   GO TO 160
120   PRINT "+-+-+-+";TAB(X2);"*"; TAB(N-6);"+-+-+-+X";
121   PRINT TAB(N+3);"(";X;",";Y;")"
125   GO TO 160
130   IF Y=0 THEN 120
135   PRINT TAB(X2);"*"; TAB(N+3);"(";X;",";Y;")"
140   GO TO 160
145   IF X2>N THEN 160
150   IF Y=0 THEN 120
155   PRINT TAB(M);"!"; TAB(X2);"*"; TAB(N+3);"(";X;",";Y;")"
160   NEXT Y
165   GO TO 175
170   PRINT "TOO WIDE"
175   END
```

```
WHAT VALUES DO YOU WANT FOR A(<>0), B(<>0), C?1,1,6

1X + 1Y = 6    Y-INTERCEPT = 6
```

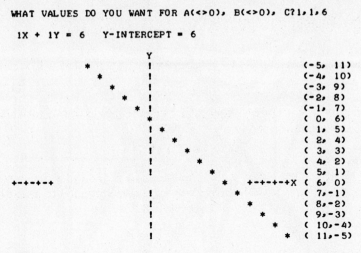

```
                        Y
            *           !                        (-5, 11)
              *         !                        (-4, 10)
                *       !                        (-3, 9)
                  *     !                        (-2, 8)
                    * ! !                        (-1, 7)
                      *                          ( 0, 6)
                      ! *                        ( 1, 5)
                      !   *                      ( 2, 4)
                      !     *                    ( 3, 3)
                      !       *                  ( 4, 2)
                      !         *                ( 5, 1)
+-+-+-+               !           *   +-+-+-+X   ( 6, 0)
                      !             *            ( 7,-1)
                      !               *          ( 8,-2)
                      !                 *        ( 9,-3)
                      !                   *      ( 10,-4)
                      !                     *    ( 11,-5)

END
```

If you input 3, −5, 7 for A, B, C the print-out should look like this:

```
WHAT VALUES DO YOU WANT FOR A(<>0), B(<>0), C? 3,-5, 7

3X +-5Y = 7    Y-INTERCEPT =-1.4

                Y
                !           *        ( 5.66667, 2)
                !         *          ( .4, 1)
+-+-+-+                 *    +-+-+-+X ( 2.33333, 0)
                !*                   ( .666667,-1)
          * !                        (-1,-2)
        *       !                    (-2.66667,-3)
      *         !                    (-4.33333,-4)
    *           !                    (-6,-5)

END
```

**1.** RUN the program with the following inputs:

$$1, -1, 6 \qquad 1, 1, -6 \qquad 1, -1, -6$$

Compare these graphs with the graph at the top of the page. What do you observe?

**2.** RUN the program with the following inputs:

$$3, 5, 7 \qquad 3, 5, -7 \qquad 3, -5, -7$$

Compare these graphs with the second graph above. What do you observe?

**3.** RUN the program to graph the following equations.
   **a.** $2X + 3Y = 8$         **b.** $.75X - .25Y = 1$

**4.** Make up some equations and plot their graphs.
   (Remember that for this program, $A \neq 0$ and $B \neq 0$.)

# 3

# Systems
# of
# Linear
# Open
# Sentences

## LINES AND LINEAR EQUATIONS

### 3–1  Open Sentences in Two Variables

Mathematical descriptions of practical situations may call for the use of two or more variables. Suppose, for example, that you are studying the motion of a particle which starts at a point $P$ on a line and moves along the line for three seconds with a certain velocity and then for five seconds with another velocity. In order to write an expression for the net displacement of the particle from $P$, you use two variables, say $x$ and $y$, to denote the first and second velocities, respectively, both in centimeters per second. Then the net displacement in centimeters is $3x + 5y$ (Figure 3–1 indicates two possibilities). The replacement set of

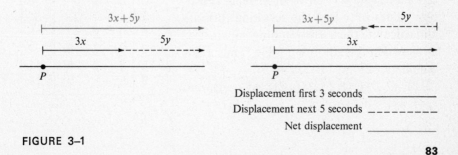

Displacement first 3 seconds _____
Displacement next 5 seconds _ _ _ _ _ _ _ _
Net displacement _____

FIGURE 3–1

each variable is $\Re$ because during each time interval the particle may travel at *any* rate in *either* direction. (Motion to the right corresponds to positive velocity and motion to the left corresponds to negative velocity.)

Suppose that you measure the net displacement of the particle to be 40 centimeters. Then

$$3x + 5y = 40.$$

To find a **solution** of such an open sentence in *two* variables, you must determine a *pair* of values, one for $x$ and a corresponding one for $y$, which make the equation true. We write these corresponding values as **ordered pairs** of the form $(x, y)$; $x$ is called the **first coordinate** and $y$ the **second coordinate** of the ordered pair. For example, $(10, 2)$ is a solution of $3x + 5y = 40$ because $3 \cdot 10 + 5 \cdot 2 = 40$ is a true statement.

Two ordered pairs are **equal** if their first coordinates are equal and their second coordinates are equal. Thus

$$(10, 2) = (2 \cdot 5, 1 + 1),$$

but

$$(10, 2) \neq (2, 10).$$

Notice that although $(10, 2)$ is a solution of $3x + 5y = 40$, $(2, 10)$ is not, because $3 \cdot 2 + 5 \cdot 10 \neq 40$.

The **solution set** of an open sentence in two variables is the set of all ordered pairs of numbers which belong to the domains of the variables and for which the sentence is true. Notice that the equation

$$3x + 5y = 40$$

has an *infinite* solution set. This means that you can find as many solutions of the equation as you please. To do this, first transform the equation into an equivalent one expressing $y$ in terms of $x$. Then substitute values of $x$ into the resulting formula and calculate the corresponding values of $y$, as shown in the adjoining table. Thus a few members of $\{(x, y): 3x + 5y = 40\}$ are $(0, 8)$, $(2, 6\frac{4}{5})$, $(5, 5)$, $(10, 2)$, $(15, -1)$, and $(-5, 11)$.

$$3x + 5y = 40$$
$$5y = 40 - 3x$$
$$y = 8 - \tfrac{3}{5}x$$

| $x$ | $8 - \tfrac{3}{5}x$ | $y$ |
|---|---|---|
| 0 | $8 - \tfrac{3}{5} \cdot 0$ | 8 |
| 2 | $8 - \tfrac{3}{5} \cdot 2$ | $6\frac{4}{5}$ |
| 5 | $8 - \tfrac{3}{5} \cdot 5$ | 5 |
| 10 | $8 - \tfrac{3}{5} \cdot 10$ | 2 |
| 15 | $8 - \tfrac{3}{5} \cdot 15$ | $-1$ |
| $-5$ | $8 - \tfrac{3}{5}(-5)$ | 11 |

## ORAL EXERCISES

State whether or not the ordered pairs are equal.

**1.** $(3, 5), (5, 3)$    **3.** $(2, 5 - 1), (6 \div 3, 4)$    **5.** $(0, 2), (0, |-2|)$

**2.** $(3, 3), (5, 5)$    **4.** $(2 \times 3, 0), (3 \times 2, 0)$    **6.** $(-3, 0), (|-3|, |0|)$

Is the given ordered pair $(x, y)$ a solution of the given equation?

**7.** $2x + 3y = 0; \ (x, y) = (0, 0)$

**8.** $7x - 5y = -7; \ (x, y) = (-1, 0)$

**9.** $40x - 20y = 10; \ (x, y) = (0, \frac{1}{2})$

**10.** $5x + 3y = 13; \ (x, y) = (2, 1)$

## WRITTEN EXERCISES

Transform each equation into an equivalent one expressing $y$ in terms of $x$. Then find the value of $y$ when $x$ is **(a)** $-1$, **(b)** 0, **(c)** 1, **(d)** 2.

**A**

**1.** $6x + 3y = 9$        **4.** $7x + 2y = 0$

**2.** $15x - 5y = 30$        **5.** $2x + 3y = 10$

**3.** $3x - 5y = 0$        **6.** $5x + 4y = 11$

Find three solutions of each sentence over the specified domain. Give coordinates in alphabetical order of the variables involved. For example, if the problem involves the variables $r$ and $s$, give coordinates in the order $(r, s)$.

**7.** $3x + 4y = 16; \ \Re$        **12.** $3r + 5s = -2; \ \{\text{integers}\}$

**8.** $5x - 2y = 12; \ \Re$        **13.** $x + 4z = 7 - 2x; \ \{\text{integers}\}$

**9.** $x + y = 6; \ \{\text{integers}\}$        **14.** $2\ell + 3m = 5 - 2\ell; \ \{\text{integers}\}$

**10.** $2x + y = 10; \ \{\text{integers}\}$        **15.** $3a + 12 = 3(a - b); \ \{\text{integers}\}$

**11.** $2x + 3y = 12; \ \{\text{integers}\}$        **16.** $2(u - v) = 6 - 2v; \ \{\text{integers}\}$

**17.** $-3h + 2(h - k) = -(h + k) - k; \ \Re$

**18.** $5p - 3(p - q) = 2q - (p - q) + 3p; \ \Re$

B **19.** $3x - 2y \leq 2$; $\mathcal{R}$

**20.** $9a - 2b \geq -3$; $\mathcal{R}$

**21.** $3(2x + y) > 0$; {integers}

**22.** $4(2u - v) < 0$; {integers}

**23.** $|x| + |y| = 6$; {negative integers}

**24.** $|x| - |y| = -2$; {negative integers}

**25.** $|a + b| < |a| + |b|$; $\mathcal{R}$

**26.** $|a + b| = |a| - |b|$; $\mathcal{R}$

**Find all real values of $r$ and $s$ for which the ordered pairs are equal.**

**27.** $(r, s) = (4, 2 - s)$

**28.** $(r, s) = (2 - r, -s)$

**29.** $(1 - r, 2 - s) = (-2r, s)$

**30.** $(r + 1, 3s - 2) = (2r - 3, 5 - 4s)$

C **31.** $(|r|, s) = (2 - |r|, -s)$

**32.** $(r, |s|) = (-r, 3 - s)$

**33.** $(|r|, |s|) = (-2 - r, 3 - |s|)$

**34.** $(|r|, |s|) = (4 - |r|, 4 + |s|)$

*PROBLEMS*

**Solve each of the following problems.**

**Sample.** The units digit in a certain two-digit numeral is 5 less than twice the tens digit. Find all positive integers represented by such numerals.

*Solution:* Recall that every nonnegative integer can be expressed as a sum of different powers of 10, each multiplied by a number belonging to the set $\{0, 1, 2, 3, 4, 5, 6, 7, 8, 9\}$.
For example, $304 = 3 \cdot 10^2 + 0 \cdot 10 + 4$.

1. Since the numeral described in the problem has 2 digits, let $t =$ tens digit and $u =$ units digit. Then the corresponding integer is $10t + u$.

2. The units digit is twice the tens digit less five.

$$u \quad = \quad 2t \quad - \quad 5$$

3. Since the domain of $t$ is $T = \{1, 2, 3, 4, 5, 6, 7, 8, 9\}$ and the domain of $u$ is $U = \{0, 1, 2, \ldots, 9\}$, you can show by substitution that

$$\{(t, u): t \in T, u \in U, \text{ and } u = 2t - 5\}$$
$$= \{(3, 1), (4, 3), (5, 5), (6, 7), (7, 9)\}.$$

Hence the required integers are 31, 43, 55, 67, and 79.

*Check:*    4. Is 31 a two digit numeral?  Yes.
Is the units digit 5 less than twice the tens digit?  Yes, because $1 = 2 \cdot 3 - 5$.

Checks for the other numbers are left to you.

---

**A**  1. The absolute value of the sum of two numbers is 10.  Find a pair of integers satisfying this condition with (**a**) both numbers positive, (**b**) both numbers negative, (**c**) one number positive and the other negative.

2. The average of Sam's scores on three tests, each having a maximum of 100 points and a minimum of 0 points, is 89.  If he scored 5 points higher on the last test than he did on the first, find two possibilities for his scores on the three tests.

3. Sara has $1.15 in change in her pocket.  If this change is composed solely of dimes and quarters, what are the possibilities for the numbers of each?

4. If the digits of a two-digit numeral are reversed, then the corresponding positive integer is 9 greater than the positive integer represented by the original numeral.  Find all possibilities for the positive integer represented by the original numeral.

5. The hundreds digit of a three-digit numeral is twice the units digit and the sum of the digits is 10.  Find all positive integers represented by such numerals.

6. The perimeter of a rectangle is 16 inches.  Find all integral possibilities for the number of inches in its length and its width.

**B**  7. The perimeter of an isosceles triangle is 16 inches.  Find all integral possibilities for the lengths (in inches) of the sides.

8. The sum of two integers is 5.  The absolute value of the sum of the integers is equal to the difference of the absolute values of the integers.  Find two such pairs of integers.

**9.** Bill has $2.50 in change in his pockets. If this change is composed solely of nickels, dimes, and quarters, and if the number of quarters is 2 less than the number of nickels, find two possibilities for the number of nickels, dimes, and quarters in his pockets. (Assume he has at least one of each coin.)

**10.** A dietician planning a meal wants the combined amounts of Vitamin A and Vitamin D per serving to be 2400 U.S.P. units, with the amount of Vitamin A between 1000 and 3000 units and the amount of Vitamin D between 300 and 500 units. Find two possibilities for the number of units of each vitamin.

**11.** Elm City recycled 8 tons of newspaper, glass, and cans last month. If there was twice as much glass as cans, find all integral possibilities for the weights of the recyclables in tons.

## 3–2　Cartesian Coordinates

Solution sets of open sentences in *one* variable can be graphed on a line, that is, a *one*-dimensional space. To graph the solution set of an open sentence in *two* variables, like $3x + 5y = 40$, you need a *two*-dimensional space, that is, a plane. To introduce a **rectangular** or **Cartesian*** **coordinate system** in a plane, choose two reference lines, called **axes**, intersecting at right angles at a point $O$, the **origin**. Using convenient units of length (ordinarily, but not necessarily, the same for both axes) make each axis into a number line with its zero point at $O$. You usually consider one axis (the $x$-axis) horizontal and the other (the $y$-axis)

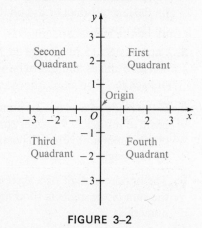

FIGURE 3–2

vertical, with positive and negative numbers assigned to points on the axes, as shown in Figure 3–2. The figure also shows the numbering of the **quadrants** into which the axes divide the plane.

　To associate a point in the plane with an ordered pair of numbers, such as $(-5, 11)$ in the solution set of $3x + 5y = 40$, draw (perhaps mentally) a vertical line through the point $-5$ on the $x$-axis and a

---

*In honor of René Descartes (1596–1650), a French mathematician and philosopher who introduced coordinates.

horizontal line through the point 11 on the $y$-axis (Figure 3–3). The point of intersection of these lines is the **graph** of $(-5, 11)$ and is called "the point $(-5, 11)$." In Figure 3–3 we locate, or **plot**, the graphs of several ordered pairs in the solution set of $3x + 5y = 40$ (see page 84).

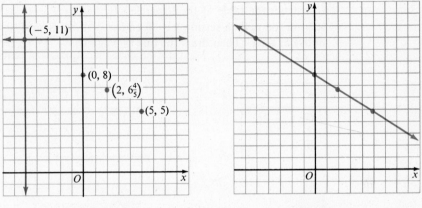

FIGURE 3–3         FIGURE 3–4

By reversing the procedure described above, you can start with any point in the plane and find an ordered pair corresponding to it. The first coordinate of the ordered pair is the **abscissa** of the point, and the second coordinate is its **ordinate**.

The **graph of an open sentence** is the set of all points whose coordinates satisfy the open sentence. Figure 3–3 suggests that the graph of $3x + 5y = 40$ is the straight line shown in Figure 3–4. That this is indeed the case follows from a theorem of analytic geometry:

---

**THEOREM.**    The graph of every equation of the form

$$Ax + By = C,$$

where not both $A$ and $B$ are zero, is a straight line. Conversely, every straight line in the plane is the graph of an equation of the form $Ax + By = C$.

---

Equations of the form $Ax + By = C$ are called **linear equations**. An **equation of a line** is one whose graph is the given line. Thus $3x + 5y = 40$ and $y = 8 - \frac{3}{5}x$ are both equations of the line pictured in Figure 3–4. You often say, for example, "the line $3x + 5y = 40$" rather than "the line whose equation is $3x + 5y = 40$."

Although you need plot only two points to determine the graph of a linear equation, it is good practice to plot a third point as a check. Points having at least one coordinate zero are easy to plot.

**Example 1.** Graph (that is, *draw the graph of*) $4x - 3y = 8$.

*Solution:* Construct the table at the lower left by letting $x = 0$ and solving for $y$, then setting $y = 0$ and $y = 4$ (to avoid fractions) in turn. Then draw the line through $(0, -2\frac{2}{3})$, $(2, 0)$, $(5, 4)$.

| x | y |
|---|---|
| 0 | $-2\frac{2}{3}$ |
| 2 | 0 |
| 5 | 4 |

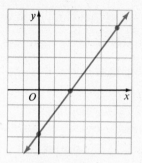

**Example 2.** Graph:
**a.** $\{(x, y): y = -3\}$      **b.** $\{(x, y): x = 4\}$

*Solution:*

**a.** The equation can be written

$$0 \cdot x + 1 \cdot y = -3.$$

The graph is the horizontal line shown below.

**b.** The equation can be written

$$1 \cdot x + 0 \cdot y = 4.$$

The graph is the vertical line shown below.

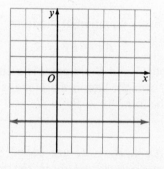 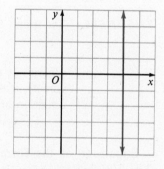

Example 2 illustrates the fact that the graph of $Ax + By = C$ is a horizontal line if $A$, the coefficient of $x$, is zero, and is a vertical line if $B$, the coefficient of $y$, is zero.

*ORAL EXERCISES*

**1.** State the coordinates of each point.

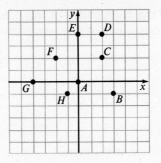

In which quadrant does each of the following points lie?

**2.** $(1, 1)$      **3.** $(-2, -1)$      **4.** $(-3, 5)$      **5.** $(3, -4)$

On which axis or axes does each of the following points lie?

**6.** $(0, 3)$      **7.** $(-7, 0)$      **8.** $(0, 0)$      **9.** $(5, 0)$

Give the coordinates of three points on each of the following lines.

**10.**           **11.**           **12.**

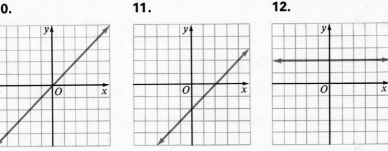

**13.**           **14.**

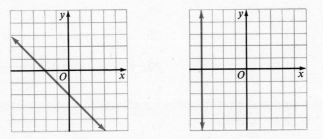

**15.** If a line passes through the points with coordinates $(3, 5)$ and $(3, 10)$, what can you conclude about the line? Why?

**16.** If a line passes through the points with coordinates $(-1, 7)$ and $(1, 7)$, what can you conclude about the line? Why?

State the coordinates of the point(s) in which the graph of each of the following equations intersects the coordinate axes.

**Sample.** $2x + 3y = 6$     *Solution:*   $(0, 2)$ and $(3, 0)$

**17.** $x + y = 4$                         **23.** $3x + 4y = 12$

**18.** $x - y = 4$                         **24.** $5x - 10y = 20$

**19.** $y - 2 = 0$                         **25.** $3x + 5y = 0$

**20.** $y + 6 = 0$                         **26.** $2x - 6y = 0$

**21.** $x + \frac{3}{2} = 0$                **27.** $\frac{1}{4}x + \frac{3}{4}y = 1$

**22.** $x - 7 = 0$                         **28.** $\frac{3}{4}x - \frac{1}{4}y = 1$

## WRITTEN EXERCISES

Write each of the following equations equivalently in the form $Ax + By = C$, where $A$, $B$, and $C$ are integers.

**A**   **1.** $y = -3x + 5$       **3.** $x = -\frac{4}{3}y + 2$       **5.** $x = 2$

**2.** $y = 2x - 3$       **4.** $x = \frac{3}{4}y - 3$       **6.** $y = -3$

**7–18.** Graph each of the equations in Oral Exercises 17–28.

Graph each equation.

**19.** $2x - 3y = 8$                     **23.** $x = \frac{1}{2}y + 1$

**20.** $3x + 2y = -10$                  **24.** $x = -\frac{1}{3}y - 1$

**21.** $\frac{x}{2} + \frac{y}{3} = 1$        **25.** $2(2 - x) + y = 4$

**22.** $\frac{x}{4} - \frac{y}{2} = 1$        **26.** $y - 2(x + y + 1) = -2$

Determine $k$ so that the point whose coordinates are given belongs to the graph of the given equation.

**B**   **27.** $9x + ky = 18; (1, 3)$

**28.** $2kx - 3y = 33; (3, -1)$

**29.** $(k + 4)x - 3ky = 2(2k + 1); (-1, -1)$

**30.** $4kx - (1 - y)k = 3(2k - 4); (2, 3)$

Graph each equation.

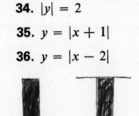

**31.** $y = |x|$     **34.** $|y| = 2$     **37.** $y = x + |x|$

**32.** $y = -|x|$     **35.** $y = |x + 1|$     **38.** $y = x - |x|$

**33.** $|x| = 1$     **36.** $y = |x - 2|$

## 3–3 Slopes of Lines

If a road rises 20 feet for each 100 feet of horizontal distance (Figure 3–5), then the steepness, or *grade*, of the road is the ratio of "rise" to "run," $\frac{20}{100}$, or 20%. To find the steepness, or *slope*, of a line, such as the graph of $3x - 4y = 6$ shown in Figure 3–6, choose two points on it, for example, $H(-2, -3)$ and $K(6, 3)$, and form a similar quotient:

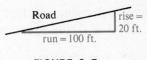

**FIGURE 3–5**

$$\text{slope} = \frac{\text{rise}}{\text{run}} = \frac{\text{ordinate of } K - \text{ordinate of } H}{\text{abscissa of } K - \text{abscissa of } H}$$

$$= \frac{3 - (-3)}{6 - (-2)} = \frac{3 + 3}{6 + 2} = \frac{6}{8} = \frac{3}{4}.$$

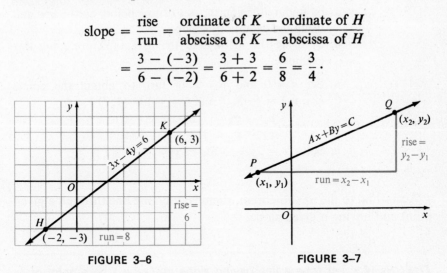

**FIGURE 3–6**     **FIGURE 3–7**

In the general case (Figure 3–7) you choose any two points $P(x_1, y_1)$ and $Q(x_2, y_2)$ on the line $Ax + By = C$ ($B \neq 0$) and compute

$$\frac{\text{ordinate of } Q - \text{ordinate of } P}{\text{abscissa of } Q - \text{abscissa of } P} = \frac{y_2 - y_1}{x_2 - x_1}.$$

(*Note:* $(x_1, y_1)$ is read "x sub one, y sub one" or "x one, y one.")

Recall that a point is on a line if and only if its coordinates satisfy the equation of the line:

$Q$ is on the line: $\qquad Ax_2 + By_2 = C$

$P$ is on the line: $\qquad Ax_1 + By_1 = C$

Subtract: $A(x_2 - x_1) + B(y_2 - y_1) = 0$

$$B(y_2 - y_1) = -A(x_2 - x_1)$$

$$\therefore \frac{y_2 - y_1}{x_2 - x_1} = -\frac{A}{B} \qquad x_2 \neq x_1.$$

Since $-\dfrac{A}{B}$ is a constant, $\dfrac{y_2 - y_1}{x_2 - x_1}$ does not depend on the particular points chosen, and this ratio is defined to be the **slope** of the line.

If $B \neq 0$, the equation $Ax + By = C$ can be put in the equivalent form $y = -\dfrac{A}{B}x + \dfrac{C}{B}$, and you can see that *the slope of a line is the coefficient of $x$ when its equation is solved for $y$.*

**Example 1.**  Find the slope $m$ of the line $x = 2y - 3$

  **a.** by finding the coordinates of two points on the line and using the definition of slope.

  **b.** by using the method of the paragraph preceding this example.

*Solution:*  **a.** Set $y = 0$ and $y = 2$ in turn to obtain the points $(x_1, y_1) = (-3, 0)$ and $(x_2, y_2) = (1, 2)$.

$$\therefore m = \frac{y_2 - y_1}{x_2 - x_1} = \frac{2 - 0}{1 - (-3)} = \frac{2}{4} = \frac{1}{2}.$$

  **b.** Solve $x = 2y - 3$ for $y$: $y = \frac{1}{2}x + \frac{3}{2}$

  $\therefore m = \frac{1}{2}.$

According to the next theorem, there is only one line through a given point and having a given slope.

---

**THEOREM.**  Let $L$ be a line having slope $m$, $(x_1, y_1)$ be a point of $L$, and $(x_2, y_2)$ be a point with $x_2 \neq x_1$.

  **1.** If $(x_2, y_2)$ is on $L$, then $\dfrac{y_2 - y_1}{x_2 - x_1} = m$.

  **2.** If $\dfrac{y_2 - y_1}{x_2 - x_1} = m$, then $(x_2, y_2)$ is on $L$.

---

Do you see that Statement **1** is just the definition of slope? A proof of the converse Statement **2** is outlined in Exercises 35–42, page **99**. The theorem is used in the next example.

**Example 2.** **a.** Find the slope $m$ of the line $3x + 2y = 2$.

**b.** Find the point of intersection of the line and the $y$-axis.

**c.** Use **a** and **b** to draw the line.

*Solution:* **a.** Compare $3x + 2y = 2$ with $Ax + By = C$ to obtain $A = 3$ and $B = 2$ for this equation.

$$\therefore m = -\frac{A}{B} = -\frac{3}{2}.$$

**b.** Substitute 0 for $x$; then

$$y = 1.$$

$\therefore$ the required point is $(0, 1)$.

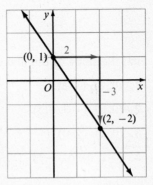

**c.** Since $m = -\dfrac{3}{2} = \dfrac{-3}{2}$, we find another point on the line by measuring 2 units to the right from $(0, 1)$, then 3 units down, and draw the line.

Notice that as a point moves from left to right along a line, the slope gives the change in its ordinate for each change of 1 unit in its abscissa. Thus if $(-2, 1)$ is a point on a line of slope 2, then $(-2 + 1, 1 + 2)$ or $(-1, 3)$ is another point on the line (Figure 3–8). But if the line through $(-2, 1)$ has slope $-2$, then another point on the line is $(-2 + 1, 1 + [-2])$ or $(-1, -1)$ (Figure 3–9). Do you see that lines which *rise* from left to right have *positive* slopes, while lines which *fall* from left to right have *negative* slopes?

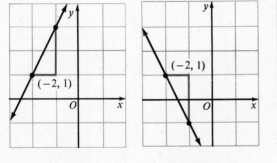

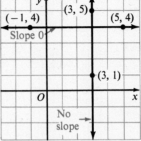

**FIGURE 3–8**          **FIGURE 3–9**          **FIGURE 3–10**

The slope of the horizontal line joining $(-1, 4)$ and $(5, 4)$ in Figure 3–10 is $\dfrac{4 - 4}{5 - (-1)} = \dfrac{0}{6} = 0$. As in this example, you can show that *the slope of every horizontal line is* 0. If, however, you tried to find the slope of the vertical line through $(3, 5)$ and $(3, 1)$ in Figure 3–10, the denominator, $x_2 - x_1$, would be 0, and division by 0 is not allowed. *Vertical lines have no slope.*

### ORAL EXERCISES

**Find the slope of the line containing the given points.**

**1.** $(0, 0)$, $(3, 5)$

**2.** $(0, 0)$, $(-3, -5)$

**3.** $(0, 1)$, $(5, 0)$

**4.** $(-2, 0)$, $(0, -7)$

**5.** $(2, 3)$, $(4, 5)$

**6.** $(6, 2)$, $(-2, -4)$

**7.** $(-1, 3)$, $(5, 3)$

**8.** $(-5, -4)$, $(9, -4)$

**Find the slope of each of the following lines.**

**9.**

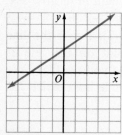

**11.**

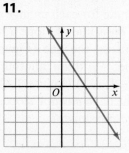

**13.**

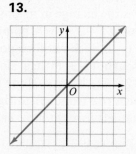

**10.**

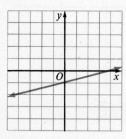

**12.**

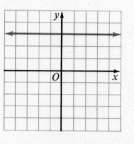

**14.**

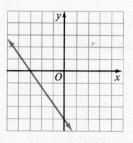

State the slope of each line and the coordinates of the point in which the line intersects the y-axis.

**15.** $y = x + 2$      **18.** $y = -\frac{1}{4}x$      **21.** $y = 5$

**16.** $y = 3x - 1$      **19.** $2y = 6x - 6$      **22.** $y = -3$

**17.** $y = \frac{1}{2}x$      **20.** $3y = 9x + 27$      **23.** $x = -1$

## WRITTEN EXERCISES

Plot each pair of points, draw a line through them, and determine the slope of the line from the graph. Then check your result by using the slope formula.

**A**

**1.** $(2, 3), (5, 6)$          **5.** $(2, 3), (5, 3)$

**2.** $(-3, 4), (-6, 1)$      **6.** $(-1, 5), (2, 5)$

**3.** $(-3, -2), (1, 6)$      **7.** $(-2, 1), (-2, 5)$

**4.** $(1, -3), (4, -9)$      **8.** $(\frac{1}{2}, 4), (\frac{1}{2}, \frac{9}{2})$

Find the slope $m$ of each line **(a)** by finding the coordinates of two points on the line and using the definition of slope and **(b)** by using the formula

$$m = -\frac{A}{B}.$$

**9.** $2x + 3y = -6$        **12.** $-2x + 13y = -52$

**10.** $x - 4y = 8$          **13.** $y = -\frac{1}{2}x + \frac{3}{2}$

**11.** $-3x - 2y = 12$      **14.** $y = \frac{1}{3}x - \frac{2}{3}$

Find three other points on the line containing the given point and having the given slope $m$.

**Sample.**    $(2, 5); \; m = -\frac{1}{2}$

*Solution:*    Since the slope $-\frac{1}{2}$ can be thought of as the change, along the line, in the $y$-direction per unit change in the $x$-direction, the point $(2 + 1, 5 + (-\frac{1}{2})) = (3, \frac{9}{2})$ is also on the line. Similarly, the points $(2 + 2, 5 + 2(-\frac{1}{2})) = (4, 4)$, $(2 + 3, 5 + 3(-\frac{1}{2})) = (5, \frac{7}{2})$, and, in general, $(2 + n, 5 + n(-\frac{1}{2})), n \in \Re$, are on the line.

Find three other points on the line containing the given point and having the given slope *m*.

**15.** $(1, 1)$; $m = 2$

**16.** $(-2, 3)$; $m = -1$

**17.** $(6, 0)$; $m = \frac{2}{3}$

**18.** $(0, -4)$; $m = \frac{4}{3}$

**19.** $(-1, -7)$; $m = -\frac{1}{4}$

**20.** $(5, -2)$; $m = -\frac{3}{2}$

Graph the line containing the given point and having the given slope *m*.

**21.** $(0, 0)$; $m = \frac{1}{2}$

**22.** $(-1, 0)$; $m = \frac{2}{3}$

**23.** $(2, 1)$; $m = 1$

**24.** $(-1, -2)$; $m = 2$

**25.** $(1, -3)$; $m = 0$

**26.** $(2, 2)$; no slope

Determine whether or not the points whose coordinates are given lie on a line. If they do, give the slope of the line.

Sample.

|   |   3 |   3 |   3 |    |
|---|-----|-----|-----|----|
| **x** | 1 | 4 | 7 | 10 |
| **y** | 8 | 7 | 6 | 5 |
|   | −1 | −1 | −1 |    |

*Solution:* As indicated in the table, each change of 3 in *x* produces a change of −1 in *y*. Thus each point lies on a line with slope $-\frac{1}{3}$.

**27.**

| **x** | 1 | 2 | 3 | 4 |
|-------|---|---|---|---|
| **y** | 1 | 3 | 5 | 7 |

**29.**

| **x** | 2 | 4 | 6 | 7 |
|-------|----|----|----|-----|
| **y** | −1 | −5 | −9 | −13 |

**28.**

| **x** | 0 | 1 | 2 | 3 |
|-------|----|----|---|---|
| **y** | −4 | −2 | 0 | 2 |

**30.**

| **x** | 3 | 4 | 5 | 7 |
|-------|----|----|----|----|
| **y** | −5 | −4 | −3 | −1 |

For each line determine the slope and the coordinates of the point of intersection of the line with the *y*-axis. Then use this information to graph the line.

**31.** $y = 2x + 1$

**32.** $y = x - 2$

**33.** $2x + 3y = 4$

**34.** $x - 4y = 2$

Prove Statement **2** of the Theorem on page 94 by letting $Ax + By = C$ be an equation of $L$ and justifying the following steps.

**B** **35.** $m = -\dfrac{A}{B}$

**36.** $\dfrac{y_2 - y_1}{x_2 - x_1} = -\dfrac{A}{B}$

**37.** $B(y_2 - y_1) = -A(x_2 - x_1)$

**38.** $By_2 - By_1 = -Ax_2 + Ax_1$

**39.** $Ax_2 + By_2 = Ax_1 + By_1$

**40.** But $Ax_1 + By_1 = C$.

**41.** Hence $Ax_2 + By_2 = C$.

**42.** $\therefore (x_2, y_2)$ is on $L$.

Determine $k$ so that the slope $m$ of the line containing the given points has the given value.

**43.** $(2, 2k), (-2, -3k); \; m = \frac{1}{10}$   **44.** $(2, 0), \left(4, \dfrac{1}{k}\right); \; m = -3$

**C** **45.** $(0, |k|), (1, k); \; m = -1$   **46.** $(1, k)(2, |k|); \; m = 1$

## 3–4 Equations of Lines

The line $L$ pictured in Figure 3–11 has slope $-\frac{2}{3}$ and passes through the point $(4, -1)$. What is its equation? You can conclude from the theorem on page 94 that a point $(x, y)$, where $x \neq 4$, is on $L$ if and only if

$$\frac{y - (-1)}{x - 4} = -\frac{2}{3}$$

$$\frac{y + 1}{x - 4} = \frac{-2}{3}$$

$$3(y + 1) = -2(x - 4)$$

$$3y + 3 = -2x + 8$$

$$2x + 3y = 5.$$

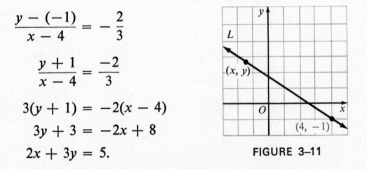

FIGURE 3–11

We can use the method illustrated above to find an equation of the line having slope $m$ and passing through a given point $(x_1, y_1)$. Any other point will be on the line if and only if $\dfrac{y - y_1}{x - x_1} = m$, or

$$y - y_1 = m(x - x_1).$$

In the derivation above we assumed $(x, y) \neq (x_1, y_1)$. However, you can verify by substitution that the pair $(x_1, y_1)$ satisfies the equation. Thus $y - y_1 = m(x - x_1)$, called the **point-slope form** of the equation of the line, is an algebraic statement of the geometric condition that the line have slope $m$ and pass through $(x_1, y_1)$.

**Example 1.** Find an equation of the line passing through the points $(3, -5)$ and $(-2, 0)$.

*Solution:*     Slope $= m = \dfrac{0 - (-5)}{-2 - 3} = \dfrac{5}{-5} = -1.$

Choose one of the points, $(-2, 0)$ say, to be $(x_1, y_1)$ and use the point-slope form:

$$y - 0 = -1(x - [-2])$$
$$y = -x - 2$$
$$x + y = -2$$

An important case of the point-slope form results when the given point is on the $y$-axis.

**Example 2.** Find an equation of the line
**a.** having slope 3 and containing $(0, -2)$.
**b.** having slope $m$ and containing $(0, b)$.

*Solution:*   **a.** $\quad y - y_1 = m(x - x_1)$   **b.** $y - y_1 = m(x - x_1)$
$\qquad\qquad y - (-2) = 3(x - 0) \qquad\qquad y - b = m(x - 0)$
$\qquad\qquad\quad y + 2 = 3x \qquad\qquad\qquad\quad y - b = mx$
$\qquad\qquad\qquad y = 3x - 2 \qquad\qquad\qquad\qquad y = mx + b$

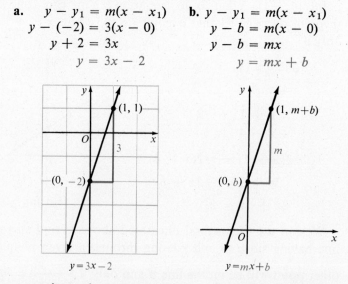

$$y = 3x - 2 \qquad\qquad\qquad y = mx + b$$

In each of these equations, the constant term ($-2$ and $b$, respectively) is the ordinate of the point where the line crosses the $y$-axis; it is called the *y*-intercept of the line. The general equation in Example 2b,

$$y = mx + b,$$

which represents the line with slope *m* and *y*-intercept *b*, is called the **slope-intercept form** of the equation of the line.

In a plane, different lines having the same slope, or having no slope, are called **parallel** lines.

Using slope-intercept form, you can show that parallel lines do not intersect. The lines

$$y = \tfrac{1}{3}x + 2 \qquad \text{and} \qquad y = \tfrac{1}{3}x - 1$$

shown in Figure 3–12 cannot have a point in common, for if $y = \tfrac{1}{3}x + 2$ and $y = \tfrac{1}{3}x - 1$ were both true statements for some $(x, y)$, then we would have

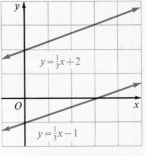

$$\tfrac{1}{3}x + 2 = \tfrac{1}{3}x - 1, \qquad \text{or} \qquad 2 = -1,$$

**FIGURE 3–12**

a false statement. It can be shown that, conversely, two nonintersecting lines in a plane either have no slope (are vertical lines) or have the same slope.

**Example 3.** Find an equation of the line which crosses the *x*-axis at $(-2, 0)$ and is parallel to the line $3x + 5y = 10$.

*Solution:* The slope of $3x + 5y = 10$ is $-\tfrac{3}{5}$. ∴ the slope-intercept form of the equation of the required line is

$$y = -\tfrac{3}{5}x + b.$$

Since $(-2, 0)$ is on this line, substitute $-2$ for *x* and 0 for *y* and solve for *b*:

$$0 = -\tfrac{3}{5}(-2) + b$$
$$0 = \tfrac{6}{5} + b$$
$$b = -\tfrac{6}{5}$$

∴ the required equation is

$$y = -\tfrac{3}{5}x - \tfrac{6}{5}$$

or

$$3x + 5y = -6.$$

The abscissa of the point where a line intersects the *x*-axis is the *x*-intercept of the line. Thus the line $3x + 5y = -6$ has *x*-intercept $-2$.

*ORAL EXERCISES*

State the point-slope form of the equation of the line containing the given point and having the given slope *m*.

**1.** $(1, 0)$; $m = 1$         **4.** $(-2, 1)$; $m = 2$

**2.** $(0, 3)$; $m = -1$        **5.** $(3, 2)$; $m = 0$

**3.** $(1, -1)$; $m = -2$       **6.** $(-4, -1)$; $m = 0$

State the slope-intercept form of the equation of the line having the given slope *m* and the given *y*-intercept *b*.

**7.** $m = 1$; $b = 2$          **10.** $m = \frac{2}{3}$; $b = 3$

**8.** $m = \frac{1}{2}$; $b = 1$   **11.** $m = 0$; $b = -2$

**9.** $m = -\frac{1}{3}$; $b = 0$  **12.** $m = 0$; $b = 0$

State the slope and the *y*-intercept of the lines whose equations are given.

**13.** $y = \frac{1}{2}x - 3$     **16.** $y - 9x = -4$

**14.** $y = -2x + 5$            **17.** $4x + 2y = 1$

**15.** $2x + y = 7$             **18.** $6x - 2y = -2$

State whether or not the lines whose equations are given are parallel.

**19.** $y = 2x + 1, y = 2x - 3$          **22.** $x = 2, x = -1$

**20.** $y = -x + \frac{1}{2}, y = x + \frac{1}{2}$   **23.** $y = x + 1, 2y = 2x + 2$

**21.** $y = 5, y = -3$                   **24.** $x + y = 3, -x - y = -3$

*WRITTEN EXERCISES*

Find an equation of the form $Ax + By = C$, where *A*, *B*, and *C* are real numbers, for the line containing the given point and having the given slope *m*.

**A**  **1.** $(1, 2)$; $m = 1$          **6.** $(0, 0)$; $m = -\frac{1}{3}$

   **2.** $(3, 2)$; $m = -2$          **7.** $(-4, 2)$; $m = 0$

   **3.** $(-1, -3)$; $m = 2$        **8.** $(3, -2)$; $m = 0$

   **4.** $(-2, 1)$; $m = -1$         **9.** $(6, -2)$; no slope

   **5.** $(0, 0)$; $m = \frac{1}{2}$    **10.** $(-3, -7)$; no slope

Find an equation of the form $Ax + By = C$, where $A$, $B$, and $C$ are real numbers, for the line containing the given points.

**11.** $(0, 4), (4, 8)$

**12.** $(1, 0), (2, 2)$

**13.** $(-2, 3), (-1, 2)$

**14.** $(4, -5), (5, -6)$

**15.** $(2, -3), (1, 1)$

**16.** $(-5, 1), (1, 3)$

**17.** $(-2, 1), (-7, 1)$

**18.** $(3, -3), (5, -3)$

**19.** $(2, -5), (2, 7)$

**20.** $(-6, 0), (-6, 4)$

Find an equation of the line satisfying the given conditions.

**B** **21.** Containing $(1, 2)$ and parallel to the graph of $x + y = 1$.

**22.** Containing $(-3, 1)$ and parallel to the graph of $x - y = 5$.

**23.** With $y$-intercept 5 and parallel to the graph of $2x + 4y = 8$.

**24.** With $y$-intercept $-3$ and parallel to the graph of $3x - 5y = 10$.

**25.** Containing $(-1, -3)$ and parallel to the graph of $y = 5$.

**26.** Containing $(1, -2)$ and parallel to the graph of $x = -4$.

**27.** With $x$-intercept 2 and parallel to the graph of $2x + 5y = 1$.

**28.** With $x$-intercept $-3$ and parallel to the graph of $3x - 2y = 9$.

**29.** With $x$-intercept 1 and $y$-intercept $-2$.

**30.** With $x$-intercept $-2$ and $y$-intercept 4.

**31.** Containing $(1, 1)$ and parallel to the line containing $(2, 3)$ and $(3, 5)$.

**32.** Containing $(-2, 3)$ and parallel to the line containing $(-1, -1)$ and $(1, 5)$.

**C** **33.** Containing $(1, 3)$ and having its $x$-intercept equal to its $y$-intercept.

**34.** Having $x$-intercept $a$ and slope $\dfrac{1}{a}$, $a \neq 0$.

**35.** Containing $(b, 2b)$ and having slope $-\dfrac{1}{b}$, $b \neq 0$.

**36.** Containing $(r, s)$ and parallel to the graph of $Ax + By = C$, where not both $A$ and $B$ are 0.

**37.** Find the $x$- and $y$-intercepts of the line with equation $Ax + By = C$, where $A \neq 0$, $B \neq 0$, and $C \neq 0$. Then show that the quotient of the $y$-intercept and the $x$-intercept of the line is equal to the negative of the slope of the line.

**38.** Show that the quadrilateral with vertexes $(3, 1)$, $(8, 2)$, $(10, 4)$, and $(5, 3)$ is a parallelogram.

**39.** Show that the quadrilateral with vertexes $(0, 0)$, $(a, b)$, $(a + c, b + d)$, and $(c, d)$ is a parallelogram.

**40.** Show that an equation of the line containing the points $(x_1, y_1)$ and $(x_2, y_2)$, with $x_1 \neq x_2$, is

$$y - y_1 = \frac{y_2 - y_1}{x_2 - x_1}(x - x_1). \qquad \text{(two-point form)}$$

**41.** Show that an equation of the line having $x$-intercept $a$ and $y$-intercept $b$, where $a \neq 0$ and $b \neq 0$, is

$$\frac{x}{a} + \frac{y}{b} = 1. \qquad \text{(intercept form)}$$

**42.** The line $L_1$ has equation $A_1 x + B_1 y = C_1$ and the line $L_2$ has equation $A_2 x + B_2 y = C_2$.
**a.** Show that if $L_1$ and $L_2$ are parallel, then $A_1 B_2 = A_2 B_1$.
**b.** Show that if $A_1 B_2 = A_2 B_1$, then $L_1$ and $L_2$ are either parallel or are the same line.

## SYSTEMS OF LINEAR SENTENCES

### 3-5 Systems of Two Linear Equations

When you graph two linear equations in two variables, the resulting lines may, as illustrated in Figure 3–13, (a) be parallel, (b) be coincident, or (c) intersect in a single point.

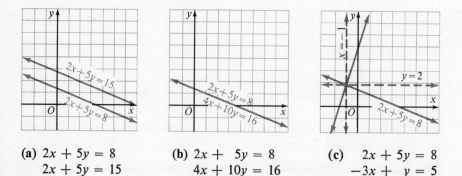

(a) $2x + 5y = 8$
$2x + 5y = 15$

(b) $2x + 5y = 8$
$4x + 10y = 16$

(c) $2x + 5y = 8$
$-3x + y = 5$

FIGURE 3–13

If $S$ and $T$ are the solution sets of the two equations in each case, then the geometrical situations above correspond, respectively, to these algebraic ones:

**(a)** $S \cap T = \emptyset$; there is no common solution. The equations are called inconsistent.

**(b)** $S = T$; every solution of either equation is a solution of the other. The equations are called consistent and dependent.

**(c)** $S \cap T$ contains a single ordered pair; there is a unique common solution. The equations are called consistent and independent.

When two equations place two conditions on the same variables *at the same time*, we sometimes say that they form a system of simultaneous equations, and we call the ordered pairs which satisfy both equations common, or simultaneous, solutions.

Two systems of equations are equivalent if they have the same solution set. For example, Figure 3–13(c) suggests that the systems

$$2x + 5y = 8 \qquad\qquad x = -1$$
$$-3x + \ y = 5 \quad \text{and} \quad y = 2$$

are equivalent. You can solve a system of equations algebraically if you can transform it into an equivalent system whose solution set can be found by inspection. The following transformations are useful in solving systems of linear equations:

---

**Transformations Which Produce an Equivalent System of Linear Equations**

1. Replacing either equation of the system by an equivalent equation in the same variables (see page 56).

2. Replacing either equation by the sum of that equation and a constant multiple\* of the other equation. (The replacement is called a *linear combination* of the given equations.)

3. In either equation substituting for one variable **(a)** an expression for it obtained from the other equation or **(b)** its value if known.

---

\*A *constant multiple* of an equation is obtained by multiplying each member of the equation by some number. To *add two equations* means to add their left members, add their right members, and then equate these sums.

You use these transformations to replace an equation of a system by one in which one of the variables has 0 as coefficient.

**Example 1.** Solve the system: $x + 3y = 9$
$4x + 5y = 1$

*Solution 1:* **(*Linear-Combination Method*)**

1. To obtain equations having the same number as coefficient of $x$, multiply each member of the first equation, $x + 3y = 9$, by 4. (Transformation 1)

$4x + 12y = 36$
$4x + 5y = 1$

2. In the new system, keep the second equation but replace the first equation by the difference of the equations in Step 1. (Transformation 2)

$7y = 35$
$4x + 5y = 1$

3. Solve $7y = 35$ for $y$. (Transformation 1)

$y = 5$

4. Substitute 5 for $y$ in $4x + 5y = 1$. (Transformation 3)

$4x + 5(5) = 1$

5. Solve for $x$. (Transformation 1)

$4x + 25 = 1$
$4x = -24$
$x = -6$

∴ the solution set is $\{(-6, 5)\}$.

*Check:* 6. Verify that $(-6, 5)$ is a solution of both the original equations.

$$-6 + 3(5) = -6 + 15 = 9$$
$$4(-6) + 5(5) = -24 + 25 = 1$$

*Solution 2:* **(*Substitution Method*)**

1. Transform the first equation to express $x$ in terms of $y$. (Transformation 1)

$x + 3y = 9$
$x = 9 - 3y$

2. Substitute this expression for $x$ in the other equation. (Transformation 3)

$4x + 5y = 1$
$4(9 - 3y) + 5y = 1$

3. Solve the result for $y$. (Transformation 1)

$36 - 12y + 5y = 1$
$y = 5$

**4.** Substitute the value of $y$ in the expression for $x$. (Transformation 3)

$$x = 9 - 3(5)$$

**5.** Solve for $x$. (Transformation 1)

$$x = -6$$

∴ the solution set is $\{(-6, 5)\}$.

*Check:*     **6.** The check is the same as Step 6 in Solution 1.

**Example 2.**   Find $\{(x, y): -2x + 4y = 1 \text{ and } 3x - 6y = 2\}$.

*Solution:*     (*Linear-Combination Method*)

**1.** To obtain equations having the coefficients of one variable (say $y$) equal in absolute value, multiply the first equation by 3 and the second by 2.

$$-6x + 12y = 3$$
$$6x - 12y = 4$$

**2.** Replace either equation, say the first, by the sum of the equations.

$$0 = 7$$
$$6x - 12y = 4$$

Since the given system is equivalent to one in which one equation has no solution ($0 = 7$ is a false statement), the given equations have no common solution and are therefore inconsistent.

$$\therefore \{(x, y): -2x + 4y = 1 \text{ and } 3x - 6y = 2\} = \emptyset.$$

Another way you can see that the equations in Example 2 are inconsistent is to notice that the first equation requires that $6x - 12y$ be equal to $-3$, while the second equation requires that the same expression, $6x - 12y$, be equal to 4.

**Example 3.**   Solve the equations $-4x + 2y = -6$ and $2x - y = 3$ simultaneously.

*Solution:*   If you multiply the second equation by $-2$, it becomes $-4x + 2y = -6$. Thus the given system is equivalent to one in which both equations are the same. The solution set of the system is therefore the same as the solution set of one of the equations, namely, the infinite set

$$\{(x, y): 2x - y = 3\}.$$

*ORAL EXERCISES*

State whether the given ordered pair is a solution of the given system of equations.

**1.** $(3, 0)$; $4x + 5y = 12$
$\quad\quad\quad\quad x - y = 3$

**4.** $(4, 3)$; $5x - 2y = 14$
$\quad\quad\quad\quad 3x + 4y = 24$

**2.** $(0, -1)$; $2x + y = -1$
$\quad\quad\quad\quad\quad 3x - 2y = -2$

**5.** $(-1, 3)$; $7x - 2y = -13$
$\quad\quad\quad\quad\quad -2x - y = -5$

**3.** $(1, 2)$; $2x + 3y = 8$
$\quad\quad\quad\quad\quad 3x + y = 5$

**6.** $(-2, -1)$; $x + 3y = -5$
$\quad\quad\quad\quad\quad\quad -2x + y = 3$

State whether the given systems of equations are equivalent.

**7.** $x = 1$, $2x + y = 4$
$\quad\quad y = 2$, $x - 2y = -3$

**9.** $2x - 3y = 6$, $2x - 3 = 6$
$\quad\quad\quad\quad y = 1$, $\quad\quad y = 1$

**8.** $4x + 2y = -12$, $x = -4$
$\quad\quad 3x - y = -14$, $y = 2$

**10.** $x + y = 9$, $x + y = 9$
$\quad\quad\quad x - y = 27$, $y = 27 - x$

Transform each equation into an equivalent one expressing the specified variable in terms of the other variable.

**Sample.** $3x - 5y = 15$; $x$ $\quad$ *Solution:* $x = 5 + \frac{5}{3}y$

**11.** $-y = 2x - 5$; $y$

**14.** $2p - q = 6$; $p$

**12.** $3y = 6x - 21$; $y$

**15.** $2m - 3n = 12$; $m$

**13.** $r + 2s = 4$; $s$

**16.** $4\ell + 5m = 20$; $m$

Tell how to combine the given equations to obtain an equation in which the variable shown in red has coefficient 0.

**Sample.** $3x - 5y = -2$
$\quad\quad\quad\quad 2x + 3y = 5$ ; $y$

*Solution:* Multiply the first equation by 3, and the second equation by 5, and then add the resulting equations.

**17.** $3x + y = 4$
$\quad\quad\quad x - 2y = -1$ ; $y$

**19.** $3u + 12v = -6$
$\quad\quad\quad\quad 5u + 3v = 7$ ; $v$

**18.** $x + 5y = 6$
$\quad\quad\quad -2x - 2y = -4$ ; $x$

**20.** $5a + 7b = -12$
$\quad\quad\quad\quad 3a - 2b = -1$ ; $b$

## WRITTEN EXERCISES

Graph each system of equations and state whether the equations are inconsistent, consistent and dependent, or consistent and independent.

**A**

**1.** $3x + 2y = 6$
$6x + 4y = 4$

**2.** $2x + 5y = 12$
$x - 3y = -5$

**3.** $8r - s = 4$
$4r - \frac{1}{2}s = 2$

**4.** $6a - 12b = 12$
$2a - 4b = 4$

**5.** $3f + 7g = -10$
$2f - 5g = 3$

**6.** $9x - 2z = 4$
$-18x + 4z = -3$

Find the solution set of each system.

**7.** $2x + y = 3$
$x - 2y = -1$

**8.** $a - 2b = 3$
$5a + 3b = 2$

**9.** $s - t = 0$
$6s + 5t = -22$

**10.** $2m - n = 0$
$3m + 5n = 78$

**11.** $x + 5z = -2$
$2x + z = 5$

**12.** $2u + v = 1$
$9u + 3v = -3$

**13.** $2x + 3y = -5$
$3x + 5y = -9$

**14.** $6x - 5y = 19$
$5x + 2y = -15$

**15.** $2r = 9 + 5s$
$18r - 45s = 72$

**16.** $21x - 24z = 27$
$7x = 10 + 8z$

**17.** $3x - 5y = -7$
$-36x + 60y = 84$

**18.** $15x + 2y = 17$
$5x + \frac{2}{3}y = 5\frac{2}{3}$

**19.** $6r - 2s = 1$
$4r + 3s = \frac{17}{6}$

**20.** $10r + 3s = 4$
$15r - 5s = -\frac{1}{3}$

**21.** $6x - 4y = 4$
$5x + 6y = 8$

**22.** $9x + 21z = -11$
$8x + 6z = -\frac{4}{3}$

Solve each system for $x$ and $y$. In Exercises 27 and 28 assume that $a^2 + b^2 \neq 0$.

**B**

**23.** $x + y = k$
$x - y = k$

**24.** $2x - y = 2s$
$2x - 2y = 4s$

**25.** $x + y = 3b - a$
$-3x + 2y = b + 3a$

**26.** $x - y = c + 3d$
$5x - 2y = 8c + 6d$

**27.** $ax + by = k_1$
$bx - ay = k_2$

**28.** $bx - ay = 0$
$ax + by = c$

**C** **29.** Solve for $x$ and $y$ assuming $r \neq 0$, $s \neq 0$, $r \neq s$.

$$rx + sy = r + s$$
$$rsx + rsy = r^2 + s^2$$

**30.** Solve for $a$ and $b$. $\left( \textit{Hint: First solve for } \dfrac{1}{a} \text{ and } \dfrac{1}{b} \cdot \right)$

$$\frac{1}{a} - \frac{1}{b} = \frac{1}{6}$$
$$\frac{3}{a} + \frac{4}{b} = \frac{17}{6}$$

**31.** Show that if two different nonvertical lines have the same slope, then they do not intersect and hence are parallel. (*Hint:* The lines can be represented by equations of the form $y = mx + b_1$ and $y = mx + b_2$, where $b_1 \neq b_2$. Show that the system

$$y = mx + b_1$$
$$y = mx + b_2$$

has no common solution.)

**32.** Show that if two nonvertical lines are parallel, then they have the same slope. (*Hint:* This is equivalent to showing that if the lines have *different* slopes, then they are *not parallel*, that is, they intersect. Let the lines be represented by equations of the form $y = m_1x + b_1$ and $y = m_2x + b_2$, where $m_1 \neq m_2$, and consider the solution of this system.)

## 3–6 Applications of Systems

Many practical problems lead to systems of linear equations. In working the examples below, we follow the four-step plan introduced in Section 2–7.

The following terms are used in problems involving the motion of aircraft:

**tail wind:**    a wind blowing in the same direction as the one in which the airplane is heading.

**head wind:**    a wind blowing in the direction opposite to the one in which the airplane is heading.

**wind speed:**    the speed of the wind in relation to the ground.

**airspeed:**    the speed of the airplane in still air.

**ground speed:**    the speed of the airplane in relation to the ground.

With a tail wind, an airplane's ground speed is the sum of its airspeed and the wind speed. With a head wind, the ground speed is the difference between airspeed and wind speed.

**Example.** With a certain tail wind a private plane took four hours to fly 800 miles. Flying back against the same wind the plane took one hour longer to make the trip. Find the wind speed and the plane's airspeed.

*Solution:*   1. Let $x$ be the wind speed (in mph) and $y$ be the plane's airspeed (in mph). We use the relationship distance = rate $\times$ time, $d = rt$, and arrange some of the given facts in a chart:

|  | Ground speed $r$ (mph) | Time (hr) $t$ | Distance (mi) $d = rt$ |
|---|---|---|---|
| **With tail wind** | $y + x$ | 4 | $4(y + x)$ |
| **With head wind** | $y - x$ | 5 | $5(y - x)$ |

2. The distance with a tail wind   is   800 miles.

$$4(y + x) \qquad = \qquad 800$$

The distance with a head wind   is   800 miles.

$$5(y - x) \qquad = \qquad 800$$

3. Solve the system

$$\begin{array}{ll} 4(y + x) = 800 \\ 5(y - x) = 800 \end{array} \quad \text{or} \quad \begin{array}{l} y + x = 200 \\ y - x = 160 \end{array}$$

Add:  $\quad 2y = 360 \quad \therefore \quad y = 180$
$\qquad 180 + x = 200 \quad \therefore \quad x = 20$

$\therefore$ the wind speed is 20 mph, and the airspeed is 180 mph.

*Check:*   4. When flying *with* the wind, the plane's ground speed is $180 + 20 = 200$ mph, and the time to travel 800 miles at this speed is $\frac{800}{200} = 4$ hours. When flying *against* the wind, the plane's ground speed is $180 - 20 = 160$ mph, and the time to travel 800 miles at this speed is $\frac{800}{160} = 5$ hours, one hour longer than for the first trip.

*PROBLEMS*

A **1.** Find two numbers whose sum is 18 and whose difference is 6.

**2.** One number is 11 more than twice another number. If the sum of the numbers is twice their difference, find the numbers.

**3.** In her purse Janet has one dollar in change composed entirely of nickels and quarters. If she has 2 more nickels than quarters, how many of each does she have?

**4.** For a vacation trip Mrs. Andrews bought $300 worth of travelers checks in ten-dollar and twenty-dollar denominations. If she had 22 travelers checks in all, how many of each denomination did she have?

**5.** If the difference in measure of two complementary angles is 10°, what are the measures of the angles?

**6.** If the measure of one of two supplementary angles is 20° more than 7 times the measure of the other, what are the measures of the angles?

**7.** Five years from now, Joel's father's age will be 3 times Joel's present age. The sum of Joel's age 5 years ago and his father's present age is 50. What is Joel's father's age?

**8.** Marissa is 3 years younger than her sister Angela. Two years from now, Marissa's age will be $\frac{5}{6}$ Angela's age. How old are the two girls now?

**9.** If each of the longer sides of a rectangle is shortened by 1 foot and each of the shorter sides is lengthened by 1 foot, then the new rectangle is a square. If each side of the original rectangle is lengthened by 1 foot, then the perimeter of the new rectangle is $\frac{5}{4}$ the perimeter of the original rectangle. Find the dimensions of the original rectangle.

**10.** The perimeter of an isosceles triangle is 15 inches. If the base of the triangle is lengthened by 2 inches and each of the congruent sides is shortened by 1 inch, then the new triangle is equilateral. Find the dimensions of the original triangle.

**11.** With a certain tail wind a jet aircraft arrives at its destination, 1890 miles away, in 3 hours. Flying against the same wind, the plane makes the return trip in $3\frac{3}{8}$ hours. Find the wind speed and the plane's airspeed.

**12.** A private plane can make a 125-mile trip in half an hour when flying against a certain head wind. Flying with the same wind, it can make the trip in 25 minutes. Find the speed of the wind and the plane's airspeed.

**13.** The diameter of the Earth is (approximately) 482 miles less than twice the diameter of the planet Mars. The sum of the radius of Mars and the radius of Earth is 6059 miles. Find the diameters of the two planets.

**14.** Mrs. Adams lives near the border of a state in which the sales tax is 2% lower than in her own state and she occasionally shops in the neighboring state to save money. She paid a total of $11.10 in sales tax when she bought $150 worth of merchandise in her own state and $120 worth of merchandise in the neighboring state. What is the rate of the sales tax in each state?

In Exercises 15 and 16, the original number is a positive integer whose numeral has 2 digits. Find the original number in each case.

**B**

**15.** The units digit is one more than twice the tens digit. The sum of the original number and the number represented when the digits are interchanged is 77.

**16.** The sum of the digits is 10. The difference between the original number and the number represented when the digits are reversed is 6 more than 3 times the sum of the digits.

**17.** On a camping trip Elsa and Janet found that it took them as long to paddle 12 miles up a certain river as it did to paddle 16 miles down the river. The rate at which the girls can paddle in still water is 6 mph greater than the rate of the current. Find both of these rates.

**18.** It takes a total of $2\frac{1}{2}$ hours for the excursion boat on the Olympic River to make the trip to a point 12 miles upstream and to return. If the rate at which the boat travels in still water is 5 times the rate of the river current, what is the rate of the current?

**19.** Find $a$ and $b$ so that the graph of $ax + by = 6$ contains the points $(1, -4)$, and $(4, 2)$.

**20.** Determine $A$ and $C$ so that the set of ordered pairs

$$\{(x, y): Ax + 2y = C\}$$

contains $(2, 1)$ and $(-4, 16)$.

**21.** Find $m$ and $b$ so that the graph of $y = mx + b$ passes through the points $(-1, -3)$, and $(8, 15)$.

**22.** Find $r$ and $s$ so that the graph of $y = rx^2 + s$ contains the points $(1, -2)$ and $(-3, 6)$.

**23.** Determine $a$ and $b$ so that $\{(x, y): ax^2 - by^2 = 1\}$ contains $(1, -2)$ and $(\frac{1}{2}, \frac{1}{2})$.

**C**

**24.** One mile upstream from her starting point, a rower passed a log floating with the current. After rowing upstream for one more hour, she rowed back and reached her starting point just as the log arrived. How fast was the current flowing?

## 3–7  Linear Inequalities

For each point *on* the line $y = x$ (Figure 3–14) the ordinate is *equal* to the abscissa. Notice that if a point is *above* this line, then its ordinate, $y$, is *greater* than its abscissa, $x$, that is, $y > x$. Similarly, the coordinates of each point below the line is a solution of the inequality $y < x$.

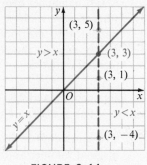

**FIGURE 3–14**

Figure 3–15 shows partial graphs of four linear inequalities related to the equation $y = x$. Each graph, pictured as a shaded region, is a *half-plane* bounded by the graph of $y = x$. If the boundary is part of the region, the line is a solid line; if not, a dashed line is used.

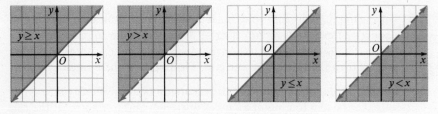

**FIGURE 3–15**

**Example 1.**  Graph the inequality $x + 2y < 6$.

*Solution:*  1. Transform the inequality into an equivalent one in which $y$ appears as one member and nowhere else.

$$x + 2y < 6$$
$$2y < 6 - x$$
$$y < 3 - \tfrac{1}{2}x$$

2. Graph $y = 3 - \tfrac{1}{2}x$ showing it as a dashed line.
3. Shade the half-plane below the line.

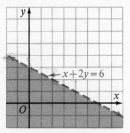

Figure 3–16 shows the graphs of both inequalities of the system:

$$x + 2y < 6$$
$$x - y \geq 2$$

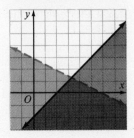

**FIGURE 3–16**

The region where the shadings overlap is the graph of the solution set of the system, that is, $\{(x, y): x + 2y < 6 \text{ and } x - y \geq 2\}$ or $\{(x, y): x + 2y < 6\} \cap \{(x, y): x - y \geq 2\}$.

**Example 2.** A food package for a space station is to contain at least 10 pounds of ingredient $A$ and 20 pounds of ingredient $B$. The combined amounts of $A$ and $B$ must be at least 40 pounds but may not exceed 80 pounds. Draw a figure showing all allowable combinations of the two ingredients.

*Solution:*  Let $x$ = number of pounds of ingredient $A$
$y$ = number of pounds of ingredient $B$.
Then

$$x \geq 10$$
$$y \geq 20$$
$$x + y \geq 40$$
$$x + y \leq 80.$$

Since *all* these conditions must be met, the set of allowable pairs $(x, y)$ is the intersection of the four sets determined by the four inequalities.

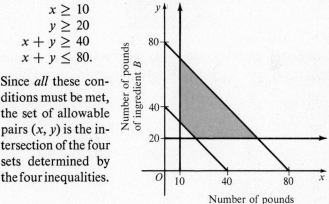

Number of pounds
of ingredient $A$

*ORAL EXERCISES*

Does the point whose coordinates are given lie above, on, or below the line with the given equation?

**1.** $(2, 4); y = x$

**2.** $(0, -1); y = x$

**3.** $(1, 3); y = 2x$

**4.** $(3, 1); y = \frac{1}{2}x$

**5.** $(4, 11); y = 3x - 1$

**6.** $(-1, 3); y = 2x + 5$

Is the point whose coordinates are given in the solution set of the given equation?

**7.** $(1, 2); y - x \leq 0$

**8.** $(4, -1); x + y \leq 0$

**9.** $(1, -2); 2x + y > 0$

**10.** $(3, 1); x - 3y < 0$

**11.** $(2, 1); 3x - 2y < 7$

**12.** $(-3, 2); 5y + 4x > -5$

**13.** $(\frac{3}{2}, 2); y > \frac{3}{2}$

**14.** $(-\frac{3}{2}, -\frac{1}{2}); x < -\frac{1}{2}$

Select the inequality whose graph is shown.

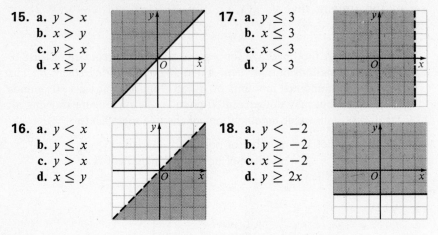

**15.** a. $y > x$
   b. $x > y$
   c. $y \geq x$
   d. $x \geq y$

**17.** a. $y \leq 3$
   b. $x \leq 3$
   c. $x < 3$
   d. $y < 3$

**16.** a. $y < x$
   b. $y \leq x$
   c. $y > x$
   d. $x \leq y$

**18.** a. $y < -2$
   b. $y \geq -2$
   c. $x \geq -2$
   d. $y \geq 2x$

Transform each inequality into one having $y$ as one member.

**19.** $3x + 6y \geq 0$

**22.** $-3x - 2y > -5$

**20.** $8y - 2x \leq 0$

**23.** $3(x + 2) < y + 6$

**21.** $2x - y < -2$

**24.** $2(x - 2) \geq 4(y - 1)$

## WRITTEN EXERCISES

Graph the solution set of each inequality or system of inequalities.

**A**

**1.** $y \leq x$

**2.** $y > x$

**3.** $y > 2x$

**4.** $y < 3x$

**5.** $y < -x$

**6.** $-x < y$

**7.** $x \leq -2$

**8.** $y \geq 2$

**9.** $y - 2x < 2$

**10.** $y - 3x > -3$

**11.** $5x + 10y \leq 20$

**12.** $2x + 4y \geq 16$

**13.** $x - 2y > -4$

**14.** $2x - y < -8$

**15.** $3x - 6y \leq 9$

**16.** $11x - 33y \geq 22$

**17.** $y \geq x$
    $x \leq 2$

**18.** $y \leq x$
    $y \geq -1$

**19.** $y > -x$
    $x < 0$

**20.** $y < -x$
    $x > 0$

**21.** $y \geq x - 2$
$\quad\;\; y \leq x + 6$

**22.** $y \geq -x + 3$
$\quad\;\; y \leq -x + 5$

**23.** $\quad 2x + 3y \leq 6$
$\quad -3x + \;\; y \leq 3$

**24.** $-3x + y > -2$
$\quad\;\;\; 2x + y < 5$

**25.** $x - 2y < 4$
$\quad x - 2y \leq -8$

**26.** $-x - 2y < -4$
$\quad -x - 2y \leq -12$

**B** **27.** A storekeeper is mixing a special blend of coffee using South American and African beans. He wants to have between 8 and 10 pounds (inclusive) of this special blend, and he wants to use at least 5 pounds of South American beans and 2 pounds of African beans in the blend. Draw a diagram showing all allowable combinations of the two kinds of beans.

**28.** A house painter wants to combine ivory paint with forest green paint to make a mixture containing at least 4 but not more than 5 cans of paint in all. She wants to use some of each color, and she wants the amount of ivory to be less than or equal to $\frac{1}{3}$ the amount of forest green. Draw a diagram showing all allowable combinations of the two colors.

Graph the solution set of each inequality or system of inequalities. Recall that a continued inequality such as "$-2 \leq y \leq 5$" is equivalent to the compound inequality "$y \geq -2$ and $y \leq 5$."

**29.** $-1 \leq x \leq 3$

**30.** $0 \leq x \leq 2$

**31.** $2 \leq y \leq 4$

**32.** $-2 \leq y \leq 1$

**33.** $x - 5 < y < x + 1$

**34.** $-2x - 1 < y < -2x + 3$

**35.** $y \leq x$ or $y \geq 1$

**36.** $y \leq -x$ or $x \leq 3$

**37.** $|x| > 2$

**38.** $|y| > 1$

**39.** $-2 \leq 2x + y \leq 3$

**40.** $-6 \leq x + 2y \leq 4$

## 3–8  Systems of Three Equations

The *ordered triple* $(2, 1, 5)$ is a solution of the equation

$$2x - 3y + z = 6$$

because $2 \cdot 2 - 3 \cdot 1 + 5 = 6$ is a true statement. (We are letting alphabetical order determine that $x$ is the first variable, $y$ is the second, and $z$ the third.) Some other solutions of this equation are $(0, 0, 6)$, $(0, -2, 0)$, $(3, 0, 0)$, and $(1, 0, 4)$.

A **linear\*** **equation** in the three variables $x$, $y$, and $z$ is one of the form

$$Ax + By + Cz = D,$$

where not all of $A$, $B$, and $C$ are zero. You can find as many solutions as you please of such an equation by assigning values to two of the variables and solving for the third.

In solving practical problems you will sometimes need to find solutions of systems of three linear equations in three variables. You can do this by using methods similar to those of Section 3–5, as illustrated below.

**Example 1.** Find the solution set of the system:

$$2y + z = 3$$
$$x - y + 2z = 1$$
$$2x - 3y + z = 3$$

*Solution:*

1. Solve the first equation for $z$ in terms of $y$.

$$2y + z = 3$$
$$z = 3 - 2y$$

2. Substitute $3 - 2y$ for $z$ in the second given equation and solve for $x$ in terms of $y$.

$$x - y + 2z = 1$$
$$x - y + 2(3 - 2y) = 1$$
$$x - y + 6 - 4y = 1$$
$$x = -5 + 5y$$

3. Substitute $-5 + 5y$ for $x$ and $3 - 2y$ for $z$ in the third given equation. Solve for $y$.

$$2x - 3y + z = 3$$
$$2(-5 + 5y) - 3y + (3 - 2y) = 3$$
$$-10 + 10y - 3y + 3 - 2y = 3$$
$$5y = 10$$
$$y = 2$$

4. Substitute 2 for $y$ in the equations obtained in Steps 1 and 2 to find the values of $x$ and $z$.

$$x = -5 + 5y \qquad z = 3 - 2y$$
$$x = -5 + 5 \cdot 2 \qquad z = 3 - 2 \cdot 2$$
$$x = 5 \qquad z = -1$$

5. Check by substituting $(5, 2, -1)$ in each of the given equations.

$$2 \cdot 2 + (-1) = 3$$
$$5 - 2 + 2(-1) = 1$$
$$2 \cdot 5 - 3 \cdot 2 + (-1) = 3$$

∴ the solution set is $\{(5, 2, -1)\}$.

---

\*Although such equations are called linear, their graphs are not lines, but planes in space.

By forming linear combinations of the given equations in a system of three linear equations in three variables, you can obtain two equations in which one of the variables does not appear.

**Example 2.** Solve simultaneously:

$$x - 2y - z = 3$$
$$2x + 3y + 2z = 4$$
$$x + 2y + 3z = 7$$

*Solution:*

1. To replace the second equation by one not involving $x$,

   (a) multiply the first equation by 2;
   (b) subtract the result from the second equation.

$$2(x - 2y - z) = 2 \cdot 3$$
$$2x - 4y - 2z = 6$$

$$2x + 3y + 2z = 4$$
$$\underline{2x - 4y - 2z = 6}$$
$$7y + 4z = -2$$

2. To replace the third equation by one not involving $x$, subtract the first equation from it.

$$x + 2y + 3z = 7$$
$$\underline{x - 2y - \ z = 3}$$
$$4y + 4z = 4$$
$$y + \ z = 1$$

3. The equations obtained in Steps 1 and 2 involve only $y$ and $z$. You can solve them simultaneously by the methods of Section 3–5.

$$y = -2$$
$$z = 3$$

4. Substitute $-2$ for $y$ and 3 for $z$ in the first of the given equations and solve for $x$. Then check as in Example 2.

$$x - 2y - z = 3$$
$$x - 2(-2) - 3 = 3$$
$$x + 1 = 3$$
$$x = 2$$

$\therefore$ the solution set is $\{(2, -2, 3)\}$.

## ORAL EXERCISES

Is the given ordered triple a solution of the given equation?

**1.** $x + y - z = 1$; $(1, 1, 1)$

**2.** $2x - y + z = 4$; $(3, 1, -1)$

**3.** $3x + 2y - 2z = -3$; $(1, -1, -2)$

**4.** $-2x + 5y + z = 6$; $(-1, 1, 3)$

**5.** $x - 2y - 2z = -2$; $(0, 1, 0)$

**6.** $x - 2y - 2z = -2$; $(0, 0, 1)$

Find two members of each of the following sets.

**7.** $\{(x, y, z): z = 2\}$

**8.** $\{(r, s, t): r = -1\}$

**9.** $\{(x, y, z): x + y = 2\}$

**10.** $\{(u, v, w); v - w = 0\}$

**11.** $\{(a, b, c): a + 2b - c = 4\}$

**12.** $\{(l, m, n): 2l - m - n = -2\}$

**13.** $\{(x, y, z): -x + 2y - z = 4\}$

**14.** $\{(x, y, z): -2x - y + z = 2\}$

## WRITTEN EXERCISES

Solve each system of equations.

**A**

**1.** $\begin{aligned} x + 5y - 2z &= 3 \\ y &= 0 \\ 3x \phantom{{}+5y} - 5z &= 8 \end{aligned}$

**2.** $\begin{aligned} x + y + z &= 2 \\ 3x - 2y &= -1 \\ z &= 4 \end{aligned}$

**3.** $\begin{aligned} x + y &= 2 \\ 2y - z &= 3 \\ 2x + y + z &= 2 \end{aligned}$

**4.** $\begin{aligned} y - z &= 2 \\ 2x - 2y + z &= 4 \\ x \phantom{{}-2y} + 5z &= 4 \end{aligned}$

**5.** $\begin{aligned} 2x - 2y + 4z &= -18 \\ x \phantom{{}-2y} + 3z &= -11 \\ -x - 3y + z &= -4 \end{aligned}$

**6.** $\begin{aligned} x - 3y &= 0 \\ 2x + y + z &= 2 \\ -x + 6y - 3z &= -6 \end{aligned}$

**7.** $\begin{aligned} 2x + 3y - z &= 2 \\ x - y + 2z &= -9 \\ 3x + 2y + 2z &= -10 \end{aligned}$

**8.** $\begin{aligned} 5x - 2y + 3z &= -13 \\ 2x - y - 4z &= -1 \\ x + 3y + 3z &= 12 \end{aligned}$

**9.** $\begin{aligned} 2a + b - 4c &= -4 \\ a - 3b + c &= -1 \\ -a + 2b + 3c &= 6 \end{aligned}$

**10.** $\begin{aligned} 3a + 9b - 4c &= 7 \\ a + b + c &= 1 \\ 5a + 5b - 3c &= 5 \end{aligned}$

**B**

**11.** $\begin{aligned} 2r + 3s + 2t &= 8 \\ 3r + 2s + 5t &= 13\tfrac{1}{2} \\ 5r - 3s - 3t &= -6\tfrac{1}{2} \end{aligned}$

**12.** $\begin{aligned} 4r + 3s - t &= 5 \\ 3r + 2s + 2t &= 4 \\ 2r - 5s + 3t &= 3\tfrac{1}{3} \end{aligned}$

**13.** When a projectile is fired from the top of a tower under certain specified conditions, its height $h$ (in feet) above the ground $t$ seconds after being fired is given by a formula of the form

$$h = a + bt + ct^2,$$

where $a$, $b$, and $c$ are constants. Determine $a$, $b$, and $c$ if the height of the projectile is 144 feet 1 second after being fired, 156 feet 2 seconds after being fired, and 84 feet 4 seconds after being fired.

Find *a*, *b*, and *c* so that the graph of the equation $y = ax^2 + bx + c$ will pass through the given points.

**14.** $(0, 3), (1, 3), (2, 1)$

**15.** $(0, 4), (1, 5), (-1, 7)$

Find the solution set of each system.

| C | **16.** | $x + 2y - z = 3$ | **18.** | $2a + b + 2c = 1$ |
|---|---------|------------------|---------|-------------------|

**16.**
$x + 2y - z = 3$
$2x - y + z = 3$
$3x - 4y + 2z = -1$

**17.**
$3x + 2y - z = 4$
$x + 3y + 2z = -1$
$2x - y + 3z = 5$

**18.**
$2a + b + 2c = 1$
$a + 2b - 3c = 4$
$3a - b + c = 0$

**19.**
$p + q + 4r = 1$
$-2p - q + r = 2$
$3p - 2q + 3r = 5$

## PROBLEMS

Solve each problem using 3 variables.

A

**1.** A chef is preparing the base for a French dressing using tomato juice, vinegar, and olive oil. His recipe calls for 3 times as much vinegar as tomato juice and 4 times as much olive oil as vinegar. If he wants to make 8 cups of dressing, how many cups of each of these ingredients should he use?

**2.** During the months of January, February, and March, Eva worked a total of 38 hours after school and on weekends shoveling snow. She worked twice as many hours in February as in March, and if she had worked 4 more hours in January, she would have worked twice as many hours in January as in February. How many hours did she work each month?

**3.** The sum of three numbers is 1. Twice the first number is the negative of the third number, and the sum of the second and third numbers is −1. Find the three numbers.

**4.** Find three numbers such that the sum of the first and second numbers is 3 less than the third number, the sum of the second number and twice the third number is 3 greater than the first number, and the sum of all three numbers is 1.

**5.** In his pocket Robert has $1.65 in quarters, dimes, and nickels. He has 11 coins in all, with twice as many nickels as dimes. How many of each coin does he have?

**6.** Miss Hanscom has $47 in one-dollar, five-dollar, and ten-dollar bills. If she has the same number of five-dollar bills as ten-dollar bills and she has 8 bills in all, how many bills of each denomination does she have?

## Chapter Summary

1. Two ordered pairs are **equal** if their first coordinates are equal and their second coordinates are equal.

2. The **solution set** of an open sentence in two variables is the set of all ordered pairs of numbers which belong to the domains of the variables and for which the sentence is true.

3. By means of a **Cartesian coordinate system**, each ordered pair can be assigned to a unique point in the plane, and vice versa. The **graph of an ordered pair** is the point in the coordinate plane corresponding to the ordered pair, and the **graph of an open sentence in two variables** is the set of points corresponding to the ordered pairs in the solution set of the open sentence.

4. The graph of every **linear equation** in two variables, this is, every equation of the form $Ax + By = C$, with $A$ and $B$ not both 0, is a *line*. The graph of a **linear inequality**, for example, $Ax + By \leq C$, is a *half plane*, which may or may not include its boundary.

5. If $P(x_1, y_1)$ and $Q(x_2, y_2)$ are any two points with $x_1 \neq x_2$ on the line with equation $Ax + By = C$ $(B \neq 0)$, then the **slope** of the line is

$$\frac{y_2 - y_1}{x_2 - x_1}, \quad \text{or} \quad -\frac{A}{B}.$$

Lines which rise from left to right have positive slope; lines which fall from left to right have negative slope. Horizontal lines have slope zero; vertical lines have no slope. If two (different) lines in the plane have the same slope or no slope, they are **parallel**.

6. The **point-slope form** of the equation of the line having slope $m$ and containing the point $(x_1, y_1)$ is $y - y_1 = m(x - x_1)$. The **slope-intercept form** of the line having slope $m$ and $y$-intercept $b$ is $y = mx + b$.

7. The **solution set of a system** of two simultaneous open sentences in two variables is the set of all ordered pairs whose coordinates satisfy *both* open sentences. The **graph of a system** of linear open sentences is the intersection of the graphs of the open sentences in the system.

8. Two systems of equations are **equivalent** if they have the same solution set. Transformations producing an equivalent system of linear equations can be used to solve systems of linear equations.

9. An equation of the form $Ax + By + Cz = D$, with $A$, $B$, and $C$ not all 0, is called a **linear equation in three variables**. Solutions of such an equation are **ordered triples** of real numbers.

**Vocabulary and Spelling**

Review each term by reference to the page listed.

first coordinate (*p. 84*)
second coordinate (*p. 84*)
solution set of open sentence
  in two variables (*p. 84*)
rectangular (Cartesian) coordinate
  system (*p. 88*)
axes (*p. 88*)
origin (*p. 88*)
quadrant (*p. 88*)
graph of ordered pair (*p. 89*)
plotting a point (*p. 89*)
abscissa (*p. 89*)
ordinate (*p. 89*)
graph of open sentence in
  two variables (*p. 89*)
linear equation (*pp. 89, 118*)
coefficient (*p. 90*)
slope (*p. 94*)
point-slope form (*p. 100*)
*y*-intercept of line (*p. 101*)

slope-intercept form (*p. 101*)
parallel lines (*p. 101*)
*x*-intercept of line (*p. 101*)
consistent equations (*p. 105*)
inconsistent equations (*p. 105*)
dependent equations (*p. 105*)
independent equations (*p. 105*)
system of simultaneous equations
  (*p. 105*)
simultaneous solutions (*p. 105*)
linear combination (*p. 105*)
tail wind (*p. 110*)
head wind (*p. 110*)
wind speed (*p. 110*)
airspeed (*p. 110*)
ground speed (*p. 110*)
linear inequality (*p. 114*)
half-plane (*p. 114*)
system of linear inequalities (*p. 114*)
ordered triple (*p. 117*)

## Chapter Test

**3–1** **1.** Find 2 solutions of the equation $4x - 3y = 12$ over $\Re$.

    **2.** The tens digit in a certain 2-digit numeral is twice the units digit. Find all positive integers represented by such numerals.

**3–2** **3.** Graph the equation $3x + 2y = -6$.

**3–3** **4.** Find the slope of the line $4x - y = 7$.

    **5.** Sketch the graph of the line containing the point $(1, 0)$ and having slope 2.

**3–4** **6.** Write an equation in (**a**) point-slope form and (**b**) slope-intercept form for the line containing the points $(2, -2)$ and $(-4, 1)$.

    **7.** Find an equation for the line which contains the point $(-5, -1)$ and is parallel to the line $x - y = 2$.

**3–5**   Find the solution set of each system.

   **8.** $2x - y = 0$
          $x + 2y = 5$

   **9.** $-3x + 12y = 15$
          $x - 4y = -5$

**3–6**   **10.** With a certain tail wind an airplane reached its destination, 630 miles away, in $1\frac{1}{2}$ hours. Flying back against the same wind, the plane took 15 minutes longer to make the trip. Find the wind speed and the plane's airspeed.

**3–7**   **11.** Graph the solution set of the system: $y - x \leq -2$
                                                      $y + x \geq 6$

**3–8**   **12.** Solve simultaneously: $2x + y - 3z = 1$
                                          $x - 2y + 4z = 3$
                                          $3x + 5y + z = -2$

---

## Chapter Review

---

**3–1**   **Open Sentences in Two Variables**   *Pages 83–88*

   **1.** Find $a$ and $b$ if $(a, -3) = (|-2|, b + 1)$.

Find a solution of the given equation over the specified domain.

   **2.** $2x + 3y = 12$; $\mathcal{R}$

   **3.** $|x| + |y| = 2$; {negative integers}

   **4.** The perimeter of a rectangle is 10 inches. Find all integral possibilities for the number of inches in its length and its width.

**3–2**   **Cartesian Coordinates**   *Pages 88–93*

   **5.** Does the point $(3, -2)$ lie on the line $2x - 2y = 2$?

   **6.** Graph the equation $2x - 3y = 12$.

   **7.** The abscissa of the point in which the line $5x + 2y = 10$ intersects the $x$-axis is __?__, and the ordinate of the point in which the line intersects the $y$-axis is __?__.

**3–3**   **Slope of Lines**   *Pages 93–99*

   **8.** Find the slope of the line $3x + 2y = 6$.

   **9.** Sketch the graph of the line containing $(-1, 2)$ and having slope $\frac{3}{2}$.

**3–4**   **Equations of Lines**     *Pages 99–104*

**10.** Write an equation in (**a**) point-slope form and (**b**) slope-intercept form for the line containing the points $(2, 1)$ and $(-2, 3)$.

Find an equation for the line satisfying the given conditions.

**11.** Containing $(2, 1)$ and having slope $\frac{2}{3}$.

**12.** Containing $(1, 0)$ and parallel to the graph of $2x - 2y = 5$.

**13.** With $x$-intercept 2 and $y$-intercept $-1$.

**3–5**   **Systems of Two Linear Equations**     *Pages 104–110*

**14.** Is $(2, -3)$ a solution of the system $\begin{aligned} 3x + 2y &= 0 \\ 2x - y &= 1 \end{aligned}$ ?

Find the solution set of each system.

**15.** $2x + 3y = -1$
  $3x - 2y = 5$

**16.** $9x - 6y = 2$
  $12x - 8y = 4$

**3–6**   **Applications of Systems**     *Pages 110–113*

**17.** Lisa has 55¢ in dimes and nickels in her wallet. If she has 2 more nickels than dimes, how many nickels and dimes does she have?

**18.** Find $a$ and $b$ so that the graph of $ax + by = 4$ contains $(-1, -1)$ and $(9, 1)$.

**3–7**   **Linear Inequalities**     *Pages 114–117*

**19.** Does the point $(1, 2)$ lie above, on, or below the line $y = 2x - 1$?

**20.** Graph the inequality $x - 2y \geq -6$.

**21.** Graph the solution set of the system: $\begin{aligned} y - 2x &\leq 4 \\ y - 2x &\geq -2 \end{aligned}$

**3–8**   **Systems of Three Equations**     *Pages 117–121*

**22.** Find the solution set of the system: $\begin{aligned} x - 2y + z &= -1 \\ 3x + y - 2z &= 7 \\ 2x - y + 3z &= -4 \end{aligned}$

**23.** The sum of three numbers is $-1$. The sum of the first and second numbers is 1 less than the third number, and the sum of the first and third numbers is 1. Find the three numbers.

# Emmy Noether

If it were necessary to find one or two words to characterize the mathematical work of the last 150 years, "rigorous," "abstract," or "axiomatic" might be likely choices. The success which men such as Abel and Dedekind found with the abstract approach encouraged many mathematicians in the latter half of the 19th century to adopt this point of view. In fact, the methods of generalization and abstraction led to such significant results during these years that by the 1930's mathematicians began to predict that they would soon run out of things to generalize and that these methods would lose their power. Strangely enough, it was during this period that the abstract approach was used most effectively by the German mathematician Emmy Noether to reach important new results in the field of algebra.

Amalie Emmy Noether (1882–1935) grew up in a world of mathematics. Her father was a professor of mathematics at the University of Erlangen, and her younger brother also studied mathematics. However, as a child, Miss Noether did not show exceptional interest in mathematics. She qualified as a teacher of French and English and in fact, did not choose to study mathematics until she was enrolled in the University of Erlangen. She received her degree from Erlangen and then in 1916 moved to the University of Göttingen, where she helped Hilbert and Klein develop the mathematical aspects of the theory of relativity. Since tradition opposed the appointment of women to university positions, only the continual persuasion of the mathematics faculty made her appointment to a minor position possible. In the following years her work gave evidence of her great creative power in abstract algebra. She lectured to classes of enthusiastic students at Göttingen until 1933 when the Nazis came to power and she and other outstanding scientists of Jewish background were denied the right to teach. She then came to the United States and was warmly received at Bryn Mawr College, where she taught until her death.

# Extra for Experts

## LINEAR PROGRAMMING

Systems of linear inequalities are often used in decision making; for example, in economics. Consider the following situation:

An automobile assembling company produces two models of sports cars, Leopards and Tigers, on which it makes profits of $550 and $650 per car, respectively. These cars are alike except for their engines, and enough bodies, frames, etc., are available to build 300 of them a month. The company can obtain as many as 250 Leopard engines and 200 Tiger engines each month. It takes 80 man-hours to assemble a Leopard and 140 man-hours to assemble a Tiger; 31,500 man-hours are available each month. How many of each model should the company build per month to make its total profit as great as possible? The given data can be arranged in a chart:

| | Input and output per car | | Total amount available |
|---|---|---|---|
| | Leopard | Tiger | |
| Input: Bodies, frames, etc. | 1 | 1 | 300 |
| Leopard engine | 1 | 0 | 250 |
| Tiger engine | 0 | 1 | 200 |
| Labor (man-hours) | 80 | 140 | 31,500 |
| Output: Profit ($) | 550 | 650 | |

Let $x$ = the number of Leopards to be built;
$y$ = the number of Tigers to be built.

If $P$ denotes the total profit in dollars, then $P = 550x + 650y$.

The company wishes to *maximize P* (cause $P$ to take on its greatest value) subject to these inequalities (*constraints*):

1. $x + y \leq 300$
2. $x \leq 250$  The total amount of each input cannot exceed the amount available.
3. $y \leq 200$
4. $80x + 140y \leq 31,500$

5. $x \geq 0$  A negative number of cars cannot be built.
6. $y \geq 0$

(Because the constraints, as well as $P$, are linear in $x$ and $y$, this is called a *linear programming* problem.)

The intersection of the solution sets of these constraint inequalities is shown as the shaded region in Figure 3–17 and is called the *feasible region* because the coordinates of each of its points satisfy all the constraints.

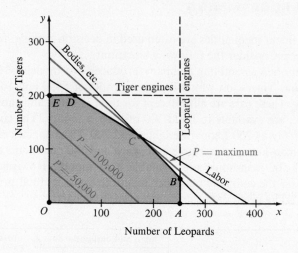

FIGURE 3–17

For any value of $P$, such as $P = 50,000$ or $P = 100,000$, the graph of the profit equation $550x + 650y = P$ is a line. You wish to find the greatest value of $P$ such that the corresponding line contains a point of the feasible region, and also to find the coordinates of this point.

In many cases, as in this example, we have a convex polygonal feasible region. It can be shown that when a linear expression like $550x + 650y$ is evaluated over such a region, it takes on its maximum value at one vertex of the polygon and its minimum value at another vertex. By substitution you find that the expression for $P$ takes on the following values at the vertexes, whose coordinates are found by solving simultaneously the equations of adjacent sides of the polygon:

| Vertex | $550x + 650y$ | $P$ |
|---|---|---|
| $O(0, 0)$ | $550(0) + 650(0)$ | 0 |
| $A(250, 0)$ | $550(250) + 650(0)$ | 137,500 |
| $B(250, 50)$ | $550(250) + 650(50)$ | 170,000 |
| $C(175, 125)$ | $550(175) + 650(125)$ | 177,500 |
| $D(50, 200)$ | $550(50) + 650(200)$ | 157,500 |
| $E(0, 200)$ | $550(0) + 650(200)$ | 130,000 |

Thus, over the feasible region, the maximum for $P$ is 177,500, and this occurs at the point $C(175, 125)$. Consequently to maximize its total profit this company should build 175 Leopards and 125 Tigers each month, and its profit would then be $177,500 per month.

# Exercises

Exercises 1 and 2 refer to the example discussed above with the constraints unchanged.

1. How many of each model should be built to maximize total profit if the profit per Leopard is $500 and the profit per Tiger is $800?

2. How many of each model car should be built to maximize total profit if the profit on each Leopard is greater than the profit on each Tiger? (*Hint:* Compare the slope of the graph of the profit equation with the slope of the graph of $x + y = 300$, and use Figure 3–17 to give a geometric argument.)

3. A convex polygonal region has vertexes $(1, 0)$, $(2, 1)$, $(2, 2)$, $(0, 3)$ and $(0, 1)$. Find the maximum and minimum values of $2x + 5y$ over this region.

4. Find the maximum and minimum values of $x - 2y$ over the intersection of the solution sets of the inequalities $1 \leq y \leq 3$, $y \leq x$, and $y \leq 7 - x$.

5. To fill part of her vitamin needs, Miss Smith intends to buy a supply of pills which must contain 1300 units of $B_1$, 1800 units of $B_2$ and 1000 units of $B_6$. For 7 cents a pill she can buy Brand 1, each pill containing 6 units of $B_1$, 4 units of $B_2$, and 2 units of $B_6$. Also available, at 12 cents a pill, is Brand 2, each pill containing 2 units of $B_1$, 6 units of $B_2$ and 4 units of $B_6$. How many pills of each brand should she buy to fill her needs at lowest cost?

6. A farmer finds that the production of 100 bushels of corn requires 2.5 acres of land, $70 in capital, 2 hours of labor in August, and 2 hours of labor in September. To produce 100 bushels of wheat he needs 5 acres of land, $50 in capital, 4 hours of labor in August, and 10 hours of labor in September. Available to him are 100 acres of land, $2100 in capital, 200 hours of labor in August, and 160 hours of labor in September. One hundred bushels of corn will bring a return of $150 and 100 bushels of wheat a return of $250. How should he divide his production between corn and wheat to make the dollar return as large as possible?

# 4

# Introduction to Functions

## FUNCTIONS AND FUNCTIONAL NOTATION

### 4–1   Relations

No doubt you are familiar with the word *relationship* as it is used in everyday life. For example, the phrase "was the husband of" describes a relationship between persons. Because the sentence "Prince Albert was the husband of Queen Victoria" is true, you can say that the *ordered pair* (Queen Victoria, Prince Albert) is in this relationship, or *relation.*

In mathematics any set of ordered pairs is a **relation**. The set of first coordinates of these pairs is the **domain** of the relation, and the set of second coordinates is its **range**. Do you see that the relation

$$\{(1, 3), (2, 3), (3, 2), (3, -1), (4, 0)\},$$

whose graph is shown in Figure 4–1, has $\{1, 2, 3, 4\}$ as its domain and $\{-1, 0, 2, 3\}$ as its range?

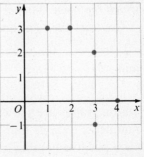

FIGURE 4–1

You can think of a relation as assigning to each member $x$ of its domain one or more members of its range, namely, the second coordinates of the ordered pairs having $x$ as first coordinate. Thus, the relation described in the preceding paragraph assigns 3 to 1, 3 to 2, both 2 *and* $-1$ to 3, and 0 to 4. Usually a relation is described by a formula or rule in the form of an open sentence which tells you how this assignment is made.

Relations are often named by single letters, as in the following example.

**Example.**   **a.** Graph the relation

$$K = \{(x, y): x + y = 6, x \in \mathcal{R} \text{ and } y \in \mathcal{R}\}.$$

**b.** Find the elements of the range which $K$ assigns to 2 and to 5.

*Solution:*   **a.** The graph of $K$ is just the graph of the equation $x + y = 6$, or $y = 6 - x$, namely, the line shown at the right.

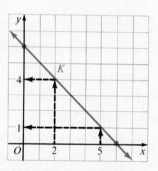

**b.** From the equation $y = 6 - x$, you can see that when $x$ is 2, $y$ is 4; that is, $K$ assigns 4 to 2. Similarly, $K$ assigns 1 to 5. (Notice the geometrical construction in the diagram at the right.)

For brevity you can omit the "$x \in \mathcal{R}$ and $y \in \mathcal{R}$" part of the description of $K$ in the example above. From now on, *whenever a relation is described by an open sentence and the domain and range are not specified, include in the domain and range those and only those real numbers for which the open sentence is true.*

Contrast the relations

$$A = \{(x, y): y = |x|\}$$

and

$$B = \{(x, y): x = |y|\}$$

whose graphs are shown in Figure 4–2. The relation $A$ assigns to each number $x$ a *single* number, its absolute value $|x|$, whereas $B$ assigns to each positive number $x$ the *two* numbers having $x$ as absolute value.

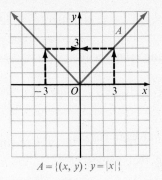

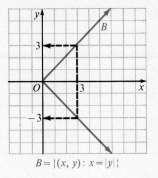

$A = \{(x, y): y = |x|\}$

$B = \{(x, y): x = |y|\}$

FIGURE 4–2

## ORAL EXERCISES

State the domain and range of each relation.

**1.** $\{(6, 6), (5, 5), (4, 4)\}$

**2.** $\{(-1, 3), (-2, 2), (-3, 1)\}$

**3.** $\{(5, 2), (0, 2), (-5, 2)\}$

**4.** $\{(3, 2), (3, 4), (3, 5)\}$

**5.** $\{(5, 2), (5, 1), (4, 0)\}$

**6.** $\{(-1, 1), (-2, 2), (-3, 3)\}$

**7.** $\{(4, 2), (5, 2), (3, 0), (2, 0)\}$

**8.** $\{(0, 0), (0, 7), (-1, -6), (-1, -7)\}$

State the domain of each relation (see page 132).

**Sample.** $\left\{(x, y): y = \dfrac{2}{(x - 2)(x + 2)}\right\}$

*Solution:* $\{x: (x - 2)(x + 2) \neq 0\}$ that is,

$\{$all real numbers except 2 and $-2\}$.

**9.** $\left\{(x, y): y = \dfrac{3}{x - 5}\right\}$

**10.** $\left\{(x, y): y = \dfrac{2}{x + 3}\right\}$

**11.** $\left\{(u, v): v = \dfrac{3}{|u| - 4}\right\}$

**12.** $\left\{(t, r): r = \dfrac{5}{|t + 1|}\right\}$

**13.** $\left\{(z, y): y = \dfrac{z}{|z| + 1}\right\}$

**14.** $\left\{(t, x): x = -\dfrac{2t}{2|t| + 3}\right\}$

**15.** $\{(x, y): |x| + |y| = 1\}$

**16.** $\left\{(x, y): y = \dfrac{4}{1 - |x|}\right\}$

*WRITTEN EXERCISES*

State the range and graph each of the following relations if
$x \in \{-3, -2, -1, 0, 1, 2, 3\}$.

**Sample.**    $\{(x, y): y \geq 2 + |x|\}$

*Solution:*    **1.** Replace $x$ with each of the values in turn.

**2.** Determine *all* corresponding real values of $y$.

**3.** Since the least value for $y$ is 2, the range is $\{y: y \geq 2\}$.

| $x$ | $|x|$ | $y \geq 2 + |x|$ |
|-----|-------|------------------|
| $-3$ | 3 | $y \geq 2 + 3 = 5$ |
| $-2$ | 2 | $y \geq 2 + 2 = 4$ |
| $-1$ | 1 | $y \geq 2 + 1 = 3$ |
| 0 | 0 | $y \geq 2 + 0 = 2$ |
| 1 | 1 | $y \geq 2 + 1 = 3$ |
| 2 | 2 | $y \geq 2 + 2 = 4$ |
| 3 | 3 | $y \geq 2 + 3 = 5$ |

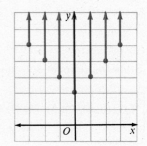

**A**  **1.** $\{(x, y): y = 2x + 2\}$

**2.** $\{(x, y): y = 4 - x\}$

**3.** $\{(x, y): y \geq x + 1\}$

**4.** $\{(x, y): y < 1 - 2x\}$

**5.** $\{(x, y): y = -4\}$

**6.** $\{(x, y): y = 3\}$

**7.** $\{(x, y): y = 2|x|\}$

**8.** $\{(x, y): y = -|x|\}$

**9.** $\{(x, y): 2x + y \leq 4\}$

**10.** $\{(x, y): 3x - 4y \leq 12\}$

**11.** $\{(x, y): y = 1 + |x|\}$

**12.** $\{(x, y): y = 2 - |x|\}$

Graph each relation over $\mathcal{R}$.

**13.** $\{(x, y): x - 3y = 6\}$

**14.** $\{(x, y): 4x + y = 8\}$

**15.** $\{(r, s): s = |r + 1|\}$

**16.** $\{(y, z): z = -|y - 2|\}$

**B**  **17.** $\left\{(x, y): y = \dfrac{1}{x}\right\}$

**18.** $\{(t, d): 0 \leq d \leq 2t + 1\}$

**19.** $\left\{(x, y): y \geq -\dfrac{1}{x}\right\}$

**20.** $\{(n, m): 0 \leq m \leq 4 - n\}$

**C**  **21.** $\{(x, y): y = x|x|\}$

**22.** $\{(x, y): y = |x| - x\}$

**23.** $\left\{(x, y): y = \dfrac{|x|}{x}\right\}$

**24.** $\{(x, y): y = |x| + x\}$

## 4–2 Functions

A **function** is a relation which assigns to each member of its domain a *single* member of its range. Thus of the two relations pictured in Figure 4–3, *F* is a function but *H* is not. (*H* associates both 1 and 3 with 2.)

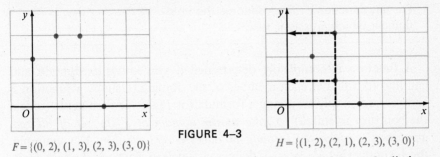

**FIGURE 4–3**

$F = \{(0, 2), (1, 3), (2, 3), (3, 0)\}$ $H = \{(1, 2), (2, 1), (2, 3), (3, 0)\}$

You can think of a function as a relation in which each distinct ordered pair has a different first coordinate. From this observation we obtain the *vertical-line test:*

> No vertical line intersects the graph of a function in more than one point.

You can see by using this test that the curve shown in color in Figure 4–4 is *not* the graph of a function. By contrast, the curve in Figure 4–5 *is* the graph of a function.

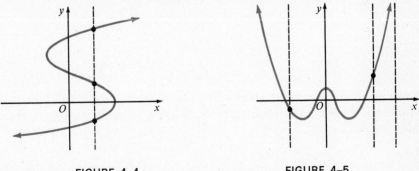

**FIGURE 4–4** **FIGURE 4–5**

The letters *f*, *F*, *g*, *G*, and *φ* (Greek *phi*, pronounced fē or fī) are often used to name functions. For example, you might write

$$f = \{(x, y): y = 2x - 1\},$$

to indicate that *f* is the function which pairs each number with its double decreased by 1. To designate the unique number associated with 3, for example, we use the symbol $f(3)$, read "$f$ of 3" or "the value of $f$ at 3." Thus $f(3) = 2 \cdot 3 - 1 = 5$. [*Warning: $f(3)$ does not mean $f$ multiplied by 3.*] Similarly, $f(1) = 2 \cdot 1 - 1 = 1$, and $f(0) = 2 \cdot 0 - 1 = -1$.

---

For any function $F$, $F(x)$ denotes the **value of $F$ at $x$**, that is, the number which $F$ assigns to $x$.

---

A function is completely determined if you know its domain and can find its value at each member of the domain. Usually a function $F$ is described simply by giving a formula for $F(x)$. We then agree (unless the contrary is stated) to *take as the domain of $F$ the set of all real numbers for which the formula produces a real number.* For example, the formula $f(x) = 2x - 1$ defines a function $f$ whose domain is $\Re$. But the domain of the function $F$ defined by $F(x) = \dfrac{1}{x}$ is the set of all *nonzero* real numbers.

For brevity we often replace a phrase like "the function $g$ defined by $g(x) = 3x - 2$," by the phrase "the function $g(x) = 3x - 2$."

**Example.** If $g(x) = 3x - 2$ and $\varphi(x) = x + 1$, find:

    **a.** $g(2)$     **b.** $g(2) \cdot \varphi(2)$     **c.** $g(\varphi(2))$     **d.** $\varphi(g(2))$

*Solution:*     **a.** $g(2) = 3 \cdot 2 - 2 = 6 - 2 = 4$

           **b.** $\varphi(2) = 2 + 1 = 3$, $\therefore g(2) \cdot \varphi(2) = 4 \cdot 3 = 12$

           **c.** $g(\varphi(2)) = g(3) = 3 \cdot 3 - 2 = 9 - 2 = 7$

           **d.** $\varphi(g(2)) = \varphi(4) = 4 + 1 = 5$

---

*ORAL EXERCISES*

State whether the relation specified by graph, roster, or rule is a function. If it is not a function, explain why not. If it is a function, give its domain and range.

**1.**         **2.**         **3.**

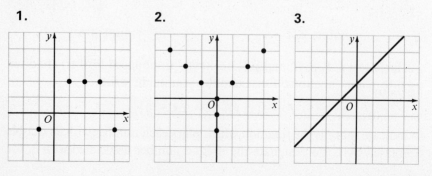

**4.**   **5.**   **6.**

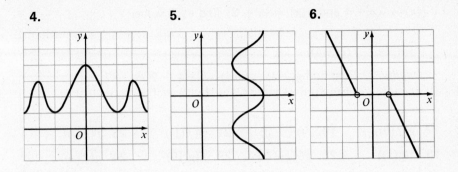

**7.** $\{(1, 3), (2, 4), (3, 5), (4, 6)\}$

**8.** $\{(1, 3), (2, 3), (3, 3), (4, 3)\}$

**9.** $\{(1, 3), (1, 4), (1, 5), (1, 6)\}$

**10.** $\{(2, 1), (2, 2), (3, 1), (3, 2)\}$

**11.** $\{(5, 1), (6, 2), (7, 0), (8, 0)\}$

**12.** $\{(0, 5), (2, 1), (1, -3), (0, 0)\}$

**13.** $\{(x, y): y = 3x\}$

**14.** $\{(t, d): d = 2t + 1\}$

**15.** $\{(m, n): n = |m|\}$

**16.** $\{(r, s): |s| = r\}$

**17.** $\{(x, y): y = 7\}$

**18.** $\{(x, y): x = -2\}$

If $f(x) = 2x + 3$, state each of the following values.

**19.** $f(0)$   **21.** $f(-3)$   **23.** $f(\frac{1}{2})$   **25.** $f(a)$

**20.** $f(3)$   **22.** $f(-10)$   **24.** $f(-\frac{1}{2})$   **26.** $f(|b|)$

## WRITTEN EXERCISES

If $f(x) = x(3x - 2)$, find:

**A**  **1.** $f(0)$   **3.** $f(-1)$   **5.** $3f(2)$   **7.** $f(a + 1)$

 **2.** $f(4)$   **4.** $f(10)$   **6.** $\frac{1}{3}f(6)$   **8.** $f(a - 1)$

If $g(x) = x + 1$ and $h(x) = 2x - 1$, find:

 **9.** $g(0)$

 **10.** $h(0)$

 **11.** $5h(0)$

 **12.** $-2g(0)$

 **13.** $g(2) - h(2)$

 **14.** $g(1) + h(1)$

 **15.** $h(3) \cdot g(3)$

 **16.** $\dfrac{h(3)}{g(2)}$

If $f(x) = 2x + 1$ and $g(x) = |x + 2|$, find each value.

**Sample.** $f(g(-4))$

*Solution :* First find $g(-4)$: $g(-4) = |-4 + 2| = |-2| = 2$.
Then $f(g(-4)) = f(2) = 2 \cdot (2) + 1 = 5$.

**B** **17.** $g(f(0))$     **19.** $f(g(-2))$     **21.** $g(f(1))$     **23.** $f(g(a))$

**18.** $f(g(1))$     **20.** $g(f(-2))$     **22.** $f(g(10))$     **24.** $g(f(a))$

If $t(x) = 3x - 12$ and $r(x) = \frac{1}{3}x + 4$, find $t(r(x))$ and $r(t(x))$ for each given value of $x$.

**25.** 0         **26.** $-1$         **27.** $a$         **28.** $-a$

**C** **29.** Let $g$ be a function for which $g(x + 2) = g(x) + g(2)$ for every real number $x$. Show that **(a)** $g(0) = 0$, and **(b)** $g(-2) = -g(2)$.

**30.** Let $h$ be a function such that $h(x + a) = h(x)$ for every real number $x$. Show that **(a)** $h(2a) = h(a)$ and **(b)** $h(2a) = h(0)$.

**31.** Let $f$ be a function such that $f(xy) = f(x) + f(y)$ for all real numbers $x$ and $y$. Show that **(a)** if $x = 0$, then $f(y) = 0$ and **(b)** $f(1) = 0$. (*Hint:* Let $x = y = 1$.)

**32.** Let $E$ be a function such that $E(x + y) = E(x) \cdot E(y)$ for all real numbers $x$ and $y$, and $E$ never has value 0. Show that **(a)** $E(0) = 1$ and **(b)** $E(-n) = \dfrac{1}{E(n)}$.

## LINEAR FUNCTIONS

### 4–3  General Linear Functions

Do you see that the function $g(x) = -2x + 3$ and the equation $y = -2x + 3$ have the same graph? It is the *line* having slope $-2$ and $y$-intercept 3 shown in Figure 4–6. For this reason the function (as well as the equation) is called *linear*.

FIGURE 4–6     $g = \{(x, y): y = -2x + 3\}$

If $f(x) = mx + b$, where $m$ and $b$ are constants, then $f$ is a **linear function**. Its graph is the same as that of the equation $y = mx + b$, namely the line having slope $m$ and $y$-intercept $b$ (Figure 4–7).

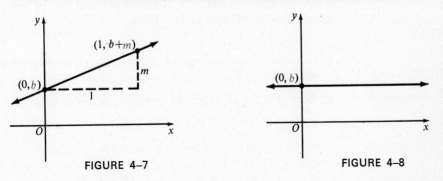

FIGURE 4–7                    FIGURE 4–8

If $m = 0$, $f$ is a **constant function**, $f(x) = b$, whose graph is a horizontal line, as illustrated in Figure 4–8.

**Example 1.**   It is known of a linear function $F$ that the slope of its graph is $\frac{1}{2}$ and that $F(4) = 3$.

 **a.** Find a formula for $F(x)$.
 **b.** Determine which, if either, of the points $(2, 1)$ and $(-2, 0)$ is on the graph of $F$.

*Solution :*   **a.** Because $F$ is linear, there are constants $m$ and $b$ such that $F(x) = mx + b$. Since $m$ is the slope of the graph of $F$,

$$F(x) = \tfrac{1}{2}x + b.$$

$F(4) = \frac{1}{2} \cdot 4 + b = 2 + b$, and since $F(4) = 3$ (given), $2 + b = 3$, or $b = 1$.

$\therefore F(x) = \frac{1}{2}x + 1.$

 **b.** The point $(2, 1)$ is on the graph of $F$ if and only if $F(2) = 1$. Since

$$F(2) = \tfrac{1}{2} \cdot 2 + 1 = 2 \neq 1,$$

$(2, 1)$ is *not* on the graph of $F$.  Since

$$F(-2) = \tfrac{1}{2}(-2) + 1 = -1 + 1 = 0,$$

the point $(-2, 0)$ *is* on the graph of $F$.

If the domain of a function of the form $f(x) = mx + b$ is a *proper* subset of $\Re$, we still continue to call the function *linear*. This situation occurs in Example 2 below, where the variable $t$ is a measure of a *physical* quantity, namely, temperature; its domain, and therefore that of the function $g$, is not all of $\Re$.

**Example 2.** Let $g(t)$ be the reading in degrees (°) on the centigrade temperature scale when the reading on the Fahrenheit scale is $t°$. The function $g$ is linear. The temperature of freezing water is 0°C (centigrade) and 32°F (Fahrenheit), and that of boiling water is 100°C and 212°F. Find:

**a.** a formula for $g(t)$;

**b.** the centigrade reading for normal body temperature, which is 98.6°F.

*Solution:* **a.** Since $g$ is linear, there are constants $m$ and $b$ such that

$$g(t) = mt + b.$$

At the temperature of freezing water, $t = 32$ and $g(t) = 0$; that is, $g(32) = 0$. Therefore $0 = g(32) = m \cdot 32 + b$. Similarly, $100 = g(212) = m \cdot 212 + b$. We now use the methods of Section 3–5 to solve the system

$$212m + b = 100$$
$$32m + b = 0$$

and obtain $m = \frac{5}{9}$, $b = -\frac{160}{9}$. Therefore

$$g(t) = \frac{5}{9}t - \frac{160}{9}.$$

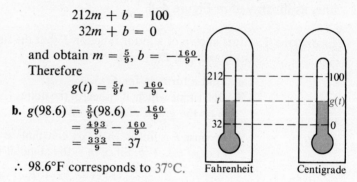

**b.** $g(98.6) = \frac{5}{9}(98.6) - \frac{160}{9}$

$= \frac{493}{9} - \frac{160}{9}$

$= \frac{333}{9} = 37$

∴ 98.6°F corresponds to 37°C.

ORAL EXERCISES

State whether or not there is a linear function containing the ordered pairs in each table. If there is, state the slope of its graph.

Sample 1.

| x | 2 | 3 | 4 | 5 |
|---|---|---|---|---|
| y | 8 | 9 | 10 | 11 |

*Solution:* The specified function is linear because for each pair of ordered pairs, $(x_1, y_1)$ and $(x_2, y_2)$, in the table,

$$\frac{y_2 - y_1}{x_2 - x_1} = 1.$$

The slope of its graph is 1.

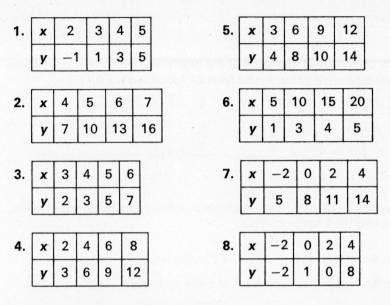

**1.**

| x | 2 | 3 | 4 | 5 |
|---|---|---|---|---|
| y | −1 | 1 | 3 | 5 |

**5.**

| x | 3 | 6 | 9 | 12 |
|---|---|---|---|---|
| y | 4 | 8 | 10 | 14 |

**2.**

| x | 4 | 5 | 6 | 7 |
|---|---|---|---|---|
| y | 7 | 10 | 13 | 16 |

**6.**

| x | 5 | 10 | 15 | 20 |
|---|---|---|---|---|
| y | 1 | 3 | 4 | 5 |

**3.**

| x | 3 | 4 | 5 | 6 |
|---|---|---|---|---|
| y | 2 | 3 | 5 | 7 |

**7.**

| x | −2 | 0 | 2 | 4 |
|---|---|---|---|---|
| y | 5 | 8 | 11 | 14 |

**4.**

| x | 2 | 4 | 6 | 8 |
|---|---|---|---|---|
| y | 3 | 6 | 9 | 12 |

**8.**

| x | −2 | 0 | 2 | 4 |
|---|---|---|---|---|
| y | −2 | 1 | 0 | 8 |

Each table specifies a linear function. In each case, supply the missing elements in the domain or range.

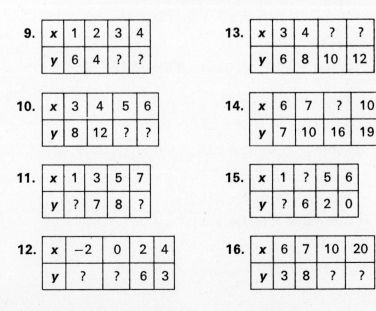

Sample 2.

| x | 2 | 4 | 6 | 8 |
|---|---|---|---|---|
| y | 3 | 7 | ? | ? |

*Solution:* For each increase of 2 in the domain, the range increases 4. Therefore, the missing elements in the range are 11 and 15.

**9.**

| x | 1 | 2 | 3 | 4 |
|---|---|---|---|---|
| y | 6 | 4 | ? | ? |

**13.**

| x | 3 | 4 | ? | ? |
|---|---|---|---|---|
| y | 6 | 8 | 10 | 12 |

**10.**

| x | 3 | 4 | 5 | 6 |
|---|---|---|---|---|
| y | 8 | 12 | ? | ? |

**14.**

| x | 6 | 7 | ? | 10 |
|---|---|---|---|---|
| y | 7 | 10 | 16 | 19 |

**11.**

| x | 1 | 3 | 5 | 7 |
|---|---|---|---|---|
| y | ? | 7 | 8 | ? |

**15.**

| x | 1 | ? | 5 | 6 |
|---|---|---|---|---|
| y | ? | 6 | 2 | 0 |

**12.**

| x | −2 | 0 | 2 | 4 |
|---|---|---|---|---|
| y | ? | ? | 6 | 3 |

**16.**

| x | 6 | 7 | 10 | 20 |
|---|---|---|---|---|
| y | 3 | 8 | ? | ? |

*WRITTEN EXERCISES*

In each Exercise 1–8, the slope of the graph of a linear function is given, and also a function value. Find a formula specifying the function.

| | | | |
|---|---|---|---|
| **A** | **1.** $3; f(2) = 1$ | **5.** | $-\frac{2}{3}; f(-2) = -1$ |
| | **2.** $4; g(3) = -1$ | **6.** | $\frac{1}{5}; h(-1) = 2$ |
| | **3.** $-3; h(0) = 4$ | **7.** | $0; f(3) = 1$ |
| | **4.** $-1; h(2) = 6$ | **8.** | $7; f(0) = 7$ |

In Exercises 9–16, two function values of a linear function are given. Find the required third function value.

**Sample.**   $f(2) = 4; f(-3) = -12; f(0) = \underline{\phantom{?}}$

*Solution:*   There are constants $m$ and $b$ such that
$$f(x) = mx + b.$$
We have
$$f(2) = m(2) + b = 4 \quad \text{and} \quad f(-3) = m(-3) + b = -12.$$
Solving the system
$$2m + b = 4$$
$$-3m + b = -12$$
we have,
$$5m = 16 \quad \text{or} \quad m = \tfrac{16}{5}.$$
Then
$$2(\tfrac{16}{5}) + b = 4 \quad \text{or} \quad 32 + 5b = 20,$$
and $b = -\tfrac{12}{5}$.
$$\therefore f(x) = \tfrac{16}{5}x - \tfrac{12}{5}, \quad \text{and} \quad f(0) = \tfrac{16}{5}(0) - \tfrac{12}{5} = -\tfrac{12}{5}.$$

**9.** $f(1) = 4; f(3) = 0; f(-1) = \underline{\phantom{?}}$

**10.** $f(-2) = -1; f(3) = 9; f(10) = \underline{\phantom{?}}$

**11.** $f(0) = 4; f(3) = -2; f(-1) = \underline{\phantom{?}}$

**12.** $f(-2) = 3; f(6) = 7; f(0) = \underline{\phantom{?}}$

**13.** $g(-5) = 7; g(-1) = 4; g(2) = \underline{\phantom{?}}$

**14.** $h(-4) = 8; h(-2) = 3; h(3) = \underline{\phantom{?}}$

**15.** $f(3) = 0; f(-3) = 4; f(7) = \underline{\phantom{?}}$

**16.** $f(5) = 3; f(-5) = 9; f(2) = \underline{\phantom{?}}$

In each Exercise 17–22, assume that the relationship described is linear, find a formula relating the specified variables, and answer the question.

**B** 17. The Thomas Car Rental Company charged $10.10 to rent a car driven 30 miles and $11.50 to rent a car driven 50 miles. What would be the charge to rent a car to drive 100 miles?

18. It costs the Acme Manufacturing Company $110 to manufacture 10 radios, and $128 to manufacture 16 radios. What will it cost them to manufacture 25 radios?

19. The taxes on a profit of $30,000 in Central City are $16,000 and on a profit of $35,000 are $17,500. What would be the taxes on a profit of $50,000?

20. If it costs $80 to ship a piece of machinery 300 miles and $92 to ship it 450 miles, what would it cost to ship the machinery 700 miles?

21. It costs the Janus Publishing Company $16,000 to print and bind 8000 books and $10,600 to print and bind 5000 of the same books. What would it cost them to print and bind 10,000 of these books?

22. A force of 40 pounds will compress a spring 12 inches in length to a length of 10.4 inches, and a 60 pound force will compress it from 12 inches to 9.6 inches. What force would it take to compress the spring to a length of 11 inches?

## 4–4  Direct Variation

The table in Figure 4–9 shows the stretch $y$, in inches, of a coil spring, caused by a load of $x$ pounds. Notice that the ratio $\dfrac{y}{x}$ is the same for each ordered pair $(x, y)$; that is, $\dfrac{y}{x} = 0.3$, or $y = 0.3x$.

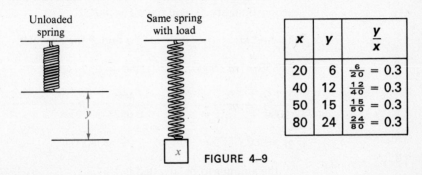

| Unloaded spring | Same spring with load | | $x$ | $y$ | $\dfrac{y}{x}$ |
|---|---|---|---|---|---|
| | | | 20 | 6 | $\frac{6}{20} = 0.3$ |
| | | | 40 | 12 | $\frac{12}{40} = 0.3$ |
| | | | 50 | 15 | $\frac{15}{50} = 0.3$ |
| | | | 80 | 24 | $\frac{24}{80} = 0.3$ |

FIGURE 4–9

The type of linear function associated with an equation of the form $y = mx$ is called a **direct variation**. We say that *y varies directly as x,* or *y varies as x,* or *y is directly proportional to* $x$; the constant $m$ is called the **constant of proportionality,** or **constant of variation.**

Since the ordered pair $(0, 0)$ is a solution of $y = mx$, the graph of a direct variation lies on a line passing through the origin. Also if $(x_1, y_1)$ and $(x_2, y_2)$ are any two ordered pairs in a direct variation with $x_1 \neq 0$ and $x_2 \neq 0$ (Figure 4–10), you have

$$y_1 = mx_1 \quad \text{and} \quad y_2 = mx_2,$$

or

$$\frac{y_1}{x_1} = m \quad \text{and} \quad \frac{y_2}{x_2} = m.$$

Therefore

$$\frac{y_1}{x_1} = \frac{y_2}{x_2}.$$

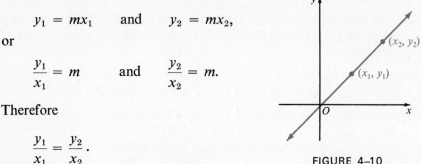

**FIGURE 4–10**

Such an equality of ratios is called a **proportion**. It can be read "$y_1$ is to $x_1$ as $y_2$ is to $x_2$," and is sometimes written $y_1 : x_1 = y_2 : x_2$. In this proportion, $y_1$ and $x_2$ are called the **extremes,** and $x_1$ and $y_2$ the **means.** Since $x_1 y_2 = x_2 y_1$, you have: *In any proportion, the product of the means equals the product of the extremes.*

**Example.**   Interest earned by money invested in a savings account varies directly as the amount invested. If $5000 earns $287.50 in a year, how much should be invested to earn $460.00 in a year?

*First Solution:*   Let $I$ denote the interest earned, in dollars, by the invested amount $P$ dollars. Then for some constant of proportionality $m$, you have $I = mP$.

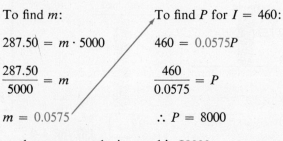

To find $m$:

$$287.50 = m \cdot 5000$$

$$\frac{287.50}{5000} = m$$

$$m = 0.0575$$

To find $P$ for $I = 460$:

$$460 = 0.0575P$$

$$\frac{460}{0.0575} = P$$

$$\therefore P = 8000$$

$\therefore$ the amount to be invested is $8000.

*Second Solution:* Let $I_1$ and $I_2$ denote the interest earned by amounts $P_1$ and $P_2$, respectively. Then

$$\frac{P_1}{I_1} = \frac{P_2}{I_2}, \quad \text{or} \quad I_1 P_2 = I_2 P_1.$$

Thus $\qquad 287.50 \times P_2 = 460 \times 5000$

or $\qquad P_2 = \dfrac{460 \times 5000}{287.50} = 8000.$

∴ the amount to be invested is $8000.

## ORAL EXERCISES

In each Exercise 1–8, an ordered pair in a direct variation is given. State the constant of proportionality.

**Sample.** (4, 12) $\qquad$ *Solution:* $\dfrac{y}{x} = \dfrac{12}{4} = 3$

**1.** (5, 20) $\qquad$ **3.** (7, 11) $\qquad$ **5.** $(1, \frac{1}{2})$ $\qquad$ **7.** $(a, 3a)$

**2.** (14, 2) $\qquad$ **4.** (18, 12) $\qquad$ **6.** $(\frac{1}{2}, 1)$ $\qquad$ **8.** $(4b, b)$

State a formula for the values of the given direct variation. Unless implied in the statement, the constant of proportionality, $m$, is given.

**9.** The circumference $C$ of a circle varies directly as the length of its radius $r$.

**10.** The distance $d$ traveled by a moving object in 10 seconds is directly proportional to its average rate $r$.

**11.** The volume $V$ of a right circular cylinder is directly proportional to the area $B$ of its base ($m = h$).

**12.** The voltage $E$ in an electric circuit of constant current is directly proportional to the resistance $R$ of the circuit ($m = I$).

## WRITTEN EXERCISES

**A**

**1.** If $y$ varies directly as $x$, and $y = 26$ when $x = 2$, find $y$ when $x = 3$.

**2.** If $y$ varies directly as $x$, and $y = 4$ when $x = 12$, find $y$ when $x = 3$.

**3.** If $t$ varies directly as $s$, and $s = 8$ when $t = 30$, find $s$ when $t = 25$.

**4.** If $r$ varies directly as $z$, and $r = 15$ when $z = 9$, find $z$ when $r = 24$.

**Sample.** If $y$ is proportional to $x + 2$, and $y = 7$ when $x = 12$, find $y$ when $x = 82$.

*Solution:* You have $y = m(x + 2)$. Substituting 12 for $x$ and 7 for $y$ produces

$$7 = m(12 + 2),$$

from which

$$7 = 14m, \quad \text{or} \quad m = \tfrac{1}{2}.$$

Using $y = \tfrac{1}{2}(x + 2)$, and replacing $x$ with 82, you find that $y = \tfrac{1}{2}(82 + 2) = \tfrac{1}{2}(84) = 42$.

**5.** If $r$ is proportional to $s - 2$, and $r = 8$ when $s = 6$, find $r$ when $s = 18$.

**6.** If $y$ is proportional to $z + 5$, and $y = 6$ when $z = 25$, find $y$ when $z = 55$.

**7.** If $m$ varies directly as $n + 2$, and $m = 35$ when $n = 5$, find $n$ when $m = 60$.

**8.** If $t$ varies directly as $s - 6$, and $t = 45$ when $s = 15$, find $s$ when $t = 20$.

**9.** If $x$ varies directly as $2y - 1$, and $x = 9$ when $y = 2$, find $y$ when $x = 15$.

**10.** If $y$ is proportional to $3x + 5$, and $y = 14$ when $x = 3$, find $x$ when $y = -7$.

If none of $x_1$, $x_2$, $y_1$, or $y_2$ denotes zero, prove each of the following properties of the proportion $\dfrac{y_1}{x_1} = \dfrac{y_2}{x_2}$ .

**B** **11.** $x_1 y_2 = x_2 y_1$    **12.** $\dfrac{x_1}{y_1} = \dfrac{x_2}{y_2}$    **13.** $\dfrac{y_1}{y_2} = \dfrac{x_1}{x_2}$    **14.** $\dfrac{x_2}{x_1} = \dfrac{y_2}{y_1}$

**15.** $\dfrac{x_1 + x_2}{x_2} = \dfrac{y_1 + y_2}{y_2}$  $\left(\textit{Hint: Add 1 to each member of } \dfrac{x_1}{x_2} = \dfrac{y_1}{y_2}.\right)$

**16.** $\dfrac{x_1 - x_2}{x_2} = \dfrac{y_1 - y_2}{y_2}$  $\left(\textit{Hint: Add } -1 \text{ to each member of } \dfrac{x_1}{x_2} = \dfrac{y_1}{y_2}.\right)$

**17.** If $y_1 \neq y_2$, $\dfrac{x_1 - x_2}{y_1 - y_2} = \dfrac{x_1}{y_1}$  (*Hint:* Use the results of Exercises 16 and 12.)

**18.** If $x_1 \neq x_2$ and $y_1 \neq y_2$, $\dfrac{x_1 + x_2}{x_1 - x_2} = \dfrac{y_1 + y_2}{y_1 - y_2}$  (*Hint:* Use the results of Exercises 15 and 16.)

**C** **19.** If $f(x) = 3x + 2$, and $a$ and $c$ are distinct real numbers, find the value of $\dfrac{f(c) - f(a)}{c - a}$ .

**20.** If $f(x) = mx + b$ $(m \neq 0)$, show that the graphs of $y = -f(x)$ and $y = f(-x)$ have the same slopes and hence are parallel lines.

**21.** Prove that if $g$ is a direct variation over $\Re$, then for every pair of real numbers $a$ and $c$, $g(a + c) = g(a) + g(c)$.

**22.** Prove that if $g$ is a linear function other than a direct variation, then for every pair of real numbers $a$ and $c$, $g(a + c) \neq g(a) + g(c)$.

**23.** Prove that if $g$ is a direct variation over $\Re$, then for each pair of real numbers $k$ and $a$, $g(ka) = kg(a)$.

**24.** Prove that if $g$ is a linear function other than a direct variation, then for every pair of real number $k \neq 1$ and $a$, $g(ka) \neq kg(a)$.

## PROBLEMS

A

**1.** In Central City, the cost of building a house varies directly as the number of square feet the house contains. If it costs $27,500 to build a home containing 1250 square feet, what would be the cost of building a home containing 1700 square feet?

**2.** If 7 gallons of premium gasoline costs $2.66, what does 11 gallons of the same gasoline cost?

**3.** Fines for speeding are frequently assessed at a fixed rate per mile an hour over the speed limit. If a driver must pay a fine of $45 for traveling 83 miles per hour and a fine of $35 for traveling 79 miles per hour, what is the legal speed limit?

**4.** If it takes 40 pounds of force to stretch a spring to a length of 18 inches, and 52 pounds of force to stretch it to $19\frac{1}{2}$ inches, what is the natural (unstretched) length of the spring?

**5.** If it takes a computer 1 second to perform 5000 multiplication operations, how many microseconds (millionths of a second) will it take the computer to perform 250 such operations?

**6.** A broker makes a commission of $1248 on a sale of $10,400. At the same rate, what would be her commission on a sale of $8000?

**7.** The dividend on 400 shares of a stock is $240 per year. At the same rate, what is the dividend on 50 shares of the stock?

**8.** If 20 pounds of copper and 48 pounds of nickel are used to form 68 pounds of an alloy, how many pounds of nickel and how many pounds of copper would be used to form 100 pounds of the same alloy?

**9.** A pump can raise the water level in a rectangular aquarium 10 inches in one half hour. How long would it take the pump to raise the water level 4 feet?

10. A manufacturing plant increases the pollution in a neighboring lake by 3 parts per million (ppm) in 8 months. How long would it take the plant to raise the pollution level by 135 ppm?

B 11. If 10 cc. of a normal specimen of human blood contains 1.2 grams of hemoglobin, how many grams of hemoglobin would 22 cc. of the same blood contain?

12. The average number of red cells in a cubic millimeter of blood is 5,000,000 for men and 4,500,000 for women. If a laboratory blood count procedure results in a dilution of 1 part of a man's blood to 199 parts of a saline solution, how many red cells should be contained in $\frac{1}{2000}$ cubic millimeter of the diluted solution?

13. The eastern plant of the Macdonald Manufacturing Company can produce 43 tractors in the same time the northern plant produces 35. If both plants work to meet a combined quota of 468 tractors, how many tractors will each plant produce?

14. In an electric circuit under a voltage of 56 volts, a meter which reads in proportion to voltage registers 77 on a scale 0–100. What is the maximum number of volts the circuit can handle in order for the meter to give a valid reading?

15. For each ounce of copper in an alloy there are 3 ounces of nickel, and for each ounce of nickel there is $\frac{1}{4}$ ounce of chromium. If there are 5 ounces of chromium in an ingot of the alloy, how many ounces of copper are in the ingot?

16. The Central City School District spends $244 per year to educate one student, and 64 cents of each dollar spent goes for teachers' pay. What is the payroll for teachers in Central City in a year when it has 4200 students enrolled in schools?

## POLYNOMIAL FUNCTIONS

### 4–5 Polynomials

From now on you will often use exponential notation in studying equations and functions. Recall that an expression such as $a \cdot a \cdot a \cdot a$ can be shortened by writing it as $a^4$, where the *exponent* 4 shows that the *base a* is used as a factor 4 times. Using this notation you can write, for example, $2 \cdot a \cdot a \cdot a \cdot a \cdot y \cdot y \cdot y$ as $2a^4y^3$. In general, if $n$ is a natural number and $a$ is a real number, the product of $n$ factors, each of which is $a$, is called the **nth power of** $a$ (also read "$a$ to the $n$th power"). In the expression $a^n$, $a$ is the **base** and $n$ is the **exponent**.

Here are some powers of $x$:

First power $\quad x^1 = x \quad\quad\quad\quad$ (ordinarily the exponent 1 is omitted)

Second power $x^2 = x \cdot x \quad\quad$ (also read "$x$ squared")

Third power $\quad x^3 = x \cdot x \cdot x \quad$ (also read "$x$ cubed")

Fourth power $x^4 = x \cdot x \cdot x \cdot x$ (also read "$x$ fourth")

A **monomial** is an expression which is either a constant or a product of a constant and one or more variables. The constant factor is called the **coefficient** of the monomial. Thus $6x^2z$, 3, $y$, and $-z^2$ are monomials having coefficients 6, 3, 1, and $-1$, respectively. Monomials which are constants, such as 4 and $-1$, or monomials which have the same variable factors, such as $3xy^2$ and $-xy^2$ are said to be **similar** or **like**. The **degree** of a nonconstant monomial is the sum of the exponents of its variables. For example, the degree of $5xz^3$ is 4 because $1 + 3 = 4$. A nonzero constant monomial has degree zero, and the monomial 0 has no degree.

A monomial or a *sum of monomials* is a **polynomial**. A polynomial is **simplified**, or in **simple form**, if no two of its terms (monomials) are similar. In the following example we simplify a polynomial by combining like terms.

**Example.** Simplify $3x^2 + 2ax - 1 - 2x^2 - 4ax$.

*Solution:* 
$$\begin{aligned}
3x^2 + 2ax - 1 - 2x^2 - 4ax &= 3x^2 - 2x^2 + 2ax - 4ax - 1 \\
&= (3 - 2)x^2 + (2 - 4)ax - 1 \\
&= x^2 - 2ax - 1
\end{aligned}$$

The **degree** of a simplified polynomial is the greatest of the degrees of its terms. Since the degrees of the terms of

$$2x^3 + 3x^2y^3 + 6$$

are, in order, 3, 5, and 0, the degree of the polynomial itself is 5.

We often wish to regard some of the letters appearing in a polynomial as constants. We would call $3h^2 - 2hx + x^2$ *a polynomial in x* if we wish to regard $h$ as a constant; if, however, both $h$ and $x$ are variables, then we refer to the expression as *a polynomial in h and x*.

A polynomial is in simple form in a given variable when no two terms are similar and its terms appear in order of either decreasing or increasing degree in that variable. Thus, $ax + a^2$ is a polynomial of two terms, a **binomial**, which is in simple form in both $x$ and $a$. The **trinomial**, or polynomial of three terms, $2y^2 + 3y^3z + yz^2$ is in simple form in $z$ but not in $y$.

*ORAL EXERCISES*

Identify each polynomial as a monomial, binomial, or trinomial, state the coefficient of each term, and give the degree of the polynomial.

**Sample.** $3x^2 - xy^3$

*Solution:* Binomial; coefficients are 3 and $-1$; degree is 4 in $x$ and $y$.

**1.** $4n^2 + 6n - 1$        **6.** $3a - 4ab + 6b^2$

**2.** $3t^2 + 4st - 3t^3$       **7.** $7 - 2x + 5x^4$

**3.** $7$                        **8.** $7x^2y + 3xy^2 - x^5y$

**4.** $-y^5$              **9.** $-9z + 3$

**5.** $4xy - 7x^2y^2 + x^3y^3$    **10.** $x^4 - 3xy^3 + 2x^2y^2 - 4xy^5$

Express each polynomial in simple form in *x* using decreasing powers of *x*.

**11.** $2x^2 - x^3 + 5$          **14.** $15x - 3 + 4ax^4 - a^3x^2$

**12.** $3y^2z + 2x^2y - 3xz$     **15.** $3xy^3 - 4y^2 + x^2y - x^3$

**13.** $6ax^5 - 4a^2x + 3ax^3$    **16.** $2a^3x^3 + 3a^5x^2 - 4x + 2$

*WRITTEN EXERCISES*

Simplify each polynomial.

**Sample.** $4z^2 - 3yz + 7y^2z + 5yz$

*Solution:* $4z^2 + (-3 + 5)yz + 7y^2z = 4z^2 + 2yz + 7y^2z$

$\boxed{A}$ **1.** $4x - 3 + 2x + 1$       **3.** $6xy^2 - 2x^2y^3 + 3x^2y - 2$

     **2.** $7y + 5 - y + 3$         **4.** $4a^2b - 3a^4b^2 + 2a^2b + 4a^3b$

**5.** $-3mn + 4m^2n - mn - 5$

**6.** $-5uv^2 + 2uv^4 + 3uv^2 + 6u$

**7.** $4x^5 - 3x^3 + x^4 - 2x^3 + x^5 - x$

**8.** $-y^4 + 3y^3 - 2y^2 + y^4 - 3y^3 + y$

**9.** $-6 + 2x^4y + 3 - 2xy^4 + x^4y$

**10.** $4 - 3a^2y + 2by + 4a^2y - 4$

**11.** $2z^5 - 3z^3 + z^5 - 4z^3 - 3z^5 + z$

**12.** $x - 2x^2 + 3 - x + 2x^2 - 3$

**13.** $d^2 - 2d + 3 - 4d - 6 + 2d^2$

**14.** $k + 2kl - 3k + kl + k - l$

**15.** $2t^3 - 4t + 6 - 8t^3 + 2t^2 + 4t$

**16.** $4a - 3a^2 + 6a - 4 + 2a^2 + a^3$

B **17.** $\frac{1}{3}x^2 - \frac{2}{5}x + 6 + \frac{7}{10}x - \frac{5}{6}x^2 + \frac{3}{8}$

**18.** $\frac{3}{10}y - \frac{4}{3}y^2 + \frac{1}{6} + \frac{5}{9}y^2 - \frac{4}{5}y + \frac{2}{3}$

**19.** $3.2x - 4.12x^2 - 1.3 + 6.1x + 2.5x^2$

**20.** $6.1z - 6.1 + 2.32z^2 - 1.2z^2 + 8.31$

**21.** $\frac{2}{5}y - 6.2y^2 + 2\frac{1}{10} - 8\frac{3}{8}y^2 - 1.62y$

**22.** $5.32x^2 + 7\frac{3}{20} + \frac{5}{4}x - 2.3 + \frac{5}{8}x^2$

## 4–6  Sums and Differences of Polynomials

The **sum** of the polynomials $2x^2 + 3x$ and $x^2 - 3x + 5$ is the polynomial $(2x^2 + 3x) + (x^2 - 3x + 5)$ and their **difference** is the polynomial $(2x^2 + 3x) - (x^2 - 3x + 5)$. The sum can be simplified as follows:

$$(2x^2 + 3x) + (x^2 - 3x + 5) = (2 + 1)x^2 + [3 + (-3)]x + (0 + 5)$$
$$= 3x^2 + 5$$

To express the difference in simple form, you use the property of the negative of a sum.

$$(2x^2 + 3x) - (x^2 - 3x + 5) = (2x^2 + 3x) + (-x^2 + 3x - 5)$$
$$= (2 - 1)x^2 + (3 + 3)x + (0 - 5)$$
$$= x^2 + 6x - 5$$

---

To express the sum (or difference) of two polynomials in simple form, add (or subtract) the coefficients of similar terms.

---

Any polynomial in one variable defines a function having domain $\Re$. For example, the function $g(x) = x^2 - 3x + 2$ is a *polynomial function* of degree 2.

**Example 1.** If $g(x) = x^2 - 3x + 2$, find each of the following:

$$g(0), \quad g(2), \quad g(a), \quad g(-a)$$

*Solution:*

$g(0) = 0^2 - 3 \cdot 0 + 2 = 0 - 0 + 2 = 2$

$g(2) = 2^2 - 3 \cdot 2 + 2 = 4 - 6 + 2 = 0$

$g(a) = a^2 - 3a + 2$

$g(-a) = (-a)^2 - 3(-a) + 2 = a^2 + 3a + 2$

[Note that $(-a)^2 = (-a)(-a) = a^2$ (see page 20).]

In Section 4–3, you studied linear functions, that is, functions of the form $f(x) = mx + b$. These are polynomial functions of degree 1 if $m \neq 0$ and of degree 0 if $m = 0$ and $b \neq 0$. Notice that the functional notation $f(x) = mx + b$ tells you to regard $x$ as the variable and $m$ and $b$ as constants. That is, since $x$ is the only variable appearing in the left member, it must be the only variable in the right member.

**Example 2.** Let $c$ be a constant. Then:

    **a.** Describe the function $F$ which assigns to each number its square decreased by $c^2$.

    **b.** Find the value of $F$ at $c$ and at $-c$.

*Solution:*

    **a.** $F(x) = x^2 - c^2$

    **b.** $F(c) = c^2 - c^2 = 0$

    $F(-c) = (-c)^2 - c^2 = c^2 - c^2 = 0$

You can add two functions, $f$ and $g$, having the same domain, to obtain their **sum function**, $f + g$. The values of $f + g$ are given by the formula

$$(f + g)(x) = f(x) + g(x).$$

The **difference function**, $f - g$, is defined by

$$(f - g)(x) = f(x) - g(x).$$

**Example 3.** Given $f(x) = x^2 + x - 3$ and $g(x) = x^2 - x + 3$, find $(f + g)(x)$ and $(f - g)(x)$.

*Solution:*

$(f + g)(x) = f(x) + g(x)$

$\qquad = (x^2 + x - 3) + (x^2 - x + 3)$

$\qquad = 2x^2$

$(f - g)(x) = f(x) - g(x)$

$\qquad = (x^2 + x - 3) - (x^2 - x + 3)$

$\qquad = 2x - 6$

## ORAL EXERCISES

In each Exercise 1–10, state (**a**) the sum and (**b**) the difference of the given polynomial when the second is subtracted from the first.

**1.** $x^2 + 3x + 2;\ x^2 - 2x + 2$

**2.** $3y^2 + 2y - 4;\ 2y^2 + 2y + 1$

**3.** $5 + 3y + 2y^2;\ 2 + 2y - y^2$

**4.** $6 - a + 3a^2;\ 2 - a + a^2$

**5.** $4t^2 - 3t + 2;\ 2 - t^2$

**6.** $5s^2 + 3s + 6;\ 7 + 3s$

**7.** $5 - z^2;\ z^2 + 3z - 2$

**8.** $4x + x^2;\ 2x^2 - 3x - 2$

**9.** $2k - 3 + k^2;\ k^2 - 5k + 5$

**10.** $u - 3u^2 + 2;\ 5 + u - 3u^2$

In each Exercise 11–20, use the distributive law to express as a sum of monomials.

   **Sample 1.** $3(2x^2 - 3x + 4)$   *Solution:* $6x^2 - 9x + 12$

**11.** $3(t^2 - 2t + 5)$

**12.** $5(4t^2 - 6t + 1)$

**13.** $-3(2s^2 + 3s - 2)$

**14.** $-2(3n^2 - 3n - 3)$

**15.** $4(x^3 - 2x^2 + 3x - 1)$

**16.** $10(2y^3 - 6y^2 + y + 1)$

**17.** $-1(6n^2 + 18n - 8)$

**18.** $0(4y^2 - 4y + 15)$

**19.** $5z - 2(2z + 1)$

**20.** $-2(x + 3) + 2x$

If $f(x)$ is as given, express $f(-x)$ as a polynomial in simple form.

   **Sample 2.** $f(x) = x^2 - 3x + 2$

   *Solution:* $f(-x) = (-x)^2 - 3(-x) + 2 = x^2 + 3x + 2$

**21.** $f(x) = x^4 + 2x^2$

**22.** $f(x) = x^3 + 2x$

**23.** $f(x) = 2x^2 + 3x + 5$

**24.** $f(x) = 3 + 2x - x^2$

**25.** $f(x) = 4 - 2x + x^3$

**26.** $f(x) = 3 - x^2 + x^4$

**27.** Let $f$ be a polynomial function. Under what circumstances will:

   **a.** $f(-x) = f(x)$ for all $x \in \mathcal{R}$?

   **b.** $f(-x) = -f(x)$ for all $x \in \mathcal{R}$?

   **c.** neither **a** nor **b** hold?

*WRITTEN EXERCISES*

Simplify each expression.

A **1.** $(5z^2 + 3z - 2) + (z^2 - 2z - 1)$

**2.** $(7a^2 + 3a - 5) + (5a^2 + 6a - 8)$

**3.** $(6t^3 - 3t^2 + 2t) - (t^3 - 3t^2 - 2t)$

**4.** $(-3b^3 + 2b^2 - 5b + 1) - (b^3 + 3b^2 - 5b + 2)$

**5.** $3(y^2 - 2y + 1) + (3 + 6y + y^2)$

**6.** $2(2x^2 - 3x + 5) + (5 + 7x - 6x^2)$

**7.** $3(z^2 - 2z) - 4(2z^2 + 6) + 2(z + 1)$

**8.** $-4(n^2 + 3n) + 5(2n^2 - 3) - (7 - 4n^2)$

**9.** $3(r^3 - 2r^2) - (4r^2 + 3) + 2(r - r^3)$

**10.** $5(a^3 - 3a^2) + 4(a^3 - 2a + 1) - 3(a + a^3 - 2)$

**11.** $3(x^2 - 3x + 5) - 2(x - x^2 + 5) + 3(2x^2 - 3x + 9)$

**12.** $4(y^2 - 2y + 5) - (3y + 5y^2 - 2) + 2(4 - y^2 + y)$

In Exercises 13–16, let $f(x) = 3x^2 - 2x + 6$ and $g(x) = 2x^2 + 5x - 4$. Express as a polynomial in simple form.

**13.** $(f + g)(x)$          **15.** $f(x) - 4g(x)$

**14.** $(f - g)(x)$          **16.** $3g(x) - 2f(x)$

In Exercises 17–20, let $f(x) = 4 + 3x - 5x^2$ and $g(x) = 6 - 3x + 6x^2$. Express as a polynomial in simple form.

**17.** $(f - g)(x)$          **19.** $(f + g)(x) - (f - g)(x)$

**18.** $(f + g)(x)$          **20.** $(f - g)(x) - 2(f + g)(x)$

In Exercises 21–26, let $f(x) = 2x^3 - 3x^2 + 2x + 1$ and $g(x) = x^3 - 5x^2 + 3x + 2$. Express as a polynomial in simple form.

B **21.** $f(-x)$       **23.** $(f + g)(-x)$       **25.** $f(x) + g(-x)$

**22.** $g(-x)$       **24.** $(f - g)(-x)$       **26.** $2g(-x) - f(x)$

C **27.** Show that if $f(x + a) = f(x) + f(a)$, and $g(x + a) = g(x) + g(a)$, then $(f + g)(x + a) = (f + g)(x) + (f + g)(a)$, and $(f - g)(x + a) = (f - g)(x) + (f - g)(a)$.

**28.** Show that if for all $a, b \in \Re$, $f(ax) = af(x)$ and $g(bx) = bg(x)$, then $(f + g)(cx) = c(f + g)(x)$.

**29.** Show that if $f(-x) = -f(x)$ and $g(-x) = -g(x)$, then $(f + g)(-x) = -(f + g)(x)$.

**30.** Show that if $f(-x) = f(x)$ and $g(-x) = g(x)$, then $(f + g)(-x) = (f + g)(x)$.

## 4–7  Products of Polynomials

The examples

$$x^2 \cdot x^4 = (x \cdot x) \cdot (x \cdot x \cdot x \cdot x) = x^6 = x^{2+4}$$
$$b^5 \cdot b^3 = (b \cdot b \cdot b \cdot b \cdot b) \cdot (b \cdot b \cdot b) = b^8 = b^{5+3}$$

suggest the following:

> **Law of Exponents:** If $m$ and $n$ are natural numbers and $a$ is a real number, then
> $$a^m \cdot a^n = a^{m+n}.$$

This law and the commutative and associative axioms are used in multiplying monomials:

$$(3x^2z)(6x^3z^2) = (3 \cdot 6) \cdot (x^2 \cdot x^3) \cdot (z \cdot z^2) = 18x^5z^3$$
$$(2ax^2)(-abx^2) = [2 \cdot (-1)] \cdot (a \cdot a) \cdot (b) \cdot (x^2 \cdot x^2) = -2a^2bx^4$$

You use the distributive law in multiplying a monomial and a polynomial. For example,

$$2ax^2(x^2 + ax - 3a^2) = 2ax^2 \cdot x^2 + 2ax^2 \cdot ax + 2ax^2(-3a^2)$$
$$= 2ax^4 + 2a^2x^3 - 6a^3x^2.$$

In the next example a binomial and a trinomial are multiplied.

$$(2x - 3)(3x^2 - 2x + 1)$$
$$= 2x(3x^2 - 2x + 1) - 3(3x^2 - 2x + 1)$$
$$= (6x^3 - 4x^2 + 2x) - (9x^2 - 6x + 3)$$
$$= (6 + 0)x^3 + (-4 - 9)x^2 + (2 + 6)x + (0 - 3)$$
$$= 6x^3 - 13x^2 + 8x - 3$$

You can also arrange this work as follows:

$$3x^2 - 2x + 1$$
$$\underline{2x - 3}$$
$$\overline{6x^3 - 4x^2 + 2x} \leftarrow \text{This is } 2x(3x^2 - 2x + 1)$$
$$\underline{- 9x^2 + 6x - 3} \leftarrow \text{This is } -3(3x^2 - 2x + 1)$$
$$\overline{6x^3 - 13x^2 + 8x - 3}$$

Notice that the product of two polynomials is the polynomial you obtain by multiplying each term of one of the given polynomials by each term of the other, and then adding these monomial products. You can usually multiply two binomials mentally by the method illustrated below.

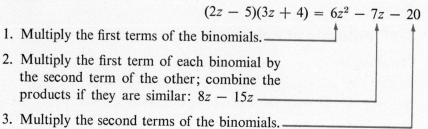

$$(2z - 5)(3z + 4) = 6z^2 - 7z - 20$$

1. Multiply the first terms of the binomials.

2. Multiply the first term of each binomial by the second term of the other; combine the products if they are similar: $8z - 15z$

3. Multiply the second terms of the binomials.

If $f$ and $g$ are two functions having the same domain, their **product** is the function $fg$ which is defined by the formula $(fg)(x) = f(x) \cdot g(x)$.

**Example.** If $f(x) = x^2 - 3x + 2$ and $g(x) = x^2 - 2$, find $(fg)(x)$.

*Solution:* $fg(x) = (x^2 - 3x + 2)(x^2 - 2)$

$$
\begin{array}{r}
x^2 - 3x + 2 \\
x^2 \qquad - 2 \\
\hline
x^4 - 3x^3 + 2x^2 \\
- 2x^2 + 6x - 4 \\
\hline
x^4 - 3x^3 \qquad + 6x - 4
\end{array}
$$

$\therefore (fg)(x) = x^4 - 3x^3 + 6x - 4$

*ORAL EXERCISES*

Express each product as a polynomial in simple form.

1. $x^2 \cdot x^5$

2. $y^3 \cdot y^7$

3. $(2x^2)(3x^5 y)$

4. $(6a^3)(2a^2 z)$

5. $(-4z^2)(-3z^{10})$

6. $(5xy^2)(-2x^3 y)$

7. $(-6r^2 s^3)(-3rs^5)$

8. $(10m^3 n^4)(5m^5 n^3)$

9. $3x(x^2 + 2x - 4)$

10. $-3y(2y^2 + 3y - 5)$

11. $-5z^2(2z^2 + 3z - 6)$

12. $6t^3(2t^2 - 3t - 4)$

13. $(x + 2)(x + 1)$

14. $(y + 3)(y + 4)$

15. $(z - 2)(z - 4)$

16. $(a - 6)(a - 3)$

17. $(t + 3)(t - 5)$

18. $(s - 4)(s + 5)$

19. $(2 + x)(3 + x)$

20. $(4 + t)(2 + t)$

21. $(2y + 1)(y + 2)$

22. $(3x - 1)(x - 5)$

23. $(2x + 3y)(3x - y)$

24. $(2a - b)(2a + 3b)$

## WRITTEN EXERCISES

Express each of the following as a polynomial in simple form. Assume that variables in exponents denote positive integers.

**A**

**1.** $-15x^3(3x^2 + 12x - 10)$

**2.** $12y^4(-11y^3 + 3y^2 - 5)$

**3.** $24b^2(8b^3 - 3b^2 + 2b)$

**4.** $-8n^4(12n^5 - 5n^3 + 3n)$

**5.** $(3a + 5)(8a - 7)$

**6.** $(6t + 7)(3t - 10)$

**7.** $(6 - 5c)(8 + 3c)$

**8.** $(9 - 2t)(5 - 8t)$

**9.** $(3x^2 - 4)(5x^2 + 7)$

**10.** $(2y^3 + 5)(3y^2 - 8)$

**11.** $(2x + 1)(x^2 - 3x + 2)$

**12.** $(5z + 3)(2z^2 - 5z + 2)$

**13.** $(3 - 4t)(2 + 7t - t^2)$

**14.** $(5 + 6z)(6 - 2z + 3z^2)$

**15.** $(2y + 1)(3y - 4)(2y + 5)$

**16.** $(5x - 2)(3x + 5)(4x - 9)$

**17.** $(2x^2 - 3x + 1)(x^2 - 5x + 2)$

**18.** $(4n^2 + 3n - 6)(2n^2 + 3n + 1)$

**Sample.** $x^{2n}(x^{n+1} + 2x^n - 3)$

*Solution:* $x^{2n}(x^{n+1} + 2x^n - 3) = x^{2n+(n+1)} + 2x^{2n+n} - 3x^{2n}$
$$= x^{3n+1} + 2x^{3n} - 3x^{2n}$$

**B**

**19.** $a^{3n}(a^{n+1} + 4a^{2n} - 3)$

**20.** $y^{3a}(y^{2a+1} - 3y^{a+2} + 5)$

**21.** $(x^n + 2)(x^n - 1)$

**22.** $(3x^a + 5)(2x^a - 4)$

**23.** $(x^n + 2)(x^{2n} + 3x^n - 4)$

**24.** $(y^b - 5)(2y^{2b} + 5y^b - 4)$

**25.** $(2z^n - 3)(z^n + 2)(z^n - 1)$

**26.** $(x^m - 2)(x^{2m} + 3)(x^m - 5)$

In Exercises 27–32, let $f(x) = x - 4$ and $g(x) = x^2 - 3x + 1$. Express as a polynomial in simple form.

**27.** $(fg)(x)$

**28.** $(fg)(-x)$

**29.** $(gf)(x + 1)$

**30.** $(gf)(1 - x)$

**31.** $3f(-x) + 2(fg)(2x)$

**32.** $2(gf)(-x) - 3g(x)$

**C**

**33.** Show that if $f(-x) = f(x)$ and $g(-x) = g(x)$, then $(fg)(-x) = (fg)(x)$.

**34.** Show that if $f(-x) = -f(x)$ and $g(-x) = -g(x)$, then $(fg)(-x) = (fg)(x)$.

**35.** Show that if $f(-x) = f(x)$ and $g(-x) = -g(x)$, then $(fg)(-x) = -(fg)(x)$.

**36.** Show that if $f(ax) = af(x)$ and $g(ax) = ag(x)$, then $(fg)(ax) = a^2(fg)(x)$.

## 4–8  Special Product Formulas

The examples

$$(ab)^3 = (ab) \cdot (ab) \cdot (ab) = (a \cdot a \cdot a) \cdot (b \cdot b \cdot b) = a^3 \cdot b^3$$

and

$$(a^3)^2 = a^3 \cdot a^3 = (a \cdot a \cdot a) \cdot (a \cdot a \cdot a) = a^6 = a^{3 \cdot 2}$$

suggest the second and third of the laws of exponents displayed below. (The first law was introduced in Section 4–7.)

---

Let *a* and *b* be real numbers and *m* and *n* be natural numbers. Then:

**1.** $a^m \cdot a^n = a^{m+n}$   **2.** $(ab)^n = a^n b^n$   **3.** $(a^m)^n = a^{mn}$

---

You can extend Law 1 and Law 2 to more than two factors, as indicated below:

$$a^m a^n a^p = (a^m a^n) \cdot a^p = a^{m+n} \cdot a^p = a^{(m+n)+p} = a^{m+n+p}$$

$$(abc)^n = [(ab)c]^n = (ab)^n c^n = a^n b^n c^n$$

**Example 1.**  Simplify, that is, transform the given expression into one in which each base appears only once, and in which there are no "powers of powers."

        **a.** $(x^2 z^3)^4$                 **b.** $(2a)^3(-3a^2 x)^2$

*Solution:*      **a.** $(x^2 z^3)^4 = (x^2)^4 (z^3)^4 = x^8 z^{12}$

           **b.** $(2a)^3 \cdot (-3a^2 x)^2 = (2^3 \cdot a^3) \cdot [(-3)^2 \cdot (a^2)^2 \cdot x^2]$
$$= 8a^3 \cdot 9a^4 x^2 = 72a^7 x^2$$

*Warning:* Law 1 applies only when the bases of the exponential expressions are the same, and Law 2 applies only when the exponents are the same. Neither law can be applied to $p^3 q^2$, for example, if $p \neq q$.

By using the methods of Section 4–7, you can derive the *special product formulas* displayed below. For example, $(a + b)^2 = (a + b) \cdot (a + b) = a \cdot a + (a \cdot b + b \cdot a) + b \cdot b = a^2 + 2ab + b^2$.

---

**1.**                $(a + b)^2 = a^2 + 2ab + b^2$
**2.**                $(a - b)^2 = a^2 - 2ab + b^2$
**3.**        $(a + b)(a - b) = a^2 - b^2$
**4.** $(a + b)(a^2 - ab + b^2) = a^3 + b^3$
**5.** $(a - b)(a^2 + ab + b^2) = a^3 - b^3$

---

The proof of Formula 5 can be arranged as follows (see page 155):

$$a^2 + ab + b^2$$
$$\underline{a\ -\ b}$$
$$a^3 + a^2b + ab^2$$
$$\underline{\quad -\ a^2b - ab^2 - b^3}$$
$$\overline{a^3 \qquad\qquad\quad -\ b^3}$$

Notice how the special product formulas are used in Example 2.

**Example 2.**   Write each of the following products as a polynomial in simple form.

a. $(x^3 - 2a)(x^3 + 2a)$

b. $(z + 3)(z^2 - 3z + 9)$

c. $(5y^2 - b^3)^2$

*Solution:*   a. $(x^3 - 2a)(x^3 + 2a) = (x^3)^2 - (2a)^2 = x^6 - 4a^2$

b. $(z + 3)(z^2 - 3z + 9) = z^3 + 3^3 = z^3 + 27$

c. $(5y^2 - b^3)^2 = (5y^2)^2 - 2(5y^2)(b^3) + (b^3)^2$
$$= 25y^4 - 10b^3y^2 + b^6$$

## ORAL EXERCISES

Simplify each expression.

1. $(xy)^3$

2. $(3z^2)^2$

3. $(a^2b^3)^3$

4. $(a^4b^3)^5$

5. $(3x^2)^3$

6. $3(x^2)^3$

7. $(-2x^2)^2$

8. $((-2x)^2)^2$

9. $(x^n)^2$

10. $(x^2)^n$

11. $(x^n)^n$

12. $(x^{2n})^{2n}$

Express as a polynomial in simple form.

13. $(x + 3)^2$

14. $(y - 5)^2$

15. $(x + 6)(x - 6)$

16. $(z + 2u)(z - 2u)$

17. $(x - 1)(x^2 + x + 1)$

18. $(z - 2)(z^2 + 2z + 4)$

19. $(2a + 1)(4a^2 - 2a + 1)$

20. $(2 - y)(4 + 2y + y^2)$

## WRITTEN EXERCISES

Simplify each expression.

**A**  **1.** $(a^2x)^4$       **5.** $(3x^2y)^3(-x^4y)^4$       **9.** $(a^{2n}x^n)^n$

  **2.** $(3n^2m^3)^2$       **6.** $(-2p^2q^2)^2(-2p^2q^2)^3$   **10.** $(b^ny^{2n})^{n+1}$

  **3.** $(3a^2n)^2(an^2)^3$   **7.** $(3x^{2n})^2$       **11.** $(a^{2n})^{n+1} \cdot (a^n)^{n-2}$

  **4.** $(-4bc^3)(-2b^2c)^3$   **8.** $(x^ny^{2n})^5$       **12.** $(b^n)^{2n} \cdot (b^2)^{2n}$

Write as a polynomial in simple form.

**13.** $(3x^2 - 2y)^2$       **18.** $(3 + 2z)(9 - 6z + 4z^2)$

**14.** $(5a^2 + 6b)^2$       **19.** $(x^2 - 2)(x^4 + 2x^2 + 4)$

**15.** $(7x^2 - 2)(7x^2 + 2)$   **20.** $(2y^2 - 3)(4y^4 + 6y^2 + 9)$

**16.** $(3t - 4s^2)(3t + 4s^2)$   **21.** $(2x - a)^2(2x + a)^2$

**17.** $(4a + 2)(16a^2 - 8a + 4)$   **22.** $(3 - 5t)^2(3 + 5t)^2$

**23.** $[(x - y) + (x + y)][(x - y) - (x + y)]$

**24.** $[(2a^2 + 1) - a][(2a^2 + 1) + a]$

**B**  **25.** $[(3z^2 + 1) - z^2][(3z^2 + 1) + z^2]$

  **26.** $[(x + y) + 2z][(x + y) - 2z]$

  **27.** $[(r + 2) + t][(r + 2)^2 - t(r + 2) + t^2]$

  **28.** $[a + (b - 2c)][a^2 - a(b - 2c) + (b - 2c)^2]$

  **29.** $(a + b + c)^2$

  **30.** $(x - y - z)^2$

  **31.** $(a + b + c + d)^2$

  **32.** $(a + b + c)^3$

  **33.** $(2x + 5)^2 + (2x + 5)(2x - 5) + (2x + 5)(4x^2 - 10x + 25)$

  **34.** $(3 - 5t)^2 - (3 - 5t)(3 + 5t) - (3 - 5t)(9 + 15t + 25t^2)$

**C**  **35.** Express $(a + b)^3$ as a polynomial in simple form, and hence develop a formula for cubing a binomial.

  **36.** Use the results of Exercise 35 to express $(x + 4)^3$ as a polynomial in simple form.

**37.** Express $(a + b)^4$ as a polynomial in simple form and hence develop a formula for raising a binomial to the fourth power.

**38.** Use the result of Exercise 37 to express $(z - 1)^4$ as a polynomial in simple form.    *Hint:* $z - 1 = z + (-1)$

## 4–9   Zeros of Functions

One of the most useful facts of algebra is that a product of real numbers is zero if and only if at least one of the factors is zero.

---

**THEOREM.**    Let $a \in \mathcal{R}$ and $b \in \mathcal{R}$. Then $ab = 0$ if and only if $a = 0$ or $b = 0$.

---

[Recall that an "if and only if" theorem is equivalent to two "if-, then-" theorems which are converses of each other (page 55).]

The "if" part is: $ab = 0$ if $a = 0$ or $b = 0$, that is:

$$\text{If } a = 0 \text{ or } b = 0, \text{ then } ab = 0.$$

*Proof:*        This follows directly from the Zero-factor Property, page 25.

The "only if" part is: $ab = 0$ only if $a = 0$ or $b = 0$, that is:

$$\text{If } ab = 0, \text{ then } a = 0 \text{ or } b = 0.$$

*Proof:*        If $a = 0$, the conclusion is true.  Hence we need only consider the case $a \neq 0$.  Then

| | |
|---|---|
| $\dfrac{1}{a} \in \mathcal{R}$ | Axiom of Reciprocals |
| $ab = 0$ | Hypothesis |
| $\dfrac{1}{a} \cdot (ab) = \dfrac{1}{a} \cdot 0$ | Multiplying each member by $\dfrac{1}{a}$ (Multiplication Property of Equality) |
| $\left(\dfrac{1}{a} \cdot a\right) b = 0$ | Associative Axiom for Multiplication and Zero-factor Property |
| $1 \cdot b = 0$ | Axiom of Reciprocals |
| $b = 0$ | Axiom of One |

$\therefore$ given $ab = 0$, it follows that either $a = 0$ or $b = 0$.

You can use this theorem to solve an equation having one member 0 and the other member a product of first-degree polynomials.

**Example 1.** Solve $(x + 2)(3x - 1) = 0$.

*Solution:* By the theorem just proved, this equation is equivalent to the compound sentence

| *Either* $x + 2 = 0$ | *or* | $3x - 1 = 0.$ |
|---|---|---|
| $x + 2 - 2 = 0 - 2$ | | $3x - 1 + 1 = 0 + 1$ |
| $x = -2$ | | $3x = 1$ |
| | | $x = \frac{1}{3}$ |

Therefore the solution set of the given equation is $\{-2\} \cup \{\frac{1}{3}\} = \{-2, \frac{1}{3}\}$.

A number $z$ is a **zero** of a function $f$ if $f(z) = 0$. For example, 3 is a zero of the function $\varphi(x) = x^2 - 9$ because $\varphi(3) = 3^2 - 9 = 9 - 9 = 0$. Notice that the *zeros* of the function $f$ are the same as the *roots* of the equation $f(x) = 0$.

**Example 2.** Find the real numbers which are zeros of the function $g(x) = x(x - 2)(x^2 + 1)$.

*Solution:* Since $g(x) = x(x - 2)(x^2 + 1)$, we need to find the roots of the equation $x(x - 2)(x^2 + 1) = 0$. By an extension of the theorem on page 161 to the case of more than two factors, this equation is equivalent to the compound sentence

| *Either* $x = 0$ *or* | $x - 2 = 0$ | *or* $x^2 + 1 = 0.$ |
|---|---|---|
| | $x - 2 + 2 = 0 + 2$ | If $x \in \mathcal{R}, x^2 \geq 0$ |
| | $x = 2$ | $x^2 + 1 \geq 1 > 0$ |
| | | $\therefore x^2 + 1 \neq 0$ |

Thus the solution set of the equation is $\{0\} \cup \{2\} \cup \emptyset = \{0, 2\}$, and therefore the zeros of $g$ are 0 and 2.

*ORAL EXERCISES*

Solve each equation over $\mathcal{R}$.

**1.** $(x + 3)(x - 5) = 0$

**2.** $(z + 2)(z + 7) = 0$

**3.** $z(z - 1) = 0$

**4.** $t(t + 2) = 0$

**5.** $2n(n - 4)(n - 5) = 0$

**6.** $3r(r + 2)(r - 8) = 0$

**7.** $(2x - 3)(x + 2) = 0$

**8.** $(4x + 1)(2x + 1) = 0$

**9.** $(y - 4)(2y + 3)(5y - 2) = 0$

**10.** $(t - 4)(3t + 1)(2t - 5) = 0$

**11.** $4t^2(2t + 3) = 0$

**12.** $-6t^3(t^2 + 1) = 0$

State all zeros of the specified function.

**13.** $f(x) = 2x - 4$

**14.** $g(x) = 6x + 2$

**15.** $h(x) = (x - 2)(2x + 3)$

**16.** $t(x) = 3x^2(x - 5)$

**17.** $f(x) = 3(x - 2a)$

**18.** $g(x) = 2(3x - 6a)$

## WRITTEN EXERCISES

Solve each equation over $\mathfrak{R}$.

**A**

**1.** $(5n - 15)(n + 2) = 0$

**2.** $(3x - 9)(5x + 30) = 0$

**3.** $(12y + 8)(6y + 32) = 0$

**4.** $(9k + 11)(3k - 48) = 0$

**5.** $-4x^2(3x - 7) = 0$

**6.** $-12x^2(15x - 70) = 0$

**7.** $2x(x^2 + 6) = 0$

**8.** $3z(2z^2 + 15) = 0$

**9.** $(5x - 7)^2(2x + 3) = 0$

**10.** $(6y - 2)^2(3y + 8) = 0$

**Sample.** $(z + 8)^2 - (z - 5)^2 = 0$

*Solution:* First, simplify the left member. Squaring $z + 8$ and $z - 5$, you have

$$(z^2 + 16z + 64) - (z^2 - 10z + 25) = 0$$
$$z^2 + 16z + 64 - z^2 + 10z - 25 = 0$$
$$26z + 39 = 0$$
$$26z = -39$$
$$z = -\tfrac{39}{26} = -\tfrac{3}{2}.$$

The solution set is $\{-\tfrac{3}{2}\}$.

**11.** $(x + 3)^2 - (x - 5)^2 = 0$

**12.** $(t - 8)^2 - (t + 6)^2 = 0$

**13.** $s(s - 11) - (s + 2)(s - 4) = 0$

**14.** $v(5 - v) + (v + 3)(v - 7) = 0$

**15.** $(x + 3)(x - 5) + (5 + x)(5 - x) = 0$

**16.** $z(12 - z) + (z - 3)(z + 8) = 0$

**B** **17.** $x(x^2 + 3x) - (x + 1)^3 = 0$  **19.** $(3x + 5)^3 - 27x^2(x + 5) = 0$

**18.** $(2y - 1)^3 + 2y^2(6 - 4y) = 0$  **20.** $27z^2(z - 4) + (4 - 3z)^3 = 0$

**C** **21.** Show that {zeros of $(fg)(x)$} = {zeros of $f(x)$} $\cup$ {zeros of $g(x)$}.

**22.** Show that {zeros of $(f^2 + g^2)(x)$} = {zeros of $f(x)$} $\cap$ {zeros of $g(x)$}.

## Chapter Summary

1. To specify a **relation**, you indicate by rule, roster, or graph the pairing of each element of one set (the domain) with one or more elements of another set (the range). For a relation to be a **function**, every element in its domain must be paired with one and only one element in its range.

2. If $m$ is a nonzero constant, then the **linear function** $\{(x, y): y = mx\}$ is a **direct variation**. If $(x_1, y_1)$ and $(x_2, y_2)$ belong to a direct variation, then $\dfrac{y_1}{x_1} = \dfrac{y_2}{x_2}$ provided $x_1 \neq 0$ and $x_2 \neq 0$.

3. A **polynomial** is a monomial or the sum of monomials, and the **degree** of a polynomial is the same as the degree of its monomial (term) of highest degree.

4. If $f(x)$ is a polynomial, then $y = f(x)$ defines a **polynomial function**.

5. If $f$ and $g$ are functions having the same domains, then the sum, difference and product of $f$ and $g$ are defined by:
$$(f + g)(x) = f(x) + g(x)$$
$$(f - g)(x) = f(x) - g(x)$$
$$(fg)(x) \quad = f(x) \cdot g(x)$$

6. If a number $c$ is such that $f(c) = 0$, then $c$ is a **zero** of $f$. If $c$ is a zero of either $f$ or $g$, then $c$ is a zero of $fg$.

### Vocabulary and Spelling

Review each term by reference to the page listed.

## Chapter Test

**4–1** **1.** State the range and graph $\{(x, y): y \geq 2x - 3\}$ over the domain $\{-2, -1, 0, 1, 2\}$.

**4–2** **2.** If $f(x) = 2x^2 - 3x + 1$, find $f(0)$ and $f(-2)$.

**3.** If $f(x) = 3x + 2$ and $g(x) = |x| - 1$, find $f(g(-1))$.

**4–3** **4.** Find a formula specifying the function $f$ whose graph has slope $-2$ and for which $f(1) = 3$.

**5.** If $f$ is a function for which $f(-1) = 6$ and $f(1) = 4$, find $f(6)$.

**4–4** **6.** If $y$ varies directly as $x$, and $y = 6$ when $x = \frac{1}{3}$, find $y$ when $x = 2$.

**7.** If 8 pounds of force stretch a spring $1\frac{1}{2}$ inches, how far will 6 pounds of force stretch the spring?

**4–5** **8.** Simplify $9x^2 - 3x + x^2 - 5 + 4x + 2$.

**4–6** **9.** Simplify $2(2x^2 - 3x + 4) - (4x^2 + 2x - 6)$.

**10.** If $f(x) = 3x^2 - 2x + 1$ and $g(x) = 3 - 2x + 5x^2$, express $(f + g)(x)$ as a polynomial in simple form.

**4–7** **11.** Express $(2x - 3)(x^2 + x - 5)$ as a polynomial in simple form.

**12.** If $f(x) = 3x - 5$, and $g(x) = 4x - 2$, express $(fg)(x)$ as a polynomial in simple form.

**4–8** Express as a polynomial in simple form.

**13.** $(z^2 - 1)(z^4 + z^2 + 1)$     **14.** $(3t^2 - 2)^2$

**4–9** Solve each equation over $\mathscr{R}$.

**15.** $z(2z - 3)(z^2 + 4) = 0$     **16.** $(x + 2)^2 + (1 - x)^2 = 0$

---

## Chapter Review

---

### 4–1  Relations  *Pages 131–134*

1. A relation is any set of __?__  __?__ .

2. The set of all first components in the ordered pairs in a relation is called the __?__ of the relation.

3. The relation defined by $y = x^2$ assigns 4 to two numbers, the numbers __?__ and __?__ .

4. The domain of the relation defined by $y = \dfrac{3}{x - 6}$ is the set of all real numbers except __?__ .

### 4–2  Functions  *Pages 135–138*

5. A function is a relation which assigns to each member of its domain a __?__ member of its range.

6. No vertical line intersects the graph of a function at more than __?__ point(s).

7. If $f$ denotes a function, then $f(2)$ denotes the unique member of the __?__ of $f$ associated with the member __?__ in the domain.

8. For $f(x) = 2x^2 - 3x + 4$, find $f(-1)$.

### 4–3  General Linear Functions  *Pages 138–143*

9. An equation of the form $f(x) = mx + b$ defines a __?__ function.

10. The domain and range of a nonconstant linear function are ordinarily taken to be the set of __?__  __?__ .

11. A constant function has only one number in its __?__ .

12. If $g$ is a linear function for which $g(2) = 4$ and $g(0) = 2$, find $g(5)$.

### 4–4  Direct Variation  *Pages 143–148*

13. An equality of ratios is called a __?__ .

14. In any proportion the __?__ of the means is equal to the __?__ of the extremes.

15. If $y$ varies directly as $x$, and $y = 12$ when $x = 3$, find $y$ when $x = 15$.

16. If $t$ varies directly as $s$, and $t = 6$ when $s = 20$, find $t$ when $s = 70$.

**4–5  Polynomials**  *Pages 148–151*

**17.** An expression that is either a constant or a product of a constant and one or more variables is called a __?__.

**18.** The degree of a polynomial is the degree of its monomial or term of __?__ degree.

**19.** The polynomial $3x + 2x^3 - 4x^2 + 5$ is of degree __?__, and its simple form is __?__.

**4–6  Sums and Differences of Polynomials**  *Pages 151–155*

**20.** If $f(x) = x^3 - 2x + 3$, then $f$ is a __?__ function of degree __?__.

**21.** If $f$ and $g$ are functions having the same domain, then $(f + g)(x) = $ __?__, and $(f - g)(x) = $ __?__.

**22.** If $f(x) = 3x^3 - 2x^2 + 2x - 1$, find $f(0)$ and $f(-1)$.

**4–7  Products of Polynomials**  *Pages 155–157*

**23.** If $m$ and $n$ are natural numbers, and $a$ is a real number, then $a^m \cdot a^n = $ __?__.

**24.** If $f$ and $g$ are functions with the same domain, then $(fg)(x) = $ __?__.

**25.** If $f(x) = x^2 - 2$ and $g(x) = x^2 - 2x + 3$, then, expressed as a polynomial in simple form, $(fg)(x) = $ __?__.

**4–8  Special Product Formulas**  *Pages 158–161*

**26.** If $a$ and $b$ are real numbers, and $m$ and $n$ are natural numbers, then $(ab)^n = $ __?__ and $(a^m)^n = $ __?__.

**27.** Expressed as polynomials in simple form, $(x - 2)^2 = $ __?__, and $(2x - 3)(2x + 3) = $ __?__.

**28.** Expressed as polynomials in simple form, $(z - 4)(z^2 + 4z + 16) = $ __?__, and $(2t + 3)(4t^2 - 6t + 9) = $ __?__.

**29.** Simplify $(-3xy)^2 (x^2 y^3)^4$.

**4–9  Zeros of Functions**  *Pages 161–163*

**30.** If $(2x - 3)(x - 5) = 0$, solve for $x$.

**31.** If $f$ is a function, and $f(c) = 0$, then $c$ is a __?__ of $f$, and a __?__ of $f(x) = 0$.

**32.** If $g(x) = 2x(x - 3)(x^2 + 4)$, find the zeros of $g$.

# Mathematics and Economics

The rising cost of goods and services has caused many consumers to brush up on their arithmetic so that they can obtain the best value for their money. In fact, it is not at all uncommon to see a shopper in a supermarket roaming the aisles with grocery cart in one hand and calculator in the other, pausing now and then to decide, for example, whether 32 ounces of diet soft drink for 49¢ or 60 ounces of diet soft drink for 99¢ is the better buy. Although unit-pricing sometimes eliminates the need for calculators in supermarkets, there are still many instances in which a knowledge of basic mathematics is essential to a consumer in solving his household economic problems.

A shopper trying to obtain the best food values within the limits of his budget is an example of an *optimization problem*, that is, a problem in which something is to be maximized or minimized subject to certain constraints. The analysis and solution of optimization problems is one of the important areas of study in the field of economics. Of course, the optimization problems that economists deal with

are usually much more complex than those of the average consumer. For example, an economist might try to determine how a certain manufacturer or group of manufacturers could maximize their profits or minimize their costs subject to given technological constraints.

Economists are interested in a wide variety of problems, from determining a minimum cost diet for livestock that satisfies certain nutritional requirements to creating a mathematical model of an expanding national economy; and in all of this work, whether at the individual, corporate, or national level, economists are coming more and more to rely on the use of mathematical techniques to reach solutions. These techniques come from many different branches of mathematics, such as set theory, calculus, matrix theory and linear algebra, differential equations, probability theory, statistical inference, and computer science. Mathematicians have even developed new areas of mathematics, such as linear programming and game theory, to aid in the solution of economic problems.

You are already familiar with some of the mathematical ideas used in economics. For example, linear programming was discussed on pages 127–129. The *function concept*, which you studied in this chapter, is also important in economics. For example, under idealized conditions, the price $p$ of a commodity, say corn, is a function $f$ of the demand $x$ for the quantity; that is, $p = f(x)$, where $x$ is measured in some convenient units and $p$ is the price in some specified monetary units for each unit of $x$. The graph of $f$, called a *demand curve*, is shown at the right. Notice that this graph illustrates the economic principle that the higher the price is, the lower the demand.

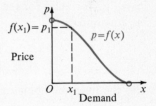

In Chapter 16 you will study another important mathematical concept used in economics: *matrices*. The following greatly simplified example illustrates one way in which matrices might be used.

**Example.** In order to meet the deadline on an important government project, an engineering firm hires 5 engineers, 3 technicians, 2 draftsmen, and 2 typists for a two-month period. If engineers receive $1800 a month, technicians $1100 a month, draftsmen $1000 a month and typists $750 a month, what is the payroll for the temporary help for the two-month period?

*Solution:* The temporary help can be represented by the matrix $[5\ 3\ 2\ 2]$ and their monthly wages can be represented by the matrix $\begin{bmatrix} 1800 \\ 1100 \\ 1000 \\ 750 \end{bmatrix}$.

Then the payroll for the two months is given by the product

$$2 \times [5\ 3\ 2\ 2] \begin{bmatrix} 1800 \\ 1100 \\ 1000 \\ 750 \end{bmatrix} = 2[5(1800) + 3(1100) + 2(1000) + 2(750)]$$

$$= 2(9000 + 3300 + 2000 + 1500)$$
$$= 2(15,800) = 31,600.$$

**169**

# Explorations with a Computer

(To be used if you have access to an electronic computer that will accept BASIC.)

In Section 3–3 you learned about the slope of a line. Use the BASIC program on page 80, to suggest graphs of the following equations. Compare the results.

1. $X + Y = 6$       4. $2X + Y = 6$

2. $X + 2Y = 6$      5. $3X + Y = 6$

3. $X + 3Y = 6$

In order to find the solution set of a pair of linear equations in two variables by computer with the methods that have been developed so far, we must find formulas for the coordinates of the solution. Consider the following equations:

$$(A1)X + (B1)Y = C1$$
$$(A2)X + (B2)Y = C2$$

Use the linear-combination method to verify that an equivalent system is:

$$X = (B2 * C1 - B1 * C2)/(A1 * B2 - A2 * B1)$$
$$Y = (A1 * C2 - A2 * C1)/(A1 * B2 - A2 * B1)$$

For convenience we let $D = A1 * B2 - A2 * B1$ in the following BASIC program:

```
10    PRINT "INPUT A1, B1, C1";
15    INPUT A1,B1,C1
20    PRINT "INPUT A2, B2, C2";
25    INPUT A2,B2,C2
30    PRINT
35    PRINT A1;"X +";B1;"Y =";C1
40    PRINT A2;"X +";B2;"Y =";C2
45    PRINT
50    LET D=A1*B2-A2*B1
55    IF D=0 THEN 80
60    LET X=(B2*C1-B1*C2)/D
65    LET Y=(A1*C2-A2*C1)/D
70    PRINT "SOLUTION IS (";X;",";Y;")"
75    GOTO 85
80    PRINT "NO UNIQUE SOLUTION"
85    END
```

**1.** Use the program for the systems given in Examples 1–3 on pages 106–107 and compare the results.

**2.** Starting with the system

$$X + 3Y = 9$$
$$4X + 5Y = 1$$

add 1 to each constant in the system, obtaining

$$2X + 4Y = 10$$
$$5X + 6Y = 2$$

and find the solution. Continue in this manner until you get to

$$8X + 10Y = 16$$
$$11X + 12Y = 8$$

and compare the solutions.

**3.** Starting with the system

$$2X + 5Y = 8$$
$$-3X + Y = 5$$

add 1 to each constant as in Exercise 2 and continue several times. Compare the solutions.

**4.** Starting with the system

$$2X + 5Y = 8$$
$$-3X + Y = 5$$

subtract 1 from each constant. Continue until D is 0. What equations do you have then?

# 5

# Factoring
# Polynomials

## METHODS OF FACTORING

### 5–1   Factors of Monomials

When you write $144 = 16 \cdot 9$, you have *factored* 144 into a product of integers, and you call 16 and 9 *integral factors* of 144. In general, if $p$, $q$, and $r$ are integers, and $r = p \cdot q$, then $p$ and $q$ are **integral factors** of $r$, and $r$ is an **integral multiple** of $p$ and of $q$. An integer greater than 1 whose only positive integral factors are itself and 1 is called a **prime**, or a **prime number**. Thus, 7 is a prime number because, although $7 = 7 \cdot 1$, 7 is not the product of any other pair of positive integers. A number which is not prime has more than one prime factor. Since $144 = 2^4 \cdot 3^2$, the prime factors of 144 are 2 and 3.

In factoring, it is important to specify the **factor set**, that is, the set from which the factors may be selected. You say that you are factoring *over* this set. Thus, over the set of integers, 144 may be factored in many ways: $144 = 6 \cdot 24$, $144 = 72 \cdot 2$, $144 = (-2) \cdot (-9) \cdot 8$. But over the *set of primes*,

$$\{2, 3, 5, 7, 11, 13, 17, 19, \ldots\},$$

there is essentially only one factorization of 144, namely $2^4 \cdot 3^2$. Other prime factorizations merely vary the order in which the factors appear: $144 = 3^2 \cdot 2^4$, $144 = 2^2 \cdot 3 \cdot 2^2 \cdot 3$, and so forth.

A systematic way of finding the prime factorization of an integer is illustrated at the right. Notice that you take the primes in increasing order of magnitude, using each prime as often as possible before going to the next. You can use prime factorization to find (1) the **greatest common factor** (GCF) of two or more integers, that is, the greatest integer which is a factor of both of them, and (2) their **least common multiple** (LCM), the least positive integer having each as a factor.

$$756 = 2 \cdot 378$$
$$= 2 \cdot 2 \cdot 189$$
$$= 2 \cdot 2 \cdot 3 \cdot 63$$
$$= 2 \cdot 2 \cdot 3 \cdot 3 \cdot 21$$
$$= 2 \cdot 2 \cdot 3 \cdot 3 \cdot 3 \cdot 7$$
$$= 2^2 \cdot 3^3 \cdot 7$$

**Example 1.** Find the greatest common factor and least common multiple of 144 and 756.

*Solution:* Factoring, you have

$$144 = 2^4 \cdot 3^2 \quad \text{and} \quad 756 = 2^2 \cdot 3^3 \cdot 7.$$

| *To find the GCF* | *To find the LCM* |
|---|---|
| The greatest power of 2 common to 144 and 756 is $2^2$, and the greatest common power of 3 is $3^2$. | Any integer having 144 and 756 as factors must have $2^4$, $3^3$, and $7^1$ as factors. |
| $\therefore$ GCF $= 2^2 \cdot 3^2 = 36$ | $\therefore$ LCM $= 2^4 \cdot 3^3 \cdot 7^1 = 3024$ |
| No greater integer is a factor of both 144 and 756. | No lesser positive integer is a multiple of both 144 and 756. |

The definitions of **factor** and **multiple** for monomials having integral coefficients are similar to those given above for integers. For example, $6ay$ is a factor of $24a^2xy$ because $24a^2xy = (4ax)(6ay)$, and $24a^2xy$ is a multiple of $4ax$ and of $6ay$.

The monomial having greatest degree and greatest coefficient that is a factor of each of several monomials is their **greatest common factor** (GCF). For example,

$2a$ is the GCF of $4ax$ and $6ay$.

The monomial with least degree and least positive coefficient that is a multiple of each of several monomials is their **least common multiple** (LCM). Thus

$12axy$ is the LCM of $4ax$ and $6ay$.

**Example 2.** Find the greatest common factor and least common multiple of the given monomials.

    **a.** $36a^2xy^2, -24a^2x^3$

    **b.** $36a^2xy^2, -24a^2x^3, 60a^3x^2y$

*Solution:*    **a.** The prime factorizations of the coefficients are:

$$\cdot 36 = 2^2 \cdot 3^2 \qquad\qquad -24 = -2^3 \cdot 3$$

| *To find the GCF* | *To find the LCM* |
|---|---|
| The coefficient is the GCF of the coefficients of the given monomials: | The coefficient is the LCM of the coefficients of the given monomials: |
| $$2^2 \cdot 3 = 12$$ | $$2^3 \cdot 3^2 = 72$$ |
| Compare the powers of each variable which occurs in *both* of the monomials and take the power with the *least* exponent: | Compare the powers of each variable which occurs in *either* of the monomials and take the power with the *greatest* exponent: |
| Compare $a^2, a^2$; take $a^2$ | Compare $a^2, a^2$; take $a^2$ |
| Compare $x, x^3$; take $x$ | Compare $x, x^3$; take $x^3$ |
| $\therefore$ GCF $= 12a^2x$ | Take $y^2$ |
| | $\therefore$ LCM $= 72a^2x^3y^2$ |

    **b.** The prime factorizations of the coefficients are:

$$36 = 2^2 \cdot 3^2, \quad -24 = -2^3 \cdot 3, \quad 60 = 2^2 \cdot 3 \cdot 5$$

| *To find the GCF* | *To find the LCM* |
|---|---|
| The coefficient is: | The coefficient is: |
| $$2^2 \cdot 3 = 12$$ | $$2^3 \cdot 3^2 \cdot 5 = 360$$ |
| Compare $a^2, a^2, a^3$; take $a^2$ | Compare $a^2, a^2, a^3$; take $a^3$ |
| Compare $x, x^3, x^2$; take $x$ | Compare $x, x^3, x^2$; take $x^3$ |
| $\therefore$ GCF $= 12a^2x$ | Compare $y^2, y$; take $y^2$ |
| | $\therefore$ LCM $= 360a^3x^3y^2$ |

*ORAL EXERCISES*

Give the prime factorization of each integer.

**1.** 12            **3.** 9            **5.** 50

**2.** 20            **4.** 25            **6.** 18

The prime factorizations of two integers are given. Find (a) the GCF of the integers and (b) the LCM of the integers.

**Sample.** $3^2 \cdot 5, \ 3 \cdot 5^2$    *Solution:*   GCF is $3 \cdot 5$, LCM is $3^2 \cdot 5^2$.

**7.** $2^2 \cdot 3^2, \ 2^3 \cdot 3$                 **9.** $5 \cdot 7^2 \cdot 11, \ 2 \cdot 3 \cdot 5^2$

**8.** $2 \cdot 3 \cdot 5, \ 2^2 \cdot 7$                **10.** $2 \cdot 3^4, \ 2 \cdot 3^2$

*WRITTEN EXERCISES*

Factor each integer over the set of primes.

**A**   **1.** 75          **3.** 120         **5.** 343         **7.** 2340

     **2.** 64          **4.** 105         **6.** 154         **8.** 2546

Find (a) the GCF and (b) the LCM of the following monomials.

   **9.** $45, 27$       **13.** $3a^2b^2, 21ab^3$       **17.** $2, 3x, x^2$

**10.** $25, 40$       **14.** $2xy^2, 10x^2y$        **18.** $3a, ab, ab^2$

**11.** $80, -240$     **15.** $-15m^2n^3, -20m^4n$    **19.** $57a^2b^2c, 38ac$

**12.** $-363, 55$     **16.** $-9ax^3, -15bx^4$      **20.** $75xy^2z^4, 150x^2y^3z^2$

**B**   **21.** List all the positive integral factors of 6, other than 6 itself, and show that the sum of the numbers in the list equals 6.

   **22.** List all the positive integral factors of 28, other than 28 itself, and show that the sum of the numbers in the list equals 28.

   **23.** Let $p$ be an *odd prime* number. List the positive integral factors of $3p$, and show that their sum is $4(p + 1)$.

   **24.** Let $p$ be a *prime greater than* 3. List the positive integral factors of $2p$, and show that their sum is $3(p + 1)$.

   **25.** The GCF of two positive integers is 12. Their LCM is 396. If one of the integers is 36, find the other integer.

   **26.** The GCF of two monomials is $a^3b^2$. Their LCM is $10a^5b^3$. If one of the monomials is $2a^3b^3$, find the other monomial.

## 5–2 Factoring Polynomials

To *factor a polynomial* you express it as a product of other polynomials belonging to a given factor set. Unless otherwise indicated, *the factor set for a polynomial having integral coefficients will be the set of all polynomials with integral coefficients.*

Some polynomials have a monomial factor other than 1 and $-1$. For example, you can factor $4ax - 6ay$ by using the distributive law: $4ax - 6ay = 2a \cdot 2x - 2a \cdot 3y = 2a(2x - 3y)$. We call $2a$ the greatest monomial factor of $4ax - 6ay$ because it is the greatest common factor of the terms of the polynomial. The first step in factoring a polynomial in simple form is to *factor out the greatest monomial factor*, that is, to express the polynomial as the product of its greatest monomial factor and another polynomial.

**Example 1.** Factor out the greatest monomial factor:

$\quad$ **a.** $20x^3 - 20x^2 + 5x$ $\qquad$ **b.** $24ax^4 + 54ax^2y^2$

*Solution:* $\quad$ **a.** $20x^3 - 20x^2 + 5x = 5x \cdot 4x^2 - 5x \cdot 4x + 5x \cdot 1$
$$= 5x(4x^2 - 4x + 1)$$

$\quad$ **b.** $24ax^4 + 54ax^2y^2 = 2^3 \cdot 3ax^4 + 2 \cdot 3^3ax^2y^2$
$$= 2 \cdot 3ax^2 \cdot 2^2x^2 + 2 \cdot 3ax^2 \cdot 3^2y^2$$
$$= 6ax^2(4x^2 + 9y^2)$$

You can factor many polynomials by recognizing that they have the form of one of the special products 1 through 5 on page 158.

---

**1.** $a^2 + 2ab + b^2 = (a + b)^2$
**2.** $a^2 - 2ab + b^2 = (a - b)^2$

---

**Example 2.** **a.** $x^2 + 6x + 9 = (x + 3)^2$

$\qquad$ **b.** $4x^2 - 4xy + y^2 = (2x)^2 - 2(2x)y + y^2$
$$= (2x - y)^2$$

---

**3.** $a^2 - b^2 = (a + b)(a - b)$

---

**Example 3.** **a.** $4a^2 - 25 = (2a)^2 - 5^2 = (2a + 5)(2a - 5)$

$\qquad$ **b.** $3x^2 - 12y^4 = 3(x^2 - 4y^4) = 3[x^2 - (2y^2)^2]$
$$= 3(x + 2y^2)(x - 2y^2)$$

$\qquad$ **c.** $1 - z^6 = 1^2 - (z^3)^2 = (1 + z^3)(1 - z^3)$

---

**4.** $a^3 + b^3 = (a + b)(a^2 - ab + b^2)$
**5.** $a^3 - b^3 = (a - b)(a^2 + ab + b^2)$

---

**Example 4.** **a.** $x^3 + 27 = x^3 + 3^3 = (x + 3)(x^2 - 3x + 9)$
**b.** $1 - z^6 = (1 + z^3)(1 - z^3)$ (by Example 3c)
$= (1 + z)(1 - z + z^2)(1 - z)(1 + z + z^2)$

You can sometimes factor a polynomial by rearranging its terms and grouping them appropriately. In the example below an appropriate grouping of terms is indicated by arrows:

$$6pq - 6 + 4q - 9p = (6pq - 9p) + (4q - 6)$$
$$= 3p(2q - 3) + 2(2q - 3)$$
$$= (3p + 2)(2q - 3)$$

In the next example, three of the terms form a *perfect square*, that is, they are the square of a binomial. When these three terms are grouped, the polynomial itself is the difference of two squares.

$$a^2 - b^2 - 2a + 1 = (a^2 - 2a + 1) - b^2$$
$$= (a - 1)^2 - b^2$$
$$= [(a - 1) + b][(a - 1) - b]$$
$$= (a + b - 1)(a - b - 1)$$

*ORAL EXERCISES*

State the factors of the following polynomials.

| | | |
|---|---|---|
| **1.** $3x + 3y$ | **9.** $4x^2 - 9$ | **17.** $s^3 - 1$ |
| **2.** $4s - 4t$ | **10.** $49x^2 - 1$ | **18.** $x^3 + 8$ |
| **3.** $a^2x - a^2y$ | **11.** $81 - a^2$ | **19.** $x^2 + 4x + 4$ |
| **4.** $my + ny$ | **12.** $1 - 36a^2$ | **20.** $a^2 - 10a + 25$ |
| **5.** $ax + 2ax^2$ | **13.** $16c^2 - 25d^2$ | **21.** $9y^2 + 6y + 1$ |
| **6.** $3ab - 6a^2b$ | **14.** $100a^2 - 144b^2$ | **22.** $36a^2 - 12a + 1$ |
| **7.** $x^2 - 25$ | **15.** $x^2 - 4y^2z^2$ | **23.** $4x^2 + 12x + 9$ |
| **8.** $a^2 - 64$ | **16.** $9a^2 - 1$ | **24.** $9x^2 - 12x + 4$ |

## WRITTEN EXERCISES

Factor the following polynomials. Variables in exponents denote positive integers.

A **1.** $2x^2 - 8$

**2.** $3y^2 - 27$

**3.** $a^3 - 9a$

**4.** $b^3 - 4b$

**5.** $x^4 - 81$

**6.** $y^4 - 1$

**7.** $3a^2 - 12a + 12$

**8.** $5a^2 + 20a + 20$

**9.** $-x^2 + 14x - 49$

**10.** $-y^2 - 20y - 100$

**11.** $3ab + 3ac + 5bd + 5cd$

**12.** $ax + by + ay + bx$

**13.** $5a^2 + 40a + 80$

**14.** $7b^2 + 70b + 175$

**15.** $2s^2 - 100s + 1250$

**16.** $3x^2 - 66x + 363$

**17.** $x^3 - y^3$

**18.** $x^3 + s^3$

**19.** $27a^3 + 1$

**20.** $s^3 - 8$

**21.** $x^3 - 64$

**22.** $8y^3 - 1$

**23.** $2ax - 3by - 6ay + bx$

**24.** $x^2 + xy - 2x - 2y$

**25.** $1000 - x^3$

**26.** $729 - y^3$

**27.** $343 - 8a^3$

**28.** $216 - 27b^3$

B **29.** $x^6 + 8$

**30.** $x^6 - 64$

**31.** $a^3 + b^3c^3$

**32.** $r^3s^6 + t^3$

**33.** $27m^6 + 1$

**34.** $1 - 1000n^6$

**35.** $x^3 - 2x^2 - x + 2$

**36.** $2a^3 - 5a^2 - 2a + 5$

**37.** $(s - 2)^2 - t^2$

**38.** $(r + 1)^2 - s^2$

**39.** $4 + 4(x + 3) + (x + 3)^2$

**40.** $25 + 10(4 - a) + (4 - a)^2$

**41.** $x^2 + 4x + 4 - y^2$

**42.** $9x^2 - 6xy - 25 + y^2$

**43.** $x^2 - y^2 + x - y$

**44.** $16 - a^2 - 2ab - b^2$

C **45.** $a^6 - x^6$

**46.** $x^{2n} - 1$

**47.** $x^{2n} + 2x^n + 1$

**48.** $4 - x^{2k}$

**49.** $(x^2 + 1)^2 - (x^2 + 1)$

**50.** $(a + b)^3 - (a + b)^5$

**51.** $(x - 3)^2 - (x - 3)^5$

**52.** $(a - 1)^6 + (1 - a)^3$

## 5-3 Factoring Quadratic Polynomials

A **quadratic polynomial** in $x$ is a polynomial of the form

$$ax^2 + bx + c, \qquad a \neq 0.$$

The term $ax^2$ is of the second degree (in $x$) and is often called the **quadratic term**, while $bx$ and $c$ are called the **linear term** and **constant term**, respectively.

If the binomials $px + r$ and $qx + s$ are factors of $ax^2 + bx + c$, then the corresponding coefficients indicated in the following identity must be equal.

$$(px + r)(qx + s) = pqx^2 + (ps + qr)x + rs = ax^2 + bx + c$$

These correspondences provide clues helpful in factoring quadratic trinomials.

**Example 1.** Factor $x^2 - 9x + 14$.

*Solution:*

*Step 1.*     *Find possibilities for the linear terms of the factors.* Since the product of these linear terms must be $x^2$, we may begin the factorization with $(x \quad)(x \quad)$.

*Step 2.*     *Find possibilities for the constant terms of the factors.* The product of the constant terms is positive $(+14)$ and their sum is negative $(-9)$. Therefore both constant terms must be negative. The only factorizations of 14 into negative integers are $(-1)(-14)$ and $(-2)(-7)$.

*Step 3.*     *Decide which of the possibilities (if any) give the desired factorization.*

| Possible Factors | Corresponding Linear Term |
|---|---|
| $(x - 1)(x - 14)$ | $-14x - x = -15x$ |
| $(x - 2)(x - 7)$ | $-7x - 2x = -9x$ |

The second factorization above gives the correct linear term. Therefore,

$$x^2 - 9x + 14 = (x - 2)(x - 7).$$

Except for trivial variations such as writing the factors in the form $(2 - x)(7 - x)$ or $(x - 7)(x - 2)$, this factorization of $x^2 - 9x + 14$ is unique; that is, it is the only one possible over the agreed-upon factor set.

**Example 2.**  Factor $10z^2 - 7z - 3$.

*Solution:*  We follow the steps used in Example 1.

*Step 1.*  The product of the linear terms is $10z^2$. The possibilities are $(2z\quad)(5z\quad)$ and $(z\quad)(10z\quad)$.

*Step 2.*  The product of the constant terms is $-3$; therefore one of the constant terms must be positive and the other negative. The possibilities are 1 and $-3$ or $-1$ and 3.

*Step 3.*

| *Possible Factors* | *Corresponding Linear Term* |
|---|---|
| $(2z + 1)(5z - 3)$ | $-6z + 5z = -z$ |
| $(2z - 3)(5z + 1)$ | $2z - 15z = -13z$ |
| $(2z - 1)(5z + 3)$ | $6z - 5z = z$ |
| $(2z + 3)(5z - 1)$ | $-2z + 15z = 13z$ |
| $(z + 1)(10z - 3)$ | $-3z + 10z = 7z$ |
| $(z - 3)(10z + 1)$ | $z - 30z = -29z$ |
| $(z - 1)(10z + 3)$ | $3z - 10z = -7z$ |
| $(z + 3)(10z - 1)$ | $-z + 30z = 29z$ |

Since the required linear term is $-7z$, the next-to-last possibility is the correct one. Therefore

$$10z^2 - 7z - 3 = (z - 1)(10z + 3).$$

**Example 3.**  Factor $t^2 + 2t + 4$.

*Solution:*

*Step 1.*  We begin with $(t\quad)(t\quad)$.

*Step 2.*  The only possibilities for the constant terms are positive numbers whose product is 4; that is, 1 and 4 or 2 and 2.

*Step 3.*

| *Possible Factors* | *Corresponding Linear Term* |
|---|---|
| $(t + 1)(t + 4)$ | $4t + t = 5t$ |
| $(t + 2)(t + 2)$ | $2t + 2t = 4t$ |

Since neither possibility gives the required linear term, $t^2 + 2t + 4$ cannot be factored over the set of polynomials with integral coefficients.

A polynomial which cannot be expressed as a product of polynomials of lower degree belonging to a given factor set is said to be **irreducible** over that set. An irreducible polynomial whose greatest monomial factor is 1 is called a **prime polynomial** over the set. Thus $t^2 + 2t + 4$ is a prime polynomial with integral coefficients, while $2t^2 + 4t + 8 = 2(t^2 + 2t + 4)$ is irreducible but not prime.

The factorization of a polynomial over a set is **complete** when each factor is either a constant, a prime polynomial, or a power of a prime polynomial. To factor $6x^7 - 96x^3$ completely over the set of polynomials with integral coefficients, you write

$$6x^7 - 96x^3 = 6x^3(x^4 - 16)$$
$$= 6x^3(x^2 + 4)(x^2 - 4)$$
$$= 6x^3(x^2 + 4)(x + 2)(x - 2).$$

*ORAL EXERCISES*

Decide whether each of the following is the proper beginning to a possible factorization. Complete those which are started correctly.

**1.** $x^2 - 2x - 3 = (x - \quad)(x - \quad)$

**2.** $x^2 - 5x + 4 = (x - \quad)(x + \quad)$

**3.** $y^2 + 7y + 12 = (y + \quad)(y + \quad)$

**4.** $y^2 - 5y - 6 = (y - \quad)(y + \quad)$

**5.** $z^2 - z - 20 = (z + \quad)(z - 10)$

**6.** $z^2 + 3z - 10 = (z + 5)(z - \quad)$

**7.** $2w^2 - w - 1 = (2w - \quad)(w + \quad)$

**8.** $2w^2 + w - 6 = (2w - \quad)(w + \quad)$

*WRITTEN EXERCISES*

Factor each of the following completely.

A

**1.** $x^2 + 12x + 35$

**2.** $x^2 + 11x + 24$

**3.** $x^2 - 4x + 3$

**4.** $y^2 - 7y + 6$

**5.** $x^2 - 9x + 20$

**6.** $x^2 + 3x - 40$

**7.** $w^2 + w - 2$

**8.** $w^2 + 8w - 20$

**9.** $k^2 + 3k - 54$

**10.** $30 - m - m^2$

**11.** $18 - 7m - m^2$

**12.** $n^2 - 18n + 81$

**13.** $2x^2 + 5x + 2$

**14.** $2x^2 - 3x - 2$

**15.** $4y^2 + 12y + 9$

**16.** $3y^2 + 10y + 8$

**17.** $3z - 4z^2 + 1$

**18.** $14 - 13p + 3p^2$

**19.** $5q + 6q^2 - 4$

**20.** $10a^2 + 49a - 5$

**21.** $14x^2 + 7x - 21$

**22.** $6y^2 - 10y - 4$

**23.** $8w^2 + 44w + 20$

**24.** $15x^2 - 35x + 10$

B **25.** $k^2 - 3kl - 10l^2$

**26.** $5a^2 + 3ab - 2b^2$

**27.** $x^4 - x^5 + x^6$

**28.** $x^5 + x^6 + x^7$

**29.** $4x^2y - 6xy - 4y$

**30.** $12w^2v - 26wv + 12v$

**31.** $y^2 + \frac{1}{6}y - \frac{1}{6}$

**32.** $2n^2 - \frac{11}{12}n + \frac{1}{12}$

**33.** $x^4 - 7x^2 - 8$

**34.** $x^4 - 2x^2y^2 + y^4$

**35.** $x^5 - 25x^3$

**36.** $6x^6 - 6x^2$

C **37.** $2y^{2n} + 3y^n + 1$

**38.** $10s^{4k} + 49s^{2k} - 5$

**39.** $(a + b)^2 - 14(a + b) + 40$

**40.** $3(p + q)^2 - 8(p + q) + 4$

**41.** $(x^2 + 3x - 10)^2 - (x^2 + 2x - 8)^2$

**42.** $y^4 - (4y - 5)^2$

## APPLICATIONS OF FACTORING

## 5-4 Common Factors and Multiples of Polynomials

A common factor of two or more polynomials is a polynomial which is a factor of each of them, while a common multiple of the polynomials is a polynomial having each of them as a factor. The common factor having greatest degree and greatest constant factor is the greatest common factor (GCF) of the given polynomials. The least common multiple (LCM) is the common multiple having least degree and least positive constant factor.

To find the GCF and LCM of a set of polynomials, you proceed as in the following example.

**Example.**   Find the greatest common factor and least common multiple of the polynomials

$$4x^5 + 8x^4 - 12x^3 \quad \text{and} \quad 6x^4 - 12x^3 + 6x^2.$$

*Solution:*   Factor the given polynomials completely:

$$4x^5 + 8x^4 - 12x^3 = 4x^3(x^2 + 2x - 3)$$
$$= 4x^3(x - 1)(x + 3)$$
$$6x^4 - 12x^3 + 6x^2 = 6x^2(x^2 - 2x + 1)$$
$$= 6x^2(x - 1)^2$$

| *To find the GCF* | *To find the LCM* |
|---|---|
| The constant factor is the GCF of the constant factors, $4 = 2^2$ and $6 = 2 \cdot 3$, of the given polynomials, namely 2. | The constant factor is the LCM of the constant factors, $4 = 2^2$ and $6 = 2 \cdot 3$ of the given polynomials, namely $2^2 \cdot 3 = 12$. |
| Compare the powers of each prime polynomial which appears in *both* polynomials and take the one with the least exponent: | Compare the powers of each prime polynomial which appears in *at least one* polynomial and take the one with the greatest exponent: |
| Compare $x^3$ and $x^2$; take $x^2$ | Compare $x^3$ and $x^2$; take $x^3$ |
| Compare $(x - 1)$ and $(x - 1)^2$; take $(x - 1)$ | Compare $(x - 1)$ and $(x - 1)^2$; take $(x - 1)^2$ |
| $\therefore$ GCF $= 2x^2(x - 1)$ | Take $(x + 3)$ |
|  | $\therefore$ LCM $= 12x^3(x - 1)^2(x + 3)$ |

*WRITTEN EXERCISES*

Find **(a)** the GCF and **(b)** the LCM of each of the following sets of polynomials. Leave answers in factored form.

**A**
1. $y^2;\ -y^3;\ y^5$
2. $5x;\ 10x^2;\ -2x^3$
3. $2a + b;\ a + 2b$
4. $6c - 10d;\ c + 10d$
5. $15x^2 - 10;\ 9x^2 - 6$
6. $10s^2 + 8t;\ 15s^2 + 12t$

7. $x^2 - 36;\ 5x + 30$
8. $y^2 - 121;\ y^2 - 9y - 22$
9. $64x^3(y + z)^2;\ 48x^5(y + z)^3$
10. $-42n^5;\ -28n^2(n - 3)(n + 2)$
11. $x^2 + x - 6;\ 2x^2 + x - 15$
12. $3x^2 + 18x + 15;\ 3x^2 + 6x + 3$

**B**
13. $x^3 + 27;\ x^2 - 9;\ x + 3$
14. $(a - b)^2;\ a^2 - b^2;\ a^3 - b^3$
15. $w^2 - 1;\ (w - 1)^2;\ w + 1$
16. $3x^3 + 6x^2;\ 4x^3 - 4x^2 - 24x;\ x^3$

C **17.** $x^5y^3 - x^2y^6$; $2x^4 - 4x^2y^2 + y^4$

**18.** $6a^2 - 33ab + 36b^2$; $6a^2 - 28ab + 16b^2$

**19.** $3r^2 - 8sr - 3s^2$; $3r^2 + 8sr - 3s^2$

**20.** $12x^4 + 10x^3y - 8x^2y^2$; $6y^2 - 30xy + 36x^2$

## 5–5 Solving Equations

You can use the extended theorem of Section 4–9 to find zeros of a polynomial function if you can find linear factors of the polynomial.

**Example 1.**  Find the zeros of $g(x) = x^3 - x^2 - 6x$.

*Solution:*   The problem is equivalent to solving the polynomial equation $x^3 - x^2 - 6x = 0$.

Factor the polynomial completely.   $x(x^2 - x - 6) = 0$
$$x(x - 3)(x + 2) = 0$$

Solve the equivalent compound sentence.

$$x = 0 \quad or \quad x - 3 = 0 \quad or \quad x + 2 = 0$$
$$x = 0 \quad \mid \quad x = 3 \quad \mid \quad x = -2$$

∴ the solution set is $\{0, 3, -2\}$.

*Check:*   $g(0) = 0^3 - 0^2 - 6 \cdot 0 = 0$
$g(3) = 3^3 - 3^2 - 6 \cdot 3 = 27 - 9 - 18 = 0$
$g(-2) = (-2)^3 - (-2)^2 - 6(-2) = -8 - 4 + 12 = 0$

A **polynomial equation** can be written with one member 0 and the other a polynomial. Sometimes the polynomial has two or more identical factors; these yield **double** or **multiple** roots of the equation.

**Example 2.**  Solve $(x - 3)^2 = 4(x - 4)$.

*Solution:*   It is important to transform the given equation to make one member 0.
$$(x - 3)^2 = 4(x - 4)$$
$$x^2 - 6x + 9 = 4x - 16$$
$$x^2 - 10x + 25 = 0$$

Factor the polynomial completely.   $(x - 5)^2 = 0$

Solve the equivalent compound sentence.
$$x - 5 = 0 \quad or \quad x - 5 = 0$$
$$x = 5 \quad \mid \quad x = 5$$

∴ 5 is a double root. $\{5\}$

*Check:*   $(x - 3)^2 = (5 - 3)^2 = 2^2 = 4$
$4(x - 4) = 4(5 - 4) = 4 \cdot 1 = 4$

As you saw in Section 4–3, the domain of a variable appearing in a practical problem is often restricted for physical or geometrical reasons.

**Example 3.** A rectangular corner lot originally had dimensions 50 yards by 60 yards, but one-third of its original area was lost when the two adjacent streets were widened by equal amounts. Find the new dimensions of the lot.

*Solution:*

1. Let $x$ = number of yards each street was widened. Note that $0 < x < 50$. The new dimensions of the lot are $(50 - x)$ yards by $(60 - x)$ yards.

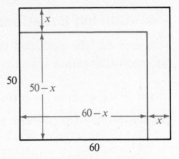

2. The present area is two-thirds of the original area.

$$(50 - x)(60 - x) = \tfrac{2}{3} \times 50 \times 60$$

3. $3000 - 110x + x^2 = 2000$
   $x^2 - 110x + 1000 = 0$
   $(x - 10)(x - 100) = 0$

   $x - 10 = 0$    *or*    $x - 100 = 0$
          $x = 10$   |          $x = 100$   Rejected as failing to satisfy $0 < x < 50$.

   $\therefore$ the new dimensions, in yards, are $50 - 10$ by $60 - 10$, that is, 40 by 50.

*Check:*

4. The new area is $40 \times 50 = 2000$ square yards. The original area was $50 \times 60 = 3000$ square yards.
   $\therefore$ area lost $= 3000 - 2000 = 1000$ square yards, which is one-third of the original area.

In Example 3 we rejected the root 100 because it is not in the domain of the variable $x$. In solving the problem we reasoned that *if* $x$ satisfies the conditions of the problem, then it must satisfy the equation in Step 2. The converse is not necessarily true; that is, a solution of this equation need not satisfy the *other* conditions of the problem (one of which is $x < 50$). The roots of the equation are *possible* solutions of the problem. By checking these possibilities in the words of the given problem, you find the *actual* solutions.

## ORAL EXERCISES

State a compound sentence equivalent to each of the following open sentences.

**1.** $x(x + 3) = 0$

**2.** $y(y - 5) = 0$

**3.** $(y + 2)(y + 3) = 0$

**4.** $(3y - 7)(y + 11) = 0$

**5.** $(10x - 7)(12x + 5) = 0$

**6.** $x(x - 9)(x + 13) = 0$

**7.** $y^2(2y - 15) = 0$

**8.** $w(3w - 2)(w + 4) = 0$

## WRITTEN EXERCISES

Solve each of the following equations over the set of real numbers.

**A**

**1.** $x^2 + 4x + 3 = 0$

**2.** $x^2 + 9x + 18 = 0$

**3.** $y^2 - 6y + 5 = 0$

**4.** $y^2 - 10y + 24 = 0$

**5.** $y^2 + 2y - 15 = 0$

**6.** $w^2 + 8w - 9 = 0$

**7.** $z^2 - 13z - 48 = 0$

**8.** $z^2 - z - 12 = 0$

**9.** $3x^2 - 12 = 0$

**10.** $n^2 = -16n$

**11.** $w^2 - 36 = 0$

**12.** $5y^2 - 125 = 0$

**13.** $2a^2 = 3 - a$

**14.** $3v = v^2 - 28$

**15.** $2c^2 = 7c - 6$

**16.** $3y^2 = 10y - 3$

**17.** $5x^2 + 11x = -2$

**18.** $3y^2 + 4 = 8y$

**19.** $2n^2 + 7 = 5(1 - n)$

**20.** $3x^2 + x = 14(x - 1)$

**21.** $3x^2 + 6x = 45$

**22.** $10x^2 = 15 - 5x$

**23.** $w^2 + 3w - 4 = 50$

**24.** $9(x^2 + x) = -2$

**25.** $(x - 5)(x - 4) = 3x$

**26.** $(a + 2)(a - 2) = 3a$

**27.** $x^2 - 13 = (5 - x)^2$

**28.** $w^2 + (5 - 2w)^2 = 10$

**B**

**29.** $y(y + 5)(y - 8) = 0$

**30.** $(y - 2)(y + 7)(y - 11) = 0$

**31.** $x^3 - 4x = 0$

**32.** $3x^3 - 10x^2 = -3x$

Solve each equation for $x$ ($a^2 \neq b^2$):

**C**

**33.** $a^2x + b = b^2x + a$

**34.** $2x^3 - 9x^2 = 2x - 9$

## PROBLEMS

Solve and check each problem.

**A**  **1.** Find two consecutive integers whose product is 462.

**2.** The difference of the cubes of two consecutive integers is 217. Find these two integers.

**3.** Find four consecutive integers such that the sum of the squares of the first two is 11 less than the square of the fourth.

**4.** Find four consecutive integers such that the sum of the squares of the first and fourth is 65.

**5.** The length of a rectangular pool is 4 yards longer than its width. The area of the pool is 60 square yards. What are the dimensions of the pool?

**6.** The sum of two adjacent sides of a rectangle is 8. Find the dimensions of this rectangle if its area is 15.

**7.** The dimensions of a rectangular field are 8 yards and 7 yards. If each dimension is increased the same number of yards, the area of the new field exceeds the area of the original field by 54 square yards. What are the dimensions of the new field?

The altitude $h$ of an object $t$ seconds after it is thrown upward with an initial speed of $r$ feet per second from an altitude of $k$ feet is given by the formula

$$h = k + rt - 16t^2.$$

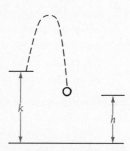

**Sample.**  A ball is thrown upward with a speed of 56 feet per second from a tower 100 feet high. When will it be 28 feet above the ground?

*Solution:*  Let $t$ be the number of seconds it takes the ball to reach the altitude of 28 feet. $h = 28, k = 100, r = 56$

$$28 = 100 + 56t - 16t^2$$
$$16t^2 - 56t - 72 = 0$$
$$2t^2 - 7t - 9 = 0$$
$$(2t - 9)(t + 1) = 0$$

$$2t - 9 = 0 \quad or \quad t + 1 = 0$$
$$t = \tfrac{9}{2} = 4\tfrac{1}{2} \quad | \quad t = -1 \text{ (rejected)}$$

∴ the ball is 28 feet above the ground $4\tfrac{1}{2}$ seconds after being thrown.

8. A missile is fired from ground level with an upward velocity of 4000 feet per second. After how many minutes will it return to the ground?

9. A baseball is thrown upward with a speed of 128 feet per second from a point opposite the base of a building 240 feet high. At what times is it opposite the top of the building?

**B** 10. How much time will elapse before a missile fired with an upward velocity of 8000 feet per second from the edge of a cliff 1800 feet high again reaches an altitude of 1800 feet?

11. From an airplane whose altitude is 2664 feet a bullet is fired with an upward muzzle velocity of 3200 feet per second. When will the bullet's altitude be 9000 feet?

12. The legs of a right triangle have lengths 4 feet and 5 feet. When the legs are decreased in length by equal amounts, the area of the resulting triangle is 4 square feet less than that of the original triangle. Find the lengths of the sides of the new triangle.

13. The hypotenuse of a right triangle is 3 feet longer than one of the legs and 6 feet longer than the other. How long are the legs? (*Hint:* First find the length of the hypotenuse.)

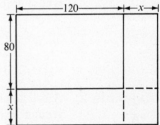

14. A rectangular field is 120 yards long and 80 yards wide. In order to double the area of the field, strips of equal width are added to one side and one end as in the diagram at the right. Find the width of these strips.

15. A rectangular recreational park with dimensions 200 yards and 300 yards is to be built on a plot of land with an area of 81,600 square yards. If there is to be a uniform strip of land surrounding the park, how wide will this strip be?

## 5–6 Solving Inequalities

Recall from page 49 that:

A product of nonzero factors is positive if the number of negative factors is even. A product of nonzero factors is negative if the number of negative factors is odd.

You can use these facts to solve certain inequalities.

**Example 1.** Graph the solution set of $x^2 > 3x + 4$.

*Solution:*
$$x^2 > 3x + 4$$
$$x^2 - 3x - 4 > 0$$
$$(x + 1)(x - 4) > 0$$

The last inequality will be satisfied if and only if the factors of the left member are either *both positive* or *both negative*.

| *Both Positive* | *or* | *Both Negative* |
|---|---|---|

$x + 1 > 0$ and $x - 4 > 0$ $\qquad$ $x + 1 < 0$ and $x - 4 < 0$

$x > -1$ and $x > 4$ $\qquad\qquad$ $x < -1$ and $x < 4$

The intersection of the solution sets of these inequalities is $\{x: x > 4\}$. $\quad$ The intersection of the solution sets of these inequalities is $\{x: x < -1\}$.

$\therefore$ the solution set of the given inequality is
$$\{x: x < -1\} \cup \{x: x > 4\},$$

whose graph is:

**Example 2.** A ball is thrown vertically upward from a point 5 feet above ground level with an initial velocity of 56 feet per second. During what period is it higher than 45 feet?

*Solution:* The height $h$ of the ball $t$ seconds after it is thrown can be found from the formula preceding the Sample, page 188, with $r = 56$ and $k = 5$:
$$h = 56t - 16t^2 + 5.$$

We wish to solve the inequality
$$56t - 16t^2 + 5 > 45.$$
$$-16t^2 + 56t - 40 > 0$$
$$16t^2 - 56t + 40 < 0$$
$$8(2t^2 - 7t + 5) < 0$$

We may divide both members by the *positive* number 8:
$$2t^2 - 7t + 5 < 0$$
$$(2t - 5)(t - 1) < 0$$

The last inequality will be satisfied if and only if one of the factors of the right member is negative and the other is positive.

$2t - 5 < 0$  and  $t - 1 > 0$  *or*  $2t - 5 > 0$  and  $t - 1 < 0$

$\qquad t < \frac{5}{2}$  and  $t > 1$ $\qquad\qquad$ $t > \frac{5}{2}$  and  $t < 1$

$\qquad\qquad 1 < t < \frac{5}{2}$ $\qquad\qquad$ There is no such number $t$.

$\therefore$ the height of the ball is greater than 45 feet during the period from 1 second to $2\frac{1}{2}$ seconds after it was thrown.

## WRITTEN EXERCISES

Graph the solution set of the following inequalities over $\mathfrak{R}$.

**A**
1. $x^2 - 7x > 0$
2. $9y - 3y^2 < 0$
3. $(x + 3)(x - 5) > 0$
4. $(x - 2)(x + 1) \leq 0$
5. $y^2 - 5y + 4 \leq 0$

6. $w^2 - 7w + 10 \geq 0$
7. $5a^2 + 4 \geq 12a$
8. $3n^2 \leq 8n - 4$
9. $9x < 10 + 2x^2$
10. $3x^2 < 5x + 12$

**B**
11. $y^3 - 16y < 0$
12. $x^3 - 81x > 0$
13. $6x^3 + 5x^2 - 4x \leq 0$

14. $3y^3 + 4y \geq 8y^2$
15. $x^3 - x^2 < 0$
16. $6w^3 + 4w^2 > 10w$

In working Exercises 17–19, use the formula on page **188** giving the altitude of a falling object.

**C**
17. A ball is tossed upward from the top of a 100-foot tower with initial speed 40 feet per second. During what period of time is it higher than the top of the tower?

18. A baseball is thrown upward from the roof of a building 120 feet high with initial speed 64 feet per second. During what period of time is the altitude of the ball greater than 168 feet?

19. During what period of time is the altitude of the ball in Exercise 18 greater than 40 feet?

## QUOTIENTS OF POLYNOMIALS

### 5–7 Dividing Polynomials

You can express an improper fraction, such as $\frac{32}{5}$, as a mixed number, $6\frac{2}{5}$, by using the identity $\frac{a+b}{c} = \frac{a}{c} + \frac{b}{c}$ (Exercise 34, page 54):

$$\frac{32}{5} = \frac{30+2}{5} = \frac{30}{5} + \frac{2}{5} = 6 + \frac{2}{5}.$$

The equation $\frac{32}{5} = 6 + \frac{2}{5}$ illustrates the *division algorithm:*

$$\frac{\text{Dividend}}{\text{Divisor}} = \text{Quotient} + \frac{\text{Remainder}}{\text{Divisor}}$$

or

$$\text{Dividend} = \text{Quotient} \times \text{Divisor} + \text{Remainder}$$

For larger numbers you can use the long-division process illustrated below for the quotient $567 \div 16$.

$$
\begin{array}{r}
35 \\
16\overline{)567} \\
480 \\ \hline
87 \\
80 \\ \hline
7
\end{array}
$$

← subtract $30 \times 16$

← subtract $5 \times 16$

$\therefore 567 - (30 \times 16) - (5 \times 16) = 7$
or $567 = (30 \times 16) + (5 \times 16) + 7$
$567 = (35 \times 16) + 7$
$\frac{567}{16} = 35 + \frac{7}{16}$

Note that the process stops when the remainder is less than the divisor.

You can use a similar process to find a quotient of polynomials. For example, to divide $x^3 + 5x^2 + 8x + 9$ by $x + 2$, you proceed as follows:

$$
\begin{array}{r}
x^2 + 3x + 2 \\
x + 2\overline{)x^3 + 5x^2 + 8x + 9} \\
x^3 + 2x^2 \\ \hline
3x^2 + 8x \\
3x^2 + 6x \\ \hline
2x + 9 \\
2x + 4 \\ \hline
5
\end{array}
$$

subtract $x^2(x+2)$

subtract $3x(x+2)$

subtract $2(x+2)$

$\therefore x^3 + 5x^2 + 8x + 9 - x^2(x+2) - 3x(x+2) - 2(x+2) = 5,$
or $x^3 + 5x^2 + 8x + 9 = x^2(x+2) + 3x(x+2) + 2(x+2) + 5.$

$$\underbrace{x^3 + 5x^2 + 8x + 9}_{\text{Dividend}} = \underbrace{(x^2 + 3x + 2)}_{\text{Quotient}}\underbrace{(x + 2)}_{\text{Divisor}} + \underbrace{5}_{\text{Remainder}}$$

or

$$\frac{\text{Dividend} \longrightarrow x^3 + 5x^2 + 8x + 9}{\text{Divisor} \xrightarrow{\hspace{2em}} x + 2} = \underbrace{x^2 + 3x + 2}_{\text{Quotient}} + \frac{5 \longleftarrow \text{Remainder}}{x + 2 \longleftarrow \text{Divisor}}$$

The process ends when the degree of the remainder is less than that of the divisor, or when the remainder is 0.

Before applying the division process, arrange the terms of both the dividend and the divisor in order of decreasing degree in the same variable. The next example shows how to insert missing terms by using 0 as a coefficient.

**Example.**   Divide $2s^3 + t^3 - 3s^2t$ by $2s^2 - t^2 - st$.

*Solution:*

$$
\begin{array}{r}
s - t \\
2s^2 - st - t^2{\overline{\smash{\big)}\,2s^3 - 3s^2t + 0st^2 + t^3}} \\
\underline{2s^3 - s^2t - st^2} \\
-2s^2t + st^2 + t^3 \\
\underline{-2s^2t + st^2 + t^3} \\
0
\end{array}
$$

*Check:* $(s - t)(2s^2 - st - t^2) = (2s^3 - s^2t - st^2) - (2s^2t - st^2 - t^3)$
$$= 2s^3 - 3s^2t + t^3$$

Notice that in the preceding example the remainder is 0, and that we have partially factored the polynomial $2s^3 - 3s^2t + t^3$.

## WRITTEN EXERCISES

Perform the indicated divisions.

[A]  **1.** $\dfrac{24x^3 - 16x^2 + 2x}{2x}$

**2.** $\dfrac{51x^5 - 34x^4 + 17x - 2}{17x^2}$

**3.** $\dfrac{x^5 - 6x^6 + 8x^8}{2x^2}$

**4.** $\dfrac{12x^6 - 16x^4 + 20x - 4}{4x^4}$

**5.** $\dfrac{15x^3 - 12x^5 + 18x^6}{3x^3}$

**6.** $\dfrac{4r^4s^4 - 8r^3s^3 + 12rs - 16}{4r^2s^2}$

**7.** $\dfrac{5x^{12} - 10x^6 - 25x^3 + x}{-5x}$

**8.** $\dfrac{36x^8 - 8x^6 + 14x^5 + x^4}{-2x^4}$

**9.** $\dfrac{w^2 - 8w + 7}{w - 1}$

**10.** $\dfrac{y^2 - 15y - 54}{y + 3}$

**11.** $\dfrac{n^2 - 7n - 60}{n - 12}$

**12.** $\dfrac{a^2 + 5a - 24}{a - 3}$

**13.** $\dfrac{4x^2 + 6x + 9}{2x - 5}$

**14.** $\dfrac{15m^2 - m - 16}{5m + 3}$

**15.** $\dfrac{5x^2 - 33x - 24}{5x + 2}$

**16.** $\dfrac{20x^2 - 37x + 15}{5x - 3}$

**17.** $\dfrac{6x^3 + x^2 - 28x - 30}{2x - 5}$

**18.** $\dfrac{6n^3 - 8n^2 - 17n - 6}{3n + 2}$

**19.** $\dfrac{6k^3 + 11k^2 - 1}{3k + 1}$

**20.** $\dfrac{y^3 - 27}{y - 3}$

**B 21.** $\dfrac{7d^2 - 32cd - 15c^2}{d - 5c}$

**22.** $\dfrac{27a^2 + 3ab - 14b^2}{9a + 7b}$

**23.** $\dfrac{10y^2 + 38xy - 8x^2}{2y + 8x}$

**24.** $\dfrac{3p^4 - 17p^2q + 10q^2}{p^2 - 5q}$

**25.** $\dfrac{x^3 - a^3}{x - a}$

**26.** $\dfrac{y^3 + b^3}{y + b}$

**27.** $\dfrac{x^4 - 8x^3y + 21x^2y^2 - 16xy^3 - 7y^4}{x^2 - 5xy + 7y^2}$

**28.** $\dfrac{2a^4 - 2a^3b - 5a^2b^2 + 4ab^3 + 5b^4}{2a^2 - 6ab + 5b^2}$

**C 29.** $\dfrac{a^5 + b^5}{a + b}$

**30.** $\dfrac{x^{12} + y^{12}}{x^4 + y^4}$

**31.** Find the value of $k$ so that $3x - 1$ is a factor of $21x^2 - 10x + k$.

**32.** Find the value of $k$ so that $6x - 5$ is a factor of $kx^2 - 61x + k$.

**33.** If the quotient obtained in dividing $y^2 - 7y + a$ by $y + b$ is $y - 2$ and the remainder is 5, find the values of $a$ and $b$.

## 5–8 Synthetic Division (Optional)

**Synthetic division** is a short process for dividing a polynomial by a binomial of the *special form* $x - r$. To divide $3x^3 - 2x^2 - 10x + 9$ by $x - 2$, for example, follow these steps:

1. Copy the *coefficients* of the dividend and put the "*r* number" (in this case, 2) on a "shelf" to the left of them:

$$\underline{2\,|} \quad 3 \quad -2 \quad -10 \quad 9$$

2. Skip a space vertically, then draw a horizontal line and repeat the first coefficient of the dividend below this line:

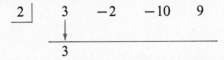

3. Multiply the number below the line by the number on the "shelf," put the product above the line in the second column, and add:

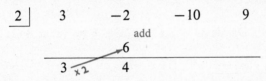

4. Multiply the last number below the line by the number on the "shelf," put the product above the line in the next column, and add:

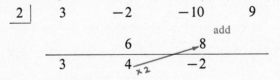

5. Continue the process described in Step 4:

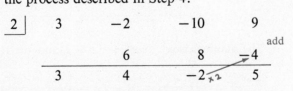

When the process is complete, the *last* number in the bottom row is the *remainder,* and the other numbers in the bottom row are the *coefficients* of the *quotient.* Thus when $3x^3 - 2x^2 - 10x + 9$ is divided by $x - 2$, the quotient is $3x^2 + 4x - 2$, and the remainder is 5.

The two division processes are compared below:

$$
\begin{array}{r|rrrr}
2 & 3 & -2 & -10 & 9 \\
  &   & 6  & 8   & -4 \\
\hline
  & 3 & 4  & -2  & 5
\end{array}
$$

$$
\begin{array}{r}
3x^2 + 4x - 2 \\
x - 2 \overline{)\, 3x^3 - 2x^2 - 10x + 9} \\
\underline{3x^3 - 6x^2} \\
4x^2 - 10x \\
\underline{4x^2 - 8x} \\
-2x + 9 \\
\underline{-2x + 4} \\
5
\end{array}
$$

When you use synthetic division, write the terms of the dividend in order of decreasing degree; and if a power is missing, supply the coefficient 0. Notice that when you divide by $x + 2$ the "shelf" number is $-2$ because $x + 2 = x - (-2)$.

**Example.** Divide $x^4 - 2x^3 - 7x - 6$ by (a) $x + 2$; (b) $x - 3$.

*Solution:*  **a.**
$$
\begin{array}{r|rrrrr}
-2 & 1 & -2 & 0  & -7  & -6 \\
   &   & -2 & 8  & -16 & 46 \\
\hline
   & 1 & -4 & 8  & -23 & 40
\end{array}
$$

The quotient is $x^3 - 4x^2 + 8x - 23$ and the remainder is 40:

$$
\frac{x^4 - 2x^3 - 7x - 6}{x + 2} = x^3 - 4x^2 + 8x - 23 + \frac{40}{x + 2} \text{ or}
$$

$$
x^4 - 2x^3 - 7x - 6 = (x + 2)(x^3 - 4x^2 + 8x - 23) + 40
$$

**b.**
$$
\begin{array}{r|rrrrr}
3 & 1 & -2 & 0 & -7 & -6 \\
  &   & 3  & 3 & 9  & 6 \\
\hline
  & 1 & 1  & 3 & 2  & 0
\end{array}
$$

The quotient is $x^3 + x^2 + 3x + 2$ and the remainder is 0. Thus

$$
x^4 - 2x^3 - 7x - 6 = (x - 3)(x^3 + x^2 + 3x + 2),
$$

and $x - 3$ is a *factor* of $x^4 - 2x^3 - 7x - 6$.

The process of synthetic division is derived from the standard division process by eliminating unneeded symbols (see Extra for Experts, page 255).

## WRITTEN EXERCISES

Use synthetic division to find the quotient and the remainder.

**A** 1. $\dfrac{x^3 - 5x^2 + 3x - 1}{x - 1}$

2. $\dfrac{x^3 + 3x^2 - 4x + 2}{x - 2}$

3. $\dfrac{x^4 - 3x^3 + 8x^2 - x - 6}{x + 1}$

4. $\dfrac{y^4 - y^3 + 2y^2 - 11y + 3}{y - 5}$

5. $\dfrac{2x^4 - 3x^3 + 4x^2 + x - 1}{x + 3}$

6. $\dfrac{3x^4 + 2x^3 - 5x^2 - 2x + 4}{x - 3}$

7. $\dfrac{x^3 - x + 5}{x + 4}$

8. $\dfrac{3x^4 - 5x^2 + 2x - 1}{x - 6}$

**Sample:** $\dfrac{6x^3 + 5x^2 - x + 2}{2x + 1}$

*Solution:* $\dfrac{6x^3 + 5x^2 - x + 2}{2x + 1} = \dfrac{1}{2} \cdot \dfrac{6x^3 + 5x^2 - x + 2}{x + \frac{1}{2}}$

$$
\begin{array}{r|rrrr}
-\frac{1}{2} & 6 & 5 & -1 & 2 \\
 & & -3 & -1 & 1 \\
\hline
 & 6 & 2 & -2 & 3
\end{array}
$$

$$\dfrac{6x^3 + 5x^2 - x + 2}{2x + 1} = \dfrac{1}{2}\left(6x^2 + 2x - 2 + \dfrac{3}{x + \frac{1}{2}}\right)$$

$$= 3x^2 + x - 1 + \dfrac{3}{2x + 1}$$

$\therefore$ quotient $= 3x^2 + x - 1$; remainder $= 3$.

**B** 9. $\dfrac{2x^3 - x^2 + 5x + 3}{2x - 1}$

10. $\dfrac{4x^3 - 3x^2 - x + 6}{2x - 3}$

11. $\dfrac{6x^3 - x^2 + 5x + 7}{2x + 3}$

12. $\dfrac{6x^3 + 7x^2 - x + 8}{3x - 2}$

13. $\dfrac{8x^4 - 5x^2 - 3x - 2}{4x - 1}$

14. $\dfrac{10x^4 + 6x^3 - 5}{5x + 2}$

Determine $k$ so that the first polynomial is a factor of the second, that is, so that when the second polynomial is divided by the first, the remainder is zero.

**C** 15. $x - 1$; $x^3 - 5x^2 + x + k$

16. $x + 2$; $x^4 - 6x^3 - 2x + k$

17. $x - 5$; $2x^3 - 30x^2 + kx - 10$

18. $x - 3$; $x^3 - 4x^2 + kx - 2k$

## 5–9  More on Factoring (Optional)

When you divide $x^3 - 3x^2 + 2x - 6$ by $x - 3$, you find that the remainder is 0:

$$
\begin{array}{r|rrrr}
3 & 1 & -3 & 2 & -6 \\
  &   & 3  & 0 & 6 \\
\hline
  & 1 & 0  & 2 & 0
\end{array}
$$

Since the first three numbers in the last row are the coefficients of the quotient, you can write

$$x^3 - 3x^2 + 2x - 6 = (x - 3)(x^2 + 2),$$

and therefore you have factored $x^3 - 3x^2 + 2x - 6$. Since both factors are prime polynomials (page 181), the factorization is complete.

**Example 1.**  Show that $x + 5$ is a factor of $x^3 + 7x^2 + 11x + 5$, and complete the factorization.

*Solution:*

$$
\begin{array}{r|rrrr}
-5 & 1 & 7 & 11 & 5 \\
   &   & -5 & -10 & -5 \\
\hline
   & 1 & 2 & 1 & 0
\end{array}
$$

$$\therefore\ x^3 + 7x^2 + 11x + 5 = (x + 5)(x^2 + 2x + 1)$$

The trinomial can be factored by the methods of Section 5–2:

$$x^2 + 2x + 1 = (x + 1)^2.$$

Hence

$$x^3 + 7x^2 + 11x + 5 = (x + 5)(x + 1)^2.$$

The following fact from Section 7–4 helps you decide which factors to try.

---

Suppose a polynomial with integral coefficients has $x - r$ as a factor, where $r$ is an integer. Then $r$ is an integral factor of the constant term of the polynomial.

---

For example, $x - 3$ and $x + 2$ *might* be factors of $x^3 + 2x + 12$ because 3 and $-2$ are factors of 12, but you would not even try $x - 5$ as a factor because 5 is not an integral factor of 12.

**Example 2.** Factor completely: $x^3 + x^2 - 4x - 4$

*Solution:* If $x - r$ is to be a factor of $x^3 + x^2 - 4x - 4$, then $r$ must be an integral factor of $-4$, that is,

$$r \in \{4, 2, 1, -1, -2, -4\}.$$

Use synthetic division to discover which, if any, of these binomials are factors.

$$\begin{array}{r|rrrr} 4 & 1 & 1 & -4 & -4 \\ & & 4 & 20 & 64 \\ \hline & 1 & 5 & 16 & 60 \end{array} \neq 0$$

$\therefore x - 4$ is not a factor.

$$\begin{array}{r|rrrr} 2 & 1 & 1 & -4 & -4 \\ & & 2 & 6 & 4 \\ \hline & 1 & 3 & 2 & 0 \end{array}$$

$\therefore x - 2$ is a factor, and

$$x^3 + x^2 - 4x - 4 = (x - 2)(x^2 + 3x + 2).$$

The trinomial can be factored using the methods of Section 5–3: $x^2 + 3x + 2 = (x + 1)(x + 2)$.

$\therefore x^3 + x^2 - 4x - 4 = (x - 2)(x + 1)(x + 2)$.

## WRITTEN EXERCISES

Factor each of the following polynomials completely.

A

**1.** $x^3 + x^2 - 9x - 9$

**2.** $x^3 + x^2 - 4x - 4$

**3.** $x^3 - 5x^2 - 13x - 7$

**4.** $x^3 - x^2 - 8x + 12$

**5.** $x^3 - 7x^2 + 8x + 16$

**6.** $x^3 - 12x - 16$

**7.** $x^3 - 19x - 30$

**8.** $x^3 - 5x^2 - 9x + 45$

**9.** $x^3 - 8x^2 + 21x - 18$

**10.** $x^3 - 12x^2 + 48x - 64$

**11.** $x^3 - 6x^2 + x - 6$

**12.** $x^3 - 2x + 4$

B

**13.** $x^4 - 3x^3 - 8x^2 + 12x + 16$

**14.** $x^4 + 3x^3 - 3x^2 - 11x - 6$

**15.** $x^4 - 3x^3 - 23x^2 - 33x - 14$

**16.** $x^4 + 4x^3 - 6x^2 - 4x + 5$

C

**17.** $x^4 + 2x^2 - 24$

**18.** $x^4 + 3x^3 - 9x - 27$

**19.** $x^5 - x^3 - 8x + 8$

**20.** $x^5 - x^4 - 8x^2 - 25x - 15$

---

# *Chapter Summary*

---

1. To **factor** an integer, you express it as a product of other integers. A **prime** number is an integer greater than 1 which has only itself and 1 as factors.

2. To **factor** a polynomial, you express it as a product of other polynomials. To do this, first factor out the greatest monomial factor. At the next step certain special product forms may be useful; for example:

$$a^2 - b^2 = (a + b)(a - b)$$
$$a^2 + 2ab + b^2 = (a + b)^2$$
$$a^3 + b^3 = (a + b)(a^2 - ab + b^2)$$

You can often factor quadratic trinomials by trying various possibilities for the linear factors.

3. To find the GCF and LCM of several numbers or of several polynomials, begin by factoring the integers or the polynomials into prime factors.

4. Factoring often is useful in solving equations or inequalities which involve polynomials.

5. Division of polynomials, like division of integers, is described by the **division algorithm**.

   In dividing one polynomial by another, the process terminates when the remainder is 0 or of lower degree than the divisor. If the remainder is 0, then the divisor and quotient are factors of the dividend.

6. **Optional**. If in a division problem the divisor is linear, then a short process, **synthetic division**, can be used.

### Vocabulary and Spelling

Review each term by reference to the page listed.

## Chapter Test

**5–1**  **1.** Find the GCF and LCM of the monomials $42x^2y^3z$, $36x^4yz^2$, and $12xz$.

**5–2**  Factor completely.

**2.** $10x^2y - 15x^3y^2z + 5x^4y^3$      **3.** $y^3 + 64$      **4.** $12x^2 - 75$

**5–3**  Factor completely.

**5.** $4x^2 + 20x + 25$          **6.** $2x^2 - 13x + 21$

**5–4**  **7.** Find the GCF and LCM of the polynomials $2x^3 + 3x^2 - 2x$, $3x^2 + 11x + 10$, and $7x^2 + 14x$.

**5–5**  Find the solution set.

**8.** $2x^2 + 2 = -7x - 4$        **9.** $3x^3 + 3x^2 - 18x = 0$

**10.** A girl, mowing a lawn 30 feet wide and 40 feet long, mows a uniformly wide strip about its border. How wide is this strip of grass when one-half of the lawn is cut?

**5–6**  **11.** Find the solution set of $3x^2 + 2x > 5$.

**5–7**  **12.** Divide $6x^3 + 5x^2 - 29x - 10$ by $3x^2 - 5x - 2$.

**5–8**  **13.** Use synthetic division to find the quotient and remainder when $2x^3 + 3x^2 - 7x + 1$ is divided by $x + 4$ (Optional).

**5–9**  **14.** Factor completely: $x^3 - 5x^2 - 8x + 12$ (Optional).

## Chapter Review

**5–1**   **Factors of Monomials**      *Pages 173–176*

**1.** The prime factors of 720 are __?__ .

Find (a) the GCF and (b) the LCM of the following monomials.

**2.** 144, 320                **3.** $36x^3y^2$, $24xy^4$, $16x^2y^3$

**5–2** **Factoring Polynomials** *Pages 177–179*

Factor completely.

**4.** $36x^2y^3z - 63x^3y^2z^2$      **6.** $9x^2 - 16$

**5.** $y^2 - 18y + 81$      **7.** $27z^3 + 8$

**5–3** **Factoring Quadratic Polynomials** *Pages 180–183*

Factor completely.

**8.** $x^2 - 6x - 7$      **10.** $15x^3 - 9x^2 - 24x$

**9.** $6x^2 - 10x - 24$      **11.** $10x^2 + 6xy - 4y^2$

**5–4** **Common Factors and Multiples** *Pages 183–185*

Find (a) the GCF and (b) the LCM of the polynomials.

**12.** $x^2 + x - 12, x^2 - 9$

**13.** $2x^3 + 5x^2 - 3x, x^3 + 3x^2, x^2 - 2x - 15$

**5–5** **Solving Equations** *Pages 185–189*

**14.** If $xy = 0$ then $x = \underline{\quad?\quad}$ or $y = \underline{\quad?\quad}$.

**15.** If $\dfrac{x}{y} = 0$ then $x = \underline{\quad?\quad}$ and $y \neq \underline{\quad?\quad}$.

Find the solution set.

**16.** $x^2 - 6x = -5$      **17.** $2y^2 + 7 = 5(1 - y)$

**18.** Find three consecutive odd integers where the sum of the square of the smallest and the product of the remaining two is 844.

**5–6** **Solving Inequalities** *Pages 189–191*

**19.** If $ab > 0$, which of the following cannot be true?
   (a) $a > 0, b > 0$      (c) $a > 0, b < 0$
   (b) $a < 0, b > 0$      (d) $a < 0, b < 0$

Find the solution set:

**20.** $3x^2 - 12 > 0$    **21.** $y^2 - 12y < -35$    **22.** $5x < 2x^2 + 2$

**5–7** **Dividing Polynomials** *Pages 192–194*

**23.** Divide $x^3 - 8x^2 + 5x + 50$ by $x - 5$.

**24.** Divide $12x^3 - 23x^2y - 8xy^2 + 12y^3$ by $4x^2 - 5xy - 6y^2$.

**5–8** **Synthetic Division (Optional)** *Pages 195–197*

Find the quotient and remainder when $5x^4 - 3x^3 + 4x - 1$ is divided by:

**25.** $x - 3$                 **27.** $2x + 6$

**26.** $x + 5$                 **28.** $3x - 2$

**5–9** **More on Factoring (Optional)** *Pages 198–199*

Factor completely.

**29.** $x^3 + 3x^2 - 4x - 12$      **31.** $x^5 - 9x^3 + 8x^2 - 72$

**30.** $x^3 - 19x + 30$          **32.** $x^4 + 2x^3 - 2x^2 - 8x - 8$

# Extra for Experts

## THE REMAINDER AND FACTOR THEOREMS

Given the polynomial

$$P(x) = x^3 - 3x^2 + 4x + 3,$$

let us first find $P(2)$ by substitution, and then divide $P(x)$ by $x - 2$.

$$
\begin{array}{ll}
P(2) = 2^3 - 3 \cdot 2^2 + 4 \cdot 2 + 3 \\
\quad\;\; = 8 - 12 + 8 + 3 \\
\quad\;\; = 7
\end{array}
\qquad
\begin{array}{r}
x^2 - x + 2 \\
x - 2 \overline{)\; x^3 - 3x^2 + 4x + 3} \\
\underline{x^3 - 2x^2} \\
-x^2 + 4x \\
\underline{-x^2 + 2x} \\
2x + 3 \\
\underline{2x - 4} \\
7
\end{array}
$$

The fact that $P(2)$ is the remainder when $P(x)$ is divided by $x - 2$ illustrates the next theorem:

---

**REMAINDER THEOREM.** When the polynomial $P(x)$ is divided by $x - r$, the remainder is $P(r)$.

---

The Remainder Theorem follows from the following general theorem about dividing polynomials, which we state without proof:

---

**THEOREM.**  Let $P(x)$ and $D(x)$ be polynomials, with $D(x)$ not the zero polynomial. Then there exist unique polynomials $Q(x)$ and $R(x)$ such that

$$P(x) = Q(x) \cdot D(x) + R(x)$$

and $R(x)$ is either the zero polynomial or a polynomial of degree less than the degree of $D(x)$.

---

If you let the divisor $D(x)$ be $x - r$, then the remainder $R(x)$ must be a constant, say $k$, and you have

$$P(x) = Q(x)(x - r) + k.$$

Substituting $r$ for $x$, you have

$$P(r) = Q(r) \cdot (r - r) + k = Q(r) \cdot 0 + k = k.$$

Thus the remainder $k$ is $P(r)$ and the theorem is proved.

Synthetic division (Section 5–8) is often called **synthetic substitution** because, by the Remainder Theorem, you can use it to find values of polynomial functions without actually substituting. Thus if $F(x) = x^4 + 5x^3 + 5x^2 - 7x + 8$, you can find $F(-3)$ as follows:

$$
\begin{array}{r|rrrrr}
-3 & 1 & 5 & 5 & -7 & 8 \\
   &   & -3 & -6 & 3 & 12 \\
\hline
   & 1 & 2 & -1 & -4 & 20 = F(-3)
\end{array}
$$

If a polynomial $P(x)$ has $x - r$ as a factor, then $P(x) = (x - r)Q(x)$ for some polynomial $Q(x)$. Hence, substituting $r$ for $x$, you have

$$P(r) = (r - r)Q(r) = 0 \cdot Q(r) = 0.$$

Conversely, if $P(r) = 0$, then by the Remainder Theorem,

$$P(x) = (x - r)Q(x) + P(r) = (x - r)Q(x) + 0 = (x - r)Q(x).$$

This proves the following:

---

**FACTOR THEOREM.**  The polynomial $P(x)$ has $x - r$ as a factor if and only if $P(r) = 0$, that is, if and only if $r$ is a root of the equation $P(x) = 0$.

---

If you know one root of a polynomial equation, you can sometimes find the other roots.

**Example.** Find the zeros of the function $P(x) = x^3 + 3x^2 - 25x - 75$.

*Solution:*   (*Plan:* Solve the equation $P(x) = 0$.)

1. By the displayed statement on page 198, the only possible integral roots are 1, 3, 5, 15, 25, 75 and their negatives.

2. Use the Factor Theorem and synthetic substitution to test each possibility. By doing the addition mentally, you can arrange the work compactly, as shown.

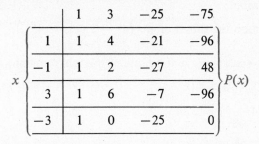

Since $P(-3) = 0$, $-3$ is a root of $P(x) = 0$.

3. Solve the *depressed equation*, $x^2 - 25 = 0$.

$$(x + 5)(x - 5) = 0$$
$$x = -5 \text{ or } 5$$

∴ the set of zeros of $P$ is $\{-3, -5, 5\}$.

# Exercises

Find the solution set of each of the following:

**1.** $x^3 - 2x^2 - 5x + 6 = 0$     **3.** $x^3 + 2x^2 - 9x - 18 = 0$

**2.** $x^3 - x^2 - 17x - 15 = 0$     **4.** $x^3 + 9x^2 + 26x + 24 = 0$

Given the indicated root, find the other real roots of the equation. If there are no real roots, so state.

**5.** $4x^3 - 24x^2 - x + 6 = 0$; $\frac{1}{2}$     **7.** $3x^3 + x^2 + 18x + 6 = 0$; $-\frac{1}{3}$

**6.** $4x^3 - 8x^2 - 29x - 12 = 0$; 4   **8.** $2x^3 + 10x^2 + x + 5 = 0$; $-5$

Find the real zeros of the following polynomial functions.

**9.** $P(x) = x^3 + 3x^2 + x + 3$     **11.** $P(x) = x^4 - 3x^2 - 4$

**10.** $P(x) = x^3 - 3x^2 + 2x - 6$   **12.** $P(x) = x^4 - x^3 - 5x^2 - x - 6$

## Cumulative Review: Chapters 1–5

**1.** Simplify $3[2(4 + 6) - 4(2 + 1)]$.

**2.** Specify {letters in the word "character"} by roster.

**3.** Write the sentence "$x$ is less than 3 or $x$ is equal to 3" using a single order symbol.

**4.** Graph the subset of $\mathcal{R}$: $\{x: -2 \leq x < 4\}$

Justify each statement by one of the real number axioms.

**5.** $-8 + 0 = -8$

**6.** $2(4 + 7) = 2 \cdot 4 + 2 \cdot 7$

**7.** $3 \cdot (5 + 8) = (5 + 8) \cdot 3$

**8.** $7 \cdot \frac{1}{7} = 1$

**9.** Graph $|n| \leq \frac{3}{2}$.

**10.** Solve over $\mathcal{R}$: $|x + 2| = 3$

Simplify.

**11.** $-8 + 7 - (10 - 2)$

**12.** $-3 - 8 + (8 - 1)$

**13.** $(-6)(5)(\frac{2}{3})(-\frac{4}{5})$

**14.** $\dfrac{(-7)(-12) + 6}{8 - 12}$

**15.** Find three solutions of $3x - 2y = 24$ over $\mathcal{R}$.

**16.** Graph $3x - y = 9$.

**17.** Find the slope and $y$-intercept of the graph of $2x - 4y = 5$.

**18.** Find an equation of the line which contains the point $(5, -2)$ and is parallel to the graph of $4x - 2y = 7$.

Solve each system.

**19.** $3x - y = -9$
$x + 4y = 10$

**20.** $x + 2y - z = -3$
$3x + z = 5$
$x - 2y + 3z = 9$

**21.** If $f(x) = 3x^2 + x - 5$, find **(a)** $f(0)$; **(b)** $f(-2)$; and **(c)** $f(2a)$.

Simplify.

**22.** $(4y^2 - 3y + 2) + (y^2 - y + 6)$

**23.** $3(2x^2 + 3x - 1) - 2(4x^2 + 3x + 2)$

**24.** $(3x - 7)(x^2 + 2x - 1)$

**25.** $(t^2 + 2)(t^3 - t + 4)$

Factor.

**26.** $x^2 + 6x + 5$     **28.** $z^3 - z^4 + z^5$

**27.** $18x^2 - 21xy - 9y^2$    **29.** $3t - 81t^4$

**30.** Find the GCF and LCM of $3x - 9$, $3x$, and $3 - x$.

**31.** The sum of three consecutive even integers is 10 more than twice the least of the three integers. Find the greatest of these integers.

**32.** At 10:00 A.M., two trains leave cities that are 200 miles apart and travel toward each other along parallel tracks. If one train averages 52 miles per hour and the other 48 miles per hour, at what time do the trains meet?

**33.** How many pounds of coffee at 72 cents per pound should be mixed with 20 pounds of coffee at 90 cents per pound to produce a mixture costing 78 cents per pound?

**34.** An angle measures at least $\frac{2}{3}$ of its supplement and at most 5 times its complement. Find the interval of possible measures for the angle.

**35.** A boat travels 6 miles upstream on a river and then returns. If the trip upstream requires one hour while the return trip requires 40 minutes, find the rate of the boat in still water and the rate of the current.

**36.** A collection consisting of 28 coins in quarters and nickels has a total value of \$4. How many coins of each kind are there?

**37.** If the length of one side of a square is increased by 1 inch and that of an adjacent side is decreased by 2 inches, a rectangle is formed whose area is 180 square inches. Find the length of a side of the square.

**38.** A rectangular garden is twice as long as it is wide, and is surrounded by a cement path 4 feet wide. If the garden and path together cover 2880 square feet, find the dimensions of the garden.

**39.** A factory can produce a bench in 24 minutes and a picnic table in 32 minutes. If it plans to turn out twice as many benches as tables, what is the maximum number of complete units of each item it can produce in an eight-hour workday?

**40.** The base of a triangle is 2 inches less than twice the altitude. If the area of the triangle is 30 square inches, find the lengths of the base and the altitude.

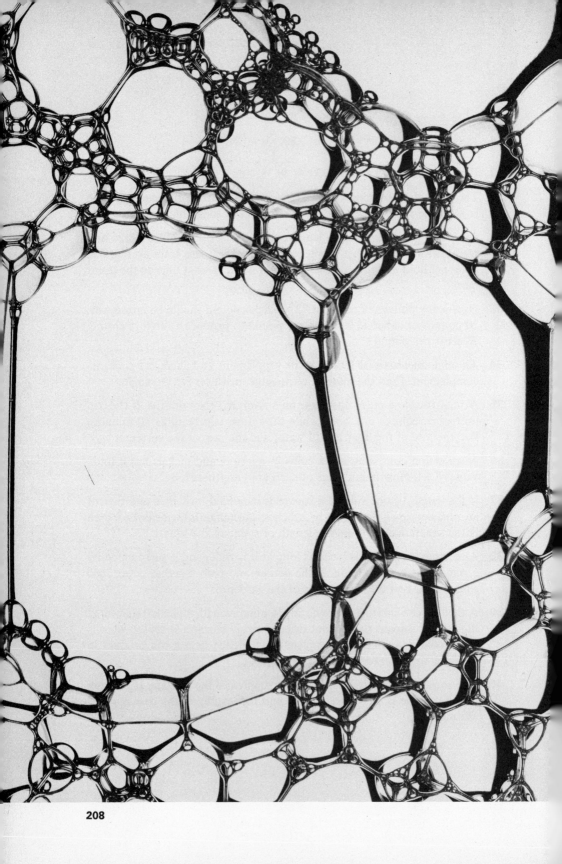

# 6

# Rational Numbers, Expressions, and Functions

## EXTENDING THE LAWS OF EXPONENTS

### 6–1 Quotients of Powers

The example $\dfrac{14 \cdot 15}{7 \cdot 3} = \dfrac{14}{7} \cdot \dfrac{15}{3} = 2 \cdot 5 = 10$ illustrates the fact that a quotient of products can be written as a product of quotients. In general we have:

> **The Basic Property of Quotients.**
>
> $$\frac{p \cdot r}{q \cdot s} = \frac{p}{q} \cdot \frac{r}{s} \qquad (q \neq 0, s \neq 0)$$

The proof of this property is outlined in Oral Exercises 17–21.

Notice that when the basic property of quotients is written as $\dfrac{p}{q} \cdot \dfrac{r}{s} = \dfrac{pr}{qs}$, it becomes a rule for multiplying fractions. Thus:

$$\frac{2}{3} \cdot \frac{5}{7} = \frac{2 \cdot 5}{3 \cdot 7} = \frac{10}{21} \quad \text{and} \quad \frac{2a}{3b} \cdot \frac{1}{b^2} = \frac{2a \cdot 1}{3b \cdot b^2} = \frac{2a}{3b^3} \qquad (b \neq 0)$$

If you replace $s$ by $r$ in the basic property of quotients, you obtain $\dfrac{pr}{qr} = \dfrac{p}{q} \cdot \dfrac{r}{r}$, or, since $\dfrac{r}{r} = 1$:

---

**The Cancellation Rule.**

$$\frac{pr}{qr} = \frac{p}{q} \qquad (q \neq 0, r \neq 0)$$

---

You can use this rule to reduce a fraction to lowest terms by taking $r$ to be the greatest common factor (GCF) of the numerator and denominator:

$$\frac{24}{40} = \frac{3 \cdot 8}{5 \cdot 8} = \frac{3}{5} \qquad\qquad \frac{3ab}{6bc} = \frac{a \cdot 3b}{2c \cdot 3b} = \frac{a}{2c}$$

(A quotient of monomials is in lowest terms when the GCF of the numerator and denominator is 1.)

By Law 1 of Exponents (page 210), $a^7 = a^3 \cdot a^4$. Therefore:

$$\frac{a^7}{a^4} = \frac{a^3 \cdot a^4}{1 \cdot a^4} = \frac{a^3}{1} = a^{7-4}$$

$$\frac{a^4}{a^7} = \frac{1 \cdot a^4}{a^3 \cdot a^4} = \frac{1}{a^3} = \frac{1}{a^{7-4}}$$

These examples illustrate Laws 4 and 5 below.

---

**The Laws of Exponents**

**1.** $a^m \cdot a^n = a^{m+n}$

**2.** $(ab)^n = a^n \cdot b^n$

**3.** $(a^m)^n = a^{mn}$

**4.** $\dfrac{a^m}{a^n} = a^{m-n} \quad (a \neq 0)$

**5.** $\dfrac{a^m}{a^n} = \dfrac{1}{a^{n-m}} \quad (a \neq 0)$

**6.** $\left(\dfrac{a}{b}\right)^n = \dfrac{a^n}{b^n} \quad (b \neq 0)$

---

At present we assume that all exponents are positive integers so that in Law 4, $m > n$, and in Law 5, $m < n$. You can derive Law 5 by using

Law 1 and the Cancellation Rule: Since $a^{n-m} \cdot a^m = a^{(n-m)+m} = a^n$, you have

$$\frac{a^m}{a^n} = \frac{1 \cdot a^m}{a^{n-m} \cdot a^m} = \frac{1}{a^{n-m}}.$$

The other laws can be derived in a similar manner.

**Example.** Simplify; that is, write as a fraction in lowest terms in which each base appears only once and in which there are no "powers of powers."

**a.** $\dfrac{2a^3b^2}{6ab^5}$     **b.** $\dfrac{x}{y}\left(\dfrac{y^2}{x}\right)^3$

*Solution:* **a.** *First Method.* Think of each variable individually and use Law 4 or 5:

$$\frac{2a^3b^2}{6ab^5} = \frac{2}{6} \cdot \frac{a^3}{a} \cdot \frac{b^2}{b^5} = \frac{1}{3} \cdot \frac{a^2}{1} \cdot \frac{1}{b^3} = \frac{a^2}{3b^3}$$

*Second Method.* Find the GCF of the numerator and the denominator and use the Cancellation Rule:

$$\frac{2a^3b^2}{6ab^5} = \frac{a^2 \cdot 2ab^2}{3b^3 \cdot 2ab^2} = \frac{a^2}{3b^3}$$

**b.** $\dfrac{x}{y}\left(\dfrac{y^2}{x}\right)^3 = \dfrac{x}{y} \cdot \dfrac{(y^2)^3}{x^3} = \dfrac{x}{y} \cdot \dfrac{y^6}{x^3} = \dfrac{xy^6}{yx^3} = \dfrac{y^5}{x^2}$

*ORAL EXERCISES*

Simplify each expression. Assume that variable expressions in exponents denote positive integers, and that no denominator is equal to 0.

**1.** $\dfrac{a^6}{a^2}$    **5.** $\left(\dfrac{x^2}{3}\right)^2$    **9.** $\dfrac{6y^3z}{3yz^3}$    **13.** $\left(\dfrac{6z^2}{3z}\right)^2$

**2.** $\dfrac{x^{12}}{x^9}$    **6.** $\left(\dfrac{y^2}{2}\right)^3$    **10.** $\dfrac{4m^3n}{8mn}$    **14.** $\left(\dfrac{4y^3}{8z}\right)^3$

**3.** $\dfrac{a^2x^3}{a^3x}$    **7.** $\dfrac{a^x}{a^2}$ $(x > 2)$    **11.** $\dfrac{-12t^4s}{4ts^5}$    **15.** $\left(\dfrac{24t^2}{-8t^3}\right)^3$

**4.** $\dfrac{by^3}{b^3y^3}$    **8.** $\dfrac{a^x}{a^2}$ $(x < 2)$    **12.** $\dfrac{15km}{-10km}$    **16.** $\left(\dfrac{-12n^4}{36n}\right)^2$

Supply the missing reasons in the following proof of the basic property of quotients.

**17.** $\dfrac{p \cdot r}{q \cdot s} = (p \cdot r) \cdot \dfrac{1}{q \cdot s}$　　　　　**17.** __?__

**18.** $(p \cdot r) \cdot \dfrac{1}{q \cdot s} = (p \cdot r) \cdot \left(\dfrac{1}{q} \cdot \dfrac{1}{s}\right)$　　**18.** __?__

**19.** $(p \cdot r) \cdot \left(\dfrac{1}{q} \cdot \dfrac{1}{s}\right) = \left(p \cdot \dfrac{1}{q}\right) \cdot \left(r \cdot \dfrac{1}{s}\right)$　　**19.** __?__

**20.** $\left(p \cdot \dfrac{1}{q}\right) \cdot \left(r \cdot \dfrac{1}{s}\right) = \dfrac{p}{q} \cdot \dfrac{r}{s}$　　　**20.** __?__

**21.** $\dfrac{p \cdot r}{q \cdot s} = \dfrac{p}{q} \cdot \dfrac{r}{s}$　　　　　**21.** __?__

## WRITTEN EXERCISES

Simplify each expression. Assume that no denominator is equal to 0.

**A**　**1.** $\dfrac{18t^4}{6t}$　　**5.** $\dfrac{-12r^3 s^5}{-36r^5 s^2}$　　**9.** $\left(\dfrac{3x^5 y}{6xy^2}\right)^2$　　**13.** $\dfrac{(-2a^2 b)^4}{(6ab^2)^2}$

**2.** $\dfrac{38r^2}{2r}$　　**6.** $\dfrac{-50t^{12}}{-15r^2 t}$　　**10.** $\left(\dfrac{12m^2 n^5}{15m^3 n^6}\right)^2$　　**14.** $\dfrac{27k^3 m^2}{(-3km)^3}$

**3.** $\dfrac{-48n^2 m^3}{64nm^5}$　　**7.** $\dfrac{6n^2}{5t} \cdot \dfrac{p}{q^2}$　　**11.** $\dfrac{(-3x^2 y)^2}{9x^3 y^2}$　　**15.** $\dfrac{(a^2 x^5 z)^6}{(ax^4 z^2)^5}$

**4.** $\dfrac{96x^3 y^4}{-24xy^6}$　　**8.** $\dfrac{4x^2 y}{z} \cdot \dfrac{3a}{5b}$　　**12.** $\dfrac{25x^4 y^6}{(5x^2 y^3)^2}$　　**16.** $\dfrac{(p^3 s^2 q^3)^4}{(p^2 s^3 q)^7}$

**B**　**17.** $\left(\dfrac{a^2}{5}\right)^3 \cdot \dfrac{ab}{c^2}$　　　　　**20.** $\left(\dfrac{-4x^2}{3y}\right)^2 \cdot \left(\dfrac{5z^2}{y^2}\right)^3$

**18.** $\dfrac{x^2 y}{z} \cdot \left(\dfrac{2u^2}{v}\right)^3$　　　　**21.** $\left(\dfrac{21a^3 b^2 c}{14ab^2 c^2}\right)^3$

**19.** $\left(\dfrac{3t^2}{5}\right)^2 \cdot \left(\dfrac{-7t}{2r^2}\right)^3$

## 6–2 Zero and Negative Exponents

Although we have needed only positive exponents up to this point, it is convenient to extend the definition of a power to permit *any* integer to be used as an exponent. We shall assign meanings to such expressions as $5^0$, $10^{-7}$, $a^0$, and $x^{-1}$ so that the laws of exponents stated on page 210 continue to hold.

If you wish the law $a^m \cdot a^n = a^{m+n}$ to hold for $m = 0$, you must have

$$a^0 \cdot a^n = a^{0+n} = a^n.$$

Therefore if $a \neq 0$,

$$a^0 = \frac{a^n}{a^n} = 1.$$

Also, if $n$ is a positive integer,

$$a^{-n} \cdot a^n = a^{-n+n} = a^0 = 1.$$

Therefore if $a \neq 0$,

$$a^{-n} = \frac{1}{a^n}.$$

We therefore make these definitions:

---

For every nonzero number *a* and every positive integer *n*,

$$a^{-n} = \frac{1}{a^n}, \quad \text{and} \quad a^0 = 1.$$

---

Thus $10^{-3} = \dfrac{1}{10^3} = \dfrac{1}{1000}$, $5^0 = 1$, $2 \cdot 5^{-2} = 2 \cdot \dfrac{1}{5^2} = \dfrac{2}{25}$, and

$(2 \cdot 5)^{-2} = 10^{-2} = \dfrac{1}{100}$. No meaning is assigned to the expression $0^0$.

It can be shown that all the laws of exponents continue to hold even though some or all of the exponents are zero or negative integers. For example, if $a \neq 0$,

$$(a^2)^{-3} = a^{2(-3)} = a^{-6}.$$

You can justify this statement as follows:

$$(a^2)^{-3} = \frac{1}{(a^2)^3} = \frac{1}{a^6} = a^{-6}.$$

Recall that division by 0 is not defined. Therefore, since $a^{-n} = \frac{1}{a^n}$, a power with a negative exponent is not defined if the base is 0. To simplify matters, we agree to restrict the domains of all variables appearing in algebraic expressions so that *neither denominators of fractions nor bases of powers with nonpositive exponents will be zero.* For example, when we write

$$\frac{x^{-2}y^{-1}}{y-3},$$

we assume that $x \neq 0$, $y \neq 0$, and $y \neq 3$.

**Example 1.** Express without using negative exponents:

        **a.** $5 \times 10^{-3}$         **b.** $\dfrac{2ax^{-2}}{2^0 \cdot y^{-1}}$

*Solution:*    **a.** $5 \times 10^{-3} = 5 \times \dfrac{1}{10^3} = \dfrac{5}{1000}$, or 0.005, or $\dfrac{1}{200}$

        **b.** $\dfrac{2ax^{-2}}{2^0 \cdot y^{-1}} = \dfrac{2ax^{-2}}{1 \cdot y^{-1}} = \dfrac{2ay}{x^2}$

Sometimes we introduce negative exponents in order to write an expression without fractions. For example, you can write $\dfrac{1}{x^2}$ as $x^{-2}$ and $\dfrac{3}{z}$ as $3z^{-1}$.

**Example 2.** Express without using fractions:

        **a.** $\dfrac{3}{1,000,000}$         **b.** $\dfrac{2a}{x^2y}$

*Solution:*    **a.** $\dfrac{3}{1,000,000} = \dfrac{3}{10^6} = 3 \times \dfrac{1}{10^6} = 3 \times 10^{-6}$

        **b.** $\dfrac{2a}{x^2y} = 2a \cdot \dfrac{1}{x^2y} = 2a \cdot \dfrac{1}{x^2} \cdot \dfrac{1}{y} = 2ax^{-2}y^{-1}$

## ORAL EXERCISES

Express each of the following as a numeral or as a power of a variable. Assume that no variable takes on a value of 0.

**1.** $3^{-2} \cdot 3^5$

**2.** $7^4 \cdot 7^{-3}$

**3.** $x^{-1} \cdot x^{-3}$

**4.** $z^5 \cdot z^{-7}$

**5.** $v^n \cdot v^{-3}$

**6.** $v^6 \cdot v^{-n}$

**7.** $6^a \cdot 6^b$

**8.** $5^r \cdot 5^r$

**9.** $7^0 \cdot 7^4$

**10.** $t^3 \cdot t^{-3}$

**11.** $v^m \cdot v^{-n}$

**12.** $a^b \cdot a^0$

**13.** $\dfrac{1}{t^{-3}}$

**14.** $\dfrac{x}{x^{-3}}$

**15.** $\dfrac{y^3}{y^{-1}}$

**16.** $\dfrac{t^0}{t^{-3}}$

**17.** $\left(\dfrac{1}{n}\right)^{-1}$

**18.** $\left(\dfrac{1}{n^{-1}}\right)^{-1}$

**19.** $\left(\dfrac{1}{t^{-2}}\right)^3$

**20.** $\left(\dfrac{1}{t^{-3}}\right)^{-2}$

**21.** $\left(\dfrac{1}{x^0}\right)^{-1}$

**22.** $\dfrac{2}{y^0 + 1}$

**23.** $\dfrac{1}{a^0 + b^0}$

**24.** $\left(\dfrac{4}{t^{-3}}\right)^0$

Express each of the following using positive exponents only.

**25.** $\dfrac{a^{-3}x}{z^{-1}}$

**26.** $\dfrac{5p^2q^{-2}}{r}$

**27.** $-6x^{-2}y$

**28.** $3a^{-1}b^{-2}c$

**29.** $\dfrac{5^{-1}r^{-1}t^{-2}}{s}$

**30.** $\dfrac{5k}{l^{-1}mn^{-2}}$

**31.** $\dfrac{5t^0z^{-1}}{3^{-1}w}$

**32.** $\dfrac{5a^2b^{-3}}{2^{-2}c^{-1}}$

## WRITTEN EXERCISES

Simplify each expression.

**Sample.** $(-2 \cdot 3)^{-2} = (-6)^{-2} = \dfrac{1}{(-6)^2} = \dfrac{1}{36}$

**A**

**1.** $\left(\frac{3}{5}\right)^{-3}$

**2.** $\left(-\frac{5}{3}\right)^{-2}$

**3.** $(6 + 3)^{-2}$

**4.** $(6 - 3)^{-4}$

**5.** $(-4^{-3})^{-1}$

**6.** $(7^{-2})^{-1}$

**7.** $(6 \cdot 4)^{-2}$

**8.** $(-1 \cdot 5)^{-3}$

**9.** $(114^0 \cdot 6^{-2})^{-1}$

**10.** $(210^0 \cdot 3^2)^{-3}$

**11.** $7.83 \times 10^{-2}$

**12.** $6.81 \times 10^{-3}$

**13.** $\left(\frac{2}{3}\right)^{-3} \cdot \left(\frac{2}{3}\right)$

**14.** $\left(\frac{5}{4}\right)^{-4} \cdot \left(\frac{5}{4}\right)^2$

**15.** $\dfrac{1}{3^{-2}} + \dfrac{1}{2^{-3}}$

**16.** $1^{-5} - \dfrac{1}{5^{-3}}$

Express each of the following without negative exponents.

**17.** $\dfrac{a^{-4}b}{c^3}$

**18.** $\dfrac{2x}{y^{-3}}$

**19.** $\dfrac{6u^4v^{-3}}{7w^{-2}}$

**20.** $\dfrac{5z^4}{6x^{-3}y^{-5}}$

**21.** $\dfrac{x^{-4}y^5z^{-5}}{xy^{-2}z^3}$

**22.** $\dfrac{r^{-1}s^{-4}t^2}{r^{-2}st^{-3}}$

**23.** $\dfrac{6s}{s^{-1}} + \dfrac{5}{s^{-2}}$

**24.** $\dfrac{3a}{b^{-1}} + \dfrac{2b}{a^{-1}}$

By assigning the indicated values to the variables, show that the following statements are *not true* for *all* real numbers.

**25.** $(x + y)^{-1} = x^{-1} + y^{-1}$;  $x = 1, y = 1$

**26.** $(x^{-1} + y^{-1})^{-1} = x + y$;  $x = 1, y = 1$

**27.** $3a^{-4} = (3a)^{-4}$;  $a = 2$

**28.** $1 + a^{-1} = \dfrac{1}{1 + a}$;  $a = 1$

In each Exercise 29–38, replace the question mark with a variable expression to make a true sentence.

>   **Sample.**  $a^{-3} + 2a^{-1} - 5 = a^{-3}( \ ? \ )$
>
>   *Solution:*  $a^{-3} + 2a^{-1} - 5 = a^{-3}(1 + 2a^2 - 5a^3)$

**B** **29.** $3t^{-2} + 5t^{-1} - 3 = t^{-2}( \ ? \ )$

**30.** $4n^{-3} - 6n^{-2} + 8n^{-1} = 2n^{-3}( \ ? \ )$

**31.** $6x - 3x^{-1} + 9x^{-3} - 3x^{-4} = 3x^{-4}( \ ? \ )$

**32.** $7 - 8z^{-2} + 5z^{-3} + z^{-5} = z^{-5}( \ ? \ )$

**33.** $\dfrac{5}{x^3} + \dfrac{7}{x^2} - \dfrac{2}{x} + 3 - x = \dfrac{1}{x^3} ( \ ? \ )$

**34.** $\dfrac{6}{n^4} - \dfrac{3}{n^2} + 6 - 9n^2 + n^4 = \dfrac{3}{n^4} ( \ ? \ )$

**C** **35.** $2z(2z + 3)^{-2} - (2z + 3)^{-1} = (2z + 3)^{-2}( \ ? \ )$

**36.** $(3 - t^2)^{-3} + t^2(3 - t^2)^{-3} = (3 - t^2)^{-3}( \ ? \ )$

**37.** $6(5 - x)(x + 2)^{-1} - 2(5 - x)^2(x + 2)^{-2} = 2(x + 2)^{-2}( \ ? \ )$

**38.** $5(3x + 1)(6 - x)^{-2} + 15(6 - x)^{-1} = 5(6 - x)^{-2}( \ ? \ )$

Given that $n$ is a positive integer, and that $a$ and $b$ are nonzero real numbers, show that:

**39.** $\dfrac{1}{b^{-n}} = b^n$

**41.** $\left(\dfrac{a}{b}\right)^{-n} = \left(\dfrac{b}{a}\right)^n$

**40.** $(ab)^{-n} = a^{-n}b^{-n}$

**42.** $\left(\dfrac{a}{b}\right)^{-n} = \dfrac{a^{-n}}{b^{-n}}$

## OPERATING WITH FRACTIONS

### 6–3  Rational Algebraic Expressions

A **rational number** is one which can be expressed as a quotient of integers, for example:

$$-\frac{7}{24} = \frac{-7}{24}, \qquad 17 \times 10^{-3} = \frac{17}{1000}, \qquad 2.64 = \frac{264}{100}$$

By analogy, we call a quotient of polynomials a **rational algebraic expression**. Some examples are:

$$\frac{2ax^2}{3x}, \quad \frac{x^2 + 2x + 1}{x^2 + 1}, \quad x^2(x - 3)^{-1} = \frac{x^2}{x - 3}$$

A rational algebraic expression is in **simple form**, or **lowest terms**, when it is expressed as a quotient of two polynomials whose greatest common factor is 1 (recall Section 5–4). You simplify a rational algebraic expression by dividing its numerator and denominator by their greatest common factor.

**Example 1.**  Write $(2x^2 - 8)(2x^2 + 5x + 2)^{-1}$ as a rational algebraic expression in lowest terms.

*Solution:*

Rewrite as a fraction:  $(2x^2 - 8)(2x^2 + 5x + 2)^{-1} = \dfrac{2x^2 - 8}{2x^2 + 5x + 2}$

Factor the numerator and denominator: $= \dfrac{2(x + 2)(x - 2)}{(2x + 1)(x + 2)}$

Divide numerator and denominator by their GCF: $= \dfrac{2(x - 2)}{2x + 1}$

Factors of the numerator and denominator may be negatives of each other, as in Example 2.

**Example 2.** Simplify: $\dfrac{z^2 - z^3}{z^3 + z^2 - 2z}$

*Solution:*
$$\frac{z^2 - z^3}{z^3 + z^2 - 2z} = \frac{z^2(1 - z)}{z(z - 1)(z + 2)}$$

Write $1 - z$ as
$(-1)(z - 1)$:
$$= \frac{z \cdot z \cdot (-1)(z - 1)}{z(z - 1)(z + 2)}$$

$$= \frac{(-1)z}{z + 2}$$

$$= \frac{-z}{z + 2}, \quad \text{or} \quad -\frac{z}{z + 2}.$$

You can see from the first line of the solution that the rational expression in Example 2 would have denominator 0 if $z \in \{0, 1, -2\}$ and would therefore not be defined. Recall that the numbers 0, 1, and $-2$ are excluded from the domain of $z$ by the agreement made on page 214.

Any rational algebraic expression defines a **rational function**, that is, one which is a quotient of two polynomial functions.

**Example 3.** Find the domain of the rational function

$$f(x) = x(2 - x - x^2)^{-1}.$$

*Solution:*  Express $f(x)$ as
a fraction.
$$f(x) = x(2 - x - x^2)^{-1}$$

$$= \frac{x}{2 - x - x^2}$$

Find the zeros (p. 162)
of the denominator.
$$2 - x - x^2 = 0$$
$$(2 + x)(1 - x) = 0$$
$$x = -2 \quad \text{or} \quad x = 1$$

∴ The domain of $f$ is $\{x: x \neq -2 \text{ and } x \neq 1\}$.

If $f$ and $g$ are two functions having the same domain $D$, the quotient function $\dfrac{f}{g}$ is defined by the formula

$$\frac{f}{g}(x) = \frac{f(x)}{g(x)}.$$

The domain of $\dfrac{f}{g}$ is $\{x: x \in D \text{ and } g(x) \neq 0\}$.

## ORAL EXERCISES

State each of the following as a fraction in lowest terms, and give all restrictions on any variable(s) involved.

**Sample.** $4 \cdot (x - 1)^{-2}$    *Solution:*   $\dfrac{4}{(x - 1)^2}$,   $x \neq 1$

**1.** $5(z + 2)^{-1}$        **5.** $x^{-2}(2x + 3)$        **9.** $n^0[n - 4]^{-1}$

**2.** $3(y - 5)^{-1}$        **6.** $n^{-1}(5n - 4)$        **10.** $3r^0[r + 5]^{-2}$

**3.** $x^2 \cdot x^{-3}$        **7.** $x[x(x - 3)]^{-1}$        **11.** $3(2n - 3)^{-1}$

**4.** $ay^3 \cdot a^{-3}$        **8.** $z[z(z + 2)^2]^{-1}$        **12.** $7(5p + 10)^{-1}$

## WRITTEN EXERCISES

Write each expression as a fraction in lowest terms, and give all restrictions on any variable(s) involved.

**A**    **1.** $(x^2 - 9)(2x^2 - 2x - 12)^{-1}$        **6.** $(3y^2 - 3)(y^2 + 4y - 5)^{-1}$

     **2.** $(m^2 - 9)(m^2 - 7m + 12)^{-1}$        **7.** $(x^2 + 7x + 10)(x^2 + 4x - 5)^{-1}$

     **3.** $(t - 1)^3(1 - t)^{-4}$        **8.** $(t^2 + t - 56)(t^2 - t - 42)^{-1}$

     **4.** $(x - 4)^4(4 - x)^{-3}$        **9.** $(9y^3 - y)(3y^2 + 8y - 3)^{-1}$

     **5.** $(4x^2 - 1)(2x^2 - 7x + 6)^{-1}$        **10.** $(1 - 6x + 9x^2)(9x^2 - 1)^{-1}$

Reduce each fraction to lowest terms. State all restrictions on any variable(s) involved.

**11.** $\dfrac{18(2a - b)(a + b)}{9(2a - b)(a^2 + 2ab + b^2)}$        **16.** $\dfrac{n^3x - 8x}{(n^2 - 4)(n - 1)}$

**12.** $\dfrac{12(a - 3)(a^2 + 6a + 9)}{4(a^2 - 9)(a + 3)}$        **17.** $\dfrac{y^2 - 16}{24 - 2y - y^2}$

**13.** $\dfrac{x^3 - 8x^2 + 12x}{5x^2 - 60x + 180}$        **18.** $\dfrac{a^2 + a - 6}{15 + 2a - a^2}$

**14.** $\dfrac{ay^2 - ay - 12a}{3y^2 + 13y + 12}$        **19.** $\dfrac{16t^4 + 4t^2 + 1}{64t^6 - 1}$

**15.** $\dfrac{t^3 - 64}{t^2s - 16s}$        **20.** $\dfrac{27x^3 + 64y^3}{9x^2 + 24xy + 16y^2}$

Find the domain of the rational function defined by the given sentence, and state all zeros of the function over $\mathcal{R}$.

**Sample.** $f(x) = \dfrac{x^2 - 9x + 18}{3x^2 + 3x - 36}$

*Solution:* Factor the numerator and denominator:

$$f(x) = \frac{(x - 6)(x - 3)}{3(x + 4)(x - 3)}$$

By inspection, the domain is $\{x: x \neq -4 \text{ and } x \neq 3\}$. Since 3 is not in the domain of $f$, the only zero is that for which $x - 6 = 0$, namely, 6.

**21.** $f(x) = \dfrac{x^3 - 25x}{2x^2 - 11x + 5}$

**22.** $g(x) = \dfrac{x^4 - 4x^2}{x^2 - 4}$

**23.** $h(x) = \dfrac{x^2 - 5x + 6}{x^2 - 4x + 4}$

**24.** $f(x) = \dfrac{x^2 - 5x - 14}{x^2 - 9x + 14}$

**25.** $g(x) = \dfrac{x^3 - 1}{x^2 - 1}$

**26.** $h(x) = \dfrac{x^3 + 1}{x^2 - x + 1}$

**27.** $f(x) = \dfrac{x^4 - 16}{x^2 + 4}$

**28.** $g(x) = \dfrac{x^4 - 16}{x^4 - x^2 - 12}$

Reduce each fraction to lowest terms.

**B**   **29.** $\dfrac{3a^4 - 3b^4}{a^4 + 2a^2b^2 + b^4}$

**30.** $\dfrac{-(x - y)(y - z)(x - z)}{(y - x)(z - y)(z - x)}$

**31.** $\dfrac{x^3 - x^2 + y^2 + y^3}{3x^2 + 6xy + 3y^2}$

**32.** $\dfrac{3x^4 - 3y^4}{6x^2 - 6y^2}$

**33.** $\dfrac{a^2 - 2ab + b^2 + a - b}{b^2 - a^2}$

**34.** $\dfrac{8r^2 - 8rs - r^3 - s^3 + 8s^2}{r^4 + r^2s^2 + s^4}$

**C**   **35.** $\dfrac{4(x^2 - 1)^4(2x - 3) - 8x(2x - 3)^2(x^2 - 1)^3}{(x^2 - 1)^8}$

**36.** $\dfrac{(2y^2 + 3)^2(6y - 1) - 8y(3y^2 - y)(2y^2 + 3)}{(2y^2 + 3)^4}$

**37.** $\dfrac{3z^2(2z^2 + 5z)^2 - 8z(z^3 - 2)(2z^2 + 5z)}{(2z^2 + 5z)^4}$

**38.** $\dfrac{9x(x^2 + 1)^2(3x - 1)^2 - 4x(x^2 + 1)(3x - 1)^3}{(x^2 + 1)^4}$

## 6–4   Products and Quotients

The basic property of quotients, $\dfrac{p}{q} \cdot \dfrac{r}{s} = \dfrac{pr}{qs}$, which we used on page 209 to multiply fractions, can be extended to more than two factors.

> The product of two or more fractions is the fraction whose numerator is the product of the numerators and whose denominator is the product of the denominators of the given fractions.

For example,

$$\frac{5}{3} \cdot \frac{6x}{y} \cdot \frac{y^2}{x+1} = \frac{5 \cdot 6x \cdot y^2}{3 \cdot y \cdot (x+1)} = \frac{5 \cdot 2 \cdot x \cdot y \cdot 3y}{(x+1) \cdot 3y} = \frac{10xy}{x+1}.$$

You can often simplify a product of rational expressions by factoring numerators and denominators.

**Example 1.**   Multiply $\dfrac{x^3+8}{3x-6}$ by $\dfrac{x-2}{x^2+4x+4}$.

*Solution:*
$$\begin{aligned}
\frac{x^3+8}{3x-6} \cdot \frac{x-2}{x^2+4x+4} &= \frac{(x+2)(x^2-2x+4)}{3(x-2)} \cdot \frac{x-2}{(x+2)^2} \\
&= \frac{(x^2-2x+4)(x+2)(x-2)}{3(x+2)(x+2)(x-2)} \\
&= \frac{x^2-2x+4}{3(x+2)}
\end{aligned}$$

Recall that a quotient is the product of the dividend and the reciprocal of the divisor (page 50). Moreover, the reciprocal of $\dfrac{c}{d}$ is $\dfrac{d}{c}$ (because $\dfrac{c}{d} \cdot \dfrac{d}{c} = 1$). Combining these facts we have:

$$\frac{a}{b} \div \frac{c}{d} = \frac{a}{b} \cdot \frac{d}{c} \qquad\qquad \frac{\dfrac{a}{b}}{\dfrac{c}{d}} = \frac{a}{b} \cdot \frac{d}{c}$$

When we **invert** the fraction $\dfrac{c}{d}$, we obtain the fraction $\dfrac{d}{c}$. Thus we can state the rule:

> To divide one fraction by another, invert the divisor and multiply by the dividend.

For example, $\dfrac{7}{9} \div \dfrac{35}{27} = \dfrac{7}{9} \cdot \dfrac{27}{35} = \dfrac{7 \cdot 3^3}{3^2 \cdot 5 \cdot 7} = \dfrac{3 \cdot 7 \cdot 3^2}{5 \cdot 7 \cdot 3^2} = \dfrac{3}{5},$   and

$$\frac{u^2}{3v}\left(\frac{u}{6v^2}\right)^{-1} = \frac{\dfrac{u^2}{3v}}{\dfrac{u}{6v^2}} = \frac{u^2}{3v} \cdot \frac{6v^2}{u} = \frac{6u^2v^2}{3uv} = 2uv.$$

It is sometimes helpful to introduce 1 as a denominator, as in Example 2.

**Example 2.**  Simplify: $(x^2 - 2ax + a^2) \div \dfrac{x^2 - a^2}{x^2 + a^2}$

*Solution:*  $(x^2 - 2ax + a^2) \div \dfrac{x^2 - a^2}{x^2 + a^2} = \dfrac{x^2 - 2ax + a^2}{1} \cdot \dfrac{x^2 + a^2}{x^2 - a^2}$

$$= \frac{(x - a)^2}{1} \cdot \frac{x^2 + a^2}{(x + a)(x - a)}$$

$$= \frac{(x - a)(x^2 + a^2)(x - a)}{(x + a)(x - a)}$$

$$= \frac{(x - a)(x^2 + a^2)}{x + a}$$

### WRITTEN EXERCISES

Express each product or quotient as a single fraction in lowest terms. Assume all variables are restricted so that no denominator or divisor is equal to zero.

**A**

**1.** $\left(-\frac{5}{14}\right) \cdot \frac{28}{20} \cdot \left(\frac{9}{10}\right)^{-1}$

**2.** $\left(-\frac{15}{18}\right)\left(-\frac{6}{30}\right)\left(\frac{7}{12}\right)^{-1}$

**3.** $\dfrac{6x^2y}{15a^3b^2} \cdot \dfrac{5ab^2}{2xy}$

**4.** $\dfrac{3rs^2}{20t^2} \cdot \dfrac{25t^3}{3r^2s^2}$

**5.** $\left(-\dfrac{12x^2y}{5z}\right) \div \left(\dfrac{24x^3y}{10y^2z}\right)$

**6.** $\left(\dfrac{3uv}{r^2s}\right) \div \left(\dfrac{9u^3v^2}{rs}\right)$

**7.** $\dfrac{5x^2y}{2xz^2} \cdot \dfrac{y^2z}{3x^2y} \div \dfrac{5y^2z}{6xz^2}$

**8.** $\dfrac{27a^3b^2}{20a^4c} \cdot \dfrac{30d}{14b^2c^2} \div \dfrac{18d^2}{7b^2c^4}$

**9.** $\dfrac{y^2 - y - 20}{y^2 + 7y + 12} \cdot \dfrac{2y^2 + 6y}{y^2 - 25}$

**10.** $\dfrac{x^2 + 3x - 18}{x^4 - 8x^3 + 12x^2} \cdot \dfrac{2x^3 + 4x^2}{x^2 - 36}$

**11.** $\dfrac{x^3 - y^3}{x^2 + xy + y^2} \div \dfrac{(x - y)^2}{4x + 4y}$

**12.** $\dfrac{a^4 - b^4}{(a - b)^2} \div \dfrac{a^2 + b^2}{a + b}$

If $f(x) = \dfrac{x^2 - 5x - 14}{x^2 + 5x - 24}$ and $g(x) = \dfrac{x^2 - 3x - 28}{x^2 - 8x + 15}$, find:

**13.** $fg(x)$

**14.** $\dfrac{f}{g}(x)$

If $h(x) = \dfrac{4x^2 + 8x + 3}{2x^2 - 5x + 3}$ and $g(x) = \dfrac{6x^2 - 9x}{1 - 4x^2}$, find:

**15.** $gh(x)$

**16.** $\dfrac{g}{h}(x)$

Write the given expression as a single fraction in lowest terms.

**17.** $\dfrac{25x^2y^2 - 16}{4xy + 1} \div \dfrac{5xy + 4}{16x^2y^2 + 16xy + 3}$

**18.** $\dfrac{y^2 + 2y - 15}{y^2 + 3y - 10} \cdot \left(\dfrac{y^2 - 9y + 14}{y^2 - 9}\right)^{-1}$

**19.** $\dfrac{ab - 3a + b - 3}{a - 2} \cdot \left(\dfrac{a + 1}{a^2 + 4}\right)^{-1}$

**20.** $\dfrac{2rs + 4r + 3s + 6}{2r + 3} \cdot \left(\dfrac{r + 2}{r - 1}\right)^{-1}$

**B** **21.** $\dfrac{a^2 + 7ab + 10b^2}{a^2 + 6ab + 5b^2} \cdot \dfrac{a + b}{a^2 + 4ab + 4b^2} \cdot (a + 2b)^{-1}$

**22.** $\dfrac{3x^2 + 10xy + 3y^2}{2x^2 + 5xy - 3y^2} \cdot \left(\dfrac{6x^2 + 11xy + 3y^2}{4x^2 + 12xy + 9y^2}\right)^{-1} \cdot \dfrac{2x - y}{x + 3y}$

**23.** $\dfrac{2a^2 + a - 3}{2a^2 - a - 3} \cdot \left(\dfrac{a^2 + 2a - 3}{6a^2 - 7a - 3}\right)^{-1} \div \left(\dfrac{a + 3}{2a + 3}\right)^{-1}$

**24.** $\dfrac{(2a - 3b)^2}{a^2 + 4ab + 4b^2} \cdot \left(\dfrac{4a^2 - 9b^2}{4a^2 - 4b^2}\right)^{-1} \div \left(\dfrac{5a^2 + 10ab}{3ab - 3b^2}\right)^{-1}$

**25.** $\dfrac{a^3 + 27}{a^2 + a - 12} \cdot \left(\dfrac{a^2 + 2a - 8}{a^2 - 3a + 9} \cdot \dfrac{3a - 9}{a^2 + a - 6}\right)^{-1}$

**26.** $\dfrac{x^2 + 2x - 15}{x^2 - 9} \cdot \left(\dfrac{9x - 6x^2 + x^3}{30 - 11x + x^2} \cdot \dfrac{25 - x^2}{x^2 - 3x}\right)^{-1}$

Let $f(x) = \dfrac{P(x)}{Q(x)}$ and $g(x) = \dfrac{R(x)}{S(x)}$, where $P(x)$, $Q(x)$, $R(x)$, and $S(x)$ are polynomials, and show that:

**C** **27.** $fg(x)$ is a rational function.

**28.** $\dfrac{f}{g}(x)$ is a rational function.

## 6–5 Sums and Differences

The meaning of division, together with the distributive law, provides a means of writing the sum of fractions with equal denominators as a single fraction:

$$\frac{a}{c} + \frac{b}{c} = a \cdot \frac{1}{c} + b \cdot \frac{1}{c} = (a + b) \cdot \frac{1}{c} = \frac{a + b}{c}.$$

> The sum of fractions with equal denominators is the fraction whose denominator is the common denominator and whose numerator is the sum of the numerators of the given fractions.

For example, $\dfrac{11}{12} + \dfrac{5}{12} - \dfrac{7}{12} = \dfrac{11 + 5 - 7}{12} = \dfrac{9}{12} = \dfrac{3}{4}$, and

$$\frac{2x}{x + 2} - \frac{1}{x + 2} + \frac{x + 7}{x + 2} = \frac{2x - 1 + x + 7}{x + 2}$$

$$= \frac{3x + 6}{x + 2} = \frac{3(x + 2)}{x + 2} = 3.$$

To add or subtract fractions with unequal denominators, use the steps:

> **1.** Find the least common denominator (LCD) of the fractions (that is, the LCM of their denominators).
>
> **2.** Express each fraction as an equivalent fraction having the LCD as denominator.
>
> **3.** Combine the resulting fractions, and simplify.

**Example 1.** Simplify, that is, express as a single fraction in lowest terms:

$$\frac{1}{15} + \frac{7}{20} - 1$$

*Solution:*

**1.** The LCM of $15 = 3 \cdot 5$ and $20 = 2^2 \cdot 5$ is $2^2 \cdot 3 \cdot 5 = 60$. This is the LCD.

**2.** $\dfrac{1}{15} = \dfrac{1 \cdot 4}{15 \cdot 4} = \dfrac{4}{60}$; $\dfrac{7}{20} = \dfrac{7 \cdot 3}{20 \cdot 3} = \dfrac{21}{60}$; $1 = \dfrac{60}{60}$

**3.** $\dfrac{1}{15} + \dfrac{7}{20} - 1 = \dfrac{4}{60} + \dfrac{21}{60} - \dfrac{60}{60} = \dfrac{4 + 21 - 60}{60}$

$$= \frac{-35}{60} = -\frac{7}{12}$$

**Example 2.** Simplify: $\dfrac{1}{a^2} + \dfrac{1}{b^2} - \dfrac{2}{ab}$.

*Solution:*

1. The LCM of $a^2$, $b^2$, and $ab$ is $a^2b^2$.

2. $\dfrac{1}{a^2} = \dfrac{1 \cdot b^2}{a^2 \cdot b^2} = \dfrac{b^2}{a^2b^2}$ ; $\dfrac{1}{b^2} = \dfrac{1 \cdot a^2}{b^2 \cdot a^2} = \dfrac{a^2}{a^2b^2}$ ; $\dfrac{2}{ab} = \dfrac{2ab}{a^2b^2}$

3. $\dfrac{1}{a^2} + \dfrac{1}{b^2} - \dfrac{2}{ab} = \dfrac{b^2}{a^2b^2} + \dfrac{a^2}{a^2b^2} - \dfrac{2ab}{a^2b^2}$

   $= \dfrac{b^2 + a^2 - 2ab}{a^2b^2} = \dfrac{(a - b)^2}{a^2b^2}$

**Example 3.** Given $f(x) = \dfrac{1}{(x - a)^2}$ and $g(x) = \dfrac{1}{x^2 - a^2}$. Express $(f + g)(x)$ in simple form.

*Solution:* $(f + g)(x) = f(x) + g(x) = \dfrac{1}{(x - a)^2} + \dfrac{1}{x^2 - a^2}$.

1. The LCM of $(x - a)^2$ and $x^2 - a^2 = (x + a)(x - a)$ is $(x - a)^2(x + a)$.

2. $\dfrac{1}{(x - a)^2} = \dfrac{x + a}{(x - a)^2(x + a)}$

   $\dfrac{1}{x^2 - a^2} = \dfrac{1 \cdot (x - a)}{(x + a)(x - a)(x - a)} = \dfrac{x - a}{(x - a)^2(x + a)}$

3. $(f + g)(x) = \dfrac{1}{(x - a)^2} + \dfrac{1}{x^2 - a^2}$

   $= \dfrac{x + a}{(x - a)^2(x + a)} + \dfrac{x - a}{(x - a)^2(x + a)}$

   $= \dfrac{x + a + x - a}{(x - a)^2(x + a)} = \dfrac{2x}{(x - a)^2(x + a)}$

## WRITTEN EXERCISES

Write each expression as a single fraction in lowest terms. Assume that the variables are restricted so that no denominator is equal to 0.

**A** 1. $\frac{5}{11} - \frac{8}{11} + \frac{13}{11}$    4. $-\frac{3}{20} + \frac{7}{5} + \frac{7}{15}$    7. $\dfrac{5}{x - 2} + \dfrac{12}{x - 2}$

2. $\frac{23}{30} + \frac{5}{30} - \frac{37}{30}$    5. $\dfrac{6x - 5}{8} + \dfrac{3x + 7}{12}$    8. $\dfrac{-6}{y + 3} + \dfrac{5}{y + 3}$

3. $\frac{5}{9} - \frac{7}{27} + \frac{1}{3}$    6. $\dfrac{2y - 8}{7} + \dfrac{3y + 5}{14}$    9. $\dfrac{2}{3x} + \dfrac{3}{4x}$

**10.** $\dfrac{7}{8y} + \dfrac{9}{5y}$

**11.** $\dfrac{2x}{5yz} + \dfrac{z}{2xy} - \dfrac{3y}{4xz}$

**12.** $\dfrac{5}{2} + \dfrac{5}{4x^2} - \dfrac{2}{x}$

**13.** $\dfrac{5x + 1}{6x} - \dfrac{3x - 2}{2x}$

**14.** $\dfrac{3x + 2y}{3y} - \dfrac{x + 2y}{6x}$

**15.** $2(x + 1)^{-1} + 2(x + 1)^{-1}$

**16.** $(2x + 1)^{-1} - 3(x - 2)^{-1}$

**17.** $\dfrac{3x}{3x - 4} + \dfrac{5x}{5x + 6}$

**18.** $\dfrac{3b}{3b - a} + \dfrac{2a}{2a - 3b}$

**19.** $\dfrac{5}{3 - x} - \dfrac{2}{x - 3}$

**20.** $\dfrac{6}{b - 3} + \dfrac{3}{3 - b}$

If $f(x) = \dfrac{x + 1}{x + 2}$ and $g(x) = \dfrac{x + 2}{x + 3}$, express in simple form:

**21.** $(f + g)(x)$

**22.** $(f - g)(x)$

If $g(x) = \dfrac{x - 4}{x + 2}$ and $h(x) = \dfrac{2x - 2}{x - 2}$, express in simple form:

**23.** $(h + g)(x)$

**24.** $(h - g)(x)$

Simplify.

**25.** $\dfrac{a + 2b}{2a - b} - \dfrac{2a + b}{a - 2b}$

**26.** $\dfrac{5a - b}{3a + b} - \dfrac{6a - 5b}{2a - b}$

**27.** $\dfrac{2y}{y^2 - 1} - \dfrac{y + 3}{y^2 + y}$

**28.** $\dfrac{2x}{x^2 - 1} - \dfrac{x - 1}{x^2 - x}$

**B** **29.** $\dfrac{1}{x + y} + \dfrac{1}{x - y} + \dfrac{2x}{x^2 - y^2}$

**30.** $\dfrac{1}{a - x} - \dfrac{3x}{a^2 - x^2} - \dfrac{a}{ax + x^2}$

**31.** $\dfrac{2}{y^2 - 5y + 6} - \dfrac{5}{y^2 + 2y - 15}$

**32.** $\dfrac{2a}{6 - a - a^2} + \dfrac{3a}{a^2 + 6a + 9}$

**33.** $x - \dfrac{2x}{x^2 - 1} + \dfrac{3}{x + 1}$

**34.** $y - 1 + \dfrac{3}{2y - 1} - \dfrac{y}{4y^2 - 1}$

**35.** $3x(1 - 2x)^{-1} - 2x(2x + 1)^{-1} - 3(4x^2 - 1)^{-1}$

**36.** $a(a - b)^{-1} + (a^2 + b^2)(b^2 - a^2)^{-1} + b(a + b)^{-1}$

**37.** $3(x + 1)^{-1} + 3(1 - x)^{-1} - 6(x^2 - 1)^{-1}$

**38.** $6(x^2 - 1)^{-1} - 5(x - 1)^{-2}$

**C** **39.** $\dfrac{x+1}{x^2-x-6} - \dfrac{x-4}{x^2-4x+3} + \dfrac{x+3}{x^2+x-2}$

**40.** $\dfrac{5x+1}{4x^2-1} + \dfrac{5x-3}{2x^2+5x+2} - \dfrac{3x+3}{2x^2+3x-2}$

**41.** $\dfrac{3y+2}{6y^2-y-1} + \dfrac{y+3}{3y^2+7y+2} - \dfrac{y-2}{2y^2+3y-2}$

**42.** $\dfrac{1}{(x-y)(y-z)} + \dfrac{1}{(y-z)(z-x)} + \dfrac{1}{(z-x)(x-y)}$

## 6–6  Complex Fractions

A **complex fraction** is a fraction whose numerator or denominator—or both—contains one or more fractions (or powers with negative exponents). For example, $\dfrac{\frac{3}{4}-\frac{1}{3}}{1-\frac{1}{6}}$ and $\dfrac{x-6+9x^{-1}}{\frac{x}{3}-1}$ are complex fractions. Either of two methods may be used to simplify such fractions.

---

**Method 1.** Multiply the numerator and denominator by the LCD of all the fractions within them.

**Method 2.** Express the fraction as a quotient of simplified fractions and proceed as in Section 6–4.

---

**Example.**  Simplify: **a.** $\dfrac{\frac{3}{4}-\frac{1}{3}}{1-\frac{1}{6}}$  **b.** $\dfrac{x-6+9x^{-1}}{\frac{x}{3}-1}$

*Solution:*

**a.**

Method 1

$$\frac{\frac{3}{4}-\frac{1}{3}}{1-\frac{1}{6}} = \frac{\left(\frac{3}{4}-\frac{1}{3}\right)\cdot 12}{\left(1-\frac{1}{6}\right)\cdot 12}$$

$$= \frac{9-4}{12-2}$$

$$= \frac{5}{10} = \frac{1}{2}$$

Method 2

$$\frac{\frac{3}{4}-\frac{1}{3}}{1-\frac{1}{6}} = \frac{\frac{9-4}{12}}{\frac{6-1}{6}}$$

$$= \frac{5}{12} \div \frac{5}{6}$$

$$= \frac{5}{12}\cdot\frac{6}{5} = \frac{1}{2}$$

**b.**

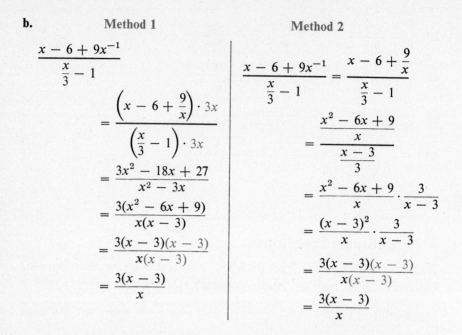

Method 1

Method 2

$$\frac{x - 6 + 9x^{-1}}{\frac{x}{3} - 1}$$

$$= \frac{\left(x - 6 + \dfrac{9}{x}\right) \cdot 3x}{\left(\dfrac{x}{3} - 1\right) \cdot 3x}$$

$$= \frac{3x^2 - 18x + 27}{x^2 - 3x}$$

$$= \frac{3(x^2 - 6x + 9)}{x(x - 3)}$$

$$= \frac{3(x - 3)(x - 3)}{x(x - 3)}$$

$$= \frac{3(x - 3)}{x}$$

$$\frac{x - 6 + 9x^{-1}}{\frac{x}{3} - 1} = \frac{x - 6 + \dfrac{9}{x}}{\dfrac{x}{3} - 1}$$

$$= \frac{\dfrac{x^2 - 6x + 9}{x}}{\dfrac{x - 3}{3}}$$

$$= \frac{x^2 - 6x + 9}{x} \cdot \frac{3}{x - 3}$$

$$= \frac{(x - 3)^2}{x} \cdot \frac{3}{x - 3}$$

$$= \frac{3(x - 3)(x - 3)}{x(x - 3)}$$

$$= \frac{3(x - 3)}{x}$$

If a fraction in the numerator or denominator of a complex fraction is itself complex, it should first be simplified independently. For example:

$$\frac{\dfrac{a}{b} + \dfrac{1 + \dfrac{1}{b}}{a}}{\dfrac{1}{a} + \dfrac{1}{b}} = \frac{\dfrac{a}{b} + \dfrac{1 + \dfrac{1}{b}}{a} \cdot \dfrac{b}{b}}{\dfrac{1}{a} + \dfrac{1}{b}} = \frac{\dfrac{a}{b} + \dfrac{b + 1}{ab}}{\dfrac{1}{a} + \dfrac{1}{b}} = \frac{\left(\dfrac{a}{b} + \dfrac{b + 1}{ab}\right) \cdot ab}{\left(\dfrac{1}{a} + \dfrac{1}{b}\right) \cdot ab}$$

$$= \frac{a^2 + b + 1}{b + a}$$

*WRITTEN EXERCISES*

Simplify. Assume that no denominator is equal to 0.

**A** **1.** $\dfrac{2 + \dfrac{3}{5}}{5 + \dfrac{1}{4}}$

**3.** $\dfrac{\dfrac{3ab}{x}}{\dfrac{6a^2b}{x^2}}$

**5.** $\dfrac{x + \dfrac{x}{y}}{1 + \dfrac{1}{y}}$

**7.** $\dfrac{r - \dfrac{1}{r^2}}{1 - \dfrac{1}{r}}$

**2.** $\dfrac{\dfrac{1}{6} + 1}{\dfrac{1}{12} - 2}$

**4.** $\dfrac{\dfrac{2x}{5y}}{\dfrac{3x}{10y^2}}$

**6.** $\dfrac{1 + \dfrac{1}{2z}}{z - \dfrac{1}{4z}}$

**8.** $\dfrac{2 - \dfrac{3}{x}}{2 + \dfrac{3}{x}}$

**9.** $\dfrac{z + 1 - \dfrac{20}{z}}{z - 2 - \dfrac{8}{z}}$

**11.** $\dfrac{1 - \dfrac{1}{x} - \dfrac{12}{x^2}}{3 + \dfrac{13}{x} + \dfrac{12}{x^2}}$

**13.** $\dfrac{\dfrac{1}{xy} + \dfrac{2}{yz} + \dfrac{3}{xz}}{\dfrac{2x + 3y + z}{xyz}}$

**10.** $\dfrac{1 + \dfrac{1}{a} - \dfrac{2}{a^2}}{2 + \dfrac{5}{a} + \dfrac{2}{a^2}}$

**12.** $\dfrac{2 - \dfrac{7}{b} + \dfrac{6}{b^2}}{2 + \dfrac{3}{b} - \dfrac{9}{b^2}}$

**14.** $\dfrac{\dfrac{x}{yz} - \dfrac{y}{xz} + \dfrac{z}{xy}}{\dfrac{1}{x^2y^2} - \dfrac{1}{x^2z^2} + \dfrac{1}{y^2z^2}}$

**B** **15.** $\dfrac{2}{x^{-1} + y^{-1}}$

**18.** $\dfrac{1 - \dfrac{8(a^2 + b^2)}{9a^2 - b^2}}{1 - \dfrac{2(a + 2b)}{3a + b}}$

**21.** $\dfrac{\dfrac{x}{y} + \dfrac{x + y}{x - y}}{\dfrac{y}{x} + \dfrac{x + y}{x - y}}$

**16.** $\dfrac{1}{x^{-2} + y^{-2}}$

**19.** $a - \dfrac{a}{a + \dfrac{1}{4}}$

**22.** $\dfrac{\dfrac{x + 1}{x - 1} + \dfrac{x - 1}{x + 1}}{\dfrac{x + 1}{x - 1} - \dfrac{x - 1}{x + 1}}$

**17.** $\dfrac{1 - \dfrac{u - v}{1 + uv}}{1 + \dfrac{u^2 - uv}{1 + uv}}$

**20.** $z - \dfrac{z}{1 - \dfrac{z}{1 - z}}$

**C** **23.** $[(x + y)(x - y)^{-1} + xy^{-1}][xy(x^2 - y^2)^{-1} + x(x + y)^{-1}]^{-1}$

**24.** $[(a + b)a^{-1} - 2b(a + b)^{-1}][(a - b)b^{-1} + 2a(a - b)^{-1}]^{-1}$

## APPLICATIONS

### 6–7  Rational Coefficients

Mathematical descriptions of practical situations often involve expressions whose numerical coefficients are fractions. In working with these, the idea of least common denominator is a useful tool.

**Example 1.**  Express $\dfrac{x^2}{3} - \dfrac{x}{6} - \dfrac{1}{2}$ as the product of a rational number and prime polynomials with integral coefficients.

*Solution:*  The LCM of 3, 6, and 2 is 6.

$$\frac{x^2}{3} - \frac{x}{6} - \frac{1}{2} = \frac{2x^2}{6} - \frac{x}{6} - \frac{3}{6}$$

$$= \tfrac{1}{6}(2x^2 - x - 3)$$

$$= \tfrac{1}{6}(2x - 3)(x + 1)$$

**Example 2.** Graph the solution set of $\dfrac{x + 9}{5} < \dfrac{x}{2} + 3$.

*Solution:* Multiply both members of the inequality by the LCM of the denominators, the *positive* number 10.

$$\frac{x + 9}{5} < \frac{x}{2} + 3$$

$$10\left(\frac{x + 9}{5}\right) < 10\left(\frac{x}{2} + 3\right)$$

$$2x + 18 < 5x + 30$$

$$-12 < 3x$$

$$-4 < x \quad (x > -4)$$

**Example 3.** A reservoir can be filled in 6 days by pipe $A$ running alone, and in 4 days by pipe $B$ alone. How long would it take to fill the reservoir if both pipes were running?

*Solution:*

1. Let $t$ = number of days required for both pipes to fill the reservoir.

   The part of the reservoir which can be filled by pipe $A$ in 1 day is $\frac{1}{6}$. The part which is filled by pipe $A$ in $t$ days is $t \cdot \dfrac{1}{6} = \dfrac{t}{6}$.

   Similarly, the part of the reservoir which can be filled by pipe $B$ in $t$ days is $t \cdot \dfrac{1}{4} = \dfrac{t}{4}$.

2. The part of the reservoir which is filled by both pipes in $t$ days is 1 (full reservoir).

   | Part filled by pipe $A$ | plus | part filled by pipe $B$ | is | part filled by both pipes. |
   |:---:|:---:|:---:|:---:|:---:|
   | $\dfrac{t}{6}$ | $+$ | $\dfrac{t}{4}$ | $=$ | $1$ |

3. To simplify the equation, multiply by the LCD, 12.

$$12\left(\frac{t}{6} + \frac{t}{4}\right) = 12 \cdot 1$$

$$2t + 3t = 12, \; 5t = 12$$

$$t = \tfrac{12}{5} = 2\tfrac{2}{5}$$

$\therefore$ it takes $2\frac{2}{5}$ days for the two pipes to fill the reservoir.

*Check:*  4. In $2\frac{2}{5}$, or $\frac{12}{5}$, days, pipe $A$ fills $\frac{12}{5} \cdot \frac{1}{6} = \frac{2}{5}$ of the reservoir, and pipe $B$ fills $\frac{12}{5} \cdot \frac{1}{4} = \frac{3}{5}$ of the reservoir. Since $\frac{2}{5} + \frac{3}{5} = 1$, the two pipes completely fill the reservoir in $2\frac{2}{5}$ days.

Fractions often enter problems through the use of *percentages*. **Percent** means *hundredths* and is denoted by $\%$. Thus $65\%$ means $\dfrac{65}{100}$, or 0.65.

**Example 4.**  Dieticians preparing a food package for a space station have available supplies of ingredient $A$ containing $65\%$ protein and $15\%$ fat and ingredient $B$ containing $30\%$ each of protein and fat. How much of each ingredient should be combined to yield a mixture containing 27 pounds of protein and 12 pounds of fat?

*Solution:*  1. Let $x$ = number of pounds of ingredient $A$;
 $y$ = number of pounds of ingredient $B$.

The facts of the problem are shown in the chart below.

|  | Ingredient $A$ | Ingredient $B$ | Mixture |
|---|---|---|---|
| **Number of pounds of protein** | 0.65x | 0.30y | 27 |
| **Number of pounds of fat** | 0.15x | 0.30y | 12 |

2. The sum of the amounts of protein in ingredients $A$ and $B$ **is** the amount of protein in the mixture.

$$0.65x + 0.30y = 27$$

The sum of the amounts of fat in ingredients $A$ and $B$ **is** the amount of fat in the mixture.

$$0.15x + 0.30y = 12$$

3. Solve the system

$$\begin{array}{ll} 0.65x + 0.30y = 27 \\ 0.15x + 0.30y = 12 \end{array} \quad \text{or} \quad \begin{array}{ll} 65x + 30y = 2700 \\ 15x + 30y = 1200 \end{array}$$

by one of the methods of Section 3–5 to find:

$$x = 30$$
$$y = 25$$

$\therefore$ 30 pounds of ingredient $A$ and 25 pounds of ingredient $B$ should be used.

*ORAL EXERCISES*

State the LCD of the terms of each open sentence, and then read the sentence after multiplying by the LCD.

**1.** $\dfrac{3z}{2} - \dfrac{z}{4} = 3$     **4.** $k + \dfrac{3k}{10} \geq -\dfrac{1}{5}$     **7.** $\dfrac{t^2}{3} - \dfrac{5t}{2} = \dfrac{3}{4}$

**2.** $\dfrac{7x}{6} - \dfrac{2x}{3} = 4$     **5.** $\dfrac{n}{4} - \dfrac{2n}{3} = \dfrac{1}{6}$     **8.** $\dfrac{x^2}{2} - \dfrac{11x}{4} = \dfrac{-5}{8}$

**3.** $\dfrac{y}{7} - 2y < \dfrac{1}{2}$     **6.** $\dfrac{5b}{6} - \dfrac{b}{3} = \dfrac{5}{12}$     **9.** $\dfrac{2y^2}{15} - \dfrac{4y}{3} < y + \dfrac{1}{2}$

*WRITTEN EXERCISES*

Write each expression as the product of a rational number and prime polynomials with integral coefficients, and solve the given sentence over $\mathcal{R}$.

**A**  **1.** $\dfrac{x^2}{5} - \dfrac{3x}{10} - \dfrac{1}{2} = 0$     **7.** $z^2 + \dfrac{19z}{15} + \dfrac{2}{5} = 0$

**2.** $y^2 + \dfrac{5y}{12} - \dfrac{1}{4} = 0$     **8.** $\dfrac{5a^2}{9} + \dfrac{a}{2} - \dfrac{1}{2} = 0$

**3.** $\dfrac{5a^2}{14} - \dfrac{a}{7} = \dfrac{1}{2}$     **9.** $\dfrac{7z^2}{4} + \dfrac{13z}{2} = 2$

**4.** $\dfrac{8z^2}{3} - \dfrac{z}{9} = \dfrac{10}{9}$     **10.** $\dfrac{2x^2}{5} - \dfrac{5x}{3} = -\dfrac{5}{3}$

**5.** $\dfrac{3x^2}{2} + \dfrac{17x}{12} = \dfrac{5}{4}$     **11.** $\dfrac{y^2}{6} + \dfrac{y}{3} = \dfrac{1}{2}$

**6.** $x^2 - \dfrac{31x}{14} = \dfrac{5}{7}$     **12.** $\dfrac{2x^2}{3} - \dfrac{4x}{3} = \dfrac{7}{2}$

Graph the solution set of each inequality over $\mathcal{R}$.

**13.** $\dfrac{x - 3}{4} < \dfrac{2}{3}$     **16.** $\dfrac{3(3y + 2)}{5} \leq 2 + \dfrac{2(2y - 1)}{3}$

**14.** $\dfrac{x + 2}{2} \geq \dfrac{2x}{3}$     **17.** $\dfrac{3z - 2}{2} < 1 + \dfrac{2z - 1}{4}$

**15.** $\dfrac{3(2x - 1)}{4} \geq \dfrac{4x + 3}{2}$     **18.** $\dfrac{2t - 1}{3} - 1 < \dfrac{2 - t}{5}$

Graph each system of inequalities in the Cartesian plane.

**Sample.** $\frac{x}{6} + \frac{y}{3} > -1$

$\frac{x}{2} - y < -3$

*Solution:* Multiply each inequality by the LCD of its fractions.

$$x + 2y > -6$$
$$x - 2y < -6$$

Graph these inequalities. The double shading represents the solution set of this system.

**B** **19.** $\frac{x}{3} + \frac{2y}{3} \leq 2$     **22.** $\frac{x}{7} - \frac{3y}{4} > 1$     **25.** $\frac{4x}{3} + \frac{7y}{6} \leq 2$

$\frac{x}{2} - \frac{y}{4} \geq 1$         $x - \frac{y}{2} < -2$          $\frac{x + 2y}{4} + \frac{2x + y}{3} \geq 1$

**20.** $\frac{x}{3} + \frac{y}{2} \geq 1$     **23.** $\frac{x}{4} + \frac{y}{5} \leq 1$     **26.** $\frac{x + y}{2} - \frac{x - y}{3} > 8$

$\frac{x}{4} - \frac{y}{3} \leq 1$         $\frac{2x}{9} + 2 \geq \frac{y}{9}$          $\frac{x + y}{3} + \frac{x - y}{4} > 11$

**21.** $\frac{2x}{3} - y > 4$     **24.** $\frac{2x}{3} + \frac{3y}{4} > -\frac{7}{2}$

$x - \frac{3y}{4} < 6$         $\frac{x}{4} - \frac{11}{2} < \frac{2y}{5}$

*PROBLEMS*

Use either one or two variables to solve these problems.

**A** **1.** Deimos, one of Mars' satellites, is approximately $\frac{4}{7}$ as wide as Phobos, another of Mars' satellites. If Phobos is about 4 kilometers longer than it is wide, and is also about twice as long as Deimos which is about 12.5 kilometers in length, about how wide is Deimos?

**2.** The ratio of Democrats to Republicans in Mayview is 10 to 9. If a sample poll is taken of 380 of the voters in Mayview, how many of each party should be interviewed if the sample is to have the same party distribution as the city's electorate?

3. The Milnor Publishing Company publishes a cookbook in a standard edition selling for $12.35 and a deluxe edition selling for $15.00. If sales for both editions one year totaled 50,000 copies and receipts were $638,700, how many copies of each edition were sold?

4. Monarch Airlines charges $80 for a first-class ticket and $64 for a coach ticket to fly from Central City to Almaden. If a flight carrying 77 passengers produces $5,168 in revenue for the airline, how many tickets of each kind were sold for the flight?

5. The average of two numbers is 28. Find the numbers if the smaller is three fifths of the larger.

6. Paul can type a page of manuscript in 6 minutes while Bill can type a page in 5 minutes. Working together, how long (exclusive of idle time) would it take them to type a 231 page book?

7. It took 6 hours for the first computer at Acme Industries to process the payroll. A new computer can process it in 3 hours. How long will it take to process the payroll if both machines operate together?

8. One pump can empty a full storage tank in 8 hours whereas it takes 12 hours for a second pump to empty the same tank. How long will it take to empty the full tank if both pumps operate together?

9. Mrs. Edmonds can travel down the White River at 12 miles per hour, but averages only 4 miles an hour going upstream. If she has $5\frac{1}{3}$ hours to sail down the river and return, how far can she go downstream?

10. A climbing party started at the 12,000 foot level and climbed Mount Behemoth. The party could average 400 feet per hour going up, and 800 feet per hour returning. If the party spent 1 hour on top of the mountain and returned to its starting position $10\frac{3}{4}$ hours after departing, how high is Mount Behemoth?

11. It takes the Mighty Mite computer 5.4 seconds to complete an algorithm. If the algorithm involves $\frac{3}{5}$ as many multiplication operations as addition operations, and if the machine needs 0.0002 seconds for one multiplication operation and 0.00015 seconds for one addition operation, how many of each type of operation are involved in the algorithm?

12. The Alpine Ski Lift travels at a steady rate of 6 miles per hour. Sandra rode the lift to the top and then, after waiting for 6 minutes, made the ski run back to the bottom of the lift at an average rate of 30 miles per hour. If the entire trip took Sandra 30 minutes, how long is the ski lift?

13. How much water must be added to 6 ounces of a 50% acid solution to produce a solution that is 25% acid?

14. How many gallons of a 25% salt solution must be mixed with 10 gallons of a 15% salt solution to produce a 17% solution?

**15.** How much water must be evaporated from a 300 gallon solution containing 2% salt to produce a 10% salt solution?

**B** **16.** A water reservoir can be filled by one input pump in 8 hours and by another one in 6 hours. During a fire, the maximum output of water took 4 hours to empty this reservoir. If, once the reservoir was empty, both input pumps were started, and if the maximum output continued, how long did it take to fill the reservoir again?

**17.** A pump can empty a tank in 18 hours. With the tank full, this pump and another pump are started and together they empty $\frac{3}{5}$ of the tank in 6 hours. If the first pump breaks at that time, how long will it take the second pump to finish emptying the tank?

**18.** Patricia has $5000 invested at 4%. What is the least amount she must invest at 6% in order that her income be at least 5% of the total she has invested?

**19.** A sum of $2000 is invested, part at 3% and part at 4% interest. What is the least amount that can be invested at 4% if the annual return from the investments is to be at least $66?

**20.** What is the greatest distance the Comptons can sail down river at 10 miles per hour, if they can return up river at 7 miles per hour and have at most 11 hours for sailing.

**21.** Mr. Thomkins rode a bus averaging 30 miles per hour from his home to the airport, and then flew to another city at 250 miles per hour. If his total trip was 765 miles and it took him 4 hours to complete, how far does he live from the airport?

**C** **22.** The Safety Moving Company van left Mr. Perkins' old house with his furniture and traveled at an average rate of 40 mph directly to a new residence. After waiting 45 minutes, Mr. Perkins left in his own car and averaged 65 miles per hour along the same route. Mr. Perkins stopped an hour and a half for lunch and 15 minutes for gasoline and arrived at his new house at the same time as the van. How far was Mr. Perkins' new house from his old house?

**23.** Mr. Edwards bought and sold a house upon which he realized a 40% profit. Using this money, he bought and sold a second house sustaining a 10% loss in the process. Using this money, in turn, he bought and sold a third house, realizing a 20% profit on a selling price of $40,000. What did Mr. Edwards pay for his first house?

**24.** A speeder going 75 miles per hour passes a state trooper parked by the side of the freeway. The trooper gives chase; within $1\frac{1}{2}$ minutes he has reached a speed of 90 mph and has gone 0.15 miles. If he continues at this speed, how long does it take him to overtake the speeder?

## 6–8 Fractional Equations

An equation such as $\dfrac{4}{x^2 - 1} - \dfrac{2x}{x - 1} + \dfrac{5}{2} = 0$ in which a variable appears in the denominator of a fraction is called a *fractional equation*. Transforming a fractional equation by multiplying each of its members by the least common denominator of its terms may *not* produce an equivalent equation.

**Example 1.** Solve $\dfrac{4}{x^2 - 1} - \dfrac{2x}{x - 1} + \dfrac{5}{2} = 0.$

*Solution:*

$$\frac{4}{x^2 - 1} - \frac{2x}{x - 1} + \frac{5}{2} = 0$$

$$2(x^2 - 1)\frac{4}{x^2 - 1} - 2(x^2 - 1)\frac{2x}{x - 1} + 2(x^2 - 1)\cdot\frac{5}{2} = 0$$

$$8 - 4x(x + 1) + 5(x^2 - 1) = 0$$

$$x^2 - 4x + 3 = 0$$

$$(x - 3)(x - 1) = 0$$

$$x = 3 \mid x = 1$$

Notice what happens when you test 3 and 1 in the original equation:

When $x = 3$,

$$\frac{4}{x^2 - 1} - \frac{2x}{x - 1} + \frac{5}{2}$$

$$= \frac{4}{3^2 - 1} - \frac{2\cdot 3}{3 - 1} + \frac{5}{2}$$

$$= \frac{4}{8} - \frac{6}{2} + \frac{5}{2}$$

$$= \frac{1}{2} - \frac{6}{2} + \frac{5}{2} = 0.$$

$\therefore$ 3 is a root.
The solution set is $\{3\}$.

When $x = 1$,

$$\frac{4}{x^2 - 1} - \frac{2x}{x - 1} + \frac{5}{2}$$

$$= \frac{4}{1^2 - 1} - \frac{2\cdot 1}{1 - 1} + \frac{5}{2}$$

$$= \frac{4}{0} - \frac{2}{0} + \frac{5}{2}.$$

The fractions $\frac{2}{0}$ and $\frac{4}{0}$ are meaningless.
$\therefore$ 1 is *not* a root.

In Example 1 the equation obtained by multiplying the given equation by $x^2 - 1$ has the extra root 1, a number for which the multiplier is zero. When you multiply an equation by a polynomial, the solution set of the resulting equation includes the roots of the original equation but may also contain numbers for which the multiplier is zero. Therefore, *always test each root of the resulting equation in the original equation.*

**Example 2.** A driver making a 200-mile trip travels the first 100 miles at 40 mph. At what speed must he travel the second 100 miles if his average speed for the entire 200 miles is to be 50 mph?

*Solution:*

1. Let $v$ = speed in mph at which he must travel the second 100 miles. In constructing the table below, we use the formula time $= \dfrac{\text{distance}}{\text{speed}}$.

|  | First 100 miles | Second 100 miles | Entire trip |
|---|---|---|---|
| Distance (miles) | 100 | 100 | 200 |
| Speed (mph) | 40 | $v$ | 50 |
| Time (hours) | $\dfrac{100}{40} = \dfrac{5}{2}$ | $\dfrac{100}{v}$ | $\dfrac{200}{50} = 4$ |

2. Time for first plus time for second is time for
100 miles        100 miles         entire trip.

$$\frac{5}{2} \quad + \quad \frac{100}{v} \quad = \quad 4$$

3. Multiply by $2v$:

$$5v + 200 = 8v$$
$$3v = 200, \quad v = \tfrac{200}{3} = 66\tfrac{2}{3}$$

∴ the driver must cover the second 100 miles at a speed of $66\tfrac{2}{3}$ mph. The check is left to you.

**Example 3.** After modernization a factory can build 6 more cars per hour than previously and the day's output of 360 cars takes 2 hours less to produce. How many cars does the modernized factory produce per hour?

*Solution:*

1. Let $n$ = number of cars per hour now;
$n - 6$ = number of cars per hour previously

2. Number of hours number of hours
to produce 360 is to produce 360 less 2.
cars now cars previously

$$\frac{360}{n} \quad = \quad \frac{360}{n-6} \quad - \quad 2$$

Solve the equation to find that 36 cars per hour are produced in the modernized factory.

*WRITTEN EXERCISES*

Solve over ℛ and check. If no solutions exist, so state.

**A**   **1.** $\dfrac{2}{x} + \dfrac{4}{3x} = \dfrac{10}{9}$

**9.** $\dfrac{5y - 4}{5} - \dfrac{10y + 9}{10} = \dfrac{51}{6}$

**2.** $\dfrac{2}{y} + \dfrac{3}{2y} = \dfrac{7}{10}$

**10.** $\dfrac{t - 3}{2t} + \dfrac{1}{3} = \dfrac{3t - 7}{2t}$

**3.** $\dfrac{12}{t - 2} = 3$

**11.** $\dfrac{3}{y + 1} + 5 = \dfrac{18}{y + 1}$

**4.** $\dfrac{10}{z - 3} = \dfrac{9}{z - 5}$

**12.** $\dfrac{y}{y + 2} - \dfrac{3}{y - 2} = \dfrac{y^2 + 8}{y^2 - 4}$

**5.** $\dfrac{4 - n}{1 - n} = \dfrac{15 - n}{3 - n}$

**13.** $\dfrac{3}{z - 3} = \dfrac{z}{z - 3} + 7$

**6.** $\dfrac{x + 2}{2x - 4} = \dfrac{3x + 5}{6x - 1}$

**14.** $\dfrac{k}{k + 2} = 8 - \dfrac{2}{k + 2}$

**7.** $\dfrac{4}{y - 3} - \dfrac{3}{y - 1} = \dfrac{10}{y^2 - 4y + 3}$

**15.** $\dfrac{2}{x + 1} + \dfrac{1}{3x + 3} = \dfrac{1}{6}$

**8.** $\dfrac{5}{y - 4} - \dfrac{3}{y - 5} = \dfrac{5}{y^2 - 9y + 20}$

**16.** $\dfrac{4}{2t - 3} + \dfrac{4t}{4t^2 - 9} = \dfrac{1}{2t + 3}$

**B**   **17.** $\dfrac{3}{x + 6} - \dfrac{2}{x - 6} = \dfrac{1}{4}$

**20.** $\left(\dfrac{t}{t - 1}\right)^2 + 4 = 5\left(\dfrac{t}{t - 1}\right)$

**18.** $\dfrac{7}{z - 3} - \dfrac{3}{z - 4} = \dfrac{1}{2}$

**21.** $\left(\dfrac{x + 2}{x - 1}\right)^2 - \dfrac{5x + 10}{x - 1} = -6$

**19.** $\dfrac{x + 2}{x + 3} = \dfrac{36}{(x + 3)^2} - 1$

**22.** $\dfrac{1}{x^4} - \dfrac{10}{x^2} + 9 = 0$

Solve each system and check.

**Sample.**   $2a - b = 5$

$\dfrac{a + 4}{b + 9} = \dfrac{3}{4}$

*Solution:*   Rewrite the second equation free of fractions.

$$\dfrac{a + 4}{b + 9} = \dfrac{3}{4}$$

$$4a + 16 = 3b + 27$$

$$4a - 3b = 11$$

Solve the system consisting of this rewritten equation and $2a - b = 5$ to obtain $a = 2$ and $b = -1$.

*Check:*   $2a + b = 2(2) - (-1) = 4 + 1 = 5$

$$\frac{a + 4}{b + 9} = \frac{2 + 4}{-1 + 9} = \frac{6}{8} = \frac{3}{4}$$

∴ the solution set is $\{(2, -1)\}$.

**C** **23.** $\dfrac{x + 5}{2} + \dfrac{y + 1}{3} = 2$

$\dfrac{x + 7}{2y + 1} = \dfrac{4}{5}$

**24.** $\dfrac{x}{3} - \dfrac{y}{2} = 1$

$\dfrac{2x + 3}{5} - \dfrac{5y + 1}{11} = 2$

**25.** $\dfrac{x + y}{2} - \dfrac{x - y}{3} = 8$

$\dfrac{x + y}{3} - \dfrac{x - y}{4} = 11$

**26.** $\dfrac{3}{a - 1} + \dfrac{4}{b - 1} = 0$

$\dfrac{5}{2a - 3} - \dfrac{7}{2b + 13} = 0$

*Hint:* In Exercises 27–30 let $\dfrac{1}{x} = a$ and $\dfrac{1}{y} = b$, solve first for $a$ and $b$ and then for $x$ and $y$.

**27.** $\dfrac{1}{x} + \dfrac{1}{y} = 5$

$\dfrac{1}{x} - \dfrac{1}{y} = 1$

**28.** $\dfrac{3}{x} - \dfrac{1}{y} = 9$

$\dfrac{4}{x} + \dfrac{3}{y} = -1$

**29.** $\dfrac{9}{x} + \dfrac{10}{y} = -1$

$\dfrac{6}{x} + \dfrac{15}{y} = 1$

**30.** $\dfrac{6}{x} + \dfrac{4}{y} = \dfrac{4}{5}$

$\dfrac{9}{x} + \dfrac{5}{y} = \dfrac{7}{10}$

## PROBLEMS

**A** **1.** The riverboat Shenandoah makes a 120 mile excursion trip upstream and returns the next day. If the total travel time both ways is $22\frac{1}{2}$ hours, and if the speed of the current is 4 miles per hour, what is the speed of the Shenandoah in still water?

**2.** Sue Zody commutes 63 miles to work each morning and returns the same distance each evening. She finds that by increasing her average speed by 8 miles per hour, she can save 1 hour in her travel time. What was her original rate?

**3.** Mr. Coven's science class went on a two-day field trip. Mr. Coven and each student contributed equally to obtain the required $360 for the trip. When six students could not attend, each of the remaining students and Mr. Coven had to pay $2 more to make up the needed $360. How many students went on the field trip?

**4.** One outlet can empty a tank in 3 hours. With a second outlet operating, the tank can be emptied in 2 hours. How long would it take the second outlet alone to empty the tank?

**5.** The numerator of a fraction is 3 less than $\frac{1}{2}$ of the denominator, and the fraction is equivalent to $\frac{1}{3}$. Find the numerator and denominator.

**6.** The sum of the reciprocals of two consecutive positive integers is $\frac{11}{30}$. Find the integers.

**7.** If a rational number is decreased by $\frac{1}{2}$ of its reciprocal, the result is $\frac{7}{30}$. Find the rational number.

**8.** When the lesser of the reciprocals of two consecutive odd integers is subtracted from the greater, the result is $\frac{2}{63}$. Find the integers.

**9.** Two numbers are in the ratio $\frac{5}{12}$. If the first number is decreased by 3 and the second increased by 6, the resulting numbers are in the ratio $\frac{2}{7}$. Find the original numbers.

**10.** A crew of three painters can paint a building in 6 hours, while a four-man crew can paint it in 4 hours. If the three-man crew has worked 1 hour and is then joined by a fourth man, how long will it take the four men to complete the job?

**11.** A mailman with a 6 mile delivery route completes the first 3 miles of his route at an average rate of $\frac{3}{4}$ mile per hour. What must be his average rate for the rest of his route if he is to finish the entire route at a rate of 1 mile per hour?

**12.** In still water, a boat travels at 15 miles per hour. If the boat can go 40 miles up a river and then return in a total of 6 hours, what is the rate of the river's current?

**13.** The Federal Savings and Loan Company raised the interest rate on savings by $\frac{1}{2}\%$. As a result Mr. Remington found that he could reduce the amount he had on deposit by $1000 and still earn the same amount of simple interest per year. If his account had been paying him $550 per year, what was the old interest rate and how much did he originally have on deposit?

**14.** Driving inside the city limits, Mr. Brown can average 25 miles per hour less than he can on the highway. If he takes a trip that consists of 14 miles of city driving and 12 miles along the highway, he finds that the trip takes him 36 minutes. What is his average rate in the city? On the highway? How far did he travel in all?

**B** **15.** A bookstore bought a shipment of books for $200. If, subsequently, 5 of the books were damaged and not sold, and if the store made a profit of $70 on a markup of $2.00 per book sold, how many were there in the original shipment?

**16.** Using two pumps, a tank can be filled in $3\frac{1}{3}$ hours. If it takes one pump 5 hours to fill the tank by itself, how long does it take the other pump to fill the tank by itself?

**17.** It takes two presses 1 hour and 12 minutes to print the morning issue of *The Times*. If one of the presses can print the issue alone in 3 hours, how long would it take the other press alone to print the issue?

**18.** A delivery man has a route 120 miles long and has 18 stops, each of which takes 10 minutes. He finds that by increasing his average speed between stops by 6 miles per hour, he can save 1 hour of time on his route. What is his usual average speed between stops?

**19.** The numerator of a fraction is 3 greater than twice the denominator. The sum of the number named by the fraction and its reciprocal is $\frac{29}{10}$. Find the fraction.

**20.** If the sum of three consecutive even integers is divided by each of the first two integers in turn, the sum of the resulting quotients is equal to the quotient when 54 is divided by 3 times the first integer. Find the integers.

**C** **21.** Going upstream a boat travels 12 miles per hour and making the return trip downstream it travels 18 miles per hour. Find the average rate for the round trip.

**22.** For the first $\frac{1}{3}$ of a trip a motorist averages 50 miles per hour and for the last $\frac{2}{3}$ of the trip she averages 60 miles per hour. What is her average rate for the whole trip?

**23.** A metal foundry has some gun metal which is 15% tin, the rest being copper. Determine the range for the number of pounds of pure tin to mix with 300 pounds of the 15% gun metal to produce a grade of gun metal containing at least 20%, but no more than 25% tin.

**24.** What are the limits on the number of ounces of copper to mix with 220 ounces of pure silver to produce an alloy that is between 81% and 90% silver?

## DECIMALS AND SCIENTIFIC NOTATION

### 6–9   Decimals for Rational Numbers

To express a rational number as a decimal numeral, write the rational number as a quotient of integers and perform the indicated division.

$$\tfrac{57}{16} = 57 \div 16 \qquad \tfrac{7}{22} = 7 \div 22$$

```
      3.5625                 0.31818
 16)57.0000            22)7.00000
    48                     66
    ──                     ──
    90                     40
    80                     22
    ───                    ───
    100                    180
     96                    176
    ───                    ───
     40                     40
     32                     22
     ──                    ───
     80                    180
     80                    176
     ──                    ───
      0                      4
```

The symbol 3.5625, called a **terminating** or **finite** decimal, is a brief way of writing $3 + \dfrac{5}{10} + \dfrac{6}{10^2} + \dfrac{2}{10^3} + \dfrac{5}{10^4}$. By using the properties of fractions, you can check that the sum does equal $\tfrac{57}{16}$.

The conversion of $\tfrac{7}{22}$ to decimal form never terminates, however, because the recurring remainders 4 and 18 produce the **repeating block** of digits 18 in the quotient. You write

$$\tfrac{7}{22} = 0.31818 \ldots \qquad \text{or} \qquad \tfrac{7}{22} = 0.3\overline{18},$$

where the dots and the bar indicate that the block is to be repeated unendingly. Such decimals are called **repeating** or **periodic** decimals. Also, because they continue without end, they are said to be **non-terminating** or **infinite** decimals.

The conversion of $\tfrac{57}{16}$ and $\tfrac{7}{22}$ to decimals illustrates what happens whenever you express a rational number by a decimal. When you divide an integer $r$ by a positive integer $s$, the remainder at each step belongs to $\{0, 1, 2, \ldots, s - 1\}$. Therefore, within $s$ steps after only zeros are left in the dividend, either 0 occurs as a remainder and the division process stops, or a nonzero remainder recurs, and the process thereafter produces a repeating sequence of remainders with a repeating block of digits in the quotient.

The decimal for any rational number $\dfrac{r}{s}$ either terminates or eventually repeats in a block of fewer than $s$ digits.

The converse of this statement is also true:

All terminating decimals and all repeating decimals represent rational numbers.

In converting a terminating decimal into a quotient of integers, you simply add fractions: $1.27 = 1 + \frac{2}{10} + \frac{7}{100} = \frac{127}{100}$. The following examples show how to convert a repeating decimal into a common fraction.

**Example 1.** Write $0.2\overline{18}$ as a common fraction.

*Solution:* Let $N = 0.2\overline{18}$

$$100\,N = 21.8\overline{18}$$
$$\text{Subtract:} \quad N = \phantom{21.}0.2\overline{18}$$
$$99\,N = 21.6$$
$$N = \tfrac{21.6}{99} = \tfrac{24}{110}$$

**Example 2.** Write $0.\overline{435}$ as a common fraction.

*Solution:* Let $N = 0.\overline{435}$

$$1000\,N = 435.\overline{435}$$
$$\text{Subtract:} \quad N = \phantom{435.}.\overline{435}$$
$$999\,N = 435$$
$$N = \tfrac{435}{999} = \tfrac{145}{333}$$

Note that any terminating decimal is equivalent to a fraction whose denominator is a power of 10. Since any power of 10 contains only powers of 2 and 5 when written in completely factored form, you can see that if a fraction in lowest terms is equivalent to a terminating decimal, its denominator can contain no factors other than powers of 2 and 5. For example, $\frac{3}{8}$, $\frac{11}{80}$, and $\frac{7}{125}$ are equivalent to terminating decimals because $\dfrac{3}{8} = \dfrac{3}{2^3}$, $\dfrac{11}{80} = \dfrac{11}{2^4 \cdot 5}$, and $\dfrac{7}{125} = \dfrac{7}{5^3}$. On the other hand, $\frac{5}{6}$, $\frac{9}{14}$, and $\frac{5}{52}$ are equivalent to nonterminating repeating decimals, because $\dfrac{5}{6} = \dfrac{5}{2 \cdot 3}$, $\dfrac{9}{14} = \dfrac{9}{2 \cdot 7}$, and $\dfrac{5}{52} = \dfrac{5}{2^2 \cdot 13}$.

*ORAL EXERCISES*

State whether the given fraction is equivalent to a terminating decimal or to a repeating decimal.

**1.** $\frac{5}{22}$     **3.** $\frac{7}{40}$     **5.** $\frac{5}{30}$     **7.** $\frac{3}{42}$     **9.** $\frac{23}{60}$     **11.** $\frac{4}{250}$

**2.** $\frac{3}{16}$     **4.** $\frac{11}{24}$     **6.** $\frac{21}{25}$     **8.** $\frac{13}{50}$     **10.** $\frac{3}{80}$     **12.** $\frac{3}{175}$

*WRITTEN EXERCISES*

Write as a terminating or repeating decimal.

**A**

**1.** $\frac{7}{8}$     **3.** $-\frac{7}{9}$     **5.** $-\frac{21}{16}$     **7.** $\frac{53}{18}$     **9.** $\frac{5}{14}$

**2.** $\frac{3}{50}$     **4.** $-\frac{3}{11}$     **6.** $-\frac{7}{40}$     **8.** $\frac{17}{15}$     **10.** $\frac{10}{21}$

Write as a common fraction in lowest terms.

**11.** 3.172     **13.** $-0.0104$     **15.** $0.\overline{9}$     **17.** $0.\overline{26}$

**12.** 6.7777     **14.** $-0.00032$     **16.** $0.\overline{5}$     **18.** $0.\overline{62}$

**B**

**19.** $-3.\overline{712}$     **21.** $2.12\overline{71}$     **23.** $-61.28\overline{125}$     **25.** $0.0006\overline{9}$

**20.** $-5.3\overline{21}$     **22.** $3.8\overline{13}$     **24.** $72.618\overline{182}$     **26.** $0.0003\overline{9}$

**C**

**27.** If the representation of $x$ as a common fraction in lowest terms has denominator 11, and the decimal representation is $0.\overline{2a}$, find $x$.

**28.** If the representation of $x$ as a common fraction in lowest terms has denominator 111, and the decimal representation is $0.\overline{08b}$, find $x$.

## 6–10   Approximations

It is often convenient to break off, or **round**, a lengthy or infinite decimal, leaving an approximation of the number represented by the original decimal. Using $\doteq$ to mean "equals approximately," you may write

$$\tfrac{7}{22} \doteq 0.3, \quad \tfrac{7}{22} \doteq 0.32, \quad \text{and} \quad \tfrac{7}{22} \doteq 0.318$$

as approximations of $\frac{7}{22}$ to the *nearest tenth, nearest hundredth,* and *nearest thousandth,* respectively. In rounding, use this rule:

> To round a decimal, add 1 to the last digit retained if the first digit dropped is 5 or more; otherwise leave the retained digits unchanged.

Under this rule the difference between a number and its approximation (the **round-off error**) is *at most* half the unit of the last digit retained. For example, the statement $s \doteq 1.32$ is equivalent to:

$$1.32 - 0.005 \leq s < 1.32 + 0.005$$

or

$$1.315 \leq s < 1.325$$

Measurements almost always produce approximations. The **precision** of an approximation is given by the unit used in making it. For example, a chemist reporting the weight of a sample of uranium to be 0.0307 grams *precise to the nearest ten-thousandth of a gram* means that the true weight $w$ satisfies the inequality $0.03065 \leq w < 0.03075$. The unit (precision) of this measurement is one ten-thousandth of a gram and the *maximum possible error* is half that unit, or 0.00005 gram. On the other hand, a rancher may list the weight of a steer as 772 pounds, *to the nearest pound*. In this measurement the precision is one pound and the maximum possible error is half a pound.

Which of the two weights given in the preceding paragraph is the more *accurate* measurement? We define the **accuracy** of a measurement to be the **relative error**, that is, the ratio of the maximum possible error in the measurement to the measurement itself. Thus the accuracy of the chemist's measurement is

$$\frac{0.00005}{0.0307} = \frac{5}{3070} \doteq 0.002, \quad \text{or} \quad 0.2\%.$$

The relative error of the rancher's measurement, however, is

$$\frac{0.5}{772} = \frac{5}{7720} \doteq 0.0006, \quad \text{or} \quad 0.06\%.$$

Thus, the rancher has made the more accurate measurement!

To indicate the accuracy of a measurement, we use the following *standard,* or *scientific, notation:* Express the measurement as a product, $a \times 10^n$, where the absolute value of $a$, a rational number in finite decimal form, is between 1 and 10, and $n$ is an integer. We use as many digits in the decimal for *a* as are justified by the accuracy of the measurement; these are called significant digits. For example:

| Chemist's measurement | Rancher's measurement |
|---|---|
| $0.03_{\wedge}07 = 3.07 \times 10^{-2}$ (grams) | $7_{\wedge}72 = 7.72 \times 10^{2}$ (pounds) |

**Significant digits**

three: 3, 0, 7 | three: 7, 7, 2

**Maximum possible error**

$0.00005 = 5 \times 10^{-5}$ (grams) | $0.5 = 5 \times 10^{-1}$ (pounds)

**Relative error (Accuracy)**

$$\frac{5 \times 10^{-5}}{3.07 \times 10^{-2}} = \frac{5}{3.07} \times 10^{-3} \qquad \frac{5 \times 10^{-1}}{7.72 \times 10^{2}} = \frac{5}{7.72} \times 10^{-3}$$

$$\doteq 0.002 \qquad\qquad\qquad \doteq 0.0006$$

Notice the red caret, $\wedge$ , after the first significant digit in each example. By counting the number of places *from the caret to the decimal point*, you obtain the exponent *n*; *n* is positive or negative according as you count to the right or to the left from the caret.

When a measurement is written in ordinary decimal form, all nonzero digits are significant and so are zeros whose purpose is not just to place the decimal point. In the following examples, the significant digits are shown in red:

$$0.502 \quad 0.006040 \quad 5006 \quad 250.05 \quad 250.00$$

In a number such as 2700 we cannot tell without additional information which, if any, of the zeros is significant. Scientific notation eliminates this problem by writing 2700 as

$$2.7 \times 10^{3}, \quad 2.70 \times 10^{3}, \quad \text{or} \quad 2.700 \times 10^{3}$$

according as none, one, or both of the zeros are significant.

Numbers written in scientific notation can be compared easily:

$$5.6 \times 10^{5} < 7.8 \times 10^{5} \qquad \text{because} \quad 5.6 < 7.8$$
$$3.26 \times 10^{-3} < 6.00 \times 10^{-3} \qquad \text{because} \quad 3.26 < 6.00$$
$$8.2 \times 10^{4} < 3.6 \times 10^{5} \qquad \text{because} \quad 10^{4} < 10^{5}$$
$$7.3 \times 10^{-4} > 8.3 \times 10^{-5} \qquad \text{because} \quad 10^{-4} > 10^{-5}$$

Scientific notation helps you make a quick "order-of-magnitude" estimate of products and quotients.

**Example.** Find a one-significant-digit estimate of

$$x = \frac{6807 \times 31.42 \times 0.07602}{0.5220}$$

*Solution:*

$$x = \frac{6807 \times 31.42 \times 0.07602}{0.5220}$$

Round each number to one significant digit.

$$x \doteq \frac{7000 \times 30 \times 0.08}{0.5}$$

Convert each approximation into scientific notation.

$$x \doteq \frac{7 \times 10^3 \times 3 \times 10 \times 8 \times 10^{-2}}{5 \times 10^{-1}}$$

Compute and round the result to one significant digit.

$$x \doteq \frac{7 \cdot 3 \cdot 8}{5} \times 10^{3+1-2-(-1)}$$

$$x \doteq 30 \times 10^3 \text{ or } 3 \times 10^4 \text{ or } 30{,}000.$$

*ORAL EXERCISES*

Approximate each number to the nearest hundredth.

**1.** 6.183    **3.** −4.8449    **5.** 31.2$\overline{7}$    **7.** $\frac{1}{4}$          **9.** 61.0029

**2.** 7.296    **4.** −3.2051    **6.** 42.2$\overline{4}$    **8.** $\frac{1}{3}$          **10.** 62.0007

Give a three-significant-digit approximation to each number.

**11.** 1.4386    **13.** 0.030821  **15.** 423,792   **17.** 0.$\overline{317}$     **19.** 0.7$\overline{6}$

**12.** 31.7215   **14.** 0.071652  **16.** 671,098   **18.** 0.$\overline{713}$     **20.** 0.7$\overline{4}$

State the number of significant digits in each of the following numerals.

**21.** $6.3 \times 10^5$                    **25.** $7.003 \times 10^6$

**22.** $7.21 \times 10^4$                   **26.** $12.07 \times 10^{-7}$

**23.** $2.07 \times 10^{-3}$                **27.** $7.00 \times 10^{10}$

**24.** $1.02 \times 10^{-4}$                **28.** $3.20 \times 10^{-6}$

## WRITTEN EXERCISES

Write each numeral in scientific notation.

**A**

**1.** 61.8      **5.** 6.92      **9.** 428,900      **13.** 0.00000213

**2.** 32.7      **6.** 8.104      **10.** 3,421,000      **14.** 0.000000721

**3.** 0.21      **7.** 0.00852      **11.** 1,000      **15.** 421,300,000

**4.** 0.38      **8.** 0.00061      **12.** 0.0001      **16.** 7,000,000,000

Write each numeral in decimal form.

**17.** $10^3$      **19.** $10^{-4}$      **21.** $3.21 \times 10^4$    **23.** $2.1 \times 10^{-3}$

**18.** $10^5$      **20.** $10^{-6}$      **22.** $7.23 \times 10^6$    **24.** $4.021 \times 10^{-1}$

Simplify each expression. Give your result in decimal form.

**25.** $\dfrac{(12 \times 10^4)(15 \times 10^{-8})}{(18 \times 10^{-3})}$      **28.** $\dfrac{(3.6 \times 10^{-4})(4.8 \times 10^{-2})}{(7.2 \times 10^{-8})}$

**26.** $\dfrac{(15 \times 10^{-5})(6 \times 10^{12})}{(45 \times 10^{14})}$      **29.** $\dfrac{0.6 \times 0.00084 \times 0.093}{0.00021 \times 0.00031}$

**27.** $\dfrac{(4.2 \times 10^3)(6.4 \times 10^{-5})}{(6.72 \times 10^8)}$      **30.** $\dfrac{0.0054 \times 0.05 \times 300}{0.0015 \times 0.27 \times 80}$

Give a one-significant-digit estimate of each of the following.

**31.** $\dfrac{261.2 \times 38.217}{(0.54)^2}$      **34.** $\dfrac{(71,200)(0.627)}{(0.0942)(20.325)}$

**32.** $\dfrac{(-21015)(0.972)}{(1.83)(0.2716)}$      **35.** $\dfrac{(41,350)(0.49)^2}{(0.082)(51.72)}$

**33.** $\dfrac{(42,000)(0.829)}{(0.473)(2.16)}$      **36.** $\dfrac{(-693.2)(1.23)^3}{(-0.032)(0.9)^2}$

Each of the following represents a measurement in meters given in scientific notation. For each give (a) the precision, (b) the maximum possible error, and (c) the relative error as a percent.

**37.** $5 \times 10^{-3}$      **41.** $8.0 \times 10^{-2}$      **45.** $5.00 \times 10^{-2}$

**38.** $1.02 \times 10^0$      **42.** $1.6 \times 10^{-4}$      **46.** $3.20 \times 10^{-4}$

**39.** $5.0 \times 10^2$      **43.** $3.75 \times 10^6$      **47.** $5.103 \times 10^8$

**40.** $2.25 \times 10^4$      **44.** $6.25 \times 10^8$      **48.** $6.1005 \times 10^6$

*PROBLEMS*

Give each result the same number of significant digits as in the least accurate item of data.

**A**  **1.** In San Diego County, the average daily emissions of air contaminants total approximately $4.378 \times 10^6$ pounds. Of these emissions, about $3.412 \times 10^6$ pounds are emitted by motor vehicles. Find the ratio of emissions by motor vehicles to total emissions and express the ratio to the nearest tenth of a percent.

**2.** The astronomical unit (AU) used in astronomy to measure distances is equal to the average distance between the Earth and the Sun. This distance is taken as 92,897,000 miles. Express this measurement in scientific notation; state the precision and accuracy of the measurement.

**3.** The light-year is a unit of distance in astronomy and is approximately equal to $6.32 \times 10^4$ AU's (see Problem 2). Use scientific notation to state the approximate number of miles in a light-year.

**4.** The parsec is an astronomical unit approximately equal to 3.258 light years. Use the results of Problem 3 to find an approximation for the number of miles in a parsec, and state your result in scientific notation.

**5.** One Ångström Unit (Å) is approximately equal to $3.937 \times 10^{-9}$ inches. The wave length of light of maximum visibility is about 5560 Å. Use scientific notation to express the wave length in inches. What is the precision and accuracy of the inch-equivalent to 1 Å?

**6.** If an electron travels about $2.99 \times 10^{10}$ centimeters per second in a vacuum, how long will it take to cross a gap of 0.42 centimeters in a vacuum tube? What is the precision and accuracy of the given speed of an electron in a vacuum?

**7.** A coulomb of electricity contains about $6.28 \times 10^{18}$ electronic charges. About how many electronic charges are there in an Ampere-hour which is equal to 3600.0 coulombs? What is the precision and accuracy of the stated measurement of a coulomb?

**8.** If a person were to receive as wages 1 cent for the first day worked, 2 cents for the second day, 4 cents for the third day, and so on, with each days wages twice the wages of the preceding day, about how many dollars would the person receive on the 31st day worked? Use the fact that $2^{10} \doteq 10^3$ to approximate $2^{30}$. Express your result as a decimal numeral.

**B**  **9.** To the nearest inch, the length of the edge of a square is 4 inches. What are the least and greatest possible values for the area of the square?

**10.** To the nearest inch, the length of the edge of a cube is 4 inches. What are the least and greatest possible values for the volume of the cube?

## Chapter Summary

1. The laws of exponents may be extended to negative integers and zero as exponents:

$$b^{-n} = \frac{1}{b^n} \text{ for } b \neq 0 \text{ and } n \text{ a positive integer; } b^0 = 1 \text{ for } b \neq 0.$$

2. A rational algebraic expression represents a real number for all real values of the variables for which the denominator is not zero.

3. To simplify a **product** or a **quotient** of **fractions**, factor each numerator and denominator to determine common factors. Then divide both the numerator and denominator by their GCF. To simplify a **sum** or a **difference** of **fractions**, find the LCD of the fractions, express each fraction equivalently having the LCD as denominator, combine the resulting fractions, and simplify. You may simplify a **complex fraction** by multiplying numerator and denominator by the LCD of all the fractions in it, or by expressing numerator and denominator as single fractions to be divided.

4. To solve an equation containing fractions, multiply each member by the LCD. In a **fractional equation**, such multiplication may introduce a value for which the multiplier is zero and which may not belong to the solution set of the original equation.

5. A number can be represented by a terminating or repeating decimal if and only if it is a rational number.

   To **round a decimal**, add 1 to the last digit to be retained if the following digit is 5 or more; otherwise, leave the retained digits unchanged.

6. The **accuracy** of a measurement is the **relative error**, that is, the maximum possible error divided by the measurement itself.

7. To compare numbers or to estimate products and quotients rapidly, express each number in **standard**, or **scientific notation**: $a \times 10^n$ where $1 \leq |a| < 10$ and $n$ is an integer.

## Vocabulary and Spelling

Review each term by reference to the page listed.

lowest terms (simple form) (*p. 210*)
rational number (*p. 217*)
rational algebraic expression
  (*p. 217*)
invert (*p. 221*)
complex fraction (*p. 227*)
percentage (*p. 231*)
percent (%) (*p. 231*)
fractional equation (*p. 236*)
terminating (finite) decimal (*p. 242*)
repeating block (*p. 242*)

repeating (periodic) decimal (*p. 242*)
nonterminating  (infinite) decimal
  (*p. 242*)
rounding a decimal (*p. 244*)
round-off error (*p. 245*)
precision (of a measurement) (*p. 245*)
maximum possible error (*p. 245*)
accuracy (of a measurement) (*p. 245*)
relative error (*p. 245*)
standard (scientific) notation (*p. 245*)
significant digits (*p. 246*)

## *Chapter Test*

**6–1**   Simplify.

**1.** $\dfrac{-18t^3s^2}{6t^4s}$

**2.** $\dfrac{(-4x^2y)^3}{16x^5y^3}$

**6–2**   Write in simple form with positive exponents only.

**3.** $\dfrac{-6x^{-3}b^0c^5}{x^{-1}b^{-1}c^2}$

**4.** $\dfrac{7t}{t-3} + \dfrac{3t^3}{t-1}$

**6–3**   **5.** Write $(t^2 - 4)(t^2 - 2t - 3)^{-1}$ as a fraction, and give all restrictions on $t$.

**6.** Reduce $\dfrac{9y^3 - y}{3y^2 + 8y - 3}$ to lowest terms and give all restrictions on $y$.

**6–4**   **7.** Simplify: $\dfrac{x^2 - 4x}{x^2 + 2x} \cdot \left(\dfrac{x^2 - 9x + 20}{x^2 - 3x - 10}\right)^{-1}$

**8.** If $f(x) = \dfrac{1 + 3x - 18x^2}{6x^2 - 17x - 3}$ and $g(x) = \dfrac{x - 3}{3x - 1}$, find an expression for $fg(x)$ in simple form.

**6–5**   **9.** Express $\dfrac{x - y}{3x + 6y} - \dfrac{x - 2y}{4x + 8y}$ as a single fraction in lowest terms.

**10.** If $f(x) = \dfrac{x}{6x - 3}$ and $g(x) = \dfrac{3}{12x - 6}$, express $(f + g)(x)$ and $(f - g)(x)$ in simple form.

6–6  **11.** Simplify $\dfrac{x - \dfrac{9}{x}}{1 + \dfrac{x}{3}}$.

6–7  **12.** Solve $\dfrac{4x}{15} - \dfrac{2x - 1}{10} = \dfrac{1}{2}$ over $\mathfrak{R}$.

**13.** One pump can empty a cistern in 18 minutes while a second pump can empty it in 21 minutes. How long will it take to empty the cistern using both pumps at the same time?

6–8  **14.** Solve and check the system: $\dfrac{2}{x} + \dfrac{1}{y} = 3$

$$\dfrac{3}{x} - \dfrac{2}{y} = 8$$

**15.** A man drives half the distance to his destination at an average speed of 50 mph. At what average speed must he drive the rest of the way if his average speed for the whole trip is to be 60 mph?

6–9  **16.** Write $\frac{4}{15}$ as a repeating decimal.

**17.** Write $0.\overline{32}$ as a common fraction in lowest terms.

6–10  **18.** Write 0.000471 in scientific notation.

**19.** Simplify $\dfrac{(6 \times 10^2)(2 \times 10^{-3})}{3 \times 10^{-4}}$ ; express your result in decimal form.

## Chapter Review

6–1  **Quotients of Powers**     *Pages 209–212*

**1.** The Basic Property of Quotients is expressed by the equation $\dfrac{pr}{qs} = \underline{\ ?\ }$. The Cancellation Rule for fractions is expressed by the equation $\dfrac{pr}{qr} = \underline{\ ?\ }$.

Simplify.

**2.** $\dfrac{24m^2 t^3}{6mt^4}$

**3.** $\dfrac{(-4n^2 y^3)^2}{8n^4 y}$

**6–2**  **Zero and Negative Exponents**  *Pages 213–216*

**4.** For every nonzero number $a$ and every positive integer $n$, $a^{-n} = \underline{\ ?\ }$ and $a^0 = \underline{\ ?\ }$.

**5.** $0^{-3}$ is not defined, because $\underline{\ ?\ }$ by 0 is not defined.

**6.** Express $\dfrac{3a^{-2}b^3}{c^{-4}d^{-1}}$ using positive exponents only.

**7.** Written without negative exponents, $x^{-1} + y^{-1} = \underline{\ ?\ }$ and $(x + y)^{-1} = \underline{\ ?\ }$.

**6–3**  **Rational Algebraic Expressions**  *Pages 217–220*

**8.** You simplify a rational algebraic expression by dividing the numerator and denominator by their $\underline{\ ?\ }\ \underline{\ ?\ }\ \underline{\ ?\ }$.

Reduce each fraction to lowest terms.

**9.** $-\dfrac{6 - 2x}{x^2 - 9}$

**10.** $\dfrac{y^2 - 3y + 2}{y^2 - 1}$

**6–4**  **Products and Quotients**  *Pages 221–223*

Express as a single fraction in lowest terms.

**11.** $\dfrac{9xy^3}{3ay} \cdot \dfrac{8a^4 x}{(2y)^2}$

**12.** $\dfrac{15 + 2x - x^2}{1 + 2x + x^2} \div \dfrac{2x^2 - 12x + 10}{x^2 - 1}$

**6–5**  **Sums and Differences**  *Pages 224–227*

Express as a fraction in simple form.

**13.** $\dfrac{3x}{2x + 2y} - \dfrac{1}{2x + 2y} - \dfrac{x - 7}{2x + 2y}$

**14.** $\dfrac{y - 4}{6y} - \dfrac{3 + 5y}{5y} + \dfrac{1}{10y}$

If $f(x) = \dfrac{5}{x^2 + 5x}$ and $g(x) = \dfrac{10}{x^2 - 25}$, express in simple form:

**15.** $(f + g)(x)$

**16.** $(f - g)(x)$

**6–6**   **Complex Fractions**   *Pages 227–229*

Simplify.

**17.** $\dfrac{x - \dfrac{9}{x}}{1 + \dfrac{x}{3}}$
    **18.** $\dfrac{y - az^{-1}}{ay^{-1} - z}$
    **19.** $\dfrac{r - \dfrac{r^2 - 1}{r}}{1 - \dfrac{r - 1}{r}}$

**6–7**   **Rational Coefficients**   *Pages 229–235*

**20.** $\dfrac{y^2}{2} - \dfrac{7}{4}y - 1 = \dfrac{1}{4}(\underline{\quad ? \quad})$

**21.** Solve $\dfrac{n}{3} - \dfrac{3n - 5}{2} - 8 = 0$ over $\mathcal{R}$.

**22.** Graph the solution set of $\dfrac{5 + x}{12} - \dfrac{3 + x}{8} \le 0$ over $\mathcal{R}$.

**6–8**   **Fractional Equations**   *Pages 236–241*

Solve over $\mathcal{R}$.

**23.** $1 + \dfrac{12}{y^2 - 4} = \dfrac{3}{y - 2}$
    **24.** $\dfrac{4}{10 + x} + \dfrac{4}{10 - x} = \dfrac{5}{6}$

**25.** A number increased by 4 times its reciprocal equals $4\frac{1}{6}$. Find the number.

**6–9**   **Decimals for Rational Numbers**   *Pages 242–244*

**26.** Write 0.234 as a common fraction in lowest terms.

**27.** The decimal for $\frac{5}{18}$ is a __?__ decimal, because 18 contains factors other than powers of __?__ and __?__.

**28.** Write $\frac{5}{18}$ as a decimal.

**6–10**   **Approximations**   *Pages 244–249*

**29.** 7.046 is precise to the nearest __?__.

**30.** The maximum possible error of a measurement of 7.046 centimeters is __?__ centimeters.

**31.** The relative error of a measurement of 7.046 inches is about __?__ %.

**32.** In scientific notation 321,000 is written __?__, and the decimal numeral for $7.3 \times 10^{-4}$ is __?__.

# Extra for Experts

## SYNTHETIC DIVISION AND SUBSTITUTION

Synthetic division (Section 5–8) results from the standard division process, with division of the form $x - r$, by eliminating unneeded symbols. This is illustrated below for the case $x^3 + 3x^2 - 6x + 5$ divided by $x - 2$. The unabbreviated process is:

$$
\begin{array}{r}
x^2 + 5x + 4 \\
x - 2\overline{)x^3 + 3x^2 - 6x + 5} \\
\underline{x^3 - 2x^2} \\
5x^2 - 6x \\
\underline{5x^2 - 10x} \\
4x + 5 \\
\underline{4x - 8} \\
13
\end{array}
$$

Since a polynomial is determined if you know its coefficients, you may eliminate all symbols of the form $x^n$, leaving only the coefficients. (Because the coefficient of $x$ in the divisor *must* be 1, you need not write it.)

$$
\begin{array}{r}
1 \quad 5 \quad 4 \\
-2\overline{)1 \quad 3 \quad -6 \quad 5} \\
\underline{1 \ -2} \\
5 \ -6 \\
\underline{5 \ -10} \\
4 \quad 5 \\
\underline{4 \ -8} \\
13
\end{array}
$$

Since each number printed in color is a duplicate of the number immediately above it, you may eliminate it to obtain:

$$
\begin{array}{r}
1 \quad 5 \quad 4 \\
-2\overline{)1 \quad 3 \quad -6 \quad 5} \\
\underline{-2} \\
5 \\
\underline{-10} \\
4 \\
\underline{-8} \\
13
\end{array}
$$

Now compress the array vertically to bring the segments shown in color into alignment:

$$
\begin{array}{r}
1 \quad\; 5 \quad\; 4 \\
-2)\overline{1 \quad 3 \;\; -6 \quad 5} \\
-2 \; -10 \; -8 \\
\hline
5 \quad\; 4 \quad 13
\end{array}
$$

Next eliminate the top member of each vertical pair of duplicates (shown in color), but put the first one into the bottom row (so that the new numbers shown in color are the coefficients of the quotient):

$$
\underline{-2} \quad\begin{array}{cccc}
1 & 3 & -6 & 5 \\
 & -2 & -10 & -8 \\
\hline
1 & 5 & 4 & 13
\end{array}
$$

The diagram below summarizes how, starting with the numbers in the first row, you can obtain the rest of the numbers.

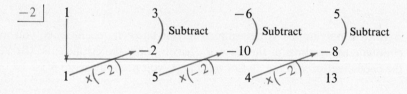

If you change the $-2$ to 2, then the subtractions are replaced by additions:

The end result is the synthetic division process used in Chapter 5:

$$
\underline{2} \quad\begin{array}{cccc}
1 & 3 & -6 & 5 \\
 & 2 & 10 & 8 \\
\hline
1 & 5 & 4 & 13
\end{array}
$$

The first three numbers in the bottom row are the coefficients of the quotient, $1x^2 + 5x + 4$, when $P(x) = x^3 + 3x^2 - 6x + 5$ is divided by $x - 2$, and the last number, 13, is the remainder. By the Factor Theorem (page 204), $P(2) = 13$.

This synthetic substitution process can be carried out in a different way, illustrated as follows:

If $x = 2$, then

$$1 \cdot x^3 + 3x^2 - 6x + 5 =$$
$$x \cdot x^2 + 3x^2 - 6x + 5 =$$
$$2 \cdot x^2 + 3x^2 - 6x + 5 =$$
$$5x^2 - 6x + 5 =$$
$$5 \cdot x \cdot x - 6x + 5 =$$
$$5 \cdot 2x - 6x + 5 =$$
$$4x + 5 =$$
$$4 \cdot 2 + 5 = 13 = P(2)$$

The process above was carried out according to the following general rule:

Beginning with the term of highest degree, write $x^k = x \cdot x^{k-1}$, substitute $r$ for the first factor, and combine with the term involving $x^{k-1}$. Continue the process until a constant is obtained. The constant is $P(r)$.

After some practice you can carry the process out mentally.

# Exercises

Use the rule stated above to find $P(r)$ in each of the following cases.

**1.** $P(x) = x^3 - x^2 - 3x + 1;\ r = 3$

**2.** $P(x) = x^3 - x^2 - 3x + 1;\ r = -2$

**3.** $P(x) = x^4 + 3x^3 - 7x^2 - 2x + 5;\ r = 2$

**4.** $P(x) = x^4 + 4x^3 + 3x^2 + 5x + 8;\ r = -2$

**5.** Apply the general rule stated above to divide the polynomial

$$P(x) = ax^3 + bx^2 + cx + d$$

by $x - r$.

**6.** Apply the general rule stated above to divide the polynomial

$$P(x) = ax^4 + bx^3 + cx^2 + dx + e$$

by $x - r$.

# Évariste Galois

Group theory is one of the most powerful ideas in modern mathematics. The study of abstract groups has clarified the structure of the number system, uncovered surprising similarities between the underlying structures of algebra and geometry, and helped to explain physical phenomena such as the behavior of electrons in an atom. The term "group" was first used in its mathematical sense by a 21-year-old French mathematician, Évariste Galois (pronounced Gal-wah), in his feverishly scribbled "will," written the night before his death. Instead of dividing up his property, Galois's "will" left to the world something far more important—ideas. It was a summary of some of Galois's mathematical discoveries which never before had been written down and which were to have a profound effect upon mathematics.

The brief life of Galois, who was born near Paris in 1811, was one of continual frustration. Twice he tried to enter the École Polytechnique, then the leading school for training mathematicians, but he was not admitted. Although he was already doing highly original work, his genius was not recognized by those who administered the examinations. By the age of seventeen, he had made a number of noteworthy mathematical discoveries, which he submitted to the Academy of Sciences, but through carelessness his paper was never presented and was later lost. Two years later he again submitted some significant work, which was lost before being read.

Discouraged with scholarly pursuits, Galois became a radical in politics and supported a party opposed to the King, Louis Philippe. He was imprisoned on trumped-up charges as a dangerous revolutionist and was later released only to be maneuvered by his political enemies into the senseless duel of honor in which he was killed. Before his death Galois wrote to a friend, "Preserve my memory, since fate has not given me life enough for my country to know my name." Today, mathematicians throughout the world know his name for his work with groups and with the conditions for the solution of algebraic equations.

# Mathematics
# and
# Medicine

A century ago, most medical discoveries were made by practicing physicians, but today, medical science is so complex that a medical practitioner spends hours each week just in reading the professional journals to keep abreast of new findings, and the bulk of medical investigation must now be carried on by full-time scientists. Some of these workers have been educated as doctors; others, however, enter medical research from other fields of science.

Medicine is, in fact, heavily dependent on the "basic sciences." Pre-med students usually take a number of science courses, and the first-year medical school curriculum is largely devoted to such subjects as anatomy, physiology, biochemistry, and biomathematics.

Because it is concerned with a very broad area of learning, medical science is prone to increasing specialization. Thus, it is not surprising that medical research often enlists the aid of specialists from fields other than medicine. A neurology laboratory, for example, might employ an electrical engineer to design equipment for measuring the charge involved in nerve impulses. Or an orthopedics laboratory might employ a mineralogist to study the formation and atomic structure of human bone tissue. Many such specialists work in cooperation with medical personnel, each investigator attacking a small part of a complex problem. With an efficient pooling of resources, research teams of this sort can probably accomplish more in a few years than individuals working independently can hope to achieve in a lifetime.

In laboratories such as the one shown, scientists study the hormones which control growth of cells. Among the various anticipated applications of this type of research are treatment of cancer, healing of wounds, lowering of blood cholesterol, and resistance to infection.

# 7

# Irrational and Complex Numbers

## POWER FUNCTIONS AND VARIATION

### 7–1 Direct Variation

Figure 7–1 shows several points of the graph of the function $g(x) = x^2$, that is, of the equation $y = x^2$. As additional points are plotted the pattern shown in Figure 7–2 emerges.

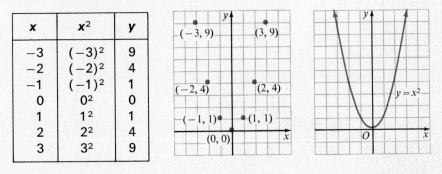

| x | x² | y |
|---|-----|---|
| −3 | $(-3)^2$ | 9 |
| −2 | $(-2)^2$ | 4 |
| −1 | $(-1)^2$ | 1 |
| 0 | $0^2$ | 0 |
| 1 | $1^2$ | 1 |
| 2 | $2^2$ | 4 |
| 3 | $3^2$ | 9 |

FIGURE 7–1          FIGURE 7–2

Functions of the form $f(x) = x^n$ are called **power functions**. Thus Figure 7–2 shows the graph of the power function of degree 2. By plotting enough points to enable you to draw smooth curves, you would find that the equations $y = x^3$, $y = x^4$, and $y = x^5$ have the graphs shown in Figure 7–3. These are, of course, the graphs of the power functions of degrees 3, 4, and 5.

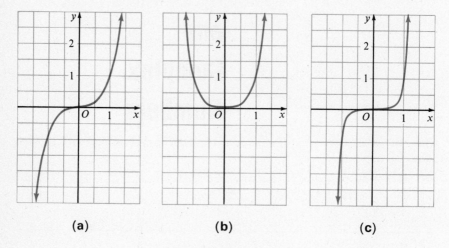

| (a) | (b) | (c) |

FIGURE 7–3

Equations of the form $y = ax^n$ define functions closely related to power functions. You can construct the graph of $y = 2x^2$, for example, by doubling the ordinate of each point on the graph of $y = x^2$ (see Figure 7–4).

Functions of the type just discussed occur in many practical problems involving *variation*. If $y = ax^n$ ($n > 0$), we say that $y$ varies directly as $x^n$, or that $y$ is directly proportional to the $n$th power of $x$, and we call $a$ the constant of proportionality.

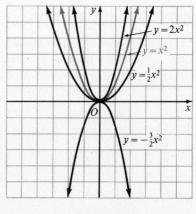

FIGURE 7–4

**Example 1.** The force exerted by the wind on a sail of fixed area varies directly as the square of the wind velocity. If a wind of 10 miles per hour exerts a force of 300 pounds, what is the force due to a 15-mile-per-hour wind?

*Solution:* Let $F$ be the force, in pounds, exerted by a wind of $v$ miles per hour. Use either method at the top of the next page.

| First Method | Second Method |
|---|---|
| $F = av^2$ | $\dfrac{F_1}{v_1^2} = \dfrac{F_2}{v_2^2}$ |
| $300 = a \cdot 10^2$ | |
| $300 = 100a$ | $\dfrac{300}{10^2} = \dfrac{F_2}{15^2}$ |
| $3 = a$ | $\dfrac{300}{100} = \dfrac{F_2}{225}$ |
| $F = 3v^2$ | |
| $F = 3 \cdot 15^2 = 675$ | $F_2 = 3 \cdot 225 = 675$ |

$\therefore$ 675 pounds force is exerted by a 15 mph wind.

## ORAL EXERCISES

State whether or not $y$ varies directly as a positive power of $x$. If it does, give the constant of proportionality.

**1.** $y = 4x^5$      **5.** $y = -5x^{15}$      **9.** $y = 3 + x^{-2}$

**2.** $y = 7x^2 - 3$      **6.** $y = \frac{1}{5}x^4$      **10.** $y = 5 + \dfrac{3}{x^{-1}}$

**3.** $y = \dfrac{6}{x^2}$      **7.** $y = 3x^{-2}$

**4.** $y = 3 - 2x$      **8.** $y = \dfrac{4}{x^{-3}}$

Given that $f(x) = x^n$, state $f(2)$ if:

**11.** $n = 2$      **12.** $n = 3$      **13.** $n = 4$      **14.** $n = 5$

Given that $f(x) = 3x^n$, state $f(-1)$ if:

**15.** $n = 2$      **16.** $n = 5$      **17.** $n = 10$      **18.** $n = 17$

Given that $f(x) = 4x^n$, state $f(\frac{1}{2})$ if:

**19.** $n = 2$      **20.** $n = 3$      **21.** $n = 4$      **22.** $n = 5$

## WRITTEN EXERCISES

By plotting selected points, graph each pair of equations on the same set of axes.

**A**

**1.** $y = 2x^2$; $y = -2x^2$      **5.** $y = 2x^2$; $y = 2x^3$

**2.** $f(x) = 3x^2$; $g(x) = -3x^2$      **6.** $y = 2x^2$; $y = 2x^4$

**3.** $g(x) = \frac{1}{4}x^2$; $h(x) = 4x^2$      **7.** $y = x^3$; $y = x^5$

**4.** $y = -\frac{1}{3}x^2$; $y = -3x^2$      **8.** $y = -x^2$; $y = -x^4$

Find the constant *a*, assuming that the given ordered pair is in the power function defined by the given equation.

**9.** $(3, 18)$; $y = ax^2$

**10.** $(3, 108)$; $y = ax^3$

**11.** $(4, 64)$; $y = 2ax^3$

**12.** $(5, 125)$; $y = ax^4$

**13.** $(4, -8)$; $y = 4ax^3$

**14.** $(2, -256)$; $y = 2ax^5$

**15.** If $y$ varies directly as $x^2$, and $y = 2$ when $x = 1$, find $y$ when $x = 10$.

**16.** If $y$ varies directly as $x^3$, and $y = 96$ when $x = 4$, find $y$ when $x = 2$.

**17.** If $r$ varies directly as $t^3$, and $r = -36$ when $t = 3$, find $r$ when $t = -2$.

**18.** If $z$ varies directly as $y^4$, and $z = 9$ when $y = 2$, find $z$ when $y = 5$.

**19.** If $A$ varies directly as $r^2$, and $A = \pi$ when $r = 1$, find $A$ when $r = 5$.

**20.** If $T$ varies directly as the cube of $h$, and $T = \frac{4}{3}$ when $h = 2$, find $T$ when $h = 3$.

**B** **21.** If $y$ varies directly as $x^2$, and $y = 36$ when $x = 2$, find $x$ when $y = \frac{4}{81}$.

**22.** If $y$ varies directly as $x^3$, and $y = 1$ when $x = \frac{1}{2}$, find $x$ when $y = 27$.

**23.** If $s$ varies as the square of $r$, and $r$ is proportional to $t$, what happens to the value of $s$ when the value of $t$ is doubled?

**24.** If $A$ varies as the square of $x$, and $x$ is proportional to $y^2$, what happens to the value of $A$ when the value of $y$ is doubled?

**25.** If $k$ varies directly as the cube of $t$, determine how $k$ changes when $t$ is halved.

**26.** If $R$ varies as the fourth power of $S$, determine how $R$ changes when $S$ is tripled.

**C** **27.** If $\left(a, \dfrac{a}{4}\right)$ is in the power function defined by $y = ax^2$, $a \neq 0$, determine all possible real-number values of $a$.

**28.** If $\left(2a, \dfrac{a}{2}\right)$ is in the power function defined by $y = ax^2$, $a \neq 0$, determine all possible real-number values of $a$.

*PROBLEMS*

**A** **1.** The wind force on a vertical surface of fixed area varies directly as the square of the wind velocity. If the pressure on a plate glass window is 30 pounds when the wind velocity is 15 miles per hour, what will be the pressure on the window when the wind velocity is 45 miles per hour?

**2.** The amount of electric energy converted into heat in a given resistor in one second varies directly as the square of the current in the resistor. If a resistor produces 18 joules of heat per second from a current of 3 amperes, how much heat per second would the resistor produce under a current of $4\frac{1}{2}$ amperes?

**3.** The kinetic energy of a moving object varies directly as the square of its speed. If a car traveling 40 miles per hour has a kinetic energy of $3 \times 10^5$ foot pounds, what will be its kinetic energy when it is traveling 60 miles per hour?

**4.** The kinetic energy of a small satellite orbiting at 5 miles per second is equivalent to $3 \times 10^4$ British Thermal Units (BTU) of heat. What is the heat equivalent of the same satellite traveling 6 miles per second? (See Problem 3.)

**5.** An object traveling in a circular path is subject to a centrifugal force that varies as the square of the velocity of the object. If a tether ball traveling at 10 feet per second exerts a centrifugal force of $\frac{1}{2}$ pound, what would be its centrifugal force if it traveled at 25 feet per second?

**6.** The rate of flow of water through a fire hose under fixed pressure is directly proportional to the square of the radius of the hose. If a two-inch hose carries 864 cubic inches of water per second, how much more water will a hose with a radius of $2\frac{1}{4}$ inches carry under the same pressure?

**7.** The amount of coal used to drive a steam turbine varies as the square of the speed of the turbine. If it takes 2 tons of coal to drive the turbine at a speed of 4 revolutions per second (rps) for 5 hours, how many tons of coal would it take to drive it at 5 rps for the same length of time?

**8.** The amount of paint needed to paint a spherical tank varies directly as the square of the radius of the tank. If it takes $2\frac{1}{2}$ gallons of paint to cover a spherical tank of radius 10 feet, how much paint would be needed to paint a tank with a radius of 12 feet?

**9.** The horsepower that can be safely transmitted by a rotating shaft at a given speed of rotation varies directly as the cube of the radius of the shaft. Compare the safe load of a shaft with a radius of 2 inches with that of a shaft of 4 inches rotating at the same speed.

**10.** The power required to propel a ship varies as the cube of its speed. Compare the power required to drive a vessel at 12 knots with that required to drive it at 18 knots.

**B** **11.** The moment of inertia of a circular disk about an axis through its center varies directly as the fourth power of the radius of the disk. Compare the moment of inertia of a disk of radius 3 inches with one of radius 9 inches.

In Exercises 12–15 use these facts:

A. The areas of similar polygons and the surface areas of similar solids are proportional to the squares of corresponding linear dimensions.
B. The volumes of similar solids are proportional to the cubes of corresponding linear dimensions.

**12.** If the length of each side of a rectangle is $\frac{2}{5}$ that of a corresponding side in a similar rectangle, compare the areas of the two rectangles.

**13.** If the length of a side of a cube is $\frac{7}{5}$ the length of a side of another cube, compare the volumes of the cubes.

**14.** A pyramid has a volume of 288 cubic meters and an edge of 6 meters. Find **(a)** the volume of a similar pyramid whose corresponding edge is 15 meters long, and **(b)** the ratio of the surface area of the first pyramid to that of the second.

**15.** The altitudes of two similar cylinders are 8 and 12 centimeters respectively. If the surface area of the smaller cylinder is 240 square centimeters, find **(a)** the surface area of the larger cylinder, and **(b)** the ratio of their volumes.

## 7–2 Inverse and Combined Variation

The volume $v$ of gas in a container is related to its pressure $p$ in such a way that the product $pv$ is constant. Figure 7–5(a) shows a glass cylinder fitted with a piston and containing 30 cubic inches of oxygen at 20 pounds per square inch (psi). In Figure 7–5(b) the pressure has been increased to 50 psi. Can you find the new volume of the gas? Since $pv = k$, a constant, you have

$$20 \cdot 30 = k,$$

or

$$k = 600.$$

Therefore,

$$pv = 600,$$

and when $p = 50$,

$$50v = 600,$$

or

$$v = 12.$$

Thus the new volume is 12 cubic inches.

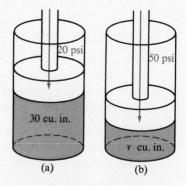

FIGURE 7–5

The expression $pv = 600$ is equivalent to $v = \dfrac{600}{p}$ or $p = \dfrac{600}{v}$ ; we say that $v$ and $p$ *vary inversely as each other*, or are *inversely proportional to each other*.

We say that $y$ **varies inversely as** $x$ if $y = \dfrac{k}{x}$ for some nonzero constant $k$, and we sometimes call a function of the form $\{(x, y): xy = k\}$ an **inverse variation**. In general, $y$ **varies inversely as** $x^n$, or $y$ **is inversely proportional to** $x^n$ if $y = \dfrac{k}{x^n}$ for some constant $k$. The case $n = 2$ is particularly important because the *inverse-square law* occurs so frequently in physical situations.

**Example.** The illumination provided by a point source of light varies inversely as the square of the distance from the source. If the illumination at a distance of 10 feet from the source is 90 foot-candles, what is the illumination at a distance of 15 feet from the source?

*Solution:* Let the illumination $d$ feet from the source be $I$ foot-candles.

| *First Method* | *Second Method* |
|---|---|
| $I = \dfrac{k}{d^2}$ | $I_1 d_1^2 = I_2 d_2^2$ |
| $90 = \dfrac{k}{10^2}$ | $90 \cdot 10^2 = I_2 \cdot 15^2$ |
| $k = 90 \cdot 10^2 = 9000$ | $9000 = 225 I_2$ |
| $I = \dfrac{9000}{d^2}$ | $I_2 = \dfrac{9000}{225} = 40$ |
| $I = \dfrac{9000}{15^2} = 40$ | |

$\therefore$ the illumination is 40 foot-candles.

The law of illumination used in Example 1 is a special case of a more general one: The illumination varies *directly* as the candlepower $P$ of the source and *inversely* as the square of the distance from the source. This means that for some constant $h$, $I = \dfrac{hP}{d^2}$ . Such variation is called **combined variation**. If $y$ varies directly with $s$ and also directly with $t$, then the equation relating these variables is of the form $y = kst$ ($k$ a constant) and you say that $y$ varies **jointly** as $s$ and $t$.

*ORAL EXERCISES*

Express each relationship in words. Assume that $k$ is the constant of proportionality.

**Sample.** $r = \dfrac{kst^2}{u}$  *Solution:* $r$ varies jointly as $s$ and the square of $t$, and inversely as $u$.

**1.** $y = kxz$   **3.** $y = \dfrac{k}{z^2}$   **5.** $V = kr^3$   **7.** $H = kri^2$

**2.** $y = \dfrac{kx}{z}$   **4.** $y = \dfrac{k}{xz^2}$   **6.** $F = \dfrac{kmn}{r^2}$   **8.** $V = \dfrac{kT}{P}$

*WRITTEN EXERCISES*

Determine the value for $C$ so that both ordered pairs belong to the same inverse variation.

**Sample.**  $(10, 25)$, $(2, C)$

*Solution:*  Since $(x_1, y_1)$ and $(x_2, y_2)$ are in the same inverse variation, you have $x_1 y_1 = x_2 y_2$; it follows that $10 \times 25 = 2C$, so that the value of $C$ is 125.

<u>A</u>  **1.** $(8, 10)$, $(C, 20)$   **5.** $(\frac{4}{7}, \frac{3}{8})$, $(C, \frac{9}{28})$

   **2.** $(15, 22)$, $(10, C)$   **6.** $(3\frac{1}{4}, 4\frac{1}{2})$, $(\frac{8}{5}, C)$

   **3.** $(-30, -14)$, $(-C, -42)$   **7.** $(6.2, 3.6)$, $(C, 4.8)$

   **4.** $(8, 18)$, $(C, C)$   **8.** $(8 \times 10^5, 3 \times 10^4)$, $(12 \times 10^7, C)$

   **9.** If $x$ varies inversely as $t$, and $x = 8$ when $t = 6$, find $t$ when $x = 12$.

   **10.** If $z$ varies inversely as $y$, and $y = 10$ when $z = 15$, find $z$ when $y = 25$.

   **11.** If $y$ varies jointly as $x$ and $z$, and $y = 8$ when $x = 2$ and $z = 2$, find $y$ when $x = 6$ and $z = 3$.

   **12.** If $r$ varies jointly as $s$ and $t$, and $r = 8$ when $s = 6$ and $t = 4$, find $s$ when $t = 16$ and $r = 24$.

   **13.** If $r$ varies directly as $s$ and inversely as $t$, and $r = 12$ when $s = 8$ and $t = 2$, find $r$ when $s = 3$ and $t = 6$.

   **14.** If $y$ varies jointly as $r$ and the square of $s$ and inversely as the square of $t$, and $y = 12$ when $r = 8$, $s = 6$ and $t = 12$, find $y$ when $r = 6$, $s = 8$, and $t = 6$.

**B** **15.** If $z$ varies directly as the square of $y$, and $y$ varies directly as $x$, what happens to $z$ if $x$ is doubled?

**16.** If $y$ varies inversely as the square of $u$, and $u$ varies directly as $t$, what happens to $y$ if $t$ is doubled?

**17.** If $r$ varies jointly as $u$ and the square of $v$, what happens to $r$ if $u$ is halved while $v$ is doubled?

**18.** If $x$ varies jointly as $u$ and $v$ and inversely as the square of $w$, what happens to $x$ if $u$, $v$, and $w$ are doubled?

**19.** In Exercise 17, what happens to $r$ if $u$ is doubled and $v$ is halved?

**20.** In Exercise 18, what happens to $x$ if $u$ and $v$ are doubled and $w$ is halved?

## PROBLEMS

**A** **1.** To produce a given amount of simple interest, the principal needed varies inversely with the interest rate. If the rate changes from $5\frac{1}{2}\%$ to $5\%$, by what amount must a principal of \$10,000 be raised to maintain the same income?

**2.** In investing in securities, a dollar-cost-averaging system dictates that the number of shares purchased be inversely proportional to the price per share. If Mrs. Thomas purchases 125 shares of a stock at \$8.00 per share one period, how many shares will be purchased the next period if the price is then \$10.00 per share?

**3.** The volume of a regular pyramid varies jointly as the area of the base and the altitude. If the volume of a pyramid with base area 24 square centimeters and altitude 8 centimeters is 64 cubic centimeters, what is the volume of a pyramid with base area 48 square centimeters and altitude 10 centimeters?

**4.** The volume of a circular cone varies jointly as the height of the cone and the square of the radius of the base. If a cone of height 24 inches and base radius 6 inches has a volume of approximately 907.2 cubic inches, what would be the approximate volume of a cone of height 20 inches and base radius 5 inches?

**5.** For belted pulleys, the speeds of the pulleys vary inversely as their radii. If a pulley with a radius of 4 inches, turning at 1452 revolutions per minute, is belted to a pulley with a radius of 6 inches, what will be the speed of the larger pulley?

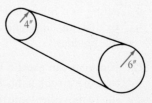

6. The speeds of meshed gears vary inversely as the numbers of teeth they have. If two meshed gears have 24 and 28 teeth, and if the gear with 24 teeth turns at 1428 revolutions per minute, what will be the speed of the gear with 28 teeth?

7. The weight of a body varies inversely as the square of its distance from the center of the earth. Assuming that the radius of the earth is about 4000 miles, about how much would a man weigh at an altitude of 150 miles above the earth's surface if he weighs 180 pounds on the earth's surface?

8. The force of repulsion of two like electrical charges is inversely proportional to the square of the distance between them. If two charges located 2.0 centimeters apart repel each other with a force of $1.5 \times 10^4$ dynes, what would be the force of repulsion if they were 5.0 centimeters apart?

**B** 9. The maximum safe uniformly distributed load for a horizontal rectangular cross-sectional beam varies jointly as the width and square of the depth of the beam, and inversely as its length. If an 8-foot beam will support up to 750 pounds when the beam is 2 inches wide and 4 inches deep, what is the maximum safe load in a similar beam 10 feet long, 2 inches wide and 6 inches deep?

10. The time required for an orbiting satellite to complete an orbit is directly proportional to the radius of the orbit, and inversely proportional to the orbital velocity. If an object requires 100 minutes to complete an orbit with a radius of 4500 miles from the center of the earth and a speed of $1.71 \times 10^4$ miles per hour, what would be the orbital time for an object traveling $1.68 \times 10^4$ miles per hour at an altitude of 4200 miles from the center of the earth?

In Exercise 11, use the fact that a lever with arms of length $d_1$ and $d_2$, supporting weights $w_1$ and $w_2$, is in balance (neglecting its own weight) if and only if

$$w_1 d_1 = w_2 d_2.$$

11. To lift a 250-pound boulder an 11-foot crowbar was placed over a stone acting as a fulcrum. How far from the boulder must the stone be placed so that an 80-pound weight on the end of the crowbar will just raise the boulder?

12. If a man can exert a force of 160 pounds, he can use a 6-foot crowbar to move a rock in his backyard. How long a bar would be needed by his son to move the same rock if the son can only exert a force of 120 pounds?

## REAL ROOTS OF REAL NUMBERS

### 7-3   Roots of Real Numbers

Does the equation $x^2 = 9$ have one or more real roots? Does $x^2 = -4$? Does $x^3 = -8$? We can use graphs to help visualize the answers to such questions. The graph of $y = x^2$, shown in Figure 7-6, is intersected by the graph of $y = 9$ in two points, $(3, 9)$ and $(-3, 9)$. When $y = 9$, $y = x^2$ becomes $9 = x^2$, and therefore the equation $x^2 = 9$ has two roots, 3 and $-3$. On the other hand, $x^2 = -4$ has no real roots because the graphs of $y = x^2$ and $y = -4$ have no points in common.

If $n$ is any *even* positive integer the graph of $y = x^n$ has the same general shape as that of $y = x^2$ (compare Figures 7-2 and 7-3(b), pages 261 and 262). Reasoning as above, we can conclude that when $n$ is *even*, $x^n = a$ has two real roots if $a > 0$ and no real roots if $a < 0$.

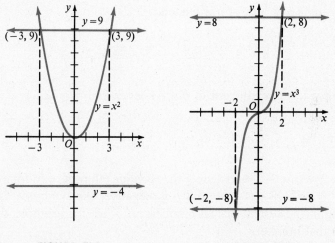

FIGURE 7-6            FIGURE 7-7

If $n$ is an *odd* positive integer, any horizontal line intersects the graph of $y = x^n$ in exactly one point. For example, Figure 7-7 shows that the graph of $y = -8$ intersects the graph of $y = x^3$ in the single point $(-2, -8)$. Thus $x^3 = -8$ has just one real root, namely $-2$. Do you see that the line $y = 8$ intersects the graph of $y = x^3$ only at $(2, 8)$ and therefore that $x^3 = 8$ has the single real root 2?

If $n$ is a positive integer, any solution of $x^n = b$ is called an **nth root of b**.

The following chart summarizes the facts about the real $n$th roots of $b$ suggested by considering points common to the graphs of $y = x^n$ and $y = b$ as we did above.

**Number and Sign of the Real $n$th Roots of $b$**

|  | $b > 0$ | $b < 0$ | $b = 0$ |
|---|---|---|---|
| **$n$ even** | One positive root<br>One negative root | No real roots | One real root,<br>namely, 0 |
| **$n$ odd** | One positive root<br>No negative roots | No positive roots<br>One negative root | One real root,<br>namely, 0 |

The symbol $\sqrt[n]{b}$ (usually read "the $n$th root of $b$") is used to denote the **principal** $n$th root of $b$, namely:

**1.** The single $n$th root of $b$ if $n$ is odd.

Examples. $\sqrt[3]{8} = 2$, $\sqrt[3]{-125} = -5$, $\sqrt[5]{32} = 2$

**2.** The *nonnegative* $n$th root of $b$ if $n$ is even and $b \geq 0$.

Examples. $\sqrt[2]{16} = 4$, $\sqrt[4]{81} = 3$, $\sqrt[6]{0} = 0$

In Case 2 we use $-\sqrt[n]{b}$ to denote the *negative* $n$th root of $b$; for example, $-\sqrt[4]{16} = -2$. If $n$ is even and $b$ is negative, the symbol $\sqrt[n]{b}$ has no meaning in the system of real numbers; $\sqrt[4]{-16}$, for example, does not name a real number.

The symbol $\sqrt[n]{b}$ is called a **radical**; $b$ is the **radicand**, and $n$ (always a positive integer) is the **index**. The index 2 is usually omitted, and the resulting symbol, $\sqrt{b}$, is read "the square root of $b$" ($\sqrt[3]{b}$ is read "the cube root of $b$"). In writing a radical be sure that the bar covers the radicand. Notice that

$$\sqrt{25 - 16} = \sqrt{9} = 3,$$

but

$$\sqrt{25} - 16 = 5 - 16 = -11.$$

Because $\sqrt[n]{b}$ is an $n$th root of $b$, you have $(\sqrt[n]{b})^n = b$. Can you simplify the expression $\sqrt[n]{b^n}$? If $n$ is odd, then $\sqrt[n]{b^n} = b$; for example, $\sqrt[3]{(-4)^3} = -4$ and $\sqrt[5]{2^5} = 2$. If $n$ is even, however, then $\sqrt[n]{b^n} = |b|$ because a radical of even index always denotes a nonnegative number (the *principal* root). Thus, $\sqrt{5^2} = |5| = 5$ and $\sqrt[4]{(-2)^4} = |-2| = 2$.

## ORAL EXERCISES

State all real roots of the given equation. If no real roots exist, so state.

**Sample.** $5x^2 = 11$

*Solution:* $\sqrt{\frac{11}{5}}$ and $-\sqrt{\frac{11}{5}}$

**1.** $3x^2 = 13$    **3.** $6x^2 = 1$    **5.** $5x^4 + 2 = 0$    **7.** $x^6 + 10 = 0$

**2.** $2x^2 = 7$    **4.** $4x^3 = -1$    **6.** $3x^4 - 7 = 0$    **8.** $x^{11} + 10 = 0$

Simplify each expression.

**Sample.** $\sqrt[3]{27}$

*Solution:* $\sqrt[3]{27} = 3$

**9.** $\sqrt[3]{8}$    **11.** $\sqrt[5]{-1}$    **13.** $\sqrt[3]{(-5)^3}$    **15.** $\sqrt[3]{-\frac{1}{125}}$

**10.** $\sqrt[4]{\frac{1}{16}}$    **12.** $\sqrt[10]{0}$    **14.** $\sqrt[4]{(-2)^4}$    **16.** $\sqrt[3]{-1000}$

State all necessary restrictions on real values of the variable for which the given sentence will be true.

**Sample.** $\sqrt{(x + 2)^2} = x + 2$

*Solution:* $x + 2 \geq 0$   or   $x \geq -2$

**17.** $\sqrt{y - 1} = 0$          **21.** $\sqrt{x^2 + 1} = x + 1$

**18.** $\sqrt{z + 2} = 0$          **22.** $\sqrt[3]{(t - 1)^2} = \sqrt{(t - 1)}$

**19.** $\sqrt[3]{(x - 1)^3} = x - 1$     **23.** $\sqrt{x^2 - 2} \in \mathcal{R}$

**20.** $\sqrt[4]{(t - 1)^4} = |t - 1|$    **24.** $\sqrt{9 - y^2} \in \mathcal{R}$

## WRITTEN EXERCISES

Use Tables 3 and 4 in the Appendix as needed to express each of the
following as a decimal to the nearest tenth.

$\boxed{A}$  **1.** $\sqrt{7}$   **4.** $\sqrt{0.49}$   **7.** $\sqrt[3]{0.038}$   **10.** $\sqrt{(3)^4}$

**2.** $-\sqrt{23}$   **5.** $\sqrt[3]{51}$   **8.** $\sqrt[3]{-0.072}$   **11.** $\sqrt[3]{(-9)^2}$

**3.** $\sqrt{0.25}$   **6.** $\sqrt[3]{92}$   **9.** $\sqrt[3]{(0.8)^2}$   **12.** $\sqrt{4^3}$

Solve over the set of rational numbers.

**Sample.** $16x^2 = 81$   *Solution:* $16x^2 = 81$
$$x^2 = \tfrac{81}{16} = \left(\tfrac{9}{4}\right)^2$$
$$x = \tfrac{9}{4} \text{ or } -\tfrac{9}{4}. \ \{\tfrac{9}{4}, -\tfrac{9}{4}\}$$

**13.** $64x^2 = 9$   **17.** $16r^4 - 1 = 0$

**14.** $225y^2 = 81$   **18.** $1 - 625y^4 = 0$

**15.** $8n^3 - 27 = 0$   **19.** $1.44n^2 - 1 = 0$

**16.** $64t^3 + 343 = 0$   **20.** $1.96t^2 - 1.21 = 0$

$\boxed{B}$  **21.** $\sqrt[2]{t^2} - t = 18$   **23.** $y - \sqrt[2]{y^2} = 46$

**22.** $\sqrt[4]{n^4} - n = 38$   **24.** $x - \sqrt[6]{x^6} + 24 = 0$

**Sample.** If $x^2 + y^2 = 10$ and $xy = 3$, find the positive value of
$x - y$.

*Solution:* 
$$\begin{array}{rcl} x^2 \phantom{{}-2xy} + y^2 &=& 10 \\ -2xy \phantom{{}+y^2} &=& -6 \\ \hline x^2 - 2xy + y^2 &=& 4 \\ (x - y)^2 &=& 4 \end{array}$$
$\therefore$ the positive value of $x - y$ is 2.

**25.** If $x^2 + y^2 = 25$ and $xy = -12$, find the positive value of $x + y$.

**26.** If $r^2 + t^2 = 32$, and $rt = -8$, find the negative value of $r + t$.

**27.** If $p^2 + q^2 = 5$ and $pq = -2$, find the positive value of $p - q$.

**28.** If $h^2 + k^2 = 12$ and $hk = 2$, find the negative value of $\dfrac{1}{h} + \dfrac{1}{k}$.

$\boxed{C}$  **29. a.** Prove: If $0 < x < 1$, then $0 < x^2 < x$ and $0 < x^3 < x$.

**b.** Use the results of **(a)** to explain why the graphs of $y = x^2$ and
$y = x^3$ are below that of $y = x$ in the interval $0 < x < 1$.

**30. a.** Prove: If $x > 1$, then $x^2 > x$ and $x^3 > x$.

    **b.** Use the result of **(a)** to explain why the graphs of $y = x^2$ and $y = x^3$ are above that of $y = x$, if $x > 1$.

**31.** Prove: If $a > b > 0$, then $\sqrt{a} > \sqrt{b}$.
    *Hint:* $a - b = (\sqrt{a} + \sqrt{b})(\sqrt{a} - \sqrt{b})$.

**32.** Prove: If $\sqrt{a} > \sqrt{b}$, then $a > b$. (*Hint:* Show that $a > \sqrt{a}\sqrt{b}$ and $\sqrt{a}\sqrt{b} > b$.) Use this result to explain why a positive number cannot have more than one positive square root.

## 7–4   Rational and Irrational Roots

When you express a rational number as a quotient of integers, you usually reduce the fraction to lowest terms; for example, you write $\frac{27}{36}$ as $\frac{3}{4}$. When you have done this, the integers in the numerator and denominator have 1 as their greatest common factor, and you call them **relatively prime**. For example, 8 and 15 are relatively prime integers, but 12 and 15 are not.

Is it possible to express $\sqrt{2}$, $\sqrt[3]{4}$, or $-\sqrt[4]{15}$ as quotients of integers? We shall answer "no" to these questions by showing that the equations $x^2 = 2$, $x^3 = 4$, and $x^4 = 15$ do not have rational roots. We shall use the following theorem, whose proof is outlined in Exercises 31 and 32.

---

**THEOREM.**   Suppose an equation has one member 0 and the other a simplified polynomial with integral coefficients. If this equation has the root $\dfrac{p}{q}$, where $p$ and $q$ are relatively prime integers, then $p$ must be an integral factor of the constant term, and $q$ must be an integral factor of the coefficient of the term of highest degree in the polynomial.

---

For example, consider the equation

$$2x^3 - 5x^2 - 4x + 3 = 0.$$

Since the numerator of a root of this equation must be a factor of $3$, and the denominator a factor of $2$, the only *possible* roots are $1$, $-1$, $3$, $-3$, $\frac{1}{2}$, $-\frac{1}{2}$, $\frac{3}{2}$, $-\frac{3}{2}$. You can easily verify that the roots are $3$, $\frac{1}{2}$, and $-1$.

Do you see that neither $\frac{3}{2}$ nor $\frac{1}{7}$ can be a root of the equation

$$4x^3 + 10x^2 + 2x + 5 = 0?$$

**Example 1.** Show that $\sqrt{2}$ is not rational.

*Solution:* Since $\sqrt{2}$ is a root of $x^2 = 2$, or

$$x^2 - 2 = 0,$$

it is sufficient to show that this equation has no rational roots. Because the constant term is $-2$ and the coefficient of the term of highest degree is 1, the only *possible* rational roots are 1, $-1$, 2, and $-2$.

If $x = 1$ or $-1$, $x^2 - 2 = 1 - 2 = -1 \neq 0$.
If $x = 2$ or $-2$, $x^2 - 2 = 4 - 2 = 2 \neq 0$.

Therefore $x^2 - 2 = 0$ has no rational roots, and since $\sqrt{2}$ *is* a root, $\sqrt{2}$ is not rational.

Real numbers which are not rational are called **irrational**. By considering the equation $x^n - k = 0$ for $n$ any integer greater than 1 and $k$ any integer, you can show that $\sqrt[n]{k}$ is irrational unless $k$ is the $n$th power of an integer. Thus, $\sqrt[4]{16} = \sqrt[4]{2^4}$ and $\sqrt[3]{-125} = \sqrt[3]{(-5)^3}$ are rational, but $\sqrt[3]{4}$ and $-\sqrt[4]{15}$ are irrational numbers.

Another method of showing a number to be irrational is illustrated below.

**Example 2.** Show that $4 + 3\sqrt{2}$ is irrational.

*Solution:* *Plan:* Show that the assumption that $4 + 3\sqrt{2}$ *is* rational leads to a contradiction and so must be false.

Suppose $4 + 3\sqrt{2}$ is rational. Then there are integers $p$ and $q$ such that

$$4 + 3\sqrt{2} = \frac{p}{q}$$

$$3\sqrt{2} = \frac{p}{q} - 4 = \frac{p - 4q}{q}$$

$$\sqrt{2} = \frac{p - 4q}{3q}$$

Since $p$, $q$, and therefore $p - 4q$ and $3q$, are integers, $\dfrac{p - 4q}{3q}$ is rational. But this contradicts the fact that $\sqrt{2}$ is irrational. Therefore, $4 + 3\sqrt{2}$ must be irrational.

## ORAL EXERCISES

State a polynomial equation in simple form that has integral coefficients and the given number as a root.

**Sample.** $\sqrt{\frac{6}{5}}$  *Solution:* $x^2 = \frac{6}{5}$; $5x^2 - 6 = 0$

**1.** $\sqrt{\frac{9}{10}}$      **4.** $\sqrt[4]{3}$      **7.** $-\sqrt{\frac{1}{3}}$      **10.** $\sqrt[3]{-\frac{3}{5}}$

**2.** $\sqrt{\frac{6}{11}}$      **5.** $\sqrt[5]{-2}$      **8.** $-\sqrt{\frac{1}{15}}$      **11.** $-\sqrt[5]{\frac{4}{5}}$

**3.** $\sqrt[3]{3}$      **6.** $\sqrt[6]{10}$      **9.** $\sqrt[3]{\frac{4}{3}}$      **12.** $-\sqrt[10]{\frac{1}{10}}$

Name all possible rational roots of the given equation.

**13.** $y^2 - 5 = 0$

**14.** $t^2 - 3 = 0$

**15.** $x^2 - 21 = 0$

**16.** $n^2 + 32 = 0$

**17.** $3y^2 - 5 = 0$

**18.** $2y^3 - 7 = 0$

**19.** $n^3 + 2n - 4 = 0$

**20.** $v^3 + 3v^2 + 6 = 0$

**21.** $2x^4 - 3x^2 + 2x - 1 = 0$

**22.** $3t^4 + 2t^3 - 3t + 5 = 0$

**23.** $6z^2 + 2z + 2 = 0$

**24.** $10k^2 - 3k - 5 = 0$

## WRITTEN EXERCISES

**A** **1–12.** Show that each of the numbers named in Oral Exercises 1–12 is irrational.

**Sample.** $\sqrt[3]{\frac{7}{10}}$

*Solution:* $\sqrt[3]{\frac{7}{10}}$ is a root of $10x^3 - 7 = 0$. The only possible rational roots of this equation are $7, -7, 1, -1, \frac{7}{2}, -\frac{7}{2}, \frac{7}{5}, -\frac{7}{5}, \frac{7}{10}$, $-\frac{7}{10}, \frac{1}{2}, -\frac{1}{2}, \frac{1}{5}, -\frac{1}{5}, \frac{1}{10}$, and $-\frac{1}{10}$. Since none of these numbers satisfies $10x^3 - 7 = 0$, none of them is equal to $\sqrt[3]{\frac{7}{10}}$ and hence $\sqrt[3]{\frac{7}{10}}$ is irrational.

Use the fact that the given number is a root of the given equation to prove that the given number is irrational.

**13.** $-3 + \sqrt{7}$; $x^2 + 6x + 2 = 0$      **15.** $2 - \sqrt{7}$; $y^2 - 4y - 3 = 0$

**14.** $1 + \sqrt{2}$; $x^2 - 2x - 1 = 0$      **16.** $4 - 2\sqrt{6}$; $t^2 - 8t - 8 = 0$

Find the rational roots of the given equation.

**B** **17.** $x^3 - 2x^2 - 5x + 6 = 0$     **20.** $3x^3 - 2x^2 + 2x + 1 = 0$

**18.** $x^3 + x^2 - 4x - 4 = 0$     **21.** $2t^4 + t^3 - 5t^2 - 3t = 0$

**19.** $4x^3 + 4x^2 - x - 1 = 0$     **22.** $3t^4 - 5t^3 - 10t^2 + 8t = 0$

Use the method of Example 2 on page 276 to prove that each of the following numbers is irrational.

**23.** $2 + \sqrt{2}$     **24.** $\frac{1}{2}(5 + \sqrt{2})$     **25.** $3\sqrt[3]{9}$     **26.** $-\dfrac{\sqrt[4]{10}}{2}$

**C** **27.** If $b$, $c$, and $d$ are integers, explain why every rational root of the equation $x^3 + bx^2 + cx + d = 0$ must be an integer.

**28.** If $b$, $c$, and $d$ are integers, explain why the only possible rational roots of $x^4 + bx^3 + cx^2 + dx + 1 = 0$ are 1 and $-1$.

**29.** Given that $\sqrt{2} + \sqrt{3}$ is a root of $x^4 - px^2 - qx + 1 = 0$, $p$ and $q$ being integers, show that $\sqrt{2} + \sqrt{3}$ is an irrational number.

**30.** Given that $\sqrt[3]{2} - \sqrt[3]{4}$ is a root of $x^3 + 6x + 2 = 0$, show that $\sqrt[3]{2} - \sqrt[3]{4}$ is an irrational number.

The proof of the theorem on page 275 is essentially the same whatever the degree of $P(x)$. The proof for degree 3 is outlined below in Exercises 31 and 32. Let

$$P(x) = ax^3 + bx^2 + cx + d,$$

where $a$, $b$, $c$, and $d$ are integers. Suppose that $p$ and $q$ are relatively prime integers and that $\dfrac{p}{q}$ is a root of $P(x) = 0$. We wish to show that $p$ is a factor of $d$ and $q$ is a factor of $a$.

**31.** Justify the following steps.

1. $a\left(\dfrac{p}{q}\right)^3 + b\left(\dfrac{p}{q}\right)^2 + c\left(\dfrac{p}{q}\right) + d = 0$
2. $ap^3 + bp^2q + cpq^2 + dq^3 = 0$
3. $ap^3 + bp^2q + cpq^2 = -dq^3$
4. $p(ap^2 + bpq + cq^2) = d(-q^3)$
5. $p$ is a factor of $d(-q^3)$.
6. $p$ is a factor of $d$.

**32.** Start with Step 2 of Exercise 31 and justify the following steps.

3'. $bp^2q + cpq^2 + dq^3 = -ap^3$
4'. $q(bp^2 + cpq + dq^2) = a(-p^3)$
5'. $q$ is a factor of $a(-p^3)$.
6'. $q$ is a factor of $a$.

## 7–5   Decimals for Real Numbers (Optional)

In Section 6–9 you studied the decimals of rational numbers. We now extend the discussion to include the decimals of all real numbers.

One method of finding successive digits in the decimal for the irrational number $\sqrt{2}$ (the length of the diagonal of the unit square) is suggested in Figure 7–8. Since the point associated with $\sqrt{2}$ on the number line is between 1 and 2, you know that $1 < \sqrt{2} < 2$. By subdividing this unit interval into ten equal parts each of length 0.1, you can see that $1.4 < \sqrt{2} < 1.5$. To verify this fact, you

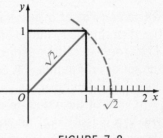

FIGURE 7–8

may compute the successive squares: $(1.1)^2$, $(1.2)^2$, and so on, until you find that $(1.4)^2 = 1.96$ and $(1.5)^2 = 2.25$.

Next divide the interval from 1.4 to 1.5 into hundredths; computing squares, you discover $(1.41)^2 = 1.9881$ and $(1.42)^2 = 2.0164$. This means

$$1.41 < \sqrt{2} < 1.42.$$

Dividing the interval from 1.41 to 1.42 into thousandths, you find that

$$1.414 < \sqrt{2} < 1.415.$$

The chart summarizes these results and indicates that each time you repeat the procedure you obtain an additional digit in the decimal for $\sqrt{2}$. Will the process lead to a terminating or periodic decimal? The answer must be "No" because a terminating or periodic decimal represents a rational number, whereas $\sqrt{2}$ is irrational.

| | |
|---|---|
| $1 < \sqrt{2} < 2$ | $\sqrt{2} \doteq 1$ |
| $1.4 < \sqrt{2} < 1.5$ | $\sqrt{2} \doteq 1.4$ |
| $1.41 < \sqrt{2} < 1.42$ | $\sqrt{2} \doteq 1.41$ |
| $1.414 < \sqrt{2} < 1.415$ | $\sqrt{2} \doteq 1.414$ |
| $\vdots$ | $\vdots$ |

The procedure used to determine successive digits in the decimal for $\sqrt{2}$ is based on the following axiom of the set of real numbers:

---

**Axiom of Completeness.** Every decimal represents a real number, and every real number has a decimal representation.

---

Do you see that the set of rational numbers is not *complete* in the sense of this axiom? For example, if you write the decimal

$$5.050050005\ldots,$$

which consists of a succession of 5's separated first by one 0, then two 0's, then three 0's, and so on, the decimal neither terminates nor repeats. Therefore, it cannot represent a rational number. However, because the set of *real* numbers is complete, this decimal does name a number, an *irrational* number.

In view of the axiom of completeness and the results of Section 6–9, you can classify real numbers according to the decimals representing them: *A rational number is a real number represented by a finite or periodic decimal; an irrational number is a real number represented by an infinite, nonperiodic decimal.*

Using decimal representations, you can illustrate the following property of the set of real numbers.

---

**Property of Density.** Between any two real numbers, there is another real number.

---

**Example.** Determine a rational number $r$ and an irrational number $s$ between $\frac{7}{5}$ and $\sqrt{2}$.

*Solution:* $\frac{7}{5} = 1.4$ and $1.41 < \sqrt{2}$. Therefore, choose $r$ and $s$ so that

$$1.4 < r \le 1.41 \quad \text{and} \quad 1.4 < s < 1.41.$$

A few choices for $r$ and $s$ appear in the table below. Can you describe the patterns in the decimals of the first two choices for $s$? The third choice is the number midway between $\frac{7}{5}$ and $\sqrt{2}$.

| $r$ | $s$ |
|---|---|
| 1.401 | 1.4040040004 . . . |
| 1.405 | 1.4071171117 . . . |
| 1.4032 | $\dfrac{\frac{7}{5} + \sqrt{2}}{2} = 0.7 + \frac{1}{2}\sqrt{2}$ |

## ORAL EXERCISES

State whether each of the following numbers is rational or irrational.

**1.** $2.4\overline{31}$

**2.** $3 - 1.0\overline{72}$

**3.** $\sqrt{2} + 1.6\overline{3}$

**4.** $\sqrt[3]{2}(0.3\overline{21})$

**5.** $\dfrac{3.0\overline{4}}{\sqrt{7}}$

**6.** $\dfrac{1 + \sqrt{2}}{0.3}$

**7.** $3.0\overline{4} + 2.\overline{40}$

**8.** $1.\overline{23} \times 3.\overline{21}$

**9.** $2.6464464446\ldots$

**10.** $1.2020020002\ldots$

**11.** The sum of the numbers in Exercises 9 and 10.

**12.** The difference of the numbers in Exercises 9 and 10.

## WRITTEN EXERCISES

Use the methods of this section to find the first three digits in the decimal for the given number.

**A**  **1.** $\sqrt{7}$  **2.** $-\sqrt{3}$  **3.** $-\sqrt[3]{-3}$  **4.** $\sqrt[3]{9}$

Insert a rational and an irrational number between the given numbers:

**5.** $\frac{1}{4}, \frac{3}{8}$

**6.** $\frac{1}{2}, \frac{2}{3}$

**7.** $\frac{9}{5}, 1.9$

**8.** $\frac{4}{3}, 1.4$

**9.** $-\frac{3}{7}, -0.43$

**10.** $-2.3, -2.\overline{3}$

**B**  **11.** $1.7, \sqrt{3}$

**12.** $\sqrt{2}, \frac{1}{2}\sqrt{10}$

**13.** $\pi, \frac{22}{7}$

**14.** $\pi, \frac{355}{113}$

(*Note:* $\pi \doteq 3.14159265\ldots$, and is irrational.)

**C**  **15.** Does the set of rational numbers have the property of density? Explain.

**16.** Does the set of integers have the property of density? Explain.

## WORKING WITH RADICALS

### 7–6  Properties of Radicals

From the facts that

$$\sqrt[3]{125 \cdot 8} = \sqrt[3]{1000} = 10 \quad \text{and} \quad \sqrt[3]{125} \cdot \sqrt[3]{8} = 5 \cdot 2 = 10,$$

you can see that

$$\sqrt[3]{125 \cdot 8} = \sqrt[3]{125} \cdot \sqrt[3]{8}.$$

Similarly, from $\sqrt{\dfrac{36}{4}} = \sqrt{9} = 3$ and $\dfrac{\sqrt{36}}{\sqrt{4}} = \dfrac{6}{2} = 3$, you can see that

$$\sqrt{\frac{36}{4}} = \frac{\sqrt{36}}{\sqrt{4}}.$$

These examples illustrate the next theorem.

---

**THEOREM.**   If $\sqrt[n]{a}$ and $\sqrt[n]{b}$ are real numbers, then,

1. $\sqrt[n]{ab} = \sqrt[n]{a} \cdot \sqrt[n]{b}$,

2. $\sqrt[n]{\dfrac{a}{b}} = \dfrac{\sqrt[n]{a}}{\sqrt[n]{b}}.$

---

*Proof of (1)*

By Law 2 of Exponents (page 210),

$$(\sqrt[n]{a} \cdot \sqrt[n]{b})^n = (\sqrt[n]{a})^n \cdot (\sqrt[n]{b})^n = a \cdot b,$$

and therefore, $\sqrt[n]{a} \cdot \sqrt[n]{b}$ is *one* of the $n$th roots of $ab$. Is it the *principal* $n$th root? It certainly is if $n$ is odd because then there is only one real $n$th root. If $n$ is even, then $\sqrt[n]{a} \geq 0$ and $\sqrt[n]{b} \geq 0$. Hence, $\sqrt[n]{a} \cdot \sqrt[n]{b} \geq 0$ and, therefore, is the principal $n$th root of $ab$ in this case as well.

The proof of (2) is similar (see Exercises 63 and 64).

**Example 1.**   Express using only one radical or no radicals:

    **a.** $\dfrac{\sqrt{30a}}{\sqrt{10a}}$   $(a > 0)$       **b.** $\sqrt[3]{9k} \cdot \sqrt[3]{3k^2}$

*Solution:*   **a.** $\dfrac{\sqrt{30a}}{\sqrt{10a}} = \sqrt{\dfrac{30a}{10a}} = \sqrt{3}$

    **b.** $\sqrt[3]{9k} \cdot \sqrt[3]{3k^2} = \sqrt[3]{9k \cdot 3k^2} = \sqrt[3]{27k^3} = \sqrt[3]{(3k)^3} = 3k$

You can sometimes use the following theorem (see Exercise 67) to replace the index of a radical by a smaller one.

---

**THEOREM.**   If each radical represents a real number, then

$$\sqrt[nq]{b} = \sqrt[n]{\sqrt[q]{b}}.$$

---

For example, $\sqrt[6]{25} = \sqrt[3 \cdot 2]{25} = \sqrt[3]{\sqrt[2]{25}} = \sqrt[3]{5}.$

---

A radical is **simplified** if its index $n$ is as small as possible and its radicand

**1.** contains no factor (other than 1) which is the $n$th power of an integer or polynomial.

**2.** contains no fraction or power with a negative exponent.

---

**Example 2.** Simplify:

$$\textbf{a. } \sqrt{\tfrac{27}{2}} \qquad\qquad \textbf{c. } \sqrt[3]{\tfrac{1}{a}}$$

$$\textbf{b. } \sqrt[4]{9z^6} \quad (z > 0) \qquad \textbf{d. } \sqrt{x^{-2} + y^{-2}} \quad (x > 0, y > 0)$$

*Solution:*

$$\textbf{a. } \sqrt{\frac{27}{2}} = \sqrt{\frac{27 \cdot 2}{2 \cdot 2}} = \sqrt{\frac{3^2 \cdot 3 \cdot 2}{2^2}} = \sqrt{\left(\frac{3}{2}\right)^2 \cdot 6} = \frac{3}{2}\sqrt{6}$$

$$\textbf{b. } \sqrt[4]{9z^6} = \sqrt{\sqrt{9z^6}} = \sqrt{\sqrt{(3z^3)^2}} = \sqrt{3z^3} = \sqrt{z^2 \cdot 3z} = z\sqrt{3z}$$

$$\textbf{c. } \sqrt[3]{\frac{1}{a}} = \sqrt[3]{\frac{1 \cdot a^2}{a \cdot a^2}} = \sqrt[3]{\frac{a^2}{a^3}} = \frac{\sqrt[3]{a^2}}{\sqrt[3]{a^3}} = \frac{\sqrt[3]{a^2}}{a} = \frac{1}{a}\sqrt[3]{a^2}$$

$$\textbf{d. } \sqrt{x^{-2} + y^{-2}} = \sqrt{\frac{1}{x^2} + \frac{1}{y^2}} = \sqrt{\frac{y^2 + x^2}{x^2y^2}} = \frac{\sqrt{y^2 + x^2}}{\sqrt{x^2y^2}}$$

$$= \frac{\sqrt{x^2 + y^2}}{xy}$$

To avoid confusing $\sqrt{2a}$ with $\sqrt{2}a$, for example, you can write $\sqrt{2}a$ as $a\sqrt{2}$. And you can write $\frac{1}{2}\sqrt{a}$ instead of $\dfrac{\sqrt{a}}{2}$ to avoid confusion with $\sqrt{\dfrac{a}{2}}$.

The next theorem (to be proved in Exercise 65) is often useful in numerical work.

---

**THEOREM.** If $m$ and $n$ are positive integers and $b$ and $\sqrt[n]{b}$ are real numbers, then

$$\sqrt[n]{b^m} = (\sqrt[n]{b})^m.$$

---

For example:
$$\sqrt[4]{16^3} = (\sqrt[4]{16})^3 = 2^3 = 8$$
$$\sqrt[3]{(-27)^2} = (\sqrt[3]{-27})^2 = (-3)^2 = 9$$

*ORAL EXERCISES*

Simplify each of the following.

**1.** $\sqrt{8}$  **4.** $\sqrt[3]{54}$  **7.** $\sqrt{\dfrac{3x^2}{4}}$  **10.** $-\sqrt{36y^3}$

**2.** $\sqrt{12}$  **5.** $\sqrt[3]{-27}$  **8.** $\sqrt{\dfrac{5r^3}{9}}$  **11.** $\sqrt[3]{8y^5}$

**3.** $\sqrt[3]{24}$  **6.** $\sqrt[3]{-125}$  **9.** $-\sqrt{25x^3}$  **12.** $\sqrt[3]{27^{-2}}$

Identify each of the following as a rational or as an irrational number.

**13.** $\dfrac{\sqrt{9}}{\sqrt{3}}$  **14.** $\dfrac{\sqrt{12}}{\sqrt{3}}$  **15.** $\dfrac{\sqrt[3]{4}}{\sqrt[3]{32}}$  **16.** $\dfrac{\sqrt[4]{16}}{\sqrt[4]{2}}$

*WRITTEN EXERCISES*

Use Tables 3 and 4 in the Appendix as needed to give an approximation correct to three significant digits of each of the following.

**Sample 1.** $\dfrac{2\sqrt{48}}{\sqrt{108}}$

*Solution:* $\dfrac{2\sqrt{48}}{\sqrt{108}} = \dfrac{2\sqrt{16}\,\sqrt{3}}{\sqrt{36}\,\sqrt{3}} = \dfrac{2\cdot 4\cdot \sqrt{3}}{6\cdot \sqrt{3}} = \dfrac{4\cdot 2\sqrt{3}}{3\cdot 2\sqrt{3}} = \dfrac{4}{3}$

$\doteq 1.33$

**Sample 2.** $\sqrt[3]{135}$

*Solution:* $\sqrt[3]{135} = \sqrt[3]{27}\cdot \sqrt[3]{5} = 3\sqrt[3]{5} \doteq 3(1.710) \doteq 5.13$

$\boxed{\text{A}}$

**1.** $5\sqrt{54}$  **6.** $\sqrt{0.61}$  **11.** $(4\sqrt{18})(2\sqrt{14})$

**2.** $2\sqrt{75}$  **7.** $\sqrt[3]{24}$  **12.** $(-5\sqrt{24})(3\sqrt{20})$

**3.** $\sqrt{\frac{25}{2}}$  **8.** $\sqrt[3]{192}$  **13.** $\dfrac{2\sqrt{162}}{3\sqrt{3}}$

**4.** $\sqrt{\frac{9}{8}}$  **9.** $\sqrt[3]{\frac{8}{5}}$  **14.** $\dfrac{5\sqrt{18}}{3\sqrt{242}}$

**5.** $\sqrt{0.7}$  **10.** $\sqrt[3]{\frac{54}{4}}$  **15.** $(3\sqrt[3]{32})(5\sqrt[3]{10})$

Express each of the following in simple radical form. You may assume that the domains of all variables in Exercises 16–60 are restricted so that the radicals denote real numbers.

**16.** $\sqrt{98}$

**17.** $\sqrt{216}$

**18.** $\sqrt[3]{108}$

**19.** $\sqrt[4]{405}$

**20.** $\sqrt{32y^4z^3}$

**21.** $\sqrt[3]{8x^4y^5}$

**22.** $\sqrt[4]{4x^5z^5}$

**23.** $\sqrt[6]{x^{-7}}$

**24.** $\sqrt[7]{z^{-8}}$

**25.** $\sqrt[6]{64x^{-9}y^{13}}$

**26.** $\sqrt[4]{162b^4c^{-5}}$

**27.** $\sqrt{8ab^2}\,\sqrt{2ab}$

**28.** $\sqrt{6x^2z^5}\,\sqrt{12x^2z^2}$

**29.** $\sqrt[3]{6ab^5}\,\sqrt[3]{12a^2b^2}$

**30.** $\sqrt[3]{-16xy^2}\,\sqrt[3]{-2x^2y}$

**31.** $\sqrt[4]{5a^4b^7}\,\sqrt[4]{125a^2b^3}$

**32.** $\dfrac{\sqrt[3]{2}}{\sqrt[3]{-108}}$

**33.** $\dfrac{8\sqrt{54a^5}}{4\sqrt{3a}}$

**34.** $\dfrac{5\sqrt{250z}}{15\sqrt{5z^2}}$

**35.** $\dfrac{\sqrt[3]{243x^2y^4}}{\sqrt[3]{36xy^2}}$

**36.** $\dfrac{\sqrt[4]{12xy^3}}{\sqrt[4]{75x^3y}}$

Simplify each radical.

**Sample.** $\sqrt[6]{a^2}$     *Solution:* $\sqrt[6]{a^2} = \sqrt[3\cdot2]{a^2} = \sqrt[3]{\sqrt[2]{a^2}} = \sqrt[3]{|a|}$

**37.** $\sqrt[6]{81}$

**38.** $\sqrt[10]{32}$

**39.** $\sqrt[4]{16x^2}$

**40.** $\sqrt[6]{x^3y^6}$

**41.** $\sqrt[8]{25y^2}$

**42.** $\sqrt[6]{125x^3y^3}$

Simplify.

**B**

**43.** $\sqrt{x^2 - 6x + 9}$

**44.** $\sqrt{y^3 + 8y^2 + 16y}$

**45.** $\sqrt{(x^2 - 4)(x + 2)}$

**46.** $\sqrt{(z^2 - 25)(z + 5)^2}$

**47.** $\sqrt[3]{\left(-\dfrac{x^{-3}}{81}\right)^{-4}}$

**48.** $\sqrt[4]{\left(\dfrac{32y^{-3}}{z^{20}}\right)^{-5}}$

**49.** $\dfrac{\sqrt[4]{8x^{-1}}}{\sqrt[4]{64x^{-4}}}$

**50.** $\dfrac{\sqrt[4]{8t^{-7}}}{\sqrt[4]{648t^{-2}}}$

**51.** $\sqrt{(y^5 - y^4)^{-1}}$

**52.** $\sqrt{(t^4 + 2t^5)^{-1}}$

**53.** $\sqrt{x^{-2} - y^{-2}}$

**54.** $\sqrt[3]{x^{-3} + y^{-3}}$

Express each of the following as a fraction with numerator free of radicals.

**55.** $\sqrt{\dfrac{5x}{7}}$

**56.** $\sqrt{\dfrac{7y}{3}}$

**57.** $\sqrt[3]{\dfrac{2(y - 2)^6}{(y - 2)}}$

**58.** $\sqrt[3]{\dfrac{2(t + 1)^9}{6(t + 1)^4}}$

**59.** $\sqrt[4]{2(x^2 + 1)}$

**60.** $\sqrt[6]{3(t^2 + 5)}$

Assuming that $a$ and $b$ denote real numbers and that $n$ represents a positive integer, prove each of the following statements, subject to the indicated additional restrictions.

C

**61.** $\sqrt[n]{ab} = \sqrt[n]{a} \cdot \sqrt[n]{b}$; $n$ odd, $a < 0$ and $b < 0$. (*Hint:* As Step 1, show that $\sqrt[n]{ab}$ and $\sqrt[n]{a} \cdot \sqrt[n]{b}$ both denote positive numbers.)

**62.** $\sqrt[n]{ab} = \sqrt[n]{a} \cdot \sqrt[n]{b}$; $n$ odd, and one of the numbers $a$, $b$ negative. (*Hint:* As Step 1, show that $\sqrt[n]{ab}$ and $\sqrt[n]{a} \cdot \sqrt[n]{b}$ both denote nonpositive numbers.)

**63.** $\sqrt[n]{\dfrac{a}{b}} = \dfrac{\sqrt[n]{a}}{\sqrt[n]{b}}$; $a \geq 0, b > 0$.

**64.** $\sqrt[n]{\dfrac{a}{b}} = \dfrac{\sqrt[n]{a}}{\sqrt[n]{b}}$; $n$ odd; $\dfrac{a}{b} < 0$.

**65.** $\sqrt[n]{b^m} = (\sqrt[n]{b})^m$; $\sqrt[n]{b}$ denotes a real number, $m$ a positive integer.

**66.** $\sqrt[n]{b^{-m}} = (\sqrt[n]{b})^{-m}$; $b \neq 0$; $\sqrt[n]{b}$ denotes a real number; $m$ a positive integer. (*Hint:* Apply the quotient property of radicals to $\sqrt[n]{b^{-m}}$ or $\sqrt[n]{\dfrac{1}{b^m}}$, and use the result of Exercise 65.)

**67.** $\sqrt[nq]{b} = \sqrt[n]{\sqrt[q]{b}}$. (*Hint:* Let $h = \sqrt[nq]{b}$ and $k = \sqrt[n]{\sqrt[q]{b}}$. Show that (1) $h^{nq} = k^{nq}$ and (2) $h$ and $k$ have the same sign. In showing (2), note that if $b < 0$, then both $n$ and $q$ are odd.)

*PROBLEMS*

Express each answer in simple radical form, and then approximate it to the nearest integer. When necessary, use $\pi \doteq \frac{22}{7}$.

A

**1.** Find the thickness $t$ in inches of a horizontal beam needed to support a weight $W$ of 1200 pounds. $\left( t = \dfrac{1}{5} \sqrt{\dfrac{W}{6}} \right)$

**2.** Find the radius $r$ of a sphere whose surface area $S$ is 308 square inches. $\left( r = \dfrac{1}{2} \sqrt{\dfrac{S}{\pi}} \right)$

**3.** Find the time $t$ in seconds required for a freely falling body to fall a distance $s$ of 300 feet. $\left( t = \frac{1}{4}\sqrt{s} \right)$

**4.** Find the vertical velocity $v$ in feet per second needed to fire a projectile to an altitude $h$ of 800 feet. $\left( v = 8\sqrt{h} \right)$

**5.** The volume $V$ of a spherical asteroid is 22 cubic miles. Find its radius $r$. $\left( V = \frac{4}{3}\pi r^3 \right)$

**6.** The surface area $S$ of a balloon requires 616 square feet of fabric for a cover. Find the diameter $d$ of the balloon. ($S = \pi d^2$)

**B**　**7.** Find the speed $v$ in meters per second of an object which, after starting from rest, has fallen a distance $s$ of 20 meters subject to an acceleration $g$ of 9.8 meters per second per second. ($v^2 = 2gs$)

**8.** Rhoda derives $a$ dollars from her business one year and $b$ dollars the next. By varying the use of her resources, she can determine $a$ and $b$ according to the formula $b = 10000 - \dfrac{a^2}{4900}$. If she wants to take $7500 next year, how much can she take this year?

**9.** A cube and a sphere have the same surface area, 60.0 square feet. What is the ratio of their volumes? (For a sphere of radius $r$, surface area $S$, and volume $V$, $S = 4\pi r^2$ and $V = \frac{4}{3}\pi r^3$.)

**10.** According to Kepler's third law of planetary motion, the cubes of the average distances of the planets from the sun are proportional to the squares of their times of one revolution around the sun. If Saturn is half as far from the sun as Uranus, find the ratio of their times of revolution.

## 7–7　Expressions Involving Radicals

The distributive law enables you to write as a single term any sum of radicals having the same index and radicand:

$$3\sqrt[3]{a} - 7\sqrt[3]{a} = (3 - 7)\sqrt[3]{a} = -4\sqrt[3]{a}$$

On the other hand, radicals with different indexes or radicands cannot be combined. Thus, in the sum $\sqrt[3]{4} + \sqrt[3]{5} + \sqrt{5}$, no two of the terms can be combined.

---

To simplify sums involving radicals:

1. Simplify each radical.
2. Combine terms in which the radicals have the same index and radicand.

---

**Example 1.**　Simplify: $3a\sqrt{8a} + \sqrt{12a^3} - 4\sqrt{2a^3}$　　$(a > 0)$

*Solution:*

$$\begin{aligned}
3a\sqrt{8a} + \sqrt{12a^3} - 4\sqrt{2a^3} &= 3a\sqrt{4 \cdot 2a} + \sqrt{4a^2 \cdot 3a} - 4\sqrt{a^2 \cdot 2a} \\
&= 6a\sqrt{2a} + 2a\sqrt{3a} - 4a\sqrt{2a} \\
&= 2a\sqrt{2a} + 2a\sqrt{3a} \\
&= 2a(\sqrt{2a} + \sqrt{3a})
\end{aligned}$$

You often use the formula $\sqrt[n]{a} \cdot \sqrt[n]{b} = \sqrt[n]{ab}$ (page 282) when you multiply expressions containing radicals.

**Example 2.** Perform the indicated multiplication and simplify the result:

$$(2 + \sqrt{5})(\sqrt{10} - \sqrt{2})$$

*Solution:* $\quad (2 + \sqrt{5})(\sqrt{10} - \sqrt{2}) = 2\sqrt{10} - 2\sqrt{2} + \sqrt{50} - \sqrt{10}$

$$= 2\sqrt{10} - 2\sqrt{2} + 5\sqrt{2} - \sqrt{10}$$

$$= \sqrt{10} + 3\sqrt{2}$$

To **rationalize the denominator** of a fraction you transform it into an equivalent fraction with denominator free of radicals. For example, to rationalize the denominator of $\dfrac{3}{2\sqrt{5}}$, you can multiply numerator and denominator by $\sqrt{5}$:

$$\frac{3}{2\sqrt{5}} = \frac{3\sqrt{5}}{2\sqrt{5}\sqrt{5}} = \frac{3\sqrt{5}}{2 \cdot 5} = \frac{3\sqrt{5}}{10}$$

It is well to rationalize denominators when obtaining decimal approximations. For example,

$$\frac{2}{\sqrt{3}} \doteq \frac{2}{1.732} \quad \text{and} \quad \frac{2}{\sqrt{3}} = \frac{2 \cdot \sqrt{3}}{\sqrt{3} \cdot \sqrt{3}} = \frac{2\sqrt{3}}{3} \doteq \frac{2 \times 1.732}{3},$$

but the second calculation is easier to do because it does not involve long division.

**Example 3.** Rationalize the denominator: $\dfrac{1 - \sqrt{3}}{5 + 2\sqrt{3}}$

*Solution:* $\quad$ Use the fact that $(a + b)(a - b) = a^2 - b^2$; in this case,

$$(5 + 2\sqrt{3})(5 - 2\sqrt{3}) = 5^2 - (2\sqrt{3})^2.$$

$$\frac{1 - \sqrt{3}}{5 + 2\sqrt{3}} = \frac{1 - \sqrt{3}}{5 + 2\sqrt{3}} \cdot \frac{5 - 2\sqrt{3}}{5 - 2\sqrt{3}}$$

$$= \frac{5 - 2\sqrt{3} - 5\sqrt{3} + 2 \cdot 3}{5^2 - (2\sqrt{3})^2}$$

$$= \frac{11 - 7\sqrt{3}}{25 - 4 \cdot 3} = \frac{11 - 7\sqrt{3}}{13}$$

Using radicals, you can factor over the set of polynomials with real coefficients certain quadratic polynomials which are irreducible over the set of polynomials with integral coefficients. For example,

$$x^2 - 8 = x^2 - (\sqrt{8})^2 = (x + \sqrt{8})(x - \sqrt{8})$$
$$= (x + 2\sqrt{2})(x - 2\sqrt{2}).$$

## ORAL EXERCISES

Express as a simplified sum.

**1.** $2\sqrt{7} - 3\sqrt{7} + 6\sqrt{7}$

**2.** $\sqrt[3]{5} + 6\sqrt[3]{5} - 3\sqrt[3]{5}$

**3.** $4\sqrt[5]{6} - 8\sqrt[5]{6} + 2\sqrt[5]{6}$

**4.** $2a\sqrt{x} - 5a\sqrt{x} + 7a\sqrt{x}$

**5.** $3(5 + \sqrt{2})$

**6.** $-4(\sqrt{3} - \sqrt{2})$

**7.** $\sqrt{2}(3 + \sqrt{2})$

**8.** $-\sqrt{5}(\sqrt{5} - \sqrt{11})$

**9.** $2\sqrt[3]{3}(\sqrt[3]{2} - 5\sqrt[3]{5})$

**10.** $-5\sqrt[4]{6}(2\sqrt[4]{7} + 5\sqrt[4]{5})$

## WRITTEN EXERCISES

Simplify.

**A**

**1.** $\sqrt{8} + 5\sqrt{2}$

**2.** $\sqrt{27} - 3\sqrt{3}$

**3.** $\sqrt{20} - 2\sqrt{45} + \sqrt{80}$

**4.** $\sqrt{54} + \sqrt{150} - 5\sqrt{24}$

**5.** $\sqrt[3]{40} + \sqrt[3]{135}$

**6.** $\sqrt[3]{54} + 2\sqrt[3]{128}$

**7.** $\sqrt{4x} - \sqrt{25x} + \sqrt{x}$

**8.** $\sqrt{8y^3} - \sqrt{18y^3} + \sqrt{2y^3}$

**9.** $3\sqrt[4]{\frac{5}{16}} + 6\sqrt[4]{80}$

**10.** $6\sqrt[5]{\frac{1}{16}} - \frac{3}{8}\sqrt[5]{486}$

**11.** $\sqrt{2}(3 + \sqrt{3})$

**12.** $3\sqrt{21}(2\sqrt{3} - 4\sqrt{7})$

**13.** $(3 - \sqrt{5})(2 + \sqrt{5})$

**14.** $(1 - \sqrt{2})(2 + \sqrt{2})$

**15.** $(2\sqrt{5} + 1)(3\sqrt{5} - 2)$

**16.** $(4\sqrt{10} + 2)(5\sqrt{10} - 3)$

**17.** $(2\sqrt[3]{9} + 2)(7\sqrt[3]{9} - 1)$

**18.** $(\sqrt[4]{8} + 2)(\sqrt[4]{8} - 2)$

**19.** $(\sqrt{6} - \sqrt{3})^2$

**20.** $(\sqrt{6} + \sqrt{2})^2$

Simplify, leaving answers with rational denominators.

**21.** $\dfrac{2}{1 + \sqrt{3}}$

**22.** $\dfrac{-3}{2 - \sqrt{2}}$

**23.** $\dfrac{1}{2 - \sqrt{y}}$

**24.** $\dfrac{x}{\sqrt{x} - 2}$

**25.** $\dfrac{\sqrt{a}}{\sqrt{a} - \sqrt{b}}$

**26.** $\dfrac{\sqrt{y}}{2 - \sqrt{y}}$

**27.** $\dfrac{\sqrt{6} - 3}{2 - \sqrt{6}}$

**28.** $\dfrac{12}{4 + 3\sqrt{6}}$

**B** **29.** $\dfrac{3x}{2} \sqrt[3]{\dfrac{24}{x^2}} - \dfrac{5}{3} \sqrt[3]{81x}$

**30.** $\sqrt[3]{\dfrac{4}{x^2}} + \dfrac{1}{x} \sqrt[3]{\dfrac{x}{2}}$

**31.** $\dfrac{1}{\sqrt{y^3}} + \dfrac{1}{y} \sqrt{\dfrac{4}{y}}$

**32.** $2a^3 \sqrt[5]{\dfrac{10}{a^3}} - \dfrac{1}{4} \sqrt[5]{320a^{12}}$

**33.** $(\sqrt{a} + \sqrt{a - b})(\sqrt{a} - \sqrt{a - b})$

**34.** $(\sqrt{b} + \sqrt{a + b})(\sqrt{b} - \sqrt{a + b})$

**35.** $\dfrac{\sqrt{x - a}}{1 - \sqrt{x - a}}$

**36.** $\dfrac{\sqrt{x + y} - \sqrt{x}}{\sqrt{x + y} + \sqrt{x}}$

**37.** $\dfrac{2}{\sqrt{a + 1} - \sqrt{a}}$

**38.** $\dfrac{2c}{-b + \sqrt{b^2 - 4ac}}$

Rationalize the numerator of each fraction.

**Sample.** $\dfrac{2 - \sqrt{x}}{1 + \sqrt{x}}$

*Solution:* $\dfrac{2 - \sqrt{x}}{1 + \sqrt{x}} = \dfrac{2 - \sqrt{x}}{1 + \sqrt{x}} \cdot \dfrac{2 + \sqrt{x}}{2 + \sqrt{x}} = \dfrac{2^2 - (\sqrt{x})^2}{2 + 3\sqrt{x} + (\sqrt{x})^2}$

$= \dfrac{4 - x}{2 + 3\sqrt{x} + x}$

**39.** $\dfrac{\sqrt{2} + 1}{\sqrt{2}}$

**40.** $\dfrac{2 - \sqrt{3}}{2 + \sqrt{3}}$

**41.** $\dfrac{1 - \sqrt{x - 1}}{1 + \sqrt{x - 1}}$

**42.** $\dfrac{\sqrt{2} + \sqrt{x}}{2 + x}$

Determine the real roots of each equation.

**43.** $y\sqrt{2} - \sqrt{6} = \sqrt{24} - y\sqrt{8}$

**44.** $r\sqrt[3]{32} - \sqrt[3]{2} = \sqrt[3]{250} - \sqrt[3]{4r^3}$

**45.** $\sqrt[3]{x} + 2\sqrt[3]{x} = 5 - \sqrt[3]{8x}$

**46.** $\sqrt[4]{z} + \sqrt[4]{81z} = 4(1 + \sqrt[4]{z})$

**47.** $(n - 2)^2 = 24$

**48.** $25(5 - y)^2 = 5$

Express each of the following as a single fraction with a numerator free of radicals.

C  **49.** $\sqrt{y+1} - \dfrac{y}{\sqrt{y+1}}$    **51.** $\dfrac{-2(r^2 - s^2)}{\sqrt{r^2 + s^2}} + \sqrt{r^2 + s^2}$

**50.** $\dfrac{a}{\sqrt{a^2+1}} - \dfrac{\sqrt{a^2+1}}{a}$    **52.** $\dfrac{n}{\sqrt{n^2-1}} + \dfrac{\sqrt{n^2-1}}{n}$

Factor each of the following completely over the set of polynomials with real coefficients.

**53.** $t^2 - 32$        **55.** $n^2 + 2\sqrt{3}\,n + 3$        **57.** $z^2 + 2\sqrt{2}\,z - 30$

**54.** $z^2 - 27$        **56.** $y^2 + 4\sqrt{3}\,y + 12$        **58.** $n^2 + 2\sqrt{3}\,n - 24$

## COMPLEX NUMBERS

## 7–8 Introducing Complex Numbers

The equation $z^2 + 1 = 0$ has no solution over the set $\mathcal{R}$ of real numbers because for any real number $z$, $z^2 + 1$ is *positive*. We wish to enlarge the replacement set of $z$ to obtain a set over which $z^2 + 1 = 0$ *does* have a solution. To do this we introduce a *new* number, $i$, called the **imaginary unit**, with the property that

$$i^2 = -1.$$

Do you see that $i$ is a solution of $z^2 + 1 = 0$?

We shall require that $i$ combine in a natural way with real numbers so that such expressions as $2i$, $-5i$, $3 + 4i$, and $4 - i$ have meaning. To do this, we introduce a set of numbers of the form $a + bi$ or $a + ib$, where $a$ and $b$ are real numbers. This set is called the set $\mathcal{C}$ of **complex numbers.**

The fact that $i^2 = -1$ suggests that you write $\sqrt{-1} = i$ and call $i$ a "square root of $-1$." Using $i$, you can find complex numbers which are square roots of negative numbers. For example, since $(2i)^2 = 2^2 \cdot i^2 = 4(-1) = -4$, you have $\sqrt{-4} = 2i$. In general, if $r$ is any positive number, then $(i\sqrt{r})^2 = i^2(\sqrt{r})^2 = (-1)r = -r$, and therefore

$$\sqrt{-r} = i\sqrt{r} \qquad (r > 0).$$

Thus

$$\sqrt{-25} = 5i \qquad \text{and} \qquad \sqrt{-12} = i\sqrt{12} = 2\sqrt{3}\,i, \text{ or } 2i\sqrt{3}.$$

---

To simplify a square root radical whose radicand is a negative number:
1. Express the radical as the product of a real number and $i$.
2. Simplify the remaining radical using the methods of Section 7–6.

---

**Example.** Simplify: **a.** $\sqrt{-2} + \sqrt{-8}$   **b.** $\sqrt{-2} \cdot \sqrt{-8}$

*Solution:*   **a.** $\sqrt{-2} + \sqrt{-8} = i\sqrt{2} + i\sqrt{8} = i\sqrt{2} + 2i\sqrt{2} = 3i\sqrt{2}$
**b.** $\sqrt{-2} \cdot \sqrt{-8} = i\sqrt{2} \cdot i\sqrt{8} = i^2\sqrt{2} \cdot \sqrt{8}$
$= (-1)\sqrt{16} = -4$

In the **b** part of the Example it would be *incorrect* to write

$$\sqrt{-2} \cdot \sqrt{-8} = \sqrt{(-2)(-8)} = \sqrt{16} = 4$$

because the rule $\sqrt{a} \cdot \sqrt{b} = \sqrt{ab}$ (page 282) does *not* hold if $a$ and $b$ are negative. We therefore avoid working with such radicals and express them as complex numbers.

Notice that if $r > 0$, $i\sqrt{r}$ is a solution of $z^2 + r = 0$ because if $z = i\sqrt{r}$, then $z^2 + r = (i\sqrt{r})^2 + r = -r + r = 0$. Of course, because $(-i\sqrt{r})^2 = i^2r = (-1)r = -r$, $-i\sqrt{r}$ is also a root of $z^2 + r = 0$. Thus by introducing $i$ we made it possible to solve all equations of the form $z^2 + r = 0$. Actually, much more is true:

---

Every polynomial equation with coefficients from $\mathcal{C}$ has a solution over $\mathcal{C}$.

---

This is the **Fundamental Theorem of Algebra**. Its proof is beyond the scope of this book.

*ORAL EXERCISES*

Express each number in the form $a$ or the form $bi$ where $a, b \in \mathcal{R}$.

**1.** $\sqrt{-4}$   **5.** $\dfrac{\sqrt{-16}}{2}$   **9.** $0\sqrt{-36}$   **13.** $i^3$

**2.** $\sqrt{-9}$   **6.** $\dfrac{\sqrt{-49}}{14}$   **10.** $-1\sqrt{-1}$   **14.** $i^4$

**3.** $\sqrt{-36}$   **7.** $3\sqrt{-100}$   **11.** $(\sqrt{-1})^2$   **15.** $i^5$

**4.** $\sqrt{-64}$   **8.** $2\sqrt{-81}$   **12.** $\sqrt{(-1)^2}$   **16.** $i^6$

## WRITTEN EXERCISES

Express in the form $a$ or the form $bi$. Simplify all radicals. Assume that all variables have domain $\Re$.

**[A]**

**1.** $\sqrt{-24}$

**2.** $\sqrt{-72}$

**3.** $\sqrt{-98}$

**4.** $\sqrt{-63}$

**5.** $\sqrt{-2} + \sqrt{-50}$

**6.** $2\sqrt{-5} - \sqrt{-500}$

**7.** $4\sqrt{-20} + 2\sqrt{-125}$

**8.** $a\sqrt{-12} + a\sqrt{-27}$

**9.** $\sqrt{-\frac{3}{8}} - \frac{1}{4}\sqrt{-6}$

**10.** $\sqrt{-\frac{5}{27}} + \frac{2}{9}\sqrt{-15}$

**11.** $(\sqrt{-2})(\sqrt{-2})$

**12.** $(\sqrt{-5})(\sqrt{-5})$

**13.** $(-2\sqrt{-8})(3\sqrt{-2})$

**14.** $(4\sqrt{-12})(-2\sqrt{-3})$

**15.** $(6\sqrt{-24})(-3\sqrt{6})$

**16.** $(2\sqrt{15})(-3\sqrt{-15})$

**17.** $(2i)(3i)^2$

**18.** $(5i)(-2i^2)$

**19.** $(2\sqrt{-3})(4i)$

**20.** $(3i)(5\sqrt{-3})$

**Sample.** $\dfrac{4}{\sqrt{-3}} = \dfrac{4}{i\sqrt{3}} = \dfrac{4 \cdot i\sqrt{3}}{i\sqrt{3} \cdot i\sqrt{3}} = \dfrac{4\sqrt{3}\,i}{(-1)(3)} = -\dfrac{4\sqrt{3}}{3}\,i$

**[B]**

**21.** $\dfrac{1}{i}$

**22.** $\dfrac{3}{i}$

**23.** $\dfrac{2}{(i^2)}$

**24.** $\dfrac{3}{i^3}$

**25.** $\left(\dfrac{2}{i}\right)\left(\dfrac{3}{i^2}\right)$

**26.** $\left(\dfrac{4}{i}\right)\left(\dfrac{6}{i^3}\right)$

**27.** $\dfrac{2}{\sqrt{-2}}$

**28.** $\dfrac{5}{\sqrt{-5}}$

**29.** $\dfrac{3}{\sqrt{-8}}$

**30.** $\dfrac{2}{\sqrt{-6}}$

**31.** $\dfrac{2\sqrt{3}}{\sqrt{-6}}$

**32.** $\dfrac{3\sqrt{2}}{\sqrt{-8}}$

**33.** $\dfrac{4\sqrt{-4}}{\sqrt{-2}}$

**34.** $\dfrac{6\sqrt{-6}}{\sqrt{-3}}$

**35.** $\dfrac{1}{i^2} - \dfrac{2}{i^4}$

**36.** $\dfrac{2}{i} - \dfrac{3}{i^3}$

**37.** $\left(\dfrac{1}{\sqrt{-3}}\right)\left(\dfrac{2}{\sqrt{-3}}\right)$

**38.** $\left(\dfrac{1}{\sqrt{-5}}\right)\left(-\dfrac{3}{\sqrt{-5}}\right)$

**39.** $2\sqrt{-\frac{4}{5}} + \sqrt{-\frac{1}{5}}$

**40.** $-3\sqrt{-\frac{9}{7}} + \sqrt{-\frac{1}{7}}$

**[C]**

**41.** $2\sqrt{-25z^4} + \sqrt{-100z^4} - 3\sqrt{-4z^4}$

**42.** $\sqrt{-\dfrac{n^2}{9}} + 3\sqrt{-\dfrac{n^2}{4}} - 5\sqrt{-\dfrac{n^2}{36}}$

**43.** $\sqrt{-2z^2} - 3\sqrt{-18z^2} + 3\sqrt{-8z^2}$

**44.** $\sqrt{-2a^6} - \sqrt{-8a^6} - 3\sqrt{-18a^6}$

## 7–9  Sums and Products of Complex Numbers

The real part of the complex number $a + bi$ is $a$ and the imaginary part is $b$ (not $bi$). We usually write such complex numbers as

$$3 + (-5)i, \quad -2 + 1i, \quad 0 + 7i, \quad \text{and} \quad 4 + 0i$$

in briefer form as

$$3 - 5i, \quad -2 + i, \quad 7i, \quad \text{and} \quad 4,$$

respectively. Do you see that the real parts of these numbers are, in order, 3, −2, 0, and 4, and that their imaginary parts are −5, 1, 7, and 0? We call a complex number imaginary if its imaginary part is not 0. An imaginary number of the form $bi$ is pure imaginary.

Two complex numbers are equal if and only if their real parts are equal and their imaginary parts are equal. Thus if $h$ and $k$ are real numbers and $h + ki = 2 - i$, then $h = 2$ and $k = -1$.

**Example 1.**  Find real numbers $x$ and $y$ such that

$$(x - y) + (x + y)i = 2 - 4i.$$

*Solution:*  Solve the system:  $x - y = 2$
$$x + y = -4$$

By addition, $2x = -2$, or $x = -1$; and, substituting into the first equation, $-1 - y = 2$, or $y = -3$.

Thus the required value of $x$ is $-1$ and of $y$ is $-3$.

To add or subtract two complex numbers, you treat them as ordinary binomials. For example:

$$(2 + 3i) + (1 + 4i) = (2 + 1) + (3i + 4i) = 3 + 7i$$
$$(2 + 3i) - (1 + 4i) = (2 - 1) + (3i - 4i) = 1 - i$$

To multiply two complex numbers, you first multiply them as ordinary binomials and then use the fact that $i^2 = -1$:

$$(2 + 3i)(1 + 4i) = 2 + 8i + 3i + 12i^2$$
$$= 2 + 11i + 12(-1)$$
$$= -10 + 11i$$

**Example 2.**  Express in $a + bi$ form the sum and product of each pair of complex numbers.

**a.** $i, 4 - 2i$          **b.** $3 + i, 3 - i$

*Solution:*
 a. $i + (4 - 2i) = 4 + (i - 2i) = 4 - i$
 $i \cdot (4 - 2i) = 4i - 2i^2 = 4i - 2(-1) = 2 + 4i$
 b. $(3 + i) + (3 - i) = (3 + 3) + (i - i) = 6$
 $(3 + i) \cdot (3 - i) = 3^2 - i^2 = 9 - (-1) = 10$

The general definitions of the sum and product of two complex numbers are as follows:

$$(a + bi) + (c + di) = (a + c) + (b + d)i$$
$$(a + bi)(c + di) = (ac - bd) + (ad + bc)i$$

The rather unnatural-looking definition of the product is motivated by the following:

$$\begin{aligned}(a + bi)(c + di) &= ac + adi + bci + bdi^2 \\ &= ac + (ad + bc)i + bd(-1) \\ &= (ac - bd) + (ad + bc)i\end{aligned}$$

It can be shown that addition and multiplication of complex numbers satisfy the commutative, associative, and distributive properties (see Exercises 33–39, pages 296 and 297).

## ORAL EXERCISES

Express each of the following in the form $a$, $bi$, or $a + bi$:

1. $(3 + 6i) + (8 + 2i)$

2. $(1 - 4i) + (3 + 3i)$

3. $(2 + 5i) - (1 + 5i)$

4. $(2 - 6i) - (2 - 6i)$

5. $(3 + 2i) + (3 - 2i)$

6. $(4 + 2i) - (4 - 2i)$

7. $6(2 + 3i)$

8. $-5(3 - 2i)$

9. $(2 + \sqrt{-3}) - (1 - \sqrt{-3})$

10. $(5 + \sqrt{-2}) + (2 + \sqrt{-2})$

State real values of $x$ and $y$ for which each equation is true.

11. $x + yi = 3 - 2i$

12. $x + yi = 6 + 5i$

13. $x - yi = 2 - 3i$

14. $x - yi = 5 + i$

15. $x + 2yi = 3$

16. $2x + yi = 5i$

## WRITTEN EXERCISES

Write each expression in the form $a$, $bi$ or $a + bi$.

**A** **1.** $(8 + 2i) + (6 - i) + 3i$     **11.** $(6 - \sqrt{-2})^2$

**2.** $(2 - 6i) + (5 + 3i) + 7$     **12.** $(2 + \sqrt{-3})^2$

**3.** $(8 - 3i) - (2 - 5i) + (6 + i)$     **13.** $(1 + i^2) - (1 - i)^2$

**4.** $(3 + 10i) - (5 - 15i) + (2 + i)$     **14.** $(1 - i)^2 + (1 + i)^2$

**5.** $2(3 - 2i) + 5(8 + 2i)$     **15.** $(2 + 3i)^2 + (2 - 3i)^2$

**6.** $6(-2 + i) - 3(-5 - 2i)$     **16.** $(4 + i)^2 + (4 - i)^2$

**7.** $-2(3 + 2i) - 4(1 + i) - 3i$     **17.** $(1 + i\sqrt{2})(1 - i\sqrt{2})$

**8.** $-6(2 - i) + 3(4 - 2i) + 1$     **18.** $(2 - \sqrt{-3})(2 + \sqrt{-3})$

**9.** $(3 - 2i)(5 + i)$     **19.** $(\sqrt{2} + i)(\sqrt{2} - i)$

**10.** $(6 + 3i)(5 - i)$     **20.** $(4 + 3i)(4 - 3i)$

Express in the form $a + bi$ the sum, difference, and product of the given pair of complex numbers.

**21.** $3i, 1 - 2i$     **23.** $2 + 7i, 2 - 7i$     **25.** $1 + i, i^3 + i^4$

**22.** $1 + i, 3 - 4i$     **24.** $4 + 2i, 4 - 2i$     **26.** $1 + i, i^2 + i^3$

Find real values of $x$ and $y$ for which each sentence is true.

**B** **27.** $(2x + y) + (x - y)i = 5 - 2i$

**28.** $(x + 2y) + (2x - y)i = 7 + 4i$

**29.** $(3x - 2y) + (x + 3y)i = -7 + 5i$

**30.** $(x + 4y) + (2x - 3y)i = 8 - 6i$

**31.** $(4 - 3i)(x + yi) = 1$

**32.** $(2 + 3i)(x + yi) = 1$

Prove that the following statements are true for all real values of $a$, $b$, $c$, $d$, $e$, and $f$.

**C** **33.** $(a + bi) + (c + di) = (c + di) + (a + bi)$

**34.** $[(a + bi) + (c + di)] + (e + fi) = (a + bi) + [(c + di) + (e + fi)]$

**35.** $(a + bi) + (0 + 0i) = a + bi$

**36.** $(a + bi) + [(-a) + (-bi)] = 0 + 0i$

**37.** $(a + bi)(c + di) = (c + di)(a + bi)$

**38.** $[(a + bi)(c + di)](e + fi) = (a + bi)[(c + di)(e + fi)]$

**39.** $(a + bi)[(c + di) + (e + fi)] = (a + bi)(c + di) + (a + bi)(e + fi)$

## 7–10  Quotients and Conjugates

The following example illustrates how to put a quotient of complex numbers into the standard $a + bi$ form.

$$(1 + 2i) \div (4 + 3i) = \frac{1 + 2i}{4 + 3i} = \frac{(1 + 2i)(4 - 3i)}{(4 + 3i)(4 - 3i)}$$

$$= \frac{4 - 3i + 8i - 6i^2}{16 - 9i^2} = \frac{4 + 5i - 6(-1)}{16 - 9(-1)}$$

$$= \frac{10 + 5i}{25} = \frac{2}{5} + \frac{1}{5}i$$

Notice that we multiplied both numerator and denominator by the *conjugate* of the denominator, that is, the number $4 - 3i$ obtained from $4 + 3i$ by changing the sign of its imaginary part. In general, the conjugate of $a + bi$ is $a - bi$. Since

$$(a + bi)(a - bi) = a^2 - b^2i^2 = a^2 - b^2(-1) = a^2 + b^2,$$

we see that the product of two conjugate complex numbers is always a nonnegative real number.

Unless told to do otherwise, you should always express complex numbers in $a + bi$ form.

**Example.**  Find the reciprocal of $3 - i$.

*Solution:*  $\dfrac{1}{3 - i} = \dfrac{1 \cdot (3 + i)}{(3 - i)(3 + i)} = \dfrac{3 + i}{9 - i^2} = \dfrac{3 + i}{9 - (-1)}$

$$= \frac{3 + i}{10} = \frac{3}{10} + \frac{1}{10}i$$

With addition and multiplication defined as in Section 7–9, the system $\mathcal{C}$ of complex numbers can be shown to satisfy all of the axioms of addition and multiplication displayed on page 29, with the symbol $\mathcal{R}$ replaced by $\mathcal{C}$. For example, the additive and multiplicative identities in $\mathcal{C}$ are 0 and 1, respectively, because

$$0 + (a + bi) = (0 + 0i) + (a + bi) = (0 + a) + (0 + b)i = a + bi$$

and $\qquad\qquad 1 \cdot (a + bi) = (1 + 0i)(a + bi) = a + bi.$

The *additive inverse*, or *negative*, of a complex number is obtained by changing the signs of its real and imaginary parts; for example, the additive inverse of $2 - 5i$ is $-(2 - 5i) = -2 + 5i$.

We can find the multiplicative inverse, or reciprocal, of a nonzero complex number such as $3 - i$ by the method of the Example on page 297, or we can reason as follows:

We seek a complex number $x + iy$ such that

$$(3 - i)(x + iy) = 1$$
$$3x + 3iy - ix - i^2y = 1$$
$$(3x + y) + (-x + 3y)i = 1 + 0i.$$

Equate real and imaginary parts to obtain the system:

$$3x + y = 1$$
$$-x + 3y = 0$$

Solve the system:

$$x = 3y, \quad 3(3y) + y = 1, \quad 10y = 1$$
$$y = \tfrac{1}{10} \quad \text{and} \quad x = \tfrac{3}{10}.$$

$$\therefore x + iy = \tfrac{3}{10} + \tfrac{1}{10}i.$$

*ORAL EXERCISES*

State the conjugate of each given complex number, and the product of the number with its conjugate.

**Sample.** $3 - 2i$

*Solution:* $3 + 2i$; $3^2 + 2^2 = 13$

**1.** $1 + i$      **4.** $3 - 1i$     **7.** $4 - i$     **10.** $8$

**2.** $1 - i$     **5.** $4i$     **8.** $5 + i$     **11.** $4 + 4i$

**3.** $1 + 2i$     **6.** $2i$     **9.** $6$     **12.** $3 - 3i$

**13–24.** State the additive inverse, or negative, of each complex number given in Oral Exercise 1–12 above.

## WRITTEN EXERCISES

Express each quotient in the form $a + bi$.

**A**  1. $\dfrac{1}{1 + i}$  5. $\dfrac{-3}{2 + i}$  9. $\dfrac{-5i}{6 + i}$  13. $\dfrac{2 + 3i}{2 + i}$

2. $\dfrac{-2}{-1 + i}$  6. $\dfrac{6}{2 + 2i}$  10. $\dfrac{-2i}{3 - 2i}$  14. $\dfrac{3 - 2i}{1 - 3i}$

3. $\dfrac{3}{3 + 2i}$  7. $\dfrac{i}{1 + 2i}$  11. $\dfrac{1 + i}{1 - i}$  15. $\dfrac{1 - 3i}{i}$

4. $\dfrac{-5}{2 - i}$  8. $\dfrac{2i}{2 - 3i}$  12. $\dfrac{1 - i}{1 + i}$  16. $\dfrac{2 + 4i}{3i}$

**B**  17. $\dfrac{2 + i}{7 - 5i}$  20. $\dfrac{1}{1 + \dfrac{1}{i}}$  23. $\dfrac{1}{1 - \dfrac{1}{i^2}}$  26. $\dfrac{(2 + 3i)^2}{(3 + i)^2}$

18. $\dfrac{-2 + 3i}{1 - i}$  21. $\dfrac{i}{\dfrac{2i}{1 + i}}$  24. $\dfrac{1}{1 + \dfrac{1}{i^3}}$  27. $\dfrac{(2 + 2i)^2}{(1 + i)^2}$

19. $\dfrac{1}{\dfrac{i}{1 - i}}$  22. $\dfrac{i}{2 + \dfrac{2}{i}}$  25. $\dfrac{(2 + i)^2}{4 - 2i}$  28. $\dfrac{(1 - i)}{(1 + i)^2}$

**C**  29. $\dfrac{1 + i}{1 - i} + \dfrac{1 - i}{1 + i}$  30. $\dfrac{i}{1 + i} + \dfrac{i}{1 - i}$

**31.** Show that if $a + bi \neq 0$ , then

$$\frac{1}{a + bi} = \frac{a}{a^2 + b^2} - \frac{b}{a^2 + b^2} i.$$

**32.** Show that the conjugate of the sum of two complex numbers is the sum of the conjugates of the numbers.

**33.** Show that the conjugate of the product of two complex numbers is the product of the conjugates of the numbers.

**34.** Show that if $b > 0$, then

$$z = \sqrt{\frac{\sqrt{a^2 + b^2} + a}{2}} + i \sqrt{\frac{\sqrt{a^2 + b^2} - a}{2}}$$

is a square root of $a + bi$. (*Hint:* Square z.)

## Chapter Summary

1. When $y = ax^n$, you say that $y$ varies directly as $x^n$. The function defined by $f(x) = x^n$ is a **power function**.

2. If $y = \dfrac{a}{x^n}$, you say that $y$ **varies inversely** as $x^n$. Other types of variation are **combined variation** and **joint variation**.

3. For every positive integer $n$, any solution of $x^n = b$ is an **$n$th root** of $b$. The **radical** $\sqrt[n]{b}$ denotes the principal $n$th root of $b$. If $n$ is odd, $\sqrt[n]{b^n} = b$; if $n$ is even, $\sqrt[n]{b^n} = |b|$. For example, $\sqrt[5]{(-2)^5} = -2$, but $\sqrt[4]{(-2)^4} = 2$.

4. If a rational root of a polynomial equation in simple form with integral coefficients is expressed in lowest terms $\dfrac{p}{q}$, with $q \neq 0$, then $p$ must be an integral factor of the constant term and $q$ an integral factor of the leading coefficient. Any other real root of the equation is an **irrational number**.

5. **The axiom of completeness** guarantees that every decimal represents a real number and that every real number has a decimal representation. Decimals representing irrational numbers are neither repeating nor terminating. The **property of density** asserts that between any two real numbers, there is always another real number.

6. In simplifying radicals, you use the theorem $\sqrt[n]{ab} = \sqrt[n]{a} \cdot \sqrt[n]{b}$ and $\sqrt[n]{\dfrac{a}{b}} = \dfrac{\sqrt[n]{a}}{\sqrt[n]{b}}$ ($n$, a positive integer, and $\sqrt[n]{a}$ and $\sqrt[n]{b}$, real numbers); and the theorem: If $m$ and $n$ are positive integers and $\sqrt[n]{b}$ denotes a real number, then $\sqrt[n]{b^m} = (\sqrt[n]{b})^m$. A term containing a radical of index $n$ is in simple form only when $n$ is as small as possible and its radicand
   1. contains no factor (other than 1) which is the $n$th power of an integer or polynomial.
   2. contains no fraction or power with a negative exponent.

7. You can write the **sum of radicals** having the same index and the same radicand as a single term by using the distributive property. You can write the **product of radicals** having the same index as a single term by applying the formula $\sqrt[n]{a} \cdot \sqrt[n]{b} = \sqrt[n]{ab}$.

8. A **complex number**, $a + bi$, where $a$ and $b$ are real numbers, is a real number if $b = 0$, is an **imaginary** number if $b \neq 0$, and is a pure imaginary number if $a = 0$, $b \neq 0$. If $a$, $b$, $c$, and $d$ are real numbers, then $a + bi = c + di$ if and only if $a = c$ and $b = d$.

**9.** Complex numbers may be added and multiplied: For real numbers $a$, $b$, $c$, and $d$, $(a + bi) + (c + di) = (a + c) + (b + d)i$; and $(a + bi) \times (c + di) = (ac - bd) + (ad + bc)i$. The closure, associative, commutative, and distributive properties hold for the set of complex numbers.

**10.** You use the fact that the product of nonzero **conjugate** complex numbers is a positive real number to express the quotient of complex numbers in standard form.

### Vocabulary and Spelling

Review each term by reference to the page listed.

power function (*p. 261*)

direct variation (*p. 262*)

inverse variation (*p. 267*)

combined variation (*p. 267*)

joint variation (*p. 267*)

$n$th root (of a number) (*p. 271*)

square root (of a number) (*p. 272*)

cube root (of a number) (*p. 272*)

principal $n$th root (*p. 272*)

radical (*p. 272*)

radicand (*p. 272*)

index (*p. 272*)

relatively prime integer (*p. 275*)

irrational number (*p. 276*)

axiom of completeness (*p. 279*)

property of density (*p. 280*)

rationalize the denominator (*p. 288*)

imaginary unit (*p. 291*)

complex numbers (*p. 291*)

Fundamental Theorem of Algebra (*p. 292*)

imaginary number (*p. 294*)

pure imaginary number (*p. 294*)

conjugate (of a complex number) (*p. 297*)

additive inverse (of a complex number) (*p. 298*)

multiplicative inverse (of a complex number) (*p. 298*)

---

## Chapter Test

---

**7–1** **1.** If $y$ varies directly as $x^3$, and $y = 18$ when $x = 6$, find $y$ when $x = 3$.

**7–2** **2.** A unit electrical charge attracts a second charge with a force directly proportional to the strength of the second charge and inversely proportional to the square of the distance between the charges. What is the effect of doubling the strength of the second charge and tripling the distance between the charges?

**7–3** **3.** Solve over the set of rational numbers: $1.21y^2 - 0.64 = 0$

7–4    **4.** Given that $3 - \sqrt{7}$ is a root of $x^2 - 6x + 2 = 0$, show that $3 - \sqrt{7}$ is irrational.

7–5    **5.** (Optional) Find **(a)** a rational number and **(b)** an irrational number between 0.7 and $\dfrac{1}{\sqrt{2}}$.

Simplify.

7–6    **6.** $\sqrt[4]{\dfrac{x^2 y^6}{8}}$    **7.** $\dfrac{\sqrt{24 x^2 y^3}}{\sqrt{3 x y^2}}$

7–7    **8.** $(\sqrt{6} + 3)(\sqrt{6} - 2)$    **9.** $\dfrac{2 - \sqrt{5}}{3 + 2\sqrt{5}}$

7–8    **10.** $(\sqrt{-6})(-\sqrt{-8})$    **11.** $\dfrac{3\sqrt{-12}}{2\sqrt{-3}}$

7–9    **12.** $2(3 - 4i) + 2(-3 + 5i)$    **13.** $(5 + 2i)(3 - i)$

7–10   **14.** $\dfrac{2}{5 + 2i}$    **15.** $\dfrac{1 + i}{2 + 3i}$

---

## Chapter Review

---

7–1    **Direct Variation**    *Pages 261–266*

**1.** Functions of the form $f(x) = x^n$ are called __?__ functions.

**2.** The equation $y = a x^n$ defines a __?__ variation.

**3.** In the equation $y = a x^n$, the number $a$ is called the constant of __?__.

**4.** If $y$ varies directly as $x^4$, and $y = 32$ when $x = 4$, find $y$ when $x = 2$.

7–2    **Inverse and Combined Variation**    *Pages 266–270*

**5.** If $y$ varies jointly as $x$ and $z^2$, and $y = 300$ when $x = 3$ and $z = 5$, find $y$ when $x = \frac{1}{2}$ and $z = 4$.

**6.** If $y$ varies directly as $x^2$ and inversely as $z^3$, and if $y = \frac{1}{2}$ when $x = 6$ and $z = 4$, find $y$ when $x = 12$ and $z = \frac{1}{2}$.

**7–3** **Roots of Real Numbers**     *Pages 271–275*

**7.** If $n$ is an even positive integer, then $x^n = b$ has __?__ real root(s) if $b > 0$, and __?__ real root(s) if $b < 0$.

**8.** If $n$ is an odd positive integer, then $x^n = b$ has __?__ real root(s).

**9.** In the symbol $\sqrt[n]{b}$, $b$ is the __?__, and $n$ is the __?__.

Simplify.

**10.** $\sqrt{54x^3y^2}$          **11.** $\sqrt[3]{16x^4y}$

**7–4** **Rational and Irrational Roots**     *Pages 275–278*

**12.** When two integers have 1 as their greatest common factor, then the integers are said to be __?__ prime.

**13.** The only possible rational roots of $x^2 + 3x - 2 = 0$ are __?__, __?__, __?__, and __?__.

**14.** The number $\sqrt{16}$ is a __?__ number, while $\sqrt{15}$ is an __?__ number.

**7–5** **Decimals for Real Numbers (Optional)**     *Pages 279–281*

**15.** Because $(2.23)^2 = 4.9729$ and $(2.24)^2 = 5.0176$, you can say that __?__ $< \sqrt{5} <$ __?__.

**16.** If $a = 0.737337333733337\ldots$ and $b = 0.151551555155551\ldots$, then $a$ is __?__, $b$ is __?__, and $a + b$ is __?__.

**17.** Find **(a)** a rational number and **(b)** an irrational number between $\frac{4}{5}$ and $\frac{5}{6}$.

**7–6** **Properties of Radicals**     *Pages 281–287*

**18.** If each radical represents a real number, then $\sqrt[nq]{b} = \sqrt[n]{\underline{\;?\;}}$.

**19.** In simple form, $\sqrt[4]{6a^2b} \cdot \sqrt[4]{8ab^3} =$ __?__, and $\dfrac{\sqrt{3a}}{\sqrt{6a}} =$ __?__.

**20.** By reducing the index on $\sqrt[8]{a^6}$   $(a > 0)$, you can write it equivalently as __?__.

**7–7** **Expressions Involving Radicals**     *Pages 287–291*

Simplify.

**21.** $\sqrt[3]{16x^4} - \sqrt[3]{54x^4} + 3x\sqrt[3]{2x}$   **22.** $(3 + \sqrt{2})(-2 + \sqrt{2})$

**23.** Rationalize the denominator of $\dfrac{2\sqrt{6}}{\sqrt{2} + \sqrt{3}}$.

**7–8** **Introducing Complex Numbers** *Pages 291–293*

**24.** Since $i^2 = -1$, $i = $ __?__.

**25.** As an expression involving $i$, $\sqrt{-24} = $ __?__.

**26.** Two solutions of $z^2 + \frac{3}{4} = 0$ are __?__ and __?__.

**27.** Every polynomial equation with complex coefficients has a solution over the set of __?__ numbers.

**7–9** **Sums and Products of Complex Numbers** *Pages 294–297*

**28.** In the complex number $a + bi$, the real part is __?__ and the imaginary part is __?__.

**29.** If its imaginary part is not 0, a complex number is called a(n) __?__ number, and if its real part is 0 it is called a __?__ __?__ number.

**30.** By definition, $(a + bi) + (c + di) = $ __?__ $+$ __?__.

**31.** By definition, $(a + bi) \cdot (c + di) = $ __?__ $+$ __?__.

**7–10** **Quotients and Conjugates** *Pages 297–299*

**32.** The conjugate of $6 - 4i$ is __?__.

**33.** In the form $a + bi$, $\dfrac{3i}{2 - i\sqrt{2}} = $ __?__.

**34.** In $a + bi$ form, the reciprocal of $1 + 2i$ is __?__.

# Extra for Experts

## CONSEQUENCES OF THE FUNDAMENTAL THEOREM

The following result, which we state without proof, is a consequence of the Fundamental Theorem of Algebra (page 292):

---

**THEOREM.** Every polynomial equation with complex coefficients and nonzero degree $n$ has exactly $n$ complex roots.

---

In applying this theorem you may have to count the same number as a root more than once. For example, we regard the *five* roots of the *fifth* degree equation $x^5 - 2x^4 + x^3 = 0$ to be 0, 0, 0, 1, 1; 0 is a *triple* root, and 1 is a *double* root. [Note that $x^5 - 2x^4 + x^3 = x^3(x - 1)^2$.]

The next theorem tells you that if a polynomial equation has *real* coefficients, then its imaginary roots occur in conjugate pairs. (The proof is outlined in Exercises 20–24, page 306.)

---

**THEOREM.** If a polynomial equation with real coefficients has $a + bi$ as a root ($a$ and $b \in \mathcal{R}$, $b \neq 0$), then $a - bi$ is also a root.

---

Suppose you are told that $1 + 2i$ is a root of

$$P(x) = x^4 - 2x^3 + 7x^2 - 4x + 10 = 0.$$

Since the equation has real coefficients, you know that $1 - 2i$ is also a root. You can now solve the equation completely: By the Factor Theorem (page 204), both $x - (1 + 2i)$ and $x - (1 - 2i)$ are factors of $P(x)$. Therefore, $[x - (1 + 2i)][x - (1 - 2i)]$, or $[(x - 1) - 2i][(x - 1) + 2i]$, or $x^2 - 2x + 5$, is a factor of $P(x)$. When you carry out the division you find that

$$P(x) \div (x^2 - 2x + 5) = x^2 + 2$$

or

$$P(x) = (x^2 - 2x + 5)(x^2 + 2).$$

The roots of the depressed equation, $x^2 + 2 = 0$, are $i\sqrt{2}$ and $-i\sqrt{2}$; hence the solution set of $P(x) = 0$ is $\{1 + 2i, 1 - 2i, i\sqrt{2}, -i\sqrt{2}\}$.

---

# Exercises

---

1. If two roots of $x^3 - 2x^2 + 4x - 8 = 0$ are 2 and $-2i$, what must be the other root? Verify your answer by synthetic substitution.

2. If two roots of $x^3 - 2x + 4 = 0$ are $-2$ and $1 + i$, what must be the other root? Verify your answer by synthetic substitution.

3. A cubic equation with real coefficient has roots $-1$ and $2 + i$. What is the third root? Find such an equation.

4. A cubic equation with real coefficients has roots 2 and $i\sqrt{3}$. What is the third root? Find such an equation.

5. A fourth-degree equation with real coefficients has $i\sqrt{2}$ and $3 + i$ as roots. What are its other roots?

**6.** A sixth-degree equation with real coefficients has $i$, $-2i$, and $3i$ as roots. What are its other roots?

**7.** Given that one root of $x^3 - 7x^2 + 17x - 15 = 0$ is $2 - i$, find the remaining roots.

**8.** Given that one root of $x^3 + 3x^2 + 4x + 2 = 0$ is $-1 + i$, find the remaining roots.

**9.** Given that one root of $x^4 + 11x^3 + 43x^2 + 67x + 34 = 0$ is $-4 + i$, find the remaining roots.

**10.** Given that one root of $x^4 - 2x^3 + 10x - 25$ is $1 - 2i$, find the remaining roots.

Factor each polynomial completely over (**a**) $\mathcal{R}$; (**b**) $\mathcal{C}$.

**11.** $x^3 - x^2 + x - 1$       **13.** $x^4 - 3x^2 - 4$

**12.** $x^3 + 2x^2 + 2x + 4$       **14.** $x^4 + x^2 - 2$

State whether each assertion is true. Justify your answer.

**15.** If $2 + 3i$ is a root of a polynomial equation, so is $2 - 3i$.

**16.** Every polynomial with real coefficients and odd degree has at least one real root.

**17.** Over the set of complex numbers every polynomial of positive degree can be written as a product of linear factors.

**18.** There is a cubic polynomial with real coefficients which has $2 + i$ and $1 - 2i$ as two of its roots.

**19.** Explain why the conjugate of a real number is the number itself.

Let $a$, $b$, $c$, $d$, and $e$ denote real numbers. Prove that each statement is true.

**20.** The conjugate of the sum of $a + bi$ and $c + di$ is the sum of the conjugates of the numbers.

**21.** The conjugate of the product of $a + bi$ and $c + di$ is the product of the conjugates of the numbers.

**22.** For each positive integer $k$, the conjugate of $e(a + bi)^k$ is $e(a - bi)^k$. (*Hint:* Use Exercises 21 and 19.)

**23.** If $P(x) = a_0x^n + a_1x^{n-1} + \cdots + a_n$ is a polynomial with *real* coefficients, then the conjugate of $P(a + bi)$ is $P(a - bi)$. (*Hint:* Use Exercises 20 and 22.)

**24.** If $P(x)$ is the polynomial described in Exercise 23, and if $P(a + bi) = 0$, then $P(a - bi) = 0$.

# Carl Friedrich Gauss

The first correct proof of the Fundamental Theorem of Algebra (see pages 292 and 304) was given by the German mathematician Carl Friedrich Gauss in his doctoral dissertation for the University of Helmstedt in 1799. Gauss felt this theorem was so significant that he gave four different proofs for it during his lifetime, the last when he was 70 years old! The area of polynomial equations was just one of the many which Gauss, generally considered to be among the most outstanding and most versatile mathematicians of all time, investigated during his lifetime. He is especially recognized for his work with number theory (including congruences), complex numbers and functions of complex variables, statistics and probability, non-Euclidean and differential geometry, and astronomy, geodesy, and electromagnetism. He even invented an electric telegraph in 1833 which he used for communication between his house and the observatory at Göttingen where he worked.

Gauss was born of poor parents in Brunswick, Germany, in 1777. Although many mathematicians make their most important discoveries relatively early in life, Gauss was unusually precocious: He corrected his father's figuring of the weekly payroll when he was not yet 3, and he amazed his mathematics teacher when he was 10 by instantaneously discovering the formula for the sum of an arithmetic progression (page 549). At 12 he was challenging the foundations of Euclidean geometry, and at 15 he gave a rigorous proof of the Binomial Theorem (page 565). Despite these and a number of other discoveries, he was still undecided when he entered the University of Göttingen at 18 whether to follow a career in mathematics or philology. In 1796, however, when he was almost 20, he discovered the unexpected result that a regular polygon of 17 sides could be constructed using straightedge and compass alone. This discovery so impressed him that he decided on a career in mathematics, and some 50 years of important contributions to mathematics followed. After Gauss' death in 1855 the people of Göttingen commemorated this great mathematician and his fateful decision in 1796 by erecting a statue of him with a pedestal in the shape of a regular 17-sided polygon.

# 8

# Quadratic Equations and Functions

## SOLVING QUADRATIC EQUATIONS

### 8–1 Completing the Square

By using radicals and complex numbers, you can find the roots of many equations you could not solve before. For example:

$$x^2 - 8 = 0 \qquad\qquad x^2 + 4 = 0$$
$$x^2 - (\sqrt{8})^2 = 0 \qquad\qquad x^2 - (\sqrt{-4})^2 = 0$$
$$(x - \sqrt{8})(x + \sqrt{8}) = 0 \qquad (x - \sqrt{-4})(x + \sqrt{-4}) = 0$$
$$x = \sqrt{8} \quad\text{or}\quad x = -\sqrt{8} \qquad x = \sqrt{-4} \quad\text{or}\quad x = -\sqrt{-4}$$
$$x = 2\sqrt{2} \quad\text{or}\quad x = -2\sqrt{2} \qquad x = 2i \quad\text{or}\quad x = -2i$$

You can see that for any real number $d$, the equation $x^2 - d = 0$, or $x^2 = d$, is equivalent to the compound sentence

$$x = \sqrt{d} \quad\text{or}\quad x = -\sqrt{d},$$

which we write in combined form as

$$x = \pm\sqrt{d}$$

(read "$x$ equals plus or minus $\sqrt{d}$"). That is, $x^2 = d$ is equivalent to $x = \pm\sqrt{d}$. Example 1 illustrates how to apply this fact when the left member is the square of a linear binomial.

**Example 1.** Solve: **a.** $(x + 3)^2 = 6$    **b.** $(x - 2)^2 = -4$

*Solution:*     **a.** $(x + 3)^2 = 6$       **b.** $(x - 2)^2 = -4$

$$x + 3 = \pm\sqrt{6} \qquad\qquad x - 2 = \pm\sqrt{-4} = \pm 2i$$

$$x = -3 \pm \sqrt{6} \qquad\qquad x = 2 \pm 2i$$

$$\{-3 + \sqrt{6}, -3 - \sqrt{6}\} \qquad \{2 + 2i, 2 - 2i\}$$

Example 1 suggests that any **quadratic equation** $ax^2 + bx + c = 0$ ($a$, $b$, and $c$ constants, $a \neq 0$) can be solved if it can be put in the form

$$(x + r)^2 = d,$$

or

$$x^2 + 2rx + r^2 = d.$$

Notice that in the left-hand member of the last equation, *the constant term, $r^2$, is the square of half the coefficient of $x$, $2r$.* We use this fact in Step 2 of the solution of Example 2.

**Example 2.** Solve $x^2 + 5x + 3 = 0$.

*Solution:*                       $x^2 + 5x + 3 \qquad\quad = 0$

1. Subtract the constant term from     $x^2 + 5x + 3 - 3 = 0 - 3$
both members.                       $x^2 + 5x \qquad\qquad = -3$

2. Add the square of half the coeffi-    $x^2 + 5x + \left(\frac{5}{2}\right)^2 \;\; = -3 + \left(\frac{5}{2}\right)^2$
cient of $x$ to both members.                       $= -3 + \frac{25}{4}$

                                             $= \frac{13}{4}$

3. Write the left member as the           $\left(x + \frac{5}{2}\right)^2 = \frac{13}{4}$
square of a binomial.

4. Proceed as in Example 1.             $x + \frac{5}{2} = \pm\sqrt{\frac{13}{4}}$

$$x + \frac{5}{2} = \pm\frac{\sqrt{13}}{2}$$

$$x = -\frac{5}{2} \pm \frac{\sqrt{13}}{2}$$

$\therefore$ the solution set is $\left\{-\dfrac{5}{2} + \dfrac{\sqrt{13}}{2}, -\dfrac{5}{2} - \dfrac{\sqrt{13}}{2}\right\}$.

The method illustrated in Example 2 is called **completing the square.** If the coefficient of the term of highest degree is not 1, another step is required.

**Example 3.** Solve $4z^2 - 8z + 5 = 0$.

*Solution:* First divide both members by 4; then proceed as in Example 2.

$$4z^2 - 8z + 5 = 0$$
$$z^2 - 2z + \tfrac{5}{4} = 0$$
$$z^2 - 2z = -\tfrac{5}{4}$$
$$z^2 - 2z + 1 = -\tfrac{5}{4} + 1 = -\tfrac{1}{4}$$
$$(z - 1)^2 = -\tfrac{1}{4}$$
$$z - 1 = \pm\sqrt{-\frac{1}{4}} = \pm i\sqrt{\frac{1}{4}} = \pm\frac{i}{2}$$
$$z = 1 \pm \frac{i}{2}$$

∴ the solution set is $\left\{1 + \dfrac{i}{2},\ 1 - \dfrac{i}{2}\right\}$.

*Check:* If $z = 1 + \dfrac{i}{2}$, then:

$$4z^2 - 8z + 5 = 4\left(1 + \frac{i}{2}\right)^2 - 8\left(1 + \frac{i}{2}\right) + 5$$
$$= 4\left(1 + i + \frac{i^2}{4}\right) - 8 - 4i + 5$$
$$= 4 + 4i + i^2 - 8 - 4i + 5 = 0$$

The check that $1 - \dfrac{i}{2}$ is a root is similar.

## ORAL EXERCISES

State the value of $k$ for which each trinomial is the square of a binomial.

**1.** $y^2 - 2y + k$

**2.** $x^2 + 4x + k$

**3.** $x^2 - 8x + k$

**4.** $z^2 + 14z + k$

**5.** $t^2 + 10t + k$

**6.** $y^2 - y + k$

**7.** $t^2 + 5t + k$

**8.** $x^2 + 9x + k$

**9.** $r^2 - \tfrac{1}{2}r + k$

**10.** $v^2 + \tfrac{1}{3}r + k$

**11.** $x^2 + \tfrac{3}{2}x + k$

**12.** $z^2 + \tfrac{1}{5}z + k$

## WRITTEN EXERCISES

In all exercises in this section, express irrational roots in simple radical form and all complex solutions in the form $a + bi$.

Solve by factoring.

**Sample.** $3x^2 - 8 = 0$    *Solution:*

$$3x^2 - 8 = 0$$
$$x^2 - \tfrac{8}{3} = 0$$
$$(x - \sqrt{\tfrac{8}{3}})(x + \sqrt{\tfrac{8}{3}}) = 0$$
$$x = \sqrt{\tfrac{8}{3}} = \tfrac{2}{3}\sqrt{6} \quad \text{or} \quad x = -\sqrt{\tfrac{8}{3}} = -\tfrac{2}{3}\sqrt{6}$$
$$\therefore \text{ the solution set is } \{\tfrac{2}{3}\sqrt{6}, -\tfrac{2}{3}\sqrt{6}\}.$$

[A]

**1.** $x^2 - 15 = 0$     **4.** $x^2 + 64 = 0$     **7.** $2x^2 + 9 = 0$

**2.** $x^2 - 50 = 0$     **5.** $4x^2 - 6 = 0$     **8.** $4x^2 + 27 = 0$

**3.** $x^2 + 49 = 0$     **6.** $3x^2 - 33 = 0$

Solve by the method of Example 1 on page 310.

**9.** $(x - 3)^2 = 16$     **12.** $(2x + 1)^2 = 25$     **15.** $(x + 2)^2 = -1$

**10.** $(x - 6)^2 = 5$     **13.** $(3x + 2)^2 = 12$     **16.** $(x - 3)^2 = -48$

**11.** $(2x + 5)^2 = 9$     **14.** $(4x - 3)^2 = 45$

Solve by completing the square.

**17.** $x^2 + 4x - 12 = 0$          **22.** $3x^2 + 9x + 1 = 0$

**18.** $y^2 - y - 6 = 0$          **23.** $15y = 15 - 3y^2$

**19.** $x^2 + 3x - 1 = 0$          **24.** $6z = 2z^2 - 54$

**20.** $z^2 - 2z - 1 = 0$          **25.** $2z^2 = 4 - 3z$

**21.** $2x^2 - 6x - 9 = 0$          **26.** $4x^2 = 9 - x$

Solve by any method.

[B]

**27.** $x^2 = \tfrac{15}{4} - x$          **31.** $1.1x^2 - 1.2x = -0.3$

**28.** $\dfrac{z^2 - 3}{2} + \dfrac{z}{4} = 1$          **32.** $6y^2 - 0.7y + 0.02 = 0$

**33.** $y^2 + \sqrt{5}\,y - 1 = 0$

**29.** $\dfrac{2y^2 - 21}{y + 3} + 3 = y + \dfrac{y}{y + 3}$          **34.** $z^2 + 2\sqrt{3}\,z + 4 = 0$

**35.** $x^2 + 2\sqrt{3}\,x - 8 = 0$

**30.** $\dfrac{y - 4}{y - 5} - \dfrac{2y - 1}{y + 4} = \dfrac{1 - 2y}{y^2 - y - 20}$          **36.** $y^2 = \sqrt{7}\,y + 1$

Solve for $x$.

C **37.** $2x^2 + x - c = 0$      **39.** $ax^2 + x + 1 = 0$

    **38.** $2x^2 + 4bx - c = 0$      **40.** $x^2 + bx + 1 = 0$

Write each equation in the form $(x - h)^2 + (y - k)^2 = c$.

**Sample.**   $x^2 + y^2 + 4x + 2y + 2 = 0$

*Solution:*   $[x^2 + 4x \quad\quad] + [y^2 + 2y \quad\quad] = -2$
$[x^2 + 4x + 4] + [y^2 + 2y + 1] = -2 + 4 + 1$
$(x + 2)^2 + (y + 1)^2 = 3$

**41.** $x^2 + y^2 + 6x + 4y - 6 = 0$    **43.** $x^2 + y^2 + 5x - 3y + 2 = 0$

**42.** $x^2 + y^2 + 8x - 6y - 1 = 0$    **44.** $x^2 + y^2 - 3x + 7y - 2 = 0$

## 8-2   The Quadratic Formula

The method we used in Section 8–1 to solve a specific quadratic equation can be applied to the general quadratic equation

$$ax^2 + bx + c = 0 \quad\quad (a \neq 0).$$

We divide by $a$ and then follow the steps of Example 2, page 310.

$$x^2 + \frac{b}{a}x + \frac{c}{a} = 0$$

**1.** $$x^2 + \frac{b}{a}x \quad\quad\quad = -\frac{c}{a}$$

**2.** $$x^2 + \frac{b}{a}x + \left(\frac{b}{2a}\right)^2 = -\frac{c}{a} + \left(\frac{b}{2a}\right)^2$$

**3.** $$\left(x + \frac{b}{2a}\right)^2 = -\frac{c}{a} + \frac{b^2}{4a^2} = \frac{-4ac + b^2}{4a^2}$$

**4.** $$x + \frac{b}{2a} = \pm\sqrt{\frac{b^2 - 4ac}{4a^2}} = \pm\frac{\sqrt{b^2 - 4ac}}{2a}$$

$$x = -\frac{b}{2a} \pm \frac{\sqrt{b^2 - 4ac}}{2a}$$

$$x = \frac{-b \pm \sqrt{b^2 - 4ac}}{2a}$$

The last equation is called the **quadratic formula**. Notice that we obtained this formula by completing the square.

**Example 1.** Solve $3x^2 - 2x + 1 = 0$.

*Solution:*      $ax^2 + bx + c = 0$ becomes $3x^2 - 2x + 1 = 0$ if

$$a = 3, \quad b = -2, \quad \text{and} \quad c = 1.$$

Substitute into the quadratic formula:

$$x = \frac{-(-2) \pm \sqrt{(-2)^2 - 4 \cdot 3 \cdot 1}}{2 \cdot 3}$$

$$= \frac{2 \pm \sqrt{4 - 12}}{6} = \frac{2 \pm \sqrt{-8}}{6}$$

$$= \frac{2 \pm 2i\sqrt{2}}{6} = \frac{1 \pm i\sqrt{2}}{3}$$

$\therefore$ the solution set is $\left\{ \dfrac{1}{3} + \dfrac{\sqrt{2}}{3}i, \dfrac{1}{3} - \dfrac{\sqrt{2}}{3}i \right\}$.

Checking these roots is left to you.

**Example 2.** A rug is to cover two-thirds of the floor area of a room 20 feet by 30 feet. The uncovered part of the floor is to form a strip of uniform width around the rug. Find the width of this strip.

*Solution:* 
1. Let $x$ = width of strip, in feet.

   Then $0 < x < 10$.

   The dimensions of the rug are $20 - 2x$ by $30 - 2x$ feet.

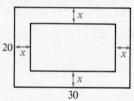

2. The area of the rug is two-thirds of the floor area.

$$\underbrace{(20 - 2x)(30 - 2x)}_{} = \underbrace{\tfrac{2}{3}}_{} \times \underbrace{(20 \times 30)}_{}$$

3. $600 - 100x + 4x^2 = \frac{2}{3} \times 600 = 400$

   $4x^2 - 100x + 200 = 0$

   $x^2 - 25x + 50 = 0$

$$x = \frac{-(-25) \pm \sqrt{(-25)^2 - 4 \cdot 1 \cdot 50}}{2 \cdot 1}$$

$$= \frac{25 \pm \sqrt{625 - 200}}{2} = \frac{25 \pm \sqrt{425}}{2}$$

$$= \frac{25 \pm 5\sqrt{17}}{2}, \text{ and } 0 < x < 10.$$

$\therefore x = \dfrac{25 - 5\sqrt{17}}{2}$.

We reject the root $\dfrac{25 + 5\sqrt{17}}{2} = 12\frac{1}{2} + \dfrac{5\sqrt{17}}{2} > 10$
since the strip must be less than 10 feet.

To determine the width to the nearest *tenth* of a foot, use Table 3, Appendix, to find $\sqrt{17}$ to *hundredths*, and substitute:

$$x \doteq \frac{25 - (5 \times 4.12)}{2} = \frac{25.00 - 20.60}{2} = \frac{4.40}{2} = 2.20$$

∴ the width of the strip is 2.2 feet to the nearest tenth of a foot.

## ORAL EXERCISES

Read each equation in the form $ax^2 + bx + c = 0$ with the value of $a$ positive.

**1.** $3x^2 - 2x = 5$

**2.** $x^2 + 6 = 3x$

**3.** $5 - 2x + x^2 = 0$

**4.** $3 - 5x - 4x^2 = 0$

**5.** $x - 4x^2 = 5$

**6.** $x - 3 = 2x^2$

Give values for $a$, $b$, and $c$ for each quadratic equation in $x$.

**7.** $x^2 - 6x + 1 = 0$

**8.** $3x^2 + 2 - 4x = 0$

**9.** $x - 3x^2 = 6$

**10.** $\sqrt{2} - x = 3x^2$

**11.** $x^2 + 2xy - 3y^2 = 0$

**12.** $x^2 + (r - 2) = (2 - r)x$

## WRITTEN EXERCISES

In all exercises in this section, express irrational roots in simple radical form and all complex roots in the form $a + bi$.

Solve by using the quadratic formula.

**A**

**1.** $x^2 - 3x + 2 = 0$

**2.** $y^2 - 5y - 6 = 0$

**3.** $3t^2 + 2t = 4$

**4.** $2z^2 - 2z = -1$

**5.** $4t^2 - 4t = 7$

**6.** $4t^2 + 4t = 1$

**7.** $3x^2 + 8x + 2 = 0$

**8.** $4y^2 + 8y + 5 = 0$

**9.** $r^2 - 8r - 1 = 0$

**10.** $n^2 - 10n + 1 = 0$

**11.** $z(z + 1) = 3(z - 1)$

**12.** $5x(x + 2) = 4(3x + 1)$

Solve by any method.

**13.** $\frac{2}{3}x^2 + \frac{1}{2}x = 4$

**14.** $\frac{1}{4}y^2 - 3y = \frac{1}{3}$

**15.** $\frac{5z + 1}{2} = \frac{3}{2 - z}$

**16.** $\frac{y - 1}{2} - \frac{5}{2} = \frac{2}{1 - y}$

**B 17.** $\frac{z + 4}{z + 3} = \frac{3z + 8}{2z + 5}$

**18.** $\frac{x^2 + x + 2}{x^2 - x + 2} = \frac{3x + 1}{3x - 1}$

**19.** $\frac{2x}{x - 1} - \frac{x + 3}{x + 1} = \frac{19}{x^2 - 1}$

**20.** $\frac{5z}{5z - 2} = \frac{1}{25z^2 - 4}$

**21.** $\frac{4}{a - 2} + \frac{2}{a + 2} = 3$

**22.** $\frac{6}{b - 4} - \frac{2}{b + 4} = 5$

**23.** $\frac{2r + 20}{r + 16} = \frac{r + 10}{r + 4} - \frac{36}{r^2 + 20r + 64}$

**24.** $\frac{5x}{5x - 4} = \frac{10x}{4 - 5x} + \frac{12}{25x^2 - 16}$

**25.** $x^2 - 4\sqrt{3}\,x + 4 = 0$

**26.** $y^2 + 4\sqrt{2}\,y + 16 = 0$

**27.** $4t^2 - 2\sqrt{6}\,t + 1 = 0$

**28.** $4x^2 + 2\sqrt{3}\,x - 2 = 0$

**29.** $x^2 - 6x = 2\sqrt{7}$

**30.** $2z^2 + 4z = \sqrt{5}$

**C 31.** $\sqrt{2}\,y^2 - 6\sqrt{5}\,y = 9\sqrt{8}$

**32.** $\sqrt{3}\,h^2 + 4\sqrt{7}\,h = 4\sqrt{27}$

**33.** $y^2 - (3 + 2i)y + (4 - i) = 0$

**34.** $z^2 - (8 - 2i)z + (23 - 14i) = 0$

**35.** $(\sqrt{3} - 1)x^2 + (4 - 2\sqrt{3})x - 4 = 0$

**36.** $(1 + \sqrt{5})y^2 - (4 + 2\sqrt{5})y + 4 = 0$

**37.** $\frac{x + 3i}{x - i} = \frac{3x + 4i}{2x + i}$

**38.** $\frac{1}{2x - i} = \frac{4}{x + 2i} - \frac{1}{2x + i}$

Solve for $x$ using the quadratic formula.

**39.** $x^2 - 4x - 8y - 4y^2 = 0$

**40.** $x^2 + x - y - y^2 = 0$

**41.** $x^2 - 2ax - 1 + 2a = 0$

**42.** $x^2 + 4bx - 1 - 4b = 0$

**43.** Show that if $c \neq 0$, an alternative form of the quadratic formula is

$$x = \frac{2c}{-b \pm \sqrt{b^2 - 4ac}}.$$

**44.** Prove that if each root of $ax^2 + bx + c = 0$, $a \neq 0$, is the reciprocal of the other, then $a = c$.

*PROBLEMS*

Express irrational results in simple radical form and also to the nearest tenth by using Table 3 in the Appendix.

**A**

1. The base of a triangle is 4 feet greater than its altitude, and the area of the triangle is 3 square feet. Find the length of the base.

2. The length of the hypotenuse of a right triangle is 7 inches, and one leg is 3 inches longer than the other. Find the area of the triangle.

3. The volume of a rectangular prism is 25 cubic cm. The prism is 5 centimeters high and 5 centimeters longer than it is wide. Find the dimensions of the prism.

4. Jill drove a new car 300 miles to deliver it to a buyer, and returned by bus. She averaged 10 miles per hour more driving the car than the bus averaged returning, and it took her one hour less to drive to her destination than it did to return. What was the average speed both ways?

5. It takes 54 feet of plywood paneling 2 feet wide to panel one wall of a room. If the width of the wall exceeds its height by 3 feet, and if the entire wall is paneled, find the dimensions of the wall.

6. A theatre seating 720 persons is remodeled. By seating 6 more persons in each row, the number of rows of seats is reduced by 4, while the capacity of the theatre remains unchanged. If the theatre is rectangular, that is, if all rows seat the same number of persons, how many persons would each row hold after the remodeling?

**B**

7. From a square piece of tin, a lidless box is constructed by cutting a 2 inch square from each corner and folding up the sides for the box. If the volume of the box is 128 cubic inches, what was the length of a side of the square piece of tin?

8. A rectangular corner lot has dimensions 50 feet by 150 feet. When two adjacent streets are widened by equal amounts, $\frac{1}{3}$ of its area is lost. What are the dimensions of the lot after the streets are widened?

9. As a result of adverse winds, the average speed of a hydrofoil on the return leg of a 180 mile round trip that took 5 hours was 15 miles per hour less than the average speed going out. Find the two speeds.

10. At 10 A.M. a long-range helicopter leaves an airport and flies due east at 200 miles per hour. At noon a cargo plane leaves the airport and flies due south at 400 miles per hour. At what time will they be 800 miles apart?

11. The sum of the squares of the first $n$ consecutive integers, i.e., $1^2 + 2^2 + 3^2 + \cdots n^2$, is given by $\dfrac{n(n + 1)(2n + 1)}{6}$. The sum of the cubes of the first $n$ integers is given by $\left[\dfrac{n(n + 1)}{2}\right]^2$. Determine all values of $n$ for which the sums of the first $n$ squares and the first $n$ cubes are equal.

12. An artist is to draw two rectangular diagrams of equal width. One is square. The height of the other is $\frac{1}{2}$ inch shorter than its width. If each dimension of each diagram is to be photo-reduced $\frac{1}{3}$ and the combined area of the two reduced diagrams is to be $7\frac{1}{3}$ square inches, what should be the dimensions of each diagram before reduction?

## ROOTS AND COEFFICIENTS

## 8–3  Relationships between Roots and Coefficients

Study these statements:

1. $2x^2 - 11x + 15 = (2x - 5)(x - 3) = 2(x - \frac{5}{2})(x - 3)$

2. $\frac{5}{2}$ and $3$ are the roots of $2x^2 - 11x + 15 = 0$.

3. $\frac{5}{2} + 3 = -\dfrac{-11}{2}$, and $\frac{5}{2} \times 3 = \dfrac{15}{2}$

Notice in Statement 3 how the sum and product of the roots, $\frac{5}{2}$ and $3$, are related to the coefficients, $2$, $-11$, and $15$, of the equation. The statements above illustrate those in the next theorem (see Exercises 29, 30, and 31).

---

*THEOREM.*   If $a \neq 0$, the following statements are equivalent (that is, any one of the statements implies each of the others):

1. $ax^2 + bx + c = a(x - r_1)(x - r_2)$

2. $r_1$ and $r_2$ are the roots of $ax^2 + bx + c = 0$.

3. $r_1 + r_2 = -\dfrac{b}{a}$, and $r_1 r_2 = \dfrac{c}{a}$

---

Since Statements 2 and 3 are equivalent, we see that $\{r_1, r_2\}$ is the solution set of $ax^2 + bx + c = 0$ if and only if $r_1 + r_2 = -\dfrac{b}{a}$ and $r_1r_2 = \dfrac{c}{a}$.

**Example 1.** Is $\left\{\dfrac{1+\sqrt{2}}{3}, \dfrac{1-\sqrt{2}}{3}\right\}$ the solution set of $9x^2 - 6x - 1 = 0$?

*Solution:*    In $9x^2 - 6x - 1$, $a = 9$, $b = -6$, $c = -1$.

$$-\frac{b}{a} = -\frac{-6}{9} = \frac{2}{3} \qquad\qquad \frac{c}{a} = \frac{-1}{9} = -\frac{1}{9}$$

$$r_1 + r_2 = \frac{1+\sqrt{2}}{3} + \frac{1-\sqrt{2}}{3} = \frac{2}{3} \qquad r_1r_2 = \frac{1+\sqrt{2}}{3} \cdot \frac{1-\sqrt{2}}{3}$$

$$= \frac{1-2}{9} = -\frac{1}{9}$$

$\therefore \left\{\dfrac{1+\sqrt{2}}{3}, \dfrac{1-\sqrt{2}}{3}\right\}$ is the solution set of $9x^2 - 6x - 1 = 0$.

**Example 2.** Find a quadratic equation whose roots are $2 + i$ and $2 - i$.

*Solution:*    $-\dfrac{b}{a} = (2 + i) + (2 - i) = 4$

$\dfrac{c}{a} = (2 + i)(2 - i) = 4 - i^2 = 4 - (-1) = 5$

$\therefore$ we may take $a = 1$, $b = -4$, $c = 5$ to obtain $x^2 - 4x + 5 = 0$.

Because Statement 2 implies Statement 1 in the theorem above, we can use the quadratic formula in factoring a quadratic polynomial.

**Example 3.** Factor $10z^2 - 7z - 3$ (see Example 2, page 181).

*Solution:*    Solve the equation $10z^2 - 7z - 3 = 0$.

$$z = \frac{-(-7) \pm \sqrt{(-7)^2 - 4(10)(-3)}}{2 \cdot 10}$$

$$= \frac{7 \pm \sqrt{49 + 120}}{20} = \frac{7 \pm \sqrt{169}}{20} = \frac{7 \pm 13}{20}$$

$$r_1 = \frac{7 + 13}{20} = \frac{20}{20} = 1, \; r_2 = \frac{7 - 13}{20} = \frac{-6}{20} = -\frac{3}{10}$$

$$\therefore 10z^2 - 7z - 3 = 10(z - r_1)(z - r_2)$$
$$= 10(z - 1)(z + \tfrac{3}{10})$$
$$= (z - 1) \cdot 10(z + \tfrac{3}{10})$$
$$= (z - 1)(10z + 3)$$

*ORAL EXERCISES*

Give the sum and the product of the roots of each equation.

**1.** $x^2 + 5x - 6 = 0$  **7.** $2z^2 - 6z + 5 = 0$

**2.** $y^2 - 3y + 2 = 0$  **8.** $4r^2 - 8r + 3 = 0$

**3.** $t^2 + 5t + 10 = 0$  **9.** $4z^2 - 1 = 0$

**4.** $z^2 - z - 1 = 0$  **10.** $2x^2 + 3x = 0$

**5.** $x^2 - 3x = 4$  **11.** $4y^2 + 3y - 2 = 0$

**6.** $y^2 - 6y = 8$  **12.** $5k^2 - 6k + 3 = 0$

State a quadratic equation whose roots have the given sum and product.

**13.** $r_1 + r_2 = 3; \ r_1r_2 = 2$  **17.** $r_1 + r_2 = \sqrt{5}; \ r_1r_2 = 0$

**14.** $r_1 + r_2 = -5; \ r_1r_2 = 4$  **18.** $r_1 + r_2 = 0; \ r_1r_2 = \sqrt{3}$

**15.** $r_1 + r_2 = 1; \ r_1r_2 = -3$  **19.** $r_1 + r_2 = \frac{3}{2}; \ r_1r_2 = \frac{5}{2}$

**16.** $r_1 + r_2 = -2; \ r_1r_2 = -5$  **20.** $r_1 + r_2 = \frac{2}{3}; \ r_1r_2 = -\frac{4}{3}$

*WRITTEN EXERCISES*

Find a quadratic equation having the given roots. Put your answer in the form $ax^2 + bx + c = 0$, where $a$, $b$, and $c$ are integers.

**Sample.**  $\frac{1}{5}, -2$

*First Solution:*

$(x - \frac{1}{5})(x + 2) = 0$
$x^2 + \frac{9}{5}x - \frac{2}{5} = 0$
$5x^2 + 9x - 2 = 0$

*Second Solution:*

$x^2 - (\frac{1}{5} + [-2])x + \frac{1}{5}(-2) = 0$
$x^2 + \frac{9}{5}x - \frac{2}{5} = 0$
$5x^2 + 9x - 2 = 0$

$\boxed{\text{A}}$

**1.** $6, -1$  **5.** $-\frac{2}{3}, \frac{5}{3}$  **9.** $1 + \sqrt{3}, 1 - \sqrt{3}$

**2.** $-2, 5$  **6.** $-\frac{6}{7}, -\frac{2}{7}$  **10.** $2 + \sqrt{5}, 2 - \sqrt{5}$

**3.** $\frac{2}{3}, -3$  **7.** $4, 0$  **11.** $7, 7$

**4.** $4, \frac{1}{4}$  **8.** $-8, 0$  **12.** $-6, -6$

**13.** $1 + i, 1 - i$

**14.** $3 + 2i, 3 - 2i$

**15.** $2 - \dfrac{i\sqrt{3}}{2}, 2 + \dfrac{i\sqrt{3}}{2}$

**16.** $\frac{2}{3} - i\sqrt{2}, \frac{2}{3} + i\sqrt{2}$

Factor using the method of Example 3.

**17.** $9x^2 - 23x + 10$

**18.** $20y^2 + 3y - 9$

**19.** $7n^2 + 20n - 3$

**20.** $14z^2 - 19z - 3$

**21.** $8n^2 - 2n - 15$

**22.** $14y^2 + 9y - 18$

**B**  **23.** If 2 is one root of $5x^2 + bx - 8 = 0$, find the other root and the value of $b$.

**24.** If $-1$ is one root of $2x^2 + x + c = 0$, find the other root and the value of $c$.

**25.** If $-3 + \sqrt{7}$ is one root of $x^2 + 6x + c = 0$, find the other root and the value of $c$.

**26.** If $2 + 2i$ is a root of $x^2 + bx + 8 = 0$, find the other root and the value of $b$.

**27.** If the difference of the roots of $y^2 + 7y + c = 0$ is 3, find both roots and the value of $c$.

**28.** If $4x^2 + 16x + c = 0$ has equal roots, find these roots and the value of $c$.

In Exercises 29–33, use the fact that $r_1$ and $r_2$, the roots of $ax^2 + bx + c = 0$, are given by

$$r_1 = \frac{-b + \sqrt{b^2 - 4ac}}{2a} \quad \text{and} \quad r_2 = \frac{-b - \sqrt{b^2 - 4ac}}{2a}.$$

**C**  **29.** Show that $r_1 + r_2 = -\dfrac{b}{a}$.

**30.** Show that $r_1 r_2 = \dfrac{c}{a}$.

**31.** Show that $ax^2 + bx + c = a(x - r_1)(x - r_2)$.

**32.** Show that $r_1^2 + r_2^2 = \dfrac{b^2 - 2ac}{a^2}$.

**33.** Show that $\dfrac{1}{r_1} + \dfrac{1}{r_2} = -\dfrac{b}{c}$.

## 8–4 The Discriminant

Using the quadratic formula, you can write the roots of the equation $ax^2 + bx + c = 0$ $(a \neq 0)$ as follows:

$$r_1 = \frac{-b + \sqrt{D}}{2a} \quad \text{and} \quad r_2 = \frac{-b - \sqrt{D}}{2a}, \quad \text{where} \quad D = b^2 - 4ac.$$

Study these examples to see how $D$ is related to the nature of the roots:

**1.** $x^2 + 2x - 5 = 0$

$D = 2^2 - 4 \cdot 1 \cdot (-5) = 4 + 20 = 24$  $D$ is positive.

$$\left. \begin{aligned} r_1 &= \frac{-2 + \sqrt{24}}{2} = -1 + \sqrt{6} \\ r_2 &= \frac{-2 - \sqrt{24}}{2} = -1 - \sqrt{6} \end{aligned} \right\}$$
$r_1$ and $r_2$ are real and unequal.

**2.** $9x^2 - 12x + 4 = 0$

$D = (-12)^2 - 4 \cdot 9 \cdot 4 = 144 - 144 = 0$  $D$ is zero.

$$\left. \begin{aligned} r_1 &= \frac{-(-12) + \sqrt{0}}{2 \cdot 9} = \frac{2}{3} \\ r_2 &= \frac{-(-12) - \sqrt{0}}{2 \cdot 9} = \frac{2}{3} \end{aligned} \right\}$$
$r_1$ and $r_2$ are real and equal.

**3.** $x^2 - 6x + 10 = 0$

$D = (-6)^2 - 4 \cdot 1 \cdot 10 = 36 - 40 = -4$  $D$ is negative.

$$\left. \begin{aligned} r_1 &= \frac{-(-6) + \sqrt{-4}}{2} = 3 + i \\ r_2 &= \frac{-(-6) - \sqrt{-4}}{2} = 3 - i \end{aligned} \right\}$$
$r_1$ and $r_2$ are conjugate imaginary.

You can see from the above examples that the following is true:

---

Let $ax^2 + bx + c = 0$ be a quadratic equation with *real* coefficients. Then:

**1.** If $b^2 - 4ac > 0$, there are two real roots.

**2.** If $b^2 - 4ac = 0$, there is one real root (called a double root).

**3.** If $b^2 - 4ac < 0$, there are two conjugate imaginary roots.

---

Because the value of $D = b^2 - 4ac$ distinguishes the three cases, $D$ is called the **discriminant** of the quadratic equation and also of the quadratic function defined by $q(x) = ax^2 + bx + c$.

The discriminant also helps you to tell whether the roots of a quadratic equation with *integral* coefficients are rational numbers. If $a$, $b$, and $c$ are *integers*, $a \neq 0$, then the numbers

$$\frac{-b \pm \sqrt{b^2 - 4ac}}{2a}$$

are rational if and only if

$$\sqrt{b^2 - 4ac}$$

is rational. But $\sqrt{b^2 - 4ac}$ is rational if and only if

$$b^2 - 4ac$$

is the square of an integer (Section 7–4). Therefore:

---

A quadratic equation with integral coefficients has rational roots if and only if its discriminant is the square of an integer.

---

**Example.** Without solving the equation, determine the nature of its roots.
  **a.** $9x^2 - 6x + 2 = 0$
  **b.** $x^2 + \frac{7}{3}x - 2 = 0$
  **c.** $x^2 + 2\sqrt{2}\,x - 2 = 0$

*Solution:* **a.** $D = (-6)^2 - 4 \cdot 9 \cdot 2 = 36 - 72 = -36 < 0$
    The roots are conjugate imaginary.

  **b.** The given equation is equivalent to $3x^2 + 7x - 6 = 0$, which has integral coefficients.

  $$D = 7^2 - 4 \cdot 3 \cdot (-6) = 49 + 72 = 121 = 11^2 > 0$$

  The roots are real, unequal, and rational.

  **c.** $D = (2\sqrt{2})^2 - 4 \cdot 1 \cdot (-2) = 8 + 8 = 16 > 0$

  The roots are real and unequal.

In part **c** of the Example, the discriminant is the square of an integer. However, we cannot conclude that the equation has rational roots because it does not have integral coefficients, and it cannot be transformed into an equation which does. Indeed, the roots are irrational numbers, namely $2 - \sqrt{2}$ and $-2 - \sqrt{2}$, as you can easily verify.

## WRITTEN EXERCISES

Without solving the equation, determine the nature of its roots.

A

1. $n^2 + 6n - 3 = 0$

2. $t^2 + 4t + 3 = 0$

3. $2z^2 + 3z + 1 = 0$

4. $5x^2 + 4x - 1 = 0$

5. $8z^2 - 9z + 6 = 0$

6. $3t^2 + 7t + 2 = 0$

7. $3y^2 - 5y = -1$

8. $4m^2 + 12m = -9$

9. $\sqrt{5}\,u^2 - 6u + \sqrt{5} = 0$

10. $n^2 - 4\sqrt{3}\,n + 12 = 0$

11. $2x^2 + 5 = 2\sqrt{10}\,x$

12. $7x^2 + 2 = 2\sqrt{14}\,x$

Find the values of $k$ for which the given equation will have the specified type of roots.

**Sample.** $2x^2 - 3x + k = 0$; two real roots

*Solution:* The equation will have two real roots provided the discriminant is positive. You have

$$D = b^2 - 4ac = (-3)^2 - 4(2)(k) = 9 - 8k.$$

Thus, for $9 - 8k > 0$ or $k < \frac{9}{8}$ the two roots are real.

B

13. $y^2 - 2ky + 7 = 0$; one real root

14. $x^2 - x + k = 0$; complex roots

15. $z^2 - 4z + k = 0$; rational roots

16. $t^2 - 12t + k = 0$; two real roots

17. $kx^2 + 3x - 1 = 0$; one real root

18. $x^2 - x + k - 2 = 0$; complex roots

In Exercises 19 and 20, $r$ is a real constant. Show that the equation has no real root.

C

19. $\dfrac{1}{x+r} = \dfrac{1}{x} + \dfrac{1}{r}$

20. $\dfrac{1}{x-r} = \dfrac{1}{x} - \dfrac{1}{r}$

In Exercises 21 and 22, $h$ and $k$ are rational numbers. Show that the roots of the equation are rational.

21. $x^2 + hx + (h - 1) = 0$

22. $hx^2 + (h + k)x + k = 0$

## EQUATIONS RELATED TO QUADRATIC EQUATIONS

### 8–5  Radical Equations

A **radical equation** is one having a variable in a radicand. You can solve the simple radical equation $\sqrt[3]{5-x} = 2$ by cubing both members:

$$\sqrt[3]{5-x} = 2$$
$$(\sqrt[3]{5-x})^3 = 2^3$$
$$5 - x = 8$$
$$x = -3$$

$$Check: \sqrt[3]{5 - (-3)} = \sqrt[3]{8} = 2$$

$\therefore$ the solution set is $\{-3\}$.

The first step in solving an equation having just one term with a variable in a radicand is to "isolate" that term in one member of the equation. Then raise each member to the power equal to the index of the radical.

**Example 1.**  Solve $x + \sqrt{x-2} = 4$.

*Solution:*

1. Isolate the radical in one member of the equation.

$$x + \sqrt{x-2} = 4$$
$$\sqrt{x-2} = 4 - x$$

2. Square both members.

$$(\sqrt{x-2})^2 = (4-x)^2$$
$$x - 2 = 16 - 8x + x^2$$

3. Solve the resulting equation.

$$x^2 - 9x + 18 = 0$$
$$(x-3)(x-6) = 0$$
$$x = 3 \quad \text{or} \quad x = 6$$

4. *Check:*  $3 + \sqrt{3-2} = 3 + \sqrt{1} = 3 + 1 = 4$
   $6 + \sqrt{6-2} = 6 + \sqrt{4} = 6 + 2 = 8 \neq 4$

$\therefore$ the solution set is $\{3\}$.

The "extraneous value" 6 appeared because Step 2 is not reversible. For example, when $x = 6$, the equation $(\sqrt{x-2})^2 = (4-x)^2$ becomes $2^2 = (-2)^2$; we certainly cannot conclude from this that $2 = -2$, that is, that $\sqrt{x-2} = 4 - x$.

If a radical equation has more than one term with a variable in a radicand, you may have to "isolate and raise to a power" more than once.

**Example 2.** Solve $\sqrt{3x - 2} - 2\sqrt{x} = 1$.

*Solution:*

$$\sqrt{3x - 2} = 2\sqrt{x} + 1$$
$$(\sqrt{3x - 2})^2 = (2\sqrt{x} + 1)^2$$
$$3x - 2 = 4x + 4\sqrt{x} + 1$$
$$-x - 3 = 4\sqrt{x}$$
$$(-x - 3)^2 = (4\sqrt{x})^2$$
$$x^2 + 6x + 9 = 16x$$
$$x^2 - 10x + 9 = 0$$
$$(x - 9)(x - 1) = 0$$
$$x = 9 \quad \text{or} \quad x = 1$$

*Check:*

$$\sqrt{3 \cdot 9 - 2} - 2\sqrt{9} = \sqrt{25} - 2 \cdot 3 = 5 - 6 = -1 \neq 1$$
$$\sqrt{3 \cdot 1 - 2} - 2\sqrt{1} = \sqrt{1} - 2 = 1 - 2 = -1 \neq 1$$

Neither 9 nor 1 is a root, and the solution set is $\emptyset$.

*WRITTEN EXERCISES*

Solve each equation.

**A**

**1.** $\sqrt{x - 3} = 5$

**2.** $\sqrt{y - 4} = 3$

**3.** $\sqrt[3]{r + 1} = 2$

**4.** $\sqrt[3]{z - 2} = 1$

**5.** $\sqrt[4]{\dfrac{x}{2}} = 2$

**6.** $\sqrt[4]{2s} = 3$

**7.** $\sqrt{2x + 3} = x$

**8.** $\sqrt{y + 2} = y$

**9.** $2z - 3 = \sqrt{7z - 3}$

**10.** $\sqrt{3u + 10} = u + 4$

**11.** $2\sqrt{y - 1} = y - 1$

**12.** $\sqrt{t - 2} = 4 - t$

**13.** $\sqrt{x - 5} - \sqrt{x} = 1$

**14.** $\sqrt{x + 4} = \sqrt{x + 20} - 2$

**15.** $\sqrt{4y + 17} = 4 - \sqrt{y + 1}$

**16.** $\sqrt{5 + r} + \sqrt{r} = 5$

**17.** $\sqrt{x + 6} - \sqrt{x} = \sqrt{2}$

**18.** $\sqrt{r - 3} + \sqrt{r} = \sqrt{2}$

**19.** $\sqrt{2x - 1} + \sqrt{x - 1} = 1$

**20.** $\sqrt{2x + 1} = 1 + \sqrt{x}$

**21.** $\sqrt{x^2 + 9} = x - 1$

**22.** $\sqrt{3y^2 + 9y - 5} = 2y + 1$

**23.** $\sqrt{x^2 - 3x + 5} = \sqrt{x + 2}$

**24.** $\sqrt{x^2 - 3} = \sqrt{x + 1}$

**B** **25.** $\sqrt[3]{(4y + 7)^2} = 9$

**28.** $\sqrt{t + 1} = \sqrt{t + 6} - \dfrac{2}{\sqrt{t + 1}}$

**26.** $\sqrt[4]{z^3 + 7} = 2$

**29.** $\sqrt{x + 4} + \sqrt{x - 1} = \sqrt{x - 4}$

**27.** $\sqrt{y} + \sqrt{y - 3} = \dfrac{3}{\sqrt{y - 3}}$

**30.** $3\sqrt{t - 1} = 2\sqrt{t + 2} - \sqrt{2t - 3}$

**31.** $\sqrt{x^2 + x - 1} + \sqrt{x^2 + 3x + 3} = 2$

**32.** $\sqrt{m^2 + 5m + 3} - \sqrt{m^2 + 3m} = 1$

Solve each equation for the variable in red.

**33.** $r = \sqrt{\dfrac{A}{\pi}}$

**36.** $y = \dfrac{1}{\sqrt{1 - x}}$

**34.** $t = \sqrt{\dfrac{2y}{g}}$

**37.** $\sqrt{ax - a^2} + \sqrt{3ax + a^2} = 2a$

**35.** $P = \pi \sqrt{\dfrac{l}{g}}$

**38.** $\sqrt{x + 3d^2} = \sqrt{2x - d^2} + d$

**C** **39.** $\sqrt{a(x - b)} + \sqrt{b(x - a)} = x$

**40.** $\sqrt{x - t} + \sqrt{x - t} = \sqrt{x + 2t}$

**41.** Find a polynomial equation with integral coefficients having $3 - 2\sqrt{5}$ as a root. Show that $3 + 2\sqrt{5}$ is also a root. (*Hint:* Start with $x = 3 - 2\sqrt{5}$, from which $x - 3 = -2\sqrt{5}$.)

**42.** Find a polynomial equation with integral coefficients having $\sqrt{3} + \sqrt{2}$ as a root. Using this equation, show that $\sqrt{3} + \sqrt{2}$ is an irrational number. (*Hint:* Start with $x = \sqrt{2} + \sqrt{3}$.)

## 8–6  Equations in Quadratic Form

The equation

$$x^4 + 3x^2 - 4 = 0$$

is not a quadratic equation, but because it can be written in the form

$$(x^2)^2 + 3x^2 - 4 = 0,$$

it is said to be *quadratic in* $x^2$.

You can first solve for $x^2$ by factoring the trinomial and then solve the resulting equation for $x$:

$$(x^2 - 1)(x^2 + 4) = 0$$

$$x^2 = 1 \quad \text{or} \quad x^2 = -4$$
$$x = \pm 1 \quad \text{or} \quad x = \pm\sqrt{-4} = \pm 2i$$

Therefore, the roots of the given equation are $1$, $-1$, $2i$, and $-2i$.

An equation is in **quadratic form** if it can be written as

$$a[f(x)]^2 + b[f(x)] + c = 0$$

where $a \neq 0$ and $f(x)$ is some expression in $x$. You can solve such an equation for $f(x)$, and sometimes the resulting equations can be solved for $x$. It is often helpful to begin the solution by making a substitution of the form $z = f(x)$.

**Example.**   Solve $\left(x + \dfrac{4}{x}\right)^2 - 9\left(x + \dfrac{4}{x}\right) + 20 = 0$.

*Solution:*    Let $z = x + \dfrac{4}{x}$. The given equation becomes:

$$z^2 - 9z + 20 = 0$$
$$(z - 4)(z - 5) = 0$$
$$z = 4 \quad \text{or} \quad z = 5$$

Replace $z$ by $x + \dfrac{4}{x}$:

$$x + \frac{4}{x} = 4 \quad \text{or} \quad x + \frac{4}{x} = 5$$
$$x^2 + 4 = 4x \qquad\qquad x^2 + 4 = 5x$$
$$x^2 - 4x + 4 = 0 \qquad x^2 - 5x + 4 = 0$$
$$(x - 2)^2 = 0 \qquad (x - 1)(x - 4) = 0$$
$$x = 2 \qquad\qquad x = 1 \quad \text{or} \quad x = 4$$

*Check:*

$(2 + \tfrac{4}{2})^2 - 9(2 + \tfrac{4}{2}) + 20 = 4^2 - 9 \cdot 4 + 20 = 16 - 36 + 20 = 0$

$(1 + \tfrac{4}{1})^2 - 9(1 + \tfrac{4}{1}) + 20 = 5^2 - 9 \cdot 5 + 20 = 25 - 45 + 20 = 0$

$(4 + \tfrac{4}{4})^2 - 9(4 + \tfrac{4}{4}) + 20 = 5^2 - 9 \cdot 5 + 20 = 25 - 45 + 20 = 0$

$\therefore$ the solution set is $\{2, 1, 4\}$.

*ORAL EXERCISES*

State the variable expression in which each equation is quadratic.

**Sample.** $x + 2\sqrt{x} - 3 = 0$

*Solution:* Since $(\sqrt{x})^2 = x$, you have $(\sqrt{x})^2 + 2\sqrt{x} - 3 = 0$. ∴ the given equation is quadratic in $\sqrt{x}$.

**1.** $x - 2\sqrt{x} - 15 = 0$

**2.** $r + 3\sqrt{r} - 10 = 0$

**3.** $x^4 - 3x^2 - 4 = 0$

**4.** $z^4 - 8z^2 + 16 = 0$

**5.** $2x^4 + 17x^2 - 9 = 0$

**6.** $r^4 - 2r^2 - 24 = 0$

**7.** $(y^2 + 1)^2 - 7(y^2 + 1) + 10 = 0$

**8.** $(z^2 + 5z)^2 - 8z(z + 5) - 84 = 0$

**9.** $\sqrt[3]{r^2} - 2\sqrt[3]{r} - 3 = 0$

**10.** $\sqrt[3]{x^2} - 2\sqrt[3]{x} - 35 = 0$

*WRITTEN EXERCISES*

**A** **1–10.** Solve each Oral Exercise 1–10.

Solve.

**11.** $2x^{-2} + 3x^{-1} = 2$

**12.** $3y^{-2} - 4y^{-1} = -1$

**13.** $8z^6 - 19z^3 - 27 = 0$

**14.** $r^6 - 82r^3 + 81 = 0$

**B** **15.** $\sqrt{x - 1} - 2\sqrt[4]{x - 1} = 15$

**16.** $z^2 - 5 - 5\sqrt{z^2 - 5} = -6$

**17.** $(2y^2 + y)^2 - 7(2y^2 + y) = -6$

**18.** $(r^2 - 2r)^2 - 5(r^2 - 2r) = 6$

**19.** $(t^2 + 1)^2 + 10 = 7(t^2 + 1)$

**20.** $(p^2 + 2)^2 - 17(p^2 + 2) = -66$

**C** **21.** $6\left(\dfrac{1}{x - 1}\right)^2 - \left(\dfrac{1}{x - 1}\right) = 1$

**22.** $\left(\dfrac{z^2 + 4}{z}\right)^2 - 9\left(\dfrac{z^2 + 4}{z}\right) + 20 = 0$

**23.** $\left(\dfrac{k - 1}{k + 3}\right)^2 + 3 = 4\left(\dfrac{1 - k}{k + 3}\right)$

**24.** $\left(\dfrac{m + 3}{m + 2}\right) - 3\left(\dfrac{m + 2}{m + 3}\right) = 2$

**25.** $\left(\dfrac{t + 1}{t + 2}\right)^2 - 8 = -2\left(\dfrac{t + 1}{t + 2}\right)$

**26.** $\left(\dfrac{z - 2}{z + 3}\right) - 2\left(\dfrac{z + 3}{z - 2}\right) = 1$

## QUADRATIC FUNCTIONS AND THEIR GRAPHS

### 8–7   Parabolas in the *xy*-plane

Curves called *parabolas* are used in studying quadratic functions. A parabola in the *xy*-plane is any curve congruent* to the graph of an equation of the form $y = ax^2$, $a \neq 0$ (recall Section 7–1). Several curves of this type are shown in Figure 8–1.

Notice that the curve $y = \frac{1}{2}x^2$, for example, can be obtained from the graph of $y = x^2$, shown in red, by multiplying the ordinate of each point by $\frac{1}{2}$. Do you see that the parabola $y = ax^2$ opens upward if $a > 0$ and opens downward if $a < 0$.

If a point $(p, q)$ lies on one of the parabolas shown in Figure 8–1, then its "mirror image" in the *y*-axis, the point $(-p, q)$ also lies on the parabola. We say that the *y*-axis is the axis of symmetry, or simply the axis, of the parabola and that the parabola is symmetric with respect to the *y*-axis. The vertex of a parabola is the point in which it intersects its axis. Thus, the origin is the vertex of each of the parabolas shown in Figure 8–1.

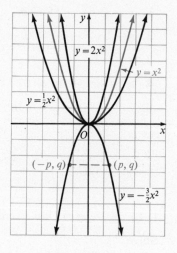

FIGURE 8–1

The graph of $y = (x - 2)^2$ is shown in Figure 8–2, along with the parabola $y = x^2$. If you move the parabola $y = x^2$ rigidly 2 units to the right, you obtain the curve $y = (x - 2)^2$. This curve is therefore a parabola having the line $x = 2$ as axis and the point $(2, 0)$ as vertex.

| $y = x^2$ | | | | $y = (x - 2)^2$ | | |
|---|---|---|---|---|---|---|
| **x** | **x²** | **y** | | **x** | **(x − 2)²** | **y** |
| −3 | $(-3)^2$ | 9 | | −1 | $(-1 - 2)^2$ | 9 |
| −2 | $(-2)^2$ | 4 | | 0 | $(0 - 2)^2$ | 4 |
| −1 | $(-1)^2$ | 1 | | 1 | $(1 - 2)^2$ | 1 |
| 0 | $0^2$ | 0 | | 2 | $(2 - 2)^2$ | 0 |
| 1 | $1^2$ | 1 | | 3 | $(3 - 2)^2$ | 1 |
| 2 | $2^2$ | 4 | | 4 | $(4 - 2)^2$ | 4 |
| 3 | $3^2$ | 9 | | 5 | $(5 - 2)^2$ | 9 |

*Two curves are congruent if one can be moved rigidly so as to coincide with the other.

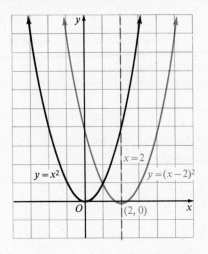

FIGURE 8–2

Do you see that the curve $y = \frac{1}{2}(x - 2)^2$, shown in Figure 8–3, is related to the parabola $y = (x - 2)^2$ in the same way the parabola $y = \frac{1}{2}x^2$ is related to the parabola $y = x^2$?

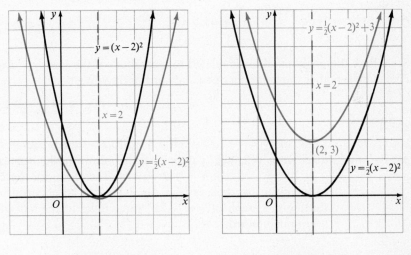

FIGURE 8–3                                    FIGURE 8–4

If the parabola $y = \frac{1}{2}(x - 2)^2$ is moved rigidly upward 3 units, the equation of the resulting parabola is $y = \frac{1}{2}(x - 2)^2 + 3$ because the *ordinate* of each point is increased 3 units. The vertex of the parabola $y = \frac{1}{2}(x - 2)^2 + 3$ is the point $(2, 3)$ and its axis is the line $x = 2$ (see Figure 8–4). In general:

The graph of

$$y = a(x - h)^2 + k \qquad (a \neq 0)$$

is a parabola having the point $(h, k)$ as vertex and the line $x = h$ as axis. The parabola opens upward when $a > 0$ and downward when $a < 0$.

You can draw a parabola after plotting only a few points because all parabolas have the same shape.

**Example.** Graph $y = -2(x + 1)^2 + 3$.

*Solution:*

Compare the given equation with the general form. The required graph is a parabola with vertical axis and having vertex $(-1, 3)$. Since $a = -2, < 0$, the parabola opens downward.

| x | y |
|---|---|
| 0 | 1 |
| −1 | 3 |
| −2 | 1 |

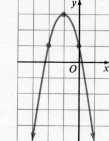

*ORAL EXERCISES*

State how the given parabola is related to a parabola of the form $y = ax^2$. First state whether the parabola opens upward or downward.

**Sample.** $y = -2(x + 3)^2 + 4$

*Solution:* Compare $y = -2(x + 3)^2 + 4$ with $y = a(x - h)^2 + k$. Since $a = -2$, the parabola opens downward. Because $x + 3 = x - (-3), h = -3; k = 4$. Therefore, the vertex of the given parabola is $(-3, 4)$, and it can be obtained by shifting the graph of $y = -2x^2$ three units to the left and four units upward.

1. $y = x^2 + 2$

2. $y = -x^2 + 4$

3. $y = -2x^2 - 8$

4. $y = 3x^2 - 12$

5. $y = (x - 2)^2 + 1$

6. $y = (x + 3)^2 - 2$

7. $y = -(x - 1)^2 + 1$

8. $y = -(x + 2)^2 + 3$

9. $y = \frac{1}{2}(x + 3)^2 + 1$

10. $y = \frac{1}{4}(x - 1)^2 + 2$

**11–20.** State an equation of the axis of symmetry for each graph in Oral Exercises 1–10.

## WRITTEN EXERCISES

<u>A</u> **1–10.** Graph each equation in Oral Exercises 1–10 in the $xy$-plane.

Find an equation for the parabola which has vertical axis, has the point $V$ as vertex, and which contains the point $Q$.

**Sample.** $V(2, 1)$; $Q(4, -1)$

*Solution :* Setting $(h, k) = (2, 1)$, we can substitute for $h$ and $k$ in $y = a(x - h)^2 + k$ to obtain

$$y = a(x - 2)^2 + 1.$$

Since the parabola is to contain $Q(4, -1)$, we can replace $x$ with 4 and $y$ with $-1$ to obtain

$$-1 = a(4 - 2)^2 + 1$$

from which

$$-1 = 4a + 1 \quad \text{or} \quad a = -\tfrac{1}{2}.$$

Therefore, the required equation is

$$y = -\tfrac{1}{2}(x - 2)^2 + 1.$$

**11.** $V(2, 0)$; $Q(0, 2)$          **14.** $V(-2, 8)$; $Q(0, 0)$

**12.** $V(0, -2)$; $Q(-2, 0)$          **15.** $V(3, 1)$; $Q(1, 37)$

**13.** $V(2, -7)$; $Q(5, 20)$          **16.** $V(-1, 1)$; $Q(1, -1)$

Find all values of $a$, $h$, or $k$ so that the graph of the given equation contains the given point.

**17.** $y = -3(x + 1)^2 + k$; $(2, -30)$     **22.** $y = a(x - 3)^2 - 5$; $(-1, -1)$

**18.** $y = 2(x - 3)^2 + k$; $(-1, 30)$     **23.** $y = -(x + 2)^2 + k$; $(0, 0)$

**19.** $y = \tfrac{1}{2}(x - h)^2 + 4$; $(2, \tfrac{9}{2})$     **24.** $y = \tfrac{2}{3}(x + 5)^2 + k$; $(-2, 2)$

**20.** $y = -\tfrac{2}{3}(x - h)^2 + 1$; $(3, 1)$     **25.** $y = -2(x - h)^2$; $(-1, -32)$

**21.** $y = a(x - 3)^2 - 2$; $(5, 0)$     **26.** $y = a(x + 4)^2$; $(-1, 12)$

Find the *x*- and *y*-intercepts of the parabola whose equation is given.

**Sample.** $y = 2(x - 3)^2 + 5$

*Solution:* To find the *y*-intercept, set $x = 0$ and solve for $y$.

$$y = 2(0 - 3)^2 + 5 = 18 + 5 = 23$$

To find the *x*-intercepts (if they exist) set $y = 0$ and solve for $x$.

$$0 = 2(x - 3)^2 + 5 \quad \text{or} \quad (x - 3)^2 = -\tfrac{5}{2}$$

Since $(x - 3)^2 \geq 0$ for all $x \in \mathcal{R}$, no *x*-intercepts exist. Thus, the *y*-intercept is 23 and there are no *x*-intercepts.

**27.** $y = 2(x + 4)^2 - 8$　　　　**30.** $y = \tfrac{2}{3}(x - 2)^2 + 3$

**28.** $y = -3(x - 2)^2 + 3$　　　　**31.** $y = -8(x - 3)^2 + 2$

**29.** $y = \tfrac{1}{2}(x + 1)^2 - 8$　　　　**32.** $y = 10(x + 5)^2 - 5$

Find *a* and *k* so that the graph of the given equation contains the given points. Write the resulting equation in *x* and *y*.

**Sample.** $y = a(x + 1)^2 + k; \ (1, 7), (-2, 1)$

*Solution:* Obtain two equations in *a* and *k* by substituting the coordinates of the given points for *x* and *y* in the given equation.

$$7 = a(1 + 1)^2 + k \qquad 1 = a(-2 + 1)^2 + k$$

Then, solve the system

$$7 = 4a + k$$
$$1 = a + k$$

to obtain $a = 2$ and $k = -1$. Thus the required equation is $y = 2(x + 1)^2 - 1$.

[B]　**33.** $y = a(x - 2)^2 + k; \ (0, -2), (2, 2)$

**34.** $y = a(x + 1)^2 + k; \ (1, 11), (0, 2)$

**35.** $y = a(x - 1)^2 + k; \ (-1, 5), (3, 5)$

**36.** $y = a(x + 3)^2 + k; \ (1, -3), (-1, -3)$

**37.** $y = ax^2 + k; \ (2, 7), (-1, -2)$

**38.** $y = ax^2 + k; \ (1, -6), (2, -24)$

C **39.** Show that the parabola $y = a(x - h)^2 + k$ contains both of the points

$$\left(h + \frac{1}{a}, \; k + \frac{1}{a}\right)$$

and

$$\left(h - \frac{1}{a}, \; k + \frac{1}{a}\right).$$

Illustrate with graphs of $y = x^2$, $y = \frac{1}{2}x^2$, and $y = 2x^2$.

**40.** Show that if the points $(r_1, s)$ and $(r_2, s)$, $r_1 \neq r_2$, belong to the parabola $y = a(x - h)^2 + k$, then

$$\frac{r_1 + r_2}{2} = h.$$

Illustrate with the graph of $y = (x - 2)^2 + 1$.

**41.** Graph the equation $y = |x^2 - 1|$.

**42.** Graph the equation $y = |x^2 - 4|$.

## 8–8 Graphs of Quadratic Functions

If $f(x) = ax^2 + bx + c$, $a \neq 0$, then $f$ is a **quadratic function**. By completing the square (Section 8–1) you can show that the graph of a quadratic function is a parabola.

**Example.** Let $f(x) = 2x^2 - 8x + 5$. Show that the graph of $f$ is a parabola, find its vertex, and draw it.

*Solution:* The graph is the same as the graph of the equation

$$
\begin{aligned}
y &= 2x^2 - 8x + 5. \\
y - 5 &= 2x^2 - 8x \\
y - 5 &= 2(x^2 - 4x \quad ) \\
y - 5 + 2 \cdot 4 &= 2(x^2 - 4x + 4) \\
y + 3 &= 2(x - 2)^2 \\
y &= 2(x - 2)^2 - 3.
\end{aligned}
$$

| x | y |
|---|---|
| 1 | −1 |
| 2 | −3 |
| 3 | −1 |

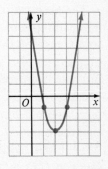

Compare this equation with the general form $y = a(x - h)^2 + k$ (page 332). The graph is a parabola with vertex $(2, -3)$.

You can apply the method of the Example to the general case by letting $y = f(x) = ax^2 + bx + c$.

$$y - c = ax^2 + bx$$

$$y - c = a\left(x^2 + \frac{b}{a}x \quad\right)$$

$$y - c + a \cdot \frac{b^2}{4a^2} = a\left(x^2 + \frac{b}{a}x + \frac{b^2}{4a^2}\right)$$

$$y - c + \frac{b^2}{4a} = a\left(x + \frac{b}{2a}\right)^2$$

$$y = a\left(x + \frac{b}{2a}\right)^2 + c - \frac{b^2}{4a}$$

$$y = f(x) = a\left(x + \frac{b}{2a}\right)^2 + \frac{4ac - b^2}{4a}$$

Comparing the last equation with $y = a(x - h)^2 + k$, you can conclude that:

---

The graph of the general quadratic function,

$$f(x) = ax^2 + bx + c \qquad (a \neq 0),$$

is a parabola with vertex $\left(-\dfrac{b}{2a}, \dfrac{4ac - b^2}{4a}\right)$. Its axis is vertical and the parabola opens upward when $a > 0$ and downward when $a < 0$.

---

The abscissa of a point where a curve intersects the *x*-axis is an **x-intercept** of the curve. For example, the *curve* $y = x^2 - 2x - 3$ has two *x*-intercepts, $-1$ and $3$, because for these values of $x$, $y$ is 0 (see Figure 8–5). You can see also that the *equation* $x^2 - 2x - 3 = 0$ has $-1$ and $3$ as *roots*, and that the *function* $f(x) = x^2 - 2x - 3$ has $-1$ and $3$ as *zeros* (page 162).

| x | y |
|----|----|
| -2 | 5 |
| -1 | 0 |
| 0 | -3 |
| 1 | -4 |
| 2 | -3 |
| 3 | 0 |
| 4 | 5 |

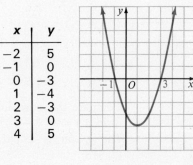

FIGURE 8–5

In general:

---

The following statements are equivalent:

**1.** *r* is a real zero of the function *f*.

**2.** *r* is a real root of the equation $f(x) = 0$.

**3.** *r* is an *x*-intercept of the graph of *f*.

---

The algebraic fact that a quadratic equation can have two, one, or no real roots corresponds to the geometric fact that a parabola can have two, one, or no *x*-intercepts, as illustrated below.

Parabola: $y = x^2 - 4x + 1$

Equation: $x^2 - 4x + 1 = 0$

Discriminant: $(-4)^2 - 4 \cdot 1 \cdot 1 = 12 > 0$

Nature of roots: Two real roots

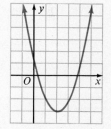

---

Parabola: $y = x^2 - 4x + 4$

Equation: $x^2 - 4x + 4 = 0$

Discriminant: $(-4)^2 - 4 \cdot 1 \cdot 4 = 0$

Nature of roots: One real (double) root

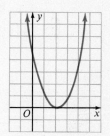

---

Parabola: $y = x^2 - 4x + 5$

Equation: $x^2 - 4x + 5 = 0$

Discriminant: $(-4)^2 - 4 \cdot 1 \cdot 5 = -4 < 0$

Nature of Roots: No real roots (Roots are imaginary)

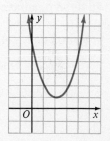

---

*WRITTEN EXERCISES*

Find the vertex of the graph of each function and then sketch the graph.

A   **1.** $f(x) = x^2 + 2x$                   **7.** $h(x) = 4x - 4 - x^2$

    **2.** $g(x) = x^2 - 4x$                   **8.** $t(x) = 2x - 1 - x^2$

    **3.** $h(x) = x^2 - 4x - 12$            **9.** $f(x) = 2x^2 + 4x + 1$

    **4.** $f(x) = x^2 + 6x + 5$             **10.** $g(x) = 2x - 2 - x^2$

    **5.** $F(x) = 6x - x^2$                 **11.** $G(x) = 7x - 2 - 3x^2$

    **6.** $G(x) = 2x - x^2$                 **12.** $H(x) = 3x^2 + 5x + 1$

Without drawing the graph of the given function, determine **(a)** how many points it has in common with the x-axis, and **(b)** whether its vertex lies above, below, or on the x-axis.

    **Sample.**   $f(x) = 2x - 3 - x^2$

    *Solution:*   **a.** The x-intercepts of the graph are solutions of the equation

$$2x - 3 - x^2 = 0, \quad \text{or} \quad -x^2 + 2x - 3 = 0.$$

Its discriminant is $(2)^2 - 4(-1)(-3) = -8$ and therefore it has no real roots. The graph has *no* points in common with the x-axis.

**b.** Since the graph of $f$ opens downward, its vertex must be *below* the x-axis (for otherwise the graph would intersect the x-axis).

    **13.** $g(x) = x^2 + 4x + 1$            **16.** $F(x) = 2x - 4 - x^2$

    **14.** $f(x) = x^2 - 4x + 1$            **17.** $G(x) = 6 + x - 2x^2$

    **15.** $h(x) = 3 - 2x - x^2$          **18.** $t(x) = 4x - 4 - x^2$

Find the quadratic function f that satisfies the given conditions.

    **Sample.**   $f(1) = 0, \; f(-1) = 10, \; f(3) = -2$

    *Solution:*   Since the function is defined by $f(x) = ax^2 + bx + c$,

$$\begin{array}{lll}
f(1) = a(1)^2 + b(1) + c = 0 & & a + b + c = 0 \\
f(-1) = a(-1)^2 + b(-1) + c = 10 & \text{or} & a - b + c = 10 \\
f(3) = a(3)^2 + b(3) + c = -2 & & 9a + 3b + c = -2
\end{array}$$

Solving this system by the method in Section 3–8, you find that $a = 1$, $b = -5$, and $c = 4$. Therefore, $f(x) = x^2 - 5x + 4$.

B **19.** $f(1) = 0$, $f(-1) = -8$, $f(2) = 7$

**20.** $f(1) = -4$, $f(-1) = -10$, $f(3) = 18$

**21.** $f(1) = 0$, $f(-1) = -14$, $f(2) = 4$

**22.** $f(1) = 4$, $f(2) = 1$, $f(3) = 0$

**23.** $f(1) = f(-1)$, $f(0) = 5$, $f(2) = 1$

**24.** $f(2) = f(-2) = 4$, $f(1) = 1$

C **25.** Show that if $f(x) = ax^2 + bx + c$, and $f(x) = f(-x)$ for all $x$, then $b = 0$.

**26.** Show that there exists no quadratic function, $f(x) = ax^2 + bx + c$, such that $f(x) = -f(-x)$ for all $x$.

## 8–9 Extrema of Quadratic Functions

Consider the functions $g$ and $h$ defined by

$$g(x) = -x^2 + 4x - 1 \quad \text{and} \quad h(x) = x^2 - 6x + 10.$$

By using the methods of Section 8–8, you can put these equations into the form

$$g(x) = -(x - 2)^2 + 3 \quad \text{and} \quad h(x) = (x - 3)^2 + 1$$

and then obtain the graphs shown in Figures 8–6 and 8–7.

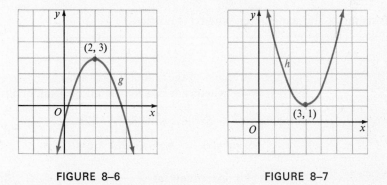

FIGURE 8–6  FIGURE 8–7

The point $(2, 3)$ is called the **maximum point** of the graph of $g$ because it is the *highest* point of the curve. Since the ordinates of points on the graph of a function are the *values* of the function, you can see from Figure 8–6 that the greatest, or **maximum**, value of $g$ is 3. Similarly, the point $(3, 1)$ is the **minimum point** of the graph of $h$ and the **minimum** value of $h$ is 1.

Do you see that you could have found the maximum value of $g$ without drawing its graph? From

$$g(x) = -(x - 2)^2 + 3 = 3 - (x - 2)^2$$

you see that the value of $g$ at 2 is 3 because

$$g(2) = 3 - (2 - 2)^2 = 3 - 0^2 = 3.$$

But if $x \neq 2$, then $(x - 2)^2$ is *positive* and therefore $g(x) < 3$. Thus $g$ has the value 3 but no value greater than 3. You can use a similar algebraic argument to show that 1 is the minimum value of $h$.

When we wish to speak of a maximum or a minimum value of a function without specifying which, we use the word **extremum** (plural: **extrema**). Thus 3 is the extremum of the quadratic function $g$ considered above and this extremum is a maximum.

By writing $f(x) = ax^2 + bx + c$ $(a \neq 0)$ in the form

$$f(x) = a\left(x + \frac{b}{2a}\right)^2 + \frac{4ac - b^2}{4a} \qquad \text{(page 336)}$$

you can see that:

---

When $x = -\dfrac{b}{2a}$, the function $f(x) = ax^2 + bx + c$ $(a \neq 0)$

has the extremum $\dfrac{4ac - b^2}{4a}$. This extremum is a maximum when $a$ is negative and a minimum when $a$ is positive.

---

**Example 1.** Find the extremum of the function $f(x) = 3x^2 - 12x + 7$ and determine whether it is a maximum or a minimum.

*Solution:* Compare $ax^2 + bx + c$ with $3x^2 - 12x + 7$:

$$a = 3, \qquad b = -12, \qquad c = 7$$

$\therefore f$ has an extremum at $-\dfrac{b}{2a} = -\dfrac{-12}{2 \cdot 3} = 2$. Since $a > 0$, the extremum is a minimum, and the minimum value is $f(2) = 3 \cdot 2^2 - 12 \cdot 2 + 7 = 12 - 24 + 7 = -5$.

Notice that it is not necessary to remember the expression $\dfrac{4ac - b^2}{4a}$ because the extreme value is $f\left(-\dfrac{b}{2a}\right)$.

**Example 2.**  A rectangular plot is to be enclosed by using part of an existing fence as one side and 20 yards of fencing for the other three sides. What is the greatest area which can be enclosed?

*Solution:*  (*Plan:* Express the area of any plot which can be so enclosed in terms of some convenient variable. Find the maximum value of this function.) Let $x =$ length, in yards, of each of the sides perpendicular to the existing fence. Then the length of the side parallel to the existing fence must be $20 - 2x$ yards since a total of 20 yards of fencing is available.

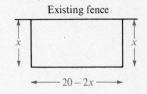

For a given $x$, the dimensions of the plot are $x$ and $20 - 2x$, and therefore its area is

$$A(x) = x(20 - 2x) = -2x^2 + 20x.$$

The maximum of $A$ occurs when $x = -\dfrac{20}{2(-2)} = 5$, and

$$A(5) = -2 \cdot 5^2 + 20 \cdot 5 = -50 + 100 = 50.$$

∴ 50 square yards is the greatest area which can be enclosed.

## ORAL EXERCISES

State the value of $x$ for which the specified function has an extremum. State whether the extremum is a maximum or a minimum.

**Sample.**  $f(x) = 2x^2 - 3x + 2$

*Solution:*  You have $-\dfrac{b}{2a} = -\dfrac{-3}{4} = \dfrac{3}{4}$. The $x$-coordinate of the extremum is $\frac{3}{4}$, and since $a > 0$, the parabola opens upward and hence the extremum is a minimum.

**1.** $f(x) = 4 - x^2$

**2.** $f(x) = 2x^2 + 4x$

**3.** $f(x) = x^2 - x + 2$

**4.** $f(x) = x^2 - 4x - 3$

**5.** $f(x) = 4 - 2x + x^2$

**6.** $f(x) = 2x - 3x^2 + 2$

**7.** $f(x) = 5x^2$

**8.** $f(x) = 3x^2$

**9.** $f(x) = 6x^2 - 3x + 2$

**10.** $f(x) = 2x^2 - 12x - 3$

*WRITTEN EXERCISES*

By the method of Example 1, page 340, find the extremum of $f$ and state whether the extremum is a maximum or a minimum. Use the resulting information to state the number of $x$-intercepts of the graph of the function.

A **1–10.** The function specified in Oral Exercises 1–10.

**11.** $f(x) = x^2 - 3x - 10$  **15.** $f(x) = 2x^2 - 5x - 3$

**12.** $f(x) = 5 + 4x - x^2$  **16.** $f(x) = 3x^2 + 7x - 1$

**13.** $f(x) = 6x - 4x^2$  **17.** $f(x) = 4 - 4x - x^2$

**14.** $f(x) = 7x^2 - 2$  **18.** $f(x) = x^2 - 10x + 25$

*PROBLEMS*

B **1.** Find two numbers whose sum is 18 and whose product is as great as possible.

**2.** Find two numbers whose difference is 10 and whose product is a minimum.

**3.** If a rifle bullet is fired vertically upward with an initial speed of 960 feet per second, its height in feet $t$ seconds later is given approximately by $h = 960t - 16t^2$. Find the maximum height attained by the bullet.

**4.** If a stone is tossed vertically upward on the moon with an initial speed of 108 feet per second, its height in feet $t$ seconds later is given by $h = 108t - 2.7t^2$. What is the maximum height attained by the stone?

**5.** In a 120-volt circuit having resistance of 16 ohms, the power $w$ in watts when a current of $I$ amperes is flowing is given by $w = 120I - 16I^2$. Find the maximum power which can be delivered in the circuit.

**6.** When an automobile dealer's markup over the factory price is $\$x$ per car, she can sell $600 - 2x$ cars each month, and so her monthly profit (before fixed expenses) is $\$(600 - 2x)x$. What should her markup be in order to maximize the profit?

C **7.** From the right triangle $ACB$, a triangle $PCQ$ is cut. If $AC = 4$, $BC = 8$, and $PC = x = \frac{1}{2}QB$:
 **a.** Express the area $z$ of the quadrilateral $APQB$ in terms of $x$.
 **b.** Find the minimum value of $z$.

**8.** In the figure, $P$ is a first-quadrant point on the line $2x + 3y = 6$, and $M$ is the foot of the perpendicular dropped from $P$ to the $x$-axis. How large an area can the triangle $OMP$ have?

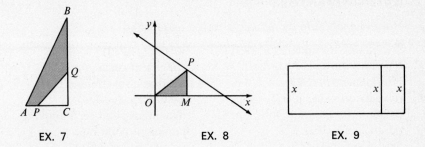

EX. 7        EX. 8        EX. 9

**9.** A rectangular field is to be enclosed by a fence and divided into two smaller rectangular fields by another fence. Find the dimensions of the field of greatest area which can thus be enclosed and partitioned with 1200 feet of fencing.

---

## *Chapter Summary*

---

**1.** The **Quadratic Formula**, $x = \dfrac{-b \pm \sqrt{b^2 - 4ac}}{2a}$ enables you to solve any quadratic equation of the form $ax^2 + bx + c = 0$, $a \neq 0$. The **sum of the roots** of such a quadratic equation is $-\dfrac{b}{a}$; the **products of the roots** is $\dfrac{c}{a}$; so that you may write $x^2 -$ (sum of roots)$x +$ (product of roots) $= 0$ as an equivalent equation. From the value of the **discriminant** $b^2 - 4ac$ of a quadratic equation, you can tell whether its real roots are rational or irrational, and equal or unequal.

**2.** To solve a **radical equation**, you isolate a radical in one member and raise each member to the power corresponding to the root index. The process is then repeated if necessary.

**3.** To solve an equation in **quadratic form**,

$$a[f(x)]^2 + b[f(x)] + c = 0,$$

you first solve for $f(x)$ and then for $x$.

**4.** The graph of the **quadratic function** $y = ax^2 + bx + c$, where $a \neq 0$, is a **parabola** which is **symmetric** with respect to a vertical line. When you transform $y = ax^2 + bx + c$ to the form $y = a(x - h)^2 + k$, $x = h$ is an equation of the **axis of symmetry** of the graph of $f$, and $(h, k)$ are the coordinates of the **vertex**. The value of $k$ is $f\left(-\dfrac{b}{2a}\right)$; it is a **maximum** value of the function when $a < 0$, and a **minimum** value when $a > 0$.

## Vocabulary and Spelling

Review each term by reference to the page listed.

completing the square (*p. 310*)
quadratic formula (*p. 313*)
discriminant (*p. 323*)
radical equation (*p. 325*)
equation in quadratic form (*p. 328*)
parabola (*p. 330*)
axis of symmetry (*p. 330*)
symmetric (*p. 330*)

vertex (*p. 330*)
congruent curves (*p. 330*)
quadratic function (*p. 335*)
*x*-intercept (*p. 336*)
maximum point on a graph (*p. 339*)
minimum point on a graph (*p. 339*)
extremum (*p. 340*)

---

## Chapter Test

8–1  **1.** Solve $x^2 - 3x - 1 = 0$ by completing the square. Express the roots in simple radical form and also to the nearest tenth.

8–2  **2.** Solve $x^2 - 4x + 5 = 0$ using the quadratic formula. Express the roots in the form $a + bi$ and $a - bi$.

**3.** The sum of a number and its reciprocal is $2\frac{1}{20}$. Find the numbers.

8–3  **4.** Find a quadratic equation in simple form whose roots are $1 + 2\sqrt{3}$ and $1 - 2\sqrt{3}$.

8–4  **5.** Without solving the equation, determine the nature of the roots of $4x^2 + 5x - 1 = 0$.

8–5  Solve each equation.

**6.** $\sqrt[3]{2x - 1} = -2$　　　　**7.** $\sqrt{x + 5} + 2\sqrt{x} = 7$

8–6  **8.** $2x^4 + 17x^2 - 9 = 0$.

8–7  **9.** Graph $y = 3(x + 2)^2 - 2$ in the *xy*-plane.

8–8  **10.** Without drawing the graph of $y = x^2 - 3x + 3$, determine how many points it has in common with the *x*-axis.

8–9  **11.** Find the extremum of $f(x) = 6x - x^2 - 5$ and state whether it is a maximum or whether it is a minimum.

**12.** Find two numbers whose sum is 8 and whose product is a maximum.

---

## Chapter Review

---

**8–1**  **Completing the Square**  *Pages 309–313*

1. The equation $x^2 = d$ is equivalent to the compound sentence $x = \underline{\;?\;}$ or $x = \underline{\;?\;}$.

2. To complete the square in $x^2 + bx$ you would add the term $\underline{\;?\;}$.

3. The equation $x^2 - 4x + 5 = 0$ is equivalent to the equation $(x - \underline{\;?\;})^2 = \underline{\;?\;}$.

4. The roots of $x^2 - 4x + 5 = 0$ are $\underline{\;?\;}$ and $\underline{\;?\;}$.

**8–2**  **The Quadratic Formula**  *Pages 313–318*

5. The quadratic formula is $x = \underline{\;?\;}$.

6. In the equation $3x^2 = 6 - 2x$, $a = \underline{\;?\;}$, $b = \underline{\;?\;}$, and $c = \underline{\;?\;}$.

7. To solve $3x^2 = 6 - 2x$ using the quadratic formula, you would first have $x = \dfrac{(\underline{\;?\;}) \pm \sqrt{\underline{\;?\;}}}{\underline{\;?\;}}$.

8. In simple radical form, the roots of $3x^2 = 6 - 2x$ are $\underline{\;?\;}$ and $\underline{\;?\;}$.

**8–3**  **Relationships between Roots and Coefficients**
*Pages 318–321*

9. If $r_1$ and $r_2$ are roots of $ax^2 + bx + c = 0$, then $r_1 + r_2 = \underline{\;?\;}$ and $r_1 r_2 = \underline{\;?\;}$.

10. The sum of the roots of $2x^2 - 4x + 5 = 0$ is $\underline{\;?\;}$, and their product is $\underline{\;?\;}$.

11. Find the quadratic equation whose roots are $2 - 3i$ and $2 + 3i$.

12. In factored form, $5x^2 - 7x - 6 = \underline{\;?\;}$.

**8–4**  **The Discriminant**  *Pages 322–324*

13. The discriminant of $ax^2 + bx + c = 0$ is $\underline{\;?\;}$.

14. The roots of $ax^2 + bx + c = 0$ are real and equal if the discriminant $D$ is $\underline{\;?\;}$, are real and unequal if $D$ is $\underline{\;?\;}$ and are conjugate imaginary if $D$ is $\underline{\;?\;}$.

15. The discriminant of $2x^2 - 4x + 5 = 0$ is $\underline{\;?\;}$.

16. The roots of $2x^2 - 4x + 5 = 0$ are $\underline{\;?\;}$ and $\underline{\;?\;}$ numbers.

**8–5 Radical Equations**     *Pages 325–327*

17. To solve $x + \sqrt{x - 4} = 10$, you would first rewrite the equation in the form __?__.

18. If $\sqrt{x - 4} = 10 - x$, then $x - 4 =$ __?__.

19. The solutions of $x^2 - 21x + 104 = 0$ are __?__ and __?__.

20. The solution set of $x + \sqrt{x - 4} = 10$ is __?__.

**8–6 Equations in Quadratic Form**     *Pages 327–329*

21. The equation $x^{-2} - 5x^{-1} + 6 = 0$ is quadratic in __?__.

22. If $x^{-2} - 5x^{-1} + 6 = 0$, then $(x^{-1} -$ __?__$)(x^{-1} -$ __?__$) = 0$.

23. The roots of $x^{-2} - 5x^{-1} + 6 = 0$ are __?__ and __?__.

24. The roots of $(x + 2)^2 + 3(x + 2) - 4 = 0$ are __?__ and __?__.

**8–7 Parabolas in the xy-plane**     *Pages 330–335*

25. The graph of $y = a(x - h)^2 + k$ $(a \neq 0)$ is a __?__ having the point __?__ as vertex and the line with equation __?__ as axis.

26. The graph of $y = a(x - h)^2 + k$ $(a \neq 0)$ opens upward if __?__ and downward if __?__.

27. The vertex of the graph of $y = -2(x + 3)^2 - 4$ is the point __?__, its axis has the equation __?__, and the curve opens __?__.

28. The graph of $y = (x - 2)^2$ can be viewed as the result of shifting the graph of $y = x^2$ __?__ units to the __?__.

**8–8 Graphs of Quadratic Functions**     *Pages 335–339*

29. If $f(x) = ax^2 + bx + c$, then $f$ is a __?__ function.

30. The vertex of $f(x) = 2x^2 - 6x + 5$ is the point __?__, and its axis of symmetry has equation __?__.

31. The graph of $g(x) = 2x^2 - 4x - 5$ has __?__ x-intercept(s).

**8–9 Extrema of Quadratic Functions**     *Pages 339–343*

32. The graph of $f(x) = 2(x - 4)^2 + 6$ has $($ __?__$,$ __?__$)$ as its __?__ (maximum/minimum) point.

33. If the sum of two numbers is 25, then the maximum value their product can have is __?__.

34. If the difference of two numbers is 7, then the minimum value their product can have is __?__.

# Extra for Experts

## DESCARTES' RULE

Throughout this section $P(x)$ will denote a simplified polynomial with real coefficients.

When the coefficients of successive terms of $P(x)$ have opposite signs, a **variation in sign** is said to occur. For example,

$$x^5 - 2x^4 - 4x^2 + 3x - 5$$

has three variations in sign. The following is known as:

### DESCARTES' RULE OF SIGNS

The equation $P(x) = 0$ has:

1. As many positive roots as $P(x)$ has variations in sign, or fewer by an even number.
2. As many negative roots as $P(-x)$ has variations in sign, or fewer by an even number.

**Example 1.** State the possible number of **(a)** positive roots, and **(b)** negative roots of $x^4 - 3x^3 - 6x^2 + 3x - 5 = 0$.

*Solution:*   **a.** The polynomial has 3 variations in sign. There are either 3 positive roots or just 1 positive root.

**b.** Replacing $x$ by $-x$, you have

$$(-x)^4 - 3(-x)^3 - 6(-x)^2 + 3(-x) - 5 =$$
$$x^4 + 3x^3 - 6x^2 - 3x - 5,$$

which has *one* variation in sign. There is 1 negative root.

Do you see that the equation in Example 1 must have at least one positive root and exactly one negative root? Since there are *four* complex roots in all, you can state the possibilities for the nature of the roots of $x^4 - 3x^3 - 6x^2 + 3x + 5 = 0$ as follows:

|  | Number of positive real roots | Number of negative real roots | Number of imaginary roots |
|---|---|---|---|
| Possibility 1 | 3 | 1 | 0 |
| Possibility 2 | 1 | 1 | 2 |

A number $b$ is an **upper bound** for the roots of $P(x) = 0$ if no root exceeds $b$. Similarly, $a$ is a **lower bound** if every root of $P(x) = 0$ is greater than or equal to $a$. The following rules are useful.

## BOUNDS FOR THE ROOTS OF $P(x) = 0$

1. The positive number $b$ is an upper bound if, when $P(x)$ is divided by $x - b$, the coefficients of the quotient and remainder are all nonnegative or all nonpositive.
2. The negative number $a$ is a lower bound if when $P(x)$ is divided by $x - a$, the coefficients of the quotient and remainder are alternately nonnegative and nonpositive.

To apply these rules, you inspect the numbers in the last line of the synthetic substitution of various numbers for $x$ in $P(x)$.

**Example 2.** Find upper and lower bounds for the real roots of

$$x^3 - 3x^2 + 3x + 8 = 0.$$

*Solution:* Use synthetic substitution of 1, 2, 3, ... until an upper bound is found, and of $-1, -2, -3, \ldots$ until a lower bound is found.

|      | 1 | $-3$ | 3  | 8    |                        |
|------|---|------|----|------|------------------------|
| 1    | 1 | $-2$ | 1  | 9    |                        |
| 2    | 1 | $-1$ | 1  | 10   |                        |
| 3    | 1 | 0    | 3  | 17   | ← all nonnegative      |
| $-1$ | 1 | $-4$ | 7  | 1    |                        |
| $-2$ | 1 | $-5$ | 13 | $-18$| ← alternating          |

$\therefore$ the upper and lower bounds are 3 and $-2$, respectively.

# Exercises

In each of the following, **(a)** use Descartes' Rule to give possibilities for the nature of the roots; **(b)** find upper and lower bounds for the real roots, as in Example 2.

**1.** $x^3 - 6x^2 + 12x - 8 = 0$    **3.** $2x^3 - 7x^2 - 4 = 0$

**2.** $z^3 + 3z^2 - 2z - 5 = 0$    **4.** $2y^3 - 3y^2 - 3y + 3 = 0$

**5.** $x^4 - 3x^3 - x + 7 = 0$      **7.** $y^5 + y^4 - 2y - 4 = 0$

**6.** $x^4 + 3x^3 + 4x^2 - 36 = 0$      **8.** $z^5 - z^4 + 2z^3 - 3 = 0$

**9.** What is the greatest number of real roots that the equation

$$x^5 + x^4 + x^3 + x + 1 = 0$$

may have? the least?

**10.** What is the greatest number of real roots that the equation

$$x^5 + 2x^2 - 3 = 0$$

may have? the least?

**11.** Explain why $x^6 + 2x^4 + x^2 + 3 = 0$ can have no real roots.

**12.** Explain why $x^5 + 2x^2 - 3 = 0$ must have two imaginary roots.

**13.** Explain why $x^5 - 2x^2 - 3 = 0$ must have four imaginary roots.

**14.** Find one real root of $x^5 - x^4 + 3x^3 - 3x^2 + 2x - 2 = 0$ and apply Descartes' Rule to the depressed equation to show that there are no other real roots.

**15.** Find one real root of $x^5 + x^4 + 5x^3 + 3x^2 + 4x + 6 = 0$ and show that there are no other real roots.

**16.** Let $r$ be a positive number. Show that $-r$ is a root of $P(x) = 0$ if and only if $r$ is a root of $P(-x) = 0$.

**17.** Assuming part 1 of Descartes' Rule, use Exercise 16 to prove part 2.

**18.** Let $P(x) = (x - b)Q(x) + P(b)$, where $b > 0$. Show that if the coefficients of $Q(x)$ and $P(b)$ are nonnegative, then $P(x) = 0$ can have no root greater than $b$. (*Hint:* For all $x > b$, $x - b > 0$ and $Q(x) > 0$.)

**19.** Let $P(x) = (x - a)Q(x) + P(a)$, where $a < 0$. Show that if the coefficients of $Q(x)$ and $P(a)$ are alternately nonnegative and nonpositive, then $P(x) = 0$ can have no roots less than $a$. (*Hint:* Observe that if $x < a$, then $x - a < 0$, and $Q(x)$ and $P(a)$ have opposite signs.)

# 9

# Quadratic Relations and Systems

## DISTANCES IN A PLANE

### 9–1  Distance between Points

To find the distance between two points *on a number line,* you calculate the absolute value of the difference between the coordinates of the points. Thus for the points pictured in Figure 9–1, you have

$$RS = |6 - 2| = 4, \quad ST = |-1 - 6| = 7, \quad \text{and} \quad TR = |2 - (-1)| = 3.$$

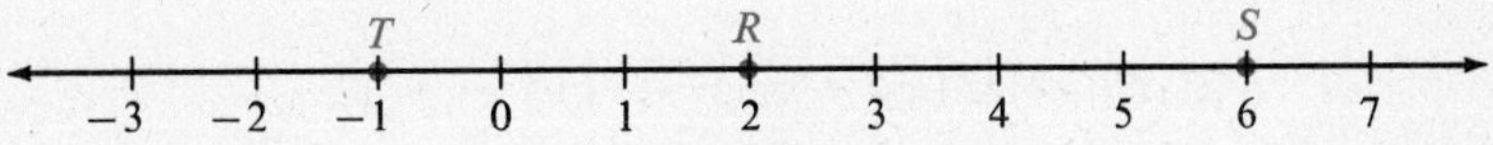

FIGURE 9–1

If two points *in a coordinate plane* lie on the same horizontal or vertical line, you can find the distance between them in a similar way. For example, the distances between the pairs of points shown in Figure 9–2 are

$$AB = |-4 - 2| = 6,$$
$$CD = |1 - (-2)| = 3,$$
$$PQ = |y_2 - y_1|.$$

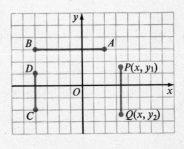

FIGURE 9–2

The formula for the distance between *any* two points in a coordinate plane depends on the following famous theorem:

---

**Pythagorean Theorem.** In a right triangle the square of the length $c$ of the hypotenuse equals the sum of the squares of the lengths $a$ and $b$ of the other two sides (the legs):

$$c^2 = a^2 + b^2$$

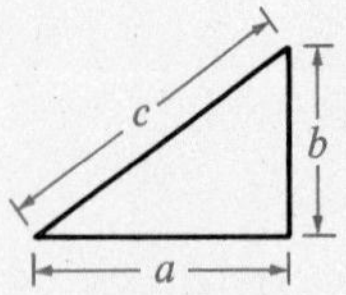

FIGURE 9–3

---

Recall that the hypotenuse is opposite the right angle and is the longest side of the triangle (see Figure 9–3).

To find the distance between any two points $P_1(x_1, y_1)$ and $P_2(x_2, y_2)$, construct a right triangle having the line segment $\overline{P_1P_2}$ as hypotenuse: Let $Q$ be the point where the horizontal line through $P_1$ intersects the vertical line through $P_2$ (see Figure 9–4). Notice that the coordinates of $Q$ are $(x_2, y_1)$. Therefore the legs of the right triangle $P_1QP_2$ have lengths $P_1Q = |x_2 - x_1|$ and $QP_2 = |y_2 - y_1|$. Now use the Pythagorean Theorem:

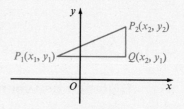

FIGURE 9–4

$$(P_1P_2)^2 = (P_1Q)^2 + (QP_2)^2$$
$$= |x_2 - x_1|^2 + |y_2 - y_1|^2$$

Since $|x_2 - x_1|^2 = (x_2 - x_1)^2$ and $|y_2 - y_1|^2 = (y_2 - y_1)^2$, you have the distance formula

$$P_1P_2 = \sqrt{(x_2 - x_1)^2 + (y_2 - y_1)^2}.$$

**Example 1.** Find the distance between $P_1(4, 0)$ and $P_2(-2, 3)$.

*Solution:* 
$$P_1P_2 = \sqrt{(-2 - 4)^2 + (3 - 0)^2}$$
$$= \sqrt{36 + 9} = \sqrt{45} = 3\sqrt{5}$$

**Example 2.** Find a formula for the distance $r$ from the origin to an arbitrary point $(x, y)$.

*Solution:* 
$$r = \sqrt{(x - 0)^2 + (y - 0)^2}$$
$$r = \sqrt{x^2 + y^2}$$

The distance formula can be used to show that the **midpoint** of the line segment joining $P_1(x_1, y_1)$ and $P_2(x_2, y_2)$ is

$$M\left(\frac{x_1 + x_2}{2}, \frac{y_1 + y_2}{2}\right)$$

(see Exercise 21, page 356).

**Example 3.** Find the midpoint of the line segment joining $(5, 3)$ and $(-1, 0)$.

*Solution:* The midpoint is $\left(\dfrac{5 + (-1)}{2}, \dfrac{3 + 0}{2}\right)$, or $(2, \frac{3}{2})$.

*WRITTEN EXERCISES*

Find **(a)** the coordinates of the midpoint and **(b)** the length of the line segment joining the points with given coordinates. Express all radicals in simple form.

A

**1.** $(6, -2), (9, -6)$

**2.** $(-4, 3), (8, 8)$

**3.** $(5, \frac{2}{3}), (3, \frac{5}{3})$

**4.** $(\frac{3}{5}, \frac{1}{4}), (-\frac{6}{5}, \frac{7}{4})$

**5.** $(4, 1), (4, -4)$

**6.** $(3, -2), (-5, -2)$

**7.** $(\sqrt{3}, 2), (2\sqrt{3}, -1)$

**8.** $(5, \sqrt{6}), (8, -2\sqrt{6})$

**9.** $(2\sqrt{3}, \sqrt{2}), (\sqrt{3}, 2\sqrt{2})$

**10.** $(\sqrt{10}, 2\sqrt{5}), (3\sqrt{10}, 5\sqrt{5})$

**11.** $(r, s), (2r, -s)$

**12.** $(-3u, 2v), (u, -2v)$

Find the coordinates of $A$ if $M$ is the midpoint of line segment $\overline{AB}$.

**Sample.** $M(3, 1)$; $B(5, 4)$

*Solution:* By inspection, the $x$-coordinate of $M$ is $5 - 3$ or 2 less than the $x$-coordinate of $B$, so the $x$-coordinate of $A$ will be 2 less than the $x$-coordinate of $M$, or 1. Similarly the $y$-coordinate of $A$ will be 3 less than the $y$-coordinate of $M$, or $-2$. Therefore, the coordinates of $A$ are $(1, -2)$.

**13.** $M(2, 6)$; $B(5, 8)$

**14.** $M(-4, 3)$; $B(-1, 7)$

**15.** $M(2, -5)$; $B(0, 6)$

**16.** $M(-3, -8)$; $B(2, 1)$

The vertices of a triangle are as specified. For each triangle, find **(a)** the perimeter, **(b)** whether it is isosceles, and **(c)** whether it is a right triangle. If it is a right triangle, find its area.

**B**

**17.** $(0, 1)$; $(8, -7)$; $(1, -6)$    **21.** $(5, 6)$; $(11, -2)$; $(-10, -2)$

**18.** $(-4, 4)$; $(-2, -4)$; $(6, -2)$    **22.** $(10, 1)$; $(3, 1)$; $(5, 9)$

**19.** $(0, 6)$; $(9, -6)$; $(-3, 0)$    **23.** $(-1, 5)$; $(8, -7)$; $(4, 1)$

**20.** $(0, 0)$; $(6, 0)$; $(3, 3)$    **24.** $(2, 2\sqrt{3})$; $(0, 0)$; $(3, -\sqrt{3})$

Show that the given points are collinear by using **(a)** the slope formula, and **(b)** the distance formula.

**25.** $(1, -4)$; $(-2, -10)$; $(4, 2)$    **27.** $(1, 0)$; $(2, 6)$; $(-2, -18)$

**26.** $(2, 0)$, $(-2, 3)$, $(\frac{2}{3}, 1)$    **28.** $(0, 2)$; $(5, 1)$; $(-10, 4)$

*PROBLEMS*

Sketch figures as needed to help solve each problem. Express irrational results in simple radical form and correct to tenths.

**A**

**1.** How long must a wire be to stretch from the top of a 40-foot telephone pole to a point on the ground 30 feet from the foot of the pole?

**2.** Find the length of the diagonal of a rectangle whose length is 12 inches and whose area is 60 square inches.

**3.** Find the length of the diagonal of a square whose area is 80 square feet.

**4.** Show that the altitude $h$ of an equilateral triangle with sides of length $s$ is given by $h = \dfrac{\sqrt{3}}{2} s$.

(*Hint:* The altitude of an equilateral triangle bisects the base.)

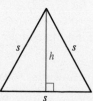

**5.** Use the result of Exercise 4 to show that the area of an equilateral triangle is given by $A = \dfrac{\sqrt{3}}{4} s^2$.

**6.** How far below the point where the wire is connected to the wall should the brace pictured at the right be placed, if the wire is 3 feet long and the lamp is to hang 15 inches below the point this wire meets the brace?

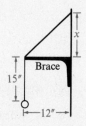

Find the area of the shaded region in each figure if $\overline{AB}$ is 10 centimeters long, and the right triangle shown is isosceles.

**7.**                                 **8.**

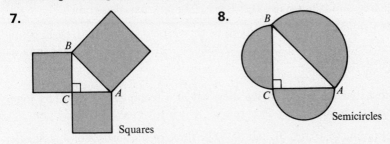

Squares                              Semicircles

**Sample.**   Find the equation expressing the condition that the point $(x, y)$ is twice as far from the point $(2, 1)$ as it is from the point $(-1, 3)$.

*Solution:*   Make a sketch. The distance to $(x, y)$ from the point $(-1, 3)$ is given by

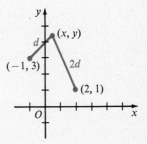

$$\sqrt{(x + 1)^2 + (y - 3)^2}$$

and the distance to $(x, y)$ from the point $(2, 1)$ is

$$\sqrt{(x - 2)^2 + (y - 1)^2}.$$

Since the second distance is twice the first,

$$\sqrt{(x - 2)^2 + (y - 1)^2} = 2\sqrt{(x + 1)^2 + (y - 3)^2}.$$

Squaring both members, you have

$$(x - 2)^2 + (y - 1)^2 = 4[(x + 1)^2 + (y - 3)^2],$$

which can be simplified as follows:

$$x^2 - 4x + 4 + y^2 - 2y + 1 = 4x^2 + 8x + 4 + 4y^2 - 24y + 36$$
$$3x^2 + 3y^2 + 12x - 22y + 35 = 0$$

Find the equation expressing the condition that point $(x, y)$ is:

**B**   **9.** Equidistant from $(0, 4)$ and $(4, 0)$.

**10.** Equidistant from $(2, 4)$ and $(0, 0)$.

**11.** Twice as far from $(0, 0)$ as it is from $(4, 0)$.

**12.** Twice as far from $(2, 2)$ as it is from $(-1, 1)$.

**13.** 6 units from $(3, 3)$.         **14.** 4 units from $(1, 4)$.

**15.** Show that the sum of the squares of the four sides of the parallelogram with vertices $(0, 0)$, $(1, 1)$, $(2, 1)$, and $(1, 0)$ is equal to the sum of the squares of its diagonals. This is true of every parallelogram.

**16.** Let $C(0, 0)$, $B(4, 0)$, and $A(0, 6)$ be the vertices of a right triangle. Show that the line joining $C$ and the midpoint of side $AB$ is half as long as that side.

C **17.** Show that the length of the diagonal of a rectangular box $a$ feet long, $b$ feet wide, and $c$ feet deep is $\sqrt{a^2 + b^2 + c^2}$ feet.

**18.** Let points $(0, 0)$, $(a, 0)$, and $(0, b)$ be three of the vertices of a rectangle. Find the fourth vertex and show that the diagonals of the rectangle bisect each other.

**19.** Let $O(0, 0)$, $A(a, 0)$, and $B(0, b)$ be the vertices of a right triangle. Show that the midpoint of side $AB$ is equidistant from the three vertices.

**20.** Let $R(-a, 0)$, $S(a, 0)$, and $T(0, b)$ be vertices of an isosceles triangle. Prove that the medians bisecting sides $RT$ and $ST$ are equal in length.

**21.** Show that the coordinates of the midpoint of the line segment joining $P_1(x_1, y_1)$ and $P_2(x_2, y_2)$ are

$$M\left(\frac{x_1 + x_2}{2}, \frac{y_1 + y_2}{2}\right).$$

(*Hint:* Show that $P_1M = \frac{1}{2}P_1P_2$ and that $MP_2 = \frac{1}{2}P_1P_2$.)

## 9–2   Perpendicular Lines

In Section 3–4 you learned that two nonvertical lines are *parallel* if and only if their *slopes are equal*. Thus, you can tell that the lines with equations $6x - 2y = 1$ and $3x = y - 1$ are parallel by writing their equations as $y = 3x - \frac{1}{2}$ and $y = 3x + 1$, respectively, and noting that the lines both have slope 3.

Two lines which intersect at right angles are called **perpendicular lines**. Thus any vertical line, such as $x = -3$ in Figure 9–5, is perpendicular to any horizontal line, as, for example, $y = 2$.

The lines shown in Figure 9–6 also are perpendicular. Their equations are $y = \frac{1}{2}x + 3$ and $y = -2x + 1$. Notice that the product of their slopes is $-1$: $\frac{1}{2} \times (-2) = -1$. This relationship is typical of perpendicular lines neither of which is vertical. The proof of the theorem on page 357 is outlined in Exercises 21–24, page 359.

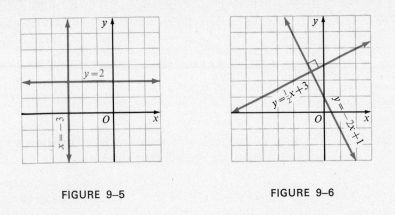

FIGURE 9–5                    FIGURE 9–6

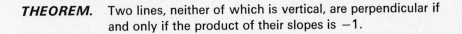

**THEOREM.**    Two lines, neither of which is vertical, are perpendicular if and only if the product of their slopes is −1.

Let us denote the lines by $L_1$ and $L_2$ and their slopes by $m_1$ and $m_2$, respectively. The theorem is equivalent to the two statements:

1. If $L_1$ and $L_2$ are perpendicular, then $m_1m_2 = -1$.
2. If $m_1m_2 = -1$, then $L_1$ and $L_2$ are perpendicular.

Notice that $m_1m_2 = -1$ means that $m_1 = -\dfrac{1}{m_2}$ and $m_2 = -\dfrac{1}{m_1}$.

Thus, nonvertical lines are perpendicular if and only if their slopes are *negative reciprocals* of each other.

**Example.**    Find an equation of the line which passes through $(1, -3)$ and is perpendicular to the line $2x + 5y = 10$.

*Solution:*     **1.** $2x + 5y = 10$
$$y = -\tfrac{2}{5}x + 2$$

∴ the slope of $2x + 5y = 10$ is $-\tfrac{2}{5}$. Any line perpendicular to it has slope $-\dfrac{1}{-\frac{2}{5}} = \dfrac{5}{2}$.

**2.** Use the point-slope form (page 100):

$$y - y_1 = m(x - x_1)$$
$$\downarrow \qquad \downarrow \qquad \downarrow$$
$$y - (-3) = \tfrac{5}{2}(x - 1)$$
$$2y + 6 = 5(x - 1), \quad \text{or} \quad 5x - 2y = 11$$

*ORAL EXERCISES*

State the slope of the line perpendicular to the line whose equation is given.
If the slope is not defined, so state.

**Sample.** $y = 3x - 4$

*Solution:* The slope of the graph of $y = 3x - 4$ is 3. Therefore, a line
perpendicular to this line will have slope $-\frac{1}{3}$.

**1.** $y = 5x - 3$      **4.** $y = 7x - 10$      **7.** $y = -\frac{1}{5}x + 10$

**2.** $y = -4x + 7$      **5.** $y = \frac{2}{3}x + \frac{1}{2}$      **8.** $y = 4$

**3.** $y = -6x$      **6.** $y = \frac{4}{3}x - \frac{1}{5}$      **9.** $x = 6$

*WRITTEN EXERCISES*

Find the slope of a line perpendicular to the line containing the given points.

**A**   **1.** $(0, 3)$; $(3, 0)$      **4.** $(6, 2)$; $(-5, 3)$      **7.** $(-2, 2)$; $(-3, 6)$

    **2.** $(5, -2)$; $(-2, 5)$      **5.** $(3, 2)$; $(-2, 3)$      **8.** $(4, -2)$; $(6, 4)$

    **3.** $(0, 0)$; $(-3, -4)$      **6.** $(3, 4)$; $(5, 8)$

Find an equation of the line satisfying the given conditions.

**9.** Passes through $(4, 5)$ and is perpendicular to the line $3x - 2y = 6$.

**10.** Passes through $(-2, 1)$ and is perpendicular to the line $x - 3y = 8$.

**11.** Passes through the origin and is perpendicular to the line $3x - y = 7$.

**12.** Passes through the origin and is perpendicular to the line $2x + 3y = 8$.

**B**   **13.** Is the perpendicular bisector of the segment with endpoints $(5, 2)$ and
$(-1, 4)$.

    **14.** Is the perpendicular bisector of the segment with endpoints $(4, 7)$ and
$(-2, 1)$.

Use **(a)** the slope relationship, and **(b)** the converse of the Pythagorean
Theorem to show that the given points are the vertices of a right triangle.

**15.** $(10, 4)$; $(3, -5)$; $(1, 1)$      **17.** $(3, -2)$; $(7, -5)$; $(5, -6)$

**16.** $(2, 4)$; $(-2, 3)$; $(8, -20)$      **18.** $(-2, 2)$; $(1, 3)$; $(2, 0)$

Show that the given points are the vertices of a rectangle.

**19.** $(8, 0)$; $(6, 6)$; $(-3, 3)$; $(-1, -3)$

**20.** $(10, 8)$; $(-3, 9)$; $(-4, -4)$; $(9, -5)$

In Exercises 21–24, $L_1$ and $L_2$ are lines having slope $m_1$ and $m_2$, respectively; $P(r, s)$ is their point of intersection; and $T_1$ and $T_2$ are the points in which the vertical line $x = r + 1$ intersects $L_1$ and $L_2$, respectively. (See the adjoining figure.) Exercises 21–24 give steps in the proof of the theorem on page 357.

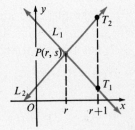

**C**  **21.** Show that the coordinates of $T_1$ and $T_2$ are $(r + 1, s + m_1)$ and $(r + 1, s + m_2)$, respectively.

**22.** Find the distances $T_1T_2$, $PT_1$, and $PT_2$ in terms of $m_1$ and $m_2$.

**23.** If $L_1$ and $L_2$ are perpendicular, then triangle $T_1PT_2$ is a right triangle. Apply the Pythagorean Theorem, and simplify to obtain $m_1m_2 = -1$.

**24.** Suppose $m_1m_2 = -1$. Prove that $(T_1T_2)^2 = (PT_1)^2 + (PT_2)^2$. It follows from the converse of the Pythagorean Theorem that triangle $T_1PT_2$ has a right angle at $P$. Hence $L_1$ is perpendicular to $L_2$.

## GRAPHING QUADRATIC RELATIONS

### 9–3  Circles

With the help of the distance formula, you can use algebra to study many important geometrical objects: circles, for example. Recall that a circle with center $C$ and radius $r$ is the set of all points $P$ such that the distance from $P$ to $C$ is $r$.

**Example 1.**  Find an equation of the circle having center $C(3, -2)$ and radius 4.

*Solution:*  A point $P(x, y)$ is on the circle if and only if $PC = 4$, that is:

$$\sqrt{(x - 3)^2 + (y - (-2))^2} = 4$$
$$\sqrt{(x - 3)^2 + (y + 2)^2} = 4$$
$$(x - 3)^2 + (y + 2)^2 = 16$$

The steps above can be reversed. Therefore, a point belongs to the circle if and only if its coordinates satisfy

$$(x - 3)^2 + (y + 2)^2 = 16.$$

By using the method of Example 1, you can show that an equation of the circle with center $(h, k)$ and radius $r$ is

$$(x - h)^2 + (y - k)^2 = r^2.$$

For example, the circle having center $(-3, 1)$ and radius 4, shown in Figure 9–7, has the equation

$$(x - [-3])^2 + (y - 1)^2 = 4^2,$$

or

$$(x + 3)^2 + (y - 1)^2 = 16,$$

which is equivalent to

$$x^2 + y^2 + 6x - 2y - 6 = 0.$$

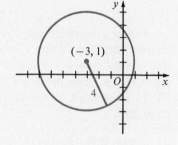

FIGURE 9–7

To transform the latter equation back into the original form (which tells directly the center and radius of the circle), you complete the square twice (recall page 309):

1. Add 6 to both members.

$$x^2 + y^2 + 6x - 2y - 6 = 0$$
$$x^2 + y^2 + 6x - 2y = 6$$

2. Group the terms involving $x$ and those involving $y$.

$$(x^2 + 6x \quad) + (y^2 - 2y \quad) = 6$$

3. Complete the square within each set of parentheses, adding $(\frac{6}{2})^2$, or 9, and $(\frac{-2}{2})^2$, or 1, to both members of the equation.

$$(x^2 + 6x + 9) + (y^2 - 2y + 1) = 6 + 9 + 1$$

4. Write in the form $(x - h)^2 + (y - k)^2 = r^2$.

$$(x + 3)^2 + (y - 1)^2 = 16$$

**Example 2.** Graph the inequality $x^2 + y^2 - 8x + 4y + 16 \leq 0$; that is, graph the relation $\{(x, y): x^2 + y^2 - 8x + 4y + 16 \leq 0\}$.

*Solution:* Transform the inequality into the form $(x - h)^2 + (y - k)^2 \leq r^2$:

1. $$x^2 + y^2 - 8x + 4y \leq -16$$

2. $$(x^2 - 8x \quad) + (y^2 + 4y \quad) \leq -16$$

3. $$(x^2 - 8x + 16) + (y^2 + 4y + 4) \leq -16 + 16 + 4$$

**4.** $(x - 4)^2 + (y + 2)^2 \leq 4$

$(x - 4)^2 + (y - (-2))^2 \leq 2^2$

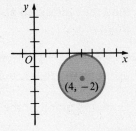

Using a compass, draw the circle with center $(4, -2)$ and radius 2. The graph is the circle and the shaded region within it.

*ORAL EXERCISES*

State an equation in the form $(x - h)^2 + (y - k)^2 = r^2$ for the circle with the given point as center and radius $r$.

**Sample.** $(5, -1)$; $r = 6$    *Solution:*    $(x - 5)^2 + (y + 1)^2 = 36$

**1.** $(4, 1)$; $r = 4$      **4.** $(0, -4)$; $r = 5$      **7.** $(-2, 5)$; $r = \frac{5}{4}$

**2.** $(3, 5)$; $r = 3$      **5.** $(-3, -7)$; $r = 2$      **8.** $(0, 0)$; $r = a$

**3.** $(-2, 0)$; $r = 1$      **6.** $(3, -1)$; $r = \frac{1}{2}$      **9.** $(h, k)$; $r = b$

*WRITTEN EXERCISES*

Write an equation in the form $x^2 + y^2 + ax + by + c = 0$ for the circle with given center and radius.

**A** **1–8.** Oral Exercises 1–8.                **9.** $(\frac{1}{2}, \frac{3}{2})$; 1

**10.** $(\frac{2}{3}, \frac{3}{5})$; $\frac{1}{3}$      **11.** $(\frac{2}{3}, \frac{3}{5})$; $\sqrt{3}$      **12.** $(\frac{1}{10}, \frac{1}{10})$; $\sqrt{10}$

Graph the given equation. Use an appropriate scale on each set of axes.

**13.** $x^2 + y^2 - 49 = 0$          **17.** $x^2 + y^2 + 2x + 4y - 4 = 0$

**14.** $x^2 + y^2 = 81$          **18.** $x^2 + y^2 - 4x + 8y - 5 = 0$

**15.** $x^2 + y^2 + 4x = 0$          **19.** $2x^2 + 2y^2 - x + y - \frac{1}{8} = 0$

**16.** $x^2 + y^2 - 6y = 0$          **20.** $4x^2 + 4y^2 + 3x - y - \frac{3}{8} = 0$

Graph the given relation.

**B** **21.** $\{(x, y): x^2 + y^2 \leq 4\}$          **23.** $\{(x, y): x^2 + y^2 - 4y < 0\}$

**22.** $\{(x, y): x^2 + y^2 \geq 49\}$          **24.** $\{(x, y): x^2 + y^2 + 10x \geq 0\}$

Find the radius and an equation of the circle with center $C$ and containing point $P$.

**25.** $C(0, 0)$; $P(3, 7)$          **27.** $C(2, 1)$; $P(-1, 4)$

**26.** $C(0, 0)$; $P(-3, -5)$        **28.** $C(-3, -4)$; $P(0, 0)$

Draw the graph of the given relation.

**29.** $\{(x, y)\colon y = \sqrt{4 - x^2}\}$        **31.** $\{(x, y)\colon x = \sqrt{1 - y^2}\}$

**30.** $\{(x, y)\colon y = \sqrt{2x - x^2}\}$      **32.** $\{(x, y)\colon 4 \le x^2 + y^2 \le 9\}$

## 9–4   Parabolas

Suppose that a point $P$ moves so that its distance from the point $F(2, 3)$ is always equal to its (perpendicular) distance from the line $L$ having equation $y = -1$ (see Figure 9–8). You can find an equation of the path of $P$ by noting that the following statements are equivalent.

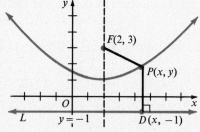

FIGURE 9–8

$$PF = PD$$
$$\sqrt{(x - 2)^2 + (y - 3)^2} = \sqrt{(x - x)^2 + (y - (-1))^2}$$
$$(x - 2)^2 + (y - 3)^2 = (y + 1)^2$$
$$(x - 2)^2 + y^2 - 6y + 9 = y^2 + 2y + 1$$
$$(x - 2)^2 + 8 = 8y$$
$$y = \tfrac{1}{8}(x - 2)^2 + 1$$

Because the last equation is of the form $y = a(x - h)^2 + k$, displayed on page 332, you see that the path of $P$ is a *parabola*.

    The definition of *parabola* given in Section 8–7 is equivalent to the following more geometric one: A parabola is the set of all points equidistant from a fixed line called the directrix of the parabola and a fixed point not on the line called the focus (plural: foci) of the parabola. Figure 9–9 shows two other features of a parabola (discussed in Section 8–7):

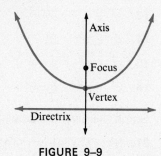

FIGURE 9–9

axis (or **axis of symmetry**), the line through the focus perpendicular to
the directrix.

**vertex,** the point in which the parabola intersects its axis.

Notice that the vertex is the midpoint of the segment of the axis
between the focus and the directrix.

**Example.**   Find an equation of the parabola whose focus is the origin
and whose directrix is the line $x = 4$. Draw the parabola and
label its focus, vertex, directrix and axis.

*Solution:*

$$PF = PD$$
$$\sqrt{x^2 + y^2} = \sqrt{(x - 4)^2 + (y - y)^2}$$
$$x^2 + y^2 = (x - 4)^2$$
$$x^2 + y^2 = x^2 - 8x + 16$$
$$y^2 = -8x + 16$$
$$x = -\tfrac{1}{8}y^2 + 2$$

To construct a table, assign values
to $y$ and then compute the corresponding values of $x$.

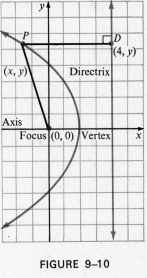

| $x$ | $-\tfrac{1}{8} \cdot y^2 + 2$ | $y$ |
|-----|-----|-----|
| $-\tfrac{5}{2}$ | $-\tfrac{1}{8} \cdot 6^2 + 2$ | 6 |
| 0 | $-\tfrac{1}{8} \cdot 4^2 + 2$ | 4 |
| $\tfrac{3}{2}$ | $-\tfrac{1}{8} \cdot 2^2 + 2$ | 2 |
| 2 | $-\tfrac{1}{8} \cdot 0^2 + 2$ | 0 |
| $\tfrac{3}{2}$ | $-\tfrac{1}{8} \cdot (-2)^2 + 2$ | $-2$ |
| 0 | $-\tfrac{1}{8} \cdot (-4)^2 + 2$ | $-4$ |
| $-\tfrac{5}{2}$ | $-\tfrac{1}{8} \cdot (-6)^2 + 2$ | $-6$ |

FIGURE 9–10

Notice that the points of the graph in the preceding example can be
paired so that the $x$-coordinates of the points in a pair are the same
but the $y$-coordinates are negatives of each other: $(\tfrac{3}{2}, 2)$ with $(\tfrac{3}{2}, -2)$;
$(0, 4)$ with $(0, -4)$; and so on. If a curve has the property that whenever
the point $(r, t)$ belongs to it, the point $(r, -t)$ also belongs to it, then
the curve is said to be **symmetric with respect to the $x$-axis.** Compare
this definition with that of *symmetry with respect to the $y$-axis,* page 330.

Parabolas occur frequently in science and technology. The optical property of parabolas indicated in Figure 9–11 makes parabolic mirrors useful for astronomical telescopes and for searchlight reflectors. The supporting cables of a suspension bridge (ideally) hang in the form of parabolic arcs (Figure 9–12) and parabolic arches are often used in concrete bridges supported from below (Figure 9–13). Unless they are falling vertically, particles under the influence of gravity alone move along parabolas; thus the jets of water in a fountain approximate parabolic arcs.

FIGURE 9–11

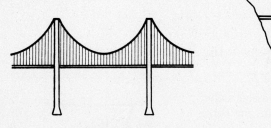

FIGURE 9–12

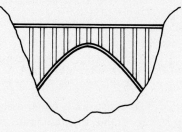

FIGURE 9–13

*WRITTEN EXERCISES*

**Graph each equation.**

Sample.     $x = 5 - 2y - y^2$

*Solution:*     **1.** To determine the vertex, complete the square in $y$.

$$x = 5 - 2y - y^2$$
$$5 - x = y^2 + 2y$$
$$5 - x + 1 = y^2 + 2y + 1$$
$$6 - x = (y + 1)^2$$
$$x = -[y - (-1)]^2 + 6$$

**2.** In plotting points use the fact that the line with equation $y = -1$ is the axis of symmetry and $(6, -1)$ is the vertex. The parabola will open to the left. See table and graph at top of next page.

| $x$ | -3 | 2 | 5 | 6 | 5 | 2 | -3 |
|---|---|---|---|---|---|---|---|
| $y$ | -4 | -3 | -2 | -1 | 0 | 1 | 2 |

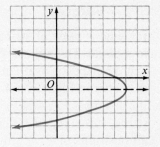

**A**  1. $y = x^2 - 2x - 3$

2. $y = x^2 + 4x - 5$

3. $y = 6 + x - x^2$

4. $y = 4 + 3x - x^2$

5. $x = \frac{1}{2}y^2$

6. $x = -3y^2$

7. $x = (y + 2)^2 - 1$

8. $x = -(y - 2)^2 + 2$

9. $x = y^2 - 6y + 7$

10. $x = 5 + 4y - y^2$

**B**  11. $y \geq x^2 - 5x - 6$

12. $y < x^2 + 2x - 15$

13. $x \geq (y - 2)^2 - 3$

14. $x \leq 4 - (y - 1)^2$

Find an equation of the parabola with given focus and directrix.

15. focus: $(2, 0)$; directrix: $y = -2$

16. focus: $(-1, 3)$; directrix: $y = 5$

17. focus: $(0, 0)$; directrix: $x = -2$

18. focus: $(1, 1)$; directrix: $x = 3$

Find an equation satisfied by all points $P(x, y)$ subject to the given conditions.

**C**  19. The slope of the line joining $P$ and $(-1, 3)$ is 2.

20. The perpendicular distance from $P$ to the line with equation $x = 6$ is 4.

21. The distance from $P$ to the point $(2, -1)$ is equal to the distance from $P$ to the $x$-axis.

22. The distance from $P$ to the point $(-1, 3)$ is equal to the distance from $P$ to the line $x = 4$.

## 9–5   Ellipses

Imagine a piece of string with ends fastened at two points $F_1$ and $F_2$ in a plane. The point $P$ of a moving pencil which holds the string taut will describe an oval curve called an *ellipse*, as shown in Figure 9–14. Because the sum of the distances $PF_1$ and $PF_2$ is the length of the string, this sum, $PF_1 + PF_2$, is the same for all positions of $P$.

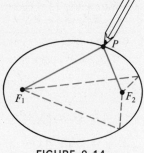

An ellipse is a set of points in a plane such that for each point of the set, the sum of its distances to two fixed points is a given constant. Each of the fixed points is called a focus. (The given constant must be greater than the distance between the foci.) For each

FIGURE 9–14

point $P$ of an ellipse the distances $PF_1$ and $PF_2$ are called the focal radii of $P$.

The ellipse having the points $F_1(-c, 0)$ and $F_2(c, 0)$ as foci and the constant $2a$ as the sum of the focal radii for each of its points has the equation

$$\frac{x^2}{a^2} + \frac{y^2}{b^2} = 1, \text{ where } b^2 = a^2 - c^2.$$

You can verify this in the special case where $c = 4$ and $a = 5$ by observing that the following equations are equivalent:

$$PF_1 \qquad + \qquad PF_2 \qquad = 10$$
$$\sqrt{(x+4)^2 + (y-0)^2} + \sqrt{(x-4)^2 + (y-0)^2} = 10$$
$$\sqrt{(x+4)^2 + y^2} \qquad\qquad = 10 - \sqrt{(x-4)^2 + y^2}$$

Square each member and simplify:

$$(x+4)^2 + y^2 = 100 - 20\sqrt{(x-4)^2 + y^2} + (x-4)^2 + y^2$$
$$5\sqrt{(x-4)^2 + y^2} = 25 - 4x$$

Square again, and simplify:

$$25(x^2 - 8x + 16 + y^2) = 625 - 200x + 16x^2$$
$$9x^2 + 25y^2 = 225$$

Divide both members by 225:

$$\frac{x^2}{25} + \frac{y^2}{9} = 1$$

You can draw an ellipse quite easily if you first study its equation as illustrated below for the case $\dfrac{x^2}{25} + \dfrac{y^2}{9} = 1$.

1. Whenever $(r, s)$ satisfies the equation, so also do $(r, -s)$ and $(-r, s)$.
   *∴ the curve is symmetric with respect to both coordinate axes.*

2. Replacing $y$ by 0, you obtain $\dfrac{x^2}{25} + \dfrac{0}{9} = 1$; $x^2 = 25$, or $x = \pm 5$.
   *∴ the x-intercepts of the ellipse are 5 and $-5$.*

3. Replacing $x$ by 0, you obtain $\dfrac{0}{25} + \dfrac{y^2}{9} = 1$; $y^2 = 9$, or $y = \pm 3$.
   *∴ the y-intercepts of the ellipse are 3 and $-3$.*

4. Transforming the equation to express each variable in terms of the other, you obtain

$$y = \pm\tfrac{3}{5}\sqrt{25 - x^2} \quad \text{and} \quad x = \pm\tfrac{5}{3}\sqrt{9 - y^2}.$$

Hence $y$ is real if and only if $|x| \le 5$, and $x$ is real if and only if $|y| \le 3$.

*∴ the ellipse lies in the rectangle formed by the lines $x = 5$, $x = -5$, $y = 3$, and $y = -3$.*

With this information about symmetry, intercepts, and extent, you can graph the equation with the help of a very short table of *first quadrant* points (Figure 9–15).

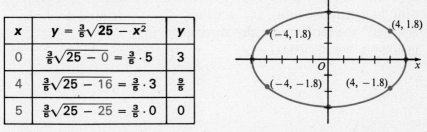

| x | $y = \tfrac{3}{5}\sqrt{25 - x^2}$ | y |
|---|---|---|
| 0 | $\tfrac{3}{5}\sqrt{25 - 0} = \tfrac{3}{5}\cdot 5$ | 3 |
| 4 | $\tfrac{3}{5}\sqrt{25 - 16} = \tfrac{3}{5}\cdot 3$ | $\tfrac{9}{5}$ |
| 5 | $\tfrac{3}{5}\sqrt{25 - 25} = \tfrac{3}{5}\cdot 0$ | 0 |

**FIGURE 9–15**

When the foci of an ellipse are on the $y$-axis at $(0, -c)$ and $(0, c)$ and the sum of the focal radii is $2a$, an equation of the ellipse is

$$\frac{x^2}{b^2} + \frac{y^2}{a^2} = 1, \text{ where } b^2 = a^2 - c^2.$$

**Example.** Graph the relation $\{(x, y): 16x^2 + y^2 = 16\}$.

*Solution:*     $16x^2 + y^2 = 16;\ \dfrac{x^2}{1} + \dfrac{y^2}{16} = 1$

1. Symmetric with respect to both axes

2. $x$-intercepts 1 and $-1$

3. $y$-intercepts 4 and $-4$

4. Extent: $-1 \le x \le 1$,
        $-4 \le y \le 4$

Table of first-quadrant points:

| $x$ | $y = 4\sqrt{1 - x^2}$ | $y$ |
|---|---|---|
| 0 | $4\sqrt{1 - 0} = 4$ | 4 |
| $\frac{1}{2}$ | $4\sqrt{1 - \frac{1}{4}} = 2\sqrt{3}$ | $\doteq 3.5$ |
| 1 | $4\sqrt{1 - 1} = 0$ | 0 |

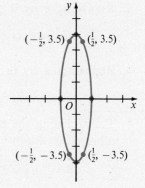

The reflective property of ellipses illustrated in Figure 9–16 has been used in the design of so-called "whispering chambers," in which a person whispering at one focus can be heard clearly at the other focus but nowhere else in the room. The planets move in ellipses with the sun at one focus, and satellites "circling" the earth are usually moving in ellipses, not in circles.

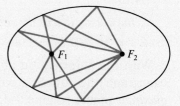

FIGURE 9–16

*WRITTEN EXERCISES*

Graph each equation or relation. Use an appropriate scale on each set of axes.

**A**   1. $\dfrac{x^2}{9} + \dfrac{y^2}{1} = 1$      3. $\dfrac{x^2}{16} + \dfrac{y^2}{9} = 1$      5. $\dfrac{x^2}{36} + \dfrac{y^2}{25} = 1$

     2. $\dfrac{x^2}{1} + \dfrac{y^2}{4} = 1$      4. $\dfrac{x^2}{4} + \dfrac{y^2}{25} = 1$      6. $\dfrac{x^2}{25} + \dfrac{y^2}{49} = 1$

**7.** $x^2 + 25y^2 = 100$

**8.** $16x^2 + y^2 = 64$

**9.** $x^2 + 9y^2 = 1$

**10.** $4x^2 + y^2 = 1$

**11.** $16x^2 + 1000y^2 = 1600$

**12.** $25x^2 + 64y^2 = 1600$

B  **13.** $\{(x, y): 4x^2 + y^2 \leq 16\}$

**14.** $\{(x, y): x^2 + 9y^2 \geq 36\}$

**15.** $\{(x, y): 36x^2 + y^2 > 144\}$

**16.** $\{(x, y): 9x^2 + 25y^2 < 225\}$

**17.** $\{(x, y): 36 \leq 4x^2 + 9y^2 \leq 144\}$

**18.** $\{(x, y): 100 \leq 25x^2 + 4y^2 \leq 400\}$

Use the definition of an ellipse to find an equation whose foci have the given coordinates, if the sum of the focal radii is as given.

**19.** $(2, 0), (-2, 0)$; 6

**20.** $(0, 3), (0, -3)$; 8

**21.** $(4, 0), (-4, 0)$; 10

**22.** $(0, 2), (0, -2)$; 5

C  **23.** $(c, 0), (-c, 0)$; $2a$

**24.** $(0, c), (0, -c)$; $2a$

Graph each equation or inequality.

**25.** $y \geq 4\sqrt{1 - x^2}$

**26.** $x \leq -3\sqrt{4 - y^2}$

**27.** $\dfrac{(x - 2)^2}{4} + \dfrac{(y - 1)^2}{9} = 1$

**28.** $\dfrac{(x - 1)^2}{16} + \dfrac{(y + 2)^2}{9} = 1$

## 9–6  Hyperbolas

Using a navigation system called *LORAN*, a pilot can guide the aircraft by maintaining a *constant difference* between the distances from two fixed points at which radio sending stations are located. The curve along which the pilot flies is called a branch of a hyperbola (see Figure 9–17).

A **hyperbola** is a set of points in a plane such that for each point of the set, the absolute value of the difference of its distances from two fixed points, called **foci**, is a given constant.

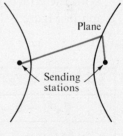

FIGURE 9–17

Consider, for example, the hyperbola for which the given constant is 6 and the foci are $F_1(-5, 0)$ and $F_2(5, 0)$. A point $P(x, y)$ belongs to the hyperbola if and only if $PF_1$ and $PF_2$, its *focal radii*, differ by 6, that is $|PF_1 - PF_2| = 6$. Thus $PF_1 - PF_2 = \pm 6$, and therefore the coordinates of $P$ satisfy the compound sentence

$$\sqrt{(x + 5)^2 + (y - 0)^2} - \sqrt{(x - 5)^2 + (y - 0)^2} = \pm 6.$$

Using the squaring process to eliminate radicals, you can show that this compound sentence is equivalent to

$$\frac{x^2}{9} - \frac{y^2}{16} = 1.$$

The equation reveals several geometric properties of the hyperbola.

1. Whenever $(r, t)$ satisfies $\dfrac{x^2}{9} - \dfrac{y^2}{16} = 1$, so also do $(r, -t)$ and $(-r, t)$.

   ∴ *the curve is symmetric with respect to both coordinate axes.*

2. Replacing $y$ by 0, you obtain $\dfrac{x^2}{9} - \dfrac{0}{16} = 1$, $x^2 = 9$, or $x = \pm 3$.

   ∴ *the x-intercepts of the hyperbola are 3 and −3.*

3. Replacing $x$ by 0, you obtain $\dfrac{0}{9} - \dfrac{y^2}{16} = 1$, or $y^2 = -16$; no real

   value of $y$ satisfies $y^2 = -16$.
   ∴ *the hyperbola has no y-intercepts.*

4. Transforming the equation you obtain

   $$y = \pm\tfrac{4}{3}\sqrt{x^2 - 9} \qquad \text{and} \qquad x = \pm\tfrac{3}{4}\sqrt{y^2 + 16}$$

   ∴ *y is real if and only if $|x| \geq 3$.*
   ∴ *no part of the hyperbola lies between the lines $x = -3$ and $x = 3$.*

With these facts and a table of first-quadrant points, you obtain the graph in Figure 9–18.

| x | $y = \tfrac{4}{3}\sqrt{x^2 - 9}$ | y |
|---|---|---|
| 3 | $\tfrac{4}{3}\sqrt{9 - 9} = \tfrac{4}{3}\sqrt{0}$ | 0 |
| 4 | $\tfrac{4}{3}\sqrt{16 - 9} = \tfrac{4}{3}\sqrt{7}$ | $\doteq 3.5$ |
| 5 | $\tfrac{4}{3}\sqrt{25 - 9} = \tfrac{4}{3}\sqrt{16}$ | $\doteq 5.3$ |
| 6 | $\tfrac{4}{3}\sqrt{36 - 9} = \tfrac{4}{3}\sqrt{27}$ | $\doteq 6.9$ |

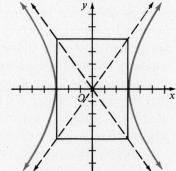

FIGURE 9–18

As you can see in Figure 9–18, the graph lies entirely within two of the regions marked out by the diagonals of the rectangle formed by the lines $x = 3$, $x = -3$, $y = 4$, $y = -4$. The equations of these diagonals, called the **asymptotes** of the hyperbola, are $y = \frac{4}{3}x$ and $y = -\frac{4}{3}x$. It can be shown that the distance between a branch of the hyperbola and the corresponding asymptote grows smaller as $|x|$ grows larger. This means that the hyperbola gets closer and closer to the diagonals, so that you can use these lines as guides in sketching the curve.

The preceding discussion suggests that a hyperbola with foci on the $x$-axis at $(c, 0)$ and $(-c, 0)$, and $2a$ as the difference of the focal radii, has an equation of the form

$$\frac{x^2}{a^2} - \frac{y^2}{b^2} = 1, \text{ where } b^2 = c^2 - a^2.$$

On the other hand, with foci at $(0, c)$ and $(0, -c)$ on the $y$-axis, the hyperbola has the equation:

$$\frac{y^2}{a^2} - \frac{x^2}{b^2} = 1, \text{ where } b^2 = c^2 - a^2.$$

**Example.**   Graph the relation $\{(x, y): 4x^2 - y^2 + 4 = 0\}$.

*Solution:*    $4x^2 - y^2 = -4;$

$$\frac{y^2}{4} - \frac{x^2}{1} = 1$$

1. Symmetric with respect to both axes.

2. No $x$-intercepts. (However, the lines $x = 1$ and $x = -1$ form the vertical sides of the rectangle whose diagonals are the asymptotes.)

3. $y$-intercepts: 2 and $-2$ (The lines $y = 2$ and $y = -2$ form the horizontal sides of the rectangle whose diagonals are the asymptotes.)

4. $y = \pm 2\sqrt{x^2 + 1}$,      $x = \pm\frac{1}{2}\sqrt{y^2 - 4}$
   No part of the curve lies between the lines $y = 2$ and $y = -2$. To graph the hyperbola, first draw the asymptotes and plot some first-quadrant points.

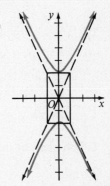

| $x$ | 0 | 1 | 2 | 3 |
|---|---|---|---|---|
| $y = 2\sqrt{x^2 + 1}$ | 2 | $\doteq 2.8$ | $\doteq 4.4$ | $\doteq 6.3$ |

The graph of an equation of the form $xy = k$ $(k \neq 0)$ can be shown to be a hyperbola having the coordinate axes as asymptotes (Exercises 28 and 29, page 373). The appearance of the curve for $k > 0$ and for $k < 0$ is shown in Figure 9–19.

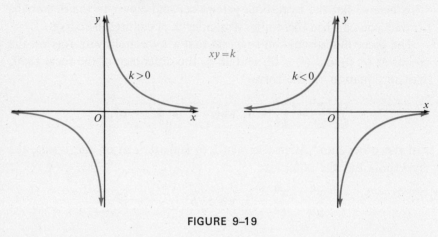

FIGURE 9–19

## WRITTEN EXERCISES

Graph each equation or inequality. Use an appropriate scale on each set of axes.

**A**

**1.** $\dfrac{x^2}{1} - \dfrac{y^2}{4} = 1$

**2.** $\dfrac{x^2}{9} - \dfrac{y^2}{1} = 1$

**3.** $\dfrac{y^2}{16} - \dfrac{x^2}{9} = 1$

**4.** $\dfrac{y^2}{25} - \dfrac{x^2}{36} = 1$

**5.** $x^2 - y^2 = 1$

**6.** $y^2 - x^2 = 1$

**7.** $4x^2 - y^2 = 16$

**8.** $9y^2 - x^2 = 36$

**9.** $3x^2 - 3y^2 + 27 = 0$

**10.** $3y^2 - 12x^2 - 48 = 0$

**B**

**11.** $x^2 - y^2 \leq 4$

**12.** $y^2 - x^2 \geq 9$

**13.** $25x^2 - 4y^2 > 100$

**14.** $25y^2 - 4x^2 < 100$

Let $A = \{(x, y): y = \sqrt{x^2 - 1}\}$, $B = \{(x, y): y = -\sqrt{x^2 - 1}\}$, $C = \left\{(x, y): y = x\sqrt{1 - \dfrac{1}{x^2}}\right\}$, and $D = \left\{(x, y): y = -x\sqrt{1 - \dfrac{1}{x^2}}\right\}$.

Graph each of the following sets.

**15.** $A$

**16.** $B$

**17.** $C$

**18.** $D$

**19.** $A \cup C$

**20.** $A \cup D$

**21.** $B \cup C$

**22.** $B \cup D$

Use the definition of a hyperbola to find an equation of the hyperbola with its given foci and given difference of focal radii.

**C**

**23.** $(4, 0), (-4, 0); 6$          **25.** $(c, 0), (-c, 0); 2a$

**24.** $(0, 3), (0, -3); 2$          **26.** $(0, c), (0, -c); 2a$

**27. a.** Show that all hyperbolas having equations of the form $x^2 - y^2 = k$ have the same asymptotes.
    **b.** In the same coordinate system, graph the hyperbolas $x^2 - y^2 = k$ for $k = 4, 1, 0, -1, -4$.

**28.** Consider the hyperbola which has the points $(-r, -r)$ and $(r, r)$ as foci and $2r$ as the difference of the focal radii of each of its points ($r > 0$).
    **a.** Show that it has the equation $xy = \frac{1}{2}r^2$.
    **b.** Use the equation in **a** to show that the hyperbola has the coordinate axes as asymptotes.

**29.** Repeat Exercise 28 with the points $(-r, r)$ and $(r, -r)$ as foci. Show that the hyperbola has equation $xy = -\frac{1}{2}r^2$.

## SOLVING QUADRATIC SYSTEMS

## 9–7   Graphical Solutions

You can sometimes use graphs to find solutions of systems of equations in two variables when one or both of the equations are quadratic.

**Example 1.**   Find the solution set of the system:

$$x^2 + y^2 = 25$$
$$x + 3y + 5 = 0$$

*Solution:*      Graph the equations on the same axes to obtain the circle and line shown at the right. Find the coordinates of their points of intersection: $(-5, 0)$ and $(4, -3)$. $\therefore$ the solution set is $\{(-5, 0), (4, -3)\}$.

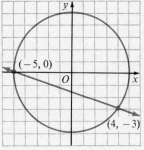

*Check:*

If $(x, y) = (-5, 0)$:
$$x^2 + y^2 = (-5)^2 + 0^2 = 25$$
$$x + 3y + 5 = (-5) + 3 \cdot 0 + 5 = 0$$

If $(x, y) = (4, -3)$:
$$x^2 + y^2 = 4^2 + (-3)^2 = 25$$
$$x + 3y + 5 = 4 + 3(-3) + 5 = 0$$

You can use graphical methods to obtain approximate solutions, as in the next example.

**Example 2.** Solve the system:

$$xy = 3$$
$$y = 8 - x^2$$

*Solution:* Draw the hyperbola and the parabola, and estimate the co-ordinates of their points of intersection. The solution set can be written

$$\{(-3, -1), (0.4, 7.8), (2.6, 1.2)\}.$$

You can check that $(-3, -1)$ is an exact solution; the other two are only approximate.

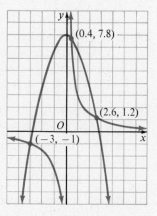

In the Exercises which follow, you will find that a system consisting of a linear and a quadratic equation may have 2, 1, or 0 real solutions; a system consisting of two quadratic equations may have 4, 3, 2, 1, or 0 real solutions.

## WRITTEN EXERCISES

Graph the equations to find the solution set of each system. Estimate non-integral coordinates of solutions to the nearest $\frac{1}{2}$ unit.

$\boxed{\text{A}}$

**1.** $y = x^2 - 5$
$y = 4x$

**5.** $x^2 + y^2 = 13$
$x + y = 5$

**9.** $x^2 + y^2 = 16$
$x^2 + y^2 = 25$

**2.** $y^2 = 9x$
$y = 3x$

**6.** $x^2 + y^2 = 9$
$y = 4$

**10.** $x^2 + y^2 = 25$
$x^2 - y^2 = 36$

**3.** $x^2 + y^2 = 16$
$y = x + 1$

**7.** $x^2 - 2x + y^2 = 3$
$2x + y = 4$

**11.** $2x^2 + y^2 = 33$
$y = x^2 - 3$

**4.** $x^2 + 4y^2 = 16$
$2x - y = 3$

**8.** $3x^2 - 2y^2 = 3$
$3x - 2y = 3$

**12.** $x^2 + y^2 = 16$
$y = x^2 - 4$

$\boxed{\text{B}}$

**13.** $3x^2 + 4y^2 = 16$
$x^2 - y^2 = 3$

**15.** $4x^2 + 3y^2 = 7$
$2x^2 - 5y^2 = 2$

**17.** $x^2 + y^2 = 8$
$xy = 4$

**14.** $4x^2 + 3y^2 = 12$
$x^2 + 3y^2 = 12$

**16.** $4x^2 + 3y^2 = 4$
$2x^2 - 5y^2 = 2$

**18.** $x^2 - y^2 = 35$
$xy = 6$

$\boxed{\text{C}}$ **19.** $y = x^2 + 4x + 2$
$\quad\quad xy = 16$

**20.** $y = x^2 - 4x + 2$
$\quad\quad xy = -16$

**21.** $x^2 + y^2 \le 25$
$\quad\quad y \ge x^2$

**22.** $x^2 + y^2 \ge 9$
$\quad\quad x^2 + 4y^2 \ge 4$

**23.** $y \le 2x^2$
$\quad\quad x \le 2y^2$

**24.** $x^2 + 9y^2 \le 9$
$\quad\quad x \ge y^2$

**25.** $y \le 2x - x^2$
$\quad\quad y \ge x^2$

**26.** $y \le 4x - x^2$
$\quad\quad xy \le 4$

## 9-8 Linear-Quadratic Systems

You may use the substitution method (Section 3–5) to solve a system consisting of a linear and a quadratic equation. Begin by solving the linear equation for one of the variables and substituting into the quadratic equation.

**Example 1.** Find the solution set of the system:

$$4x^2 + y^2 = 13$$
$$2x - y = 1$$

*Solution:*

**1.** Solve the linear equation for one of the variables, say $y$.

$\quad\quad 2x - y = 1$
$\quad\quad\quad y = 2x - 1$

**2.** Substitute the expression $2x - 1$ for $y$ in the quadratic equation and solve the resulting equation in one variable.

$\quad\quad\quad 4x^2 + y^2 = 13$
$\quad\quad\quad 4x^2 + (2x - 1)^2 = 13$
$\quad\quad 4x^2 + 4x^2 - 4x + 1 = 13$
$\quad\quad\quad 8x^2 - 4x - 12 = 0$
$\quad\quad\quad\quad 2x^2 - x - 3 = 0$
$\quad\quad\quad (2x - 3)(x + 1) = 0$
$\quad\quad x = \tfrac{3}{2}, \quad x = -1$

**3.** Substitute $\tfrac{3}{2}$ and $-1$ in turn for $x$ in the linear equation and determine the corresponding values of $y$.

$\quad\quad y = 2x - 1$
$\quad\quad y = 2(\tfrac{3}{2}) - 1, \quad y = 2(-1) - 1$
$\quad\quad y = 2 \quad\quad\quad\quad y = -3$

*Check:*

**4.** Check each ordered pair in both equations.

$4(\tfrac{3}{2})^2 + 2^2 = 4 \cdot \tfrac{9}{4} + 4 = 13 \quad | \quad 4(-1)^2 + (-3)^2 = 4 + 9 = 13$
$2 \cdot \tfrac{3}{2} - 2 = 3 - 2 = 1 \quad\quad\quad | \quad 2(-1) - (-3) = -2 + 3 = 1$

$\therefore$ the solution set is $\{(\tfrac{3}{2}, 2), (-1, -3)\}$.

**Example 2.** Find the points of intersection, if any, of the line $x + y = 9$ and the parabola $2x + y^2 = 16$.

*Solution:*

$$x + y = 9$$
$$x = 9 - y$$
$$2x + y^2 = 16$$
$$2(9 - y) + y^2 = 16$$
$$y^2 - 2y + 2 = 0$$
$$y = \frac{2 \pm \sqrt{2^2 - 4 \cdot 1 \cdot 2}}{2 \cdot 1}$$
$$= \frac{2 \pm \sqrt{-4}}{2}$$
$$y = 1 + i, \; y = 1 - i$$

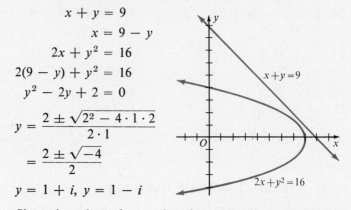

Since the values of $y$ are imaginary, the line and the parabola have no point in common. Their graphs are shown above.

## WRITTEN EXERCISES

Find the solution set of each system.

**A**

**1.** $y = x^2 + 3$
$y = 3x + 1$

**2.** $y = x^2 - 4$
$y = 2x$

**3.** $a + b = 1$
$ab = -12$

**4.** $c - 3d = 6$
$cd = 24$

**5.** $x^2 + y^2 = 13$
$2x - 3y = -5$

**6.** $2x^2 + y^2 = 12$
$2x + y = 2$

**7.** $x^2 - y^2 = 5$
$2x - y = 4$

**8.** $x^2 - y^2 = 8$
$x - y = 2$

**9.** $n^2 + 4m^2 = 36$
$n - 2m = 6$

**10.** $4r^2 + s^2 = 17$
$2r - s = 3$

**11.** $16r^2 + 9s^2 = 13$
$8r - 3s = 4$

**12.** $6x^2 - 2y^2 = 37$
$x - 3y = 1$

**13.** $2x^2 - 2xy + y^2 = 10$
$2x - y = -2$

**14.** $2r^2 - rt = 30$
$r - 2t = -11$

**15.** $p^2 - 3p - 2q = 2$
$2p + 5q = 1$

**16.** $r^2 + 2r - s = 3$
$3r + s = 3$

**17.** $n^2 - 2n + m^2 = 3$
$2n + m = 4$

**18.** $t^2 - tu - 2u^2 = 4$
$t - u = 2$

**B**

**19.** $a^2 - 4a - b^2 + 5b = -1$
$4a - 3b = 2$

**20.** $2c^2 - 7c - d^2 - 2d = 1$
$6c + 5d = 2$

**21.** $4u^2 - 5u + v^2 - 4v = 3$
$2u - v = 1$

**22.** $r^2 + 3r + 2s^2 - s = 7$
$2r + s = 1$

**23.** $\dfrac{1}{x} + \dfrac{3}{y} = \dfrac{2}{3}$
$y = 3x$

**25.** $\dfrac{4}{r} + \dfrac{3}{s} = 1$
$r = s - 3$

**24.** $\dfrac{3}{x} + \dfrac{3}{y} = 2$
$y = x + 4$

**26.** $\dfrac{6}{n} + \dfrac{3}{m} = 2$
$n = m + 5$

Solve for $x$ in terms of $a$ and $b$.

C **27.** $ax^2 - by^2 = ab^2 - a^2b$
$ax - by = 0$

**28.** $ax^2 + y^2 = b^2$
$y = ax + b$

## PROBLEMS

A **1.** Find the dimensions of a rectangle whose area is 6 square feet and whose perimeter is 10 feet.

**2.** Find the dimensions of a rectangle whose perimeter is 34 inches and which has a diagonal of length 13 inches.

**3.** The length of the hypotenuse of a right triangle is 25 inches and its perimeter is 56 inches. Find the lengths of the legs.

**4.** Find two numbers whose sum is 18 and whose product is 72.

**5.** Find two numbers whose difference is 3 and the difference of whose squares is 21.

**6.** Find two numbers whose sum is 10 and the sum of whose reciprocals is $\frac{5}{12}$.

**7.** Find two numbers whose sum is 8 and the difference of whose reciprocals is $\frac{1}{3}$.

**8.** The sum of a number and twice another number is 3, and the sum of their squares is 5. Find the numbers.

B **9.** The difference of the areas of two squares is 24. The length of the side of one square is 3 less than twice the length of the other square. Find the length of a side of each square.

**10.** A rectangular lot with one side along a straight river contains 8 square miles. To fence the three sides of the lot not on the river requires 10 miles of fencing. Find the dimensions of the lot (two sets of answers).

**11.** Find the value of $k$ for which the line $y = 2x + k$ is tangent to the parabola $y = x^2$. (*Hint:* Determine $k$ so that the system will have exactly one solution.)

**12.** Find the values of $k$ for which the line $y = 3x + k$ is tangent to the circle $x^2 + y^2 = 16$.

## 9–9 Quadratic-Quadratic Systems

You can often solve a system of two quadratic equations in two variables using the substitution method.

**Example 1.** Find the solution set of the system:

$$y^2 - x^2 = 3$$
$$xy = 2$$

*Solution:*

The second equation has no solution for $x = 0$; we may therefore assume that $x \neq 0$ throughout.

**1.** Solve the second equation for $y$.    $xy = 2$

$$y = \frac{2}{x}$$

**2.** Substitute $\frac{2}{x}$ for $y$ in the first equation.    $\left(\frac{2}{x}\right)^2 - x^2 = 3$

$$\frac{4}{x^2} - x^2 = 3$$

$$4 - x^4 = 3x^2$$

Solve the resulting equation, which is quadratic in $x^2$.

$$x^4 + 3x^2 - 4 = 0$$

$$(x^2 + 4)(x^2 - 1) = 0$$

Since $x^2 + 4 = 0$ has no real roots, we consider only the equation $x^2 - 1 = 0$.

$$x^2 + 4 = 0; \; x^2 - 1 = 0$$

$$(x - 1)(x + 1) = 0$$

$$x = 1, \; x = -1$$

**3.** Substitute 1 and $-1$ in turn for $x$ in $y = \frac{2}{x}$:

$$y = \frac{2}{1} \, ; \, y = \frac{2}{-1}$$

$$y = 2; \; y = -2$$

*Check:*

**4.** Check each ordered pair in both equations.

$$2^2 - 1^2 = 4 - 1 = 3; \; (1)(2) = 2$$
$$(-2)^2 - (-1)^2 = 4 - 1 = 3; \; (-1)(-2) = 2$$

∴ the solution set is $\{(1, 2), (-1, -2)\}$.

The linear-combination method (Section 3–5) can often be used to solve a system of quadratic equations.

**Example 2.** Find the solution set of the system: $9x^2 - 4y^2 = 36$
$$x^2 + y^2 = 43$$

*Solution:*

1. To obtain equations having coefficients of $y^2$ equal in absolute
value, multiply the second equation by 4.

$$9x^2 - 4y^2 = 36$$
$$4x^2 + 4y^2 = 172$$

2. Replace either equation by the
*sum* of the equations. Use this replacement to find the values of $x$.

$$13x^2 = 208$$
$$x^2 = 16$$
$$x = 4, \ x = -4$$

3. Substitute 4 and $-4$ in turn into
$x^2 + y^2 = 43$, or $y^2 = 43 - x^2$.

$$y^2 = 43 - 4^2,$$
$$y^2 = 43 - (-4)^2$$
$$y^2 = 27$$
$$y = 3\sqrt{3} \quad \text{or} \quad y = -3\sqrt{3}$$

*Check:*

4. Check for $(4, 3\sqrt{3})$:

$$9 \cdot 4^2 - 4(3\sqrt{3})^2 = 9 \cdot 16 - 4 \cdot 27$$
$$= 144 - 108$$
$$= 36$$
$$4^2 + (3\sqrt{3})^2 = 16 + 27 = 43$$

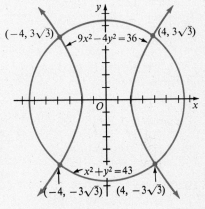

After checking this solution, you
can convince yourself by considera-
tions of symmetry that the remaining
three solutions are also correct. (See
the figure at the right.)

$\therefore$ the solution set is $\{(4, 3\sqrt{3}), (4, -3\sqrt{3}), (-4, 3\sqrt{3}), (-4, -3\sqrt{3})\}$.

*WRITTEN EXERCISES*

Find the solution set of each system.

**A** 1. $x^2 + y^2 = 10$
$9x^2 + y^2 = 18$

2. $x^2 + y^2 = 16$
$9x^2 + 25y^2 = 225$

3. $x^2 + 3y^2 = 7$
$2x^2 + y^2 = 9$

4. $n^2 + 2m^2 = 17$
$2n^2 - 3m^2 = 6$

**Sample.** $x^2 + y^2 = 34$
$xy = 15$

*Solution:* This system can be solved by the method of Example 1, page 378. Large numbers can be avoided by using the following method.

$$
\begin{array}{ll}
x^2 \qquad + y^2 = 34 & \text{If } x - y = 2, \text{ then } y = x - 2. \\
\text{Subtract:} \quad \underline{\quad 2xy \qquad = 30} & \qquad xy = x(x - 2) = 15 \\
x^2 - 2xy + y^2 = \quad 4 & \qquad x^2 - 2x - 15 = 0 \\
(x - y)^2 = \quad 4 & \qquad (x - 5)(x + 3) = 0 \\
x - y \;\; = \pm 2 & \qquad x = 5, \qquad y = 3 \\
& \qquad x = -3, \quad y = -5
\end{array}
$$

The case where $x - y = -2$ is similar. The solution set is $\{(5, 3), (-3, -5), (-5, -3), (3, 5)\}$.

**5.** $x^2 + y^2 = 25$
$xy = 12$

**11.** $r^2 - 3s = 46$
$2r^2 + 3s = 101$

**6.** $t^2 + s^2 = 12$
$ts = 4$

**12.** $5a^2 - 4b^2 = 4$
$3a^2 - b^2 = 8$

**7.** $a^2 + 4b^2 = 17$
$ab = 2$

**13.** $t^2 + 4s^2 + 2s = 11$
$t^2 - 4s^2 - 4s = 6$

**8.** $u^2 + 4v^2 = 25$
$uv = 6$

**14.** $2c^2 + d^2 + d = 4$
$3c^2 - d^2 + d = -3$

**9.** $y = x^2 + x - 3$
$y = x^2 + 2x - 6$

**15.** $x^2 - 2 = 2 - y^2$
$x^2 - 2 = 2y^2 - 1$

**10.** $y = 3x^2 - x - 8$
$y = -x^2 + 3x + 12$

**16.** $4x^2 - 5 = y^2 + 2$
$2x^2 - 3 = 6 - 5y^2$

**B** **17.** $\dfrac{1}{a} - \dfrac{1}{b} = \dfrac{1}{5}$
$ab = 80$

**20.** $ab = -2$
$(a + 1)(2b - 1) = -1$

**18.** $\dfrac{1}{x} - \dfrac{2}{y} = \dfrac{1}{2}$
$xy = 4$

**21.** $y = 2x^2 - 3x + 5$
$xy = 14$

**19.** $xy = 8$
$(x - 2)(y + 3) = -20$

**22.** $x = 4 - 2y - y^2$
$xy = -5$

**C** **23.** $x^2 + 3xy - y^2 = -3$
$x^2 - xy - y^2 = 1$

**24.** $2x^2 + xy - 2y^2 = 16$
$x^2 + 2xy - y^2 = 17$

(*Hint:* First eliminate the $x^2$ and $y^2$ terms.)

## PROBLEMS

A  1. Find two positive numbers such that the sum of their squares is 208 and the difference of their squares is 80.

2. Find two numbers whose product is 156 and the sum of whose squares is 313 (see Sample, page 380).

3. The length of the diagonal of a rectangle is 13 inches, while its area is 60 square inches. Find the dimensions of the rectangle.

4. The area of a rectangular carpet is 300 square feet, and the distance from one corner diagonally to another is 25 feet. Find the dimensions of the carpet.

5. When a 6 centimeter strip is trimmed from all four sides of a rectangular sheet of paper, the area is decreased from 1170 square centimeters to 486 square centimeters. Find the dimensions of the new rectangular sheet.

6. A rectangular garden has an area of 960 square feet. Around it is a path 5 feet wide with an area of 780 square feet. What are the dimensions of the garden?

7. Mr. Edwards drove 100 miles from Dayton to Pottsville and returned. If his return trip was made at an average speed 10 miles per hour greater than that of his trip going, and if he returned in 30 minutes less time than it took him to go, what was his rate each way?

8. When the jet stream caused an increase of 100 miles per hour over the usual speed of his plane, the pilot made the 3000-mile trip between two cities in one hour less than the usual time. Find the usual speed of the plane.

9. Ellen receives $150 per year in simple interest on a loan. If the interest rate were 1% higher, she would receive the same amount of interest on a loan of $500 less. Find the interest rate and the amount she has loaned.

10. A merchant bought some portable radios for $360. He sold all but two of them for $380, averaging a profit of $8 on each one. How many radios did he buy?

B 11. A circle with center on the $y$-axis contains the points $(5, 3)$ and $(4, 6)$. Find an equation for the circle.

12. An ellipse with equation of the form $\dfrac{x^2}{a^2} + \dfrac{y^2}{b^2} = 1$ contains the points $(4, \frac{9}{5})$ and $(-3, \frac{12}{5})$. Find $a^2$ and $b^2$.

**13.** A hyperbola with equation of the form $\dfrac{y^2}{a^2} - \dfrac{x^2}{b^2} = 1$ contains the points $\left(3, \dfrac{3\sqrt{5}}{2}\right)$ and $(-4, 3\sqrt{2})$. Find $a^2 + b^2$.

**14.** An ellipse whose equation is of the form $\dfrac{x^2}{4a^2} + \dfrac{y^2}{b^2} = 1$ and a hyperbola whose equation is of the form $\dfrac{x^2}{a^2} - \dfrac{y^2}{b^2} = 1$ are drawn on the same axes. If one of the points of intersection is $(\frac{6}{5}\sqrt{10}, \frac{3}{5}\sqrt{15})$, write each equation.

---

## *Chapter Summary*

---

**1.** The formula for the **distance between points** $P_1(x_1, y_1)$ and $P_2(x_2, y_2)$ is $d = \sqrt{(x_2 - x_1)^2 + (y_2 - y_1)^2}$. The **midpoint** of line segment $P_1P_2$ is $M\left(\dfrac{x_1 + x_2}{2}, \dfrac{y_1 + y_2}{2}\right)$.

**2.** By applying the distance formula, you can derive the equation of a **circle** with center $(h, k)$ and radius $r$; namely, $(x - h)^2 + (y - k)^2 = r^2$.

**3.** An equation of a **parabola** can be derived if you know its **focus** and its **directrix**.

**4.** When an equation of an **ellipse** is of the form $\dfrac{x^2}{a^2} + \dfrac{y^2}{b^2} = 1$ $(a \neq 0, b \neq 0)$, the ellipse intersects the coordinate axes in the points $(a, 0)$, $(-a, 0)$, $(0, b)$, $(0, -b)$ and is symmetric with respect to both coordinate axes.

**5.** When an equation of a **hyperbola** is of the form $\dfrac{x^2}{a^2} - \dfrac{y^2}{b^2} = 1$ or $\dfrac{y^2}{b^2} - \dfrac{x^2}{a^2} = 1$ $(a \neq 0, b \neq 0)$, the hyperbola is symmetric with respect to both coordinate axes. The **asymptotes**, $y = \dfrac{b}{a}x$ and $y = -\dfrac{b}{a}x$, may be used as guides in drawing the hyperbola.

**6.** The points of intersection of the graphs of the equations of a system represent the real solutions of the system. Linear-quadratic and quadratic-quadratic systems may be solved algebraically by substitution or by linear combination.

## Vocabulary and Spelling

Review each term by reference to the page listed.

Pythagorean Theorem (*p. 352*)
hypotenuse (*p. 352*)
distance formula (*p. 352*)
midpoint of a segment (*p. 353*)
perpendicular lines (*p. 356*)
parabola (*p. 362*)
  focus (*p. 362*)
  directrix (*p. 362*)

axis (of symmetry) (*p. 362*)
ellipse (*p. 366*)
  foci (*p. 366*)
  focal radii (*p. 366*)
hyperbola (*p. 369*)
  foci (*p. 369*)
  focal radii (*p. 369*)
  asymptotes (*p. 371*)

## Chapter Test

9–1  **1.** Find the length of the line segment with endpoints $(-2, 6)$ and $(8, 10)$, and specify its midpoint.

9–2  **2.** Write an equation for the line perpendicular to the line with equation $3x - 2y = 6$ and passing through $(-2, 5)$.

9–3  **3.** Find the center and radius of the circle with equation

$$x^2 + y^2 - 2x + 4y - 4 = 0.$$

Graph each equation.

9–4  **4.** $y = 2x^2 + 4x + 5$

9–5  **5.** $x^2 + 16y^2 = 16$

9–6  **6.** $x^2 - 4y^2 = 16$

9–7  **7.** Solve graphically: $x^2 - y^2 = 16$
                        $x + y = 3$

Find the solution set of each system.

9–8  **8.** $2x^2 + y^2 = 9$
         $2x - 3y = 7$

9–9  **9.** $2x^2 + y^2 = 17$
         $x^2 - 2y^2 = -14$

**10.** The area of a rectangle is 18 square centimeters, and a diagonal has length $3\sqrt{5}$. Find the dimensions of the rectangle.

---

## Chapter Review

---

### 9–1 Distance between Points    *Pages 351–356*

1. The Pythagorean Theorem asserts that the sum of the squares of the lengths of the __?__ of a right triangle is equal to the __?__ of the length of the __?__.

2. Find the distance between the points $(-5, 2)$ and $(0, 4)$.

3. Find the midpoint of the segment with endpoints $(4, 6)$ and $(-2, 2)$.

### 9–2 Perpendicular Lines    *Pages 356–359*

4. Two lines, neither of which is vertical, are perpendicular if and only if the __?__ of their slopes is __?__.

5. Find an equation for the line perpendicular to the line $3x + 4y = 2$ and containing the point $(-2, 3)$.

### 9–3 Circles    *Pages 359–362*

6. Find an equation in the form $x^2 + y^2 + ax + by + c = 0$ for the circle with center at $(-1, 2)$ and radius $\sqrt{3}$.

7. The graph of $x^2 + y^2 - 2x - 2 \leq 0$ is a __?__ and its __?__.

8. The graph of $x^2 + y^2 - 12x + 14y = 0$ is a circle with center __?__ and radius __?__.

### 9–4 Parabolas    *Pages 362–365*

9. If a curve has the property that whenever the point $(r, t)$ belongs to it, the point $(r, -t)$ also belongs to it, then the curve is __?__ with respect to the __?__-axis.

10. Write the equation $y = 2x^2 - 4x + 3$ in the form
$$y = a(x - h)^2 + k.$$

11. Find the equation for the axis of symmetry of the graph of $x = y^2 - 4y - 12$.

### 9–5 Ellipses    *Pages 366–369*

12. An ellipse is a set of points in a plane such that for each point in the set, the __?__ of its distances to two fixed points is a __?__.

13. The $x$-intercepts of the graph of $16x^2 + 25y^2 = 400$ are __?__ and __?__, while the $y$-intercepts are __?__ and __?__.

**9–6** **Hyperbolas**  *Pages 369–373*

14. A hyperbola is a set of points such that for each point of the set, the absolute value of the __?__ of its distances from two fixed points called __?__ is a __?__.

15. The $y$-intercepts of $\dfrac{y^2}{25} - \dfrac{x^2}{16} = 1$ are __?__ and __?__.

**9–7** **Graphical Solutions**  *Pages 373–375*

16. The graphs of $2x^2 - 4y^2 = 6$ and $3x + 2y = 5$ are a(n) __?__ and a(n) __?__ respectively, and these graphs can have at most __?__ points of intersection.

17. Solve graphically: $y = x^2 + 4$
$$x + y = 6$$

**9–8** **Linear-Quadratic Systems**  *Pages 375–377*

Solve each system.

18. $x^2 - 3y^2 = -11$
    $2x - y = 0$

19. $x^2 - 2x + y = 9$
    $2x - y = 5$

**9–9** **Quadratic-Quadratic Systems**  *Pages 378–382*

Find the solution set of each system.

20. $x^2 + 3y^2 = 28$
    $xy = -8$

21. $x^2 + y^2 = 20$
    $2x^2 - y^2 = 28$

22. Find two positive numbers such that the sum of their squares is 170 and the difference of their squares is 72.

# Extra for Experts

## THE CONIC SECTIONS

The diagrams show that the shape of the curve in which a plane intersects a cone whose base is circular and whose axis is perpendicular to the base depends on the angle between the plane and the axis of the cone and on whether the plane contains the vertex of the cone. But in every case the intersection is a figure you have studied: a circle, an ellipse, a parabola, a hyperbola, a point, a line, or a pair of intersecting lines. As a result, these sets of points are called **conic sections**.

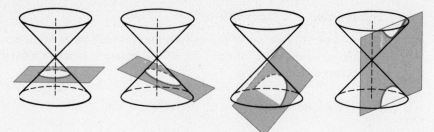

Circle      Ellipse      Parabola      Hyperbola

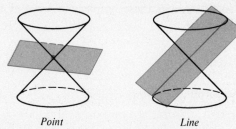

Point      Line      Intersecting Lines

Conic sections also share a common definition as sets of points in a plane. A set of points each of whose distances from a given point $F$ (the focus) is a constant $e$ times its distance from a given line $d$ (the directrix) is a conic section. Do you see that the parabola satisfies this definition with $e = 1$? The diagram below illustrates the effect of $e$, called the eccentricity, and suggests a means of classifying conics:

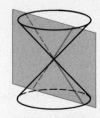

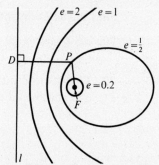

| $e$ | Conic |
|---|---|
| $0 < e < 1$ | ellipse |
| $e = 1$ | parabola |
| $e > 1$ | hyperbola |

# Exercises

Find its equation and sketch the conic with given focus, directrix, and eccentricity.

**1.** $F(0, 1)$; $y = -3$; $e = 1$

**2.** $F(3, 0)$; $x = 12$; $e = \frac{1}{2}$

**3.** $F(4, 0)$; $x = 1$; $e = 2$

**4.** $F(-\sqrt{2}, 0)$; $x = -\dfrac{\sqrt{2}}{2}$; $e = \sqrt{2}$

# Pierre de Fermat

The equation $x^2 + y^2 = z^2$ has many solutions where $x$, $y$, and $z$ are positive integers; for example, $3^2 + 4^2 = 5^2$ and $5^2 + 12^2 = 13^2$. But is there an integer $n \geq 3$ such that

$$x^n + y^n = z^n$$

has a solution for positive integers $x$, $y$, and $z$? This seemingly simple problem has baffled and intrigued mathematicians for centuries since it was proposed by the French mathematician Pierre de Fermat (1601–1665).

Fermat (pronounced Fer-mah) played an important part in establishing many branches of mathematics: analytic geometry, the theory of probability, differential calculus, and the modern theory of numbers. Yet he was an amateur mathematician. Law was his profession; mathematics, his hobby.

Fermat spent his entire life in government service and did not discover his mathematical aptitude until he was about thirty. While reading in the evenings, he recorded many of his mathematical discoveries in the margins of his books. There was usually not enough space to include proofs, but with one exception each statement that he claimed to have proved has been proved by later mathematicians. This exception is now known as "Fermat's last theorem." Fermat answered the question raised in the first paragraph as follows: "If $n$ is a number greater than two, there are no positive integers $x$, $y$, $z$ such that $x^n + y^n = z^n$. I have found a truly wonderful proof which this margin is too small to contain." Although mathematicians have shown that for many specific values of $n$ the equation has no solution, no general proof of this has been found. The problem, however, has had significance far beyond its intrinsic importance because many of the most useful techniques of mathematics were introduced in an effort to solve it.

# 10

# Exponential Functions and Logarithms

## 10–1   Rational Exponents

As you know, any *integer* can be used as an exponent in a power with a nonzero base. For example, $3^1 = 3, 2^{-3} = \frac{1}{8}, 5^0 = 1$, and $(-3)^2 = 9$. Let us see how to extend the definition of a power so that any *rational number* may serve as an exponent.

Look first at a special case, say $3^{\frac{1}{2}}$. If $3^{\frac{1}{2}}$ is to have a meaning consistent with the familiar laws of exponents, page 210, then it must be true that $(3^{\frac{1}{2}})^2 = 3^{\frac{1}{2} \cdot 2} = 3^1 = 3$. Since $3^{\frac{1}{2}}$ is to denote a number whose square is 3, we will define it to be $\sqrt{3}$. (We choose $\sqrt{3}$ rather than $-\sqrt{3}$ so that the inequality $3^0 < 3^{\frac{1}{2}} < 3^1$ will be true.) Similar reasoning requires that $3^{\frac{5}{2}} = (3^{\frac{1}{2}})^5 = (\sqrt{3})^5$, and $3^{-\frac{5}{2}} = (3^{\frac{1}{2}})^{-5} = (\sqrt{3})^{-5}$. These examples suggest the following definition:

> If $p$ is an integer, $q$ is a positive integer, and $b$ is a positive real number, then
> $$b^{\frac{p}{q}} = (\sqrt[q]{b})^p.$$

Because $b > 0$, the theorem on page 283 implies that

$$b^{\frac{p}{q}} = \sqrt[q]{b^p}.$$

Using powers with rational exponents you can write radical expressions in **exponential form**, that is, as powers or products of powers. Then, because the laws of exponents apply to these powers (Exercises 42, 43, and 44), you can use the laws to simplify the exponential expressions.

**Example 1.** $\sqrt{16a^3x^2y^{-4}} = 16^{\frac{1}{2}}(a^3)^{\frac{1}{2}}(x^2)^{\frac{1}{2}}(y^{-4})^{\frac{1}{2}} = 4a^{\frac{3}{2}}xy^{-2}$

**Example 2.** $\sqrt[3]{a^2} \cdot \sqrt[6]{\dfrac{1}{a}} = \sqrt[3]{a^2} \cdot \sqrt[6]{a^{-1}} = a^{\frac{2}{3}} \cdot a^{-\frac{1}{6}} = a^{\frac{2}{3}-\frac{1}{6}} = a^{\frac{1}{2}},$ or $\sqrt{a}$

**Example 3.** $(3^{-2} + 4^{-2})^{\frac{1}{2}} = \left(\dfrac{1}{9} + \dfrac{1}{16}\right)^{\frac{1}{2}} = \left(\dfrac{16 + 9}{9 \cdot 16}\right)^{\frac{1}{2}} = \left(\dfrac{25}{9 \cdot 16}\right)^{\frac{1}{2}}$

$$= \dfrac{5}{3 \cdot 4} = \dfrac{5}{12}$$

With the help of the theorem on page 282, we have

$$(\sqrt[kq]{b})^{kp} = \left[\left(\sqrt[k]{\sqrt[q]{b}}\right)^k\right]^p = [\sqrt[q]{b}]^p$$

provided $k$, $p$, and $q$ are integers, and $k > 0, q > 0$, and $b > 0$. From this we have

$$b^{\frac{kp}{kq}} = b^{\frac{p}{q}} \quad \text{and} \quad \sqrt[kq]{b^{kp}} = \sqrt[q]{b^p}.$$

For example, $36^{\frac{3}{6}} = 36^{\frac{1}{2}} = \sqrt{36} = 6$, and $\sqrt[6]{8} = \sqrt[6]{2^3} = 2^{\frac{3}{6}} = 2^{\frac{1}{2}} = \sqrt{2}$.

Notice that in extending powers to include rational exponents, we have restricted the base $b$ to be a positive real number. Without this restriction some of the familiar laws of exponents would not hold (see Exercise 41).

*ORAL EXERCISES*

State the value of each symbol.

**1.** $9^{\frac{1}{2}}$  **3.** $(-27)^{\frac{1}{3}}$  **5.** $(16)^{-\frac{1}{4}}$  **7.** $(32)^{\frac{3}{5}}$

**2.** $27^{\frac{1}{3}}$  **4.** $(-8)^{\frac{4}{3}}$  **6.** $(\frac{3}{5})^{-2}$  **8.** $(-27)^{-\frac{2}{3}}$

State the value of $b$.

**9.** $b^{\frac{1}{4}} = 2$  **10.** $b^{\frac{2}{3}} = 25$  **11.** $b^{\frac{3}{2}} = 27$  **12.** $b^{\frac{4}{3}} = 16$

## WRITTEN EXERCISES

Write in exponential form.

**A**

**1.** $\sqrt[5]{x^2y}$

**2.** $\sqrt[6]{a^5b^5}$

**3.** $-5\sqrt[4]{a^3b}$

**4.** $-2\sqrt[6]{ab^5}$

**5.** $\sqrt[3]{27x^{-6}y^2z^{-2}}$

**6.** $\sqrt[4]{81r^{-6}t^5s^{-3}}$

Express in simple radical form.

**7.** $\sqrt[3]{4} \cdot \sqrt[3]{2^4}$

**8.** $\sqrt[6]{9} \cdot \sqrt[3]{3^5}$

**9.** $\sqrt[6]{81} \cdot \sqrt[4]{32}$

**10.** $\sqrt[10]{32} \cdot \sqrt[6]{2^2}$

**11.** $\sqrt[6]{8} \div \sqrt[9]{16}$

**12.** $\sqrt[4]{125} \div \sqrt[6]{25}$

**13.** $\sqrt[6]{27} \div \sqrt[12]{9}$

**14.** $\sqrt[8]{32} \div \sqrt[10]{256}$

Evaluate.

**15.** $32^{-\frac{3}{5}}$

**16.** $243^{-\frac{4}{5}}$

**17.** $(-64)^{\frac{4}{3}}$

**18.** $(-32)^{\frac{6}{5}}$

**19.** $81^{-\frac{1}{2}} + 3^2$

**20.** $25^{-\frac{1}{2}} + 25^0$

**21.** $16^{-\frac{1}{2}} - 16^{\frac{1}{4}}$

**22.** $49^{-\frac{1}{2}} - 64^{\frac{1}{6}}$

Simplify.

**23.** $10^{3.1} \times 10^{2.7} \div 10^{1.3}$

**24.** $10^{5.1} \times 10^{-1.2} \div 10^{0.7}$

**25.** $\sqrt[4]{10^{3.64} \div 10^{1.28}}$

**26.** $\sqrt[3]{10^{9.3} \div 10^{-1.2}}$

**B**

**27.** $\sqrt[4]{\sqrt{125}}$

**28.** $\sqrt{\sqrt[4]{8}}$

**29.** $\sqrt{\sqrt{\sqrt{a^9}}} \cdot \sqrt{\sqrt{\sqrt{a^5}}}$

**30.** $\sqrt[3]{\sqrt{\sqrt{a^3}}} \cdot \sqrt{\sqrt{\sqrt{a^{30}}}}$

**31.** $\sqrt{a^{-1}} \cdot \sqrt[3]{a^2}$

**32.** $\sqrt{4x^3} \cdot \sqrt[3]{x^{-2}}$

**33.** $(x + 2x^{\frac{1}{2}}y^{\frac{1}{2}} + y)^{\frac{1}{2}}$

**34.** $(t - 2t^{\frac{1}{2}}s^{\frac{1}{2}} + s)^{\frac{1}{2}}$

Solve over $\Re$.

**35.** $x^{-\frac{2}{3}} = 4$

**36.** $y^{-\frac{3}{4}} = \frac{1}{27}$

**37.** $(n + 2)^{\frac{2}{5}} = 1$

**38.** $(a - 3)^{\frac{3}{2}} = 216$

**39.** $m = m^{\frac{1}{2}} + 12$
  (*Hint:* Let $x = m^{\frac{1}{2}}$ and solve for $x$.)

**40.** $m - 3m^{\frac{1}{2}} = 10$

**C** **41.** If $(-8)^{\frac{1}{3}} = \sqrt[3]{-8}$, show that $(-8)^{\frac{1}{3}} \neq (-8)^{\frac{2}{6}}$, even though $\frac{1}{3} = \frac{2}{6}$.

Let $a$ and $b$ denote positive real numbers, $r$ and $s$ positive integers, and $p$ and $q$ integers. Prove each of the following statements:

**42.** $b^{\frac{p}{r}} \cdot b^{\frac{q}{s}} = b^{\frac{ps+rq}{rs}}$

**43.** $(b^{\frac{p}{r}})^{\frac{q}{s}} = b^{\frac{pq}{rs}}$

**44.** $(ab)^{\frac{p}{r}} = a^{\frac{p}{r}}b^{\frac{p}{r}}$

  (*Hint:* Raise each member of the equation to an appropriate power.)

## 10–2   Real Numbers as Exponents

We have assigned a meaning to the symbol $2^x$ for every rational value of $x$. For example, $2^{-3} = \dfrac{1}{2^3} = \dfrac{1}{8}$, $2^{\frac{3}{2}} = \sqrt{2^3}$, and $2^{1.7} = 2^{\frac{17}{10}} = \sqrt[10]{2^{17}}$. Now we shall extend the meaning to include such expressions as $2^{\sqrt{3}}$, where the exponent is an irrational number.

Using the table of values below you can plot points of the graph of $\{(x, y): y = 2^x\}$ for selected rational values of $x$ (Figure 10–1):

| $x$ | $-3$ | $-2$ | $-1$ | $-\frac{1}{2}$ | 0 | $\frac{1}{2}$ | 1 | $\frac{3}{2}$ | 2 |
|---|---|---|---|---|---|---|---|---|---|
| $y = 2^x$ | $\frac{1}{8}$ | $\frac{1}{4}$ | $\frac{1}{2}$ | $\doteq 0.7$ | 1 | $\doteq 1.4$ | 2 | $\doteq 2.8$ | 4 |

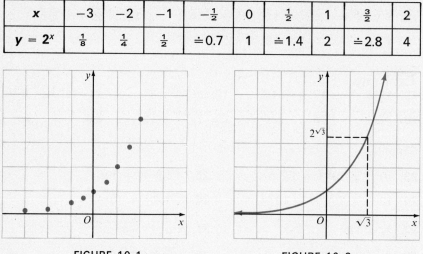

FIGURE 10–1                    FIGURE 10–2

If the complete graph is to be a smooth curve as shown in Figure 10–2, we must be able to approximate a power such as $2^{\sqrt{3}}$ as follows: Evaluate the successive powers

$$2^1, \quad 2^{1.7}, \quad 2^{1.73}, \quad 2^{1.732}, \quad \ldots$$

in which the exponents are the rational numbers obtained by taking more and more places in the decimal representing $\sqrt{3}$. Since these powers increase steadily but remain less than $2^2$, it can be proved, using advanced mathematics, that they get closer and closer to a certain positive real number, which we call $2^{\sqrt{3}}$ (to five significant digits, $2^{\sqrt{3}} = 3.3220$).

The method discussed above can be used to define $b^x$, where $b > 0$, and $x$ is any real number. Moreover, it can be shown that the laws of exponents continue to hold for these powers. For example,

$$(2^{\sqrt{3}})^{\sqrt{3}} = 2^{\sqrt{3} \cdot \sqrt{3}} = 2^3 = 8,$$

and        $$5^{1+\pi} \cdot 5^{1-\pi} = 5^{(1+\pi)+(1-\pi)} = 5^2 = 25.$$

The function defined by $y = b^x$ ($b > 0$, $b \neq 1$) is called the **exponential function with base** $b$. For $b > 1$, its graph has the general appearance of the curve in Figure 10–2. In particular, it rises continuously with increasing $x$. The curve in Figure 10–3 is typical of the graph of $y = b^x$ for $0 < b < 1$; it falls as $x$ increases. In each case, any vertical line or any horizontal line above the $x$-axis intersects the graph in just one point. This means that

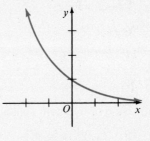

**FIGURE 10–3**

$$b^{x_1} = b^{x_2} \quad \text{if and only if} \quad x_1 = x_2.$$

**Example.** Solve for $x$: $100^{x-1} = (\frac{1}{10})^{4x-1}$

*Solution:*

1. Express each member of the equation as a power of the same base, 10, and simplify.

$$100^{x-1} = (\tfrac{1}{10})^{4x-1}$$
$$(10^2)^{x-1} = (10^{-1})^{4x-1}$$
$$10^{2x-2} = 10^{-4x+1}$$

2. Equate exponents and solve for $x$.

$$2x - 2 = -4x + 1$$
$$6x = 3$$
$$x = \tfrac{1}{2}$$

*Check:*

3. $100^{\frac{1}{2}-1} = 100^{-\frac{1}{2}} = \dfrac{1}{100^{\frac{1}{2}}} = \dfrac{1}{10}$

$$\left(\frac{1}{10}\right)^{4(\frac{1}{2})-1} = \left(\frac{1}{10}\right)^{2-1} = \frac{1}{10}$$

*WRITTEN EXERCISES*

Simplify.

**Sample.** $(2^{\sqrt[3]{6}})^{\sqrt[3]{\frac{6}{3}}}$

*Solution:* $(2^{\sqrt[3]{6}})^{\sqrt[3]{\frac{6}{3}}} = 2^{\sqrt[3]{6} \cdot \sqrt[3]{\frac{6}{3}}} = 2^{\sqrt[3]{6 \cdot \frac{6}{3}}} = 2^{\sqrt[3]{12}}$

Ⓐ **1.** $(4^{\pi})^{\sqrt{\frac{4}{\pi^2}}}$

**2.** $(2^{\sqrt[3]{5}})^{\sqrt[3]{25}}$

**3.** $(7^0)^{\pi}$

**4.** $(2^{\sqrt{3}})^{\sqrt{27}}$

**5.** $(2^{3-\sqrt{5}}) \cdot (2^{3-\sqrt{5}})$

**6.** $(3^{1-\sqrt{2}}) \cdot (3^{1+\sqrt{2}})$

**7.** $3^{1-\sqrt{2}} \div 3^{1-\sqrt{3}}$

**8.** $4^{\pi} \div 4^{\pi-2}$

**9.** $3^{\sqrt{80}} \div 9^{\sqrt{5}}$

Solve for $x$.

**10.** $4^{2x-1} = 4^5$      **13.** $2^{2x-1} = 8^{x+7}$      **16.** $4^x = 8 \cdot 2^{1-x}$

**11.** $6^{x-3} = 6^5$      **14.** $(\frac{1}{3})^x = 3^{x-6}$      **17.** $(\sqrt{3})^{x+1} = 9^{x-1}$

**12.** $3^{-x} = 9^{6-2x}$      **15.** $81^x = (\frac{1}{27})^{x-2}$      **18.** $8^{x+2} = 16^{2x-1}$

Using convenient scales on the axes, graph each function.

**19.** $\{(x, y): y = 4^x\}$      **21.** $\{(x, y): y = (\frac{1}{2})^x\}$

**20.** $\{(x, y): y = 5^x\}$      **22.** $\{(x, y): y = (\frac{1}{3})^x\}$

B **23.** $\{(x, y): 2^x + 2\}$      **25.** $\{(x, y): y = 2 \cdot 3^{x+1}\}$

**24.** $\{(x, y): 3^x - 1\}$      **26.** $\{(x, y): y = 3 \cdot 2^{-x}\}$

Sketch the graphs of the equations in the system and use the graphs to estimate the solutions of the system to the nearest $\frac{1}{2}$ unit.

C **27.** $y = 2^x$      **28.** $y = 3^x$      **29.** $y = 2^x + 2$      **30.** $y = 2^x$
    $y = 2 - x$          $y = x^2$          $y = 3^x$          $y = 10^{x-1}$

## LOGARITHMS

## 10–3  Logarithmic Functions

You can use the graph of an exponential function to find an approximation of any power of the base. For example, in Figure 10–4, you find $10^{0.9} \doteq 8$ by drawing first the vertical line, then the horizontal line shown in red. On the other hand, by reversing the construction (Figure 10–5) you find that to express 8 as a power of 10, you use an exponent approximately equal to 0.9; that is, $8 \doteq 10^{0.9}$. The figure also shows that $3 \doteq 10^{0.5}$.

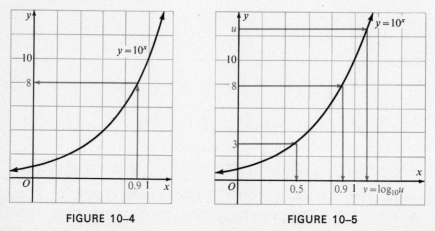

FIGURE 10–4                    FIGURE 10–5

As Figure 10–5 indicates, given any positive number $u$, there is a unique real number $v$ such that $u = 10^v$. We call $v$ the **logarithm** of $u$ to the base 10 and write $\log_{10} u = v$. Thus:

$$\log_{10} 1000 = 3 \qquad \text{because} \qquad 1000 = 10^3$$

$$\log_{10} 0.01 = -2 \qquad \text{because} \qquad 0.01 = 10^{-2}$$
$$\log_{10} 8 \doteq 0.9 \qquad \text{because} \qquad 8 \doteq 10^{0.9}$$
$$\log_{10} 3 \doteq 0.5 \qquad \text{because} \qquad 3 \doteq 10^{0.5}$$

In Sections 10–6 and 10–7 you will learn how to find the logarithm to the base 10 of any positive number.

In general, let $b$ be any positive number except 1. Then for any positive number $u$ there is a unique real number $v$, called the **logarithm of $u$ to the base $b$,** such that $u = b^v$; we write $\log_b u = v$. Thus,

$$\log_b u = v \quad \text{if and only if} \quad u = b^v.$$

You can see from this definition that

$$\log_b b^v = v \qquad \text{and} \qquad b^{\log_b u} = u.$$

**Example 1.** Find: **a.** $\log_3 9$  **b.** $\log_{\frac{1}{2}} 16$  **c.** $\log_{10} 0.001$  **d.** $2^{\log_2 6}$

*Solution:*  **a.** $\log_3 9 = \log_3 3^2 = 2$

**b.** Let $\log_{\frac{1}{2}} 16 = v$. Then $(\frac{1}{2})^v = 16$, or $(2^{-1})^v = 2^4$, or $2^{-v} = 2^4$. $\therefore -v = 4$, or $v = -4$, and $\log_{\frac{1}{2}} 16 = -4$.

**c.** $\log_{10} 0.001 = \log_{10} 10^{-3} = -3$

**d.** $2^{\log_2 6} = 6$

The function defined for $x > 0$ by the equation $y = \log_b x$ ($b > 0$, $b \neq 1$) is the **logarithmic function with base $b$.** It has the property, suggested by Figure 10–5, that

$$\log_b u_1 = \log_b u_2 \quad \text{if and only if} \quad u_1 = u_2.$$

**Example 2.** Solve: **a.** $\log_{10}(x^2 - 1) = \log_{10} 8$  **b.** $\log_9 x = \frac{3}{2}$

*Solution:*  **a.** $\log_{10}(x^2 - 1) = \log_{10} 8$
$$x^2 - 1 = 8$$
$$x^2 = 9, \qquad x = \pm 3 \qquad \{-3, 3\}$$

**b.** $\log_9 x = \frac{3}{2}$
$$x = 9^{\frac{3}{2}} = (\sqrt{9})^3 = 27 \qquad \{27\}$$

*ORAL EXERCISES*

Express using logarithmic notation.

**Sample.**  $10^2 = 100$      *Solution:*  $2 = \log_{10} 100$

**1.** $3^2 = 9$  **3.** $2^{-3} = \frac{1}{8}$  **5.** $10^{-2} = 0.01$

**2.** $4^3 = 64$  **4.** $3^{-4} = \frac{1}{81}$  **6.** $10^{-1} = 0.1$

Express using exponential notation.

**Sample.**  $\log_{25} 5 = \frac{1}{2}$     *Solution:*  $25^{\frac{1}{2}} = 5$

**7.** $\log_4 16 = 2$  **9.** $\log_{16} 8 = \frac{3}{4}$  **11.** $\log_4 \frac{1}{2} = -\frac{1}{2}$

**8.** $\log_9 27 = \frac{3}{2}$  **10.** $\log_{10} 1000 = 3$  **12.** $\log_{10} 0.001 = -3$

*WRITTEN EXERCISES*

Find each logarithm.

**A**  **1.** $\log_2 64$  **5.** $\log_4 \frac{1}{16}$  **9.** $\log_{\frac{1}{2}} \frac{1}{8}$

**2.** $\log_3 81$  **6.** $\log_2 \frac{1}{16}$  **10.** $\log_{\frac{1}{3}} \frac{1}{81}$

**3.** $\log_4 2$  **7.** $\log_{10} 10^4$  **11.** $\log_{\frac{1}{5}} 25$

**4.** $\log_{27} 3$  **8.** $\log_{10} 10^{-4}$  **12.** $\log_{\frac{1}{4}} 64$

Find the value of *x* in each equation.

**Sample.**  $\log_x 64 = 3$

*Solution:*  Write in exponential form: $x^3 = 64$  $\therefore x = \sqrt[3]{64} = 4$

**13.** $\log_x 625 = 4$  **17.** $\log_{25} x = \frac{1}{2}$  **21.** $\log_3 \frac{1}{27} = x$

**14.** $\log_x 0.1 = -1$  **18.** $\log_{64} x = -\frac{1}{3}$  **22.** $\log_7 49 = x$

**15.** $\log_3 81 = x$  **19.** $\log_{\frac{1}{2}} x = -6$  **23.** $\log_x 125 = -3$

**16.** $\log_{\frac{1}{2}} 16 = x$  **20.** $\log_4 x = -\frac{1}{2}$  **24.** $\log_x 16 = -4$

**B** **25.** $\log_{\sqrt{3}} x = 6$       **29.** $\log_{\sqrt{3}} 27 = x$

**26.** $\log_{\sqrt{2}} x = -4$       **30.** $\log_{\sqrt{5}} \frac{1}{5} = x$

**27.** $\log_x \sqrt[3]{4} = \frac{1}{6}$       **31.** $\log_{10} (x^2 + 9x) = 1$

**28.** $\log_x \sqrt{6} = \frac{1}{4}$       **32.** $\log_{10} (4x - 4) = 2$

Graph both functions using the same set of axes.

**33.** $\{(x, y): y = 2^x\}$; $\{(x, y): y = \log_2 x\}$

**34.** $\{(x, y): y = 10^x\}$; $\{(x, y): y = \log_{10} x\}$

Show that each statement is true.

**C** **35.** $\log_{10} [\log_3 (\log_5 125)] = 0$       **39.** $\log_b 1 = 0$   $(b > 0)$

**36.** $\log_{10} [\log_2 (\log_3 9)] = 0$       **40.** $\log_b b = 1$   $(b > 0)$

**37.** $\log_4 [\log_2 (\log_3 81)] = \frac{1}{2}$       **41.** $\log_b b^b = b$

**38.** $\log_2 [\log_2 (\log_2 16)] = 1$       **42.** $\log_b b^n = n$

## 10–4   Inverse Functions

You saw in Section 10–3 that

$$\log_{10} u = v \quad \text{if and only if} \quad u = 10^v.$$

If you let $\log_{10} u = f(u)$ and $10^v = g(v)$, then this fact can be written in the form

$$f(u) = v \quad \text{if and only if} \quad u = g(v).$$

Any two functions related in this way are **inverses** of each other.

**Example 1.**   Let $F(x) = 2x + 2$ and $G(x) = \frac{1}{2}x - 1$. Show that the functions $F$ and $G$ are inverses of each other.

*Solution:*   Suppose $F(u) = v$. Then

$$2u + 2 = v.$$

Solve for $u$:       $2u = v - 2$

$$u = \tfrac{1}{2}v - 1$$

Hence       $u = G(v).$

The steps above can be reversed to show that if $G(v) = u$, then $v = F(u)$.

For the inverse functions of Example 1 you can see that

$$F(1) = 2 \cdot 1 + 2 = 4$$

and

$$G(4) = \tfrac{1}{2} \cdot 4 - 1 = 1.$$

Therefore $(1, 4)$ is on the graph of $F$, while $(4, 1)$ is on the graph of $G$. If equal scales are used on the coordinate axes, these points are symmetrically located with respect to the line $y = x$ (see Figure 10–6). In general, the graphs of two inverse functions are "mirror images" of each other, with the line $y = x$ acting as the "mirror" (Exercise 19, page 400).

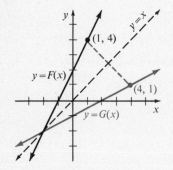

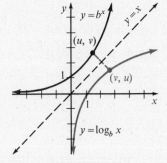

FIGURE 10–6                    FIGURE 10–7

We can show that $y = \log_b x$ and $y = b^x$ are inverse functions. Therefore, we can obtain the graph of $y = \log_b x$ by reflecting the graph of its inverse function, $y = b^x$, in the line $y = x$ (Figure 10–7).

If $f$ and $g$ are inverse functions, we write

$$g = f^{-1} \quad \text{and} \quad f = g^{-1},$$

read "$g$ is the inverse of $f$ and $f$ is the inverse of $g$." (Notice that the superscript $^{-1}$ is *not* an exponent; that is, $f^{-1}$ does *not* mean $\dfrac{1}{f}$.)

Do you see that the domain of $f^{-1}$ is the range of $f$, and that the range of $f^{-1}$ is the domain of $f$?

Recall that if an ordered pair $(x, y)$ belongs to a function $f$, then $x$ denotes a member of the domain, and $y$ the corresponding member of the range. You can use this fact to find a formula for $f^{-1}$, given a formula for $f$: If $y = f(x)$ defines $f$, then $x = f(y)$, when solved for $y$, gives a formula for $f^{-1}$, $y = f^{-1}(x)$. That is, you simply interchange the roles of $x$ and $y$ in the formula for $f$ to obtain a formula for $f^{-1}$.

**Example 2.** If $f(x) = x^3 + 2$, find a formula for $f^{-1}(x)$.

*Solution:* Let $y = f(x) = x^3 + 2$. Then $x = y^3 + 2$ defines the inverse of $f$. Solving for $y$, you have

$$y^3 = x - 2$$

or

$$y = \sqrt[3]{x - 2}.$$

The required formula is then $f^{-1}(x) = \sqrt[3]{x - 2}$.

The graph of the function $f$ in Example 2 is shown in Figure 10–8. (Since its equation is $y = x^3 + 2$, it can be obtained by raising the curve $y = x^3$ shown in Figure 7–3(b), page 262, 2 units vertically upward.) You can obtain the graph of $f^{-1}$ (shown in red) by reflecting the graph of $f$ in the line $y = x$.

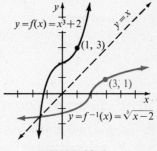

FIGURE 10–8

## WRITTEN EXERCISES

If the given function has an inverse, find a formula for it and graph it. If the given function has no inverse function, so state.

**A**
**1.** $f(x) = \frac{1}{3}x - 2$

**2.** $f(x) = 4x + 1$

**3.** $f(x) = x^2 - 2$

**4.** $f(x) = 1 - x^2$

**5.** $f(x) = 2$

**6.** $\{(x, y): y = x\}$

**7.** $\left\{(x, y): y = \frac{1}{x}\right\}$

**8.** $\{(x, y): y = \sqrt{x^2 + 4}\}$

**9.** $\{(x, y): y = \sqrt{16 - x^2}\}$

**10.** $\{(x, y): y = \sqrt{x}\}$

**B**
**11.** $f(x) = |x|$

**12.** $f(x) = |x| - 1$

**13.** $f(x) = 10^x$

**14.** $f(x) = \log_2 x$

**C**
**15.** Recall that a function is a set of ordered pairs no two of which have the same first coordinates (Section 4–2). Show that if a function $f$ has an inverse, $f^{-1}$, then $f^{-1} = \{(v, u): (u, v) \in f\}$.

**16.** Suppose the function $f$ has the inverse $f^{-1}$. Show that:
   **a.** For each $x$ in the domain of $f$, $f^{-1}(f(x)) = x$.
   **b.** For each $x$ in the domain of $f^{-1}$, $f(f^{-1}(x)) = x$.

**17.** A function $f$ is one-to-one if $x_1 \neq x_2$ implies $f(x_1) \neq f(x_2)$, or, in other words, if $f(x_1) = f(x_2)$ implies $x_1 = x_2$. Show that a function which has an inverse must be one-to-one.

**18.** Show that every one-to-one function (see Exercise 17) has an inverse.

**19.** Let the functions $f$ and $g$ be inverses of each other. Show that if $(u, v)$ is on the graph of $y = f(x)$, then $(v, u)$ is on the graph of $y = g(x)$ by justifying these steps:
1. $(u, v)$ is on the graph of $y = f(x)$.
2. $v = f(u)$
3. $u = g(v)$
4. $(v, u)$ is on the graph of $y = g(x)$.

## 10–5   Laws of Logarithms

By using a law of exponents and the fact that $v = \log_b u$ if and only if $u = b^v$, you can prove that *the logarithm of the product of two numbers equals the sum of their logarithms:*

From page 396,

$$x = b^{\log_b x} \quad \text{and} \quad y = b^{\log_b y}.$$

$$\therefore \; xy = b^{\log_b x} \cdot b^{\log_b y} = b^{\log_b x \,+\, \log_b y}.$$

Hence,

$$\log_b xy = \log_b x + \log_b y.$$

This proves the first of the *laws of logarithms* stated in the following theorem.

---

***THEOREM.***   Let $b$ be a positive number not equal to 1, let $x$ and $y$ be positive numbers and let $n$ be any real number. Then:

**1.** $\log_b xy = \log_b x + \log_b y$

**2.** $\log_b \dfrac{x}{y} = \log_b x - \log_b y$

**3.** $\log_b x^n = n \log_b x$

---

The proofs of the second and third laws of logarithms use the laws of exponents $b^p \div b^q = b^{p-q}$ and $(b^p)^n = b^{np}$, respectively.

**Example 1.** Given that $\log_3 5 = 1.465$, find:

    **a.** $\log_3 15$    **b.** $\log_3 \frac{5}{3}$    **c.** $\log_3 1.8$    **d.** $\log_3 25$

*Solution:*    Use the fact that $\log_3 3 = 1$ and $\log_3 9 = \log_3 3^2 = 2$.

    **a.** $\log_3 15 = \log_3 3 \cdot 5 = \log_3 3 + \log_3 5 = 1 + 1.465$
          $= 2.465$

    **b.** $\log_3 \frac{5}{3} = \log_3 5 - \log_3 3 = 1.465 - 1 = 0.465$

    **c.** $\log_3 1.8 = \log_3 \frac{9}{5} = \log_3 9 - \log_3 5 = 2 - 1.465$
          $= 0.535$

    **d.** $\log_3 25 = \log_3 5^2 = 2 \log_3 5 = 2 \times 1.465 = 2.930$

Recall that $b^{\log_b x} = x$ (page 395).

**Example 2.** Find the value of:

    **a.** $10^{\log_{10} 6 - \log_{10} 3}$    **b.** $4^{2 \log_4 3 + 3 \log_4 2}$    **c.** $25^{\log_5 6}$

*Solution:*    **a.** $10^{\log_{10} 6 - \log_{10} 3} = 10^{\log_{10} 6 \div 3} = 10^{\log_{10} 2} = 2$

    **b.** $4^{2 \log_4 3 + 3 \log_4 2} = 4^{\log_4 3^2 + \log_4 2^3} = 4^{\log_4 3^2 \cdot 2^3}$
          $= 4^{\log_4 72} = 72$

    **c.** $25^{\log_5 6} = (5^2)^{\log_5 6} = 5^{2 \log_5 6} = 5^{\log_5 6^2} = 6^2 = 36$

Figure 10–9 illustrates the law

$$\log_b pq = \log_b p + \log_b q.$$

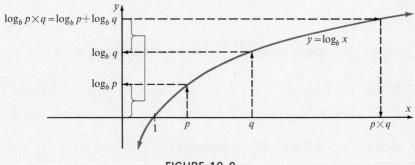

FIGURE 10–9

Notice that the figure suggests a graphical method to perform multiplication. The method is not used in practice because the slide rule, a mechanical device based on logarithms, does this and other operations more quickly and accurately.

*ORAL EXERCISES*

Express each logarithm as the sum or difference of simpler logarithmic expressions. Assume that all variables denote positive real numbers.

**Sample.** $\log_2 7x^2$

*Solution:* $\log_2 7x^2 = \log_2 7 + \log_2 x^2 = \log_2 7 + 2\log_2 x$

**1.** $\log_2 (xy)$    **4.** $\log_a x^4 y$    **7.** $\log_5 3\sqrt[3]{x}$    **10.** $\log_a x^2 y^3$

**2.** $\log_2 (abc)$    **5.** $\log_a 2x^{\frac{1}{2}}$    **8.** $\log_a (bc)^2$    **11.** $\log_b \dfrac{\sqrt{x}}{y}$

**3.** $\log_a \dfrac{x}{b}$    **6.** $\log_4 \dfrac{xy}{z}$    **9.** $\log_a \sqrt{bc}$    **12.** $\log_b \dfrac{x}{\sqrt{y}}$

Express as a single logarithm with coefficient 1.

**13.** $\log_a x + \log_a y + \log_a z$      **16.** $2\log_a x - \frac{1}{2}\log_a y$

**14.** $4\log_a x$      **17.** $2(\log_a z - \log_a 3)$

**15.** $\log_a x + \log_a y - \log_a z$      **18.** $\frac{1}{5}(2\log_a x + 3\log_a y)$

*WRITTEN EXERCISES*

Solve for *x*.

**Sample.** $\log_2 x = \log_2 15 + \log_2 3 - \log_2 5$

*Solution:* $\log_2 x = \log_2 \dfrac{15 \times 3}{5} = \log_2 9$; therefore, $x = 9$.

[A] **1.** $\log_5 x = 3\log_5 7$      **3.** $\log_3 x = 3\log_3 2 + 2\log_3 4$

**2.** $\log_2 x = \frac{1}{2}\log_2 81$      **4.** $\log_5 x = 2\log_5 7 + 2\log_5 3$

**5.** $\log_{10} x = \frac{1}{2}\log_{10} 144 - \frac{1}{3}\log_{10} 8$

**6.** $\log_{10} x = \frac{1}{3}\log_{10} 64 + \frac{1}{2}\log_{10} 121$

**7.** $\log_{10} x = 4\log_{10} 4 - 2\log_{10} 16$

**8.** $\log_{10} x = 3\log_{10} 5 - \frac{1}{2}\log_{10} 25$

**9.** $\log_{10} x = \frac{1}{5}(2\log_{10} 3 + 3\log_{10} 2)$

**10.** $\log_{10} x = \frac{1}{6}(2\log_{10} 4 + 2\log_{10} 2)$

Given that $\log_{10} 3 \doteq 0.4771$, find:

**B**  **11.** $\log_{10} 9$  **12.** $\log_{10} 30$  **13.** $\log_{10} 0.3$  **14.** $\log_{10} 300$

Use the value of $\log_{10} 3$ given above together with the fact that $\log_{10} 5 = 0.6990$ to find:

**15.** $\log_{10} 15$  **16.** $\log_{10} \frac{5}{3}$  **17.** $\log_{10} \frac{3}{5}$  **18.** $\log_{10} 45$

Find the value of each expression.

**19.** $5^{\log_5 3 + \log_5 2}$  **21.** $4^{\log_2 8}$  **23.** $3^{\log_3 5 - 2 \log_3 2}$

**20.** $10^{\log_{10} 8 - \log_{10} 4}$  **22.** $3^{\log_3 (3^{\log_3 3})}$  **24.** $8^{2 \log_8 \sqrt[3]{5}}$

Solve for $x$:

**Sample.**  $\log_{10} x + \log_{10} (x + 21) = 2$

*Solution:*  By the first law of logarithms, you have:

$$\log_{10} x(x + 21) = 2$$
$$\log_{10} (x^2 + 21x) = 2$$

Writing this equation in exponential form gives

$$x^2 + 21x = 10^2$$

from which  $\qquad x^2 + 21x - 100 = 0$
$$(x + 25)(x - 4) = 0$$

$$x = -25 \text{ or } x = 4.$$

*Check:*  $\log_{10} (-25) + \log_{10} (-25 + 21) \stackrel{?}{=} 2$.  Since $\log_{10} (-25)$ is not defined, $-25$ is not a root.

$$\log_{10} 4 + \log_{10} (4 + 21) = \log_{10} 4 + \log_{10} 25$$
$$= \log_{10} 100 = 2$$

Therefore, the solution set is $\{4\}$.

**C**  **25.** $\log_3 (x + 4) + \log_3 (x - 4) = 2$

**26.** $\log_2 (x + 2) - 1 = \log_2 (x - 2)$

**27.** $\log_7 (x + 1) + \log_7 (x - 5) = 1$

**28.** $\log_{10} (x + 3) + \log_{10} x = 1$

**29.** $\log_{10} (x + 2) + \log_{10} (x - 1) = 1$

**30.** $\log_{10} (x + 3) - \log_{10} (x - 1) = 1$

**31.** Prove the second law of logarithms.

**32.** Prove the third law of logarithms.

## 10–6   Common Logarithms

Logarithms to the base 10 are called common logarithms and are useful in simplifying computations. Since it is customary to omit writing the base 10, we will agree that $\log x = \log_{10} x$, and we will often use the word "logarithm" to mean "common logarithm."

Because $\log 10^x = \log_{10} 10^x = x$ (page 395), it is easy to find common logarithms of integral powers of 10. For example:

$$\log 0.01 = \log 10^{-2} = -2 \qquad \log 10 \quad = \log 10^1 = 1$$
$$\log 0.1 \, = \log 10^{-1} = -1 \qquad \log 100 \, = \log 10^2 = 2$$
$$\log 1 \quad = \log 10^0 \, = \quad 0 \qquad \log 1000 = \log 10^3 = 3$$

To find logarithms of numbers which are not integral powers of 10, use a "table of logarithms" such as Table 5 of the Appendix. Table 5, a part of which appears below, gives the first four decimal places of the logarithms of numbers between 1 and 10. (The decimal point which should precede each entry is usually omitted in tables.) To find an approximation for $\log x$ for $1 < x < 10$, find the first two digits of $x$ in the column headed $x$ and the third digit in the row to the right of $x$.

| x | 0 | 1 | 2 | 3 | 4 | 5 | 6 | 7 | 8 | 9 |
|----|------|------|------|------|------|------|------|------|------|------|
| **20** | 3010 | 3032 | 3054 | 3075 | 3096 | 3118 | 3139 | 3160 | 3181 | 3201 |
| **21** | 3222 | 3243 | 3263 | 3284 | 3304 | 3324 | 3345 | 3365 | 3385 | 3404 |
| **22** | 3424 | 3444 | 3464 | 3483 | 3502 | 3522 | 3541 | 3560 | 3579 | 3598 |
| **23** | 3617 | 3636 | 3655 | 3674 | 3692 | 3711 | 3729 | 3747 | 3766 | 3784 |
| **24** | 3802 | 3820 | 3838 | 3856 | 3874 | 3892 | 3909 | 3927 | 3945 | 3962 |

To find $\log 2.17$, look for 21 under $x$ and move along row 21 to the column headed 7, where you find 3365. Therefore $\log 2.17 \doteq 0.3365$. Although entries in tables of logarithms are only approximations (except for $\log 1 = 0$), it is customary to use the symbol $=$, rather than $\doteq$, in such statements as $\log 2.17 = 0.3365$.

To find the logarithm of a positive number which is not between 1 and 10, you can use standard notation (page 245) and the product law of logarithms (page 400). For example:

$$2{\scriptstyle\wedge}17 = 2.17 \times 10^2 \qquad\qquad 0.002{\scriptstyle\wedge}17 = 2.17 \times 10^{-3}$$
$$\log 2{\scriptstyle\wedge}17 = \log 2.17 + \log 10^2 \qquad \log 0.002{\scriptstyle\wedge}17 = \log 2.17 + \log 10^{-3}$$
$$= 0.3365 + 2 \qquad\qquad\qquad = 0.3365 + (-3)$$
$$= 2.3365 \qquad\qquad\qquad\qquad = -3 + 0.3365$$

The common logarithm of a number is the sum of (1) an integer, called the characteristic, and (2) a nonnegative number less than 1, called the mantissa, which is found in a table of logarithms. In practice you obtain the characteristic by inspection (using carets if you like) without actually putting the number in standard notation. In working with logarithms we usually wish to keep the "fractional part" positive, so we do *not* simplify $-3 + 0.3365$ to $-2.6635$. Instead, we ordinarily write $-3 = 7.0000 - 10$, so that $\log 0.00217 = 7.3365 - 10$. In some cases it may be more convenient to use some other difference, such as $5.000 - 8$, $6.0000 - 9$, or $17.0000 - 20$, to represent $-3$.

**Example 1.** Find $\log 61300$.

*Solution:* The characteristic of $\log 6{}_\wedge 1300$ is 4. Row 61 column 3 of Table 5 contains the mantissa .7875. Therefore $\log 61300 = 4.7875$.

**Example 2.** If the logarithm of a number is $9.3892 - 10$, find the number. That is, if $\log x = 9.3892 - 10$, find $x$.

*Solution:* Reverse the order of the steps in Example 1. The mantissa is .3892 so look for 3892 *among the entries* of Table 5. Since 3892 is located in row 24 and column 5, the digits in $x$ are 245. If the characteristic were 0, $x$ would equal 2.45. Since the characteristic is $9 - 10 = -1$, $x = 2.45 \times 10^{-1} = 0.245$.

If $\log x = a$, then $x$ is called the antilogarithm of $a$, written antilog $a$.

**Example 3.** Find antilog $1.3565$.

*Solution:* Looking among the entries in Table 5, we do not find 3565 and therefore use 3560, the entry closest to 3565. Because 3560 is in row 22 and column 7, the first three digits in the antilogarithm are 227. Since the characteristic is 1, we have antilog $1.3565 = 22.7$. (Of course, $\log 22.7 = 1.3560$, not 1.3565, but 22.7 is the best *three*-significant-digit answer. In Section 10–7 you will learn to get a *four*-significant-digit answer.)

*WRITTEN EXERCISES*

Use Table 5 in the Appendix to find each logarithm.

| A | | | | |
|---|---|---|---|---|
| **1.** $\log 3.41$ | **3.** $\log 21.6$ | **5.** $\log 0.812$ | **7.** $\log 342$ |
| **2.** $\log 9.72$ | **4.** $\log 42.8$ | **6.** $\log 0.0315$ | **8.** $\log 21{,}600$ |

**9.** $\log 0.0012$    **11.** $\log 700,000$    **13.** $\log 3010$    **15.** $\log 0.0669$

**10.** $\log 0.00999$    **12.** $\log 806,000$    **14.** $\log 1020$    **16.** $\log 3.98$

Find each antilogarithm.

**17.** antilog $0.8727$                **25.** antilog $6.9917 - 10$

**18.** antilog $1.4942$                **26.** antilog $9.0086 - 10$

**19.** antilog $3.7067$                **27.** antilog $3.9987$

**20.** antilog $9.0969 - 10$        **28.** antilog $6.9562$

**21.** antilog $7.9619 - 10$        **29.** antilog $6.3711 - 10$

**22.** antilog $2.7566$                **30.** antilog $8.7774 - 10$

**23.** antilog $7.7093 - 10$        **31.** antilog $6.9031$

**24.** antilog $8.8686 - 10$        **32.** antilog $7.6021$

Find each of the following using Table 5.

> **Sample.**   $10^{2.8069}$
>
> *Solution:*   If $x = 10^{2.8069}$, then
>
> $$\log_{10} x = 2.8069 \text{ or } x = \text{antilog } 2.8069.$$
> $$\text{From Table 5, antilog } 2.8069 = 641.$$
> $$\therefore 10^{2.8069} = 641.$$

B **33.** $10^{0.4871}$              **35.** $10^{3.9708}$              **37.** $10^{7.5977-10}$

**34.** $10^{1.6839}$              **36.** $10^{2.7459}$              **38.** $10^{9.0128-10}$

Solve each inequality.

C **39.** $\dfrac{(x-2)\log x}{x^2} > 0$                **40.** $\dfrac{\log(x-1)}{x+3} \le 0$

## 10–7   Linear Interpolation

Can you find $\log 2.354$? Because Table 5 gives direct reading for the logarithms of numbers known to at most *three* significant digits, you can find entries for $\log 2.350$ and $\log 2.360$, but not for $\log 2.354$. However, you can reason as follows: Since $2.354$ is $\frac{4}{10}$ of the way from $2.350$ to $2.360$, the number which is $\frac{4}{10}$ of the way from $\log 2.350$ to $\log 2.360$ is an acceptable approximation to $\log 2.354$.

You find this approximation to log 2.354 by adding $\frac{4}{10}$ of the differ-ence log 2.360 − log 2.350 to log 2.350. If you let

$$d = \log 2.354 - \log 2.350,$$

you can arrange the work as follows:

| $x$ | $\log x$ |
|---|---|
| 2.360 | 0.3729 |
| 2.354 | log 2.354 |
| 2.350 | 0.3711 |

0.010 [ 0.004 [ ] $d$ ] 0.0018

$d \doteq \frac{4}{10} \times 0.0018 \doteq 0.0007$ (rounded to four places because the entries in Table 5 are reliable only to four places)

∴ log 2.354 ≐ 0.3711 + 0.0007 = 0.3718.

The process described above is called **linear interpolation**. It is illus-trated geometrically in Figure 10–10 where the red curve is a schematic representation of part of the graph of $y = \log x$.

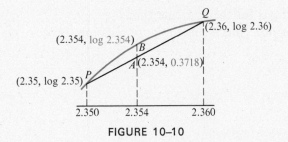

(2.354, log 2.354)

$Q$ (2.36, log 2.36)

$B$

$A$ (2.354, 0.3718)

(2.35, log 2.35) $P$

| 2.350 | 2.354 | 2.360 |

FIGURE 10–10

Linear interpolation is equivalent geometrically to replacing the arc $\overset{\frown}{PQ}$ by the line segment $\overline{PQ}$. That is, we approximate the ordinate of the point $B$ on the arc by the ordinate of the point $A$ on the line segment and on the same vertical line as $B$.

You can use linear interpolation to find the *mantissa* of the logarithm of any four-significant-digit number. For example, the work done above tells you that the mantissa of log 23,540 is 0.3718, and therefore log 23,540 = 4.3718. To find the logarithm of a number having more than four significant digits, first round the given number to four sig-nificant digits (page 245) and then proceed as above.

You can interpolate in reverse to find $x$ if $\log x$ is known but its mantissa is not an entry in the table.

**Example.**   Find antilog 1.3572.

*Solution:*   You are asked to find $x$, given that $\log x = 1.3572$. In Table 5, find the consecutive entries, 3560 and 3579, between which 3572 lies, and note the corresponding four-digit sequences, 2270 and 2280. Make the following chart, putting 22.70 and 22.80 in the $x$-column because the characteristic 1 tells you that each antilog is between 10 and 100.

$$
\begin{array}{c|c}
 & x & \log x \\
\hline
 & 22.80 & 1.3579 \\
0.10\ \Big[ & \text{antilog } 1.3572 & 1.3572 \\
 c\ \Big[ & & \\
 & 22.70 & 1.3560 \\
\end{array}
\quad 0.0012 \quad \Big] \; 0.0019
$$

$$\frac{0.0012}{0.0019} = \frac{12}{19}; \qquad c = \frac{12}{19} \times 0.10 \doteq 0.06$$

$$\therefore \ \text{antilog } 1.3572 = 22.70 + 0.06 = 22.76$$

Notice that the value of $c$ was rounded to one significant digit because reverse interpolation in a four-place table yields at most four significant digits for the antilogarithm.

With practice, much of the interpolation process can be done mentally.

*WRITTEN EXERCISES*

Find each logarithm.

**A**

**1.** $\log 5.816$     **4.** $\log 92.68$     **7.** $\log 32,110$     **10.** $\log 6.219$

**2.** $\log 3.415$     **5.** $\log 0.6144$     **8.** $\log 50,280$     **11.** $\log 0.001974$

**3.** $\log 12.72$     **6.** $\log 0.08152$     **9.** $\log 8.184$     **12.** $\log 0.1007$

Find each antilogarithm.

**13.** antilog 2.3932     **17.** antilog 4.7820     **21.** antilog 6.6975 − 10

**14.** antilog 1.8770     **18.** antilog 2.1220     **22.** antilog 7.8780 − 10

**15.** antilog 3.6948     **19.** antilog 0.4803     **23.** antilog 9.0759 − 10

**16.** antilog 0.1978     **20.** antilog 1.2440     **24.** antilog 9.9979 − 10

**COMPUTING WITH LOGARITHMS**

## 10–8 Computing Products and Quotients

For common logarithms the first two laws stated in the theorem on page 400 become

$$\log (x \times y) = \log x + \log y$$
$$\log (x \div y) = \log x - \log y$$

These laws enable you to replace the operations of $\times$ and $\div$ by $+$ and $-$, respectively.

For example, to compute $348 \times 0.509$, you can write:

$$\log (348 \times 0.509) = \log 348 + \log 0.509$$

The two logarithms in the right-hand member can be found with the help of Table 5: $\log 348 = 2.5416$ and $\log 0.509 = 9.7067 - 10$. Hence:

$$\log (348 \times 0.509) = 2.5416 + (9.7067 - 10)$$
$$= 12.2483 - 10$$
$$= 2.2483$$

Therefore,     $348 \times 0.509 = $ antilog $2.2483$
$$= 177$$

to three significant digits. Notice that addition is the only arithmetic operation used in working this multiplication problem.

To avoid mistakes in computing with logarithms, you should arrange your work neatly and systematically. It is a good plan to list numbers to be added or subtracted in vertical columns, aligning equality signs and decimal points vertically. You can indicate the operations $(+)$, $(-)$, or $(\times)$ and put in all characteristics before you use the table to find the mantissas. Label each step so that if you must check back, you will know what you are checking. The example worked above can be arranged as follows.

$$\text{Let } N = 348 \times 0.509.$$
$$\log N = \log 348 + \log 0.509$$
$$\log 348 = \quad 2.5416$$
$$\log 0.509 = \quad 9.7067 - 10 \qquad (+)$$
$$\log N = 12.2483 - 10 = 2.2483$$
$$N = \text{antilog } 2.2483 = 177$$
$$\therefore 348 \times 0.509 = 177$$

To guard against misplacing a decimal point, you can make an "order of magnitude" estimate (page 247) either before or after the actual computation. In the foregoing example, the estimate

$$348 \times 0.509 \doteq (3 \times 10^2) \times (5 \times 10^{-1}) = 15 \times 10 = 150$$

suggests that no errors of magnitude have been made.

Numbers involved in computations often are measurements and therefore approximations. In the example worked above, we assumed that the numbers 348 and 0.509 were accurate only to three significant digits and therefore expressed the answer, 177, only to three significant digits rather than use reverse interpolation to obtain 177.1. We use the following rules to decide how to round answers.

---

**1.** Give products, quotients, and powers to the same number of *significant digits* as appear in the least accurate approximation involved.

**2.** Give sums and differences to the same number of *decimal places,* as appear in the approximation with the least number of decimal places.

---

**Example 1.**    Compute $-\dfrac{7.435}{24.40}$ to four significant digits.

*Solution :*    Let $N = -\dfrac{7.435}{24.40}$. Since $N$ is negative and negative numbers do not have logarithms, work with $|N| = \dfrac{7.435}{24.40}$.

$$\log |N| = \log 7.435 - \log 24.40$$
$$\log 7.435 = 0.8713$$
$$\log 24.40 = \underline{1.3874} \qquad (-)$$
$$?$$

To obtain a difference in which the mantissa is positive, write $\log 7.435$ as $10.8713 - 10$.

$$
\begin{aligned}
\log 7.435 &= 10.8713 - 10 \\
\log 24.40 &= \underline{\ \ 1.3874} \qquad (-) \\
\log |N| &= \ \ 9.4839 - 10 \\
|N| &= \text{antilog } (9.4839 - 10) = 0.3047 \\
N &= -0.3047 \\
\therefore \ -\frac{7.435}{24.40} &= -0.3047
\end{aligned}
$$

*Estimate :*    $|N| \doteq (7) \div (2 \times 10) = 3.5 \times 10^{-1} = 0.35$

**Example 2.** Compute $\dfrac{34.60}{0.04855 \times 8.520}$.

*Solution:*  Let $N = \dfrac{34.60}{0.04855 \times 8.520}$ and $M = 0.04855 \times 8.520$.

Then $\log N = \log 34.60 - \log M$.

$$
\begin{array}{rl}
\log 0.04855 = & 8.6862 - 10 \\
\log 8.520 = & 0.9304 \qquad (+) \\
\hline
\log M = & 9.6166 - 10
\end{array}
$$

$$
\begin{array}{rl}
\log 34.60 = & 11.5391 - 10 \\
\log M = & 9.6166 - 10 \qquad (-) \\
\hline
\log N = & 1.9225 \\
N = & \text{antilog } 1.9225 = 83.66
\end{array}
$$

$$\therefore \frac{34.60}{0.04855 \times 8.520} = 83.66$$

*Estimate:*  $N \doteq \dfrac{3 \times 10}{(5 \times 10^{-2})(9)} = \dfrac{3 \times 10^{3}}{45} = \dfrac{10^{3}}{15} = \dfrac{2}{3} \times 10^{2} \doteq 67$

## *ORAL EXERCISES*

State the logarithmic equation you would use to compute the value of each of the following expressions.

**Sample.**  $\dfrac{3.81 \times 21.7}{6.29}$

*Solution:*  $\log N = \log 3.81 + \log 21.7 - \log 6.29$

**1.** $(2.45)(71.3)$

**2.** $(61.9)(427)$

**3.** $\dfrac{38.2}{7.5}$

**4.** $\dfrac{24.3}{6.7}$

**5.** $\dfrac{(812)(41.5)}{4300}$

**6.** $\dfrac{(41.6)(67)}{8200}$

**7.** $\dfrac{(0.713)(0.0214)}{0.00816}$

**8.** $\dfrac{(0.913)(0.0178)}{0.0503}$

**9.** $(612)(9.17)(-0.0216)$

**10.** $(82.4)(-3.18)(-0.176)$

**11.** $\dfrac{(6.38)(2.17)}{(72.4)(0.513)}$

**12.** $\dfrac{(71.6)(314)}{(5230)(0.816)}$

**13.** $(2138)(6.717)$

**14.** $(83.91)(5.826)$

**15.** $\dfrac{-612.8}{22.37}$

**16.** $\dfrac{82.93}{-605.3}$

**17.** $\dfrac{71.63}{(8.024)(21.63)}$

**18.** $\dfrac{82460}{(213.7)(62.81)}$

*WRITTEN EXERCISES*

A **1–18.** Compute the value of each expression in Oral Exercises 1–18.

Compute each of the following.

B **19.** $\dfrac{(0.2134)(0.01678)}{(0.09143)(0.0213)}$

**23.** $\dfrac{(26.23)(406)(0.7231)}{(6.41)(27.63)}$

**20.** $\dfrac{(2.184)(0.01763)}{(21.62)(0.0719)}$

**24.** $\dfrac{(482)(6.093)(-4.176)}{(7.126)(0.5023)}$

**21.** $\dfrac{(21.3)(6.176)(0.9821)}{(0.7132)(214.6)}$

**25.** $\dfrac{(12700)(62400)(2176)}{(78540)(21200)(0.876)}$

**22.** $\dfrac{(823)(0.3721)(1.023)}{(72.14)(4.21)}$

**26.** $\dfrac{(0.0183)(0.2177)(1.395)}{(0.0812)(0.6217)(0.0821)}$

## 10–9 Computing Powers and Roots

For common logarithms the third law stated in the theorem on page 400 becomes

$$\log x^n = n \log x.$$

You can use this law to evaluate powers and roots.

**Example 1.** Compute: **a.** $(0.408)^5$ **b.** $\sqrt[3]{0.0250}$

*Solution:*

**a.**  Let $H = (0.408)^5$
  $\log H = 5 \times \log 0.408$
  $\log 0.408 = \quad 9.6107 - 10$
  $\underline{\qquad\qquad\quad 5} \;\; (\times)$
  $\log H = 48.0535 - 50$
  $\quad\; = \quad 8.0535 - 10$
  $H = \text{antilog}\,(8.0535 - 10) = 0.0113$

  $\therefore \; (0.408)^5 = 0.0113$

**b.**  Let $K = \sqrt[3]{0.0250} = (0.0250)^{\frac{1}{3}}$
  $\log K = \frac{1}{3} \times \log 0.0250$
  $\log 0.0250 = \quad 8.3979 - 10$
  $\qquad\qquad\; = 28.3979 - 30$
  $\underline{\qquad\qquad\qquad\; \frac{1}{3} \;\; (\times)}$
  $\log K = \quad 9.4660 - 10$
  $K = \text{antilog}\,(9.4660 - 10) = 0.292$

  $\therefore \; \sqrt[3]{0.0250} = 0.292$

*Estimate:*

$H \doteq (4 \times 10^{-1})^5 = 4^5 \times 10^{-5}$
$\quad = 16 \times 64 \times 10^{-5}$
$\quad = (2 \times 10) \times (6 \times 10) \times 10^{-5}$
$\quad = 12 \times 10^2 \times 10^{-5}$
$\quad = 12 \times 10^{-3} = 0.012$

$K \doteq (3 \times 10^{-2})^{\frac{1}{3}} \doteq (30 \times 10^{-3})^{\frac{1}{3}}$
$\quad = 30^{\frac{1}{3}} \times 10^{-1} \doteq 3 \times 10^{-1}$
$\quad = 0.3$

Note that in Example 1(b) the characteristic $8.0000 - 10$ was replaced by $28.0000 - 30$ (the same number, of course) so that the negative part would be evenly divisible by 3.

In more complicated computations you may need to use all three laws of logarithms.

**Example 2.** Evaluate: $4.11 \left( \dfrac{431}{3.92 \times 0.868} \right)^{\frac{3}{2}}$

*Solution:*  Let $M = 4.11 \left( \dfrac{431}{3.92 \times 0.868} \right)^{\frac{3}{2}}$.

$$\log M = \log 4.11 + \tfrac{3}{2}[\log 431 - (\log 3.92 + \log 0.868)]$$
$$= [\log 431 - \log 3.92 - \log 0.868] \times \tfrac{3}{2} + \log 4.11$$

$$
\begin{array}{rll}
\log 431 = & 2.6345 & \\
\log 3.92 = & 0.5933 & (-) \\
\hline
 & 12.0412 - 10 & \\
\log 0.868 = & 9.9385 - 10 & (-) \\
\hline
 & 2.1027 & \\
 & 3 & (\times) \\
\hline
 & 6.3081 & \\
 & \tfrac{1}{2} & (\times) \\
\hline
 & 3.1540 & \\
\log 4.11 = & 0.6138 & (+) \\
\hline
\log M = & 3.7678 & \\
M = & \text{antilog } 3.7678 = 5860
\end{array}
$$

$$\therefore\ 4.11 \left( \frac{431}{3.92 \times 0.868} \right)^{\frac{3}{2}} = 5860$$

*Estimate:*  $M \doteq 4 \left( \dfrac{4 \times 10^2}{4 \times 1} \right)^{\frac{3}{2}} = 4 \cdot (10^2)^{\frac{3}{2}} = 4 \times 10^3 = 4000$

Many formulas can be evaluated with the help of logarithms. Among these is the compound-interest formula: If a principal $P$ is invested at $r\%$ annual interest compounded $n$ times a year, then the accumulated amount after $t$ years is

$$A = P \left( 1 + \frac{0.01r}{n} \right)^{nt}.$$

**Example 3.** If \$2500 is invested in a bank paying 6% interest compounded quarterly, how much will be in the account at the end of 5 years?

*(Solution on page 414.)*

*Solution:*     In the formula $A = P\left(1 + \dfrac{0.01r}{n}\right)^{nt}$, substitute $P = 2500$, $r = 6$, $n = 4$, and $t = 5$. Then:

$$A = 2500\left(1 + \frac{0.01 \times 6}{4}\right)^{4 \times 5} = 2500(1.015)^{20}$$

$$\log A = \log 2500 + 20 \log 1.015$$

$$\begin{aligned} \log 1.015 &= 0.0065 \\ &\underline{\phantom{=} 20} \quad (\times) \\ \log (1.015)^{20} &= 0.1300 \\ \log 2500 &= \underline{3.3979} \quad (+) \\ \log A &= 3.5279 \\ A &= \text{antilog } 3.5279 = 3372 \end{aligned}$$

$\therefore$ about \$3372 will be in the account.

Modern computers have to a great extent replaced logarithms in doing numerical calculations. The ideas involved, however, are useful in higher mathematics.

## ORAL EXERCISES

State the logarithmic equation you would use to find the value of each of the following expressions.

Sample.     $\dfrac{(21.3)(2.42)^2}{\sqrt{1.87}}$

*Solution:*     $\log N = \log 21.3 + 2 \log 2.42 - \tfrac{1}{2} \log 1.87$

**1.** $(6.14)^3$

**2.** $(21.7)^4$

**3.** $\sqrt{31.8}$

**4.** $\sqrt[4]{8.36}$

**5.** $(0.618)^2$

**6.** $(0.0925)^2$

**7.** $\sqrt{\dfrac{2.18}{0.173}}$

**8.** $\sqrt[3]{\dfrac{62.5}{38.2}}$

**9.** $\sqrt{(49.2)(7.15)^3}$

**10.** $\sqrt{(2.76)(9.38)^{\frac{1}{2}}}$

**11.** $\dfrac{(53.9)^2\sqrt{6.08}}{(7.32)^4}$

**12.** $\dfrac{(83.9)^3}{(235)^2\sqrt{7061}}$

**13.** $\dfrac{\sqrt{3.61}\,(2.147)^2}{4\sqrt[3]{1318}}$

**14.** $\dfrac{(6.49)^2\sqrt[3]{8.215}}{17.93}$

**15.** $3.28\sqrt[3]{\dfrac{690}{(0.417)(1.84)}}$

**16.** $\dfrac{28}{4.1}\sqrt{\dfrac{(32.1)^3}{(6.2)^2(1.31)^2}}$

**17.** $\sqrt{\dfrac{(4.71)(0.00482)}{(0.0432)^2}}$

**19.** $\dfrac{(0.0813)^2(8157)(2.918)}{5.62\sqrt{4.183}}$

**18.** $\sqrt{\dfrac{(3.19)^3(0.973)}{(0.0826)}}$

**20.** $\dfrac{(6.148)^2(6.192)\sqrt[3]{17.9}}{(312.6)(0.821)^3}$

## WRITTEN EXERCISES

A **1–10.** Find the value of each expression in Oral Exercises 1–10.

B **11–20.** Find the value of each expression in Oral Exercises 11–20.

C **21.** Show that $\log \dfrac{x + \sqrt{x^2 - 1}}{x - \sqrt{x^2 - 1}} = 2 \log (x + \sqrt{x^2 - 1})$.

**22.** Show that $\dfrac{\log (x + h) - \log x}{h} = \log \left(1 + \dfrac{h}{x}\right)^{\frac{1}{h}}$.

**23.** Show that $\log \dfrac{y}{a + \sqrt{a^2 + y^2}} = \log \dfrac{\sqrt{a^2 + y^2} - a}{y}$.

**24.** Solve for $x$: $\log_4 |2x + 2| - \log_4 |3x + 1| = \frac{1}{2}$.

## PROBLEMS

A **1.** A principal of $5000 is invested at 4% interest compounded annually. How much will the investment amount to after 10 years?

**2.** How much will the investment in Problem 1 amount to after 10 years if the interest is compounded semiannually?

**3.** The Monarch Savings and Loan Company pays 5% interest on regular passbook accounts, compounded every 4 months. How much will an investment of $6000 be worth after 20 years?

**4.** If $4000 is invested at 6% compounded quarterly, what will the investment be worth after 12 years?

**5.** How much principal needs to be invested now, at 4% interest compounded quarterly, to be worth $8042 after 12 years?

**6.** How much money must Miss Brown invest at 4% compounded annually if she is to receive $2500 in 20 years?

In Exercises 7, 8, and 9 use $\pi = 3.142$.

**7.** The volume $V$ of a sphere is given by $V = \frac{4}{3}\pi r^3$, where $r$ is the radius. Find $V$ when $r = 3.167$.

**8.** A pendulum of length $l$ feet makes a single oscillation in $t$ seconds where

$$t = \pi \sqrt{\frac{l}{32.16}}.$$

Find $t$ when $l = 4.138$.

**9.** A brass wire of radius $r$ and length $l$ centimeters, stretches $s$ centimeters under a weight of $m$ grams according to the relationship

$$s = \frac{mgl}{\pi r^2 k}$$

where $g = 980$ and $k$ is a constant. If a wire 219 centimeters long has a radius of 0.319 centimeters, and if it stretches 0.059 centimeters under a weight of 944 grams, find $k$.

**10.** Using the relationship given in Problem 9, find the length $l$ of a wire of radius 0.032 centimeters if it stretches 5.92 centimeters under a weight of 1820 grams and if $k = 2.1 \times 10^{12}$.

**B** **11.** A certain radioactive element decays according to the formula $Q = Q_0 e^{-0.4t}$, where $Q$ is the amount remaining of an original amount $Q_0$ after $t$ seconds. If $e \doteq 2.72$, how much would remain of 40 grams of the element after 4 seconds?

**12.** Using the formula given in Problem 11, if there were 40 grams of the element remaining after 2 seconds, how much of the element was there originally?

**13.** The atmospheric pressure $p$ is given approximately by $p = 30.0(10)^{-0.09a}$, where $p$ is in inches of mercury and $a$ is the altitude above sea level in miles. What is the atmospheric pressure at an altitude of 2 miles?

**14.** Assuming that the formula in Problem 13 is valid below sea level, what would be the atmospheric pressure at the bottom of a mine shaft $\frac{1}{2}$ mile deep?

**15.** The number of bacteria in a certain culture is related to time by the formula $N = N_0 e^{-0.4t}$, where $N_0$ is the number present initially (at time $t = 0$), and $t$ is the elapsed time in hours. If $9 \times 10^4$ bacteria are present 10 hours after the start of the experiment, how many were present initially? (use $e \doteq 2.72$)

**16.** In Problem 15, how many bacteria are present 24 hours after the start of the experiment?

## 10–10   Solving Equations

You can use logarithms to solve certain types of equations.

**Example 1.**   Find $x$ to three significant digits if $x^{\frac{2}{3}} = 150$.

*Solution:*

**1.** Equate the common logarithms of the two members.

$$x^{\frac{2}{3}} = 150$$
$$\log x^{\frac{2}{3}} = \log 150$$

**2.** Use a law of logarithms to simplify the left member.

$$\tfrac{2}{3} \log x = \log 150$$

**3.** Solve for $\log x$.

$$\log x = \tfrac{3}{2} \log 150$$
$$= \tfrac{3}{2} \times 2.1761$$
$$= 3.2642$$

**4.** Find antilog 3.2642

$$x = 1840$$

**Example 2.**   Express $\log_5 8$ in terms of common logarithms.

*Solution:*   Let $x = \log_5 8$.

**1.** Write the equation in exponential form.

$$5^x = 8$$
$$\log 5^x = \log 8$$

**2.** Equate the common logarithms of the members, simplify, and solve for $x$.

$$x \log 5 = \log 8$$

$$x = \frac{\log 8}{\log 5}$$

$$\therefore \log_5 8 = \frac{\log_{10} 8}{\log_{10} 5}$$

The result of Example 2 suggests the general relationship which holds between the logarithms of a number $n$ to two different bases, $a$ and $b$:

$$\log_b n = \frac{\log_a n}{\log_a b}$$

(See Exercise 27, page 418, for an outline of a proof.)

To express $\log_5 8$ as a decimal, start with the result of Example 2, and find the common logarithms:

$$\log_5 8 = \frac{\log_{10} 8}{\log_{10} 5} = \frac{0.9031}{0.6990}$$

You can express this quotient in decimal form either by using "long division" or by using logarithms as follows (but do *not* subtract 0.6990 from 0.9031).

$$\text{Let } x = \log_5 8 = \frac{0.9031}{0.6990}$$

$$\log x = \log \frac{0.9031}{0.6990}$$

$$= \log 0.9031 - \log 0.6990$$

$$\log 0.9031 = 9.9557 - 10$$
$$\log 0.6990 = 9.8445 - 10 \quad (-)$$
$$\overline{\log x = 0.1112}$$

$$x = 1.292, \text{ or } \log_5 8 = 1.292.$$

## WRITTEN EXERCISES

Solve each equation using logarithm tables as necessary.

A

**1.** $6^x = 72$      **5.** $x^{\frac{1}{5}} = 3.21$      **9.** $3a^{\frac{3}{5}} = 7.21$

**2.** $5^x = 104$      **6.** $y^{\frac{2}{3}} = 6.82$      **10.** $2z^{\frac{3}{2}} = 18.6$

**3.** $4^x = 2.17$      **7.** $(\frac{1}{9})^{x-1} = 27^{4-x}$      **11.** $3.12^{-x} = 7.3$

**4.** $12^x = 6.32$      **8.** $8^{2y-1} = (\frac{1}{4})^{y-1}$      **12.** $6^{x^2} = 39$

Approximate each logarithm to three significant digits.

**13.** $\log_3 5$      **15.** $\log_{12} 7$      **17.** $\log_3 18$      **19.** $\log_{3.21} 10$

**14.** $\log_7 21$      **16.** $\log_{16} 11$      **18.** $\log_4 24$      **20.** $\log_{3.21} 3$

Solve using logarithm tables as necessary.

B

**21.** $10^{x+3} = 6.18$      **23.** $9^{x-4} = 6.28$      **25.** $2.1^{x-5} = 9.32$

**22.** $10^{2y-1} = 42.6$      **24.** $5^{s+2} = 15.3$      **26.** $7.6^{z-2} = 41.7$

C

**27.** Justify each step in the following proof of $\log_a n = \dfrac{\log_b n}{\log_b a}$.

$$\text{Let } x = \log_a n.$$

$$a^x = n$$

$$\log_b a^x = \log_b n$$

$$x \log_b a = \log_b n$$

$$x = \frac{\log_b n}{\log_b a}, \quad \text{or} \quad \log_a n = \frac{\log_b n}{\log_b a}.$$

## PROBLEMS

**A**

1. How many years will it take $1200 to increase to $1500 if it is invested at 5% compounded semi-annually?

2. In how many years will an investment of $2000 amount to $3000 if the investment draws 6% interest compounded quarterly?

3. How long will it take $4000 invested at 4% compounded quarterly to grow to $6500?

4. How long will it take $3000 invested at 5% compounded semi-annually to grow to $5000?

5. How long will it take a given principal to double if it is invested at 5% compounded quarterly? (*Hint:* Let $A = 2P$ in the formula on page 413.)

6. How long will it take a given principal to double if it is invested at 5% compounded semi-annually?

**B**

7. A certain radioactive material decays at a rate given by $Q = Q_0 \cdot 10^{-kt}$ where $Q$ is in grams and $t$ is in years. If $Q_0 = 500$ grams, find $k$ if $Q = 400$ grams when $t = 1000$ years.

8. Using the results of Problem 7, find $Q$ when $t = 2000$ years.

---

## Chapter Summary

---

1. A radical may be written in **exponential form**: $(\sqrt[r]{b})^p = b^{\frac{p}{r}}$, if $\sqrt[r]{b}$ and $p$ are real numbers. The laws of exponents apply also to exponents that are real numbers.

2. By interchanging the coordinates in each of the ordered pairs belonging to the **exponential function** $\{(x, y): y = b^x, b > 0, b \neq 1\}$, you obtain the **logarithmic function** $\{(x, y): y = \log_b x\}$. If $a = b^n$, then $\log_b a = n$.

3. The **characteristic** of the common logarithm of a number is found by inspecting the number in standard form; the **mantissa** is determined from the table. In using **linear interpolation**, you assume that small portions of the graph of $y = \log x$ are straight lines.

4. The laws of exponents are the basis for the laws of logarithms:

$$\log ab = \log a + \log b; \log \frac{a}{b} = \log a - \log b; \log a^n = n \log a.$$

**5.** To find the logarithm of a number to another base, use the relationship:
$$\log_b n = \frac{\log_a n}{\log_a b}.$$

**6.** Logarithms are used also in solving equations.

### Vocabulary and Spelling

Review the meaning of each term by reference to the page listed.

exponential form (*p. 390*)

exponential function (*p. 393*)

logarithm (*p. 395*)

logarithmic function (*p. 395*)

inverse (of a function) (*p. 397*)

laws of logarithms (*p. 400*)

common logarithms (*p. 404*)

characteristic (*p. 405*)

mantissa (*p. 405*)

antilogarithm (*p. 405*)

linear interpolation (*p. 407*)

compound interest (*p. 413*)

## Chapter Test and Review

10–1   **1.** Write $3\sqrt[4]{x^3 y^5}$ in exponential form.

10–2   **2.** Simplify: $(2^{\sqrt{2}})^{\sqrt{32}}$

10–3   **3.** Find $\log_3 \frac{1}{27}$.

**4.** Solve for $x$: $\log_8 x = 2$

10–4   **5.** Find the inverse of $\{(x, y): y = 6x - 4\}$.

10–5   **6.** Solve for $x$: $\log_{10} x = 3 \log_{10} 2 - 2 \log_{10} 5$

Use Table 5 to find each of the following.

10–6   **7.** log 4.16          **8.** antilog $9.8876 - 10$

10–7   **9.** log 0.02314          **10.** antilog 2.4716

10–8   **11.** Simplify: $\dfrac{(3.12)(6.15)}{(1.24)}$

10–9   **12.** Simplify: $\sqrt[3]{(3.2)^2}$

10–10   **13.** Find $\log_6 15$.

**14.** Solve for $x$: $3^x = 14$.

# *Cumulative Review: Chapters 1–10*

Solve for $x$.

**1.** $2x - 3(x + 7) = 6 - 3x$

**2.** $\dfrac{4x}{15} - \dfrac{2x - 1}{10} = \dfrac{1}{2}$

**3.** $(5 - x)^2 + x^2 = 13$

**4.** $c^2x + d = d^2x + c$

Solve each system.

**5.** $2x + y = -2$
$\quad 6x - 5y = 18$

**6.** $x + 2y = 8$
$\quad 2x + y + z = 11$
$\quad x + y + 2z = 13$

Simplify each expression.

**7.** $(x^2 + x - 6)(x^2 + 2x - 3)^{-1}$

**8.** $-(x - y)(y^2 - 5xy + 4x^2)^{-1}$

**9.** $\dfrac{a}{6a - 3b} - \dfrac{b}{12a - 6b}$

**10.** $\dfrac{x - \dfrac{1}{x}}{1 + \dfrac{4}{x} - \dfrac{5}{x^2}}$

**11.** If $y$ varies directly as $x^2$ and inversely as $z$, and if $y = 8$ when $x = 2$ and $z = 3$, find $z$ when $x = 6$ and $y = 9$.

Simplify each expression.

**12.** $\sqrt[4]{5x^4z^7} \cdot \sqrt[4]{125x^2z^3}$

**13.** $(\sqrt{12} + \sqrt{20})(\sqrt{12} - \sqrt{20})$

**14.** $\dfrac{5}{1 - \sqrt{3}}$

**15.** $\dfrac{3 + 2i}{1 - 2i}$

**16.** If $f(x) = 3x^2 - 2x + 5$, and $g(x) = x^2 + 3x - 6$, find:

    **a.** $f(5) - g(3)$     **b.** $f(-2) - g(-1)$     **c.** $f(g(3))$

**17.** Find a quadratic equation with integral coefficients whose roots are 3 and $-\frac{1}{5}$.

**18.** Solve $2 - 3\sqrt[3]{2x + 1} = 0$ over $\Re$.

Sketch the graph of each equation.

**19.** $x^2 + y^2 - 2x + 4y - 4 = 0$

**20.** $3x^2 - 5y^2 = 15$

**21.** $4x^2 + 9y^2 = 36$

**22.** $y = 3 - 2x - x^{2}$

**23.** Evaluate $\log_3 \left(-\log_5 \left(\tfrac{1}{125}\right)\right)$

**24.** Write $2 \log_{10} x - \frac{1}{3} \log_{10} y + \log_{10} z$ as the logarithm of a single expression.

**25.** Solve $|2y - 1| = 4y + 5$ for $y$.

**26.** Graph the solution set of $5x - 2 < x + 6 < 3x + 8$.

**27.** The sum of the reciprocals of two consecutive even negative integers is $-\frac{7}{24}$. Find the integers.

**28.** A tank can be filled in 8 hours by one pump working alone and in 10 hours by another pump working alone. An outlet to the tank will empty it in 6 hours. If the outlet is left open and both pumps are started, how long will it take to fill the empty tank?

**29.** How many pounds of candy at 89 cents a pound should be mixed with 18 pounds of candy at 39 cents a pound to make a blend worth 59 cents a pound?

**30.** Mr. Donaldson made a trip to another city in 3 hours, while his return trip over the same road took $3\frac{1}{2}$ hours. If his speed returning was 5 miles per hour less than his speed going, how many miles was the round trip?

**31.** The Transtate Telephone Company has a computer that takes 28 hours to process the weekly payroll of the company. By renting another computer and using both machines at the same time, the company can process the payroll in 12 hours. How long would it take the rented computer to do the job alone?

**32.** The frequency of a electromagnetic wave is inversely proportional to the wave length. If a wave 250 meters long has a frequency of 1200 kilocycles (1000 cycles) per second, what would be the wave length of a wave with a frequency of 800 kilocycles per second?

**33.** Find the lengths of the legs of a right triangle whose hypotenuse measures 17 yards and whose area is 60 square yards.

**34.** The lengths of the side of three cubes are in the ratio $2 : 3 : 4$. If the sum of the volumes of the cubes is 7040 cubic centimeters, find the length of a side of each cube to the nearest tenth of a centimeter.

# Extra for Experts

## POLYNOMIALS AND LINEAR INTERPOLATION

In locating the real zeros of polynomial functions you can use the following fact, which is proven in higher mathematics: Graphs of polynomial functions are smooth unbroken curves. Some samples are shown in Figure 10–11.

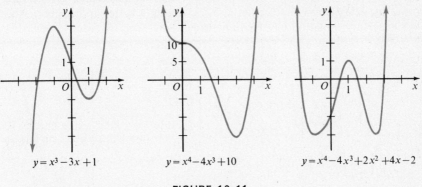

$y = x^3 - 3x + 1$        $y = x^4 - 4x^3 + 10$        $y = x^4 - 4x^3 + 2x^2 + 4x - 2$

FIGURE 10–11

As an example, consider the function $P(x) = x^3 - x^2 + 3x - 7$. Since $P(1) = -4$ and $P(2) = 3$, you see that the points $A(1, -4)$ and $B(2, 3)$ are on the graph of $P$ (Figure 10–12). Consider these facts: (1) The graph of $P$

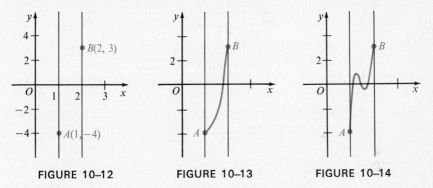

FIGURE 10–12        FIGURE 10–13        FIGURE 10–14

is an unbroken curve containing $A$ and $B$. (2) $A$ and $B$ lie on opposite sides of the $x$-axis. You can conclude that the part of the graph lying between the lines $x = 1$ and $x = 2$ must cross the $x$-axis at least once. (Two possibilities are shown in Figures 10–13 and 10–14.) It follows that $P(x)$ has at least one root between 1 and 2. This discussion illustrates the Location Principle.

If $P(x)$ is a polynomial with real coefficients and the numbers $P(a)$ and $P(b)$ have opposite signs, then the equation $P(x) = 0$ has at least one root between $a$ and $b$.

(*Warning:* Do not assume that if $P(a)$ and $P(b)$ have the same sign, then there are no roots between $a$ and $b$. The situation might be as pictured in Figure 10–15.)

The Location Principle asserts that the equation $P(x) = x^3 - 2x - 10 = 0$ has at least one root between 2 and 3 because $P(2) = -6$ and $P(3) = 11$.

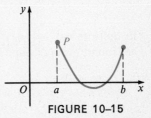

FIGURE 10–15

By Descartes' Rule, on the other hand, this equation has exactly one positive root, say $r$. We know that $2 < r < 3$. To obtain a better approximation we use linear interpolation (see Section 10–7).

$$1 \left[ c \left[ \begin{array}{c|c} x & P(x) \\ \hline 3 & 11 \\ r & P(r) = 0 \\ 2 & -6 \end{array} \right] 6 \right] 17$$

(3, 11)

(2, −6)

$$\frac{c}{1} = \frac{6}{17}; \; c \doteq 0.4$$

$$\therefore r \doteq 2 + 0.4 = 2.4$$

Using synthetic substitution we find:

$$\begin{array}{c|cccc}
2.4 & 1 & 0 & -2 & -10 \\
& & 2.4 & 5.76 & 9.024 \\
\hline
& 1 & 2.4 & 3.76 & -0.976 \doteq -0.98
\end{array}$$

$$\begin{array}{c|cccc}
2.5 & 1 & 0 & -2 & -10 \\
& & 2.5 & 6.25 & 10.625 \\
\hline
& 1 & 2.5 & 4.25 & 0.625 \doteq 0.62
\end{array}$$

Therefore the root $r$ is between 2.4 and 2.5. To obtain an even better approximation we use interpolation again, over a shorter interval:

$$0.10 \left[ c \left[ \begin{array}{c|c} x & P(x) \\ \hline 2.5 & 0.62 \\ r & P(r) = 0 \\ 2.4 & -0.98 \end{array} \right] 0.98 \right] 1.60$$

$$\frac{c}{0.1} = \frac{0.98}{1.60}$$
$$c \doteq 0.06$$
$$\therefore r \doteq 2.4 + 0.06$$
$$r \doteq 2.46$$

# Exercises

In the interval $-4 < x < 4$, locate consecutive integers between which are found roots of each of the following equations:

Sample. $12x^3 - 8x^2 - 23x + 12 = 0$

*Solution:*  Let $y = 12x^3 - 8x^2 - 23x + 12$. Use synthetic substitution to find values for $y$ when $x = -3, -2, -1, 0, 1, 2, 3$, and construct a table of values.

| $x$ | $-3$ | $-2$ | $-1$ | 0 | 1 | 2 | 3 |
|---|---|---|---|---|---|---|---|
| $y$ | $-315$ | $-70$ | 15 | 12 | $-7$ | 30 | 195 |

By examining the values of $y$, you will find that there are roots between $-2$ and $-1$, 0 and 1, and 1 and 2.

**1.** $12x^3 + 16x^2 - 95x - 50 = 0$

**2.** $8x^3 - 12x^2 - 66x + 35 = 0$

**3.** $2x^3 + x^2 + x - 1 = 0$

**4.** $2x^3 - x^2 + x + 1 = 0$

**5.** $6x^4 - x^3 - 6x^2 - 12x - 5 = 0$

**6.** $4x^4 + 8x^3 - x^2 + 8x - 5 = 0$

Each of the following functions has one real zero between $-4$ and $4$. Locate it between consecutive integers.

**7.** $\{(x, y): y = 2x^3 - x^2 + 5x - 3\}$

**8.** $\{(x, y): y = 2x^3 + 7x^2 + 7x + 5\}$

**9.** $\{(x, y): y = 2x^5 + 5x^4 + 4x + 10\}$

**10.** $\{(x, y): y = 2x^5 - 3x^4 + 8x^3 - 12x^2 + 8x - 3\}$

Find the indicated root to the nearest hundredth.

**11.** $x^3 + 3x^2 - 9x + 4 = 0$ between 0 and 1

**12.** $x^3 - 3x + 1 = 0$ between 1 and 2

**13.** $x^3 - 3x^2 - 2x + 5 = 0$ between 1 and 2

**14.** $x^3 - 4x^2 - 6x + 8 = 0$ between 0 and 1

**15.** $x^3 + 2x + 20 = 0$ between $-2$ and $-3$

**16.** $2x^3 - x^2 + 3x + 1 = 0$ between $-1$ and 0

**17.** Find all real zeros of the function $\{(x, y): y = x^3 + x - 1\}$.

**18.** Find all real zeros of the function $\{(x, y): y = x^3 + 7x + 7\}$.

# 11

# Trigonometric Functions and Vectors

## COORDINATES AND TRIGONOMETRY

### 11–1  Rays, Angles, and Points

Any point of a line, such as point *P* on line *l* in Figure 11–1, separates the line into two **opposite rays**, such as *r* and *s*, each having the point as **vertex**. You can specify a ray by indicating first its vertex and then another of its points. For example, in Figure 11–1, *r* is ray *PM*, and *s* is ray *PN*.

*In a coordinate system* a ray can also be specified as the graph of a linear relation. For example, $\{(x, y): y = x \text{ and } x \geq 0\}$ has as its graph the ray in the first quadrant with slope 1 and the origin as vertex (Figure 11–2). The opposite ray is the graph of the relation $\{(x, y): y = x \text{ and } x \leq 0\}$.

The set of points composing two rays with a common vertex, together with a *rotation* that sends one ray into the other, is called a

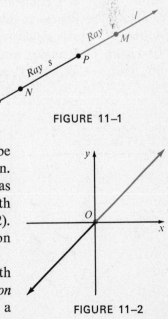

FIGURE 11–1

FIGURE 11–2

directed angle, or simply an angle. Counterclockwise rotation yields a positive angle; clockwise rotation yields a negative angle.

In Figure 11–3 ray *p* is the initial side (ray) of the angle and ray *q* is the terminal side (ray). The point *O* is the vertex of the angle.

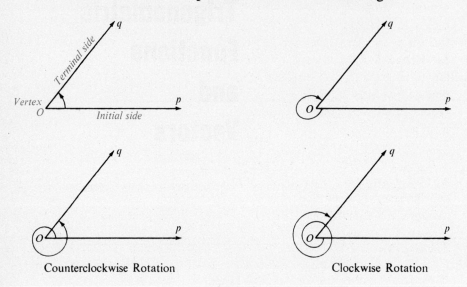

Counterclockwise Rotation          Clockwise Rotation

FIGURE 11–3

A common unit of angle measure is a *degree*, written as 1°. One degree is $\frac{1}{360}$ of a complete counterclockwise rotation of a ray about a point. The degree is divided into *minutes* and *seconds:* 1 minute, written 1′, equals $\frac{1}{60}$ degree; 1 second, written 1″, equals $\frac{1}{60}$ minute. In Figure 11–4 angles whose measures are 30° (read "30 degrees"), 90°, 180°, −180°, −405°, and −360° are shown.

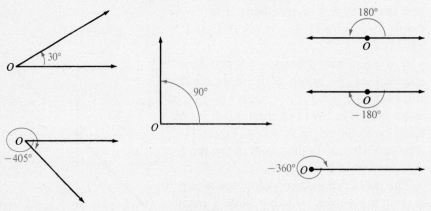

FIGURE 11–4

Angles having the same initial and terminal sides are called co-terminal angles. As illustrated in Figure 11–5, in a coordinate plane every point $P$ other than the origin determines an infinite set of co-terminal angles, each having the origin as vertex, the positive half of the $x$-axis as initial side, and ray $OP$ as terminal side. Each of these angles is said to be in standard position and to be a position angle of $P$. Notice that the measures of co-terminal angles in standard position differ by integral multiples of 360°.

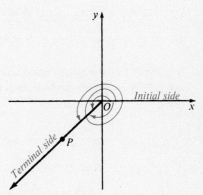

Coterminal Angles in Standard Position

If you know both the measure of a position angle of $P$ and the distance $OP$ between $P$ and the origin, you can locate $P$ in the plane.

**FIGURE 11–5**

**Example.**  Locate point $P$ if $OP = 4$ and the measure of one of its position angles is $-110°$. Indicate and give the measure of one of its positive position angles.

*Solution:*

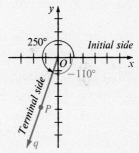

1. Draw ray $q$ forming with the positive half of the $x$-axis an angle whose measure is $-110°$.
2. On $q$ measure 4 units from $O$. The point reached is $P$.
3. A positive position angle of $P$ has measure $(-110 + 360)°$, or 250°.

## WRITTEN EXERCISES

At the given distance from the origin locate the point having a position angle with the stated measure. Indicate and give the measures of two other angles, one positive and one negative, that are also position angles of the point.

**A**

1. 3 (distance from origin);  $-40°$ (angle measure)

2. 5; $-60°$   5. 4; 280°   8. $\frac{5}{3}$; $-270°$   11. 0; 120°

3. 2; 40°   6. 5; 230°   9. $2\frac{1}{2}$; 460°   12. 0; $-210°$

4. 6; 85°   7. $\frac{5}{4}$; $-180°$   10. $1\frac{3}{4}$; $-520°$

Draw the ray which is the graph of each relation, and indicate a positive and a negative angle in standard position having the ray as terminal side.

**13.** $\{(x, y): y = 3x \text{ and } x \geq 0\}$     **16.** $\{(x, y): y = -\frac{5}{4}x \text{ and } x \geq 0\}$

**14.** $\{(x, y): y = 2x \text{ and } x \leq 0\}$     **17.** $\{(x, y): y = 0 \text{ and } x \leq 0\}$

**15.** $\{(x, y): y = -\frac{1}{2}x \text{ and } x \leq 0\}$     **18.** $\{(x, y): x = 0 \text{ and } y \geq 0\}$

If point *P* has the given coordinates, draw ray *OP*, specify a relation having ray *OP* as its graph, and show a positive angle in standard position having ray *OP* as its terminal side.

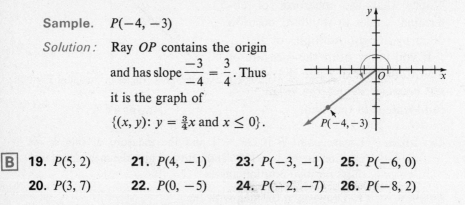

> Sample.   $P(-4, -3)$
>
> *Solution:*   Ray *OP* contains the origin and has slope $\dfrac{-3}{-4} = \dfrac{3}{4}$. Thus it is the graph of
>
> $\{(x, y): y = \frac{3}{4}x \text{ and } x \leq 0\}$.

**B**   **19.** $P(5, 2)$     **21.** $P(4, -1)$     **23.** $P(-3, -1)$   **25.** $P(-6, 0)$

**20.** $P(3, 7)$     **22.** $P(0, -5)$     **24.** $P(-2, -7)$   **26.** $P(-8, 2)$

**27.** Explain why the measures of coterminal angles in standard position differ by an integral multiple of 360°.

**28.** Explain why angles in standard position whose measures differ by an integral multiple of 360° are coterminal.

## 11–2   Sine and Cosine Functions

Consider a point *P* moving on the circle with radius 1 and center at the origin (the **unit circle**), starting at the point (1, 0). As *P* moves counterclockwise around the circle until it reaches its starting point again, the measure of its position angle runs through the set of values from 0° to 360° (Figure 11–6). By allowing *P* to repeat its counterclockwise motion on the unit circle, and also allowing *P* to travel in a clockwise direction, you can obtain a position angle of *P* whose measure in degrees is any prescribed real number.

Let the Greek letter $\theta$ (theta) be a variable whose domain is the set of all angles in standard position in the plane. As *P* moves on the circle,

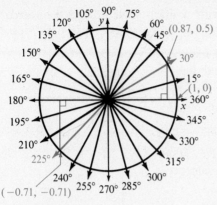

| Measure of θ | a | b |
|---|---|---|
| 0° | 1 | 0 |
| 15° | 0.97 | 0.26 |
| 30° | 0.87 | 0.50 |
| 45° | 0.71 | 0.71 |
| 90° | 0 | 1 |
| 180° | −1 | 0 |
| 225° | −0.71 | −0.71 |
| 270° | 0 | −1 |
| 360° | 1 | 0 |

FIGURE 11–6

not only does its position angle $\theta$ change, but its coordinates $(a, b)$ also change subject to the condition that the distance between $P$ and the origin is 1; that is, $\sqrt{a^2 + b^2} = 1$. Indeed, for each angle $\theta$ there is a unique corresponding pair $(a, b)$ whose coordinates can be estimated as shown for the angles of 30° and 225° in Figure 11–6. Use the figure to check the entries for the other angles given in the table.

The set of all ordered pairs $(\theta, a)$ determined by this procedure is called the cosine function; the set of all pairs $(\theta, b)$ is called the sine function. Do you see that the domain of each of these functions is the set of *angles in standard position*, and that the range is the set of *real numbers* between $-1$ and 1 inclusive? For any particular angle $\theta$ we call the values of these functions "the cosine of angle $\theta$" and "the sine of angle $\theta$" according to the following definition:

**Let $\theta$ denote any angle in standard position. If $(a, b)$ denotes the coordinates of the point one unit from the origin on the terminal side of $\theta$, then**

$$\text{cosine of angle } \theta = a \qquad \text{and} \qquad \text{sine of angle } \theta = b.$$

We usually write $\cos \theta$ as an abbreviation for "cosine of angle $\theta$" and $\sin \theta$ for "sine of angle $\theta$." *Notice that $(\cos \theta, \sin \theta)$ are the coordinates of the point of intersection of the unit circle and the terminal side of $\theta$.*

Given *any* point $T$, other than the origin, on the terminal side of an angle $\theta$ in standard position, you can determine the coordinates of the point $P$ in which the ray $OT$ intersects the unit circle $x^2 + y^2 = 1$. This means that you can find $\cos \theta$ and $\sin \theta$, as illustrated in the following example.

**Example 1.** $T(3, -4)$ is a point with position angle $\theta$. Find $\cos \theta$ and $\sin \theta$.

*Solution:* Let $P$ be the point of intersection of ray $OT$ and the
unit circle $x^2 + y^2 = 1$.

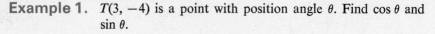

1. Ray $OT$ is the graph of

$$\{(x, y): y = -\tfrac{4}{3}x \text{ and } x \geq 0\}.$$

2. The coordinates of $P$ must satisfy the three open sentences:

$$x^2 + y^2 = 1$$
$$y = -\tfrac{4}{3}x$$
$$x \geq 0$$

3. Solving $x^2 + y^2 = 1$ and $y = -\tfrac{4}{3}x$ simultaneously, you
obtain two solutions, $(-\tfrac{3}{5}, \tfrac{4}{5})$ and $(\tfrac{3}{5}, -\tfrac{4}{5})$. Only the
second solution satisfies the condition $x \geq 0$.

∴ the coordinates of $P$ are $(\tfrac{3}{5}, -\tfrac{4}{5})$, so that

$$\cos \theta = \tfrac{3}{5} \text{ and } \sin \theta = -\tfrac{4}{5}.$$

Notice in the example above that the denominator, 5, of each of
the fractions $\tfrac{3}{5}$ and $-\tfrac{4}{5}$ is just the distance $r$ between $O$ and the point
$T(3, -4)$, since $r = \sqrt{3^2 + (-4)^2} = 5$. Thus, you can write the
results of this example as

$$\cos \theta = \frac{\text{abscissa of } T}{r} \quad \text{and} \quad \sin \theta = \frac{\text{ordinate of } T}{r}.$$

This suggests the following theorem:

---

**THEOREM.** If $(a, b)$ are the coordinates of any point other than the
origin on the terminal side of $\theta$, an angle in standard
position, then

$$\cos \theta = \frac{a}{r} \text{ and } \sin \theta = \frac{b}{r}, \text{ where } r = \sqrt{a^2 + b^2}.$$

---

This theorem can be proved by carrying through the steps of the
solution of Example 1 with $(a, b)$ in place of $(3, -4)$.

You can use the theorem above to find the sine and cosine of an
angle in standard position given the coordinates of any point other
than the origin on its terminal side, as illustrated in the next example.

**Example 2.** Determine $\cos \theta$ and $\sin \theta$ for the angle shown.

*Solution:*

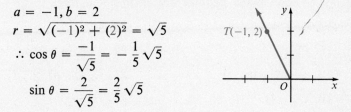

$$a = -1, b = 2$$
$$r = \sqrt{(-1)^2 + (2)^2} = \sqrt{5}$$
$$\therefore \cos \theta = \frac{-1}{\sqrt{5}} = -\frac{1}{5}\sqrt{5}$$
$$\sin \theta = \frac{2}{\sqrt{5}} = \frac{2}{5}\sqrt{5}$$

*ORAL EXERCISES*

Referring to Figure 11–6, page 431, describe the variation of (a) $\cos \theta$ and (b) $\sin \theta$ as the measure of $\theta$ increases from the first to the second value.

**Sample.** 0° to 90°     *Solution:*   **a.** $\cos \theta$ decreases from 1 to 0.
                                                       **b.** $\sin \theta$ increases from 0 to 1.

**1.** 90° to 180°       **3.** 270° to 360°       **5.** −180° to −90°

**2.** 180° to 270°      **4.** −90° to 0°         **6.** 360° to 450°

For what measures of $\theta$ between 0° and 360° are the given statements true?

**7.** $\sin \theta = 1$          **9.** $\cos \theta = -1$        **11.** $\sin \theta > 1$

**8.** $\cos \theta = 0$         **10.** $\sin \theta = 0$        **12.** $-1 \le \cos \theta \le 1$

*WRITTEN EXERCISES*

Using a large scale on the coordinate axes, copy Figure 11–6, page 431, and use it to find approximations to the nearest tenth for $\cos \theta$ and $\sin \theta$ if $\theta$ has the given measure.

**A**

| | | | |
|---|---|---|---|
| **1.** 45° | **5.** 150° | **9.** −60° | **13.** 520° |
| **2.** 60° | **6.** 120° | **10.** −75° | **14.** 750° |
| **3.** 110° | **7.** 240° | **11.** −135° | **15.** −810° |
| **4.** 135° | **8.** 300° | **12.** −225° | **16.** −135° |

Determine $\cos \theta$ and $\sin \theta$ for any position angle $\theta$ of the point with the given coordinates. Express radicals in simple form.

**17.** $(-8, 15)$             **21.** $(0, -3)$

**18.** $(2, 1)$                **22.** $(-4, 0)$

**19.** $(-30, -16)$        **23.** $(-2, -4)$

**20.** $(4, -4)$            **24.** $(6, -3)$

Determine $\cos \theta$ and $\sin \theta$ if $\theta$ is an angle in standard position whose terminal side is the graph of the given relation.

**B**   **25.** $\{(x, y): 2x + 5y = 0, x \geq 0\}$    **28.** $\{(x, y): x - y = 0, y \leq 0\}$

       **26.** $\{(x, y): 3x - 4y = 0, x \leq 0\}$    **29.** $\{(x, y): y = 0, x \leq 0\}$

       **27.** $\{(x, y): x + y = 0, y \geq 0\}$      **30.** $\{(x, y): x = 0, y \geq 0\}$

In the following exercises, $\theta$ is an angle in standard position with measure between $0°$ and $360°$. Draw $\theta$ and determine all numerical replacements of the question mark for which the given statement is true.

**31.** $\sin \theta = \frac{1}{5}$ and $\cos \theta = \underline{\ ?\ }$      **33.** $\cos \theta = -\dfrac{\sqrt{2}}{2}$ and $\sin \theta = \underline{\ ?\ }$

**32.** $\sin \theta = -\frac{2}{7}$ and $\cos \theta = \underline{\ ?\ }$      **34.** $\cos \theta = \dfrac{\sqrt{3}}{2}$ and $\sin \theta = \underline{\ ?\ }$

## 11–3   The Trigonometric Functions

Certain combinations of sine and cosine values occur so often that they are given special names, as indicated below.

If $\theta$ is the angle shown in Figure 11–7, then $\cos \theta = \frac{8}{17}$ and $\sin \theta = -\frac{15}{17}$. The quotient

$$\frac{\sin \theta}{\cos \theta} = \frac{-\frac{15}{17}}{\frac{8}{17}} = -\frac{15}{8} \text{ is called the } tangent \text{ of angle } \theta \text{ (tan } \theta\text{),}$$

$$\frac{\cos \theta}{\sin \theta} = \frac{\frac{8}{17}}{-\frac{15}{17}} = -\frac{8}{15} \text{ is called the } cotangent \text{ of angle } \theta \text{ (cot } \theta\text{),}$$

$$\frac{1}{\cos \theta} = \frac{1}{\frac{8}{17}} = \frac{17}{8} \text{ is called the } secant \text{ of angle } \theta \text{ (sec } \theta\text{), and}$$

$$\frac{1}{\sin \theta} = \frac{1}{-\frac{15}{17}} = -\frac{17}{15} \text{ is called the } cosecant \text{ of angle } \theta \text{ (csc } \theta\text{).}$$

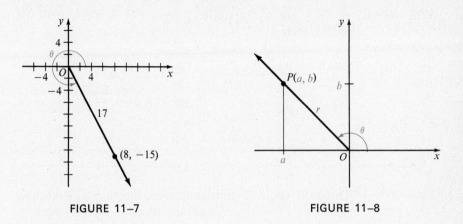

FIGURE 11–7                                    FIGURE 11–8

We can now state the following definitions of four functions whose domains are subsets of the set of angles $\theta$ in standard position:

**the tangent function** $= \left\{ (\theta, \tan \theta): \tan \theta = \dfrac{\sin \theta}{\cos \theta}, \text{ provided } \cos \theta \neq 0 \right\}$

**the cotangent function** $= \left\{ (\theta, \cot \theta): \cot \theta = \dfrac{\cos \theta}{\sin \theta}, \text{ provided } \sin \theta \neq 0 \right\}$

**the secant function** $= \left\{ (\theta, \sec \theta): \sec \theta = \dfrac{1}{\cos \theta}, \text{ provided } \cos \theta \neq 0 \right\}$

**the cosecant function** $= \left\{ (\theta, \csc \theta): \csc \theta = \dfrac{1}{\sin \theta}, \text{ provided } \sin \theta \neq 0 \right\}$

The sine, cosine, tangent, cotangent, secant, and cosecant functions are called the trigonometric functions. Using their definitions and the theorem stated on page 432, you can compute their values in terms of the coordinates $(a, b)$ of any point $P$, other than the origin, on the terminal side of the angle $\theta$ (Figure 11–8). For example, if $\cos \theta \neq 0$,

$$\tan \theta = \frac{\sin \theta}{\cos \theta} = \frac{\dfrac{b}{\sqrt{a^2 + b^2}}}{\dfrac{a}{\sqrt{a^2 + b^2}}} = \frac{b}{a}, \text{ provided } a \neq 0.$$

Similar reasoning leads to the expressions for $\cot \theta$, $\sec \theta$, and $\csc \theta$ listed in the following theorem.

---

**THEOREM.**   If $\theta$ is a position angle of a point $P(a, b)$ other than the origin, then

$$\sin \theta = \frac{b}{\sqrt{a^2 + b^2}} \qquad\qquad \csc \theta = \frac{\sqrt{a^2 + b^2}}{b}, \quad (b \neq 0)$$

$$\cos \theta = \frac{a}{\sqrt{a^2 + b^2}} \qquad\qquad \sec \theta = \frac{\sqrt{a^2 + b^2}}{a}, \quad (a \neq 0)$$

$$\tan \theta = \frac{b}{a}, \quad (a \neq 0) \qquad \cot \theta = \frac{a}{b}, \quad (b \neq 0)$$

---

**Example 1.**   Find the values of the trigonometric functions of a position angle of $P(-8, 15)$. Sketch the smallest positive position angle of $P$.

*Solution:*   $a = -8, b = 15$

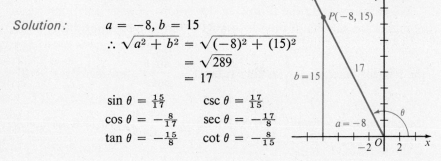

$$\therefore \sqrt{a^2 + b^2} = \sqrt{(-8)^2 + (15)^2}$$
$$= \sqrt{289}$$
$$= 17$$

$$\sin \theta = \tfrac{15}{17} \qquad \csc \theta = \tfrac{17}{15}$$
$$\cos \theta = -\tfrac{8}{17} \qquad \sec \theta = -\tfrac{17}{8}$$
$$\tan \theta = -\tfrac{15}{8} \qquad \cot \theta = -\tfrac{8}{15}$$

Notice that the values of the trigonometric functions depend only on the position of the terminal side of the angle. Because the measure of an angle in standard position determines the position of its terminal side, you can refer to such an angle simply by giving its measure. For example, you can write "sin 30°" in place of "sine of the angle in standard position whose measure is 30°."

An angle in standard position is frequently classified according to the quadrant in which its terminal side lies. Thus, the angle shown in Figure 11–7 is called a *fourth quadrant angle*, and the angle shown in Figure 11–8 is called a *second quadrant angle*. When the terminal side of the angle lies on a coordinate axis, as in the case of $\alpha$ (alpha) in Figure 11–9, we call the angle a quadrantal angle.

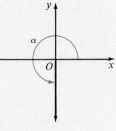

FIGURE 11–9

The following table classifies the values of the trigonometric functions of a nonquadrantal angle in standard position as positive or negative numbers according to the quadrant of the angle.

| Value | Quadrant of $\theta$ | | | |
|---|---|---|---|---|
| | **I** | **II** | **III** | **IV** |
| $\sin \theta$ and $\csc \theta$ <br> $\cos \theta$ and $\sec \theta$ <br> $\tan \theta$ and $\cot \theta$ | positive <br> positive <br> positive | positive <br> negative <br> negative | negative <br> negative <br> positive | negative <br> positive <br> negative |

If you know the quadrant in which the terminal side of $\theta$ lies and the value of one of the trigonometric functions of $\theta$, you can determine its other trigonometric function values.

**Example 2.** If $\theta$ is the third-quadrant angle of least positive measure for which $\sin \theta = -\frac{3}{4}$, draw $\theta$ and find the values of its other trigonometric functions.

*Solution:* Let $P(a, b)$ be a point other than $O$ on the terminal side of $\theta$.

1. Since $\sin \theta = \dfrac{b}{r} = -\dfrac{3}{4}$, you can chose $b = -3$ and $r = 4$. Then $P$ is the point in the third quadrant in which the horizontal line 3 units below the $x$-axis intersects the circle with center at the origin and radius 4.

2. Show $\theta$ with terminal ray $OP$.

3. To find $a$, use $a^2 + b^2 = r^2$.

$$a^2 + (-3)^2 = 4^2$$
$$a^2 + 9 = 16$$
$$a^2 = 7$$
Since $a < 0$ (why?),
$$a = -\sqrt{7}.$$

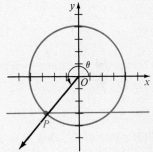

4. Use the theorem on page 436.

$$\sin \theta = -\tfrac{3}{4} \qquad\qquad \csc \theta = -\tfrac{4}{3}$$

$$\cos \theta = -\frac{\sqrt{7}}{4} \qquad\qquad \sec \theta = -\frac{4}{\sqrt{7}} = -\frac{4}{7}\sqrt{7}$$

$$\tan \theta = \frac{3}{\sqrt{7}} = \frac{3}{7}\sqrt{7} \qquad\qquad \cot \theta = \frac{\sqrt{7}}{3}$$

We define the trigonometric functions of any angle $\theta$ not in standard position to have the same values as the corresponding functions of the angle $\varphi$ which is in standard position and has the same measure as $\theta$. Moreover, the quadrant of $\theta$ is defined to be that of $\varphi$.

*ORAL EXERCISES*

State the value of the indicated trigonometric function of the angle $\theta$ in standard position.

**Sample.** $\cos \theta = \dfrac{5}{\sqrt{34}}$, $\sin \theta = -\dfrac{3}{\sqrt{34}}$, $\tan \theta = \underline{\ ?\ }$

*Solution:* $\tan \theta = \dfrac{\sin \theta}{\cos \theta} = -\frac{3}{5}$

**1.** $\cos \theta = \frac{3}{4}$, $\sec \theta = \underline{\ ?\ }$

**2.** $\sin \theta = -\frac{2}{3}$, $\csc \theta = \underline{\ ?\ }$

**3.** $\csc \theta = \frac{5}{3}$, $\sin \theta = \underline{\ ?\ }$

**4.** $\sin \theta = \frac{3}{7}$, $\cos \theta = \dfrac{-2\sqrt{10}}{7}$, $\tan \theta = \underline{\ ?\ }$

**5.** $\cos \theta = \dfrac{3\sqrt{10}}{10}$, $\sin \theta = \dfrac{\sqrt{10}}{10}$, $\cot \theta = \underline{\ ?\ }$

**6.** $\sec \theta = \frac{12}{5}$, $\cos \theta = \underline{\ ?\ }$

**7.** $\tan \theta = -\frac{1}{2}$, $\cot \theta = \underline{\ ?\ }$

**8.** $\cot \theta = 3$, $\tan \theta = \underline{\ ?\ }$

Name two quadrants in which the terminal side of $\theta$ may lie.

**9.** $\sin \theta > 0$      **11.** $\tan \theta < 0$      **13.** $\sec \theta < 0$

**10.** $\cos \theta < 0$      **12.** $\csc \theta > 0$      **14.** $\cot \theta > 0$

Name the quadrant in which the terminal side of $\theta$ must lie.

**15.** $\sin \theta > 0$, $\cos \theta < 0$      **19.** $\csc \theta > 0$, $\cos \theta > 0$

**16.** $\sin \theta < 0$, $\cos \theta < 0$      **20.** $\sec \theta < 0$, $\sin \theta > 0$

**17.** $\cos \theta > 0$, $\tan \theta < 0$      **21.** $\cot \theta < 0$, $\sec \theta > 0$

**18.** $\sin \theta < 0$, $\tan \theta > 0$      **22.** $\csc \theta < 0$, $\tan \theta < 0$

*WRITTEN EXERCISES*

Draw the smallest positive angle $\theta$ in standard position having the point with the given coordinates on its terminal side; evaluate the trigonometric functions of $\theta$.

**A**

**1.** $(3, 4)$      **4.** $(12, 5)$      **7.** $(-5, 2)$      **10.** $(-6, -8)$

**2.** $(4, 3)$      **5.** $(-3, 5)$      **8.** $(3, -6)$      **11.** $(-1, \sqrt{3})$

**3.** $(3, 3)$      **6.** $(-4, 6)$      **9.** $(-8, -6)$      **12.** $(-\sqrt{3}, 1)$

Draw in standard position the negative angle $\theta$, with measure of least absolute value, which terminates in the given quadrant; state the values of the trigonometric functions of $\theta$.

**B**

**13.** $\sin \theta = -\frac{1}{5}$; IV     **16.** $\cot \theta = \frac{3}{5}$; III     **19.** $\cos \theta = 0.7$; IV

**14.** $\cos \theta = -\frac{1}{6}$; II     **17.** $\csc \theta = -2$; III     **20.** $\csc \theta = -1.5$; III

**15.** $\tan \theta = \sqrt{3}$; I     **18.** $\tan \theta = 1$; I

Find the values of all the trigonometric functions of $\theta$, given that:

**21.** $\cos \theta = \frac{3}{8}$, and $\theta$ is not a first-quadrant angle.

**22.** $\sin \theta = -\frac{1}{2}$, and $\theta$ is not a third-quadrant angle.

**23.** $\cot \theta = \sqrt{7}$, and $\cos \theta > 0$.

**24.** $\tan \theta = -1$, and $\sin \theta < 0$.

Determine the rectangular coordinates of the point at the given distance from the origin in the stated quadrant if $\theta$ is its position angle.

**25.** 10; II; $\sin \theta = \frac{4}{5}$        **27.** 4; IV; $\tan \theta = -1$

**26.** $\sqrt{2}$; III; $\cos \theta = -\dfrac{\sqrt{2}}{2}$        **28.** 5; I; $\cot \theta = 1$

Determine all numerical replacements for the question mark for which the given statement is true.

**C**

**29.** $\sin \theta = \cos \theta = $ ___?___        **30.** $\sin \theta \sec \theta = 1$, $\cos \theta = $ ___?___

**31.** $1 + \cos \theta = 3 \cos \theta$, $\tan \theta < 0$, and $\sin \theta = $ ___?___

## 11-4   Special Angles

Referring to Figure 11-6, page 431, and the definitions of the trigonometric functions, you can verify the values given in the following table for quadrantal angles. A dash (—) means that no value exists. (Recall that division by 0 is undefined.)

| $\theta$ | $\sin \theta$ | $\cos \theta$ | $\tan \theta$ | $\csc \theta$ | $\sec \theta$ | $\cot \theta$ |
|----------|---------------|---------------|---------------|---------------|---------------|---------------|
| 0°   | 0  | 1  | 0  | —  | 1  | —  |
| 90°  | 1  | 0  | —  | 1  | —  | 0  |
| 180° | 0  | -1 | 0  | —  | -1 | —  |
| 270° | -1 | 0  | —  | -1 | —  | 0  |

An angle whose measure is between 0° and 90° is called a **positive acute angle.** Using the origin and any other point $P$ on the terminal side of a positive acute angle $\theta$ in standard position, you can form a right triangle whose third vertex is the point $S$ of intersection of the $x$-axis and the vertical line through $P$ (Figure 11–10). The values of the trigonometric functions of $\theta$ can then be expressed in terms of the lengths of the sides of the right triangle $OSP$:

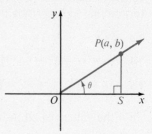

**FIGURE 11–10**

$$\sin \theta = \frac{b}{\sqrt{a^2 + b^2}} = \frac{\text{length of side opposite } \theta}{\text{length of hypotenuse}}$$

$$\cos \theta = \frac{a}{\sqrt{a^2 + b^2}} = \frac{\text{length of side adjacent to } \theta}{\text{length of hypotenuse}}$$

Similarly, in terms of the sides located relative to the angle $\theta$, you have:

$$\tan \theta = \frac{\text{length of opposite side}}{\text{length of adjacent side}} \qquad \csc \theta = \frac{\text{length of hypotenuse}}{\text{length of opposite side}}$$

$$\cot \theta = \frac{\text{length of adjacent side}}{\text{length of opposite side}} \qquad \sec \theta = \frac{\text{length of hypotenuse}}{\text{length of adjacent side}}$$

Because these statements do not involve a coordinate system, you can use them to define values for trigonometric functions of any acute angle of a right triangle.

**Example.** In triangle $ABC$, let $a$, $b$, and $c$ denote the lengths of the sides opposite angles $A$, $B$, and $C$, respectively. Find the trigonometric function values for $\angle A$ (read "angle $A$") and $\angle B$ if $a = 3$, $c = \sqrt{34}$, and $\angle C$ is a right angle.

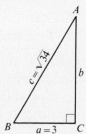

*Solution:*

**1.** Use the Pythagorean Theorem to determine $b$.

$$b = \sqrt{c^2 - a^2}$$

$$b = \sqrt{(\sqrt{34})^2 - 3^2} = \sqrt{25} = 5$$

| | Length of | | |
|---|---|---|---|
| **Angle** | **Opposite side** | **Adjacent side** | **Hypotenuse** |
| $A$ | 3 | 5 | $\sqrt{34}$ |
| $B$ | 5 | 3 | $\sqrt{34}$ |

**2.** Using the data in the chart, you find the following values:

$$\sin A = \frac{3\sqrt{34}}{34}, \quad \cos A = \frac{5\sqrt{34}}{34}, \quad \tan A = \frac{3}{5}, \quad \cot A = \frac{5}{3},$$

$$\csc A = \frac{\sqrt{34}}{3}, \quad \sec A = \frac{\sqrt{34}}{5}$$

$$\sin B = \frac{5\sqrt{34}}{34}, \quad \cos B = \frac{3\sqrt{34}}{34}, \quad \tan B = \frac{5}{3}, \quad \cot B = \frac{3}{5},$$

$$\csc B = \frac{\sqrt{34}}{5}, \quad \sec B = \frac{\sqrt{34}}{3}$$

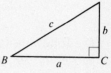

In right triangle $ABC$, angles $A$ and $B$ are complementary angles because the sum of their measures is 90°. Notice that

$$\sin B = \frac{b}{c} = \cos A, \qquad \cos B = \frac{a}{c} = \sin A, \qquad \tan B = \frac{b}{a} = \cot A.$$

Also, $\csc B = \sec A$, $\qquad \sec B = \csc A$, $\qquad \cot B = \tan A$.

Therefore, if we couple the trigonometric functions into the following pairs of **cofunctions**: the *sine and cosine* functions, the *tangent and cotangent* functions, the *secant and cosecant* functions, we can state the preceding facts as a theorem.

---

**THEOREM.** Any trigonometric function of a positive acute angle is equal to the cofunction of the complementary angle.

---

For example, $\sin 60° = \cos 30°$, $\tan 20° = \cot 70°$, $\csc 45° = \sec 45°$.

Figure 11–11 shows an equilateral triangle $ABD$ with sides of length 2 and angles of 60°. Because $AC$, the perpendicular bisector of side $BD$, bisects angle $A$, triangle $ABC$ is a right triangle whose acute angles and sides have the measures indicated in the figure. Thus you have:

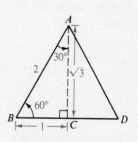

**FIGURE 11–11**

$$\sin 60° = \frac{\sqrt{3}}{2} = \cos 30° \qquad \csc 60° = \frac{2}{\sqrt{3}} = \frac{2\sqrt{3}}{3} = \sec 30°$$

$$\cos 60° = \tfrac{1}{2} = \sin 30° \qquad \sec 60° = 2 = \csc 30°$$

$$\tan 60° = \sqrt{3} = \cot 30° \qquad \cot 60° = \frac{1}{\sqrt{3}} = \frac{\sqrt{3}}{3} = \tan 30°$$

To determine the values of the trigonometric functions of an angle of 45°, consider right triangle *BCD* formed by drawing the diagonal *BD* of the unit square in Figure 11–12. Since the hypotenuse *BD* has length $\sqrt{2}$, you have

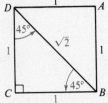

$$\sin 45° = \frac{1}{\sqrt{2}} = \frac{\sqrt{2}}{2} = \cos 45°$$

$$\tan 45° = 1 = \cot 45°$$

$$\sec 45° = \sqrt{2} = \csc 45°$$

FIGURE 11–12

The following table gives the values of the trigonometric functions of angles of 30°, 60°, and 45°. Picturing Figures 11–11, 11–12, and 11–6 in your mind will enable you to recall the entries in this table as well as those in the table for quadrantal angles on page 439.

| $\theta$ | $\sin \theta$ | $\cos \theta$ | $\tan \theta$ | $\csc \theta$ | $\sec \theta$ | $\cot \theta$ |
|---|---|---|---|---|---|---|
| 30° | $\dfrac{1}{2}$ | $\dfrac{\sqrt{3}}{2}$ | $\dfrac{\sqrt{3}}{3}$ | 2 | $\dfrac{2\sqrt{3}}{3}$ | $\sqrt{3}$ |
| 45° | $\dfrac{\sqrt{2}}{2}$ | $\dfrac{\sqrt{2}}{2}$ | 1 | $\sqrt{2}$ | $\sqrt{2}$ | 1 |
| 60° | $\dfrac{\sqrt{3}}{2}$ | $\dfrac{1}{2}$ | $\sqrt{3}$ | $\dfrac{2\sqrt{3}}{3}$ | 2 | $\dfrac{\sqrt{3}}{3}$ |

It is not necessary to remember all six of the function values for these angles, since the last three are, in order, the reciprocals of the first three.

*WRITTEN EXERCISES*

Evaluate.

**A**  **1.** $\cos 60° - \cot 45° + \sec 60°$      **4.** $\cos 180° - 2 \sec 60° + \sin 270°$

**2.** $\csc 30° + \tan 45° - \sin 30°$      **5.** $\cos 30° \sec 30° + \sin 60° \csc 60°$

**3.** $\sin 90° + 2 \cot 45° - \cos 0°$      **6.** $\sin 60° \cos 30° + 2 \sin 45° \cos 45°$

Verify that each of the following statements is true.

**7.** $\sin^2 60° + \cos^2 60° = 1$ (*Note:* $\sin^2 60° = (\sin 60°)^2$)

**8.** $\cos^2 30° - \sin^2 30° = \cos 60°$

**9.** $2 \cos^2 45° - 1 = \cos 90°$

**11.** $1 - 2 \sin^2 45° = \cos 90°$

**10.** $1 + \tan^2 30° = \sec^2 30°$

**12.** $\csc^2 60° - 1 = \cot^2 60°$

**13.** $\sin 90° \cos 30° - \cos 90° \sin 30° = \sin 60°$

**14.** $\tan 60° \tan 30° = \sec 60° - 1$

**15.** $\sin 30° = \sqrt{\dfrac{1 - \cos 60°}{2}}$

**16.** $\cos 45° = \sqrt{\dfrac{1 + \cos 90°}{2}}$

Exercises 17–20 refer to right triangle *ABC* shown below.

**17.** Find $a$ if $b = 4$ and $A$ is $30°$.

**18.** Find $c$ if $a = 9$ and $B$ is $45°$.

**19.** Find $\sin A$ if $a = 5$ and $b = 9$.

**20.** Find $\tan B$ if $a = 3$ and $c = 7$.

## EVALUATING AND APPLYING TRIGONOMETRIC FUNCTIONS

### 11–5   Using Tables

Approximations of the values of the trigonometric functions of positive acute angles have been calculated and are available in tables. In Table 6 at the end of the book the measures of angles from $0°00'$ to $45°00'$ at intervals of ten minutes are listed in the first column; angle measures from $45°00'$ to $90°00'$ are listed in the last column, reading *from bottom to top*.

To find a *four-significant-digit* approximation of the value of any trigonometric function of an angle with measure between $0°$ and $90°$:

1. Reading in the first or last column, locate the row in which the measure of the angle is listed.
2. In this row find the entry in the column
   (a) at the *top* of which the function is named if the angle is read at the *left*,
   (b) at the *bottom* of which the function is named if the angle is read at the *right*.

**Examples.**   **a.** $\sin 25°00' \doteq 0.4226;$   $\tan 40°50' \doteq 0.8642$

**b.** $\cos 65°00' \doteq 0.4226;$   $\cot 49°10' \doteq 0.8642$

Notice that angles whose measures are given at the beginning and end of any one row are complementary angles, and that functions named at the top and bottom of a column are cofunctions.

Notice also that as $\theta$ varies in measure from $0°$ to $90°$ the values $\sin\theta$, $\tan\theta$, and $\sec\theta$ increase, but $\cos\theta$, $\cot\theta$, and $\csc\theta$ decrease. This fact is important in using linear interpolation (Section 10–7) to approximate the values of the trigonometric functions of an angle whose measure is between consecutive entries in the table.

**Example 1.** Find $\sin 34°27'$ and $\cos 34°27'$.

*Solution:*    Referring to Table 6, you have

| | $\theta$ | $\sin\theta$ |
|---|---|---|
| | 34°30′ | 0.5664 |
| | 34°27′ | ? |
| | 34°20′ | 0.5640 |

$10'\left[\ 7'\left[\begin{array}{c}34°30' \\ 34°27' \\ 34°20'\end{array}\ \middle|\ \begin{array}{c}0.5664 \\ ? \\ 0.5640\end{array}\right]\ e\ \right]\ 0.0024$

(a positive number because $\sin 34°30' > \sin 34°20'$)

$$\frac{7}{10} = \frac{e}{0.0024}, \ e = \frac{7}{10}(0.0024) \doteq 0.0017.$$

Then $\sin 34°27' = 0.5640 + 0.0017 = 0.5657$.

*Note:* In working with approximations of values of trigonometric functions, it is customary to write $=$ instead of $\doteq$.

$10'\left[\ 7'\left[\begin{array}{c}34°30' \\ 34°27' \\ 34°20'\end{array}\ \middle|\ \begin{array}{c}0.8241 \\ ? \\ 0.8258\end{array}\right]\ e\ \right]\ -0.0017$

| | $\theta$ | $\cos\theta$ |
|---|---|---|
| | 34°30′ | 0.8241 |
| | 34°27′ | ? |
| | 34°20′ | 0.8258 |

(a negative number because $\cos 34°30' < 34°20'$)

$$\frac{7}{10} = \frac{e}{-0.0017}, \ e = \frac{7}{10}(-0.0017) \doteq -0.0012.$$

Then $\cos 34°27' = 0.8258 + (-0.0012) = 0.8246$.

Table 6 also enables you to approximate the measure of a positive acute angle to the nearest minute if you know a four-significant-digit approximation of the value of one of its trigonometric functions.

**Example 2.** Find the measures of the positive acute angles $\alpha$ and $\beta$ (beta), if $\tan\alpha = 1.740$ and $\cot\beta = 0.0490$.

*Solution:*   In the column labeled **Tan** locate the two consecutive entries between which 1.740 lies:

| $\alpha$ | $\tan \alpha$ |
|---|---|
| 60°10′ | 1.744 |
| ? | 1.740 |
| 60°00′ | 1.732 |

$$10' \left[ n' \left[ \begin{array}{c} 60°10' \\ ? \\ 60°00' \end{array} \right| \begin{array}{c} 1.744 \\ 1.740 \\ 1.732 \end{array} \right] 0.008 \right] 0.012$$

$$\frac{n}{10} = \frac{0.008}{0.012}, \; n = 10(\tfrac{2}{3}) \doteq 7.$$

∴ the measure of $\alpha$ is approximately 60°00′ + 7′, or 60°7′.

To determine $\beta$, work in the column **Cot**.

| $\beta$ | $\cot \beta$ |
|---|---|
| 87°20′ | 0.0466 |
| ? | 0.0490 |
| 87°10′ | 0.0495 |

$$10' \left[ n' \left[ \begin{array}{c} 87°20' \\ ? \\ 87°10' \end{array} \right| \begin{array}{c} 0.0466 \\ 0.0490 \\ 0.0495 \end{array} \right] -0.0005 \right] -0.0029$$

$$\frac{n}{10} = \frac{-0.0005}{-0.0029}, \; n = 10(\tfrac{5}{29}) \doteq 2.$$

∴ $\beta = 87°10′ + 2′$, or 87°12′.

Following common usage, we write = in place of $\doteq$ even though $\beta$ is given only to the nearest minute.

In many practical problems involving trigonometry, an angle is described as an *angle of elevation* or an *angle of depression* (see Figure 11–13). Since the point $B$ is elevated with respect to the observer at $A$, $\angle CAB$, the angle between the line of sight and the horizontal ray $AC$ through $A$, is an angle of elevation. The point $Q$ is depressed with respect to the observer at $R$; therefore $\angle SRQ$, the angle between the line of sight and the horizontal ray $RS$ through $R$, is an angle of depression.

**FIGURE 11–13**

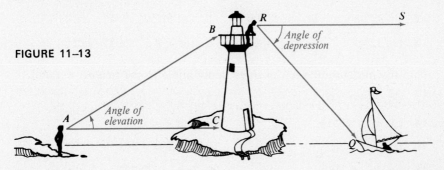

**Example 3.** The angle of elevation from a point 225 feet from the base of a building to the top of the building is 60°, while, from the same point, the angle of elevation to the top of a flagpole on the building is 65°. How tall is the flagpole if the data given are assumed to be **(a)** exact and **(b)** approximate?

*Solution:* In right triangle $ABC$,

$$\tan 60° = \frac{BC}{225}$$
$$BC = 225 \tan 60°$$
$$= 225(1.732)$$
$$= 389.7.$$

In right triangle $ADC$,

$$\tan 65° = \frac{DC}{225}$$
$$DC = 225 \tan 65°$$
$$= 225(2.145)$$
$$= 482.6.$$

To find $DB$, the height of the flagpole, you compute $DB = DC - BC$.

$$DB = 482.6 - 389.7 = 92.9$$

∴ **(a)** 92.9 feet; **(b)** 93 feet.

*WRITTEN EXERCISES*

Find a four-significant-digit approximation of the given trigonometric function value.

**A**

**1.** $\cos 23°17'$  **4.** $\cos 44°22'$  **7.** $\csc 63°18'$  **10.** $\sec 35°12'$

**2.** $\sin 38°48'$  **5.** $\sin 48°36'$  **8.** $\cot 51°59'$  **11.** $\csc 41°23'$

**3.** $\tan 12°6'$  **6.** $\tan 75°18'$  **9.** $\sec 4°4'$  **12.** $\cot 32°35'$

Find the measure of each positive acute angle to the nearest minute.

**13.** $\sin \alpha = 0.3600$    **17.** $\cos \alpha = 0.5752$

**14.** $\cos \alpha = 0.8516$    **18.** $\tan \alpha = 0.0075$

**15.** $\tan \alpha = 3.542$    **19.** $\sec \alpha = 1.413$

**16.** $\sin \alpha = 0.7538$    **20.** $\csc \alpha = 3.222$

With the data given for right triangle *ABC* (in each case *C* is the right angle) find to the nearest unit the measures of its other sides and angles.

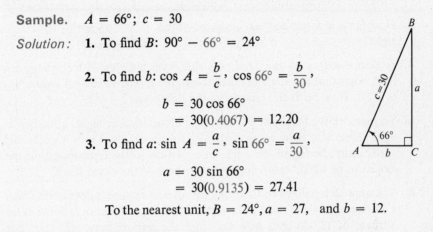

**Sample.**   $A = 66°$; $c = 30$

*Solution:*   **1.** To find $B$: $90° - 66° = 24°$

**2.** To find $b$: $\cos A = \dfrac{b}{c}$, $\cos 66° = \dfrac{b}{30}$,

$$b = 30 \cos 66°$$
$$= 30(0.4067) = 12.20$$

**3.** To find $a$: $\sin A = \dfrac{a}{c}$, $\sin 66° = \dfrac{a}{30}$,

$$a = 30 \sin 66°$$
$$= 30(0.9135) = 27.41$$

To the nearest unit, $B = 24°$, $a = 27$,  and $b = 12$.

**B**  **21.** $B = 32°$, $c = 24$    **23.** $A = 48°$, $b = 35$    **25.** $a = 6$, $b = 8$

**22.** $A = 22°$, $c = 16$    **24.** $B = 54°$, $b = 120$   **26.** $a = 5$, $b = 16$

Since most of the entries in Table 6 are approximations, lengths (not angle measures) computed using them should generally not be given to more than four significant digits, even though the data involved are assumed to be exact. The following rule comparing the accuracy of angle measure and length is used with the rules on page 410 to decide the accuracy of a computation involving approximations.

| Angles measured to | correspond to | lengths measured to |
|:---:|:---:|:---:|
| 1° | | 2 significant digits |
| 10′ | | 3 significant digits |
| 1′ | | 4 significant digits |

## PROBLEMS

Find the results if the data given are **(a)** exact and **(b)** approximate.

**A**  **1.** An aircraft takes off and climbs steadily at an airspeed of 360 miles per hour along a straight path making an angle of 15° with the ground. How high (in feet) is the airplane after 10 minutes?

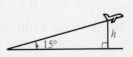

**2.** After 15 minutes, how far (in miles) along the ground has the airplane in Problem 1 traveled? (Neglect any effects of wind.)

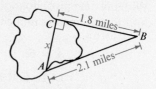

**3.** How far is it across the lake pictured at the right?

**4.** A man standing on the deck of a ship with his eyes 25 feet above sea level notes that the angle of elevation to the top of a 265-foot lighthouse is 3°27′. How far is the boat from the lighthouse?

**5.** To measure the height of the base of a cloud level at night, a meteorologist directed a light beam vertically to the clouds. From a point 800 yards away, he found the angle of elevation to the light image on the clouds to be 62°20′. How high was the base of the cloud level?

**6.** The approach pattern to Central City Airport requires pilots to establish a 12-degree angle of descent toward the runway. If a pilot is flying at an altitude of 12,000 feet, how far from the airport must she start her landing approach?

**7.** The bottom end of a 20-foot ladder rests 5 feet from the sides of a building. What angle does the ladder make with the wall of the building?

**8.** At an airport, cars drive down a ramp 96 feet long to reach the lower level baggage-claim area 13 feet below the main level. What angle does the ramp make with the ground at the lower level?

**B** **9.** The White River is 600 feet wide at a point where a 40-foot tree stands on the north bank. From the south bank, the angle of elevation to the top of the tree is 3°20′. Find whether the south bank is higher or lower than the north bank, and by how much.

**10.** From a window in the Adams Building the angle of elevation to the top of the hospital across the street is 70°, while, from the same window, the angle of depression to the foot of the hospital is 67°. If the buildings are 125 feet apart, find the height of the hospital.

**11.** Charlesville and Newton are at sea level and in line with 5000-foot Mount Rush, and the angles of depression of the two towns are 8°36′ and 5°44′, respectively. How far apart are the towns?

**12.** From the deck of a ship 25 feet above water level, the angles of elevation to the top and bottom of a lighthouse on a cliff are 47°28′ and 43°36′, respectively. If the cliff is 678 feet high, what is the height of the lighthouse?

**13.** A pendulum 40 inches long is moved 30° from the vertical. How much is the lower end of the pendulum lifted?

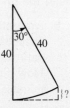

**14.** The top of a vertical tree broken by the wind hits the ground 25 feet from the foot of the tree. If the upper portion makes an angle of 30° with the horizontal ground, what was the original height of the tree?

**15.** The angle of depression of the top of the Billings Building from the roof of the Wolcott Building (in the same vertical plane) is 33°10′, and from the 15th floor it is 21°50′. If the distance between the roof and the 15th floor is 101 feet, how far apart are the buildings?

**16.** From the point on the ground 75.0 feet from the base of a building, the angle of elevation of the top of a flagpole on the edge of the roof of the building is 45°20′ and the angle of elevation of the bottom of the flagpole is 38°40′. Find the height of the pole.

## 11–6  Logarithms of the Values of Trigonometric Functions

Since each value of a trigonometric function of a positive acute angle is a positive number, it has a logarithm. For example:

$$\sin 31° = 0.5150 \quad \text{(Table 6)}$$

$$\log 0.5150 = 9.7118 - 10 \quad \text{(Table 5)}$$

∴ log sin 31° which is read "logarithm of the sine of an angle of 31°"
  = 9.7118 − 10.

In Table 7 the characteristics and four-significant-digit approximations of the mantissas of these logarithms can be read directly for angles measured to the nearest ten minutes, and by interpolation for angles measured to the nearest minute. Note that, to conserve space, Table 7 gives the logarithms increased by 10. Thus, log sin 31° is listed as 9.7118; you must add −10 to complete this and every other logarithm in the table.

To facilitate interpolation in Table 7, the difference between successive entries is given in the columns headed "*d*" or "*cd*." The numbers in the "*cd*" column are differences common to "L Tan" and "L Cot" and are used with either one. Notice that a difference listed as "43" stands for "0.0043" or for "−0.0043" according as values to which the difference refers increase (L Sin $\theta$, L Tan $\theta$) or decrease (L Cos $\theta$, L Cot $\theta$) with increasing measure of $\theta$.

**Example 1.** Use Table 7 to find log sin 22°23′ and log csc 22°23′.

*Solution:*

| | $\theta$ | L sin $\theta$ | |
|---|---|---|---|
| | 22°30′ | 9.5828 − 10 | |
| 22°23′ | ? | |
| | 22°20′ | 9.5798 − 10 | |

$10' \left[ 3' \left[ \begin{matrix} 22°30' & 9.5828 - 10 \\ 22°23' & ? \\ 22°20' & 9.5798 - 10 \end{matrix} \right] e \right] 0.0030$

$e = \tfrac{3}{10}(0.0030) = 0.0009.$

$\therefore \log \sin 22°23' = 9.5798 - 10 + 0.0009 = 9.5807 - 10.$

$\log \csc 22°23' = \log 1 - \log \sin 22°23'$ (why?)
$= (10.0000 - 10) - (9.5807 - 10) = 0.4193.$

**Example 2.** An escalator in a department store makes an angle of 43°30′ with the floor. How far does a person ride on the escalator if he travels from the first to the second floor through a vertical distance of 18.3 feet?

*Solution:* Make a sketch.

$\sin 43°30' = \dfrac{BC}{AB} = \dfrac{18.3}{AB}$

$AB = \dfrac{18.3}{\sin 43°30'}$

$\log AB = \log 18.3 - \log \sin 43°30'$

From Table 5, log 18.3 = 1.2625 = 11.2625 − 10
From Table 7,
$\log \sin 43°30' = 9.8378 - 10 = \quad \underline{9.8378 - 10}$ (−)
$\qquad\qquad\qquad\qquad\qquad\qquad 1.4247$

$AB = $ antilog 1.4247 = 26.59

$\therefore$ a person rides about 26.6 feet in going between the first and second floors.

*WRITTEN EXERCISES*

Using logarithms, find the value of each expression to four significant digits.

A **1.** 213 sin 61°10′      **3.** sin 47°40′ cos 10°20′      **5.** $\dfrac{\sin 68°10'}{\sin 41°40'}$

**2.** 8.13 cos 14°30′      **4.** tan 72°00′ sin 71°50′      **6.** $\dfrac{\tan 51°00'}{\cot 41°00'}$

**7.** $83 \dfrac{\sin 8° \sin 71°}{\sin 79°}$     **9.** $\dfrac{(\sin 18°)^3}{22 \cos 41°}$     **11.** $\sqrt{\dfrac{12 \sec 58°}{5 \sin 58°}}$

**8.** $6.1 \dfrac{\sin 18° \sin 76°}{\sin 58°}$     **10.** $\dfrac{310 \cos 50°}{\sin^2 50°}$     **12.** $\sqrt[3]{10 \csc 18° \cos 18°}$

Use logarithms to find the measure to the nearest minute of the positive acute angle specified.

**13.** $\cos \theta = \dfrac{4.930}{8.750}$       **17.** $\cos \alpha = \dfrac{\cos 15° \sin 40°}{\cos 20°}$

**14.** $\tan \theta = \dfrac{7.540}{3.290}$       **18.** $\sin \beta = \sqrt{\dfrac{13 \times 34}{52 \times 35}}$

**15.** $\sin \alpha = \dfrac{24.30 \sin 32°}{28.9}$       **19.** $\tan \theta = \dfrac{14.32 \tan 16°24'}{76.9}$

**16.** $\sin \beta = \dfrac{17.31 \sin 54°}{52.00}$       **20.** $\cot \theta = \dfrac{15.21 \cot 39°14'}{7.30}$

*PROBLEMS*

Use logarithms in solving these problems. Assume linear measures are correct to four significant digits and angle measures to the nearest minute.

**A**

1. How long a shadow will a 73.20 foot tree cast when the angle of elevation of the sun is 57°10′?

2. What angle does the diagonal of a rectangle form with the longer side if the dimensions of the rectangle are 243.0 centimeters and 167.0 centimeters?

3. A guy wire holding a television transmitter antenna is 72.40 feet long and is attached to the antenna at a point 53.70 feet from the ground. What angle does the wire make with the ground?

4. How tall is the Cooper Building if the angle of elevation to its top from a point 653.0 feet from its base is 61°20′?

5. A beam of light traveling 39.50 inches from a source to a horizontal surface hits the surface at an angle of 37°25′. How far above the surface is the light source?

6. A conveyor belt carries bales of hay from the ground to the loft of a barn. If the belt makes an angle of 52°15′ with the vertical wall of the barn, and the loft door is 27.50 feet above the ground, how far does the belt travel from the ground to the loft?

7. The State Highway Commission plans to take a triangular corner plot from the square lot at the intersection of Routes 3 and 102 for the new expressway. If the plot to be taken runs 521.5 feet along Route 3 and 636.4 feet along Route 102, what angles does the third side of this plot make with Routes 3 and 102?

8. An ore boat headed due east on Lake Superior. After it had traveled 17.25 miles, it had been blown 0.8375 miles north of this course by a southwest wind. By what angle did the boat's actual course differ from its intended course?

B 9. The lens of a camera forms on the film an inverted image of the object being photographed, as indicated in the sketch. The object is 10.3 feet away from the lens, the lens is 4.9 inches from the film, and angle $A$ is 27°31′. How high is the image on the film?

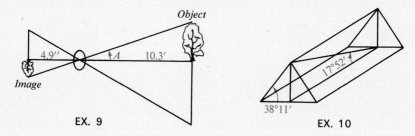

EX. 9          EX. 10

10. Ingots of an alloy are cast in the shape of a prism with isosceles triangles as ends. The base of a triangular end is 8 inches long, and the sides form a 38°11′ angle with the base. The line connecting the apex of a triangular end with the midpoint of the base of the opposite end forms an angle of 17°52′ with the line joining the midpoints of the bases of either end. Find the volume of the ingot. (Volume = area of cross section × depth.)

11. From the top of a mountain 12,350 feet high, the angles of depression of two ships lying in a vertical plane with the mountain top measure 18°24′ and 13°39′. Find the distance between the ships.

12. A right triangle whose legs are 36.25 and 84.30 centimeters long is inscribed in a circle. Find the length of the diameter of the circle and the measures of the acute angles of the triangle.

## 11–7   Reference Angles

To see how to use Tables 6 and 7 for angles of any measure, we associate with each angle $\theta$ in standard position an angle with measure between 0° and 90°, inclusive, called the *reference angle* of $\theta$. Let $P(a, b)$ be any point other than the origin on the terminal side of $\theta$, and let $T$ be the point whose coordinates are $(|a|, |b|)$. The angle in standard position with ray $OT$ as terminal side, and measure between $0°$ and $90°$, inclusive, is the reference angle of $\theta$ and is labeled $\alpha$ in Figure 11–14.

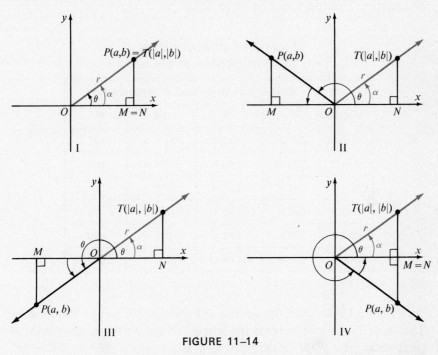

**FIGURE 11–14**

To determine the relationship between the values of the trigonometric functions of $\theta$ and $\alpha$, you must analyze the relationship between $a$ and $|a|$, and $b$ and $|b|$ for each quadrant. Notice that in all quadrants $r = \sqrt{a^2 + b^2} = \sqrt{|a|^2 + |b|^2}$.

**$\theta$ in Quadrant I:** $a = |a|$ and $b = |b|$.

This means $\sin \theta = \sin \alpha$, $\cos \theta = \cos \alpha$, and so on.

**$\theta$ in Quadrant II:** $a = -|a|$ and $b = |b|$.

$$\sin \theta = \frac{b}{r} = \frac{|b|}{r} = \sin \alpha,$$

$$\cos \theta = \frac{a}{r} = \frac{-|a|}{r} = -\cos \alpha,$$

$$\tan \theta = \frac{b}{a} = \frac{|b|}{-|a|} = -\tan \alpha.$$

Similarly, $\csc \theta = \csc \alpha$, $\sec \theta = -\sec \alpha$, and $\cot \theta = -\cot \alpha$.

By considering $\theta$ in the other quadrants, you can verify the other entries in the table below. Notice that you can obtain the value of any function of $\theta$ by multiplying the value of the corresponding function of $\alpha$ by 1 or $-1$ according as the value of the function of $\theta$ is positive or negative. (See the chart on page 437.)

**Functions of $\theta$ in terms of its reference angle $\alpha$.**

| Value | Quadrant in which $\theta$ terminates | | | |
|---|---|---|---|---|
| | I | II | III | IV |
| $\sin\ \theta$ | $\sin\ \alpha$ | $\sin\ \alpha$ | $-\sin\ \alpha$ | $-\sin\ \alpha$ |
| $\cos\ \theta$ | $\cos\ \alpha$ | $-\cos\ \alpha$ | $-\cos\ \alpha$ | $\cos\ \alpha$ |
| $\tan\ \theta$ | $\tan\ \alpha$ | $-\tan\ \alpha$ | $\tan\ \alpha$ | $-\tan\ \alpha$ |
| $\csc\ \theta$ | $\csc\ \alpha$ | $\csc\ \alpha$ | $-\csc\ \alpha$ | $-\csc\ \alpha$ |
| $\sec\ \theta$ | $\sec\ \alpha$ | $-\sec\ \alpha$ | $-\sec\ \alpha$ | $\sec\ \alpha$ |
| $\cot\ \theta$ | $\cot\ \alpha$ | $-\cot\ \alpha$ | $\cot\ \alpha$ | $-\cot\ \alpha$ |

In Figure 11–14, you can see that $\angle POM$ has the same measure as $\alpha$. This assertion is true because the lengths of the corresponding sides of right triangles $POM$ and $TON$ are equal since $MP = NT = |b|$, $OM = ON = |a|$, and $OT = OP = \sqrt{a^2 + b^2}$. Hence, the measures of the corresponding angles must be equal. This means that you can determine the measure of $\alpha$ by finding the measure of the acute angle formed by the terminal side of $\theta$ and the $x$-axis.

**Example 1.** Find $\cos 150°$ and $\tan 150°$.

*Solution:*

1. Sketch the angle in standard position.

2. $\alpha$ measures

$$180° - 150° = 30°.$$

3. Since the terminal side of the angle is in Quadrant II,

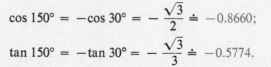

$$\cos 150° = -\cos 30° = -\frac{\sqrt{3}}{2} \doteq -0.8660;$$

$$\tan 150° = -\tan 30° = -\frac{\sqrt{3}}{3} \doteq -0.5774.$$

   To determine the measure of the reference angle when the measure of $\theta$ is greater than 360° or less than 0°, you use the fact that *angles in standard position whose measures differ by an integral multiple of 360° are coterminal.*

**Example 2.** Find four-significant-digit approximations of cos 845°50′ and sin 845°50′.

*Solution:*      **1.** 845°50′ − 2 · 360° = 125°50′

             **2.** 180° − 125°50′ = 179°60′ − 125°50′ = 54°10′

$$\left.\begin{array}{l} \cos 845°50′ = -\cos 54°10′ = -0.5854 \\ \sin 845°50′ = \sin 54°10′ = 0.8107 \end{array}\right\} \text{Table 6}$$

## WRITTEN EXERCISES

Draw the angle in standard position having the given measure. Label and give the measure of its reference angle.

**A**   **1.** 100°        **4.** −25°        **7.** 740°        **10.** 510°

      **2.** 240°        **5.** −140°       **8.** 1020°       **11.** −470°

      **3.** 290°        **6.** −225°       **9.** 410°        **12.** −800°

Copy each table and replace the question marks with exact values.

**13.**

| $\theta$ | 30° | 150° | 210° | 330° |
|---|---|---|---|---|
| sin $\theta$ | ? | ? | ? | ? |
| cos $\theta$ | ? | ? | ? | ? |
| tan $\theta$ | ? | ? | ? | ? |

**14.**

| $\theta$ | −45° | −135° | −225° | −315° |
|---|---|---|---|---|
| sin $\theta$ | ? | ? | ? | ? |
| cos $\theta$ | ? | ? | ? | ? |
| tan $\theta$ | ? | ? | ? | ? |

**15.**

| $\theta$ | 45° | 135° | 225° | 315° |
|---|---|---|---|---|
| sin $\theta$ | ? | ? | ? | ? |
| cos $\theta$ | ? | ? | ? | ? |
| tan $\theta$ | ? | ? | ? | ? |

**16.**

| $\theta$ | 60° | 120° | 240° | 300° |
|---|---|---|---|---|
| sin $\theta$ | ? | ? | ? | ? |
| cos $\theta$ | ? | ? | ? | ? |
| tan $\theta$ | ? | ? | ? | ? |

Express as a function of a positive acute angle.

**17.** $\sin 150°$    **20.** $\cot 325°$    **23.** $\cos 650°$    **26.** $\tan(-970°)$

**18.** $\cos 170°$    **21.** $\sec(-110°)$    **24.** $\sin 850°$    **27.** $\csc(-15°12')$

**19.** $\tan 210°$    **22.** $\csc(-289°)$    **25.** $\cot(-1045°)$ **28.** $\sec(-62°28')$

Find each value to four significant digits.

**29.** $\cos 118°$           **31.** $\sin(-38°)$           **33.** $\log|\cos 127°10'|$

**30.** $\cot 216°$           **32.** $\sec(-137°)$          **34.** $\log|\sin 211°50'|$

Determine the measure of $\theta$ to the nearest 10 minutes if $\theta$ is a positive angle measuring less than 360°.

**35.** $\cos \theta = 0.9462$, $\sin \theta < 0$       **37.** $\tan \theta = 6.500$, $\sin \theta < 0$

**36.** $\sin \theta = -0.1937$, $\tan \theta < 0$      **38.** $\sec \theta = 2.715$, $\sin \theta > 0$

**39.** $\log|\sin \theta| = 9.7861$, $\theta$ terminates in Quadrant III

**40.** $\log|\cos \theta| = 9.9453$, $\theta$ terminates in Quadrant II

Determine, correct to tenths, the rectangular coordinates of the point $P$ if $OP$ has the given length and the position angle of $P$ has the given measure.

B **41.** 5; 125°           **43.** 8.3; 267°           **45.** 10; −15°

**42.** 12; −40°           **44.** 12.5; 140°          **46.** 52; 224°

For the given point $P$, determine the distance $OP$ to tenths and the measure of the smallest positive position angle of $P$ to the nearest degree.

**47.** $P(4, 3)$            **49.** $P(-7, -2)$           **51.** $P(7.1, -0.5)$

**48.** $P(5, 12)$           **50.** $P(-3.4, -8.2)$        **52.** $P(-0.8, 4.2)$

## VECTORS

## 11–8  Adding Vectors

To describe the velocity of an airplane you must give both the direction and the magnitude of the velocity. Quantities whose description requires both magnitude and direction are called **vector quantities** and are represented by directed line segments, or arrows, called **vectors**.

Units of length on the arrow indicate the units of magnitude of the vector quantity, and the direction of the arrow indicates the direction of the vector quantity. Vector quantities are often denoted by a symbol with an arrow over it, as illustrated below.

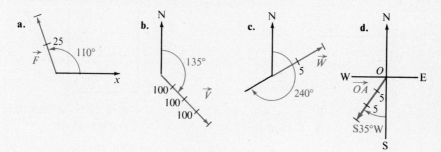

**FIGURE 11–15**

In Figure 11–15, vectors represent:

**a.** A force $\vec{F}$ of 50 pounds making an angle of 110° with the *x*-axis (the angle is measured counterclockwise from the positive *x*-axis to the vector).

**b.** A velocity $\vec{V}$ of 400 mph on a heading of 135° (the angle, called the *heading, bearing,* or *course,* is measured clockwise from north to the vector).

**c.** Wind velocity $\vec{W}$ of 10 knots from 240° (the angle is measured clockwise from north to the direction *from* which the wind blows).

**d.** A displacement $\overrightarrow{OA}$ from *O* to *A* of 15 miles S35°W (the 35° angle is measured *from* south *toward* west).

Examples **b**, **c**, and **d** above illustrate the use of the north-south line as reference in specifying direction angles in navigation and surveying. Figure 11–16 below shows other examples of this technique.

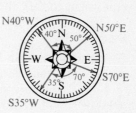

*Bearing N50°E*　　　*Heading 320°*　　　*Wind from 30°*

**FIGURE 11–16**

Vectors such as $\overrightarrow{BR}$ and $\overrightarrow{CD}$ in Figure 11–17 are equivalent vectors; that is, they have the same magnitude and direction. We may think of a point as representing a vector of length 0, a *zero vector*. All zero vectors are considered equivalent.

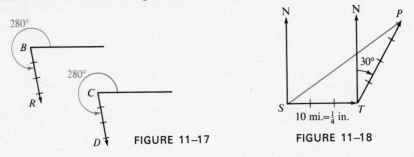

FIGURE 11–17　　　　　　FIGURE 11–18

The scale drawing in Figure 11–18 pictures the motion of a ship which sailed 30 miles due east $(\overrightarrow{ST})$, turned, and then sailed 40 miles N30°E $(\overrightarrow{TP})$. Because the vector $\overrightarrow{SP}$ represents the single motion that would bring the ship to the same final position as the combination of the given displacements, we call $\overrightarrow{SP}$ the *sum* or *resultant* of $\overrightarrow{ST}$ and $\overrightarrow{TP}$ and write $\overrightarrow{SP} = \overrightarrow{ST} + \overrightarrow{TP}$. Measure $\overrightarrow{SP}$ and $\angle TSP$ on the scale drawing; you can estimate that this resultant displacement is 60 miles N55°E.

To define the resultant of any two vectors, we use the fact that given a vector $\overrightarrow{CD}$ and a point $B$, then there is a unique vector $\overrightarrow{BR}$ at $B$ equivalent to $\overrightarrow{CD}$ (Figure 11–19).

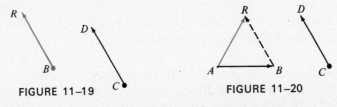

FIGURE 11–19　　　　　　FIGURE 11–20

If $\overrightarrow{AB}$ and $\overrightarrow{CD}$ are any two vectors, and if $\overrightarrow{BR}$ is the vector at $B$ equivalent to $\overrightarrow{CD}$, then the sum or resultant of $\overrightarrow{AB}$ and $\overrightarrow{CD}$ is $\overrightarrow{AR}$; that is, $\overrightarrow{AB} + \overrightarrow{CD} = \overrightarrow{AR}$ (Figure 11–20).

**Example.** A ship leaves harbor, travels 4 hours on a heading of 80°, and then turns to a heading of 170° and travels another 3 hours until it has an engine failure. (a) What heading should a repair vessel take from the harbor in order to reach the stranded ship if the ship traveled at a (constant) speed of 22 knots? Give the result to the nearest 10 minutes. (b) How fast should the repair vessel travel to reach the ship in 4 hours? Give the result to the nearest knot.

*Solution:*   **a.** Make a sketch showing the ships' distances and headings. At 22 knots, the distance traveled was:

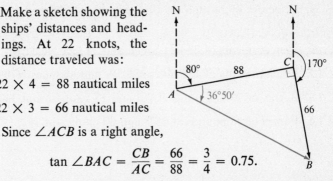

$AC = 22 \times 4 = 88$ nautical miles

$CB = 22 \times 3 = 66$ nautical miles

Since $\angle ACB$ is a right angle,

$$\tan \angle BAC = \frac{CB}{AC} = \frac{66}{88} = \frac{3}{4} = 0.75.$$

From Table 6, to the nearest 10 minutes, $\angle BAC = 36°50'$. Therefore, the repair vessel should take a heading of $80° + 36°50'$, or $116°50'$.

**b.** To find the distance from $A$ to $B$, you can use the relationship

$$\sec \angle BAC = \frac{AB}{AC},$$

$$\sec 36°50' = \frac{AB}{88},$$

$$AB = 88 \sec 36°50' = 88 \times 1.249$$
$$= 109.912 \doteq 110.$$

Since the stranded ship is 110 nautical miles from the harbor, in order to reach the ship in 4 hours, the repair vessel must average $\frac{110}{4} = 27.5$ or $28$ knots.

## WRITTEN EXERCISES

Find the resultant of the given vectors. Express angle measures to the nearest degree, magnitudes to the nearest unit.

**A**   **1.** Velocities of 32 mph west and 20 mph north.

**2.** Velocities of 84 knots south and 40 knots east.

**3.** Displacements of 35 miles west and 45 miles north.

**4.** Displacements of 18 centimeters south and 26 centimeters east.

**5.** Forces of 150 pounds to the right and 40 pounds down.

**6.** Forces of 4.2 tons to the left and 2.1 tons up.

**B**   **7.** Velocities of 500 knots on a heading of 230° and 16 knots on a heading of 140°.

**8.** Velocities of 24 mph on a bearing of N75°W and 10 mph on a bearing of N15°E.

**9.** Displacements of 28 feet S25°E and 34 feet N65°E.

**10.** Displacements of 130 miles N20°W and 40 miles N70°E.

**11.** Forces of 75 kilograms at 70° and 25 kilograms at 340°.

**12.** Forces of 48 pounds at 252° and 60 pounds at 342°.


*PROBLEMS*

Assume data given are exact.

**A**  **1.** A radar beacon bears 84° from a ship 140 miles away. The ship then sails on a heading of 174° until the beacon bears due north, or 0°. How far did the ship sail, and how far is it from the beacon?

**2.** An aircraft flies on a heading of 47° at 520 mph. If a wind is blowing at 70 mph from 317°, what is the course of the aircraft over the ground, and what is its ground speed?

**3.** Two ships start from the same harbor, one sails N22°W and the other S68°W. If the first ship averages 12 knots and the second ship 14 knots, how far apart are the ships after 8 hours, and what is the bearing from the second ship to the first?

**4.** In a naval maneuver two ships rendezvous at Point A. One then proceeds east 10 miles and north 14 miles to Point B. At what bearing should the second ship head to meet the first ship at Point B?

**B**  **5.** At what bearing and speed would a pilot head if she wanted to fly due north at 345 mph when a 40 mph west wind is blowing?

**6.** At what bearing and speed should the navigator direct the captain of a ship to head if the captain wants to steam straight ahead at 18.2 knots when seas of 7.4 knots are hitting the ship broadside?

**7.** Ida can swim at a rate of 3 mph. If she heads for a point directly across a river in which the current is 10 mph, by how many degrees does the direction in which she actually swims differ from her intended direction? If the river is 32 yards wide, will she make it across before reaching the falls 112 yards downstream?

**8.** In a military test, a ballistic missile has a target 250 miles east and 280 miles south of its launching site. How far must it travel to the target? If a second missile is sent from the target site to intercept the first missile, at what heading should it be sent?

## 11–9    Resolving Vectors

Two vectors such as $\vec{V_1}$ and $\vec{V_2}$ in Figure 11–21 whose sum is $\vec{V}$ are called **components** of $\vec{V}$. Expressing a vector in a plane as a sum of two components in definite directions (usually at right angles to each other) is called **resolving the vector**. Of particular interest are the horizontal and vertical components of a vector.

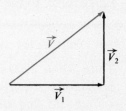

**FIGURE 11–21**

**Example 1.**    To pull a boat into a dock, a man pulls on a rope with a force of 60 pounds. To the nearest pound, what part of the man's force moves the boat toward the dock if the rope makes an angle of 34° with the horizontal?

*Solution:*    1. Draw a right triangle and label it to represent the given force and angle.

2. $\cos \angle ABC = \dfrac{BC}{AB}$,

$\cos 34° = \dfrac{BC}{60}$,

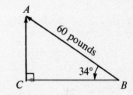

$BC = 60 \cos 34° = 60(0.8290) = 49.74.$

To the nearest pound, 50 pounds of force is being applied to move the boat horizontally.

Figure 11–22 illustrates the forces acting on an object at $A$ on a plane inclined at an angle $\theta$ with the horizontal. The vector $\overrightarrow{AW}$ represents the force of gravity pulling the object towards the center of the earth with a pull equal to the weight of the object. Its components are $\overrightarrow{AB}$ parallel to the plane and $\overrightarrow{BW}$ perpendicular to the plane. Because $\overrightarrow{AB}$ denotes

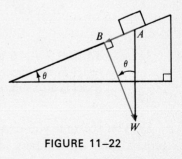

**FIGURE 11–22**

the force tending to pull the object down the plane, the length of $\overrightarrow{AB}$ tells the magnitude of the smallest braking force that will keep the object from moving. If $\theta$ measures 10° and $\overrightarrow{AW}$ represents the weight of a 2700-pound automobile, you find the magnitude of this minimum brake force, to the nearest pound, as follows:

$$\sin 10° = \frac{AB}{AW} = \frac{AB}{2700}.$$

Thus $AB = 2700 \sin 10° = 2700 \, (0.1736) \doteq 468.7.$ $\therefore$ the minimum brake force required is 469 pounds.

Can you find the resultant of two vectors whose horizontal and vertical components are known? You can if you assume the fact that *the component of the resultant in any given direction is the sum of the components of the vectors in that direction.*

**Example 2.** Two tugboats on headings of N30°W and N40°E are pulling a disabled vessel through the water. If the first tug exerts a force of 4 tons and the second a force of 6 tons on the tow rope, what is the force acting to move the towed vessel due north?

*Solution:* Let $\overrightarrow{DE}$ = the north component of the first boat's force,

$\overrightarrow{CB}$ = the north component of the second boat's force,

$\overrightarrow{AN}$ = the total north component acting on towed vessel.

Then $\overrightarrow{AN} = \overrightarrow{DE} + \overrightarrow{CB}$.

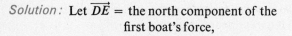

In $\triangle ADE$: $\sin 60° = \dfrac{DE}{4}$,

$DE = 4 \sin 60° \doteq 4(0.8660) = 3.464 \doteq 3.5.$

In $\triangle ABC$: $\sin 50° = \dfrac{CB}{6}$,

$CB = 6 \sin 50° \doteq 6(0.7660) = 4.596 \doteq 4.6.$

$\therefore AN \doteq 3.5 + 4.6 = 8.1$, and the force exerted on the towed vessel in the direction of due north is about 8 tons. (The towed vessel is also subject to a force of about 2 tons due east.)

Do you see that you can describe a vector at a point by giving an ordered pair of numbers indicating its horizontal and vertical components? For example, at $T$, $[2, -3]$ denotes a vector whose horizontal component is a vector of *length 2 to the right* and whose vertical component *is a vector of length 3 down* (Figure 11–23). When the initial point is the origin in a coordinate plane, the vector is said to be in standard position; the components of the vector are then the co-ordinates of the terminal point of the vector (Figure 11–24).

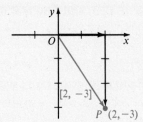

FIGURE 11–23  FIGURE 11–24

## WRITTEN EXERCISES

Determine the horizontal and vertical components of the vector having the given length and direction.

**A** 
1. 15; S45°E
3. 32; course 120°
5. 1000; heading 90°

2. 70; N62°W
4. 56; bearing 350°
6. 260; course 270°

If $P$ is a point at the given distance from the origin in a coordinate plane and $P$ has a position angle of given measure, approximate to tenths the lengths of the horizontal and vertical components of $\overrightarrow{OP}$.

7. 18; 54°
9. 72; 148°
11. 250; −310°

8. 10; 76°
10. 30; 245°
12. 68; −110°

Each ordered pair describes a vector at the origin of a coordinate system. Draw the vectors, determine the components of their resultant, and draw the resultant.

**Sample.** [−2, 3], [1, 4]

*Solution:*

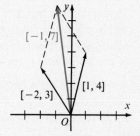

$$[-2, 3] + [1, 4] = [-2 + 1, 3 + 4]$$
$$= [-1, 7].$$

**B** 
13. [4, 2], [3, −2]
16. [4, −2], [−1, 0]
19. [−2, −3], [2, 3]

14. [6, 2], [−6, 4]
17. [−3, 5], [−1, 1]
20. [5, 0], [0, 5]

15. [0, 7], [−2, −2]
18. [3, −4], [−3, 6]

Let $[a, b]$, $[c, d]$, and $[e, f]$ denote vectors at the origin of a coordinate system. **(a)** Verify each statement if $a = -5$, $b = 3$, $c = 7$, $d = -2$, $e = 4$, $f = 1$; **(b)** prove that each statement is true for every replacement of $a$, $b$, $c$, $d$, $e$, and $f$ by real numbers.

C

**21.** $[a, b] + [c, d] = [c, d] + [a, b]$        **22.** $[a, b] + [0, 0] = [a, b]$

**23.** $([a, b] + [c, d]) + [e, f] = [a, b] + ([c, d] + [e, f])$

**24.** There is a vector $[x, y]$ such that $[a, b] + [x, y] = [0, 0]$.


*PROBLEMS*


Assume given data are exact.

A

**1.** Four guy wires making angles of 60° with the ground brace a vertical pole which weighs 55 pounds. If each wire is under a tension of 18 pounds, what is the total force pushing the pole downward against the ground?

**2.** Three girls are pushing a stalled car. If one pushes directly from the rear with a force of 60 pounds, and if each of the other two push against the car with equal forces of 50 pounds from left and right at angles of 25° to the sides of the car, what is the total force tending to move the car forward?

**3.** A ship sails 25 miles on a course N75°W, then turns to a heading of N20°W and sails 30 miles on this course. How far north has the ship sailed in all? How far west?

**4.** How much force is wasted when a man pushes with a force of 90 pounds on the handle of a lawn mower while holding it at **(a)** an angle of 30° with the ground and **(b)** an angle of 45° with the ground?

B

**5.** The string to a box kite makes a 60° angle with the ground. If the string will break when subjected to a 30-pound force and if the kite requires a minimum horizontal force of 5 pounds to fly, what is the possible range of the horizontal force of the wind in which the kite can be flown?

**6.** A wire cable on a crane can withstand a 20-ton tension. If the crane is extended at an angle of 50° from the vertical, what is the greatest weight it can lift to leave a safety margin of two tons in the wire?

**7.** A river flows from north to south. To cross from the east bank directly to the west bank a boat captain finds that he must keep on a course of 284°. If the trip takes fifteen minutes when the speed of the boat relative to the water is 12.4 mph, **(a)** how wide is the river and **(b)** how swift is the current?

**8.** A theater marquee is supported by two steel cables of equal length which make angles of 40° with the building. If the marquee weighs 600 pounds, what is the tension in each cable?

## 11–10 Vectors and Trigonometric Form for Complex Numbers

In addition to representing forces, displacements, velocities, and other physical quantities, vectors can also be used to represent complex numbers. If you associate each complex number $a + bi$ with the vector $[a, b]$ in standard position (recall Section 11–9), then each complex number corresponds to a unique vector in standard form, and conversely (see Figure 11–25). Thus it is convenient to represent complex numbers geometrically in this manner. Figure 11–26 shows the geometric representation of $4 + 2i$. Note that 4, the real part of $4 + 2i$, is the length of the horizontal component of $[4, 2]$ and that 2, the imaginary part of $4 + 2i$, is the length of the vertical component of $[4, 2]$.

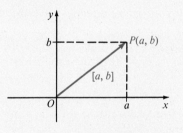

FIGURE 11–25

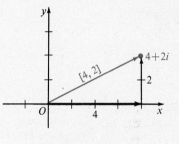

FIGURE 11–26

When vectors in standard position are considered to represent complex numbers, the $xy$-plane is called the **complex plane**. To *plot a complex number in the complex plane* means to draw the associated vector (or, in some cases, mark the terminal point of the vector) in the plane. Can you suggest why, as shown in Figure 11–27, the $x$-axis is called the *axis of reals* (or the *real axis*) and the $y$-axis is called the *axis of imaginaries* (or the *imaginary axis*) in the complex plane?

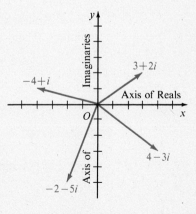

FIGURE 11–27

If $P(a, b)$ is the terminal point of the vector representing the complex number $a + bi$ (see Figure 11–28), then a position angle $\theta$ of $P$ is called an **amplitude** of $a + bi$, and the distance $OP$, or $r$, is called the **absolute value** or **modulus** of $a + bi$ and is denoted by $|a + bi|$. Since $r = \sqrt{a^2 + b^2}$, $a = r \cos \theta$ and $b = r \sin \theta$, it follows that

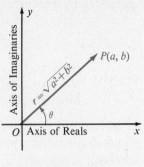

$$a + bi = r \cos \theta + ir \sin \theta$$
$$= r (\cos \theta + i \sin \theta).$$

**FIGURE 11–28**

This latter expression is called the **trigonometric form** for the complex number $a + bi$. In this book, we will use the *least nonnegative value* for $\theta$. That is, we will choose $0° \leq \theta < 360°$.

**Example.**  Express: **a.** $5 - 2i$ in trigonometric form.

**b.** $\sqrt{3} (\cos 60° + i \sin 60°)$ in the form $a + bi$.

*Solution:*  **a.** First compute

$$r = \sqrt{a^2 + b^2} = \sqrt{5^2 + (-2)^2}$$
$$= \sqrt{25 + 4}$$
$$= \sqrt{29}.$$

To determine $\theta$, note that the vector for $5 - 2i$ lies in Quadrant IV. Thus the reference angle $\alpha$ for $\theta$ is given by

$$\cos \alpha = \frac{a}{\sqrt{a^2 + b^2}} = \frac{5}{\sqrt{29}}.$$

From Tables 4 and 6, $\alpha = 21°50'$ to the nearest 10 minutes. Then

$$\theta = 360° - 21°50' = 338°10'.$$

$$\therefore 5 - 2i = \sqrt{29} (\cos 338°10' + i \sin 338°10').$$

**b.** Since $\cos 60° = \frac{1}{2}$ and $\sin 60° = \frac{\sqrt{3}}{2}$,

$$\sqrt{3} (\cos 60° + i \sin 60°) = \sqrt{3} \left( \frac{1}{2} + \frac{\sqrt{3}}{2} i \right)$$

$$= \frac{\sqrt{3}}{2} + \frac{3}{2} i.$$

*WRITTEN EXERCISES*

Plot each set of complex numbers in the complex plane.

**A**

1. $\{4 + 3i, -2 + 3i, -3i, 6 - 2i\}$
2. $\{5, -3 + i, -2 - 2i, 1 - 3i\}$
3. $\{1 + 4i, -1 + i, -4 - i, 2 - i\}$
4. $\{4, 3i, -2 - 5i, 5 - 3i\}$
5. $\{2 + i, 1 + 2i, -2 - i, 1 - 2i\}$
6. $\{3, 3i, -3, -3i\}$

Express each complex number in trigonometric form. When necessary, leave the modulus in simple radical form and express $\theta$ to the nearest 10 minutes. (Remember: $0° \leq \theta < 360°$)

7. $3 + 3i$     9. $-1 + \sqrt{3}i$    11. $4 + 3i$      13. $-5 - 12i$

8. $\sqrt{3} - i$     10. $-1 - i$      12. $6 - 3i$      14. $-7 + 15i$

Express each complex number in the form $a + bi$. Leave $a$ and $b$ in simple radical form when necessary.

15. $\sqrt{3}(\cos 30° + i \sin 30°)$      18. $3(\cos 120° + i \sin 120°)$

16. $\sqrt{2}(\cos 45° + i \sin 45°)$      19. $2\sqrt{2}(\cos 225° + i \sin 225°)$

17. $2(\cos 90° + i \sin 90°)$      20. $\sqrt{3}(\cos 330° + i \sin 330°)$

In each of the following, draw in the same complex plane (a) the vectors for each of the given complex numbers and (b) the vector for the sum of these numbers. (Notice that the vector in (b) is the sum of the vectors in (a).)

**Sample.**    $1 + 3i, 2 - i$

*Solution:*    **a.** The vectors are shown as solid arrows.

         **b.** $(1 + 3i) + (2 - i) = 3 + 2i$.

         The vector is shown as a dotted arrow.

21. $2i, 3 - i$        23. $3 + 2i, 3 - 2i$      25. $3, 2i$

22. $-2 - 2i, 3$      24. $-2 + i, 2 + i$      26. $2 - 3i, 3 - 2i$

---

## Chapter Summary

---

1. If $(a, b)$ are the coordinates of a point other than the origin on the terminal side of a position angle $\theta$, then $\cos \boldsymbol{\theta} = \dfrac{a}{\sqrt{a^2 + b^2}}$ and $\sin \boldsymbol{\theta} = \dfrac{b}{\sqrt{a^2 + b^2}}$. Certain combinations of the sine and cosine of an angle $\theta$ produce values of the other trigonometric functions: $\tan \boldsymbol{\theta} = \dfrac{\sin \theta}{\cos \theta}$, and $\cot \boldsymbol{\theta}$, $\sec \boldsymbol{\theta}$, $\csc \boldsymbol{\theta}$ are the reciprocals of $\tan \theta$, $\cos \theta$, $\sin \theta$, respectively. The exact values of certain **special angles** may be found.

2. From the table on page 454 you can find a value for a trigonometric function of any angle $\theta$ in terms of the value of a trigonometric function of the **reference angle** of $\theta$.

3. You can solve many practical problems by applying trigonometric function values or their logarithms. The **resultant** of vectors, parallel or perpendicular to each other, and the horizontal and vertical **components** of a given vector can be found similarly.

4. Complex numbers can be represented by vectors in the complex plane. The **trigonometric form** of $a + bi$ (with $a^2 + b^2 \neq 0$) is given by

$$a + bi = r(\cos \theta + i \sin \theta),$$

where $r = \sqrt{a^2 + b^2}$, $\cos \theta = \dfrac{a}{\sqrt{a^2 + b^2}}$, and $\sin \theta = \dfrac{b}{\sqrt{a^2 + b^2}}$.

### Vocabulary and Spelling

Review each term by reference to the page listed.

ray (*p. 427*)

vertex (*p. 427*)

(directed) angle (*p. 428*)

positive angle (*p. 428*)

negative angle (*p. 428*)

initial side (*p. 428*)

terminal side (*p. 428*)

degree, minute,
   second (*p. 428*)

coterminal angles (*p. 429*)

standard position (*p. 429*)

position angle (*p. 429*)

unit circle (*p. 430*)

sine (sin $\theta$) (*p. 431*)

cosine (cos $\theta$) (*p. 431*)

tangent (tan $\theta$) (*p. 435*)

cotangent (cot $\theta$) (*p. 435*)

secant (sec $\theta$) (*p. 435*)

cosecant (csc $\theta$) (*p. 435*)

trigonometric functions (*p. 435*)

first (second, third, fourth) quadrant
   angle (*p. 436*)

quadrantal angle (*p. 436*)

special angles (*p. 439*)

cofunction (*p. 441*)
angle of elevation (*p. 445*)
angle of depression (*p. 445*)
reference angle (*p. 453*)
vector quantity (*p. 456*)
vector (*p. 456*)
heading (bearing, course)
  (*p. 457*)
equivalent vectors (*p. 458*)

zero vector (*p. 458*)
resultant (*p. 458*)
component vectors (*p. 461*)
resolving a vector (*p. 461*)
complex plane (*p. 465*)
axis of reals (*p. 465*)
axis of imaginaries (*p. 465*)
amplitude of a complex number (*p. 466*)
modulus of a complex number (*p. 466*)

## *Chapter Test and Review*

**11–1**  **1.** Sketch an angle in standard position with terminal side containing $P(-3, 5)$. Show both a positive and a negative angle satisfying these conditions.

**11–2**  **2.** If the terminal side of the position angle $\theta$ contains $P(8, -15)$, find $\sin \theta$ and $\cos \theta$.

**11–3**  **3.** Draw the smallest positive position angle $\theta$ whose terminal side contains the point $P(-5, 3)$, and find the values of all trigonometric functions of $\theta$.

**11–4**  **4.** Evaluate $\tan 60° - 2 \cot 30° + 2 \tan 45°$.

**5.** Solve for $x$: $\sin x = \cos 70°$    $(0° < x < 90°)$

**11–5**  **6.** Using Table 6, find $\sin 17°44'$.

**11–6**  **7.** Use logarithms to find the measure of $\theta$ to the nearest minute if

$$\cos \theta = \frac{2.17 \sin 23°10'}{\tan 81°10'}.$$

**11–7**  **8.** Express $\sec 318°$ as a function of a positive acute angle.

**11–8**  **9.** Find the magnitude and direction of the resultant of a displacement of 16 meters south and 10 meters west.

**11–9**  **10.** In opening a vertical window from the top, a man pulls on a window pole with a force of 50 pounds. To the nearest pound, what part of the force lowers the window if the pole makes an angle of 25° with the window?

**11–10**  **11.** Write $-3 + 3i$ in trigonometric form.

**12.** Write $2(\cos 135° + i \sin 135°)$ in the form $a + bi$.

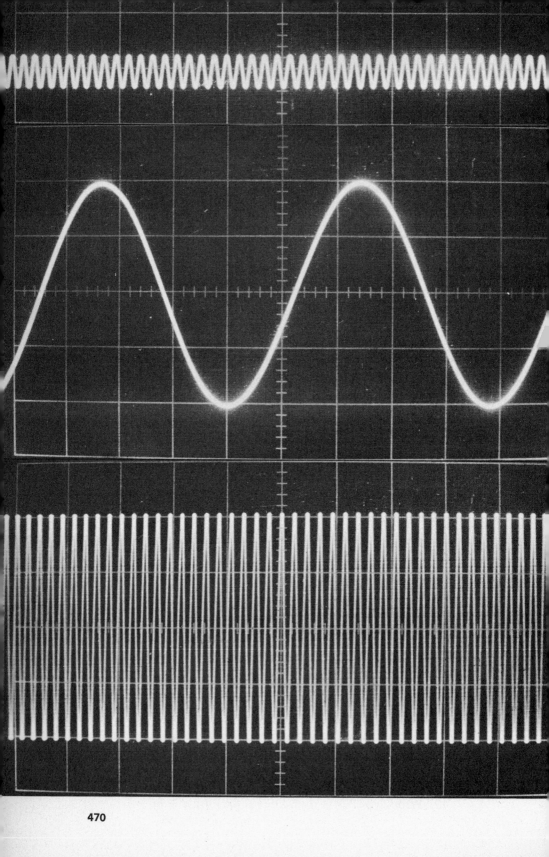

# 12

# Trigonometric Identities and Formulas

## IDENTITIES INVOLVING ONE ANGLE

### 12–1 The Fundamental Identities

Since the statement $(2x + 1)(2x - 1) = 4x^2 - 1$ is true for every replacement of $x$ by a real number, it is true in particular when $x$ is replaced by the real number $\cos 21°$; that is,

$$(2\cos 21° + 1)(2\cos 21° - 1) = 4(\cos 21°)^2 - 1.$$

Indeed, for every angle $\theta$ you may write the true statement:

$$(2\cos\theta + 1)(2\cos\theta - 1) = 4\cos^2\theta - 1,$$

where "$\cos^2\theta$" means "$(\cos\theta)^2$" and is read, "the square of $\cos\theta$," or "cosine squared $\theta$."

An equation which involves at least one trigonometric function is called a trigonometric equation. A trigonometric equation, such as $(2\cos\theta + 1)(2\cos\theta - 1) = 4\cos^2\theta - 1$, that is true for all values of the variables for which both of its members are defined, is called a trigonometric identity.

Trigonometric identities depend on the definitions of the trigonometric functions, as well as on the algebra of real numbers. Can you explain why each of the following statements is true for every angle $\theta$ for which the functions involved are defined?

**1.** $\tan \theta = \dfrac{\sin \theta}{\cos \theta}$  **3.** $\sec \theta = \dfrac{1}{\cos \theta}$  **5.** $\cot \theta = \dfrac{1}{\tan \theta}$

**2.** $\cot \theta = \dfrac{\cos \theta}{\sin \theta}$  **4.** $\csc \theta = \dfrac{1}{\sin \theta}$

Identities 1–4 follow directly from the definitions of the trigonometric functions (page 435). To see that Identity 5 follows from Identities 1 and 2, notice that if $\sin \theta \neq 0$ and $\cos \theta \neq 0$, you have:

$$\frac{1}{\tan \theta} = \frac{1}{\dfrac{\sin \theta}{\cos \theta}} \quad \text{(using Identity 1)}$$

$$= \frac{\cos \theta}{\sin \theta} \quad \begin{array}{l}\text{(using the properties}\\ \text{of real numbers)}\end{array}$$

$$= \cot \theta \quad \text{(using Identity 2)}$$

$$\therefore \cot \theta = \frac{1}{\tan \theta}$$

By recalling that for every angle $\theta$ in standard position, $(\cos \theta, \sin \theta)$ are the coordinates of a point on the unit circle $x^2 + y^2 = 1$ (Section 11–2), you can discover the identity:

$$(\cos \theta)^2 + (\sin \theta)^2 = 1, \text{ or}$$

**6.** $\sin^2 \theta + \cos^2 \theta = 1.$

If you divide each member of Identity 6 by $\cos^2 \theta$, you can derive another identity,

$$\frac{\sin^2 \theta}{\cos^2 \theta} + 1 = \frac{1}{\cos^2 \theta}, \text{ or } 1 + \left(\frac{\sin \theta}{\cos \theta}\right)^2 = \left(\frac{1}{\cos \theta}\right)^2, \qquad \cos \theta \neq 0.$$

Using Identities 1 and 3, you then find that

**7.** $1 + \tan^2 \theta = \sec^2 \theta.$

Can you suggest how to derive the following identity?

---

**8.** $1 + \cot^2 \theta = \csc^2 \theta.$

---

Identities 1–8 are called the **fundamental trigonometric identities.** Using them and the properties of real numbers, you can write any expression involving values of the trigonometric functions of an angle $\theta$ in terms of the value of $\sin \theta$ or any other trigonometric function of $\theta$, as illustrated in the following examples.

**Example 1.** Express $\cos \theta$ in terms of $\sin \theta$.

*Solution:*

$$\sin^2 \theta + \cos^2 \theta = 1$$
$$\cos^2 \theta = 1 - \sin^2 \theta$$
$$\therefore \cos \theta = \sqrt{1 - \sin^2 \theta}, \quad \theta \text{ in Quadrant I or IV}$$
$$\text{or } \cos \theta = -\sqrt{1 - \sin^2 \theta}, \quad \theta \text{ in Quadrant II or III}$$

The result of Example 1 suggests the following method for finding $\cos \theta$ given that $\theta$ is a second-quadrant angle with $\sin \theta = \frac{3}{5}$.

$$\cos \theta = -\sqrt{1 - \sin^2 \theta} = -\sqrt{1 - (\tfrac{3}{5})^2} = -\sqrt{\tfrac{16}{25}} = -\tfrac{4}{5}$$

**Example 2.** Express in terms of $\sin \theta$,

$$\tan^2 \theta \, (\csc^2 \theta - 1) + \tan \theta \cos \theta.$$

*Solution:* The expression denotes a real number provided $\sin \theta \neq 0$ and $\cos \theta \neq 0$. Since Identities 1 and 4 introduce no additional restrictions, you can use them to write

$$\tan^2 \theta \, (\csc^2 \theta - 1) + \tan \theta \cos \theta = \frac{\sin^2 \theta}{\cos^2 \theta} \left( \frac{1}{\sin^2 \theta} - 1 \right) + \frac{\sin \theta}{\cos \theta} \cos \theta$$

$$= \frac{1}{\cos^2 \theta} - \frac{\sin^2 \theta}{\cos^2 \theta} + \sin \theta$$

$$= \frac{1 - \sin^2 \theta}{\cos^2 \theta} + \sin \theta$$

$$= \frac{\cos^2 \theta}{\cos^2 \theta} + \sin \theta$$

$$= 1 + \sin \theta.$$

$$\therefore \tan^2 \theta \, (\csc^2 \theta - 1) + \tan \theta \cos \theta = 1 + \sin \theta, \text{ if}$$
$$\sin \theta \neq 0, \cos \theta \neq 0.$$

*ORAL EXERCISES*

Express each of the following in terms of a single trigonometric function.

**1.** $\dfrac{\cos \theta}{\sin \theta}$
**3.** $\dfrac{\sin^2 \theta}{\cos^2 \theta}$
**5.** $\sec \theta \sin \theta \cos \theta$

**2.** $1 - \sin^2 \theta$
**4.** $\tan \theta \cos \theta$
**6.** $\csc \theta \cot \theta \cos \theta$

**7.** $1 - \cos^2 \theta + \sin^2 \theta$
**10.** $\cos^2 \theta + \sin^2 \theta + \cot^2 \theta$

**8.** $1 - \sin^2 \theta + \cos^2 \theta$
**11.** $\cos \theta \sec \theta - \sin^2 \theta$

**9.** $\sin^2 \theta + \cos^2 \theta + \tan^2 \theta$
**12.** $\sin \theta \csc \theta - \cos^2 \theta$

*WRITTEN EXERCISES*

Write the first expression in terms of the given function.

**Sample.** $\sec \alpha \sin \alpha;\ \cot \alpha$

*Solution:* $\sec \alpha \sin \alpha = \dfrac{1}{\cos \alpha} \sin \alpha = \dfrac{\sin \alpha}{\cos \alpha} = \tan \alpha = \dfrac{1}{\cot \alpha}.$

[A] **1.** $\sec A \sin A;\ \tan A$
**9.** $\sec \alpha;\ \tan \alpha$

**2.** $\csc B \cos B;\ \cot B$
**10.** $\csc \alpha;\ \tan \alpha$

**3.** $\dfrac{\sin \alpha}{\tan \alpha};\ \cos \alpha$
**11.** $\cos^2 \beta;\ \csc \beta$

**4.** $\dfrac{\cos \theta}{\cot \theta};\ \sin \theta$
**12.** $\sec^2 C;\ \csc C$

**5.** $\cot^2 \alpha \sin^2 \alpha;\ \sin \alpha$
**13.** $\cot^2 \theta(\sec^2 \theta - 1) + \cot \theta \sin \theta;\ \cos \theta$

**6.** $\tan^2 \beta \csc^2 \beta;\ \cos \beta$
**14.** $\cos \theta \sec \theta + \dfrac{\cos \theta}{\sin \theta \tan \theta};\ \csc \theta$

**7.** $\tan \theta;\ \cos \theta$
**15.** $\dfrac{1 + \tan^2 \alpha}{\csc^2 \alpha};\ \cos \alpha$

**8.** $\csc \theta;\ \cos \theta$
**16.** $\dfrac{1 - \tan^2 \beta}{\tan \beta};\ \cos \beta$

Express in terms of sine and cosine functions only, and simplify.

**Sample.** $\dfrac{\cos \theta}{\sec \theta - \tan \theta}$

*Solution:* $\dfrac{\cos \theta}{\sec \theta - \tan \theta} = \dfrac{\cos \theta}{\dfrac{1}{\cos \theta} - \dfrac{\sin \theta}{\cos \theta}}$

$$= \frac{\cos^2 \theta}{1 - \sin \theta} = \frac{1 - \sin^2 \theta}{1 - \sin \theta}$$

$$= \frac{(1 - \sin \theta)(1 + \sin \theta)}{1 - \sin \theta}$$

$$= 1 + \sin \theta \quad (\sec \theta \neq \tan \theta)$$

**B** **17.** $\left(\dfrac{\cos \alpha - \sec \alpha}{\sec \alpha} + \cos^2 \alpha \tan^2 \alpha\right)\left(\dfrac{\tan \alpha - \sin \alpha}{\tan \alpha}\right)$

**18.** $(\tan u + \sin u)(1 - \cos u) + \dfrac{\cos u}{\csc u}$

**19.** $\dfrac{\sqrt{\cot^2 \beta + 1}}{\csc \beta}\left(\dfrac{\cot^2 \beta \sec^2 \beta - 1}{\csc \beta \cot^2 \beta \sin \beta}\right)$

**20.** $\sin \alpha \sec \alpha \left(\cos \alpha + \dfrac{\csc \alpha}{\sec^2 \alpha}\right) + (\csc \alpha + \sec \alpha)$

Find the value of the following expressions if $\sin \theta = \frac{13}{85}$ and $\cos \theta = \frac{84}{85}$.

**21.** $\cos \theta \left(\dfrac{\sin \theta \sec \theta + \tan \theta}{\sec \theta \tan \theta}\right)$

**22.** $\dfrac{\cot \theta + \cos \theta}{\sec \theta + \tan \theta}$

**23.** $\sin \theta + \dfrac{\sqrt{1 + \cos \theta}}{\sqrt{\sec \theta - 1}} \cdot \dfrac{\sqrt{1 - \cos \theta}}{\sqrt{\sec \theta + 1}}$

**24.** $(\cos \theta - \sin \theta)\left(\dfrac{\sqrt{\csc \theta + 1}}{\sqrt{1 - \sin \theta}} \cdot \dfrac{\sqrt{\csc \theta - 1}}{\sqrt{\sin \theta + 1}}\right)$

## 12–2 Proving Identities

You can sometimes verify that a trigonometric equation is an identity by using the properties of real numbers and the fundamental identities to transform the more complicated member of the equation to the form of the simpler member.

**Example 1.** Prove that $\csc \alpha = \dfrac{\sec \alpha + \csc \alpha}{1 + \tan \alpha}$ is an identity.

*Proof:* Notice that the given equation is meaningful only if $\sin \alpha \neq 0$, $\cos \alpha \neq 0$, and $\tan \alpha \neq -1$. (Why?)

**1.** Choose the right-hand member to be transformed.

$$\dfrac{\sec \alpha + \csc \alpha}{1 + \tan \alpha}$$

**2.** Express all function values in terms of $\sin \alpha$ and $\cos \alpha$.

$$= \dfrac{\dfrac{1}{\cos \alpha} + \dfrac{1}{\sin \alpha}}{1 + \dfrac{\sin \alpha}{\cos \alpha}}$$

**3.** Simplify the complex fraction by multiplying numerator and denominator by $\cos \alpha \sin \alpha$.
($\cos \alpha \neq 0$, $\sin \alpha \neq 0$)

$$= \left( \dfrac{\dfrac{1}{\cos \alpha} + \dfrac{1}{\sin \alpha}}{1 + \dfrac{\sin \alpha}{\cos \alpha}} \right) \cdot \dfrac{\cos \alpha \sin \alpha}{\cos \alpha \sin \alpha}$$

$$= \dfrac{\sin \alpha + \cos \alpha}{\cos \alpha \sin \alpha + \sin^2 \alpha}$$

**4.** Factor $\sin \alpha$ from terms in the denominator.

$$= \dfrac{\sin \alpha + \cos \alpha}{\sin \alpha(\cos \alpha + \sin \alpha)}$$

**5.** Divide numerator and denominator by $\sin \alpha + \cos \alpha$.
($\sin \alpha + \cos \alpha \neq 0$ because $1 + \tan \alpha \neq 0$)

$$= \dfrac{1}{\sin \alpha}$$

**6.** Replace $\dfrac{1}{\sin \alpha}$ with $\csc \alpha$.

$$= \csc \alpha$$

**Example 2.** Verify: $\dfrac{\cos \beta}{1 - \sin \beta} = \dfrac{1 + \sin \beta}{\cos \beta}$.

*Proof:*

**1.** Choose either member, say the left-hand one, to be transformed.

$$\dfrac{\cos \beta}{1 - \sin \beta}$$

**2.** Multiply numerator and denominator by $1 + \sin \beta$, which is the numerator of the right-hand member (valid if $\sin \beta \neq -1$). Simplify.

$$= \dfrac{\cos \beta(1 + \sin \beta)}{(1 - \sin \beta)(1 + \sin \beta)}$$

$$= \dfrac{\cos \beta(1 + \sin \beta)}{1 - \sin^2 \beta}$$

**3.** Substitute $\cos^2 \beta$ for $1 - \sin^2 \beta$.

$$= \dfrac{\cos \beta(1 + \sin \beta)}{\cos^2 \beta}$$

**4.** Divide numerator and denominator by $\cos \beta$ (valid if $\cos \beta \neq 0$).

$$= \dfrac{1 + \sin \beta}{\cos \beta}$$

Are the steps in Example 2 consistent with the conditions implied by the given equation? The members of that equation are both defined if and only if $\sin \beta \neq 1$ and $\cos \beta \neq 0$. But $\cos \beta \neq 0$ implies that $\beta$ is not coterminal with either 90° or 270° and, therefore, that $\sin \beta \neq 1$ or $-1$. Thus, the transformations introduced no additional restrictions.

Often it is convenient to transform each member of a proposed identity to the same expression in terms of sine and cosine or other functions. Checking to see that the steps in the following example are consistent with the restrictions in the equation is left to you.

**Example 3.** Verify: $(1 + \csc \theta)(1 - \sin \theta) = \cot \theta \cos \theta$.

*Proof:*

$$(1 + \csc \theta)(1 - \sin \theta) \qquad\qquad \cot \theta \cos \theta$$

$$= 1 + \csc \theta - \sin \theta - \csc \theta \sin \theta \qquad = \frac{\cos \theta}{\sin \theta} \cdot \cos \theta$$

$$= 1 + \csc \theta - \sin \theta - 1 \qquad = \frac{\cos^2 \theta}{\sin \theta}$$

$$= \csc \theta - \sin \theta \qquad = \frac{1 - \sin^2 \theta}{\sin \theta}$$

$$= \frac{1}{\sin \theta} - \sin \theta \qquad = \frac{1}{\sin \theta} - \sin \theta$$

*WRITTEN EXERCISES*

Prove each identity.

**A**

**1.** $\sin \beta \sec \beta = \tan \beta$

**2.** $\sin \alpha + \cos \alpha \cot \alpha = \csc \alpha$

**3.** $\cos \theta \csc \theta = \cot \theta$

**4.** $\sec^2 A(1 - \sin^2 A) = 1$

**5.** $\dfrac{\sin^2 \theta + \cos^2 \theta}{\cos \theta} = \sec \theta$

**6.** $\dfrac{1 + \sin \theta}{\sin \theta} = 1 + \csc \theta$

**7.** $\cos^2 \alpha(1 + \tan^2 \alpha) = 1$

**8.** $1 - \sin \theta \cos \theta \tan \theta = \cos^2 \theta$

**9.** $\tan A(\sin A + \cot A \cos A) = \sec A$

**10.** $2 \cos^2 B - \sin^2 B + 1 = 3 \cos^2 B$

**11.** $1 - 2 \sin^2 A = 2 \cos^2 A - 1$

**12.** $\sin B \tan B + \cos B = \sec B$

**13.** $\sin \theta(\sec \theta - \csc \theta) = \tan \theta - 1$

**14.** $\csc \alpha \, (\csc \alpha + \cot \alpha) = \dfrac{1}{1 - \cos \alpha}$

**15.** $\dfrac{\sin \beta \cot \beta + \cos \beta}{\sin \beta} = 2 \cot \beta$

**16.** $\dfrac{1 + 2 \sin \alpha \cos \alpha}{\sin \alpha + \cos \alpha} = \sin \alpha + \cos \alpha$

**17.** $\sec \beta + \tan \beta = \dfrac{\cos \beta}{1 - \sin \beta}$

**18.** $\cos \alpha \csc \alpha \, (1 + \sec \alpha) = \dfrac{\sin \alpha \sec \alpha}{\sec \alpha - 1}$

**19.** $\dfrac{\sin \theta \cos \theta}{1 - 2 \sin^2 \theta} = \dfrac{1}{\cot \theta - \tan \theta}$

**20.** $(1 - \tan \theta)(1 + \cot \theta) = (\cot \theta - 1)(\tan \theta + 1)$

**21.** $\dfrac{\sec \theta}{1 - \cos \theta} = \dfrac{\sec \theta + 1}{\sin^2 \theta}$

**22.** $\dfrac{\sin^2 \alpha}{\tan \alpha + \sin \alpha} = \dfrac{1 - \cos \alpha}{\tan \alpha}$

**23.** $\dfrac{\cos \beta - 1}{\sec \beta - 1} = -\dfrac{1 + \cos \beta}{1 + \sec \beta}$

**24.** $(1 + \tan \alpha + \sec \alpha)^2 = 2(1 + \sec \alpha)(\tan \alpha + \sec \alpha)$

**B** **25.** $\dfrac{\sin \theta + \cos \theta}{\sec \theta + \tan \theta} + \dfrac{\cos \theta - \sin \theta}{\sec \theta - \tan \theta} = 2 - 2 \sin^2 \theta \sec \theta$

**26.** $\dfrac{\sin \alpha \cos \alpha}{1 + \cos \alpha} - \dfrac{\sin \alpha}{1 - \cos \alpha} + (\cot \alpha \cos \alpha + \csc \alpha) = 0$

**27.** $\dfrac{\csc C + \cot C}{\csc C - \cot C} = \csc^2 C(1 + 2 \cos C + \cos^2 C)$

**28.** $\dfrac{\sin C + \cos C - 1}{\sin C - \cos C + 1} = \dfrac{\cos C}{\sin C + 1}$

**29.** $\dfrac{\sin^3 B + \cos^3 B}{\sin^2 B + 2 \sin B \cos B + \cos^2 B} = \dfrac{1}{\sin B + \cos B} - \dfrac{\cos B}{1 + \cot B}$

**30.** $\dfrac{\cos B - \sin B}{\cos^3 B - \sin^3 B} = \dfrac{1}{\tan B \cos^2 B + 1}$

**C** **31.** $\dfrac{1 - \cos x}{\sin x} = \pm \sqrt{\dfrac{\csc x - \cot x}{\csc x + \cot x}}$

**32.** $\tan x(\csc x + 1) = \pm \sqrt{\dfrac{\sin x + 1}{1 - \sin x}}$

**33.** $\dfrac{\sin^2 \theta + 2 \cos \theta - 1}{2 + \cos \theta - \cos^2 \theta} = \dfrac{1}{1 + \sec \theta}$

**34.** $\dfrac{2 \tan \alpha \sec \alpha + \sec \alpha}{3 + \tan \alpha - 2 \sec^2 \alpha} = \dfrac{1}{\cos \alpha - \sin \alpha}$

## IDENTITIES INVOLVING TWO ANGLES

### 12-3 A Distance Formula

Suppose $P$ and $Q$ are the points shown in Figure 12-1. To find the distance $PQ$, you take the steps described below:

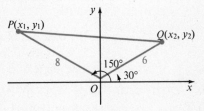

**FIGURE 12-1**

1. Use the theorem on page 432 to determine the rectangular coordinates $(x_1, y_1)$ of $P$ and $(x_2, y_2)$ of $Q$.

$P$: $\quad x_1 = 8 \cos 150° = 8(-\tfrac{1}{2}\sqrt{3}) = -4\sqrt{3}$;
$\quad\ \ y_1 = 8 \sin 150° = 8(\tfrac{1}{2}) = 4$.

$Q$: $\quad x_2 = 6 \cos 30° = 6(\tfrac{1}{2}\sqrt{3}) = 3\sqrt{3}$;
$\quad\ \ y_2 = 6 \sin 30° = 6(\tfrac{1}{2}) = 3$.

2. Use the distance formula:

$$(PQ)^2 = (x_1 - x_2)^2 + (y_1 - y_2)^2$$
$$(PQ)^2 = (-4\sqrt{3} - 3\sqrt{3})^2 + (4 - 3)^2 = 49(3) + 1 = 148$$
$$\therefore PQ = \sqrt{148} = 2\sqrt{37}$$

By following the steps above, you can derive a formula for the square of the distance between any points $P$ and $Q$ in terms of their respective position angles, $\alpha$ and $\beta$, and their distances from the origin, $p$ and $q$ (Figure 12-2).

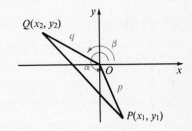

**FIGURE 12-2**

1. $P$: $\quad x_1 = p \cos\alpha; \ y_1 = p \sin\alpha$
$\ \ Q$: $\quad x_2 = q \cos\beta; \ y_2 = q \sin\beta$

2. $(PQ)^2 = (x_1 - x_2)^2 + (y_1 - y_2)^2$
$\quad = (p \cos\alpha - q \cos\beta)^2 + (p \sin\alpha - q \sin\beta)^2$
$\quad = p^2 \cos^2\alpha - 2pq \cos\alpha \cos\beta + q^2 \cos^2\beta$
$\qquad\qquad + p^2 \sin^2\alpha - 2pq \sin\alpha \sin\beta + q^2 \sin^2\beta$
$\quad = p^2(\cos^2\alpha + \sin^2\alpha) + q^2(\cos^2\beta + \sin^2\beta)$
$\qquad\qquad\qquad - 2pq(\cos\alpha \cos\beta + \sin\alpha \sin\beta)$
$\quad = p^2 \cdot 1 + q^2 \cdot 1 - 2pq(\cos\alpha \cos\beta + \sin\alpha \sin\beta)$

$\therefore (PQ)^2 = p^2 + q^2 - 2pq(\cos\alpha \cos\beta + \sin\alpha \sin\beta)$

**Example.** Approximate $(PQ)^2$ if $P$ is the point $(-10, 0)$ and $Q$ is 5 units from the origin and has position angle 305°.

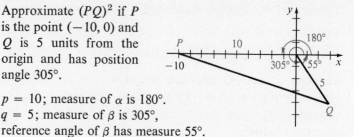

*Solution:* $p = 10$; measure of $\alpha$ is 180°.
$q = 5$; measure of $\beta$ is 305°, reference angle of $\beta$ has measure 55°.

$(PQ)^2 = 10^2 + 5^2 - 2 \cdot 10 \cdot 5(\cos 180° \cos 305° + \sin 180° \sin 305°)$
$= 100 + 25 - 100(-1 \cdot \cos 55° + 0 \cdot -\sin 55°)$
$= 125 - 100(-0.5736 + 0) = 125 + 57.36$
$\therefore (PQ)^2 \doteq 182.4$

## WRITTEN EXERCISES

Find to the nearest tenth the square of the distance between the points.

A  **1.** $P$, 10 units from the origin with position angle 50°, and $Q$, 15 units from the origin with position angle 0°.

**2.** $P$, 6 units from the origin with position angle 74°, and $Q$, 8 units from the origin with position angle 312°.

**3.** $R(0, 5)$ and $S$, 4 units from the origin with position angle 372°.

**4.** $R(-4, 0)$ and $S$, 6 units from the origin with position angle 470°.

**5.** $X$, 8 units from the origin with position angle 234°, and $Y$, 12 units from the origin with position angle $-82°$.

**6.** $X$, 10 units from the origin with position angle 327°, and $Y$, 15 units from the origin with position angle $-208°$.

**7.** $S$, with $x$-coordinate 3 and position angle 435°, and $T$, 15 units from the origin with position angle $-265°$.

**8.** $S$, $\frac{9}{2}$ units from the origin with position angle 9°, and $T$, with $y$-coordinate $-\frac{15}{2}$ and position angle $-23°$.

**9.** $C(5, 7)$ and $D$, 2 units from the origin with position angle 261°.

**10.** $C$, 21 units from the origin with position angle 912°, and $D(10, -1)$.

**11.** $G$, with $x$-coordinate $-5$ and position angle 141°, and $H$, with $y$-coordinate $-12$ and position angle $-112°$.

**12.** $G$, with $y$-coordinate $-\frac{5}{2}$ and position angle 680°, and $H$, with $x$-coordinate $-\frac{1}{2}$ and position angle 100°.

## 12–4   The Cosine of the Difference of Two Angles

In Figure 12–3, the angle $\alpha - \beta$ has the terminal side of $\beta$ as its initial side and the terminal side of $\alpha$ as its terminal side. Its measure is the difference between 150°, the measure of $\alpha$, and 60°, the measure of $\beta$: $150° - 60° = 90°$.

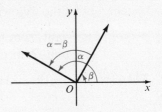

Does $\cos(\alpha - \beta) = \cos\alpha - \cos\beta$?

FIGURE 12–3

$$\cos(\alpha - \beta) = \cos 90° = 0$$

$$\cos\alpha = \cos 150° = -\frac{\sqrt{3}}{2}, \cos\beta = \cos 60° = \tfrac{1}{2}$$

$$\cos\alpha - \cos\beta = -\frac{\sqrt{3}}{2} - \frac{1}{2} = -\frac{\sqrt{3}+1}{2}$$

You see that, in general, $\cos(\alpha - \beta)$ is *not* equal to $\cos\alpha - \cos\beta$.

To discover the relationships among the trigonometric functions of $\alpha$, $\beta$, and $\alpha - \beta$, let $\alpha$ and $\beta$ be any angles in standard position, and let point $P$ (on the terminal side of $\alpha$) and $Q$ (on the terminal side of $\beta$) be 1 unit from the origin (Figure 12–4).

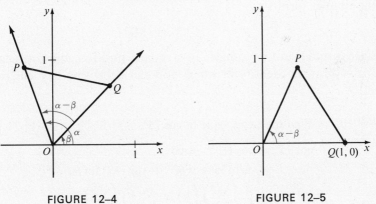

FIGURE 12–4

FIGURE 12–5

Using the distance formula of Section 12–3 with 1 in place of $p$ and $q$,

$$(PQ)^2 = 1^2 + 1^2 - 2(1)(1)(\cos\alpha \cos\beta + \sin\alpha \sin\beta)$$
$$\therefore (PQ)^2 = 2 - 2(\cos\alpha \cos\beta + \sin\alpha \sin\beta) \quad (*)$$

Now choose a new coordinate system in which $Q$ is the point $(1, 0)$; ray $OQ$, the terminal side of $\beta$, is the positive $x$-axis; and the angle $\alpha - \beta$ is in standard position (Figure 12–5, drawn to show the new *x*-axis as a horizontal line). To compute $(PQ)^2$ in this coordinate system, notice the following:

$Q$ is $1$ unit from the origin with a position angle of $0°$;
$P$ is $1$ unit from the origin with $\alpha - \beta$ as position angle.

Therefore, using the distance formula again,

$$
\begin{aligned}
(PQ)^2 &= 1^2 + 1^2 - 2(1)(1)[\cos{(\alpha - \beta)}\cos{0°} + \sin{(\alpha - \beta)}\sin{0°}] \\
&= 1 + 1 - 2[1 \cdot \cos{(\alpha - \beta)} + 0 \cdot \sin{(\alpha - \beta)}] \\
\therefore (PQ)^2 &= 2 - 2\cos{(\alpha - \beta)} \quad (*)
\end{aligned}
$$

Equate the expressions for $(PQ)^2$ in the starred (*) equations:

$$
2 - 2\cos{(\alpha - \beta)} = 2 - 2(\cos\alpha\cos\beta + \sin\alpha\sin\beta).
$$

This leads to the *formula for the cosine of the difference of two angles:*

$$
\cos{(\alpha - \beta)} = \cos\alpha\cos\beta + \sin\alpha\sin\beta.
$$

This result now enables you to write the distance formula on page 479 in the simpler form

$$
(PQ)^2 = p^2 + q^2 - 2pq\cos{(\alpha - \beta)}.
$$

Since $\alpha$ and $\beta$ are any angles in standard position, the formula for $\cos{(\alpha - \beta)}$ is an identity in which you can substitute any angles for $\alpha$ and $\beta$.

**Example.** Find the exact value of $\cos 75°$.

*Solution:*

$$
\begin{aligned}
\cos 75° &= \cos{(135° - 60°)} \\
&= \cos 135° \cos 60° + \sin 135° \sin 60° \\
&= \left(-\frac{\sqrt{2}}{2}\right)\left(\frac{1}{2}\right) + \left(\frac{\sqrt{2}}{2}\right)\left(\frac{\sqrt{3}}{2}\right) \\
\cos 75° &= -\frac{\sqrt{2} + \sqrt{6}}{4}
\end{aligned}
$$

Since the formula for $\cos{(\alpha - \beta)}$ is valid for any angle $\alpha$, it is true in particular when the measure of $\alpha$ is $90°$. Thus,

$$
\begin{aligned}
\cos{(90° - \beta)} &= \cos 90° \cos\beta + \sin 90° \sin\beta \\
&= 0 \cdot \cos\beta + 1 \cdot \sin\beta = 0 + \sin\beta \\
\cos{(90° - \beta)} &= \sin\beta.
\end{aligned}
$$

Because $\cos(90° - \beta) = \sin\beta$ is an identity in $\beta$, it remains valid when you replace $\beta$ by $90° - \beta$. In this way you obtain

$$\cos[90° - (90° - \beta)] = \sin(90° - \beta)$$
$$\cos\beta = \sin(90° - \beta), \text{ or}$$
$$\sin(90° - \beta) = \cos\beta.$$

Furthermore, since $\tan(90° - \beta) = \dfrac{\sin(90° - \beta)}{\cos(90° - \beta)} = \dfrac{\cos\beta}{\sin\beta} = \cot\beta,$

$$\tan(90° - \beta) = \cot\beta,$$

provided the function values are defined. The proofs of the following identities are left as Exercises 13, 14, 15, below:

$$\cot(90° - \beta) = \tan\beta, \ \sec(90° - \beta) = \csc\beta, \ \csc(90° - \beta) = \sec\beta$$

Compare these results, showing that *cofunctions of complementary angles are equal*, with the theorem stated on page 441.

## WRITTEN EXERCISES

Express each of the following in the form $\cos\theta$ for suitable $\theta$ and then evaluate $\cos\theta$ to four decimal places.

A

**1.** $\cos 275° \cos 175° + \sin 275° \sin 175°$

**2.** $\cos 430° \cos 220° + \sin 430° \sin 220°$

**3.** $\dfrac{\sqrt{3}}{2} \cos 220° - \tfrac{1}{2} \sin 220°$   **5.** $\dfrac{\sqrt{2}}{2} (\cos 10° + \sin 10°)$

**4.** $0.9135 \cos 12° + 0.4067 \sin 12°$   **6.** $0.5736 \cos 80° - 0.8192 \sin 80°$

Express as the cosine of the difference of two angles and evaluate.

**7.** $\cos 15°$   **8.** $\cos 105°$   **9.** $\cos 195°$   **10.** $\cos 255°$

Verify as identities.

**11.** $\cos(45° - \beta) = \dfrac{\sqrt{2}}{2} (\cos\beta + \sin\beta)$

**12.** $\cos(150° - \beta) = -\tfrac{1}{2}(\sqrt{3} \cos\beta - \sin\beta)$

Prove that the following are identities.

**13.** $\cot(90° - \beta) = \tan\beta$   **15.** $\csc(90° - \beta) = \sec\beta$

**14.** $\csc\beta = \sec(90° - \beta)$   **16.** $\cos(180° - \beta) = -\cos\beta$

**Simplify.**

B

**17.** $\cos (90° - \alpha) \sin (180° - \beta) + \cos (360° - \alpha) \sin (90° - \beta)$

**18.** $\cos (\alpha - 90°) \sin (90° - \beta) + \sin (\beta - 270°) \cos (90° - \alpha)$

**19.** $\tan (90° - \beta) \tan (180° - \beta) \sec \beta + \csc \alpha \sin (90° - \alpha) \csc (90° - \alpha)$

**20.** $\csc (90° - \theta) \sec (360° - \theta) - \tan (720° + \theta) \cot (450° - \theta)$

## 12–5   Functions of Sums and Differences of Angles

Several important identities are consequences of the formula

$$\cos (\alpha - \beta) = \cos \alpha \cos \beta + \sin \alpha \sin \beta.$$

For example, if $\alpha$ has measure 0°, you find that

$$\cos (0° - \beta) = \cos 0° \cos \beta + \sin 0° \sin \beta,$$
$$\cos (-\beta) = 1 \cdot \cos \beta + 0 \cdot \sin \beta,$$
$$\cos (-\beta) = \cos \beta.$$

To find an expression for $\sin (-\beta)$ in terms of $\sin \beta$, replace $\beta$ by $-\beta$ in the identity $\sin \beta = \cos (90° - \beta)$ derived in Section 12–4:

$$\sin (-\beta) = \cos [90° - (-\beta)],$$
$$\sin (-\beta) = \cos (90° + \beta)$$
$$= \cos [\beta - (-90°)]$$
$$= \cos \beta \cos (-90°) + \sin \beta \sin (-90°)$$
$$= \cos \beta \cdot 0 + \sin \beta \cdot (-1)$$
$$\sin (-\beta) = -\sin \beta.$$

It is left as an exercise for you to prove the following identities:

$$\tan (-\beta) = -\tan \beta \qquad \cot (-\beta) = -\cot \beta$$
$$\sec (-\beta) = \sec \beta \qquad \csc (-\beta) = -\csc \beta$$

Using the fact that $\alpha + \beta = \alpha - (-\beta)$, you can now derive a *formula for the cosine of the sum of two angles:*

$$\cos (\alpha + \beta) = \cos [\alpha - (-\beta)]$$
$$= \cos \alpha \cos (-\beta) + \sin \alpha \sin (-\beta)$$
$$= \cos \alpha \cos \beta + \sin \alpha (-\sin \beta)$$

$$\cos (\alpha + \beta) = \cos \alpha \cos \beta - \sin \alpha \sin \beta$$

Because the sine of an angle equals the cosine of its complement, you can also obtain a formula for $\sin(\alpha + \beta)$:

$$\begin{aligned}
\sin(\alpha + \beta) &= \cos[90° - (\alpha + \beta)] \\
&= \cos[(90° - \alpha) - \beta] \\
&= \cos(90° - \alpha)\cos\beta + \sin(90° - \alpha)\sin\beta
\end{aligned}$$

---

$\sin(\alpha + \beta) = \sin\alpha\cos\beta + \cos\alpha\sin\beta$

---

If you replace $\beta$ by $-\beta$ in the formula just derived, you obtain:

$$\begin{aligned}
\sin[\alpha + (-\beta)] &= \sin\alpha\cos(-\beta) + \cos\alpha\sin(-\beta) \\
\sin(\alpha - \beta) &= \sin\alpha\cos\beta + \cos\alpha(-\sin\beta)
\end{aligned}$$

---

$\sin(\alpha - \beta) = \sin\alpha\cos\beta - \cos\alpha\sin\beta$

---

**Example 1.**  Simplify $\sin 20° \cos 130° + \cos 20° \sin 130°$.

*Solution:*  $\begin{aligned}[t]
\sin 20° \cos 130° + \cos 20° \sin 130° &= \sin(20° + 130°) \\
&= \sin 150° \\
&= \tfrac{1}{2}
\end{aligned}$

**Example 2.**  Find $\sin(\alpha - \beta)$ if $\alpha$ is a fourth-quadrant angle for which $\cos\alpha = \frac{5}{13}$ and $\beta$ is a third-quadrant angle for which $\sin\beta = -\frac{4}{5}$.

*Solution:*  $\cos\alpha = \frac{5}{13}$ and $\sin\beta = -\frac{4}{5}$. Using the method of Section 11–3 or Section 12–1, you find that $\sin\alpha = -\frac{12}{13}$ and $\cos\beta = -\frac{3}{5}$. Then,

$$\begin{aligned}
\sin(\alpha - \beta) &= \sin\alpha\cos\beta - \cos\alpha\sin\beta \\
&= (-\tfrac{12}{13})(-\tfrac{3}{5}) - (\tfrac{5}{13})(-\tfrac{4}{5}) \\
&= \tfrac{36}{65} + \tfrac{20}{65} = \tfrac{56}{65}
\end{aligned}$$

$$\therefore \sin(\alpha - \beta) = \tfrac{56}{65}$$

Can you derive a formula for $\tan(\alpha + \beta)$? If $\cos(\alpha + \beta) \neq 0$,

$$\tan(\alpha + \beta) = \frac{\sin(\alpha + \beta)}{\cos(\alpha + \beta)},$$

$$\tan(\alpha + \beta) = \frac{\sin\alpha\cos\beta + \cos\alpha\sin\beta}{\cos\alpha\cos\beta - \sin\alpha\sin\beta}.$$

Assuming $\cos \alpha \neq 0$ and $\cos \beta \neq 0$, you can transform the fraction in the right-hand member of the identity above into an equivalent fraction by dividing numerator and denominator by $\cos \alpha \cos \beta$:

$$\tan (\alpha + \beta) = \frac{\dfrac{\sin \alpha \cos \beta}{\cos \alpha \cos \beta} + \dfrac{\cos \alpha \sin \beta}{\cos \alpha \cos \beta}}{\dfrac{\cos \alpha \cos \beta}{\cos \alpha \cos \beta} - \dfrac{\sin \alpha \sin \beta}{\cos \alpha \cos \beta}}$$

$$\tan (\alpha + \beta) = \frac{\tan \alpha + \tan \beta}{1 - \tan \alpha \tan \beta}$$

Rewriting this identity with $-\beta$ in place of $\beta$, and simplifying the result, you obtain for all angles $\alpha$ and $\beta$ for which the functions involved are defined:

$$\tan (\alpha - \beta) = \frac{\tan \alpha - \tan \beta}{1 + \tan \alpha \tan \beta}$$

*WRITTEN EXERCISES*

Simplify.

**A**

**1.** $\cos 90° \cos 120° + \sin 90° \sin 120°$

**2.** $\cos 120° \cos 30° - \sin 120° \sin 30°$

**3.** $\cos 180° \cos 45° - \sin 180° \sin 45°$

**4.** $\cos 135° \cos 45° + \sin 135° \sin 45°$

**5.** $\sin 42° \cos 78° + \cos 42° \sin 78°$

**6.** $\cos 543° \sin 423° - \sin 543° \cos 423°$

**7.** $\dfrac{\tan 154° - \tan 34°}{1 + \tan 154° \tan 34°}$

**8.** $\dfrac{\tan 215° + \tan 100°}{1 - \tan 215° \tan 100°}$

Apply the formulas involving sums and differences of angles to find the exact value of each of the following.

**9.** $\tan 15°$

**10.** $\sin 15°$

**11.** $\tan 75°$

**12.** $\sin 75°$

**13.** $\cos 195°$

**14.** $\tan 195°$

**15.** $\cos 165°$

**16.** $\sin 165°$

**17.** $\tan 285°$

**18.** $\cos 285°$

**19.** $\sin 345°$

**20.** $\tan 345°$

**21.** If $\alpha$ is a first-quadrant angle with $\sin \alpha = \frac{4}{5}$, and $\beta$ is a second-quadrant angle with $\cos \beta = -\frac{51}{149}$, find **(a)** $\sin (\alpha + \beta)$; **(b)** $\cos (\alpha + \beta)$; **(c)** $\sin (\alpha - \beta)$; **(d)** $\cos (\alpha - \beta)$; **(e)** $\tan (\alpha + \beta)$; **(f)** $\tan (\alpha - \beta)$.

**22.** If $\alpha$ is a third-quadrant angle with $\csc \alpha = -\frac{13}{5}$, and $\beta$ is a fourth-quadrant angle with $\sec \beta = \frac{25}{7}$, find **(a)** $\sin (\alpha + \beta)$; **(b)** $\cos (\alpha + \beta)$; **(c)** $\sin (\alpha - \beta)$; **(d)** $\cos (\alpha - \beta)$; **(e)** $\tan (\alpha + \beta)$; **(f)** $\tan (\alpha - \beta)$.

If $\theta$ is a third-quadrant angle with $\sin \theta = -\frac{8}{17}$, and $\alpha$ is a first-quadrant angle with $\sec \alpha = \frac{5}{3}$, find each of the following.

**23. a.** $\tan (\theta + \alpha)$;    **b.** $\cot (\theta + \alpha)$;    **c.** $\cos (\alpha + \theta)$;    **d.** $\sin (\alpha - \theta)$

**24. a.** $\tan (\theta - \alpha)$;    **b.** $\cot (\theta - \alpha)$;    **c.** $\cos (\alpha - \theta)$;    **d.** $\sin (\alpha + \theta)$

Prove that each of the following is an identity.

**B**

**25.** $\cos (180° - \alpha) = -\cos \alpha$      **29.** $\tan (180° + \alpha) = \tan \alpha$

**26.** $\cos (180° + \alpha) = -\cos \alpha$      **30.** $\tan (180° - \alpha) = -\tan \alpha$

**27.** $\sin (180° + \alpha) = -\sin \alpha$      **31.** $\cos (360° - \alpha) = \cos \alpha$

**28.** $\sin (180° - \alpha) = \sin \alpha$      **32.** $\sin (360° - \alpha) = -\sin \alpha$

**C**

**33.** $\tan (135° - x) = \dfrac{\sin x + \cos x}{\sin x - \cos x}$

**34.** $\tan (x + 60°) = \dfrac{4 \tan x + \sqrt{3} \sec^2 x}{\sec^2 x - 4 \tan^2 x}$

**35.** $\dfrac{\cos (A + B)}{\cos (A - B)} = \dfrac{1 - \tan A \tan B}{1 + \tan A \tan B}$

## 12–6   Double- and Half-Angle Identities

When you replace $\alpha$ by $\beta$ in the formulas for $\sin (\alpha + \beta)$, $\cos (\alpha + \beta)$, and $\tan (\alpha + \beta)$, you obtain identities known as the *double-angle formulas:*

$$\sin (\beta + \beta) = \sin \beta \cos \beta + \cos \beta \sin \beta,$$
$$\sin 2\beta = 2 \sin \beta \cos \beta.$$

$$\cos (\beta + \beta) = \cos \beta \cos \beta - \sin \beta \sin \beta,$$
$$\cos 2\beta = \cos^2 \beta - \sin^2 \beta.$$

$$\tan (\beta + \beta) = \frac{\tan \beta + \tan \beta}{1 - \tan \beta \tan \beta},$$

$$\tan 2\beta = \frac{2 \tan \beta}{1 - \tan^2 \beta}.$$

Alternative forms of the formula for $\cos 2\beta$ result when you use the identity $\sin^2\beta + \cos^2\beta = 1$ to transform the right-hand member of the identity $\cos 2\beta = \cos^2\beta - \sin^2\beta$.

$$\cos 2\beta = \cos^2\beta - \sin^2\beta$$
$$= \cos^2\beta - (1 - \cos^2\beta)$$
$$\cos 2\beta = 2\cos^2\beta - 1$$

$$\cos 2\beta = \cos^2\beta - \sin^2\beta$$
$$= (1 - \sin^2\beta) - \sin^2\beta$$
$$\cos 2\beta = 1 - 2\sin^2\beta$$

**Example 1.** Evaluate $\sin 67\frac{1}{2}° \cos 67\frac{1}{2}°$.

*Solution:* Because $\sin\beta \cos\beta = \frac{1}{2}\sin 2\beta$, you have

$$\sin 67\tfrac{1}{2}° \cos 67\tfrac{1}{2}° = \tfrac{1}{2}\sin 2(67\tfrac{1}{2})°$$

$$= \tfrac{1}{2}\sin 135° = \frac{1}{2}\left(\frac{\sqrt{2}}{2}\right)$$

$$\therefore \ \sin 67\tfrac{1}{2}° \cos 67\tfrac{1}{2}° = \frac{\sqrt{2}}{4}.$$

From the formulas for $\cos 2\beta$, you can derive the *half-angle formulas:*

$$\sin^2\frac{\theta}{2} = \frac{1 - \cos\theta}{2} \qquad\qquad \cos^2\frac{\theta}{2} = \frac{1 + \cos\theta}{2}$$

$$\tan^2\frac{\theta}{2} = \frac{1 - \cos\theta}{1 + \cos\theta}$$

To obtain the first formula, transform $\cos 2\beta = 1 - 2\sin^2\beta$ into the equivalent identity

$$\sin^2\beta = \frac{1 - \cos 2\beta}{2}.$$

Then, replace $2\beta$ by $\theta$ and $\beta$ by $\dfrac{\theta}{2}$ to obtain

$$\sin^2\frac{\theta}{2} = \frac{1 - \cos\theta}{2}.$$

You can derive the second half-angle formula similarly from the identity $\cos 2\beta = 2\cos^2\beta - 1$. Then, you have

$$\tan^2\frac{\theta}{2} = \frac{\sin^2\dfrac{\theta}{2}}{\cos^2\dfrac{\theta}{2}} = \frac{\dfrac{1 - \cos\theta}{2}}{\dfrac{1 + \cos\theta}{2}} = \frac{1 - \cos\theta}{1 + \cos\theta}.$$

Notice that the half-angle formulas give expressions for the *squares*

$\sin^2\dfrac{\theta}{2}$, $\cos^2\dfrac{\theta}{2}$, $\tan^2\dfrac{\theta}{2}$. Of course, knowing the value of $\cos\theta$, you can use these formulas to determine $\sin\dfrac{\theta}{2}$, $\cos\dfrac{\theta}{2}$, or $\tan\dfrac{\theta}{2}$ *provided that you also know the quadrant in which* $\dfrac{\theta}{2}$ *terminates*. Thus, depending on the quadrant of $\theta$, you have

$$\sin\frac{\theta}{2} = \sqrt{\frac{1-\cos\theta}{2}} \quad \text{or} \quad -\sqrt{\frac{1-\cos\theta}{2}}$$

$$\cos\frac{\theta}{2} = \sqrt{\frac{1+\cos\theta}{2}} \quad \text{or} \quad -\sqrt{\frac{1+\cos\theta}{2}}$$

$$\tan\frac{\theta}{2} = \sqrt{\frac{1-\cos\theta}{1+\cos\theta}} \quad \text{or} \quad -\sqrt{\frac{1-\cos\theta}{1+\cos\theta}}$$

**Example 2.** Find the exact values of $\sin 157\frac{1}{2}°$ and $\tan 157\frac{1}{2}°$.

*Solution:* Note first that $157\frac{1}{2} = \frac{1}{2}(315)$.

$$\sin^2 157\tfrac{1}{2}° = \sin^2 \tfrac{1}{2}(315)° \qquad \tan^2 157\tfrac{1}{2}° = \tan^2 \tfrac{1}{2}(315)°$$

$$= \frac{1-\cos 315°}{2} \qquad\qquad = \frac{1-\cos 315°}{1+\cos 315°}$$

$$= \frac{1-\dfrac{\sqrt{2}}{2}}{2} \qquad\qquad = \frac{1-\dfrac{\sqrt{2}}{2}}{1+\dfrac{\sqrt{2}}{2}}$$

$$= \frac{2-\sqrt{2}}{4} \qquad\qquad = \frac{2-\sqrt{2}}{2+\sqrt{2}} = \frac{6-4\sqrt{2}}{2}$$

$$\qquad\qquad\qquad\qquad\qquad = 3 - 2\sqrt{2}$$

Since an angle measuring $157\frac{1}{2}°$ terminates in the second quadrant, $\sin 157\frac{1}{2}° > 0$ and $\tan 157\frac{1}{2}° < 0$. Therefore,

$$\sin 157\tfrac{1}{2}° = \tfrac{1}{2}\sqrt{2-\sqrt{2}}, \ \tan 157\tfrac{1}{2}° = -\sqrt{3-2\sqrt{2}}.$$

Another formula for $\tan\dfrac{\theta}{2}$ can be derived as follows:

$$\tan\beta = \frac{\sin\beta}{\cos\beta} = \frac{2\sin\beta\cos\beta}{2\cos\beta\cos\beta} = \frac{\sin 2\beta}{1+\cos 2\beta}.$$

Replace $\beta$ by $\dfrac{\theta}{2}$ and $2\beta$ by $\theta$:

$$\tan \frac{\theta}{2} = \frac{\sin \theta}{1 + \cos \theta}$$

For example, $\qquad \tan 157\tfrac{1}{2}° = \dfrac{\sin 315°}{1 + \cos 315°}$

$$= \frac{-\dfrac{\sqrt{2}}{2}}{1 + \dfrac{\sqrt{2}}{2}} = \frac{-\sqrt{2}}{2 + \sqrt{2}} \cdot \frac{2 - \sqrt{2}}{2 - \sqrt{2}}$$

$$= \frac{2 - 2\sqrt{2}}{2} = 1 - \sqrt{2}.$$

(To show that this number is the same as the one obtained in Example 2, use the fact that two negative numbers are equal if their squares are equal.)

## WRITTEN EXERCISES

Find the exact value of each expression.

**A**

**1.** $2 \sin 15° \cos 15°$         **3.** $1 - 2 \sin^2 67\tfrac{1}{2}°$      **5.** $1 - 2 \sin^2 105°$

**2.** $\cos^2 22\tfrac{1}{2}° - \sin^2 22\tfrac{1}{2}°$     **4.** $2 \cos^2 157\tfrac{1}{2}° - 1$

**6.** $\dfrac{2 \tan 165°}{1 - \tan^2 165°}$

Using a half-angle formula, find the exact value of each of the following.

**7.** $\tan 112\tfrac{1}{2}°$     **8.** $\cos 195°$     **9.** $\sin (-15°)$     **10.** $\cot (-105°)$

If $0 < \alpha < \dfrac{\pi}{2}$, and $\sin \alpha = \tfrac{3}{5}$, find each function value.

**11.** $\cos 2\alpha$        **13.** $\tan 2\alpha$        **15.** $\sin 4\alpha$

**12.** $\sin 2\alpha$        **14.** $\sec 2\alpha$        **16.** $\cos 4\alpha$

If $90° < \theta < 180°$, and $\sin \theta = \tfrac{4}{5}$, find each function value.

**17.** $\sin \dfrac{\theta}{2}$        **19.** $\tan \dfrac{\theta}{2}$        **21.** $\cos \dfrac{\theta}{4}$

**18.** $\cos \dfrac{\theta}{2}$        **20.** $\csc \dfrac{\theta}{2}$        **22.** $\sin \dfrac{\theta}{4}$

Prove each identity.

**B**

**23.** $\cos 2t = \dfrac{\csc^2 t - 2}{\csc^2 t}$

**27.** $\dfrac{1 - \sin 2\theta}{\cos 2\theta} = \dfrac{1 - \tan \theta}{1 + \tan \theta}$

**24.** $\dfrac{\sin 2\beta}{2 \sin^2 \beta} = \cot \beta$

**28.** $\tan 3\beta = \csc 6\beta - \cot 6\beta$

**25.** $\tan 2\theta = \dfrac{\sin 4\theta}{1 + \cos 4\theta}$

**29.** $\sin 3\theta = 3 \sin \theta - 4 \sin^3 \theta$

**26.** $\tan \dfrac{x}{2} = \dfrac{1 - \cos x}{\sin x}$

**30.** $\cos 3\theta = 4 \cos^3 \theta - 3 \cos \theta$

In the rest of the exercises, we consider the **sum and product identities**. In Exercises 31 and 32, **(a)** prove the identity and **(b)** use it to write the given expression as a sum or difference.

**Sample.** $-2 \sin \alpha \sin \beta = \cos (\alpha + \beta) - \cos (\alpha - \beta)$; $\sin 36° \sin 24°$

*Solution:* **a.**

$$\cos (\alpha + \beta) = \cos \alpha \cos \beta - \sin \alpha \sin \beta$$

Subtract: $\cos (\alpha - \beta) = \cos \alpha \cos \beta + \sin \alpha \sin \beta$

$$\overline{\cos (\alpha + \beta) - \cos (\alpha - \beta) = -2 \sin \alpha \sin \beta}$$

**b.** $-2 \sin 36° \sin 24° = \cos (36° + 24°) - \cos (36° - 24°)$
$$= \cos 60° - \cos 12°$$
$$\sin 36° \sin 24° = \tfrac{1}{2} \cos 12° - \tfrac{1}{2} \cos 60°$$

**31.** $2 \cos \alpha \cos \beta = \cos (\alpha + \beta) + \cos (\alpha - \beta)$; $\cos 110° \cos 10°$

**32.** $2 \sin \alpha \cos \beta = \sin (\alpha + \beta) + \sin (\alpha - \beta)$; $\sin 70° \cos 20°$

In Exercises 33–35, **(a)** prove the identity and **(b)** use it to write the given expression as a product.

**Sample.** $\cos A + \cos B = 2 \cos \tfrac{1}{2}(A + B) \cos \tfrac{1}{2}(A - B)$; $\cos 36° + \cos 24°$.

*Solution:* **a.** *Plan:* Use the identity of Exercise 31 with

$$\alpha + \beta = A \quad \text{and} \quad \alpha - \beta = B.$$

Solve this system for $\alpha$ and $\beta$:

$$\alpha = \tfrac{1}{2}(A + B) \quad \text{and} \quad \beta = \tfrac{1}{2}(A - B).$$

Then $\cos (\alpha + \beta) + \cos (\alpha - \beta) = 2 \cos \alpha \cos \beta$ becomes

$$\cos A + \cos B = 2 \cos \tfrac{1}{2}(A + B) \cos \tfrac{1}{2}(A - B).$$

**b.** $\cos 36° + \cos 24° = 2 \cos \tfrac{1}{2}(36° + 24°) \cos \tfrac{1}{2}(36° - 24°)$
$$= 2 \cos 30° \cos 6°$$

C **33.** $\sin A + \sin B = 2 \sin \frac{1}{2}(A + B) \cos \frac{1}{2}(A - B)$; $\sin 36° + \sin 24°$

**34.** $\sin A - \sin B = 2 \cos \frac{1}{2}(A + B) \sin \frac{1}{2}(A - B)$; $\sin 130° - \sin 40°$

**35.** $\cos A - \cos B = -2 \sin \frac{1}{2}(A + B) \sin \frac{1}{2}(A - B)$; $\cos 130° - \cos 40°$

## SUMMARY OF BASIC IDENTITIES

### Reciprocal Identities

$$\cos \theta \sec \theta = 1 \qquad \sin \theta \csc \theta = 1 \qquad \tan \theta \cot \theta = 1$$

### Quotient Identities

$$\tan \theta = \frac{\sin \theta}{\cos \theta} \qquad \cot \theta = \frac{\cos \theta}{\sin \theta}$$

### Pythagorean Identities

$$\sin^2 \theta + \cos^2 \theta = 1 \qquad 1 + \tan^2 \theta = \sec^2 \theta \qquad 1 + \cot^2 \theta = \csc^2 \theta$$

### Sum and Difference Identities

$$\cos (\alpha + \beta) = \cos \alpha \cos \beta - \sin \alpha \sin \beta$$
$$\cos (\alpha - \beta) = \cos \alpha \cos \beta + \sin \alpha \sin \beta$$
$$\sin (\alpha + \beta) = \sin \alpha \cos \beta + \cos \alpha \sin \beta$$
$$\sin (\alpha - \beta) = \sin \alpha \cos \beta - \cos \alpha \sin \beta$$

$$\tan (\alpha + \beta) = \frac{\tan \alpha + \tan \beta}{1 - \tan \alpha \tan \beta} \qquad \tan (\alpha - \beta) = \frac{\tan \alpha - \tan \beta}{1 + \tan \alpha \tan \beta}$$

### Double-Angle Identities

$$\cos 2\theta = \cos^2 \theta - \sin^2 \theta \qquad \sin 2\theta = 2 \sin \theta \cos \theta$$
$$= 2 \cos^2 \theta - 1 \qquad \tan 2\theta = \frac{2 \tan \theta}{1 - \tan^2 \theta}$$
$$= 1 - 2 \sin^2 \theta$$

### Half-Angle Identities

$$\left| \cos \frac{\theta}{2} \right| = \sqrt{\frac{1 + \cos \theta}{2}} ; \qquad \left| \sin \frac{\theta}{2} \right| = \sqrt{\frac{1 - \cos \theta}{2}} ;$$

$$\left| \tan \frac{\theta}{2} \right| = \sqrt{\frac{1 - \cos \theta}{1 + \cos \theta}}, \qquad \tan \frac{\theta}{2} = \frac{\sin \theta}{1 + \cos \theta} .$$

## MISCELLANEOUS IDENTITIES

Prove each identity.

**A**

**1.** $(\csc \theta - \cot \theta)(\csc \theta + \cot \theta) = 1$

**2.** $\dfrac{1 - \tan^2 \theta}{1 + \tan^2 \theta} = 1 - 2 \sin^2 \theta$

**3.** $\dfrac{\sin \alpha \, \cos \alpha}{1 - 2 \cos^2 \alpha} = \dfrac{1}{\tan \alpha - \cot \alpha}$

**4.** $\dfrac{\tan \beta - \sin \beta}{\tan \beta \sin \beta} = \dfrac{\tan \beta \sin \beta}{\tan \beta + \sin \beta}$

**5.** $\sin 2\alpha = 2 \sin^2 \alpha \cot \alpha$

**6.** $\sin 2\alpha = 2 \cos^2 \alpha \tan \alpha$

**7.** $\dfrac{1 - \tan^2 \alpha}{1 + \tan^2 \alpha} = \cos 2\alpha$

**8.** $2 \sin^2 2\theta = 1 - \cos 4\theta$

**9.** $\sec 2\theta = \dfrac{\sec^2 \theta}{2 - \sec^2 \theta}$

**10.** $\dfrac{1 - \cos 2\alpha}{\sin 2\alpha} = \tan \alpha$

**B**

**11.** $\cos^3 \beta - \sin^3 \beta = (\cos \beta - \sin \beta)(1 + \tfrac{1}{2} \sin 2\beta)$

**12.** $2 \cos^2 \dfrac{\theta}{2} = \dfrac{1 + \sec \theta}{\sec \theta}$

**13.** $\tan 2\theta = \dfrac{2}{\cot \theta - \tan \theta}$

**14.** $\sec \dfrac{\alpha}{2} = \pm \dfrac{\sqrt{2 + 2 \cos \alpha}}{1 + \cos \alpha}$

**15.** $\dfrac{1 - (\sin 2\beta + \cos 2\beta)^2}{2 \sin 2\beta \cos 2\beta} = -1$

**16.** $\dfrac{\sin 4\theta \cos 2\theta - \cos 4\theta \sin 2\theta}{\cos^2 \theta - \sin^2 \theta} = \tan 2\theta$

**17.** $\dfrac{\sin 4\beta - \sin 2\beta}{\cos 2\beta + \cos 4\beta} = \tan \beta$

**18.** $\dfrac{\cos 2\beta - \cos 4\beta}{\sin 2\beta - \sin 4\beta} = -\tan 3\beta$

**19.** $\dfrac{\cos 3\beta + \cos \beta}{\cos \beta - 2 \sin^2 \beta \cos \beta} = 2$

**C**

**20.** $\dfrac{\cos \beta + \cos 2\beta + \cos 3\beta}{\sin 3\beta + \sin 2\beta + \sin \beta} = \cot 2\beta$

**21.** $\dfrac{\sin (\alpha + \beta + \gamma) + \sin (\alpha - \beta - \gamma)}{\cos (\alpha + \beta + \gamma) - \cos (\alpha - \beta - \gamma)} = \dfrac{\tan \beta \tan \gamma - 1}{\tan \beta + \tan \gamma}$

**22.** $\dfrac{\cos 10\beta + \cos 2\beta}{-\cos 10\beta + \cos 2\beta} = \dfrac{(1 - \tan^2 3\beta)(1 - \tan^2 2\beta)}{4 \tan 2\beta \tan 3\beta}$

**23.** $\dfrac{1 - \sin 2\beta}{1 - 4 \sin^2 \beta + 4 \sin^4 \beta} = \dfrac{1}{1 + 2 \sin \beta \cos \beta}$

**24.** $\sec^6 p - \tan^6 p = 1 + 3 \tan^2 p \sec^2 p$

**25.** $\cos 20° + \cos 100° + \cos 140° = 0$

**26.** $1 + \tan C \tan \dfrac{C}{2} = \sec C$

## TRIANGLE APPLICATIONS

### 12–7  The Law of Cosines

The formula $c^2 = a^2 + b^2$ is true for every *right* triangle (Figure 12–6). The distance formula on page 482 enables you to derive a similar formula true for *every* triangle.

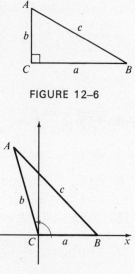

Given any triangle *ABC*, choose a coordinate system with origin at *C* and coordinate axes oriented so that angle *C* is in standard position (Figure 12–7). Then, vertex *B* is *a* units from the origin and has a position angle of 0°; vertex *A* is *b* units from the origin and has angle *C* as position angle; and the distance between *A* and *B* is *c*.

**FIGURE 12–6**

Applying the distance formula (page 482), you find

$$c^2 = (AB)^2 = a^2 + b^2 - 2ab \cos (C - 0°),$$

$$c^2 = a^2 + b^2 - 2ab \cos C.$$

**FIGURE 12–7**

If *C* is a right angle, $\cos C = 0$, so that you have the special case of the Pythagorean Theorem, $c^2 = a^2 + b^2$. Can you explain why $c^2 < a^2 + b^2$ if the measure of angle *C* is between 0° and 90° and $c^2 > a^2 + b^2$ if the measure of *C* is between 90° and 180°?

By choosing coordinate systems first with angle *A* and then with angle *B* in standard position, you can show similarly that

$$a^2 = b^2 + c^2 - 2bc \cos A; \qquad b^2 = a^2 + c^2 - 2ac \cos B.$$

These results are summarized in the following theorem called the **Law of Cosines**.

---

***THEOREM.***  In a triangle the square of the length of any side equals the sum of the squares of the lengths of the other two sides decreased by twice the product of these two sides and the cosine of their included angle.

---

**Example 1.**  In $\triangle ABC$, $a = 7$, $b = 9$, and $C = 60°$. Find $c$.

*Solution:*   **1.** Make a sketch.

**2.** Using $c^2 = a^2 + b^2 - 2ab \cos C$, you have

$$c^2 = 7^2 + 9^2 - 2(7)(9) \cos 60°$$
$$c^2 = 49 + 81 - 126(0.5)$$
$$= 130 - 63$$
$$= 67$$
$$\therefore c = \sqrt{67} \doteq 8.2$$

**Example 2.**   A mine shaft is sunk into the face of a hill that slopes upward from the horizontal by 12°. If the shaft is 50 feet long, and if it penetrates the earth at an angle of 12° with the horizontal, how far is the end of the shaft vertically from the surface?

*Solution:*   **1.** Draw a sketch showing the situation. Since $\angle BAC$ is bisected by an altitude of $\triangle ABC$, $\triangle ABC$ is isosceles and $AB = 50$. Also $\angle BAC$ measures 24°.

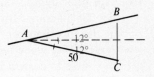

**2.** To determine $BC$:

$$(BC)^2 = b^2 + c^2 - 2bc \cos A$$
$$= 50^2 + 50^2 - 2(50)(50) \cos 24°$$
$$= 2500 + 2500 - 5000(0.9135)$$
$$= 5000 - 4567.5 \doteq 433$$
$$\therefore BC = \sqrt{433} \doteq 21, \text{ and the end of the shaft is about}$$
21 feet vertically from the surface.

## WRITTEN EXERCISES

Find the required part of $\triangle ABC$, in red, either to the nearest unit or to the nearest degree.

**A**

**1.** $a = 4, b = 5, C = 30°$; $c$

**2.** $a = 10, b = 40, C = 30°$; $c$

**3.** $b = 20, c = 30, A = 60°$; $a$

**4.** $a = 7, c = 9, B = 60°$; $b$

**5.** $b = 7, c = \sqrt{2}, A = 135°$; $a$

**6.** $a = 24, b = 30, C = 150°$; $c$

**7.** $a = 12, b = 17, C = 120°$; $c$

**8.** $b = 24, c = 18, A = 120°$; $a$

**9.** $a = 3, b = 5, c = 6$; $A$
(*Hint:* Solve $a^2 = b^2 + c^2 - 2bc \cos A$ for $\cos A$.)

**10.** $a = 5, b = 6, c = 8$; $B$

**11.** $a = 5, b = 6, c = 8$; $C$          **12.** $a = 3, b = 5, c = 6$; $B$

Solve each problem, assuming that linear measures are correct to two significant figures and angle measures to the nearest degree.

**13.** Two planes starting from the same point at the same time fly on courses diverging by 40°. If one plane averages 320 miles per hour over the ground and the other plane averages 400 miles per hour over the ground, how far apart are the planes after 15 minutes?

**14.** Two ships sail from the same harbor at the same time on courses diverging by 97°. If the ships sail at constant speeds of 12 knots and 14 knots, how many nautical miles apart are the ships after $\frac{1}{2}$ hour?

**15.** As a pilot flying from Midville to Central City is leaving Midville, he flies 20° off course for 50 miles before discovering his error. If the direct air distance between the cities is 200 miles, how far is the pilot from Central City when he finds his mistake? By how much must he change his course to correct his error?

**16.** Each of two legs of a stepladder is 14 feet long. If the angle formed by the legs measures 10°, how far apart are the feet of the stepladder?

**17.** Two forces of 30 pounds and 50 pounds act on an object at an angle of 120°. Find the magnitude of the resultant of the two forces.

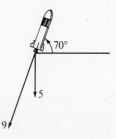

**18.** On take-off a rocket exerts on its launch pad a force of 9 tons in the direction shown, while the weight of the pad exerts a 5-ton force vertically downward. Find the magnitude of the resultant of these forces to the nearest ton.

**B** **19.** Two forces of 40 pounds and 50 pounds, respectively, acting on an object exert a resultant force of 70 pounds. Find the measure of the angle between the two forces.

**20.** Two planes, one flying at 300 mph and the other at 450 mph left an airport at the same time. Three hours later they were 1200 miles apart. What was the measure of the angle between their flight paths?

**21.** Find the measure of the angle between two forces of 20 pounds and 15 pounds if the magnitude of their resultant is 26 pounds.

**22.** The measure of two sides of a parallelogram are 50 and 80 inches, and one diagonal is 90 inches long. How long is the other diagonal?

Prove that each formula holds in triangle $ABC$.

C 23. $1 + \cos A = \dfrac{(b + c + a)(b + c - a)}{2bc}$

24. $1 - \cos A = \dfrac{(a - b + c)(a + b - c)}{2bc}$

25. $\tan\dfrac{A}{2} = \sqrt{\dfrac{(a - b + c)(a + b - c)}{(b + c + a)(b + c - a)}}$   (Half-angle Law)

(*Hint:* Use the *half-angle formula* on page 489 and the results of Exercises 23 and 24.)

26. $\tan\dfrac{A}{2} = \sqrt{\dfrac{(s - c)(s - b)}{s(s - a)}} = \dfrac{r}{s - a}$,   where   $s = \dfrac{a + b + c}{2}$  and

$r = \sqrt{\dfrac{(s - a)(s - b)(s - c)}{s}}.$    (*Hint:* Use the result of Exercise 25.)

Using logarithms, apply the result of Exercise 26 to determine to the nearest minute the measure of each angle of the triangles with sides as follows.

27. 15.3, 19.2, 13.9            29. 17.30, 25.40, 35.52

28. 8.16, 9.44, 11.0            30. 39.01, 14.00, 43.40

## 12–8   The Law of Sines

Noting that the coordinates of point $A$ in Figure 12–7, page 494, are $(b \cos C, b \sin C)$, you can deduce a useful formula for the area $K$ of triangle $ABC$. Because the $y$-coordinate of $A$ is the length of the *altitude* $AD$ drawn from $A$ perpendicular to the line containing *base* $BC$ whose length is $a$ (Figure 12–8), you have:

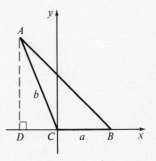

$$K = \tfrac{1}{2}(BC)(AD) = \tfrac{1}{2}a(b \sin C),$$

$$K = \tfrac{1}{2}ab \sin C.$$

By choosing the coordinate system appropriately, you can also obtain

**FIGURE 12–8**

$$K = \tfrac{1}{2}ac \sin B \quad \text{and} \quad K = \tfrac{1}{2}bc \sin A.$$

These results are summarized in the following theorem.

---

**THEOREM.** The area of a triangle equals one-half the product of the lengths of two sides and the sine of their included angle.

---

This theorem implies that the following compound sentence is true: $\frac{1}{2}bc \sin A = \frac{1}{2}ac \sin B = \frac{1}{2}ab \sin C$. Division of each member by $\frac{1}{2}abc$, yields

$$\frac{\sin A}{a} = \frac{\sin B}{b} = \frac{\sin C}{c}.$$

This relationship, called the Law of Sines, is stated in the following theorem.

---

**THEOREM.** The sines of the angles of a triangle are proportional to the lengths of the opposite sides.

---

Notice that if $C$ is a right angle, then $\sin C = 1$ and the Law of Sines yields the familiar right triangle relationships $\sin A = \frac{a}{c}$ and $\sin B = \frac{b}{c}$.

**Example 1.** If $A = 30°$ and $B = 60°$, what is the ratio of $a$ to $b$ to the nearest tenth?

*Solution:* Use $\dfrac{\sin A}{a} = \dfrac{\sin B}{b}$. Multiplying each member by $\dfrac{a}{\sin B}$, you have

$$\frac{\sin A}{\sin B} = \frac{a}{b}.$$

Then since $\sin 30° = \frac{1}{2}$ and $\sin 60° = \frac{\sqrt{3}}{2}$, you have

$$\frac{a}{b} = \frac{1}{\sqrt{3}} = \frac{\sqrt{3}}{3} \doteq \frac{1.73}{3} \doteq 0.6.$$

**Example 2.** While on opposite ends of a 7.4-mile-long beach which runs due north and south, two men observe a ship on fire. Their lines of sight to the ship have bearings N40°E and S59°E. How far is the ship from the nearer man?

*Solution:* Make a sketch, and notice that $A = 59°$ and $B = 40°$. Then

$$C = 180° - 59° - 40° = 81°.$$

To solve for $b$, use

$$\frac{b}{\sin B} = \frac{c}{\sin C}.$$

You have

$$\frac{b}{\sin 40°} = \frac{7.4}{\sin 81°},$$

or

$$b = \frac{7.4 \sin 40°}{\sin 81°} = \frac{7.4(0.6428)}{0.9877} \doteq 4.8.$$

∴ the ship is about 4.8 miles from the nearer man.

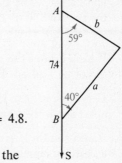

Note that a solution of the foregoing example could also have been accomplished using logarithms. To solve

$$b = \frac{7.4 \sin 40°}{\sin 81°},$$

you would use the logarithmic equation

$$\log b = \log 7.4 + \log \sin 40° - \log \sin 81°.$$

## WRITTEN EXERCISES

Find the required values for $\triangle ABC$, in red, to the nearest tenth or nearest 10 minutes.

**A**

**1.** $A = 60°$, $B = 45°$; $a : b$

**2.** $C = 30°$, $B = 45°$; $c : b$

**3.** $A = 45°$, $B = 60°$, $a = 10$; $b$

**4.** $B = 45°$, $C = 30°$, $b = 20$; $c$

**5.** $A = 30°$, $a = 15$, $b = 20$; $B$

**6.** $C = 120°$, $a = 10$, $c = 15$; $A$

**7.** $a = 16$, $b = 13$, $\sin A = \frac{2}{3}$; $B$

**8.** $b = 34$, $c = 27$, $\sin B = 0.9$; $C$

**9.** $b = 1.5$, $\sin A = \frac{2}{5}$, $\sin B = \frac{4}{7}$; $a$

**10.** $c = 7.5$, $\sin C = 0.7$, $\sin B = 0.4$; $b$

Solve each problem, assuming that angle measures are correct to the nearest degree and linear measures to two significant figures.

**11.** An isosceles triangle has base of length 20 inches. If the vertex angle of the triangle measures 30°, find the perimeter of the triangle.

**12.** Two fire towers are located 10 miles apart. If each locates the same fire and finds it to be at an angle of 62° and 71°, respectively, from the other tower, how far is the fire from the tower nearest to it?

**13.** Two radar stations located 18 miles apart spot the same thunderstorm. If each locates the storm at an angle of 110° and 52°, respectively, from the other station, how far is the storm from the station nearest to it?

**14.** How long are the legs of a stepladder if the feet of the ladder are 3 feet apart when the angle between the legs measures 32°?

**B** **15.** Radio station *A* is 120 miles due north of station *B*. Station *A* receives a distress message from a ship at a bearing of 130°, while station *B* receives the same message at a bearing of 47°. How long would a helicopter flying at 110 mph take to reach the ship from station *A*?

**16.** A surveyor laying a road due west from *A* encounters a swamp at *B*. She changes her direction to N28°W for 2500 yards to *C* and then turns S37°W. How far must she continue in this direction to reach point *D* on the east-west line through *A*?

**17.** *ABC* is an equilateral triangle whose sides are 18″ long. Lines *AD* and *AE* are drawn trisecting angle *A* and intersecting side *BC* in points *D* and *E*. Find the lengths of segments *BD*, *DE*, and *EC*.

**18.** In triangle *ABC*, prove that if $\dfrac{\cos A}{b} = \dfrac{\cos B}{a}$, then the triangle is either an isosceles or a right triangle. (*Hint:* Use Law of Sines.)

Prove that each formula holds in triangle *ABC*.

**19.** $\dfrac{a - b}{b} = \dfrac{\sin A - \sin B}{\sin B}$

**20.** $\dfrac{a + b}{b} = \dfrac{\sin A + \sin B}{\sin B}$

**C** **21.** $\dfrac{a - b}{a + b} = \dfrac{\sin A - \sin B}{\sin A + \sin B}$ (*Hint:* Use the results of Exercises 19 and 20.)

**22.** $\dfrac{a - b}{a + b} = \dfrac{\cos \frac{1}{2}(A + B) \sin \frac{1}{2}(A - B)}{\sin \frac{1}{2}(A + B) \cos \frac{1}{2}(A - B)} = \dfrac{\tan \frac{1}{2}(A - B)}{\tan \frac{1}{2}(A + B)}$

(*Hint:* Use the results of Exercise 21 and Exercises 33 and 34 page 492.) This identity is called the **Law of Tangents**.

Using logarithms, apply the result of Exercise 22 to determine to the nearest minute the measures of the other angles of the triangle in which the two sides whose lengths are given form an angle of the indicated measure.

**23.** 71.30, 36.50; 47°15′            **24.** 1240, 2413; 24°10′

## 12–9 Solving Triangles

In Sections 12–7 and 12–8, you have used one relationship, either the Law of Cosines or the Law of Sines, to solve problems involving triangles in which you were given the measures of (1) two sides and the angle included between them, or (2) three sides, or (3) two angles and one side. The process of determining the measures of the remaining sides and angles is called **solving the triangle.** By using both laws, you can sometimes shorten the solution.

**Example 1.** Solve $\triangle ABC$ if $a = 10$, $b = 16$, and $C = 28°40'$.

*Solution:*   1. Make a sketch and label all known parts.

2. Use the Law of Cosines to find $c$.

$$c^2 = a^2 + b^2 - 2ab \cos C$$
$$= 10^2 + 16^2 - 2(10)(16) \cos 28°40'$$
$$= 100 + 256 - 320(0.8774) = 356 - 280.768$$
$$c^2 = 75.232$$

Then $\log c = \frac{1}{2} \log 75.232 = \frac{1}{2}(1.8764) = 0.9382$, so that $c = $ antilog $0.9382 = 8.674$.

3. Use the Law of Sines to find the measure of the *smaller* of the remaining angles, namely $A$.

$$\frac{\sin A}{a} = \frac{\sin C}{c}$$
$$\sin A = \frac{a \sin C}{c} = \frac{10 \sin 28°40'}{8.674} = \frac{10(0.4797)}{8.674} = 0.5530$$
$$\therefore \ A = 33°34'$$

4. Finally,
$$B = 180° - A - C = 180° - 33°34' - 28°40' = 117°46'.$$

$$\therefore c = 8.674, \ A = 33°34', \text{ and } B = 117°46'.$$

*Check:*   5. Compare your results with the sketch and by substituting the results in a relationship not employed in the solution.

For example, $\dfrac{\sin A}{a} = \dfrac{\sin B}{b}$.

$$\frac{\sin A}{a} = \frac{\sin 33°34'}{10} = \frac{0.5530}{10} = 0.055$$

$$\frac{\sin B}{b} = \frac{\sin 117°46'}{16} = \frac{0.8849}{16} = 0.055$$

**Example 2.** Solve $\triangle ABC$ if $a = 5$, $b = 8$, and $c = 7$.

*Solution:*

1. Make a sketch and label all known parts.

2. Use the Law of Cosines to determine the measure of the *largest* angle, namely $B$.

$$\cos B = \frac{a^2 + c^2 - b^2}{2ac}$$

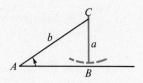

3. Since $B$ is the largest angle, the remaining angles must be acute. (Why?) Use the Law of Sines to find the measure of one of them, say $A$.

$$\sin A = \frac{a \sin B}{b}$$

4. Use $C = 180° - A - B$ to find $C$.

5. Check by using $\dfrac{\sin C}{c} = \dfrac{\sin A}{a}$.

Carrying through the steps of the calculations is left to you.

When the measures of two angles and one side of a triangle are known, you use the fact that the sum of the measures of the angles is 180° to find the measure of the other angle, and the Law of Sines to find the lengths of the other sides. You also use the Law of Sines in solving a triangle when you know the *lengths of two of its sides and the measure of an angle opposite one of them*. In this case there may be no, one, or two solutions.

Suppose, for example, that $A$, $a$, and $b$ are known. By sketching $\angle A$ and measuring the distance $b$ on one of its sides, you have the possibilities suggested in Figures 12–9 and 12–10 ($A < 90°$) and Figure 12–11 ($A \geq 90°$).

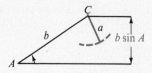

$a < b \sin A$
No solution

$a = b \sin A$
One solution

**FIGURE 12–9**

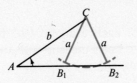

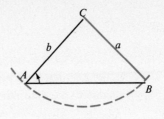

$b \sin A < a < b$
Two solutions

$a \geq b$
One solution

**FIGURE 12–10**

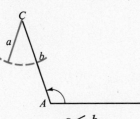

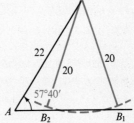

$a \leq b$
No solution

$a > b$
One solution

**FIGURE 12–11**

**Example 3.** Solve $\triangle ABC$, if $a = 20$, $b = 22$, and $A = 57°40'$.

*Solution :*

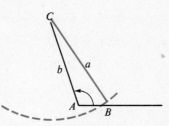

1. Sketching the triangle suggests that there are two solutions.

2. To determine $B$, use

$$\sin B = \frac{b \sin A}{a} \doteq \frac{18.59}{20} = 0.9295 \quad \text{so that either}$$

$B = 68°22'$ or $B = 180° - 68°22' = 111°38'$.

3. To determine $C$, use $C = 180° - A - B$.

| | |
|---|---|
| $C = 180° - 57°40' - 68°22'$ | $C = 180° - 57°40' - 111°38'$ |
| $C = 53°58'$ | $C = 10°42'$ |

4. To determine $c$, use $c = \dfrac{a \sin C}{\sin A}$.

Completing Step 4 to find $c = 19.14$ or $c = 4.40$, and checking, are left to you. Thus you have

$B = 68°22'$, $C = 53°58'$, $c = 19.14$; **or**

$B = 111°38'$, $C = 10°42'$, $c = 4.40$.

## WRITTEN EXERCISES

Find the measures of the other parts of each triangle *ABC* fitting the following data. Assume that angles have been measured to the nearest minute and sides to four significant figures.

A

**1.** $a = 4.000, b = 5.000, B = 30°00'$

**2.** $a = 5.000, b = 8.000, A = 150°00'$

**3.** $a = 16.00, b = 18.00, A = 58°00'$

**4.** $a = 65.00, c = 83.00, C = 99°00'$

**5.** $a = 15.00, c = 25.00, B = 45°00'$

**6.** $a = 15.00, c = 20.00, A = 30°00'$

**7.** $a = 20.00, b = 22.00, A = 57°40'$

**8.** $a = 25.00, b = 4.000, B = 37°40'$

**9.** $a = 14.00, b = 18.00, A = 34°50'$

**10.** $a = \sqrt{2}, b = \sqrt{8}, c = 5$

**11.** $c = 152.0, b = 135.0, B = 50°25'$

**12.** $a = 4900, c = 2200, C = 30°19'$

**13.** $a = 7.200, b = 5.400, c = 13.90$

If $b, B,$ and $A$ are to be parts of triangle *ABC*, determine the values of $b$ for which $A$ has **(a)** no value; **(b)** one value; **(c)** two values.

B

**14.** $a = 70, B = 120°$                       **15.** $a = 20, B = 45°$

## PROBLEMS

A

**1.** A string is tied into a loop and then rigidly stretched from three points in the string to make a triangle with sides 75 inches, 85 inches, and 100 inches. What are the measures of the angles of the triangle?

**2.** La Linda Avenue and Pasqual Street meet at an angle of 40°. From a point 1000 feet from the intersection along La Linda Avenue, Ann Evans wants to run an 800-foot fence over to Pasqual Street to establish a lot for a house. If a city ordinance specifies that no lot with less than a 2500-foot perimeter can be fenced off from adjoining land, how far down Pasqual Avenue from the intersection does Mrs. Evans' fence meet the street?

**3.** The angles of elevation from a point on a hill to the top and bottom of a 50-foot tower are 56° and 12°, respectively. How far is the point from the bottom of the tower?

**4.** Two planes leave an airport at the same time and fly courses which diverge by 38°. When one plane has flown 800 miles, it receives a message from the other that it has made an emergency landing, and although its directional radar is inoperative, it finds that the distance between the planes is 900 miles at the time. What change in heading should the first plane make to fly a course to the downed plane?

**5.** An aircraft flying at an airspeed of 420 miles per hour on a heading of 140° is affected by a wind blowing from a direction of 200°. If the velocity of the wind is 60 miles per hour, find the course of the plane over the ground and its ground speed.

**6.** From two successive milestones on a straight stretch of the Central City Expressway, the angles of elevation to a stationary balloon are 48° and 37°, respectively. If the balloon is directly over the road, find its altitude.

**7.** One minute after passing over checkpoint *A*, a plane flying due east at 600 miles per hour passes over checkpoint *B*. As it passes *B*, what is its bearing from checkpoint *C*, which is 20 miles from *A*, 12 miles from *B*, and north of the line *AB*?

**8.** A vertical television mast is mounted on the roof of a building. From a point 750 feet from and on a level with the base of the building, the angles of elevation of the base and top of the mast measure 34° and 50°, respectively. How tall is the mast?

**9.** From a point on level ground the angle of elevation to the top of a nearby building is 46°. From a point on the ground 83 feet closer to the building on a line with the first point, the angle of elevation to the top of the building is 68°. How tall is the building?

**10.** A hill slopes steadily upward at an angle of 15°. A tree that grows vertically casts a shadow down the hill of length 40 feet when the angle of elevation of the sun is 64°. How tall is the tree?

**B** **11.** A ship sailed 25 miles on course 32° and then 12 miles on course 150°. What course should the ship set and what distance must it travel to return to its starting point by the shortest route?

**12.** A plane heading due north with an airspeed of 392 mph is subject to a 29 mph wind from 159°20′. What are the plane's ground speed and course?

**13.** A pilot wants to maintain a course of 31° and ground speed of 400 mph against a 41 mph headwind from 343°. What should his heading and airspeed be?

14. From the top of a lighthouse 150 feet above the ocean, the angle of depression of a buoy due west of the lighthouse measures 36°. The angle of depression of another buoy S65°W of the lighthouse measures 24°. How far apart are the buoys?

15. From two points 6000 feet apart in the plane of the base of a hill, the angles of elevation of the summit measure 19°10′ and 20°30′. If the points are on opposite sides of the mountain but in the same vertical plane with the summit, what is the height of the mountain?

C 16. A fighter plane has a cruising speed of 600 mph. In what direction should the plane head, and how long will it take to intercept in the shortest time a bomber that is 400 miles due north and flying on course 60° at 350 mph?

17. A freighter, steaming on course 140° at 20 knots, is 40 nautical miles N20°E of a submarine with a cruising speed of 25 knots. Find the course to be set by the sub to overtake the freighter in the least amount of time, and find this minimum time.

## 12–10 Areas of Triangles

Knowing the lengths of two sides, say $a$ and $b$, of triangle $ABC$ and the measure of their included angle $C$, you can find the area of the triangle by using the formula $K = \frac{1}{2}ab \sin C$ (see page 497). From the Law of Sines you have $b = \dfrac{a \sin B}{\sin A}$. Substituting for $b$ in the area formula, you obtain

$$K = \tfrac{1}{2}a^2 \frac{\sin B \sin C}{\sin A}.$$

You can similarly derive the formulas

$$K = \tfrac{1}{2}b^2 \frac{\sin A \sin C}{\sin B} \quad \text{and} \quad K = \tfrac{1}{2}c^2 \frac{\sin A \sin B}{\sin C}.$$

With the formulas just derived you can find the area of a triangle if you know the measures of one of its sides and two of its angles.

Given the lengths of three sides of a triangle, you find the area by using the following formula, whose proof is indicated in Exercises 19 and 20 on page 507:

$$K = \sqrt{s(s - a)(s - b)(s - c)},$$

where $s$ is $\dfrac{a + b + c}{2}$, the *semiperimeter* of the triangle.

**Example.** Find the area of $\triangle ABC$, if

    **a.** $b = 12$, $c = 7$, $A = 42°$;

    **b.** $b = 4$, $A = 40°$, $C = 60°$;

    **c.** $a = 6$, $b = 8$, $c = 12$.

*Solution:*   **a.** $K = \frac{1}{2}bc \sin A = \frac{1}{2}(12)(7) \sin 42° = 42(0.6691) = 28.10.$

    **b.** $B = 180° - A - C = 180° - 40° - 60° = 80°$;

$$K = \frac{b^2 \sin A \sin C}{2 \sin B} = \frac{16 \sin 40° \sin 60°}{2 \sin 80°}$$

$$= \frac{8(0.6428)(0.8660)}{(0.9848)} = 4.52.$$

    **c.** $s = \dfrac{a + b + c}{2} = \dfrac{6 + 8 + 12}{2} = 13$;

$$K = \sqrt{s(s - a)(s - b)(s - c)} = \sqrt{13(7)(5)(1)}$$

$$= \sqrt{455} = 21.3.$$

## WRITTEN EXERCISES

A  **1–13.** Find the area of the triangle or triangles (if any) fitting the data in Exercises 1–13, page 504.

Find the measure of the indicated part of $\triangle ABC$ determined by the given data.

B  **14.** $K = 1420$, $A = 42°$, $B = 55°$; $c = \underline{\quad ? \quad}$

   **15.** $K = 16.2$, $c = 6.42$, $a = 42.30$, $A > 90°$; $B = \underline{\quad ? \quad}$

   **16.** Find the area of a parallelogram with adjacent sides of 23 inches and 14 inches if one angle measures 72°.

   **17.** Find the area of a parallelogram with diagonals of length 90 centimeters and 110 centimeters if these diagonals intersect at an angle of 48°.

Prove that each formula holds in triangle $ABC$.

C  **18.** $K = \frac{1}{2}c^2 \dfrac{\sin A \sin B}{\sin (A + B)}$        **19.** $K = bc \sin \frac{1}{2}A \cos \frac{1}{2}A$

   **20.** $K = \sqrt{s(s - a)(s - b)(s - c)}$, where $s = \dfrac{a + b + c}{2}$.

    (*Hint:* Use Exercise 19, the half-angle formulas, page 489, and the results of Exercises 23 and 24, page 497.)

# Chapter Summary

1. The **fundamental trigonometric identities**, together with other derived identities, allow you to transform trigonometric expressions into simpler ones and also to prove that certain equations are identities.

2. The **distance** between points $P$ and $Q$ can be expressed in terms of position angles of the points and their distances from the origin:

$$(PQ)^2 = p^2 + q^2 - 2pq(\cos \alpha \cos \beta + \sin \alpha \sin \beta)$$

From this formula you can derive formulas for values of functions of the **difference** and the **sum of two angles**, and then for functions of **double** and of **half a given angle**.

3. When the measures of three parts, including one side, of a triangle are known, you can use the **Law of Cosines** (page 494) or the **Law of Sines** (page 498) to find the remaining parts. When the known parts of the triangle are the lengths of **two sides** and the measure of an **angle opposite** one of them, there may be no, one, or two solutions.

4. The area of a triangle is given by $K = \frac{1}{2}ab \sin C$; by $K = \frac{1}{2}a^2 \dfrac{\sin B \sin C}{\sin A}$; by $K = \sqrt{s(s-a)(s-b)(s-c)}$, where $s$ denotes the semiperimeter.

### Vocabulary and Spelling

Review the meaning of each term by reference to the page listed.

trigonometric equation (*p. 471*)      Law of Cosines (*p. 494*)
trigonometric identity (*p. 471*)      Law of Sines (*p. 498*)
fundamental identities (*p. 473*)      solving a triangle (*p. 501*)

# Chapter Test and Review

**12–1**    **1.** Express $\sec^2 \theta$ in terms of $\sin \theta$.

**12–2**    **2.** Prove: $\cos \alpha(\csc \alpha - \sec \alpha) = \cot \alpha - 1$.

**12–3**    **3.** Find $(PQ)^2$ for $P$, 5 units from the origin with position angle 150°, and $Q$, 4 units from the origin with position angle 30°.

**12–4**  **4.** Express $\cos (x - 90°)$ as a function of $x$.

**12–5**  **5.** Given that $\sin \alpha = \frac{3}{5}$ with $\alpha$ in Quadrant II, and $\csc \beta = -\frac{4}{3}$ with $\beta$ in Quadrant III, find $\sin (\alpha + \beta)$.

**12–6**  **6.** Prove: $\tan \theta = \dfrac{1 - \cos 2\theta}{\sin 2\theta}$ .

**12–7**  **7.** In $\triangle ABC$, if $a = 7$, $b = 9$, and $C = 60°$, find $c$.

**12–8**  **8.** A parcel of land is in the shape of an isosceles triangle. The base fronts on a road and has a length of 560 feet. If the legs of the triangle meet at an angle of 24°, how long are they?

**12–9**  **9.** In $\triangle ABC$, $a = 14$, $b = 18$, and $A = 35°$. How many different triangles are possible with these measurements?

**12–10**  **10.** Find the area of $\triangle ABC$ if $b = 16$, $c = 5$, and $A = 30°$.

# Extra for Experts

## DE MOIVRE'S THEOREM

In Section 11–10, you saw that any complex number $a + bi$ can be expressed in the trigonometric form $r(\cos \theta + i \sin \theta)$, where $r = \sqrt{a^2 + b^2}$ and $\theta$ is determined by $\cos \theta = \dfrac{a}{\sqrt{a^2 + b^2}}$ and $\sin \theta = \dfrac{b}{\sqrt{a^2 + b^2}}$. Computing products and quotients of complex numbers is particularly easy when the numbers are expressed in trigonometric form.

If $z_1 = r_1(\cos \theta_1 + i \sin \theta_1)$ and $z_2 = r_2(\cos \theta_2 + i \sin \theta_2)$ are complex numbers in trigonometric form, you can verify that

$$z_1 z_2 = r_1 r_2[(\cos \theta_1 \cos \theta_2 - \sin \theta_1 \sin \theta_2) + i(\sin \theta_1 \cos \theta_2 + \cos \theta_1 \sin \theta_2)],$$

$$z_1 z_2 = r_1 r_2[\cos (\theta_1 + \theta_2) + i \sin (\theta_1 + \theta_2)].$$

Similarly, if $z_2 \neq 0$, you have

$$\frac{z_1}{z_2} = \frac{r_1}{r_2} [\cos (\theta_1 - \theta_2) + i \sin (\theta_1 - \theta_2)].$$

Hence, to multiply complex numbers in trigonometric form, you multiply their absolute values and add their amplitudes; to divide complex numbers, you divide their absolute values and find the difference of their amplitudes. Thus, if $z_1 = 6(\cos 150° + i \sin 150°)$ and $z_2 = 2(\cos 30° + i \sin 30°)$, you have the following:

$$z_1 z_2 = 6 \cdot 2[\cos(150° + 30°) + i \sin(150° + 30°)]$$
$$= 12(\cos 180° + i \sin 180°) = -12$$
$$z_1 \div z_2 = (6 \div 2)[\cos(150° - 30°) + i \sin(150° - 30°)]$$
$$= 3(\cos 120° + i \sin 120°) = -\frac{3}{2} + \frac{3\sqrt{3}}{2} i.$$

Squaring the complex number $r(\cos \theta + i \sin \theta)$, you obtain

$$[r(\cos \theta + i \sin \theta)]^2 = r^2(\cos 2\theta + i \sin 2\theta).$$

(Why?) In general, for any integer $n$, it is true that

$$[r(\cos \theta + i \sin \theta)]^n = r^n(\cos n\theta + i \sin n\theta).$$

When $r = 1$, this reduces to

$$(\cos \theta + i \sin \theta)^n = \cos n\theta + i \sin n\theta.$$

This latter result is called De Moivre's Theorem after the French and English mathematician Abraham De Moivre (1667–1754) who first proved it.

# Exercises

Write each of the numbers in the following pairs in trigonometric form. Then, using that form, find the product and the quotient of the numbers and write the results in the form $a + bi$.

**1.** $-2 + 2i, -1 + i$

**3.** $-2i, 4 - 4i$

**2.** $-1 + \sqrt{3}i, 1 + i$

**4.** $5i, -1 - \sqrt{3}i$

Use trigonometric form in computing the indicated powers. Write the result in the form $a + bi$.

**5.** $(1 + \sqrt{3}i)^4$

**7.** $(-\frac{1}{3} - \frac{1}{3}i)^3$

**6.** $(2 - 2\sqrt{3}i)^4$

**8.** $(2\sqrt{3} - 2i)^4$

**9.** Prove: Two complex numbers are equal if and only if their absolute values are equal and their amplitudes are coterminal angles.

**10.** Find the three roots of $z^3 = 27$. (*Hint:* If $z = r(\cos \theta + i \sin \theta)$, $r^3(\cos 3\theta + i \sin 3\theta) = 27(\cos 0° + i \sin 0°)$. Use Exercise 9.)

**11.** In De Moivre's Theorem for $n = 3$, expand the left member and then, by equating real and imaginary parts, deduce formulas for $\cos 3\theta$ and $\sin 3\theta$.

# Explorations with a Computer

(To be used if you have access to an electronic computer that will accept BASIC.)

The following program will print out approximations to some points on the graph of:

$$Y = SIN\ X$$

Notice that 5 vertical line feeds make 1 unit on the X-axis, while 10 horizontal spaces make 1 unit on the Y-axis. By studying the printed coordinates and noting that the TAB function acts similarly to the INT function in counting the number of spaces, you can see why the graphs have "flat" portions and why $-.9$, for example, is graphed as $-1$, while $.9$ is graphed as $.9$.

```
10    PRINT "Y = SIN X"
20    PRINT
30    LET L=20
40    LET M=L+(66-L)/2
50    PRINT "  X      Y"
60    FOR X1=-66 TO 66 STEP 2
70    LET X=.1*X1
80    LET Y=SIN(X)
90    LET Y2=10*Y+M
100   IF X1=0 THEN 210
110   IF Y2<L THEN 250
120   IF Y2>66 THEN 250
130   IF Y2<M THEN 190
140   IF Y2=M THEN 170
150   PRINT X;
151   PRINT TAB(6);Y;
152   PRINT TAB(M);"!";TAB(Y2);"*"
160   GO TO 220
170   PRINT X;
171   PRINT TAB(6);Y;
172   PRINT TAB(Y2);"*"
180   GO TO 220
190   PRINT X;
191   PRINT TAB(6);Y;
192   PRINT TAB(Y2);"*";TAB(M);"!"
200   GO TO 220
210   PRINT X;
211   PRINT TAB(6);Y;
212   PRINT TAB(Y2);"*";TAB(67);"---Y"
220   NEXT X1
230   PRINT TAB(M);"X"
240   GO TO 260
250   PRINT "TOO WIDE"
260   END
```

Change lines 10 and 80 and try the following:

$$Y = 2\ COS\ 2X \quad Y = SIN\ 2X + 1 \qquad Y = SIN\ X + COS\ X$$

$$Y = 2\ SIN\ 3X \quad Y = SIN\ X + SIN\ 2X \quad Y = SIN\ X + COS\ 2X + 1$$

In trying other graphs, you may find some "TOO WIDE." You can obtain somewhat greater width by changing 30 LET L = 6, 50 PRINT "X", and deleting lines 151, 171, 191, and 211.

# 13

# Circular Functions and Their Inverses

## VARIATION AND GRAPHS

### 13–1 Measuring Arcs and Angles

Figure 13–1 shows the point $T(1, 0)$ on the unit circle $x^2 + y^2 = 1$, whose circumference is $2\pi$. A point starting at $T$ and moving around the circle to a point $P$ travels along a circular arc, $\overset{\frown}{TP}$. If you know the length of the arc, $[m(\overset{\frown}{TP})]$, and the direction of motion, then you can locate $P$.

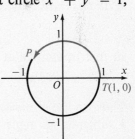

FIGURE 13–1

Figure 13–2 shows various positions of $P$ corresponding to different arc lengths. Notice that positive measures have been assigned to arcs generated by counterclockwise motion

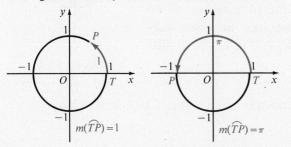

FIGURE 13–2

and negative measures to arcs generated by clockwise motion. The terminal point of an arc with measure 0 is the initial point *T* itself.

Can you explain why arcs whose measures differ by integral multiples of $2\pi$ will have the same terminal point? Figure 13–3 shows three co-terminal arcs; note that arcs of more than one revolution are indicated by multiple curves.

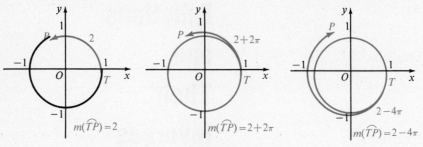

**FIGURE 13–3**

In geometry you learned that the length *s* of a circular arc is proportional to the radius *r and* the measure of the central angle $\theta$ subtended by the arc (Figure 13–4); that is,

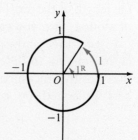

$$s = kr\theta.$$

By choosing an appropriate unit to measure the angle, you can make the constant of proportionality *k* be 1. *With this measurement system,*

**FIGURE 13–4**

$$s = r\theta;$$

so that in a circle of radius 1, you have $s = 1 \cdot \theta$, or

$$s = \theta \quad \text{and} \quad \theta = s.$$

Thus, an arc of measure 1 on a circle of radius 1 subtends an angle whose measure is 1. We call this unit of angular measure, **1 radian,** written $1^R$.

Can you find the radian measure of an angle of 180°? Because such an angle subtends half the unit circle, that is, an arc $\pi$ units long,

**FIGURE 13–5**

$$180° = \pi^R;$$

this means the radian measure of an angle of 180° is $\pi^R$. Thus,

$$1^R = \frac{180°}{\pi} \quad \text{and} \quad 1° = \frac{\pi^R}{180}.$$

Using these relationships, you can convert radian measure to degree measure, and degrees to radians. Because $\pi \doteq 3.14159$, you have $1^R \doteq 57°18'$ and $1° \doteq 0.01745^R$.

**Example 1.** Convert $\frac{2}{3}^R$ to degree measure.

*Solution:* $\dfrac{2^R}{3} = \left(\dfrac{2}{3} \cdot \dfrac{180}{\pi}\right)° \doteq \left(\dfrac{120}{3.1416}\right)° \doteq 38.20°$ or $38°12'$.

**Example 2.** Convert $60°$ to radian measure.

*Solution:* $60° = \left(60 \cdot \dfrac{\pi}{180}\right)^R = \dfrac{\pi^R}{3} \doteq \dfrac{3.1416^R}{3}$, or $1.0472^R$.

**Example 3.** A wheel revolves at an angular speed of 80 rpm (revolutions per minute). Express this speed in radians per second and find the distance, to the nearest inch, that a point 8 inches from the center of the wheel will travel in 6 seconds.

*Solution:* 80 rpm $= 80(2\pi)$, or $160\pi$ radians per minute.

$\therefore$ the angular speed is $\dfrac{160\pi}{60}$, or $\dfrac{8\pi}{3}$, radians per second.

In 6 seconds, the point has traveled an arc with central angle $\theta$ of $6 \cdot \dfrac{8\pi}{3}$, or $16\pi$, radians. Then, using the relationship $s = r\theta$, you find that the point has traveled $s = 8(16\pi) = 128\pi$ inches, or about 402 inches.

*WRITTEN EXERCISES*

Convert each degree measure to a radian measure. Leave your result as a multiple of $\pi$.

[A]

| | | | |
|---|---|---|---|
| **1.** 30° | **4.** 110° | **7.** 210° | **10.** −690° |
| **2.** 45° | **5.** −180° | **8.** 330° | **11.** 900° |
| **3.** 90° | **6.** −270° | **9.** −480° | **12.** 1050° |

Using $\pi \doteq 3.1416$, convert each degree measure to the nearest hundredth of a radian.

**13.** 80°  **14.** 142°  **15.** −218°  **16.** −341°

Convert each radian measure to a degree measure. Express your result to the nearest degree.

**17.** $\dfrac{\pi}{5}^{R}$

**18.** $\dfrac{3\pi}{4}^{R}$

**19.** $-\dfrac{5\pi}{12}^{R}$

**20.** $-\dfrac{7\pi}{18}^{R}$

**21.** $\dfrac{18\pi}{5}^{R}$

**22.** $\dfrac{16\pi}{3}^{R}$

**23.** $-\dfrac{7\pi}{12}^{R}$

**24.** $-\dfrac{11\pi}{6}^{R}$

**25.** $2^{R}$

**26.** $3^{R}$

**27.** $\frac{1}{2}^{R}$

**28.** $0^{R}$

State the exact value of each expression.

**29.** $\sin \dfrac{\pi}{6}^{R}$

**30.** $\sin \dfrac{3\pi}{4}^{R}$

**31.** $\cos \left(-\dfrac{\pi}{3}\right)^{R}$

**32.** $\cos \left(-\dfrac{\pi}{4}\right)^{R}$

**33.** $\tan 2\pi^{R}$

**34.** $\cot 3\pi^{R}$

**35.** $\sec \dfrac{7\pi}{6}^{R}$

**36.** $\csc \dfrac{5\pi}{3}^{R}$

Find the length of the arc on a circle with the given radius that is intercepted by a central angle having the given measure. Use $\pi \doteq 3.14$ and state your result to the nearest tenth.

**37.** 8 cm.; 270°

**38.** 3 ft.; 150°

**39.** 10 in.; $\dfrac{2\pi}{3}^{R}$

**40.** 24 km.; $\dfrac{7\pi}{4}^{R}$

**41.** 4 yd.; 900°

**42.** 8 m.; 1050°

Find the radius of a circle on which an arc of the given length subtends a central angle of the given measure. Use $\pi \doteq 3.14$ and state your result to the nearest tenth.

**43.** $\pi$ in.; 45°

**44.** $\dfrac{7\pi}{3}$ in.; 35°

**45.** 18 ft.; $\frac{3}{4}^{R}$

**46.** 18 yd.; $\pi^{R}$

**47.** 36 ft.; 120°

**48.** 10 yd.; 360°

B **49.** What is the speed in feet per minute of a point on the rim of a wheel of diameter 6 feet, turning at the rate of 5 rpm?

**50.** An automobile is traveling at the rate of 60 mph (88 feet per second). If a tire is 15 inches in radius, find the radian measure of the angle through which the wheel turns in 3 seconds.

**51.** To the nearest hundred miles per minute, what is the speed of an earth satellite traveling in a circular orbit 600 miles high if it completes 1 orbit every 80 minutes? (Use 4000 miles for the earth's radius.)

**52.** If the eye can detect a movement of 0.1 inches per second, how long must the minute hand of a clock be for the movement of its tip to be detected?

$\boxed{\text{C}}$ **53.** Show that the length of an arc subtended by a central angle of $d°$ in a circle of radius $r$ is given by $s = \dfrac{\pi r d}{180}$.

**54.** Show that the velocity $v$ of a point on the edge of a circular disk of radius $r$ feet turning at $p$ rpm is $2\pi p r$ feet per minute.

## 13–2 The Circular Functions

Every real number $t$ is the measure of one and only one arc with initial point $T$ on the circle $x^2 + y^2 = 1$. Hence each real number $t$ determines a unique ordered pair of numbers $(a, b)$ which are the coordinates of the terminal point $P$ of the arc (Figure 13–6). The set of all ordered pairs $(t, a)$ is called the **cosine function over the real numbers**; the set of all pairs $(t, b)$ is the **sine function over the real numbers**. We write: $\cos t = a$ and $\sin t = b$.

What is the relationship between these new functions and the cosine and sine functions whose domain is the set of angles in standard position (Chapter 11)? Notice that an arc of measure $t$ subtends at $O$ an angle in standard position with measure $t^R$. Since $P(a, b)$ is a point one unit from the origin on the terminal side of this angle, you have $\cos t^R = a$ and $\sin t^R = b$. Thus:

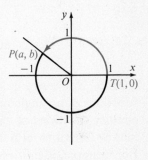

$$\cos t^R = \cos t \quad \text{and} \quad \sin t^R = \sin t.$$

**FIGURE 13–6**

Because of these relationships, you need not distinguish between $\cos t^R$ and $\cos t$, or between $\sin t^R$ and $\sin t$. For example, you can write $\cos 2$ to mean either "cosine of the number 2" or "cosine of the angle whose measure is $2^R$." On the other hand, in referring to an angle whose measure is given in degrees, you *always use the degree symbol.* Notice that $\cos 2° \neq \cos 2$, since $\cos 2° \doteq 0.9994$ and $\cos 2$ or $\cos 2^R \doteq \cos 114°35' \doteq -0.4160$.

In terms of the cosine and sine functions, you can define the tangent, cotangent, secant, and cosecant functions over the real numbers just as we did in Section 11–3 for angles. For example, the tangent function is the following set of ordered pairs of real numbers:

$$\left\{ (t, \tan t) \colon \tan t = \frac{\sin t}{\cos t}, \ \cos t \neq 0 \right\}.$$

The six "trigonometric" functions over the real numbers are sometimes called the circular functions, to distinguish them from the trigonometric functions for angles. However, to find the value of a circular function for any given real number, you can consider the real number to be the radian measure of an angle and then evaluate the corresponding trigonometric function for that angle.

**Example 1.** Find the value of (a) $\sin \frac{2}{3}$; (b) $\cos \frac{5\pi}{6}$; (c) $\tan \left(-\frac{3\pi}{4}\right)$.

*Solution:*

a. $\frac{2^R}{3} = \left(\frac{2}{3} \cdot \frac{180}{\pi}\right)^{\circ} \doteq 38°12'$;

$\therefore \sin \frac{2}{3} \doteq \sin 38°12' \doteq 0.6184.$

b. $\frac{5\pi^R}{6} = \left(\frac{5\pi}{6} \cdot \frac{180}{\pi}\right)^{\circ} = 150°$;

$\therefore \cos \frac{5\pi}{6} = \cos 150° \doteq -0.8660.$

c. $\frac{-3\pi^R}{4} = \left(-\frac{3\pi}{4} \cdot \frac{180}{\pi}\right)^{\circ} = -135°$;

$\therefore \tan \left(-\frac{3\pi}{4}\right) = \tan (-135°) = 1.$

Since, for every real number $t$, the values of the trigonometric functions of an angle of $t$ radians equal the corresponding values of the circular functions of the real number $t$, the properties of the trigonometric functions also hold for the corresponding circular functions. For example, identities such as those in the first column below, which were proved for the set of angles, can be restated as valid identities over the real numbers (column 2).

$$\sin^2 \theta + \cos^2 \theta = 1 \qquad \sin^2 t + \cos^2 t = 1$$

$$\cos (90° - \beta) = \sin \beta \qquad \cos \left(\frac{\pi}{2} - x\right) = \sin x$$

$$\sin (-\alpha) = -\sin \alpha \qquad \sin (-u) = -\sin u$$

**Example 2.** Find the value of $\cos \left(\pi - \frac{\pi}{3}\right)$.

*Solution:* $\cos \left(\pi - \frac{\pi}{3}\right) = \cos \pi \cos \frac{\pi}{3} + \sin \pi \sin \frac{\pi}{3}$

$$= (-1)(\tfrac{1}{2}) + (0)\left(\frac{\sqrt{3}}{2}\right) = -\tfrac{1}{2}$$

## WRITTEN EXERCISES

Find the exact value of each of the six circular functions for the given number.

**A**

1. $\dfrac{\pi}{6}$     3. $\dfrac{\pi}{3}$     5. $-\dfrac{2\pi}{3}$     7. $-\dfrac{3\pi}{4}$     9. $\dfrac{5\pi}{4}$

2. $\dfrac{\pi}{4}$     4. $\dfrac{\pi}{2}$     6. $-\dfrac{5\pi}{6}$     8. $-\pi$     10. $\dfrac{11\pi}{6}$

Find each value to the nearest hundredth. Use $1^R \doteq 57°18'$ and $0.1^R \doteq 5°44'$.

11. $\cos 1.4$        13. $\tan 2.3$        15. $\sec (-1.5)$

12. $\sin 0.8$        14. $\cos 1.7$        16. $\csc (-3)$

At time $t$ (in sec.) the displacement $d$ (in ft.), velocity $v$ (in ft. per. sec.), and acceleration $a$ (in ft. per. sec.²) of a weight hanging on a certain vibrating spring are given by $d = \frac{1}{3} \cos 3t$, $v = -\sin 3t$, and $a = -3 \cos 3t$. Find $d$, $v$, and $a$ for the following values of $t$.

17. $0$        18. $\frac{1}{12}\pi$        19. $\frac{1}{2}\pi$        20. $\frac{1}{3}\pi$

At time $t$ (in sec.) the voltage $V$ and the current $I$ (in amperes) in a circuit are given by $V = 100 \sin 260t$ and $I = 25 \sin 260(t - 0.001)$. Find $V$ and $I$ at the following times.

21. $0$ sec.      22. $0.001$ sec.      23. $0.005$ sec.      24. $0.01$ sec.

Prove each identity over the set of real numbers.

**B**

25. $\dfrac{\sin^2 x + \cos (\pi + x) \cos (\pi - x)}{\sec (\pi - x)} = -\cos x$

26. $\csc^2 (\pi - x) + \sec (\pi - x) \sec (\pi + x) = \csc^2 x + \sec^2 (\pi - x)$

27. $\sec^2 t - \sec^3 t \sin^3 t \cot^3 t = \tan^2 t$

28. $\csc^2 u - \cos^3 u \csc^3 u \tan^3 u = \cot^2 u$

29. $\dfrac{\tan v + \sin v}{\tan v - \sin v} = \dfrac{\sec v + 1}{\sec v - 1}$

30. $\dfrac{2 \tan r}{1 + \tan^2 r} = \sin 2r$

31. $\dfrac{2 \sin x(1 - \sin^2 x)}{\sin 2x} = \cos x$

**32.** $\dfrac{\sin 2t}{1 - \sin^2 t} - \tan t = \tan t$

**33.** $\dfrac{\sin 4r}{\sin 2r} = 2 \cos 2r$

**34.** $\dfrac{2 \cos 3x}{\sin 2x} + \dfrac{\sin 2x}{\cos x} = \dfrac{\cos 2x}{\sin x}$

**35.** $\cos \left(\dfrac{\pi}{3} + y\right) + \cos \left(\dfrac{\pi}{3} - y\right) = \cos y$

**36.** $\dfrac{\tan (r + s) - \tan s}{1 + \tan (r + s) \tan s} = \tan r$

C **37.** $\dfrac{\cos 12y - \sin 6y + \cos 6y - \sin 12y}{\cos 12y + \cos 6y} = 1 - \tan 9y$

**38.** $\dfrac{2 \cos k}{\cos k \sin 2k + \sin k \cos 2k - \sin k} = \cot k + \tan 2k$

**39.** If $\sin u = \sin v$, prove that there is an integer $n$ for which $u = v + 2n\pi$ or $u = (\pi - v) + 2n\pi$.

**40.** If $\cos u = \cos v$, prove that there is an integer $n$ for which $u = v + 2n\pi$ or $u = (2\pi - v) + 2n\pi$.

## 13–3 Graphs of Cosine and Sine Functions

In Figure 13–7, the terminal points of arcs of length $t$ for various convenient values of $t$ between 0 and $2\pi$ have been indicated on the unit circle. Recall that for each value of $t$ the abscissa of the terminal point is $\cos t$, and the ordinate is $\sin t$. Therefore, by noting the pattern of change in the coordinates of the terminal points, you can see the variation in the values of the sine and cosine functions as $t$ varies from 0 to $2\pi$.

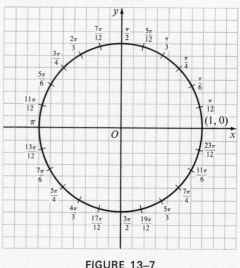

FIGURE 13–7

| When $t$ increases from | $\cos t$ | $\sin t$ |
|---|---|---|
| 0 to $\dfrac{\pi}{2}$ | decreases from 1 to 0 | increases from 0 to 1 |
| $\dfrac{\pi}{2}$ to $\pi$ | decreases from 0 to $-1$ | decreases from 1 to 0 |
| $\pi$ to $\dfrac{3\pi}{2}$ | increases from $-1$ to 0 | decreases from 0 to $-1$ |
| $\dfrac{3\pi}{2}$ to $2\pi$ | increases from 0 to 1 | increases from $-1$ to 0 |

For $t$ greater than $2\pi$ or less than 0, the values of $\cos t$ and $\sin t$ repeat the patterns of the interval from 0 to $2\pi$, because for every integer $n$, $\cos (t + n \cdot 2\pi) = \cos t$ and $\sin (t + n \cdot 2\pi) = \sin t$. Whenever $f$ is a function such that for a nonzero constant $p$, $f(x + p) = f(x)$ for each $x$ in the domain of $f$, then $f$ is called a **periodic function**. The smallest such *positive* constant $p$ is called the **period** of $f$. Thus, the cosine and sine functions are periodic with period $2\pi$.

You can picture the variation and periodicity of these functions by drawing their graphs. The graph of $\{(x, y): y = \sin x\}$ is shown in Figure 13–8. It is a smooth curve obtained by joining the points whose coordinates are given in the table and then repeating the pattern in both directions along the $x$-axis. Note that the scales on the axes are equal. For convenience, 3.14 units have been marked off on the $x$-axis to denote $\pi$, and the numbers $\dfrac{\pi}{6}, \dfrac{\pi}{3}, \dfrac{\pi}{2}$, and so forth, have been shown. Note also that $\dfrac{\sqrt{3}}{2} \doteq 0.87$.

| $x$ | 0 | $\dfrac{\pi}{6}$ | $\dfrac{\pi}{3}$ | $\dfrac{\pi}{2}$ | $\dfrac{2\pi}{3}$ | $\dfrac{5\pi}{6}$ | $\pi$ | $\dfrac{7\pi}{6}$ | $\dfrac{4\pi}{3}$ | $\dfrac{3\pi}{2}$ | $\dfrac{5\pi}{3}$ | $\dfrac{11\pi}{6}$ | $2\pi$ |
|---|---|---|---|---|---|---|---|---|---|---|---|---|---|
| $y = \sin x$ | 0 | $\dfrac{1}{2}$ | $\dfrac{\sqrt{3}}{2}$ | 1 | $\dfrac{\sqrt{3}}{2}$ | $\dfrac{1}{2}$ | 0 | $-\dfrac{1}{2}$ | $-\dfrac{\sqrt{3}}{2}$ | $-1$ | $-\dfrac{\sqrt{3}}{2}$ | $-\dfrac{1}{2}$ | 0 |

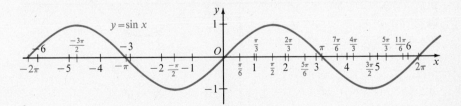

FIGURE 13–8

Figure 13–9 shows the graph of $\{(x, y): y = \cos x\}$.

| $x$ | 0 | $\dfrac{\pi}{6}$ | $\dfrac{\pi}{3}$ | $\dfrac{\pi}{2}$ | $\dfrac{2\pi}{3}$ | $\dfrac{5\pi}{6}$ | $\pi$ | $\dfrac{7\pi}{6}$ | $\dfrac{4\pi}{3}$ | $\dfrac{3\pi}{2}$ | $\dfrac{5\pi}{3}$ | $\dfrac{11\pi}{6}$ | $2\pi$ |
|---|---|---|---|---|---|---|---|---|---|---|---|---|---|
| $y = \cos x$ | 1 | $\dfrac{\sqrt{3}}{2}$ | $\dfrac{1}{2}$ | 0 | $-\dfrac{1}{2}$ | $-\dfrac{\sqrt{3}}{2}$ | $-1$ | $-\dfrac{\sqrt{3}}{2}$ | $-\dfrac{1}{2}$ | 0 | $\dfrac{1}{2}$ | $\dfrac{\sqrt{3}}{2}$ | 1 |

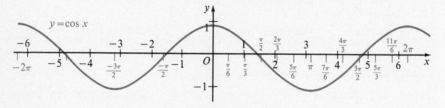

FIGURE 13–9

The curves in Figures 13–8 and 13–9 are called *sinusoidal waves*. Both curves have 1 as maximum ordinate and $-1$ as minimum ordinate. When a periodic function has a maximum value $M$ and a minimum value $m$, the **amplitude** of the function is $\dfrac{M - m}{2}$. Thus, the amplitude of the sine and cosine functions is $\dfrac{1 - (-1)}{2} = 1$.

To draw the graph of $\{(x, y): y = 3 \cos x\}$, notice that the ordinate of each point on the graph of this function is $3$ times the ordinate of the corresponding point on the curve in Figure 13–9. Therefore, as shown in Figure 13–10, this function has period $2\pi$ and amplitude 3.

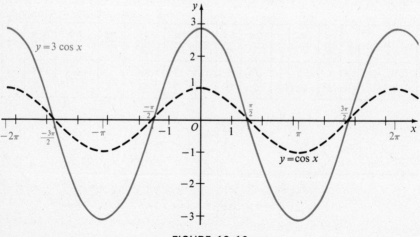

FIGURE 13–10

In general, the graph of any function of the form $\{(x, y): y = a\cos x\}$, or $\{(x, y): y = a\sin x\}$, $a \neq 0$, is a sinusoidal curve with period $2\pi$ and amplitude $|a|$.

To draw the graph of $\{(x, y): y = \cos 3x\}$, notice that as $x$ varies from 0 to $\dfrac{2\pi}{3}$, $3x$ varies from 0 to $2\pi$. Therefore, $\cos 3x$ will run through a complete cycle of cosine values as $x$ varies from 0 to $\dfrac{2\pi}{3}$. This fact also shows up in the following table. Thus, as the graph in Figure 13–11 indicates, the function has period $\dfrac{2\pi}{3}$ and amplitude 1.

| $x$ | 0 | $\dfrac{\pi}{18}$ | $\dfrac{\pi}{9}$ | $\dfrac{\pi}{6}$ | $\dfrac{2\pi}{9}$ | $\dfrac{5\pi}{18}$ | $\dfrac{\pi}{3}$ | $\dfrac{7\pi}{18}$ | $\dfrac{4\pi}{9}$ | $\dfrac{\pi}{2}$ | $\dfrac{5\pi}{9}$ | $\dfrac{11\pi}{18}$ | $\dfrac{2\pi}{3}$ |
|---|---|---|---|---|---|---|---|---|---|---|---|---|---|
| $3x$ | 0 | $\dfrac{\pi}{6}$ | $\dfrac{\pi}{3}$ | $\dfrac{\pi}{2}$ | $\dfrac{2\pi}{3}$ | $\dfrac{5\pi}{6}$ | $\pi$ | $\dfrac{7\pi}{6}$ | $\dfrac{4\pi}{3}$ | $\dfrac{3\pi}{2}$ | $\dfrac{5\pi}{3}$ | $\dfrac{11\pi}{6}$ | $2\pi$ |
| $y = \cos 3x$ | 1 | $\dfrac{\sqrt{3}}{2}$ | $\dfrac{1}{2}$ | 0 | $-\dfrac{1}{2}$ | $-\dfrac{\sqrt{3}}{2}$ | $-1$ | $-\dfrac{\sqrt{3}}{2}$ | $-\dfrac{1}{2}$ | 0 | $\dfrac{1}{2}$ | $\dfrac{\sqrt{3}}{2}$ | 1 |

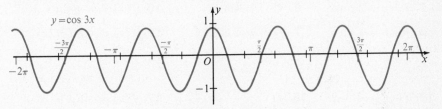

FIGURE 13–11

In general, the graph of a function of the form $\{(x, y): y = a\cos bx\}$, or $\{(x, y): y = a\sin bx\}$, $a \neq 0$, $b \neq 0$, is a sinusoidal curve with amplitude $|a|$ and period $\dfrac{2\pi}{|b|}$. By using this fact, you can quickly sketch the curve by locating its maximum and minimum points and the points where it crosses the $x$-axis.

**Example.** Sketch the graph of $y = -3\sin\frac{1}{2}x$.

*Solution:* Amplitude $= |-3| = 3$; period $= \dfrac{2\pi}{\frac{1}{2}} = 4\pi$.

Because the coefficient $-3$ is negative, the curve will be the reflection in the $x$-axis of an ordinary sine curve. Use the following information to sketch the curve over the interval $0 \leq x \leq 4\pi$:

1. The maximum point occurs at the third quarter-point of the interval: $(3\pi, 3)$.

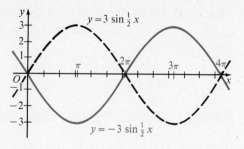

2. The minimum point occurs at the first quarter-point of the interval: $(\pi, -3)$.

3. The curve crosses the $x$-axis at the two endpoints and the midpoint: $(0, 0)$, $(2\pi, 0)$, $(4\pi, 0)$.

With this information, the curve is sketched as shown.

## WRITTEN EXERCISES

Determine the amplitude and period $p$ of each function $f$ whose values are indicated, and then sketch the graph over the following interval of two periods: $-p \leq x \leq p$.

**A**

1. $f(x) = 3 \cos x$
5. $f(x) = \cos 3x$
9. $f(x) = 2 \cos x$

2. $f(x) = 2 \sin x$
6. $f(x) = \sin 4x$
10. $f(x) = -\frac{1}{3} \sin x$

3. $f(x) = \frac{1}{2} \sin x$
7. $f(x) = 2 \cos \frac{1}{3}x$
11. $f(x) = \cos (-2x)$

4. $f(x) = \frac{2}{3} \sin x$
8. $f(x) = 3 \sin \frac{1}{2}x$
12. $f(x) = -2 \sin (-\frac{1}{2}x)$

**B**

13. $f(x) = \sin \pi x$
15. $f(x) = -\cos \frac{\pi x}{2}$
17. $f(x) = \sin x - 2$

14. $f(x) = \cos \frac{\pi}{3} x$
16. $f(x) = -\sin \frac{\pi x}{3}$
18. $f(x) = \cos x + 1$

**C**

19. $f(x) = \sin \left(x - \frac{\pi}{6}\right)$
20. $f(x) = \cos \left(x + \frac{\pi}{3}\right)$

Over the given interval draw the graphs of $f$ and $g$ in the same coordinate plane. From the graphs find to the nearest tenth the values of $x$ for which $f(x) = g(x)$.

21. $f(x) = \sin x$
$g(x) = 2 \cos x, 0 \leq x < 2\pi$

22. $f(x) = \sin 2x$
$g(x) = 3 \cos x, 0 \leq x < 2\pi$

23. $f(x) = \cos \pi x$
$g(x) = x, 0 \leq x < 2$

24. $f(x) = \sin \pi x$
$g(x) = x - 1, 0 \leq x < 2$

25. $f(x) = \sin \left(\frac{\pi}{2} - x\right)$
$g(x) = x, 0 \leq x \leq 2\pi$

## 13–4   Graphs of Other Circular Functions

Can you find the coordinates of the point of intersection of ray $\overrightarrow{OP}$ (Figure 13–12) and the line $x = 1$ which is tangent to the unit circle?

Since ray $\overrightarrow{OP}$ has slope $\dfrac{\sin t}{\cos t}$, or $\tan t$,

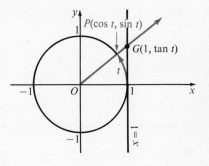

it is contained in the line $y = (\tan t)x$. Therefore, for $P$ in the first or fourth quadrant, the ray intersects the line $x = 1$ at the point $G(1, \tan t)$.

This means that you can visualize the variation in the value of $\tan t$ as

$t$ varies from $-\dfrac{\pi}{2}$ to $\dfrac{\pi}{2}$ by noting the

change in the ordinate of $G$ as $P$

**FIGURE 13–12**

moves counterclockwise on the circle from $S(0, -1)$ to $T(0, 1)$. For $P$ at $S$, that is, for $t = -\dfrac{\pi}{2}$, there is no value of $\tan t$, since the ray $\overrightarrow{OS}$ does not intersect the line $x = 1$. As $t$ varies from $-\dfrac{\pi}{2}$ to 0, $\tan t$ increases through all negative numbers to zero. As $t$ varies from 0 to $\dfrac{\pi}{2}$, $\tan t$ increases from zero through all positive numbers.

You can see from the identity

$$\tan (t + \pi) = \frac{\tan t + \tan \pi}{1 - \tan t \tan \pi} = \frac{\tan t + 0}{1 - \tan t \cdot 0} = \tan t$$

that *the tangent function has period $\pi$.* This implies that as $t$ varies from $\dfrac{\pi}{2}$ to $\dfrac{3\pi}{2}$, $\tan t$ takes on the same values it did when $t$ varied from $-\dfrac{\pi}{2}$ to $\dfrac{\pi}{2}$.

The graph of $\{(x, y): y = \tan x\}$ pictures the variation described above and is obtained by repeating the pattern shown by plotting points over one period (Figure 13–13 on the next page).

| $x$ | $-\dfrac{\pi}{2}$ | $-\dfrac{\pi}{3}$ | $-\dfrac{\pi}{4}$ | $-\dfrac{\pi}{6}$ | $0$ | $\dfrac{\pi}{6}$ | $\dfrac{\pi}{4}$ | $\dfrac{\pi}{3}$ | $\dfrac{\pi}{2}$ |
|---|---|---|---|---|---|---|---|---|---|
| $y = \tan x$ | — | $-\sqrt{3}$ | $-1$ | $-\dfrac{\sqrt{3}}{3}$ | $0$ | $\dfrac{\sqrt{3}}{3}$ | $1$ | $\sqrt{3}$ | — |

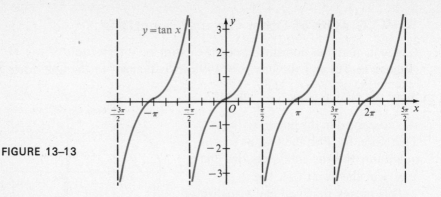

**FIGURE 13–13**

Because tan $x$ does not exist when $x$ is replaced by any odd multiple of $\dfrac{\pi}{2}$, the equations of the dashed lines (the asymptotes) in the figure are all of the form $x = (2n + 1)\dfrac{\pi}{2}$, where $n$ is any integer.

You can discuss the variations and draw the graphs of the cotangent, secant, and cosecant functions by referring to a table of values or to the known graphs of their reciprocal functions. Figures 13–14, 13–15, and 13–16 picture the graphs of each pair of reciprocal functions.

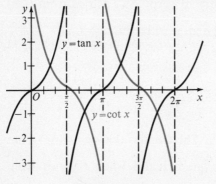

**FIGURE 13–14**

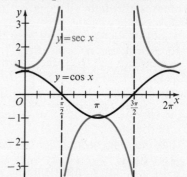

**FIGURE 13–15**

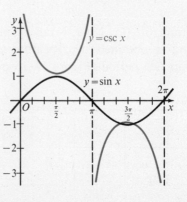

**FIGURE 13–16**

## WRITTEN EXERCISES

Determine the period of each function $f$ whose values are indicated, and sketch the graph over an interval two periods in length.

**A**

**1.** $f(x) = \frac{1}{2}\tan x$      **5.** $f(x) = -\tan x$      **9.** $f(x) = \tan \frac{1}{2}x$

**2.** $f(x) = \frac{1}{2}\sec x$      **6.** $f(x) = -\cot x$      **10.** $f(x) = \sec 2x$

**3.** $f(x) = 2\cot x$      **7.** $f(x) = -\frac{1}{2}\sec x$      **11.** $f(x) = \csc \frac{1}{2}x$

**4.** $f(x) = 2\csc x$      **8.** $f(x) = -\frac{1}{2}\csc x$      **12.** $f(x) = \cot 2x$

Describe the variation of each value as $t$ varies from **(a)** 0 to $\pi$; **(b)** $\pi$ to $2\pi$; **(c)** 0 to $-\pi$; **(d)** $-\pi$ to $-\dfrac{3\pi}{2}$.

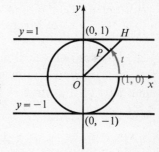

**13.** $\tan t$ and $\cot t$      **15.** $\cos t$ and $\sec t$

**14.** $\sin t$ and $\csc t$      **16.** $\sin^2 t$ and $\cos^2 t$

Refer to the figure to discuss the variation of each function in terms of the $x$-coordinate of $H$ (Ex. 17) and the distance $OH$ (Ex. 18).

**B**

**17.** cotangent function    **18.** cosecant function

From the graphs of $f$ and $g$, find to the nearest tenth the values of $x$ for which $f(x) = g(x)$.

**19.** $f(x) = x$
$\phantom{19.\ }g(x) = \tan x,\ -\pi \le x < \pi$

**20.** $f(x) = \tan\left(x + \dfrac{\pi}{4}\right)$
$\phantom{20.\ }g(x) = \frac{1}{2}$

## INVERSE FUNCTIONS AND GRAPHS

## 13–5   Inverse Sines and Inverse Cosines

The equation $\cos x = \frac{1}{2}$ has infinitely many solutions; namely, $\dfrac{\pi}{3}$, $-\dfrac{\pi}{3}$, and all the numbers which differ from either of these by an integral multiple of $2\pi$ (see Figure 13–17). However, if you require that a so-

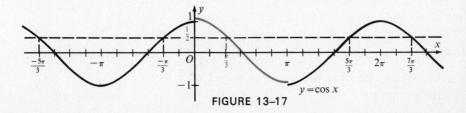

FIGURE 13–17

lution of cos $x = \frac{1}{2}$ *also* satisfy the condition

$0 \leq x \leq \pi$, then there is only one solution, $\frac{\pi}{3}$.

Indeed, for any number $u$ such that $-1 \leq u \leq 1$, there is a unique number $v$ which satisfies

$\qquad$ cos $v = u$ $\qquad$ and $\qquad$ $0 \leq v \leq \pi$

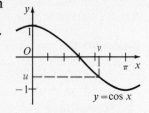

(Figure 13–18). This number $v$ is denoted by **Arccos** $u$, read "Arc cosine of $u$."

**FIGURE 13–18**

**Example 1.** Find (a) Arccos 0; (b) Arccos $(-\frac{1}{2})$; (c) Arccos 1.

*Solution:* $\qquad$ (a) Arccos 0 $= \frac{\pi}{2}$ ; (b) Arccos $(-\frac{1}{2}) = \frac{2\pi}{3}$ ;

$\qquad$ (c) Arccos 1 $= 0$.

If you restrict the domain of the cosine function to the interval $0 \leq x \leq \pi$, you obtain a new function, Cosine, whose graph is the solid part of the curve in Figure 13–19. Since $y = $ Cos $x$ if and only if

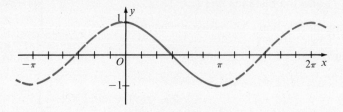

**FIGURE 13–19**

$y = \cos x$ *and* $0 \leq x \leq \pi$, you have

$\qquad$ Arccos $u = v$ $\qquad$ if and only if $\qquad$ $u = $ Cos $v$.

Thus Arccos and Cos are inverse functions (recall Section 10–4). The **Arc cosine** function is also called the **inverse cosine function** and is often denoted by Cos$^{-1}$. Do you see that the following four statements are equivalent?

$y = $ **Arccos** $x$ $\qquad$ cos $y = x$ **and** $0 \leq y \leq \pi$

$y = $ Cos$^{-1}$ $x$ $\qquad$ Cos $y = x$

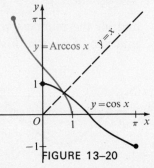

By Section 10–4, you can obtain the graph of $y = $ Arccos $x$ (shown in red in Figure 13–20) by reflecting the graph of $y = $ Cos $x$ in the line $y = x$.

**FIGURE 13–20**

In defining the **Arc sine function,** or **inverse sine function,** denoted Arcsin, you proceed as above, except that you restrict the sine function to the interval $-\dfrac{\pi}{2} \le x \le \dfrac{\pi}{2}$ instead of $0 \le x \le \pi$. (See Figure 13–21.) Thus you have

$$y = \textbf{Arcsin } x \quad \textbf{if and only if} \quad \sin y = x \textbf{ and } -\dfrac{\pi}{2} \le y \le \dfrac{\pi}{2}.$$

The notation $\mathrm{Sin}^{-1} x$ is often used in place of Arcsin $x$.

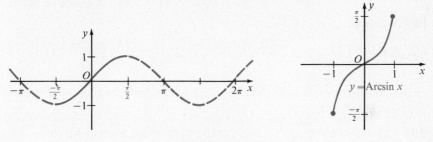

**FIGURE 13–21**

**Example 2.** Find **(a)** Arcsin $\left( \sin \dfrac{2\pi}{3} \right)$ and **(b)** cos [Arcsin $(-\tfrac{3}{5})$].

*Solution:*  **a.** Let $y = $ Arcsin $\left( \sin \dfrac{2\pi}{3} \right)$.

Then $\sin y = \sin \dfrac{2\pi}{3}$ and $-\dfrac{\pi}{2} \le y \le \dfrac{\pi}{2}$.

Since $\sin \dfrac{\pi}{3} = \sin \dfrac{2\pi}{3}$ and $-\dfrac{\pi}{2} \le \dfrac{\pi}{3} \le \dfrac{\pi}{2}$,

$$\text{Arcsin} \left( \sin \dfrac{2\pi}{3} \right) = \dfrac{\pi}{3}.$$

**b.** Let $v = $ Arcsin $(-\tfrac{3}{5})$.

Then $\sin v = -\tfrac{3}{5}$ and

$$-\dfrac{\pi}{2} \le v \le \dfrac{\pi}{2}.$$

$\therefore \cos v = +\sqrt{1 - \sin^2 v}$

$= \sqrt{1 - (-\tfrac{3}{5})^2} = \tfrac{4}{5}.$

A sketch such as the one shown may help in solving problems of this type.

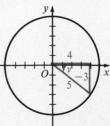

*ORAL EXERCISES*

State the value of each of the following.

**1.** Arcsin 1

**2.** Arcsin 0

**3.** Arcsin (−1)

**4.** Arccos (−1)

**5.** Arcsin $\frac{1}{2}$

**6.** Arcsin (−$\frac{1}{2}$)

**7.** Arccos $\frac{\sqrt{2}}{2}$

**8.** Arcsin $\frac{\sqrt{2}}{2}$

**9.** Arccos $\left(-\frac{\sqrt{2}}{2}\right)$

**10.** Arcsin $\left(-\frac{\sqrt{2}}{2}\right)$

**11.** Arcsin $\frac{\sqrt{3}}{2}$

**12.** Arccos $\frac{\sqrt{3}}{2}$

**13.** Arcsin $\left(-\frac{\sqrt{3}}{2}\right)$

**14.** Arccos $\left(-\frac{\sqrt{3}}{2}\right)$

State **(a)** the domain and **(b)** the range of the functions defined by the following formulas.

**15.** $f(x) =$ Arccos $x$

**16.** $f(x) =$ Arcsin $x$

**17.** $f(x) =$ Arcsin (sin $x$)

**18.** $f(x) =$ Arccos (cos $x$)

**19.** $f(x) =$ sin (Arcsin $x$)

**20.** $f(x) =$ cos (Arccos $x$)

*WRITTEN EXERCISES*

Evaluate each of the following.

**A**

**1.** Arcsin $\left[ \sin \left( -\frac{\pi}{2} \right) \right]$

**2.** Arccos [cos $\pi$]

**3.** Arccos $\left[ \cos \left( -\frac{\pi}{2} \right) \right]$

**4.** Arcsin $\left( \sin \frac{3\pi}{2} \right)$

**5.** Arcsin $\left( \sin \frac{5\pi}{6} \right)$

**6.** Arccos $\left[ \cos \left( -\frac{\pi}{3} \right) \right]$

**7.** Arcsin $\left[ \cos \left( -\frac{\pi}{4} \right) \right]$

**8.** Arccos $\left[ \sin \left( -\frac{\pi}{4} \right) \right]$

**9.** sin [Arccos (−$\frac{4}{5}$)]

**10.** cos [Arcsin $\frac{5}{13}$]

**11.** cos [Arcsin (−$\frac{4}{5}$)]

**12.** sin [Arccos (−$\frac{12}{13}$)]

**13.** cos $\left(\text{Arcsin} \frac{\sqrt{2}}{2} + \text{Arccos} \frac{\sqrt{2}}{2}\right)$

**14.** sin $\left[ \text{Arccos} \left(-\frac{\sqrt{2}}{2}\right) + \text{Arcsin} \frac{\sqrt{2}}{2} \right]$

**15.** cos (2 Arccos 0)

**16.** cos [2 Arcsin (−1)]

For each exercise, **(a)** express without using circular or inverse circular functions and **(b)** draw the graph.

**Sample.** $y = \sin (\text{Arccos } x)$

*Solution:* Note that the expression in the right member is defined only when $-1 \leq x \leq 1$.

**a.** Let $t = \text{Arccos } x$.
Then $\cos t = x$ and $0 \leq t \leq \pi$.
$$y = \sin t = +\sqrt{1 - \cos^2 t}$$
$$\therefore \ y = \sqrt{1 - x^2}$$

**b.** $y^2 = 1 - x^2$, and $y \geq 0$
$$x^2 + y^2 = 1$$

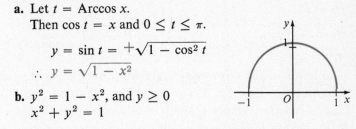

The graph is the part of the unit circle for which $y \geq 0$.

**B**

**17.** $y = \cos (\text{Arcsin } x)$

**18.** $y = \sin (-\text{Arccos } x)$

**19.** $y = \sin (\text{Arcsin } x)$

**20.** $y = \cos (\text{Arccos } x)$

**21.** $y = \text{Arcsin } (\sin x)$, $-\dfrac{\pi}{2} \leq x \leq \dfrac{\pi}{2}$

**22.** $y = \text{Arccos } (\cos x)$, $0 \leq x \leq \pi$

**23.** $y = \text{Arcsin } (\cos x)$, $0 \leq x \leq \pi$
$\left( Hint: \text{Use } \cos x = \sin \left( \dfrac{\pi}{2} - x \right). \right)$

**24.** $y = \text{Arccos } (\sin x)$, $-\dfrac{\pi}{2} \leq x \leq \dfrac{\pi}{2}$

**C**

**25.** $y = \text{Arccos } (\cos x)$, $\pi \leq x \leq 2\pi$
(*Hint:* Use $\cos x = \cos (2\pi - x)$ because $0 \leq 2\pi - x \leq \pi$.)

**26.** $y = \text{Arcsin } (\sin x)$, $\dfrac{\pi}{2} \leq x \leq \dfrac{3\pi}{2}$
$\left( Hint: \text{Use } \sin x = \sin (\pi - x) \text{ because } -\dfrac{\pi}{2} \leq \pi - x \leq \dfrac{\pi}{2}. \right)$

## 13–6  More on Inverse Functions

The graph of the tangent function in Figure 13–22, page 532, indicates that, given any real number $u$, there is a unique number $v$ in the interval $-\dfrac{\pi}{2} < v < \dfrac{\pi}{2}$ such that $\tan v = u$. We denote this number by **Arctan** $u$.

Do you see that

$$\text{Arctan } 1 = \frac{\pi}{4}, \ \text{Arctan } (-1) = -\frac{\pi}{4}, \ \text{and } \text{Arctan } \sqrt{3} = \frac{\pi}{3} ?$$

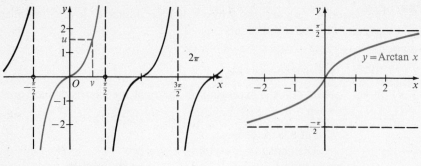

FIGURE 13–22                    FIGURE 13–23

In terms of the customary variables $x$ and $y$, we have the following definition for the **Arc tangent function** (denoted Arctan), or **inverse tangent function**, whose graph is shown in Figure 13–23:

$$y = \textbf{Arctan } x \quad \textbf{if and only if} \quad \tan y = x \quad \textbf{and} \quad -\frac{\pi}{2} < y < \frac{\pi}{2}.$$

**Example 1.**  Evaluate $\tan [\text{Arctan } \frac{1}{4} + \text{Arccos } (-\frac{4}{5})]$.

*Solution:*      Let $a = \text{Arctan } \frac{1}{4}$ and $b = \text{Arccos } (-\frac{4}{5})$.

Then $\tan a = \frac{1}{4}$ and $\cos b = -\frac{4}{5}$.

Since $b$ is restricted to the interval $0 \leq b \leq \pi$ and $\cos b$ is negative, you have $\frac{\pi}{2} < b < \pi$.

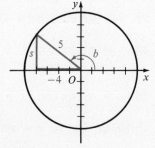

Sketch an appropriate right triangle, as shown, and determine $s$ using the Pythagorean Theorem. You have $s = 3$, and hence $\tan b = -\frac{3}{4}$.

Since $\tan (a + b) = \dfrac{\tan a + \tan b}{1 - \tan a \tan b}$,

$$\tan [\text{Arctan } \tfrac{1}{4} + \text{Arccos } (-\tfrac{4}{5})] = \frac{\tfrac{1}{4} + (-\tfrac{3}{4})}{1 - \tfrac{1}{4}(-\tfrac{3}{4})} = -\tfrac{8}{19}.$$

The **Arc cotangent function**, or **inverse cotangent function**, is defined as follows:

$$y = \textbf{Arccot } x \quad \textbf{if and only if} \quad \cot y = x \text{ and } 0 < y < \pi.$$

**Example 2.** Write $\cos(2\,\text{Arccot}\,x)$ without using circular or inverse circular functions.

*Solution:*  *Plan:* Use the formula $\cos 2y = 2\cos^2 y - 1$ with $y = \text{Arccot}\,x$.

Then $\cot y = x$ and $0 < y < \pi$.

$\sec^2 y = 1 + \cot^2 y = 1 + x^2$, and thus $\cos^2 y = \dfrac{1}{1 + x^2}$.

$\therefore \cos(2\,\text{Arccot}\,x) = \cos 2y = 2\cos^2 y - 1$

$$= 2\left(\frac{1}{1+x^2}\right) - 1 = \frac{1 - x^2}{1 + x^2}.$$

We can define the remaining inverse circular functions, **Arc secant** and **Arc cosecant**, in terms of those already discussed:

**If $|x| \geq 1$,  Arcsec $x = \text{Arccos}\,\dfrac{1}{x}$,  and  Arccsc $x = \text{Arcsin}\,\dfrac{1}{x}$.**

The material in this section and the preceding one can be restated to define the inverse *trigonometric* functions. For example, Arctan $\sqrt{3}$ can be interpreted as the *angle* whose tangent is $\sqrt{3}$ and which lies between $-90°$ and $90°$, that is, Arctan $\sqrt{3} = 60°$.

*ORAL EXERCISES*

State the domain and range of each function as **(a)** an inverse circular function and **(b)** an inverse trigonometric function.

**1.** Arc tangent  **3.** Arc sine  **5.** Arc secant

**2.** Arc cotangent  **4.** Arc cosine  **6.** Arc cosecant

State the value of each of the following **(a)** as a real number and **(b)** as an angle.

**7.** Arctan 0   **10.** Arccot $(-1)$   **13.** Arctan $(\cos \pi)$

**8.** Arcsin $\left(\dfrac{\sqrt{3}}{2}\right)$   **11.** Arctan $(-\sqrt{3})$   **14.** Arcsin $(\cos \pi)$

**9.** Arccos $\left(-\dfrac{\sqrt{3}}{2}\right)$   **12.** Arccot 0   **15.** Arccos $\left[\cot\left(-\dfrac{\pi}{4}\right)\right]$

*WRITTEN EXERCISES*

Evaluate each of the following.

**A**  **1.** cot (Arctan $\frac{2}{3}$)

**2.** cos (Arcsec 2)

**3.** sin (Arctan 3)

**4.** tan [Arccos $(-\frac{1}{3})$]

**5.** 2 sin (Arctan $\frac{5}{12}$)

**6.** 2 cos [Arcsin $(-\frac{3}{5})$]

**7.** sin (2 Arctan $\frac{5}{12}$)

**8.** cos [2 Arcsin $(-\frac{3}{5})$]

**9.** tan [2 Arccos $(-\frac{2}{3})$]

**10.** sec (2 Arccos $\frac{1}{4}$)

**11.** tan ($\frac{1}{2}$ Arcsin $\frac{2}{3}$)

**12.** cos [$\frac{1}{2}$ Arccos $(-\frac{1}{3})$]

**13.** sin (Arctan $\frac{3}{4}$ + Arctan $\frac{4}{3}$)

**14.** cos [Arccos $\frac{3}{5}$ − Arctan $(-\frac{3}{4})$]

**15.** tan [Arccot $\frac{12}{5}$ − Arccos $(-\frac{3}{5})$]

**16.** cot [Arcsin $\frac{1}{3}$ + Arccos $\frac{1}{3}$]

Write without using circular or inverse circular functions.

**B**  **17.** sin (Arccos $2x$)

**18.** tan $\left(\text{Arcsin } \dfrac{x}{2}\right)$

**19.** cos (2 Arcsin $x$)

**20.** sin (2 Arctan $x$)

**21.** tan (Arctan $x$ + Arctan $y$)

**22.** cot (Arctan $x$ − Arctan $y$)

**23.** sin (Arccos $x$ − Arccos $y$)

**24.** cos (Arcsin $x$ − Arcsin $y$)

Prove each statement.

**25.** Arctan $\dfrac{2}{3}$ + Arctan $\dfrac{1}{5}$ = $\dfrac{\pi}{4}$     **26.** Arctan $\dfrac{4}{3}$ − Arctan $\dfrac{1}{7}$ = $\dfrac{\pi}{4}$

**27.** Arcsin $\frac{12}{13}$ + Arccos $\frac{3}{5}$ = Arccos $(-\frac{33}{65})$

**28.** Arcsin $\frac{1}{3}$ − Arccos $\frac{7}{11}$ = Arcsin $(-\frac{17}{33})$

**29.** 2 Arcsin $\frac{3}{5}$ + Arccos $\frac{3}{5}$ = $\pi$ − Arcsin $\frac{4}{5}$

**30.** 2 Arctan 2 − Arctan $\left(-\dfrac{1}{7}\right)$ = $\dfrac{3\pi}{4}$

**31.** Arcsin $a$ = $\dfrac{\pi}{2}$ − Arccos $a$     **33.** Arcsin $\dfrac{2b}{1+b^2}$ = 2 Arctan $b$

**32.** Arctan $c^2$ + Arctan $\dfrac{1}{c^2}$ = $\dfrac{\pi}{2}$     **34.** 2 Arccos $x^2$ = Arccos $(2x^4 − 1)$

Solve each equation.

**35.** Arccos $(2x^2 − 1)$ = Arccos 1

**36.** Arctan $(y^2 − 1)$ = Arctan $(y + 1)$

**37.** Arctan $x$ + Arctan $3x$ = Arctan 2

**38.** Arccos $x$ + 2 Arcsin $x$ = Arccos $(-\frac{1}{2})$, $x > 0$

**39.** Sketch the graph of $y = $ Arccot $x$.

**40.** Sketch the graph of $y = $ Arcsec $x$.

**41.** Sketch the graph of $y = $ Arccsc $x$.

## OPEN SENTENCES

## 13–7  General and Particular Solutions

You can employ algebraic transformations (Chapters 2–8), trigonometric identities, and properties of the inverse functions to solve open
sentences involving trigonometric and circular functions.

**Example 1.**  Solve $\tan \theta(\csc \theta + 2) = 0$.

*Solution:*  The equation is equivalent to the compound sentence

$$\tan \theta = 0 \qquad or \qquad \csc \theta + 2 = 0$$

In the interval $-\dfrac{\pi}{2} < \theta < \dfrac{\pi}{2}$,
the only solution is $\theta = 0$.
Since tangent has period $\pi$,
the solution set is

$$\{\theta: \theta = 0 + n\pi, n \in J\}, \text{ or}$$
$$\{\theta: \theta = n\pi, n \in J\}.$$

$$\csc \theta = -2,$$
$$\sin \theta = -\tfrac{1}{2}.$$

In the interval $0 \le \theta < 2\pi$, the only solu-
tions are $\theta = \dfrac{7\pi}{6}$ and $\theta = \dfrac{11\pi}{6}$. Since sine has
period $2\pi$, the solution set is

$$\left\{\theta: \theta = \frac{7\pi}{6} + 2n\pi \text{ or } \theta = \frac{11\pi}{6} + 2n\pi, n \in J\right\}.$$

Checking these values in the original equation is left to you. Thus the solution

set is $\left\{\theta: \theta = n\pi, \dfrac{7\pi}{6} + 2n\pi, \dfrac{11\pi}{6} + 2n\pi, n \in J\right\}$.

The answer given in Example 1 is called the **general solution** of the
open sentence because it indicates all the members in the solution set.
The subset consisting of the solutions in a particular interval, usually
taken to be $0 \le \theta < 2\pi$, is called the **particular solution**. The particular

solution in Example 1 is $\left\{0, \pi, \dfrac{7\pi}{6}, \dfrac{11\pi}{6}\right\}$.

**Example 2.** Find the particular solution of $\sin^2 x + \cos 2x = \cos x$ in the interval $0 \leq x < 2\pi$.

*Solution:* First replace $\sin^2 x$ with $1 - \cos^2 x$ and $\cos 2x$ with $2 \cos^2 x - 1$ to obtain an equation in $\cos x$ only.

$$(1 - \cos^2 x) + (2 \cos^2 x - 1) = \cos x,$$
$$\cos^2 x - \cos x = 0.$$

Solve by factoring: $\cos x(\cos x - 1) = 0$

$$\cos x = 0 \quad or \quad \cos x - 1 = 0$$

$$\left\{\frac{\pi}{2}, \frac{3\pi}{2}\right\} \quad\bigg|\quad \begin{array}{c}\cos x = 1 \\ \{0\}\end{array}$$

Over the specified interval, the solution set is $\left\{0, \dfrac{\pi}{2}, \dfrac{3\pi}{2}\right\}$.

**Example 3.** Solve $\sin \theta - \cos^2 \theta = 0$ over the interval $0° \leq \theta < 360°$.

*Solution:* Replace $\cos^2 \theta$ with $1 - \sin^2 \theta$: $\sin \theta - (1 - \sin^2 \theta) = 0$,

$$\sin^2 \theta + \sin \theta - 1 = 0.$$

Use the quadratic formula to solve for $\sin \theta$:

$$\sin \theta = \frac{-1 \pm \sqrt{1 + 4}}{2} = \frac{-1 \pm \sqrt{5}}{2},$$

$$\sin \theta \doteq 0.618 \text{ or } \sin \theta \doteq -1.618.$$

Since $|\sin \theta| \leq 1$, there is no angle $\theta$ for which $\sin \theta = -1.618$. From Table 6, if $\sin \theta \doteq 0.618$, then $\theta \doteq 38°10'$. Thus, over the specified interval, the solution set is

$$\{38°10', 141°50'\}.$$

*WRITTEN EXERCISES*

Solve the following open sentences for $0 \leq x < 2\pi$, or for $0° \leq \theta < 360°$. Give approximations of values of $x$ to the nearest tenth and of $\theta$ to the nearest degree.

**A**

**1.** $\sec x - 1 = 0$

**2.** $\tan x + 1 = 0$

**3.** $2 \cos x - 1 = 0$

**4.** $2 \sec x - 2\sqrt{3} = 0$

**5.** $2 \sin \theta + \sqrt{3} = 0$

**6.** $4 \cos^2 \theta - 1 = 0$

**7.** $\tan^2 \theta = 0$

**8.** $2 \sin^2 \theta - 1 = 0$

**9.** $\sqrt{3} \csc^2 x + 2 \csc x = 0$

**10.** $3 \sec^2 x + 4 \sec x = 0$

**11.** $2 \sin^2 x + 3 \sin x = -1$    **16.** $\cos^2 x - \sin 2x = 0$

**12.** $2 \cos^2 x - \cos x = 1$    **17.** $\cos 2\theta + \cos \theta + 1 = 0$

**13.** $\cos 2\theta = \frac{1}{2}$    **18.** $\tan^2 \theta - \frac{4}{3} \tan \theta = \frac{4}{3}$

**14.** $\sin 2\theta = 1$    **19.** $2 \cos^2 \theta - 1 > 0$

**15.** $\sin^2 x + \sin 2x = 0$    **20.** $4 \sin^2 \theta - 3 < 0$

State the general solution of each of the following equations over the set of (a) real numbers; (b) angles.

**B** **21.** $\sin x = \cos x$    **22.** $\tan x = \cot x$    **23.** $4 \sin^2 x = 3 \tan^2 x - 1$

Solve each open sentence over the real numbers $x$ for which $-\pi < x \le \pi$.

**C** **24.** $0 \le \sin x \le 1$    **26.** $\tan x \le \sin x$    **28.** $\tan^2 x - 1 \le 0$

**25.** $\cot x \ge 1$    **27.** $\sin x \le \cos x$    **29.** $2 \sin^2 x - 1 \le 0$

## 13–8  Solving Additional Open Sentences

You can often find the solution set of an open sentence involving values of circular functions of an algebraic expression such as $\frac{1}{2}x$, $3x$, or $2x - \dfrac{\pi}{3}$ by *first* solving the open sentence for the given expression and then finding the corresponding solutions for $x$.

**Example 1.**    Solve $4 \sin 2x + \csc 2x = 4$ for $0 \le x < 2\pi$.

*Solution:*    $4 \sin 2x + \csc 2x = 4$

$$4 \sin 2x + \frac{1}{\sin 2x} = 4$$

$4 \sin^2 2x + 1 = 4 \sin 2x$

$4 \sin^2 2x - 4 \sin 2x + 1 = 0$

$(2 \sin 2x - 1)^2 = 0$

$2 \sin 2x - 1 = 0$

$\sin 2x = \frac{1}{2}$

Since $0 \le x < 2\pi$, you have $0 \le 2x < 4\pi$.

$$\therefore 2x = \frac{\pi}{6}, \frac{5\pi}{6}, \frac{13\pi}{6}, \frac{17\pi}{6}, \text{ and } x \in \left\{ \frac{\pi}{12}, \frac{5\pi}{12}, \frac{13\pi}{12}, \frac{17\pi}{12} \right\}.$$

Multiplying by $\sin 2x$ could have introduced extraneous solutions. You can check to see that this did not occur.

**Example 2.** Find the general solution of $\sin \frac{1}{3}z + \cos \frac{1}{3}z = 1$.

*Solution:*

1. Transform the equation to one having $\sin \frac{1}{3}z$ as its left member.

$$\sin \frac{1}{3}z + \cos \frac{1}{3}z = 1$$
$$\sin \frac{1}{3}z = 1 - \cos \frac{1}{3}z$$

2. Square each member of the equation obtained in Step 1 and substitute $1 - \cos^2 \frac{1}{3}z$ for $\sin^2 \frac{1}{3}z$.

$$\sin^2 \frac{1}{3}z = 1 - 2 \cos \frac{1}{3}z + \cos^2 \frac{1}{3}z$$
$$1 - \cos^2 \frac{1}{3}z = 1 - 2 \cos \frac{1}{3}z + \cos^2 \frac{1}{3}z$$

3. Solve the resulting equation by factoring.

$$2 \cos^2 \frac{1}{3}z - 2 \cos \frac{1}{3}z = 0$$
$$2 \cos \frac{1}{3}z(\cos \frac{1}{3}z - 1) = 0$$

$\cos \frac{1}{3}z = 0$

$$\frac{1}{3}z = \frac{\pi}{2} + 2n\pi \text{ or } \frac{3\pi}{2} + 2n\pi$$

$$\therefore z = \frac{3\pi}{2} + 6n\pi \text{ or } \frac{9\pi}{2} + 6n\pi$$

$\cos \frac{1}{3}z = 1$

$$\frac{1}{3}z = 0 + 2n\pi$$

$$z = 6n\pi$$

4. Because the squaring transformation in Step 2 may not produce an equivalent equation, it is essential to check in the original equation.

$$\sin \frac{1}{3}\left(\frac{3\pi}{2} + 6n\pi\right) + \cos \frac{1}{3}\left(\frac{3\pi}{2} + 6n\pi\right) = 1 + 0 = 1$$

$$\sin \frac{1}{3}\left(\frac{9\pi}{2} + 6n\pi\right) + \cos \frac{1}{3}\left(\frac{9\pi}{2} + 6n\pi\right) = -1 + 0 = -1$$

$$\sin \frac{1}{3}(6n\pi) + \cos \frac{1}{3}(6n\pi) = 0 + 1 = 1$$

$\therefore$ the solution set is $\left\{z: z = 6n\pi \text{ or } \frac{3\pi}{2} + 6n\pi, n \in J\right\}$.

*WRITTEN EXERCISES*

Solve the following equations for $0 \leq x < 2\pi$ or for $0° \leq \theta < 360°$.

**A** 1. $\cos 2\theta = \frac{\sqrt{2}}{2}$

7. $\sin 2x(3 - 4 \cos^2 x) = 0$

2. $\sin \frac{1}{2}\theta = \frac{1}{2}$

8. $\tan 2x(1 - \csc^2 x) = 0$

3. $\tan 3\theta = 0$

9. $\cos 3x + \sec 3x = 2$

4. $\cos 2\theta + \sin 2\theta = 0$

10. $\cos 3x \cos x - \sin 3x \sin x = 1$

5. $\cos 2x = \cos^2 x - 1$

11. $2 \sin x \cos 2x + 2 \cos x \sin 2x = 1$

6. $\sin 4x - 2 \sin 2x = 0$

12. $\sin (\pi + x) = \frac{1}{2}$

**B** **13.** $\sin \frac{1}{2}x + \cos \frac{1}{2}x = 1$      **15.** $\sin 2x + 2 \sin x - \cos x - 1 = 0$

**14.** $\cos 2x = \sin x$      **16.** $\sin 2x + \sin x + 2 \cos x + 1 = 0$

Give the general solution of each equation. If the equation is an identity, state this fact and prove it. In certain of these exercises use the sum and product formulas, Exercises 31–35, page 491.

**17.** $\sin\left(2x + \frac{2\pi}{3}\right) + \sin\left(2x - \frac{2\pi}{3}\right) = \sin\frac{\pi}{4}$

**18.** $\sin\left(\frac{\pi}{3} - \frac{1}{2}x\right) + \sin\left(\frac{\pi}{3} + \frac{1}{2}x\right) = \sin\frac{\pi}{3}$

**19.** $\cos^4 3\theta - \sin^4 3\theta = \cos 6\theta$

**20.** $|\sin \frac{1}{2}\theta| = |\cos \theta|$

**C** **21.** $1 - \cos^6 5x = \sin^2 5x(1 + \cos^2 5x + \cos^4 5x)$

**22.** $\sin 2\theta \tan^2 2\theta - \tan 2\theta = \sin 2\theta$

Solve each system for $0° \leq \theta < 360°$.

**23.** $\begin{aligned} r &= 3 \\ r^2 &= 18 \cos 2\theta \end{aligned}$      **24.** $\begin{aligned} r &= \sin 2\theta \\ r &= \cos 4\theta \end{aligned}$      **25.** $\begin{aligned} r &= \cos 3\theta \\ r &= 1 - \cos 3\theta \end{aligned}$

---

## Chapter Summary

---

1. A **radian** is the central angle subtended in a circle of radius 1 by an arc of length 1. Therefore $180° = \pi^R$. In a circle of radius $r$, a central angle $\theta$ *measured in radians* intercepts an arc of length $r\theta$.

2. The value of a **circular function** of a real number $t$ equals the value of the corresponding trigonometric function of an angle of $t$ radians.

3. Because the circular functions are **periodic functions**, their graphs consist of basic patterns repeated over each interval of length one **period**. Therefore, there is an infinite set of numbers for which a given circular function has any one of its values.

4. The **inverse circular** (or **trigonometric) functions** are the inverses of the corresponding circular (or trigonometric) functions restricted to an appropriate interval.

5. Using trigonometric identities and the usual algebraic transformations, you can solve equations involving circular or trigonometric functions.

**Vocabulary and Spelling**

Review each term by reference to the page listed.

radian (*p. 514*)
circular functions (*p. 518*)
periodic functions (*p. 521*)
period (*p. 521*)
sinusoidal wave (*p. 522*)

amplitude (*p. 522*)
Arccos *u*, etc. (*p. 528*)
inverse circular functions (*pp. 528–533*)
general solution (*p. 535*)
particular solution (*p. 535*)

---

## *Chapter Test and Review*

---

**13–1**　**1.** Convert $-135°$ to radians.　　　**2.** Convert $\dfrac{7\pi^{R}}{12}$ to degrees.

**13–2**　**3.** Give the exact value of **(a)** $\sin\left(\dfrac{7\pi}{6}\right)$ and **(b)** $\tan\left(-\dfrac{2\pi}{3}\right)$.

**13–3**　**4.** Graph $\{(x, y): y = 2 \cos \tfrac{1}{2}x\}$ over one period.

**13–4**　**5.** Graph $\{(x, y): y = \tfrac{1}{2} \sec 2x\}$ over one period.

**13–5**　**6.** Evaluate $\text{Arccos}\left(\cos\dfrac{3\pi}{2}\right)$.

**13–6**　**7.** Evaluate $\sin (\text{Arctan } 1)$.

**13–7**　**8.** Solve $2 \cos x - \sqrt{3} = 0$ over the interval $0 \le x < 2\pi$.

　　　**9.** Find the general solution of $2 \cos^2 \theta - \cos \theta = 1$.

**13–8**　**10.** Solve $\cos 2\theta(3 - 4 \sin^2 \theta) = 0$ over the interval $0° \le \theta < 360°$.

---

## *Cumulative Review: Chapters 11–13*

---

**1.** Draw a positive and negative angle in standard position whose terminal ray $\overrightarrow{OP}$ contains $P(\sqrt{5}, -2)$.

**2.** If the terminal side of an angle $\theta$ in standard position contains $P(5, -4)$, find all trigonometric functions of $\theta$.

**3.** If $\sin \theta = -\tfrac{5}{7}$, and $\theta$ terminates in Quadrant III, find $\cos \theta$.

**4.** Find $\cos 27°38'$.

**5.** If $\sec A = 1.445$, and $0° \leq A < 90°$, find $A$.

**6.** If $\sin B = \dfrac{38.2 \cos 38°10'}{51.5}$, use logarithms to find the measure of the positive acute angle $B$.

**7.** Find, to the nearest tenth, the horizontal and vertical components of a vector of magnitude 15 and bearing 82°.

**8.** From the top of a tower 94 feet tall, the angle of depression to a boulder is $11°20'$. How far is the boulder from the foot of the tower if the boulder is on a level with the foot of the tower?

**9.** Express $\cot^2 x$ in terms of $\sin x$.

**10.** Prove the identity: $\sec^2 B + \csc^2 B = (\tan B + \cot B)^2$

**11.** If $\sec A = \frac{4}{3}$ and $A$ is in Quadrant IV, while $\csc B = \frac{13}{5}$ and $B$ is in Quadrant II, find **(a)** $\sin (A + B)$; **(b)** $\tan (A - B)$.

**12.** If $\cot \theta = \dfrac{\sqrt{7}}{21}$, and $\theta$ is in Quadrant III, find $\cos \dfrac{\theta}{2}$.

**13.** Solve right triangle $ABC$, if $a = 6$ and $b = 9$.

**14.** Solve the oblique triangle $ABC$, if $a = 8$, $b = 10$, and $C = 120°$.

**15.** How many radians are equivalent to 765°?

**16.** Find the exact value of $\cos \left( -\dfrac{8\pi}{3} \right)$.

**17.** If two points on the equator differ by 10° in longitude, find the distance between them. (Use 4000 miles as the radius of the earth.)

**18.** Graph $\{(x, y): y = 3 \sin 2x\}$, for $-\pi \leq x \leq \pi$.

**19.** Graph $\{(x, y): y = \sec \frac{1}{2}x\}$, for $0 \leq x \leq 4\pi$.

**20.** If $t = \text{Arctan } (\frac{5}{12})$, find $\sin t$ and $\cos t$.

**21.** Evaluate $\cos \left[ \text{Arccos } \dfrac{\sqrt{2}}{2} - \text{Arcsin } (-\frac{3}{5}) \right]$

**22.** Solve $\cos 2\theta = 4 \cos \theta$ over $0 \leq \theta < 360°$.

**23.** Solve $1 + \cos^2 x = 2 \sin^2 x - \frac{5}{2} \sin x$ over $0 \leq x < 2\pi$.

**24.** Find a general solution of $\sin 2x - \sin 4x = 2 \sin x$.

**25.** Find the particular solution of $\cos 2x + 2 = \sin x$ for $0 \leq x < 2\pi$.

**26.** Prove that $\text{Arcsin } b = \text{Arctan } \dfrac{b}{\sqrt{1 - b^2}}$.

# 14

# Sequences, Series and Binomial Expansions

## ARITHMETIC SEQUENCES AND SERIES

### 14–1 Arithmetic Sequences

A rocket fired vertically traveled 5 meters during the first second and (as long as its engines provided thrust) traveled 35 meters farther in each succeeding second than in the preceding one. The table below shows $s(n)$, the number of meters it traveled during the $n$th second.

| $n$ | 1 | 2 | 3 | 4 | ... |
|-----|---|----|----|-----|-----|
| $s(n)$ | 5 | 40 | 75 | 110 | ... |

The values of a function whose domain is a set of consecutive natural numbers are said to form a **sequence**, and each value is a **term** of the sequence. Noting that the function $s(n)$ described in the table defines a sequence, 5, 40, 75, 110, ..., in which consecutive terms differ by 35, you can write the table as follows:

| $n$ | 1 | 2 | 3 | ... | $n$ |
|-----|---|---|---|-----|-----|
| $s(n)$ | $5 + 0 \cdot 35$ | $5 + 1 \cdot 35$ | $5 + 2 \cdot 35$ | ... | $5 + (n-1) \cdot 35$ |

The last entry in the table provides a formula for the **general**, or *nth*, **term** of this sequence:

$$s(n) = 5 + (n - 1) \cdot 35.$$

Thus the rocket traveled $s(20) = 5 + (20 - 1) \cdot 35 = 670$ meters during the 20th second.

The sequence studied above is an example of an *arithmetic\* sequence.* An **arithmetic sequence**, or **arithmetic progression (AP)** is one in which the difference obtained by subtracting any term from the next term is always the same number $d$.

If $a$ is the first term of an arithmetic sequence, and $d$ is the common difference between successive terms, then the following table suggests a formula for the $n$th term $t(n)$, also denoted by $t_n$.

| $n$ | 1 | 2 | 3 | . . . | $n$ |
|---|---|---|---|---|---|
| $t_n$ | $a + 0d$ | $a + 1d$ | $a + 2d$ | . . . | $a + (n - 1)d$ |

$$t_n = a + (n - 1)d$$

**Example.**  Which term of the arithmetic sequence 2, 9, 16, . . . is 142?

*Solution:*  Find the common difference (any term minus the preceding term):

$$d = 9 - 2 = 7.$$

Substitute 7 for $d$, 2 for $a$, and 142 for $t_n$ in the formula

$$t_n = a + (n - 1)d$$
$$142 = 2 + (n - 1) \cdot 7$$
$$140 = (n - 1) \cdot 7$$
$$20 = n - 1$$
$$n = 21$$

∴ 142 is the 21st term.

Be careful not to draw conclusions from just a few terms of a sequence unless it is *known* to be an arithmetic sequence or to have some other specified law of formation. For example, the sequence $t$ beginning 7, 4, 1, will have $-2$ as its fourth term if it is an arithmetic sequence; but if its law of formation is $t_n = |3n - 10|$, then its fourth term will be 2.

---

\*When the word "arithmetic" is used as an adjective, it is pronounced ar-ith-**met**-ic.

## WRITTEN EXERCISES

Write the first four terms in an AP using the given values for *a* and *d*.

**A**

**1.** $a = 3; d = 2$       **3.** $a = 4; d = -5$      **5.** $a = a; d = d$

**2.** $a = -3; d = 5$      **4.** $a = 1; d = -\frac{1}{2}$      **6.** $a = a; d = 3k$

Find the *n*th term of the AP having the given values for *a, d,* and *n*.

**7.** $a = 7, d = 4, n = 7$      **11.** $a = 3, d = -5, n = 20$

**8.** $a = -3, d = -9, n = 10$      **12.** $a = \frac{3}{4}, d = -\frac{5}{4}, n = 10$

**9.** $a = 2, d = \frac{1}{2}, n = 12$      **13.** $a = 2i, d = -3i, n = 9$

**10.** $a = -5, d = 3, n = 17$      **14.** $a = \sqrt{3}, d = -\sqrt{2}, n = 12$

Find the specified term in each AP.

**15.** The twelfth term of $\frac{3}{4}, \frac{3}{2}, \frac{9}{4}, \ldots$

**16.** The tenth term of $\frac{5}{6}, \frac{7}{6}, \frac{3}{2}, \ldots$

**17.** The twelfth term of $-17, -13, -9, \ldots$

**18.** The twentieth term of $10, 7, 4, \ldots$

**19.** The eighteenth term of $x, x + 3a, x + 6a, \ldots$

**20.** The fifteenth term of $z, z - \dfrac{a}{2}, z - a, \ldots$

**21.** Which term of $-2, 5, 12, \ldots$ is 138?

**22.** Which term of $3, -2, -7, \ldots$ is $-142$?

Draw the graph of the function *f* whose domain is {1, 2, 3, 4, 5} and whose values are given by the indicated formula.

**23.** $f(n) = 1 + 3n$    **25.** $f(n) = 1 - \frac{1}{3}n$      **27.** $f(n) = 4 + n(\sin 30°)$

**24.** $f(n) = 2n + 5$    **26.** $f(n) = 0.75n - 0.5$    **28.** $f(n) = 4 - n(\cos 60°)$

**B**    **29.** Show that each term of an arithmetic sequence after the first is obtained from the preceding term by adding the same number to it.

## PROBLEMS

**A**    **1.** A long-term labor contract provides annual salary increases of $700 for a position that pays $6200 per year to start. Under this contract, what would be the salary of an employee during the sixth year in this position?

**2.** If an employee hired at $4800 per year is guaranteed annual salary increases of $300 per year, how many years will it take the employee to achieve a salary of $8000 or more?

**3.** A salesman received $20 for selling a calculator, with an arrangement that for each successive machine he sold he would receive an additional $1.80 until he was paid $56 per machine sale. How many machines did he have to sell to reach the maximum commission?

**4.** A pile of bricks one brick thick has 43 bricks in the bottom row, 41 in the second row, 39 in the third row, etc., and 1 brick in the top row. How many bricks are in the 12th row?

**5.** In the pile of bricks in Problem 4, how many rows are there in the pile?

**6.** To take care of increasing population, each year Central City added 20,000 acre-feet of water to its reservoir capacity. If it had a reservoir capacity of 2,000,000 acre-feet in 1965, what was its capacity in 1972?

B **7.** At the end of one year, the trade-in value of a certain automobile is $500 less than the original cost. Each year thereafter the trade-in value decreases by $200. If the original cost of the automobile is $4000, what is the trade-in value at the end of $n$ years? When is the trade-in value 0?

**8.** The third term of an AP is 14; the ninth term is $-1$. Find the first three terms. (*Hint:* $a + 2d = 14$, etc.)

**9.** The fifth term of an AP is 9; the fourteenth term is 45. Write the first three terms.

## 14-2  Arithmetic Means

Equal numbers of passengers board a commuter train at each station after the first. When the train leaves the third station, there are 55 passengers aboard; when it leaves the seventh station there are 115 passengers aboard. How many passengers boarded at the first station?

To answer this question, notice that the numbers of passengers form an arithmetic sequence which you can represent schematically as follows:

$$\underline{\ ?\ }, \underline{\ ?\ }, 55, \underline{\ ?\ }, \underline{\ ?\ }, \underline{\ ?\ }, 115$$

You can find the common difference by considering the sequence whose first term is 55 and whose fifth term is 115. Using the equation $t_n = a + (n-1)d$, you have

$$115 = 55 + (5-1)d,$$

and you find that $d = 15$. Now consider the sequence whose third term is 55 and whose common difference is 15. Again you use the equation $t_n = a + (n - 1)d$ and obtain

$$55 = a + (3 - 1)15,$$

from which you find that $a = 25$. Therefore 25 passengers boarded the train at the first station.

Since you know the first term and the common difference, you can complete the sequence considered above to obtain

$$25, 40, 55, 70, 85, 100, 115.$$

The terms between two given terms of an arithmetic sequence are called **arithmetic means** between the given terms. Thus $\{70, 85, 100\}$ is the set of *three* arithmetic means between 55 and 115.

**Example.**   Insert four arithmetic means between 10 and $-5$.

*Solution:*   Represent the sequence schematically:

$$10, \underline{\ ?\ }, \underline{\ ?\ }, \underline{\ ?\ }, \underline{\ ?\ }, -5$$

and notice that it has 6 terms. Use $t_n = a + (n - 1)d$ with $t_n = -5$, $a = 10$, and $n = 6$ to obtain

$$-5 = 10 + (6 - 1)d, \text{ or } d = -3.$$

Hence the sequence is 10, $\underline{7}$, $\underline{4}$, $\underline{1}$, $\underline{-2}$, $-5$ and the required four arithmetic means are 7, 4, 1, $-2$.

A single arithmetic mean between two numbers is the **average**, or *the* arithmetic mean of the two numbers. For example, the arithmetic mean of 1 and 5 is half their sum, or 3.

*WRITTEN EXERCISES*

Insert the given number of arithmetic means in each case.

**A**
**1.** One, between 2 and 63   **5.** Three, between $-6$ and $-4$

**2.** Five, between 2 and 20   **6.** Four, between $-10$ and 2

**3.** Seven, between 1 and 3   **7.** Eight, between $-6$ and 22

**4.** One, between $-7$ and 5   **8.** Two, between $-8$ and 3

Find the arithmetic mean of the given numbers.

**9.** 7, 107   **10.** 9, 23   **11.** $-42$, 12   **12.** $-23$, $-6$

Find the specified term in red in an AP with the given terms.

**B**  **13.** $t_3 = 7, t_8 = 17; t_1$    **16.** $t_2 = 8, t_5 = -7; t_7$

**14.** $t_5 = -16, t_{20} = -46; t_{12}$    **17.** $t_2 = x, t_4 = -x; t_6$

**15.** $t_3 = 14, t_9 = -1; t_2$    **18.** $t_3 = a, t_5 = b; t_1$

## PROBLEMS

**A**  **1.** To fence part of one side of a 100-foot lot, Mr. Davis set a fence post 30 feet from the front of his lot and another one 5 feet from the back of his lot. If he wished to place 25 posts between these two, how far apart should he set them?

**2.** Mrs. Green received an annual salary increase of the same amount each year from 1965 to 1972. If her salary in 1965 was $8800 and in 1972 was $11,740, what was the annual increase in her salary?

**3.** Each one of a set of 8 standard weights differs from its successor and predecessor by the same amount. If the lightest weight is 2 grams and the heaviest is 30 grams, what are the other weights?

**4.** For a garden party, 15 lanterns are to be equally spaced along a 42-foot wire. If a lantern is hung at each end of this wire, where will the remaining 13 lanterns be hung?

**B**  **5.** If the arithmetic mean of two positive integers is 11 and if twice the greater of the two integers is 4 more than three times the lesser of the integers, find the integers.

**6.** The greater of two integers is 1 more than three times the arithmetic mean of the numbers, and is 2 less than the square of the smaller integer. Find the integers.

**7.** The product of two nonzero numbers is $\frac{2}{5}$ the larger number and $\frac{4}{9}$ their arithmetic mean. Find the smaller number.

**8.** The reciprocal of one number is 5 times the reciprocal of another. Seven times their arithmetic mean is 4 greater than their product. Find all such pairs of numbers.

**9.** In the equation $ax^2 + bx + c = 0$, the product of the roots is $-7$, $a = 1$, and the sum of the roots is the arithmetic mean of $a$ and $c$. Find $b$ and $c$.

**10.** A circle whose radius is 10 has its center at the origin. The arithmetic mean of the coordinates of a point on the circle is 7. Find the coordinates of all such points.

## 14–3  Arithmetic Series

A series is the indicated sum of the terms of a sequence. Thus, associated with the arithmetic sequence 1, 4, 7, 10, 13 is the arithmetic series $1 + 4 + 7 + 10 + 13$.

You can find a formula for the sum $S_n$ of the first $n$ terms of an arithmetic sequence: First write the terms of the series in the usual order, then in reversed order. Use $l$ to denote the $n$th or last term.

$$S_n = a + (a + d) + (a + 2d) + \cdots + (l - 2d) + (l - d) + l$$
$$S_n = l + (l - d) + (l - 2d) + \cdots + (a + 2d) + (a + d) + a$$

If you add these equations, the sum of each pair of corresponding terms is $a + l$, and you have

$$2S_n = (a + l) + (a + l) + (a + l) + \cdots$$
$$+ (a + l) + (a + l) + (a + l).$$

Since there are $n$ terms in the series, $2S_n = n(a + l)$, or

$$S_n = \frac{n}{2}(a + l).$$

If you write this as $S_n = n\left(\dfrac{a + l}{2}\right)$ you can see that the sum is just $n$ times the average of the first and last terms. Because $l$ is the $n$th term, you have $l = a + (n - 1)d$. Therefore you can replace $l$ by $a + (n - 1)d$ in the formula displayed above to obtain

$$S_n = \frac{n}{2}[a + a + (n - 1)d]$$

or

$$S_n = \frac{n}{2}[2a + (n - 1)d].$$

**Example 1.**  Find the sum of the first 15 terms of the arithmetic series $5 + 9 + 13 + \ldots$.

*Solution:*  The common difference is 4. Use $S_n = \dfrac{n}{2}[2a + (n - 1)d]$, replacing $n$ with 15, $a$ with 5, and $d$ with 4:

$$S_{15} = \tfrac{15}{2}[2 \cdot 5 + (15 - 1) \cdot 4]$$
$$= \tfrac{15}{2}[10 + 14 \cdot 4] = \tfrac{15}{2} \cdot 66 = 495$$

You can use the Greek letter $\Sigma$ (sigma) to write a series in brief form. For example, to abbreviate the series

$$2 \quad + \quad 5 \quad + \quad 8 \quad + \quad 11 \quad + \quad 14, \quad \text{or}$$
$$(3 \cdot 1 - 1) + (3 \cdot 2 - 1) + (3 \cdot 3 - 1) + (3 \cdot 4 - 1) + (3 \cdot 5 - 1),$$

notice that each term is of the *form* $(3 \cdot k - 1)$, and write

$$\sum_{k=1}^{5} (3k - 1) \quad \text{(read "summation from 1 to 5 of } 3k - 1\text{").}$$

The symbol $\sum$ is called the summation sign, the expression $(3k - 1)$ is the summand, and the letter $k$ is the index. Do you see that changing the index to some other letter would not change the series?

**Example 2.** Write $\displaystyle\sum_{n=1}^{6} (4n - 3)$ in expanded form.

*Solution:* Replace $n$ with the integers 1 to 6 in turn.

$$\sum_{n=1}^{6} (4n - 3) = (4 \cdot 1 - 3) + (4 \cdot 2 - 3) + (4 \cdot 3 - 3)$$
$$+ (4 \cdot 4 - 3) + (4 \cdot 5 - 3) + (4 \cdot 6 - 3)$$
$$= 1 + 5 + 9 + 13 + 17 + 21.$$

**Example 3.** Write the arithmetic series $3 + 5 + 7 + \cdots + 25$ using summation notation.

*Solution:* Each term is of the form $2j + 1$. The first term $3 = 2 \cdot 1 + 1$, and the last term $25 = 2 \cdot 12 + 1$.

$$\therefore \sum_{j=1}^{12} (2j + 1)$$

*ORAL EXERCISES*

State each series in expanded form.

**1.** $\displaystyle\sum_{j=1}^{4} j + 3$

**2.** $\displaystyle\sum_{i=1}^{3} (3i + 2)$

**3.** $\displaystyle\sum_{k=1}^{4} (k - 1)$

**4.** $\displaystyle\sum_{i=2}^{5} (i + 10)$

**5.** $\displaystyle\sum_{j=3}^{6} (4 - j)$

**6.** $\displaystyle\sum_{i=2}^{5} -3i$

**7.** $\displaystyle\sum_{k=1}^{4} -2k$

**8.** $\displaystyle\sum_{i=1}^{4} (6 - i)$

**9.** $\displaystyle\sum_{k=2}^{5} (10 - 2k)$

**10.** $\displaystyle\sum_{j=2}^{4} (2 - j)$

**11.** $\displaystyle\sum_{i=1}^{4} \frac{1}{2}i$

**12.** $\displaystyle\sum_{k=2}^{5} \frac{1}{3}k$

## WRITTEN EXERCISES

Write each series using summation notation.

**A**
**1.** $2 + 5 + 8 + 11$

**2.** $3 + 10 + 17 + 24 + 31$

**3.** $\frac{1}{2} + \frac{3}{2} + \frac{5}{2} + \frac{7}{2} + \frac{9}{2}$

**4.** $\frac{4}{5} + \frac{7}{5} + 2 + \frac{13}{5} + \frac{16}{5}$

**5.** $6 + 3 + 0 + (-3)$

**6.** $12 + 8 + 4 + 0 + (-4)$

**7.** $0.82 + 0.76 + 0.70 + 0.64$

**8.** $0.13 + 0.18 + 0.23 + 0.28$

Find the sum of the arithmetic series.

**9.** $\displaystyle\sum_{i=1}^{7} (2i + 3)$

**10.** $\displaystyle\sum_{j=1}^{6} (4j - 2)$

**11.** $\displaystyle\sum_{k=2}^{8} (5k - 3)$

**12.** $\displaystyle\sum_{j=2}^{10} (4j + 7)$

**13.** $\displaystyle\sum_{h=10}^{20} (2h - 3)$

**14.** $\displaystyle\sum_{k=1}^{31} (3k - 2)$

Find the sum of the arithmetic series having the given data.

**15.** $a = 17, l = 173, d = 3$

**16.** $a = -5, l = 2, d = \frac{1}{3}$

**17.** $a = -1, l = -35, d = -2$

**18.** $a = 8, l = 161, d = 3$

**19.** $a = -2, n = 361, d = 3$

**20.** $a = -7, n = 19, d = 9$

**Sample.** If $\displaystyle\sum_{i=1}^{5} (ik + 2) = 20$, find $k$.

*Solution:* The series is:

$$(k + 2) + (2k + 2) + (3k + 2) + (4k + 2) + (5k + 2)$$

Use $S_n = \dfrac{n}{2}[2a + (n - 1)d]$ with $S_n = 20$, $n = 5$, $a = k + 2$

and $d = k$ to obtain

$$20 = \tfrac{5}{2}[2(k + 2) + (5 - 1)k]$$
$$= \tfrac{5}{2}[6k + 4] = 5(3k + 2)$$

$$\therefore 3k + 2 = 4, \ 3k = 2, \ k = \tfrac{2}{3}$$

**B**
**21.** If $\displaystyle\sum_{j=1}^{4} aj = 40$, find $a$.

**22.** If $\displaystyle\sum_{i=1}^{5} bi = 30$, find $b$.

**23.** If $\displaystyle\sum_{i=1}^{6} (it - 4) = 25$, find $t$.

**24.** If $\displaystyle\sum_{j=1}^{5} (4 - jb) = -10$, find $b$.

C  **25.** If $\sum\limits_{i=1}^{4} (ai + b) = 28$ and $\sum\limits_{i=2}^{5} (ai + b) = 44$, find $a$ and $b$.

**26.** If $\sum\limits_{j=2}^{4} (aj + b) = 27$ and $\sum\limits_{j=1}^{3} (aj + b) = 15$, find $a$ and $b$.

**27.** Show that the sum of the first $n$ odd positive integers is $n^2$.

**28.** Show that the sum of the first $n$ even positive integers is $n^2 + n$.

## PROBLEMS

A  **1.** Mr. Jones charges $3.00 for the first foot of any well he digs and 2¢ more than the preceding one for each foot thereafter. What will he charge to dig a 100-foot well?

**2.** A set of 7 weights for a balance are in AP with the lightest one weighing 1 gram and the heaviest weighing 13 grams. How much does the heaviest object weigh that can be weighed using this set?

**3.** How many bricks will there be in a pile one brick thick, if there are 29 bricks in the first row, 27 in the second row, etc., and 1 in the top row?

**4.** If 256 bricks are piled in the same fashion as those described in Problem 3, how many bricks will there be in the fourth row from the bottom of the pile?

**5.** In flight testing a new airplane, the test pilot flew the plane for 40 minutes on the first flight and then 8 minutes longer each successive day until he had flown the airplane a total of 4 hours and 40 minutes. How many test flights had he made by that time?

**6.** Insurance charges on a TV set increase at a fixed rate of $6 per year. If Amy Scott paid $40 for her first years premium, how much money had she paid for insurance after 6 years?

B  **7.** The Midville Library charges 2 cents for the first day a book is overdue, and each day thereafter each day's fine is 2¢ more than that of the preceding day. How many days must a $5.50 book be overdue before the fine totals more than the value of the book?

**8.** The largest integer in an arithmetic progression of consecutive even integers is nine times the smallest integer. The sum of the arithmetic progression is 90. Find the numbers.

**9.** At the end of each month a credit union charges interest equal to one percent of the unpaid balance of a loan. Elise borrows $100 from the credit union and pays back $10 at the end of each month for ten months. If her first payment is made one month after the date of the loan, what is the total amount of interest charged?

## GEOMETRIC SEQUENCES AND SERIES

### 14–4 Geometric Sequences

Under conditions favorable to the growth of certain bacteria, each organism will divide into two every hour. Thus, if three organisms are present originally, there will be six at the beginning of the second hour, twelve present at the beginning of the third hour, and so on. The number of bacteria present at the beginning of each hour is a term of the sequence 3, 6, 12, 24, . . . .

The situation described above can be summarized in a table:

| Hour | 1 | 2 | 3 | 4 | . . . | $n$ |
|---|---|---|---|---|---|---|
| Number of bacteria | $3 = 3 \cdot 2^0$ | $6 = 3 \cdot 2^1$ | $12 = 3 \cdot 2^2$ | $24 = 3 \cdot 2^3$ | . . . | $3 \cdot 2^{n-1}$ |

Notice that the $n$th term of the sequence is given by

$$t_n = 3 \cdot 2^{n-1}.$$

Thus, the sixth term of the sequence is $t_6 = 3 \cdot 2^{6-1} = 3 \cdot 2^5 = 96$.

The sequence studied above is an example of a *geometric* sequence. A **geometric sequence**, or **geometric progression (GP)**, is one in which the ratio obtained by dividing any term by the preceding term is always the same number $r$.

If $a$ is the first term of a geometric sequence, and $r$ is the common ratio, the following table suggests a formula for the $n$th term.

| $n$ | 1 | 2 | 3 | 4 | . . . | $n$ |
|---|---|---|---|---|---|---|
| $t_n$ | $a \cdot r^0$ | $a \cdot r^1$ | $a \cdot r^2$ | $a \cdot r^3$ | . . . | $a \cdot r^{n-1}$ |

$$t_n = ar^{n-1}$$

**Example 1.** Find the fifth term of the GP 18, 12, 8, . . .

*Solution:* Find the common ratio $r$ by dividing any term by the preceding term: $r = \frac{12}{18} = \frac{2}{3}$. Use $t_n = ar^{n-1}$, replacing $a$ by 18, $r$ by $\frac{2}{3}$, and $n$ by 5.

$$t_5 = 18 \cdot \left(\tfrac{2}{3}\right)^{5-1}$$
$$t_5 = 18 \cdot \left(\tfrac{2}{3}\right)^4 = 18 \cdot \tfrac{16}{81} = \tfrac{32}{9}$$

The terms between two given terms of a geometric sequence are called **geometric means** between the given terms.

**Example 2.**   Insert three geometric means between 7 and 112.

*Solution:*   Schematically, the sequence looks like this:

$$7, \underline{\quad?\quad}, \underline{\quad?\quad}, \underline{\quad?\quad}, 112$$

Note that the sequence has five terms. To find $r$, use $t_n = ar^{n-1}$ with $n = 5$, $a = 7$, and $t_5 = 112$:

$$112 = 7 \cdot r^4$$
$$r^4 = \tfrac{112}{7} = 16, \quad \text{or} \quad r = \pm 2.$$

If $r = 2$, the sequence is 7, 14, 28, 56, 112, and the required means are 14, 28, 56.

If $r = -2$, the sequence is 7, $-14$, 28, $-56$, 112, and the required means are $-14$, 28, $-56$.

A single geometric mean between two numbers is called a **mean proportional** of the numbers. If $m$ is a mean proportional of $a$ and $b$, then $a$, $m$, $b$ form a geometric sequence and therefore

$$\frac{m}{a} = \frac{b}{m}, \quad \text{or} \quad m^2 = ab.$$

If $a$ and $b$ are positive, then the *positive* number

$$m = \sqrt{ab}$$

is *the* geometric mean or *the* mean proportional of $a$ and $b$.

*ORAL EXERCISES*

State whether the given sequence forms a geometric progression, an arithmetic progression, or neither. If the sequence is an arithmetic progression, give the common difference; if it is a geometric progression, give the common ratio.

**1.** 3, 6, 9          **4.** 12, 6, 1          **7.** $-4, -7, -10$

**2.** 3, 6, 12          **5.** $-1, 1, -1$          **8.** $-4, -2, -\tfrac{1}{2}$

**3.** 12, 6, 3          **6.** $-1, 0, 1$          **9.** $a, abc, ab^2c^2$

State the geometric mean of the given two numbers.

**10.** 3, 12          **12.** 1, 25          **14.** $\tfrac{1}{2}, 8$          **16.** 2, 7

**11.** 2, 8          **13.** 1, 49          **15.** $\tfrac{1}{3}, 48$          **17.** 5, 7

## WRITTEN EXERCISES

Write the next three terms in the given geometric progression; then write an equation for the $n$th term.

**A**

**1.** $2, 6, 18, \ldots$

**2.** $\frac{1}{2}, 2, 8, \ldots$

**3.** $\frac{2}{3}, \frac{4}{3}, \frac{8}{3}, \frac{16}{3}, \frac{32}{3}, \frac{64}{3}$

**4.** $\frac{1}{2}, -\frac{3}{2}, \frac{9}{2}, \ldots$

**5.** $32, 16, 8, 4, 2, 1$

**6.** $\frac{7}{2}, \frac{7}{10}, \frac{7}{50}, \ldots$

**7.** $x, ax, a^2x, \ldots$

**8.** $\dfrac{n}{a}, \dfrac{n}{ab}, \dfrac{n}{ab^2}, \ldots$

Insert the given number of geometric means and write the resulting finite geometric progression.

**9.** Three, between 3 and 48

**10.** Two, between 1 and 8

**11.** Four, between $\frac{1}{4}$ and 8

**12.** Three, between $\frac{1}{27}$ and 3

**13.** One, positive, between 3 and 75

**14.** One, positive, between $\frac{1}{2}$ and $\frac{9}{32}$

**15.** One, negative, between $a^2$ and $a^8$ $(a > 0)$

**16.** One, negative, between $x$ and $x^7$, $(x > 0)$

**17.** There are two geometric progressions of real numbers with first term of 7 and fifth term of 112. Find the two values for $r$.

**18.** There are two geometric progressions of real numbers having $-1$ as the first term and $-625$ as the fifth term. Find the two values for $r$.

**19.** In a geometric progression whose first term is 5 and whose common ratio is $-5$, there is a term $t_n = 3125$. What is the value of $n$?

**20.** The first term of a geometric progression is 27 and the common ratio is $\frac{1}{3}$. For what value of $n$ is $t_n = \frac{1}{3}$?

**21.** Which term of the geometric progression $2, -6, 18, \ldots$ is 162?

**22.** Which term of the geometric progression $243, -81, 27, \ldots$ is $-\frac{1}{9}$?

**23.** Find the first term in a geometric progression whose common ratio is 2 and whose sixth term is 96.

**24.** Find the first term in a geometric progression whose common ratio is 3 and whose fifth term is 324.

**25.** If $-\frac{64}{9}$ is the sixth term of a geometric progression whose common ratio is $-\frac{2}{3}$, what is the first term of the progression?

**26.** If $\frac{81}{16}$ is the fifth term of a geometric progression whose common ratio is $-\frac{3}{4}$, what is the first term?

**27.** Show that each term of a geometric sequence after the first is obtained from the preceding term by multiplying it by the same number.

C **28.** Prove that the logarithms of *n* positive numbers in geometric progression are in arithmetic progression.

**29.** In a geometric progression of positive numbers, each term after the second is the sum of the two preceding terms. Find the common ratio.

**30.** Prove that the arithmetic mean of two different positive numbers *a* and *b* is greater than their positive geometric mean. (*Hint:* $(a - b)^2 > 0$.)

**31.** Use the result of Exercise 30 to prove that for any three different positive numbers *a*, *b*, and *c*, $(a + b)(b + c)(c + a) > 8abc$.

*PROBLEMS*

A **1.** A home bought three years ago for $15,000 was sold today for $25,920. Assuming that the value of the home increased geometrically, find the average rate per year at which its value increased.

**2.** A hamburger stand purchased at $64,000 was sold three years later for $42,875. Assuming that the value of the stand decreased geometrically, find the average rate per year at which its value decreased.

**3.** Suppose you were to tear a sheet of newspaper in half, place the halves together and tear the two sheets, place the resulting four sheets together and tear those, etc. How many sheets of paper would you be tearing at the 12th tear (if you could tear them 11 times)?

**4.** On her income tax, Mrs. Evans depreciated a cleaning plant at the rate of 12% per year. Thus, her $72,000 plant was worth $63,360 the second year, $55,756.80 the third year, and so on. What was the value of her plant after 6 years, to the nearest dollar?

**5.** The population of Centerville increases by 10% per year. If the population is now 20,000, what, to the nearest hundred, will be the expected population 5 years from now?

**6.** Assuming that there is no duplication of ancestors, how many ancestors does a person have in the eighth generation back?

B **7.** A side of an equilateral triangle is 20 inches long. A second equilateral triangle is inscribed in it by joining the midpoints of the sides of the first triangle. The process is continued, as shown in the accompanying diagram. Find the perimeter of the fifth inscribed equilateral triangle.

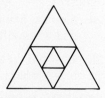

C **8.** The length of each oscillation of a pendulum is 90% of the length of the preceding oscillation. How many oscillations are required for the pendulum to damp down to one which has less than half the length of the initial oscillation?

**9.** Filled to capacity, a tank contains 10 gallons of pure antifreeze. One gallon of liquid is drawn out and the tank is filled with water. If this operation is repeated several times, after how many operations will there be less than 1 gallon of pure antifreeze left in the tank?

## 14–5   Geometric Series

A **geometric series** is the indicated sum of the terms of a geometric sequence; for example, $12 + 6 + 3 + \frac{3}{2}$ is such a series.

To find the sum $S_n$ of $n$ terms of the general geometric series, first write it in expanded form, and then, beneath this, write the result of multiplying each term by $r$:

$$S_n = a + ar + ar^2 + \cdots + ar^{n-2} + ar^{n-1}$$
$$rS_n = ar + ar^2 + ar^3 + \cdots + ar^{n-1} + ar^n$$

When you subtract the second equation from the first, the differences indicated by red lines are all zero, and therefore you obtain

$$S_n - rS_n = a - ar^n.$$

Factoring the left member, you have

$$(1 - r)S_n = a - ar^n,$$

or

$$S_n = \frac{a - ar^n}{1 - r} \qquad (r \neq 1).$$

**Example 1.**   How many ancestors does a person have in the ten generations preceding him, assuming that there have been no intermarriages?

*Solution:*   In the first preceding generation, a person has 2 parents, in the second generation 4 grandparents, and so on. Thus you wish to find the sum of the first 10 terms of the geometric series $2 + 4 + 8 + \cdots$.

In $S_n = \dfrac{a - ar^n}{1 - r}$ replace $a$ by 2, $r$ by 2, and $n$ by 10:

$$S_{10} = \frac{2 - 2 \cdot 2^{10}}{1 - 2}$$

$$= \frac{2 - 2048}{-1} = 2046.$$

If $l$ denotes the last term of a geometric series of $n$ terms, then $l = ar^{n-1}$ (page 553), and therefore $ar^n = ar^{n-1} \cdot r = lr$. Hence you can rewrite $S_n = \dfrac{a - ar^n}{1 - r}$ as follows:

$$S_n = \frac{a - rl}{1 - r} \qquad (r \neq 1)$$

**Example 2.** A geometric series has first term 48, last term 3, and common ratio $-\frac{1}{2}$. **(a)** Find its sum. **(b)** Write the series.

*Solution:*     **a.** In $S_n = \dfrac{a - rl}{1 - r}$ substitute 48 for $a$, 3 for $l$ and $-\frac{1}{2}$ for $r$:

$$S_n = \frac{48 - (-\frac{1}{2})3}{1 - (-\frac{1}{2})}$$

$$= \frac{48 + \frac{3}{2}}{1 + \frac{1}{2}} = \frac{96 + 3}{2 + 1} = \frac{99}{3} = 33$$

       **b.** $48 + (-24) + 12 + (-6) + 3$, or

         $48 - 24 + 12 - 6 + 3$.

*ORAL EXERCISES*

Give the first two terms of the expanded form of each geometric series.

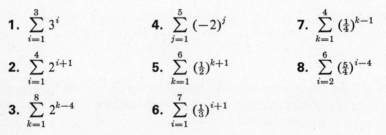

**1.** $\displaystyle\sum_{i=1}^{3} 3^i$        **4.** $\displaystyle\sum_{j=1}^{5} (-2)^j$        **7.** $\displaystyle\sum_{k=1}^{4} (\tfrac{1}{4})^{k-1}$

**2.** $\displaystyle\sum_{i=1}^{4} 2^{i+1}$        **5.** $\displaystyle\sum_{k=1}^{6} (\tfrac{1}{2})^{k+1}$        **8.** $\displaystyle\sum_{i=2}^{6} (\tfrac{5}{4})^{i-4}$

**3.** $\displaystyle\sum_{k=1}^{8} 2^{k-4}$        **6.** $\displaystyle\sum_{i=1}^{7} (\tfrac{1}{3})^{i+1}$

*WRITTEN EXERCISES*

Find the sum of the geometric series with data as given.

A    **1–8.** The series in Oral Exercises 1–8.

   **9.** $a = 24, r = \frac{1}{2}, n = 6$            **12.** $a = 27, r = -\frac{2}{3}, n = 5$

   **10.** $a = 3, r = -1, n = 40$        **13.** $a = 3, r = -1, n = 49$

   **11.** $a = 20, r = \frac{1}{10}, n = 6$        **14.** $a = 39, r = \frac{1}{3}, n = 8$

Of the values $a$, $l$, $n$, $r$, and $S_n$, three are given. Find the other two.

**B** **15.** $a = \frac{5}{8}, l = 40, S_n = \frac{215}{8}$     **19.** $a = 1, n = 3, S_n = 21$

**16.** $a = 7, l = 896, S_n = 1785$     **20.** $a = 4, n = 4, S_n = 160$

**17.** $a = 243, l = \frac{1}{32}, n = 6$     **21.** $n = 4, r = 3, S_n = 200$

**18.** $a = \frac{4}{9}, l = \frac{9}{4}, n = 5$     **22.** $n = 5, r = -3, S_n = 244$

Evaluate each expression.

**C** **23.** $\displaystyle\sum_{i=1}^{5} (3^i - 2^i)$     **24.** $\displaystyle\sum_{i=1}^{5} (2^i + 2^{-i})$     **25.** $\displaystyle\sum_{j=1}^{4} (3 + 2^j)$

(*Hint:* $\sum(a - b) = \sum a - \sum b$.)

## PROBLEMS

**A** **1.** An ocean floor drill extracts a core 20 feet in length. Each successive core however, is only $\frac{4}{5}$ as long as its predecessor. How deep will the drill penetrate in extracting 6 cores?

**2.** A "chain letter" is a letter asking the recipient to write similar letters to additional people. If a letter asks each recipient to write 5 copies of the letter and send them to a friend, how many persons will receive such letters after 10 steps in the chain, providing the chain is not broken and there are no duplicates in the chain?

**3.** A jar contains 500 cubic inches of air. On the first stroke, an air pump removes 20% of the air. On the second stroke, it removes 20% of the remaining air, and so on. How much air has been removed from the jar after 6 strokes of the pump?

**4.** A ball dropped from a height of 10 feet rebounded $\frac{2}{3}$ of the distance it was dropped, and thereafter $\frac{2}{3}$ of the distance from which it last fell. How far had the ball traveled after 8 bounces?

**5.** A swinging pendulum is brought to a stop by frictional losses. If the pendulum swings 60 centimeters on its first swing, and the length of each succeeding swing is $\frac{5}{6}$ that of the preceding swing, how far has the bob traveled in 5 swings?

**6.** If there have been no intermarriages, how many ancestors have you had in the twelve generations preceding you?

**7.** A side of a square is 10 inches. The midpoint of its sides are joined to form an inscribed square, and the process is continued as shown in the diagram. Find the sum of the perimeters of the first five squares formed in this way.

8. A side of an equilateral triangle is 12 inches. The midpoints of its sides are joined to form an inscribed equilateral triangle, and the process is continued. Find the sum of the perimeters of the first five equilateral triangles in the resulting figure.

9. If the half-life of the Uranium 230 isotope is 20.8 days, how much of a given amount of the isotope will be left after 104 days?

10. A certain lathe makes a total of 211 revolutions in the first five seconds after the motor is turned off. In any one second, its speed is two-thirds of its speed during the preceding second. What was its speed in revolutions per second at the time the motor was turned off?

**B** 11. In a certain credit union, money left on deposit for one year earns 4% interest at the end of the year. If you invested $100 at the beginning of each year in this credit union and did not withdraw the interest due at the end of the year, how much would you have on deposit at the end of the 10th year?

12. At the end of each quarter a bank pays 1% interest on money left on deposit from the beginning of the quarter. If you wished to save $1000 over a two-year period, how much would you have to deposit in this bank at the beginning of each quarter? Assume that you do not withdraw the interest due at the end of each quarter.

## 14–6 Infinite Geometric Series

Suppose that an ideal rubber ball* is thrown to a height of 45 feet and that it rebounds on each bounce to two-thirds the height from which it fell. Thus on the first up-and-down round trip the ball travels 90 feet, on the second, 60 feet, on the third, 40 feet, and so on without end. The total distance the ball travels is given by the *infinite* geometric series having first term 90 and common ratio $\frac{2}{3}$:

$$90 + 60 + 40 + \tfrac{80}{3} + \cdots$$

To find the sum of this series you first find the sum of the first $n$ terms using $S_n = \dfrac{a - ar^n}{1 - r}$ :

$$S_n = \frac{90 - 90(\tfrac{2}{3})^n}{1 - \tfrac{2}{3}}$$

---

*An "ideal" rubber ball is really a "mathematical model" of an imaginary ball which can bounce infinitely many times. Applied mathematicians study real-life situations by approximating them by such models and then using the powerful tools of pure mathematics.

The table shown below suggests that as $n$ gets larger and larger, $(\frac{2}{3})^n$ gets closer and closer to $0$.

| $n$ | 1 | 2 | 3 | 4 | 5 | $\ldots$ |
|---|---|---|---|---|---|---|
| $(\frac{2}{3})^n$ | $\frac{2}{3}$ | $\frac{4}{9}$ | $\frac{8}{27}$ | $\frac{16}{81}$ | $\frac{32}{243}$ | $\ldots$ |

Therefore, as $n$ increases indefinitely, $(\frac{2}{3})^n$ approaches 0 and $S_n$ approaches $S$, where

$$S = \frac{90 - 90 \cdot 0}{1 - \frac{2}{3}} = \frac{90}{\frac{1}{3}} = 270.$$

The number 270 is called the sum of the infinite geometric series $90 + 60 + 40 + \ldots$, and is theoretically the distance traveled by the ball before it comes to rest.

The general infinite geometric series is

$$a + ar + ar^2 + \ldots + ar^{n-1} + \ldots.$$

The sum of the first $n$ terms is given by $S_n = \dfrac{a - ar^n}{1 - r}$. If $|r| < 1$, it can be shown that $r^n$ approaches $0$ as $n$ increases indefinitely, and therefore $S_n$ approaches $S$, where $S = \dfrac{a - a \cdot 0}{1 - r} = \dfrac{a}{1 - r}$. Therefore we *define* the sum $S$ of the infinite geometric series to be

$$S = \frac{a}{1 - r} \qquad (|r| < 1).$$

**Example.** Find the sum of the infinite geometric series

$$\tfrac{4}{3} - \tfrac{2}{3} + \tfrac{1}{3} - \tfrac{1}{6} + \ldots.$$

*Solution:* The series can be written $\frac{4}{3} + (-\frac{2}{3}) + \frac{1}{3} + (-\frac{1}{6}) + \ldots$. Hence $a = \frac{4}{3}$ and $r = -\frac{1}{2}$. Using $S = \dfrac{a}{1 - r}$, we have

$$S = \frac{\frac{4}{3}}{1 - (-\frac{1}{2})} = \frac{\frac{4}{3}}{\frac{3}{2}} = \frac{4}{3} \cdot \frac{2}{3} = \frac{8}{9}.$$

Infinite repeating decimals (Section 6–9) provide other examples of infinite geometric series. Thus:

$$0.272727\ldots = 0.27 + 0.0027 + 0.000027 + \ldots$$
$$= \tfrac{27}{100} + \tfrac{27}{100} \cdot \tfrac{1}{100} + \tfrac{27}{100} \cdot (\tfrac{1}{100})^2 + \ldots$$

To find the sum of this series, that is, the number which the infinite decimal represents, you can use $S = \dfrac{a}{1-r}$ with $a = \frac{27}{100}$ and $r = \frac{1}{100}$:

$$S = \frac{\frac{27}{100}}{1 - \frac{1}{100}} = \frac{27}{99} = \frac{3}{11}$$

## WRITTEN EXERCISES

Find the sum of the given infinite geometric series. If the series has no sum, so state.

A

**1.** $12 + 6 + 3 + \cdots$

**2.** $4 + 2 + 1 + \cdots$

**3.** $6 - 4 + \frac{8}{3} - \cdots$

**4.** $\frac{3}{4} - \frac{1}{2} + \frac{1}{3} - \cdots$

**5.** $\frac{1}{49} + \frac{1}{56} + \frac{1}{64} + \cdots$

**6.** $\frac{1}{36} + \frac{1}{30} + \frac{1}{25} + \cdots$

**7.** $\frac{1}{81} - \frac{1}{54} + \frac{1}{36} - \cdots$

**8.** $3 + 2 + \frac{4}{3} + \cdots$

**9.** $2 + 0.8 + 0.32 + \cdots$

**10.** $5 + 1.5 + 0.45 + \cdots$

**11.** $\frac{2}{3} - \frac{1}{3} + \frac{1}{6} - \cdots$

**12.** $\frac{5}{6} - \frac{1}{6} + \frac{1}{30} - \cdots$

Write the first three terms of the infinite geometric series having the given data.

**13.** $a = 1$, $S = \frac{3}{2}$

**14.** $a = 6$, $S = 7\frac{1}{2}$

**15.** $r = -\frac{1}{11}$, $S = 10\frac{1}{12}$

**16.** $r = \frac{1}{3}$, $S = 9$

Find a common fraction equivalent to the given repeating decimal.

**17.** $0.3\overline{3}$

**18.** $0.6\overline{6}$

**19.** $0.3\overline{1}$

**20.** $0.4\overline{5}$

**21.** $0.1\overline{2}$

**22.** $0.3\overline{4}$

**23.** $0.0\overline{27}$

**24.** $0.4\overline{10}$

## PROBLEMS

A

**1.** The arc length through which the bob of a swinging pendulum moves is $\frac{9}{10}$ of its preceding arc length. How far will the bob move before coming to rest if the first swing covers an arc length of 20 centimeters?

**2.** A ball dropped from a height of 6 feet rebounds on each bounce to a height of 90% of the distance from which it fell. How far does it travel before coming to rest?

**3.** A side of a square is 12 inches. The midpoints of its sides are joined to form an inscribed square, and this process is continued as shown in the diagram. Find the sum of the perimeters of the squares if this process is continued without end.

**4.** A side of an equilateral triangle is 10 inches. The midpoints of its sides are joined to form an inscribed equilateral triangle and the process is continued. Find the sum of the perimeters of the triangles if the process is continued without end.

**5.** Find the sum of the areas of the squares in Problem 3.

**6.** Find the sum of the areas of the triangles in Problem 4.

**B** | **7.** A "snowflake" curve is constructed as follows: The sides of an equilateral triangle are trisected, and the middle third of the trisection serves as a base for a new equilateral triangle, following which this segment is deleted from the figure. The process is continued. If the side of the initial equilateral triangle is of length 1, what is the area enclosed by the snowflake curve if the process is continued without end?

**8.** Show that the perimeter of the figure described in Problem 7 is of unbounded length.

**C** | **9.** If $|x| < 1$, find the sum $S$ of the series $1 + 3x + 5x^2 + 7x^3 + \ldots$. (*Hint:* Consider $S - xS$.)

**10.** Find the fallacy in the following argument.

$$\text{Let } S = 1 + 2 + 2^2 + 2^3 + \ldots$$

$$= 1 + 2(1 + 2 + 2^2 + \ldots) = 1 + 2 \cdot \frac{1}{1 - 2} = 1 - 2$$

$$\therefore S = -1$$

## BINOMIAL EXPANSIONS

## 14–7  Powers of Binomials

When you expand natural number powers of binomials, you discover an interesting pattern. Notice the coefficients and the exponents of the successive powers of $(a + b)$ on the next page.

$$(a + b)^1 = a^1 + b^1$$
$$(a + b)^2 = a^2 + 2ab + b^2$$
$$(a + b)^3 = a^3 + 3a^2b + 3ab^2 + b^3$$
$$(a + b)^4 = a^4 + 4a^3b + 6a^2b^2 + 4ab^3 + b^4$$
$$(a + b)^5 = a^5 + 5a^4b + 10a^3b^2 + 10a^2b^3 + 5ab^4 + b^5$$

These examples suggest the following rules for expanding $(a + b)^n$, where $n$ is a natural number.

---

1. The first term is $a^n$, or $1 \cdot a^n \cdot b^0$.
2. For each term after the first:
   a. The coefficient is the product of the coefficient of the preceding term and the exponent of $a$ in the preceding term divided by the number of the preceding term.
   b. The exponent of $a$ is one less than the exponent of $a$ in the preceding term.
   c. The exponent of $b$ is one greater than the exponent of $b$ in the preceding term.
3. There are $n + 1$ terms, the last being $b^n$.

---

**Example.** Use the foregoing rules to expand $(x^2 - 2y)^4$.

*Solution:* $\quad (x^2 - 2y)^4 = [(x^2) + (-2y)]^4$.

Coefficients

$$1(x^2)^4 + 4(x^2)^3(-2y) + 6(x^2)^2(-2y)^2 + 4(x^2)^1(-2y)^3 + 1(-2y)^4$$

Term numbers

Simplifying,

$$x^8 + 4x^6(-2y) + 6x^4 \cdot 4y^2 + 4x^2 \cdot (-8y^3) + 16y^4,$$

or

$$x^8 - 8x^6y + 24x^4y^2 - 32x^2y^3 + 16y^4.$$

Notice that when the binomial to be expanded is a difference, the signs of the terms of the simplified expansion alternate. When the binomial is a sum, all the terms have positive signs.

## WRITTEN EXERCISES

Write each binomial in expanded form and simplify.

**A**

**1.** $(r + s)^4$      **5.** $(x - 2)^4$      **9.** $(2z - a^2)^3$

**2.** $(t + q)^5$      **6.** $(y + 2)^5$      **10.** $(x - 2y^2)^4$

**3.** $(1 - t)^6$      **7.** $\left(\dfrac{x}{3} + 3\right)^5$      **11.** $\left(2x - \dfrac{y}{2}\right)^3$

**4.** $(1 - s)^5$      **8.** $\left(\dfrac{y}{2} + 2\right)^6$      **12.** $\left(3a + \dfrac{y}{3}\right)^4$

Find the first three terms of each expansion and simplify.

**13.** $(a - 2b)^{12}$      **15.** $\left(\dfrac{z}{2} + 8\right)^8$      **17.** $(x + y)^{18}$

**14.** $(r + \frac{1}{2}s)^{10}$      **16.** $\left(\dfrac{r}{3} - 27\right)^7$      **18.** $(a + b)^{20}$

Find the value of each expression to the nearest hundredth.

**B**   **19.** $(1 + 0.02)^{10}$   **20.** $(1 + 0.01)^{15}$   **21.** $(0.99)^8$   **22.** $(0.98)^8$

## 14–8   The General Binomial Expansion

If you apply the rules stated in the preceding section to $(a + b)^n$, where $n$ is a positive integer, you will obtain the **Binomial Theorem**: $(a + b)^n =$

$$a^n + \frac{n}{1} a^{n-1}b^1 + \frac{n}{1} \cdot \frac{n-1}{2} a^{n-2}b^2 + \frac{n}{1} \cdot \frac{n-1}{2} \cdot \frac{n-2}{3} a^{n-3}b^3 + \cdots$$

$$+ \frac{n}{1} \cdot \frac{n-1}{2} \cdot \frac{n-2}{3} \cdots \frac{n - (r - 1)}{r} a^{n-r}b^r + \cdots + b^n.$$

The term of the expansion which involves $b^r$ is

$$\frac{n(n - 1)(n - 2) \cdots (n - [r - 1])}{1 \cdot 2 \cdot 3 \cdots r} a^{n-r}b^r.$$

**Factorial notation** is used to abbreviate the product in the denominator: $1 \cdot 2 \cdot 3 \cdots r = r!$ ($r!$ is read "r factorial" or "factorial r"). For example,

$$3! = 1 \cdot 2 \cdot 3 = 6 \qquad 5! = 1 \cdot 2 \cdot 3 \cdot 4 \cdot 5 = 120$$

$$1! = 1 \qquad 7! = 1 \cdot 2 \cdot 3 \cdot 4 \cdot 5 \cdot 6 \cdot 7 = 5040$$

Therefore, **the term of the expansion which involves** $b^r$ **can be written**

$$\frac{n(n-1)(n-2)\cdots(n-[r-1])}{r!}\, a^{n-r}b^r.$$

Note that the numerator of the coefficient is the product of $r$ consecutive integers decreasing from $n$ and that the numerator and denominator contain exactly the same number of factors.

**Example 1.** Find the term which involves $v^4$ in the expansion of $(u+2v)^{10}$.

*Solution:*  Here $n=10$, $a=u$, and $b=2v$.

The term involving $b^4$, or $(2v)^4$, is the one which involves $v^4$.

$\therefore\ r=4.$  $\hfill (2v)^4$

The exponent of $a$, or $u$, is $n-r$: $10-4=6$.  $\hfill u^6(2v)^4$

The denominator of the coefficient is $r!=4!$.
4! has four factors.  $\hfill \dfrac{1}{1\cdot2\cdot3\cdot4}u^6(2v)^4$

The numerator is the product of four consecutive
integers descending from 10.  $\hfill \dfrac{10\cdot9\cdot8\cdot7}{1\cdot2\cdot3\cdot4}u^6(2v)^4$

Simplify.  $\hfill 3360u^6v^4$

The **$k$th term** of the expansion $(a+b)^n$ is the one which involves $b^{k-1}$. It can be obtained from the general expression above by replacing $r$ by $k-1$.

$$\frac{n(n-1)(n-2)\cdots(n-[k-2])}{(k-1)!}\, a^{n-(k-1)}b^{k-1}.$$

**Example 2.** Find the sixth term in the expansion of $(x^3-y)^{12}$.

*Solution:*  Here $n=12$, $k=6$, $a=x^3$, and $b=-y$.

The exponent of $b$, or $-y$, is $k-1$, or 5.  $\hfill (-y)^5$

The exponent of $a$, or $x^3$, is 7 because when it is added
to 5, the exponent of $b$, the sum must be 12.  $\hfill (x^3)^7(-y)^5$

The denominator of the coefficient is
$(k-1)!=5!$. 5! has five factors.  $\hfill \dfrac{1}{1\cdot2\cdot3\cdot4\cdot5}(x^3)^7(-y)^5$

The numerator is the product of five
consecutive integers descending from 12.  $\hfill \dfrac{12\cdot11\cdot10\cdot9\cdot8}{1\cdot2\cdot3\cdot4\cdot5}(x^3)^7(-y)^5$

Simplify.  $\hfill -792x^{21}y^5$

## WRITTEN EXERCISES

Simplify each expression.

**Sample.** $\dfrac{10!}{(10 - 3)!}$

*Solution:* $\dfrac{10!}{(10 - 3)!} = \dfrac{10!}{7!} = \dfrac{1 \cdot 2 \cdot 3 \cdot 4 \cdot 5 \cdot 6 \cdot 7 \cdot 8 \cdot 9 \cdot 10}{1 \cdot 2 \cdot 3 \cdot 4 \cdot 5 \cdot 6 \cdot 7}$

$$= 8 \cdot 9 \cdot 10 = 720$$

A  **1.** $\dfrac{9!}{8!}$  **3.** $\dfrac{(8 - 2)!}{(4 + 1)!}$  **5.** $\dfrac{6!}{7! - 6!}$  **7.** $\dfrac{n!}{(n - 1)!}$

**2.** $\dfrac{12!}{9!}$  **4.** $\dfrac{4!6!}{7!}$  **6.** $\dfrac{5!}{4! - 3!}$  **8.** $\dfrac{(n + 2)!}{n!}$

Find and simplify the specified term in each expression.

**9.** $(a + t)^6$ involving $t^4$

**10.** $(1 + z)^5$ involving $z^3$

**11.** $(2 - x)^6$ involving $x^5$

**12.** $(3 - s)^5$ involving $s^3$

**13.** $(1 + t)^{12}$ involving $t^7$

**14.** $(2 + m)^{10}$ involving $m^4$

**15.** $(x^2 - 2y)^{10}$; the fourth term

**16.** $\left(t + \dfrac{y^2}{2}\right)^8$; the sixth term

**17.** $\left(r^2 - \dfrac{x^2}{2}\right)^7$; the fifth term

**18.** $\left(6 - \dfrac{x^2}{3}\right)^9$; the third term

---

## Chapter Summary

---

**1.** The **general term** of an **AP** is given by the formula $t_n = a + (n - 1)d$.

The **sum** of an AP, $\sum\limits_{i=1}^{n} [a + (i - 1)d]$, can be found from the formula

$$S_n = \frac{n}{2}(a + l), \text{ or } S_n = \frac{n}{2}[2a + (n - 1)d.]$$

**2.** The general term of a **GP** can be found from the formula $t_n = ar^{n-1}$.

The **sum**, $\sum\limits_{i=1}^{n} ar^{i-1}$, is found from $S_n = \dfrac{a - ar^n}{1 - r}$, or $S_n = \dfrac{a - rl}{1 - r}$.

3. You can determine any number of arithmetic or of geometric **means** by using the formulas for $t_n$.

4. When $|r| < 1$, the sum of an **infinite GP** having $r$ as common ratio and $a$ as first term is $\dfrac{a}{1 - r}$. Using this formula, you can express a repeating decimal as an equivalent common fraction.

5. To find any positive integral power of a binomial, use the **Binomial Theorem**:

$$(a + b)^n = a^n + na^{n-1}b + \frac{n(n - 1)}{2!} a^{n-2}b^2$$

$$+ \frac{n(n - 1)(n - 2)}{3!} a^{n-3}b^3 + \cdots + b^n$$

where the $k$th term of the expansion is given by

$$\frac{n(n - 1)(n - 2) \cdots (n - k + 2)}{(k - 1)!} a^{n-k+1}b^{k-1}$$

### Vocabulary and Spelling

Review the meaning of each term by reference to the page listed.

sequence (*p. 543*)  
term (of a sequence) (*p. 543*)  
arithmetic progression (AP) (*p. 544*)  
arithmetic means (*p. 547*)  
series (*pp. 549, 557*)  
summation $\Sigma$ (*p. 550*)

geometric progression (GP) (*p. 553*)  
geometric means (*p. 553*)  
mean proportional (*p. 554*)  
Binomial Theorem (*p. 565*)  
factorial notation ($n!$) (*p. 565*)

---

## Chapter Test and Review

---

**14–1**  1. Find the 40th term of the AP: 10, 8, 6, . . .

**14–2**  2. Insert two arithmetic means between $-3$ and 9.

3. If a position pays $5500 per year to start, and if yearly increases of $300 are awarded, in how many years will a person holding this position reach a salary of $10,000 per year?

**14–3**  4. Evaluate $\displaystyle\sum_{j=2}^{7} (3j - 5)$.

5. The bell in a church tower chimes as many times as the hour. How many times will it chime from 1:00 P.M. to 11:00 P.M., inclusive?

14–4    **6.** Find the 6th term of the geometric progression whose first two terms are 5 and $-10$.

14–5    **7.** Evaluate $\displaystyle\sum_{i=2}^{7} 3(2)^i$.

14–6    **8.** Find a common fraction equivalent to $0.\overline{41}$.

14–7    **9.** Expand $(z + \frac{1}{4})^4$.

14–8    **10.** Find the term involving $z^4$ in the expansion of $(\frac{1}{2} - z)^7$.

# Extra for Experts

## MATHEMATICAL INDUCTION

The set $N$ of natural numbers, $\{1, 2, 3, 4, \ldots\}$, is the smallest set of real numbers which (1) contains the number 1, and (2) is closed with respect to adding 1 to its members. That is, if $S$ is any *subset* of $N$ which satisfies (1) and (2), then $S$ must be $N$. Note that $1 \in S$; $1 + 1$, or 2, $\in S$; $2 + 1$, or 3, $\in S$; and, in general, if $k \in S$, then $k + 1 \in S$.

   To determine whether a set $S$ of *natural numbers* is actually the set $N$ of *all* natural numbers, you subject $S$ to two tests:
   1. Does 1 belong to $S$?
   2. For each number $k$ in $S$, is it true that $k + 1$ is in $S$?
If the answer to both questions is "Yes," then $S$ is $N$.

**Example 1.**    Let $S$ be the solution set of $2^n \geq n + 1$ over $N$. Prove that $S = N$.

*Proof:*    **1.** Is $1 \in S$? That is, does $2^n \geq n + 1$ become a true statement when $n$ is replaced by 1?

$$2^1 \geq 1 + 1 \text{ is true. } \therefore 1 \in S.$$

**2.** If $k \in S$, is $k + 1 \in S$? That is, if $2^k \geq k + 1$ is a true statement, is $2^{k+1} \geq (k + 1) + 1$ true?

Suppose        $2^k \geq k + 1$.
Then        $2 \cdot 2^k \geq 2(k + 1)$,
or        $2^{k+1} \geq 2k + 2 > k + 2 = (k + 1) + 1$.
Hence        $2^{k+1} \geq (k + 1) + 1$.

$\therefore$ if $k \in S$, then $k + 1 \in S$.

Because the answers to questions 1 and 2 are both "Yes," we conclude that $S = N$.

**Example 2.** Prove that the sum of the first $n$ even integers is $n(n + 1)$: $2 + 4 + \cdots + 2n = n(n + 1)$.

*Proof:* Let $S$ be the set of all natural numbers $n$ such that $2 + 4 + \cdots + 2n = n(n + 1)$ is true.

**1.** Is $1 \in S$? For $n = 1$, the equation becomes $2 \cdot 1 = 1(1 + 1)$, which is true. $\therefore 1 \in S$.

**2.** If $k \in S$, is $k + 1 \in S$?
Suppose $2 + 4 + \cdots + 2k = k(k + 1)$
Add the next even integer $2(k + 1)$ to both members:

$$2 + 4 + \cdots + 2k + 2(k + 1) = k(k + 1) + 2(k + 1)$$
$$= k^2 + 3k + 2$$
$$= (k + 1)(k + 2)$$

Hence $2 + 4 + \cdots + 2(k + 1) = (k + 1)((k + 1) + 1)$. That is, if the given equation is true when $n$ is replaced by $k$, it is also true when $n$ is replaced by $k + 1$; if $k \in S$, then $k + 1 \in S$.

Since $S$ passed both tests, $S = N$. Thus, we have proved that for every natural number $n$, $2 + 4 + \cdots + 2n = n(n + 1)$.

The argument used in Examples 1 and 2 is called *proof by mathematical induction.* You use the method to prove propositions which involve natural numbers.

# Exercises

Prove that each statement is true for all natural numbers $n$.

**1.** $3^n > 2n$

**2.** $1 + 2 + 3 + \cdots + n = \frac{1}{2}n(n + 1)$

**3.** $1 + 3 + 5 + \cdots + (2n - 1) = n^2$

**4.** $\dfrac{1}{1 \cdot 2} + \dfrac{1}{2 \cdot 3} + \dfrac{1}{3 \cdot 4} + \cdots + \dfrac{1}{n(n + 1)} = \dfrac{n}{n + 1}$

**5.** $(1 + a)^n \geq 1 + na \quad (a > -1)$

**6.** $1^2 + 2^2 + 3^2 + \cdots + n^2 > \frac{1}{3}n^3$

# Mathematics and Music

Knowledge of mathematics is basic to understanding music. Musical pitch, tempo, timbre (the quality given to a sound by its overtones), and the acoustics of music all can be expressed in mathematical terms. The intangible quality of art is what distinguishes a great piece of music.

Numbers enter curiously in describing what sounds are pleasing to the ear. In the 12-note tempered scale (our usual musical scale), each note is spaced in pitch (frequency) by a ratio of $\sqrt[12]{2}$. This is a compromise of several objectives. The best sounding pitch scale has notes related to each other by the set of small integer ratios such as $1:2$, $2:3$, and $3:4$. The tempered scale comes closest to including these ratios in a *uniformly* spaced set of notes.

Using a special mathematical series, called a Fourier Series, it is possible to write a mathematical formula for the harmonies generated by an instrument when a single note is sounded. Thus the general form for a Fourier Series,

$$\phi_n(x) = a_0 + a_1 \cos x + b_1 \sin x + a_2 \cos 2x + b_2 \sin 2x + \cdots$$
$$+ a_n \cos nx + b_n \sin nx + \cdots$$

is used to express an arbitrary periodic waveform and can therefore differentiate between, say, a clarinet and a bassoon, both playing at the same pitch. Note that this series is a sum of simple sine and cosine functions.

These mathematical relationships are particularly important in the development of music synthesizers such as the one shown below. The "voices" of various musical instruments such as the cello, French horn, and harpsicord are simulated within the synthesizer giving the performer incredible flexibility in producing innumerable combinations of interesting sounds.

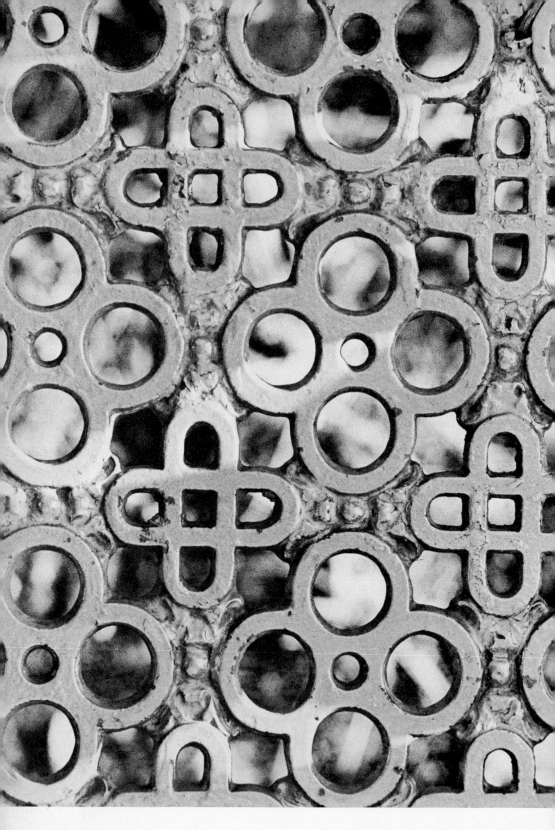

# 15

# Permutations, Combinations, and Probability

## ARRANGEMENTS OF OBJECTS

### 15–1    A Fundamental Counting Principle

The map in Figure 15–1 shows that there are two roads from city $A$ to city $B$ and three roads from city $B$ to city $C$. In how many ways can you drive from $A$ to $C$ by way of $B$? First you choose a road from $A$ to $B$; you can do this in any one of $2$ ways. Then you choose a road from $B$ to $C$ in any one of $3$ ways. Since each $A$-to-$B$ road can be paired with each $B$-to-$C$ road, there are $2 \times 3$, or 6, different ways of driving from $A$ to $C$ by way of $B$. This example illustrates the following:

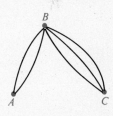

FIGURE 15–1

> **Fundamental Counting Principle.** If one action can be performed in $m$ ways, and for each of these ways, a second action can be performed in $n$ ways, then the number of ways the two actions can be performed in the order named is $m \times n$.

This principle can be extended to three or more actions performed in succession.

**Example 1.** How many three-digit numerals for even natural numbers can be formed from the digits 0, 1, 3, 7, 8, 9?

*Solution:* Think of performing three actions: choosing the hundreds digit, the tens digit, and the units digit. A diagram such as

$$\underline{\phantom{xx}},\ \underline{\phantom{xx}},\ \underline{\phantom{xx}},\qquad \text{or}\qquad \boxed{\phantom{xx}\,|\,\phantom{xx}\,|\,\phantom{xx}}$$

will help you find the *number of ways* the choices can be made.

Because the numeral may not have its first digit 0, you must choose the hundreds digit from the set $\{1, 3, 7, 8, 9\}$. Since you can do this in 5 ways, you write $\underline{\ 5\ }$, $\underline{\phantom{xx}}$, $\underline{\phantom{xx}}$. The tens digit is any member of the six-element set $\{0, 1, 3, 7, 8, 9\}$, so write 6 in the second place: $\underline{\ 5\ }$, $\underline{\ 6\ }$, $\underline{\phantom{xx}}$. The units digit must be chosen from $\{0, 8\}$, so write 2 in the third place: $\underline{\ 5\ }$, $\underline{\ 6\ }$, $\underline{\ 2\ }$. The fundamental principle tells you that there are $5 \times 6$ ways of choosing both the hundreds and tens digits; therefore there are $(5 \times 6) \times 2$, or 60, numerals of the type described.

Sometimes you must divide a problem into cases, count the objects represented in each case, and add the results.

**Example 2.** Auto license plates in a certain state consist either of one letter followed by four digits, or of two letters followed by three digits. How many plates are possible?

*Solution:* For plates having just one letter, you have the diagram $\boxed{26\,|\,10\,|\,10\,|\,10\,|\,10}$; the number of such plates is $26 \times 10^4$, or 260,000. For plates having two letters you have $\boxed{26\,|\,26\,|\,10\,|\,10\,|\,10}$; the number of plates of this type is $26 \times 26 \times 10^3$, or 676,000. Therefore the total number of plates possible is 260,000 + 676,000, or 936,000.

*WRITTEN EXERCISES*

**A**

1. A red die and a green die are tossed together. In how many ways can they fall?

2. In how many ways can a ten-question true-false test be answered if each question must be marked *true* or *false?*

3. In how many ways can a five-question true-false test be answered if it is permitted to leave blanks?

4. A debating club consists of 12 boys and 8 girls. How many teams of one boy and one girl can be formed?

5. Repeat Example 1 (page 574) with the change that the number must be odd.

6. Repeat Example 1 (page 574) with the change that the number must be a multiple of 5.

7. How many three-letter sequences can be formed from the letters A, B, C, D, and E if repetitions are allowed?

8. Repeat Exercise 7 with the change that the middle letter must be a vowel.

9. Repeat Exercise 7 with the change that the first and last letters must be consonants.

10. How many three-digit numbers can be formed from the digits 2, 3, 4, 5, 6 if repetitions are not allowed?

**B** 11. Repeat Exercise 10 with the change that the number must be even. (*Hint:* Fill the units place first.)

12. Repeat Example 2 (page 574) with the change that repetition of letters is not allowed.

13. Repeat Example 2 (page 574) with the change that repetitions of neither letters nor digits is allowed.

14. In how many ways can a five-question multiple-choice test be answered if there are four choices *a*, *b*, *c*, *d* for each question and consecutive choices must be different?

15. How many numbers of three or fewer digits can be formed from the digits 3, 4, 5, 6, 7?

16. Repeat Exercise 15 with the change that repetitions are not allowed.

17. A test consists of six true-false questions and four multiple-choice questions with three possible answers for each question. In how many ways can the test be answered (no blanks allowed)?

18. A basketball club consists of two centers, four forwards, and three guards. How many different starting teams (consisting of one center, two forwards, and two guards) are possible?

19. How many three-letter code words can be made if at least one of the letters must be a vowel (A, E, I, O, U, or Y)?

20. How many auto licenses of four symbols can be made if at least two of the symbols are letters and the rest are digits?

## 15–2   Permutations

You can arrange the members of the set $\{x, y, z\}$ in six different orders:

| | | |
|---|---|---|
| $x\ y\ z$ | $y\ z\ x$ | $z\ x\ y$ |
| $x\ z\ y$ | $y\ x\ z$ | $z\ y\ x$ |

Each of these arrangements is called a *permutation* of the letters $x$, $y$, $z$. A **permutation** of objects is an arrangement of the objects in a definite order.

Let us find the *number* of permutations of the four members of the set $\{a, b, c, d\}$. We can choose any of the 4 letters to be first in the arrangement, so we write $\boxed{4\ \ |\ \ |\ \ }$. After a letter has been chosen for first place, the choice for second place must be made from the 3 remaining letters, so we have $\boxed{4\ |\ 3\ |\ \ |\ \ }$. The third letter is chosen from the 2 now remaining, so we write $\boxed{4\ |\ 3\ |\ 2\ |\ \ }$. There is only 1 choice for the last place; thus the completed diagram is $\boxed{4\ |\ 3\ |\ 2\ |\ 1}$. Hence the number of permutations of $\{a, b, c, d\}$ is $4 \cdot 3 \cdot 2 \cdot 1 = 4! = 24$.

Some of these 24 permutations are *cabd*, *dcba*, *abcd*, and *dabc*. The method discussed above can be used to show the following fact:

---

The number of permutations of $n$ objects is $n!$.

---

**Example 1.**   Five different signal flags are available.

a. How many signals can be made by displaying all five on a vertical flagpole?

b. How many signals can be made by displaying the flags three at a time?

*Solution:*

a. The problem asks for the number of permutations of five things (taken five at a time). This is $5! = 5 \cdot 4 \cdot 3 \cdot 2 \cdot 1 = 120$.

b. In the diagram $\boxed{\ \ |\ \ |\ \ }$, the first space can be filled in 5 ways, the second in 4, and the third in 3. Thus $\boxed{5\ |\ 4\ |\ 3}$ represents the situation. By the fundamental principle (page 573) there are $5 \cdot 4 \cdot 3$, or 60, ways of displaying the flags three at a time.

The number of permutations of $n$ things taken $r$ at a time is denoted by $_nP_r$. Thus, in Example 1(b) we found $_5P_3$ to be $5 \cdot 4 \cdot 3$, or 60. In the general case, the diagram has $r$ spaces to be filled, as shown:

$$\boxed{\; n \;|\; n-1 \;|\; n-2 \;|\; \cdots \;|\; n-(r-1) \;}$$

Thus

$$_nP_r = n(n-1)(n-2)\cdots(n-[r-1]).$$

Notice that the product on the right has $r$ factors.

**Example 2.** In how many ways can four different prizes be awarded in a group of ten people if no person may win more than one prize?

*Solution:* $\quad _{10}P_4 = 10 \cdot 9 \cdot 8 \cdot 7 = 5040$

We can express $_nP_r$ in another way:

$$_nP_r = n(n-1)\cdots(n-r+1)$$
$$= \frac{n(n-1)\cdots(n-r+1)(n-r)(n-r-1)\cdots 3 \cdot 2 \cdot 1}{(n-r)(n-r-1)\cdots 3 \cdot 2 \cdot 1}$$

Therefore: $\qquad\qquad _nP_r = \dfrac{n!}{(n-r)!}$

**Example 3.** In how many ways can a president, a secretary and a treasurer be chosen in a club containing ten members?

*Solution:* $\quad _{10}P_3 = \dfrac{10!}{(10-3)!} = \dfrac{10!}{7!} = \dfrac{10 \cdot 9 \cdot 8 \cdot 7!}{7!} = 10 \cdot 9 \cdot 8 = 720$

For convenience 0! is *defined* to be 1, hence if $r = n$, you have

$$_nP_r = \frac{n!}{(n-n)!} = \frac{n!}{0!} = n!.$$

## WRITTEN EXERCISES

**A**

**1.** In how many ways can the letters in the word NUMBER be arranged if each letter is used exactly once in each arrangement?

**2.** In how many ways can seven bus drivers be assigned to seven busses?

**3.** In how many ways can a president, a vice-president, a secretary, and a treasurer be chosen in a club containing twelve members?

**4.** In how many ways can seven people be seated in a room containing ten chairs?

5. The Greek alphabet contains 24 letters. How many sorority names may be formed using three different letters?

6. How many permutations of the letters in the word NUMBER end in a vowel? end in a consonant?

7. How many permutations of the letters A, B, C, D, E, F have a vowel in the first and fifth positions?

8. How many numerals for natural numbers less than 500 can be formed by using three of the five digits 1, 3, 5, 7, 9?

Exercises 9–11 refer to a set of seven different signal flags. How many signals can be made by displaying them on a vertical flagpole:

9. Four at a time?    10. Two at a time?

11. Four or fewer (but at least one) at a time?

Exercises 12–14 refer to a club consisting of three men and three women. In how many ways can they be arranged for a photograph if:

12. they stand in a row?

13. they stand in two rows, with the women in front?

14. they stand in one row, but men and women occupy alternate positions?

Exercises 15–20 refer to a set of eight different books, four red and four green. In how many ways can the eight books be arranged if:

15. the first four books are red?    16. the first three books are red?

17. the first two books are red?

18. the first two and the last two books are red?

19. the first two books are red and the last two are green?

20. the books alternate in color?

Show that each of the following is true.

**B** 21. $_6P_4 = 6(_5P_3)$                24. $_5P_3 - _5P_2 = 2(_5P_2)$

22. $_nP_4 = n(_{n-1}P_3)$              25. $_nP_4 - _nP_3 = (n-4)(_nP_3)$

23. $_5P_r = 5(_4P_{r-1})$              26. $_nP_r - _nP_{r-1} = (n-r)(_nP_{r-1})$

Solve for *n*.

**C** 27. $_nP_3 = 3(_{n-1}P_2)$          29. $_nP_4 = 4(_nP_3)$

28. $_nP_4 = 8(_{n-1}P_3)$              30. $_nP_r = k(_{n-1}P_{r-1})$

## 15–3   Other Arrangements

Can you find the number of permutations of the five letters in the word ERROR? If you replaced the three R's by X, Y, and Z, you would have five *different* letters and the number of permutations would be 5! or 120. Notice, however, that both of the arrangements XEZYO and ZEYXO would become RERRO when you restored the R's. Indeed, there are 3! different arrangements of the form __, E, __, __, O when X, Y, and Z fill the blanks, but only 1 when R, R, and R do. Thus you multiplied the number of *distinguishable* arrangements by 3! when you replaced the R's by X, Y, and Z. Therefore, you must divide by this factor to find the number of permutations of 5 letters, 3 of which are alike.

$$\frac{5!}{3!} = \frac{5 \cdot 4 \cdot 3 \cdot 2 \cdot 1}{3 \cdot 2 \cdot 1} = 5 \cdot 4 = 20$$

In general:

> The number of permutations of $n$ objects, with $n_1$ objects alike, $n_2$ of another kind alike, and so on, is:
>
> $$\frac{n!}{n_1! \, n_2! \cdots}$$

**Example 1.**   A real-estate development contains ten identical houses in a row. The developer decides to paint three of them green, three brown, and four white. In how many ways can this be done?

*Solution:*   Imagine ten *movable* houses already painted as described. The problem is equivalent to finding the number of permutations of 10 objects (the painted houses) 3 of which are alike, 3 of a different kind are alike, and 4 of a third kind are alike:

$$\frac{10!}{3!3!4!} = \frac{10 \cdot 9 \cdot 8 \cdot 7 \cdot 6 \cdot 5 \cdot 4!}{3 \cdot 2 \cdot 1 \cdot 3 \cdot 2 \cdot 1 \cdot 4!} = 4,200$$

A **circular permutation** is an arrangement of objects around a circle. (Until now we have considered only *linear* permutations.) A common example is the seating of people around a circular table. In such an arrangement there is no first place, so that if each person shifts his position one place counterclockwise the relative positions are not changed. In fact, each person can shift position $n$ times and return to his original position without disturbing the arrangement. Therefore, if you use the formula for a linear permutation to find the number of possible arrangements, you will have counted each different arrangement $n$ times. Thus, there are $n! \div n = (n - 1)!$ distinguishable permutations.

**Example 2.**   In how many ways can five persons be seated around a circular table?

*Solution:*   Since this is a circular permutation of 5 things, there are $(5 - 1)! = 4! = 24$ possible different seating arrangements. You may think of this in a slightly different manner: Since a rotation of any permutation does not produce a new permutation, one of the positions can be considered fixed, and $\boxed{1 \mid 4 \mid 3 \mid 2 \mid 1}$ describes the situation. We see again that there are 24 different arrangements.

Figure 15–2 shows the 6 circular permutations of the letters $a$, $b$, $c$, $d$. If these diagrams represent seating arrangements about a table, it is natural to consider them as being all different. Suppose, however, that the diagrams represent arrangements of keys on a key ring. If any arrangement in the top row is flipped over, the corresponding arrangement in the second row results, and we would wish to consider these two as being the same. In this sort of situation, the number of circular permutations of $n$ objects is $\frac{1}{2}(n - 1)!$, provided $n > 2$.

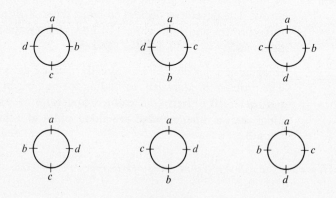

FIGURE 15–2

**Example 3.**   In how many ways can seven different charms be arranged on a charm bracelet?

*Solution:*   The situation here is analogous to that of a key ring. Hence the number of arrangements is $\frac{1}{2} \cdot 6! = 360$.

## WRITTEN EXERCISES

Find the number of (distinguishable) permutations of all the letters of each word.

**A**  **1.** ELEMENT | **3.** LETTER | **5.** INFINITE

**2.** BANANA | **4.** BANDANA | **6.** MISSISSIPPI

**7.** How many signals can be made by displaying seven flags at one time on a vertical flagpole if the flags differ only in color, and three are red, two are white, and two are blue?

**8.** In how many ways can 4 nickels, 3 dimes, and 2 quarters be distributed among 9 boys seated in a row if each receives a coin?

How many numerals can be formed using all the digits in the given numeral?

**9.** 5775    **10.** 1066    **11.** 23232    **12.** 11211    **13.** 646646

**14.** A "wheel of fortune" is to be made by dividing a circular disk into six equal pie-shaped wedges and numbering them from 1 to 6. In how many ways can this numbering be done?

**15.** In how many ways can six charms be arranged on a charm bracelet?

**16.** In how many ways can seven beads of different colors be strung on a necklace?

**B**  **17.** In how many ways can four young couples be seated at a round table so that men and women are seated alternately?

**18.** In how many ways can five men and five women be seated at a round table so that each woman is between two men?

**19.** In how many ways can a committee of six be seated at a round table if the chairman and the secretary must sit together?

## COMBINATIONS AND BINOMIAL COEFFICIENTS

### 15–4  Counting Subsets

You can check that the list below contains all of the three-element subsets of the five-element set $S = \{a, b, c, d, e\}$:

| | | | | |
|---|---|---|---|---|
| $\{a, b, c\}$ | $\{a, b, d\}$ | $\{a, b, e\}$ | $\{a, c, d\}$ | $\{a, c, e\}$ |
| $\{a, d, e\}$ | $\{b, c, d\}$ | $\{b, c, e\}$ | $\{b, d, e\}$ | $\{c, d, e\}$ |

The number of $3$-element subsets of a $5$-element set is denoted by $_5C_3$; by counting the sets in the list you see that $_5C_3 = 10$. Here is a way to find $_5C_3$ without writing out a list:

You can obtain a *permutation* of 3 of the 5 objects in $S$ in two steps: (1) choose a subset of 3 objects, and then (2) arrange these three objects. The number of ways of performing step (1) is $_5C_3$, and the number of ways of performing step (2) is $3!$. Thus the number of ways of obtaining a permutation of 3 of the 5 objects is $_5C_3 \times 3!$; that is,

$$_5C_3 \times 3! = {}_5P_3 = \frac{5!}{2!}.$$

Therefore, $\qquad _5C_3 = \dfrac{5!}{3!2!} = \dfrac{5 \cdot 4 \cdot 3 \cdot 2 \cdot 1}{3 \cdot 2 \cdot 1 \cdot 2 \cdot 1} = 10.$

An *r*-element subset of an *n*-element set is often called a **combination** of *n* objects taken *r* at a time. The *number* of combinations of *n* objects taken *r* at a time is denoted by $_nC_r$. The method of the preceding paragraph gives

$$_nC_r = \frac{{}_nP_r}{r!}.$$

Using expressions for $_nP_r$ from Section 15–2, you have

$$_nC_r = \frac{n!}{r!(n-r)!}$$

and

$$_nC_r = \frac{n(n-1)(n-2)\cdots(n-[r-1])}{r!}.$$

In the second formula note that the products in the numerator and the denominator both have *r* factors.

**Example 1.** How many different five-girl basketball teams can be formed from a group of twelve girls each of whom can play any position?

*Solution:*

$$_{12}C_5 = \frac{12 \cdot 11 \cdot 10 \cdot 9 \cdot 8}{5 \cdot 4 \cdot 3 \cdot 2 \cdot 1} = 792$$

**Example 2.** How many committees of two Republicans and three Democrats can be chosen from a group of seven Republicans and ten Democrats?

*Solution:* The number of ways of choosing 2 Republicans from a group of 7 is $_7C_2 = \dfrac{7 \cdot 6}{2 \cdot 1} = 21$, and the number of ways of choosing 3 Democrats from a group of 10 is $_{10}C_3 = \dfrac{10 \cdot 9 \cdot 8}{3 \cdot 2 \cdot 1} = 120$. By the Fundamental Counting Principle (page 573), the number of ways of selecting the committee is $21 \times 120 = 2520$.

You can sometimes use the fact that $_nC_r = {}_nC_{n-r}$. (Exercise 21, page 584) to simplify computations. For example,

$$_{40}C_{38} = {}_{40}C_2 = \frac{40 \cdot 39}{2 \cdot 1} = 780.$$

**Example 3.** Suppose you have six mathematics books and eight history books. In how many orders can you arrange four mathematics books and two history books on a shelf which holds six books?

*Solution:* You can choose the mathematics books in $_6C_4 = {}_6C_2 = \dfrac{6 \cdot 5}{2 \cdot 1} = 15$ ways and the history books in $_8C_2 = \dfrac{8 \cdot 7}{2 \cdot 1} = 28$ ways. Thus the number of ways you can select the six books is $15 \times 28 = 420$. Each of these selections can be arranged in $_6P_6 = 6! = 720$ ways. Thus the total number of arrangements is $420 \times 720 = 302{,}400$.

## WRITTEN EXERCISES

A

1. How many combinations can be formed from the letters in the word COUNT, taking them two at a time? List the combinations.

2. How many combinations can be formed from the letters in the word PRIME, taking them three at a time? List the combinations.

3. In how many ways can a committee of four be chosen from a club whose membership is 50?

4. In how many ways can a student choose to answer ten questions out of twelve on a test if the order of his answers is of no importance?

5. In how many ways can ten eggs be selected from a dozen?

6. How many different two-card hands can be drawn from a 52-card deck? How many five-card hands can be drawn?

7. A quality-control engineer must inspect a sample of three fuses from a box of 100. How many different samples can he choose?

8. Three marbles are drawn from a bag containing ten marbles. In how many ways can this be done? In how many ways can an additional three marbles be drawn?

9. How many straight lines can be formed by joining any two of six points, no three of which are in a straight line?

10. Eight points lie on the circumference of a circle. How many chords can be drawn joining them?

11. Five points lie on the circumference of a circle. How many inscribed triangles can be drawn having those points as vertices?

12. Five lines are drawn in a plane. If no three of these pass through the same point, how many points of intersection are there?

B 13. How many subcommittees of three Democrats and two Republicans can be formed from a committee whose membership is eight Democrats and six Republicans?

14. Seven men and seven girls were nominated for homecoming king and queen. How many ways can a king, a queen, and her court of two girls be chosen?

15. How many five-letter arrangements of the letters in the word LOGARITHMS consisting of three consonants and two vowels can be formed if no letter is repeated?

Exercises 16–18 deal with a bridge deck having four suits of 13 cards each. Set up a solution and estimate the answer.

16. How many 13-card hands having exactly 10 spades are there? (*Hint:* The 10 spades must come from the set of 13 spades, while the remaining 3 cards must come from the set of 39 remaining cards.)

17. How many 13-card hands are there which have exactly 10 cards of some one suit?

18. How many 13-card hands having seven spades are there? How many hands having exactly seven spades, three hearts, two diamonds, and one club are there?

19. Find $n$, given $_nC_3 = {}_{90}C_{87}$.

20. Find $n$, given $_nC_7 = {}_nC_3$.

C 21. Use the second formula for $_nC_r$ on page 582 to prove that $_nC_r = {}_nC_{n-r}$.

22. Six social workers are to form themselves into three teams of two. In how many ways can this be done?

## 15–5   Binomial Coefficients and Pascal's Triangle

The numbers $_nC_r$ occur as the coefficients in the expansion of the binomial $(a + b)^n$. To see why this is so, consider the following expansion:

$$(a + b)^3 = (a + b)(a + b)(a + b)$$
$$= aaa + aab + aba + abb + baa + bab + bba + bbb$$

To obtain each term of the sum on the second line, you choose either an $a$ or a $b$ from each factor on the first line and multiply them together. How many such terms can be simplified to $ab^2$, for instance? That is, how many terms contain exactly two $b$'s? To obtain such a term you choose 2 $b$'s from a set of 3 $b$'s (the $b$'s in the three factors on the first line). Since you can do this in $_3C_2$ ways, there are $_3C_2$ terms containing exactly two $b$'s. Thus $ab^2$ has $_3C_2$ as coefficient. Similarly, the coefficient of $a^3$ is $_3C_0$, of $a^2b$ is $_3C_1$, and of $b^3$ is $_3C_3$. Thus

$$(a + b)^3 = {}_3C_0a^3 + {}_3C_1a^2b + {}_3C_2ab^2 + {}_3C_3b^3$$
$$= 1a^3 + 3a^2b + 3ab^2 + 1b^3$$

By applying the reasoning used above to the general case, you can write the **Binomial Theorem** in the form

$$(a + b)^n = {}_nC_0a^n + {}_nC_1a^{n-1}b + {}_nC_2a^{n-2}b^2 + \cdots + {}_nC_nb^n.$$

If you set $a = 1$ and $b = 1$ you obtain:

$$(1 + 1)^n = {}_nC_0 + {}_nC_1 + {}_nC_2 + \cdots + {}_nC_n = 2^n$$

Recalling that $_nC_r$ is the number of $r$-element subsets of an $n$-element set, we see that:

**An $n$-element set has $2^n$ subsets.**

For example, the 3-element set $\{a, b, c\}$ has the following $2^3$, or 8, subsets: $\emptyset, \{a\}, \{b\}, \{c\}, \{a, b\}, \{a, c\}, \{b, c\}, \{a, b, c\}$.

**Example.**   In how many ways can a committee of two or more be selected from a group of seven people?

*Solution:*   The problem asks for the number of subsets of a 7-element set which have two or more members. There is $_7C_0 = 1$ subset with no members and there are $_7C_1 = 7$ subsets with one member each. Since the total number of subsets is $2^7$, the required number is $2^7 - 1 - 7 = 128 - 8 = 120$.

If at the top of the array of expansions on page 564, you add $(a + b)^0 = 1$, and then write only the coefficients of the right-hand members, you obtain **Pascal's Triangle:**

In this array each number except the 1's is the sum of the two numbers to the right and left of it in the row directly above. Thus you can obtain the sixth row from the fifth as illustrated below:

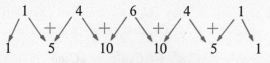

The relationship $_nC_r = {_{n-1}C_{r-1}} + {_{n-1}C_r}$ (Exercise 10, below) justifies the procedure.

*WRITTEN EXERCISES*

Expand the binomial and simplify.

**A**  **1.** $\left(x + \dfrac{1}{x}\right)^4$    **2.** $\left(x - \dfrac{1}{x}\right)^4$    **3.** $\left(x - \dfrac{1}{x}\right)^5$    **4.** $\left(x + \dfrac{1}{x}\right)^5$

**B**  **5.** How many numbers are there in the first 4 rows of Pascal's Triangle? In the first 5 rows? In the first $n$ rows? (Recall Section 14–3.)

**6.** In the United States six kinds of coins are in circulation, their values in cents being 1, 5, 10, 25, 50, and 100. How many different amounts of money can be obtained by using no more than one of each kind of coin?

**7.** There are 10 men at the base camp of an antarctic expedition. An exploring party of two or more men is to be organized. How many such parties are possible if at least two men must remain at base camp?

**8.** Show that for any natural number $n$,

$$_nC_0 - {_nC_1} + {_nC_2} - {_nC_3} + \cdots \pm {_nC_n} = 0.$$

**C**  **9.** Prove: $_nC_2 = {_{n-1}C_1} + {_{n-1}C_2}$ if $n \geq 3$.

**10.** Prove: $_nC_r = {_{n-1}C_{r-1}} + {_{n-1}C_r}$ if $n \geq r + 1$.

## PROBABILITY

## 15–6   Sample Spaces and Events

Suppose you perform the experiment of tossing three coins—a dime, a nickel, and a quarter. One possible outcome is that the dime, nickel and quarter land heads, heads, and tails, respectively. You can represent this outcome by the ordered triple $(H, H, T)$, or, more simply, $HHT$. The set of all possible outcomes of the experiment can be represented by $\{HHH, HHT, HTH, THH, HTT, THT, TTH, TTT\}$, and this set is called a *sample space* for the experiment. A **sample space** for an experiment is any set whose members correspond one-to-one with the outcomes of the experiment.

**Example 1.**   Give a sample space for the experiment of throwing a pair of dice.

*Solution:*   Since the six faces of each die bear symbols representing the numbers in

$$S = \{1, 2, 3, 4, 5, 6\},$$

the set of ordered pairs of $S$ serves as a sample space. It can be portrayed as a set in the Cartesian plane (see the adjacent figure). For example, the point $(3, 5)$ represents the outcome in which the first die

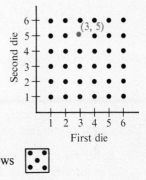

shows ⚁ and the second shows ⚃

The experiment of choosing at random one letter from the word COMPUTING has $\{$C, O, M, P, U, T, I, N, G$\}$ as a sample space. You might be interested in whether the letter chosen is a vowel. You can call the choosing of a vowel an *event* and list the outcomes which produce this event: $\{$O, U, I$\}$. An **event** is any subset of a sample space.

**Example 2.**   In the experiment of throwing a pair of dice, let $A$ be the event that the sum of the numbers thrown is eight and let $B$ be the event that at least one five is thrown. Picture these events geometrically.

*Solution:*   $A$ is the set enclosed in black. $B$ is the set enclosed in red.

**Example 3.** A bag contains five marbles: two green, one red, one white, and one blue.
  **a.** List a sample space for the experiment of drawing two marbles from the bag simultaneously.
  **b.** List the event that exactly one of the two marbles is green.
  **c.** List the event that at least one of the marbles is green.

*Solution:*  **a.** {gg, gr, gw, gb, rw, rb, wb}
  **b.** {gr, gw, gb}
  **c.** {gg, gr, gw, gb}

The experiment of drawing a letter at random from the word PUPPY has {P, U, Y} as a sample space. Not all three outcomes are equally likely, however, because if you are unbiased, you are three times as likely to choose a P as a U or a Y. You can "tag" the P's with subscripts to obtain a sample space {$P_1$, U, $P_2$, $P_3$, Y} in which all outcomes are equally likely. Similarly, if the sample space in Example 3 is written so that the outcomes are equally likely, it becomes

$$\{g_1g_2, g_1r, g_2r, g_1w, g_2w, g_1b, g_2b, rw, rb, wb\}.$$

*WRITTEN EXERCISES*

Describe a sample space for each of the following experiments, either by roster or by drawing a figure. Then list or picture the event or events described. (All drawings and choices are assumed to be random.)

A  **1.** Experiment: A number is drawn from the digits 1 through 9. Event 1: The number is odd. Event 2: The number is a multiple of 3.

  **2.** Experiment: A letter is drawn from TRIANGLE. Event: The letter is a vowel.

  **3.** Experiment: A letter is drawn from SQUARE. Event: The letter is a consonant.

  **4.** Experiment: A number is chosen from the integers whose absolute values are less than 5. Event 1: The number is positive. Event 2: The number is odd.

  **5.** Experiment: Two numbers are chosen from the set {−2, −1, 0, 1, 2}. Event: The sum of the numbers is zero.

  **6.** Experiment: Two numbers are drawn from the set {1, 2, 3, 4}. Event: The sum of the numbers is even.

  **7.** Experiment: Two dice are thrown. Event 1: Exactly one 2 is thrown. Event 2: A total of either 7 or 11 is thrown.

8. Experiment: A die is thrown and a coin is tossed. Event 1: The number on the die is even. Event 2: The number on the die is even and a head comes up.

9. Experiment: Two letters are chosen, one from MY and the other from EAR. Event 1: Both letters are vowels. Event 2: Both letters are consonants.

10. Experiment: Two letters are chosen, one from LET and the other from MAIN. Event 1: Both letters are vowels. Event 2: One letter is a vowel and the other a consonant.

11. Experiment: Two letters are drawn, one from ME and the other from TOO. (Give a sample space of equally likely outcomes.) Event 1: Both letters are vowels. Event 2: One letter is a vowel and the other a consonant.

12. Experiment: Two marbles are drawn simultaneously from a bag containing two black marbles and two white marbles. (Give a sample space of equally likely outcomes.) Event 1: The marbles have the same color. Event 2: The marbles have different colors.

13. Experiment: A marble is drawn from a bag containing two black marbles and two white marbles, returned to the bag, and a marble is drawn again. (Give a sample space of equally likely outcomes.) Event 1: The marbles drawn have the same color. Event 2: The first marble drawn is black and the second is white.

14. Experiment: Two letters are drawn from ROTOR. (Give a sample space of equally likely outcomes.) Event 1: Two consonants are drawn. Event 2: Two vowels are drawn.

## 15–7 Mathematical Probability

Consider the following experiment: From a bag containing 11 black and 7 white balls, draw one ball at random, record its color, and return it to the bag. The experiment has 18 equally likely outcomes. The event that a black ball is drawn consists of 11 outcomes. Therefore if you repeat the experiment a great many times, it seems reasonable to expect that about $\frac{11}{18}$ of the time you will draw a black ball. The ratio $\frac{11}{18}$ is called the *probability* that in any single experiment, the event of drawing a black ball will occur. This example suggests the following definition: In an experiment which has $n$ equally likely outcomes, let $E$ be an event which has $h$ outcomes. The **probability of the event $E$,** denoted by $P(E)$, is defined by

$$P(E) = \frac{h}{n}.$$

**Example 1.** A pair of dice is to be thrown. Find the probability of each of the events.

*A*: The sum of the numbers thrown is eight.
*B*: At least one five is thrown.

*Solution:* (See Example 2, page 587.) The experiment has 36 equally likely outcomes, and since the number of outcomes in *A* is 5, we have $P(A) = \frac{5}{36}$. *B* has 11 outcomes; hence $P(B) = \frac{11}{36}$.

In Example 1, the answer $P(A) = \frac{5}{36}$ does *not* tell you that if the dice are rolled 36 times, you will get a total of eight exactly 5 times. You might get 3 totals of eight, or 10 such totals, or 36. You *can* expect, however, to get a total of eight about $\frac{5}{36}$ of the time if the experiment is repeated a large number of times.

In calculating the probability of an event *E*, it is not necessary to exhibit the sample space and the set *E*. You need only find how many members each of these sets has.

**Example 2.** If there are 2 red, 3 white, and 5 blue marbles in a bag, what is the probability that two marbles drawn from the bag will both be blue?

*Solution:* Let *E* be the event of drawing two blue marbles. Since there are 10 marbles altogether, there are $_{10}C_2$ ways of drawing 2 marbles of any color (that is, the sample space has $_{10}C_2$ members). There are 5 blue marbles and therefore $_5C_2$ ways of drawing 2 blue marbles, (the event *E* has $_5C_2$ members).

Therefore, $P(E) = \dfrac{_5C_2}{_{10}C_2} = \dfrac{10}{45} = \dfrac{2}{9}$.

An event *E* in a sample space *S* is called *certain* if $E = S$ and is called *impossible* if $E = \emptyset$. Do you see that the probability of a certain event is 1 and of an impossible one is 0? If *E* is an event which is neither impossible nor certain, then $0 < P(E) < 1$.

By the **complement** of *E*, denoted by $\tilde{E}$, we mean the set of all members of *S* which do *not* belong to *E*. If *S* has *n* members and *E* has *h* members, then $\tilde{E}$ has $n - h$ members. $P(\tilde{E})$ is the probability that *E* does *not* occur, and

$$P(\tilde{E}) = \frac{n - h}{n} = 1 - \frac{h}{n} = 1 - P(E).$$

The **odds** that the event *E* will occur, or the **odds favoring** *E*, are

$$\frac{P(E)}{P(\tilde{E})}, \quad \text{or} \quad \frac{h}{n - h}, \quad \text{or} \quad \text{``}h \text{ to } n - h.\text{''}$$

The **odds against** $E$ occurring are "$n - h$ to $h$." Thus, in the original experiment (page 589), the odds are 11 to 7 *in favor of* drawing a black ball, or 7 to 11 *against* drawing a black ball.

**Example 3.**  Find the probability that, when two dice are thrown, neither shows a five. Find the odds that no five will appear.

*Solution:*  Let $E$ be the event that neither die shows a five. Then $\tilde{E}$ is the event that at least one five is thrown. From Example 1, $P(\tilde{E}) = \frac{11}{36}$. Hence $P(E) = 1 - P(\tilde{E}) = 1 - \frac{11}{36} = \frac{25}{36}$. The odds that no five will appear are 25 to 11.

## WRITTEN EXERCISES

A

1. One card is drawn from a bridge deck of 52 cards. (a) What is the probability that it is an ace? (b) What are the odds against its being an ace? (c) What is the probability that it is a spade? (d) What are the odds that is is a red card?

2. The probability that it will rain on a certain day is 40%. (a) What is the probability that it will not rain? (b) What are the odds that it will rain? (c) What are the odds that it will not rain?

3. A die is thrown. Find the probability that the number thrown is (a) even; (b) less than 5; (c) neither 1 nor 6.

4. At a charity event, a door prize is given to the person whose ticket stub is drawn from a bowl. Tickets numbered 1 through 60 have been sold. (a) Find the probability that the number of the winning ticket is $\leq 40$; is $> 30$; is both $\leq 40$ and $> 30$. (b) What are the odds against the winning number being even? being a multiple of 3?

5. If the name of one of the fifty states in the United States is drawn at random, what is the probability that (a) the name begins with $A$? (b) the name consists of two words? (c) the state has a common border with Mexico?

6. A letter is drawn at random from TRIANGLE. (a) What is the probability that it is a vowel? (b) What are the odds against it being a vowel?

7. Two numbers are drawn from the set $\{1, 2, 3, 4\}$. (a) What is the probability that their sum is even? (b) What are the odds that their sum is odd?

8. Two dice are thrown. (a) What is the probability that a total of either 7 or 11 is thrown? (b) What are the odds against this event?

9. Three coins are tossed. What is the probability that they all land heads?

10. A bag contains 2 red marbles, 3 white marbles, and 5 blue marbles. If a marble is drawn from the bag, find the probability that **(a)** it is red; **(b)** it is white; **(c)** it is not blue; **(d)** it is blue; **(e)** it is not white; **(f)** it is not red.

11. In Exercise 10, two marbles are drawn from the bag. Find the probability that: **(a)** both are red; **(b)** both are white; **(c)** one is red and the other white; **(d)** one is white and the other blue.

12. In Exercise 10, three marbles are drawn from the bag. Find the probability that **(a)** all are red; **(b)** all are white; **(c)** all are blue; **(d)** no two have the same color.

B 13. In an ESP experiment two subjects in separate rooms each have 5 cards labeled *A*, *B*, *C*, *D*, and *E*. The first subject selects a card at random and concentrates on its label, while the other subject tries to choose from his cards the one with the same label. If neither subject has ESP, what is the probability that the two cards will match?

14. Ten people comprising five married couples draw cards numbered from 1 to 10 to determine the order in which they shall stand in a row for a group photograph. What is the probability that each person stands next to his or her spouse?

15. The letters in the word MOTOR are placed at random in a row. What are the odds against the two O's being adjacent?

16. The letters in the word AMOUNT are placed at random in a row. What is the probability that three successive letters are consonants?

## 15–8 Mutually Exclusive Events

Figure 15–3 shows a sample space for the experiment of drawing one number from the set

$$S = \{1, 2, 3, 4, 5, 6, 7, 8\}.$$

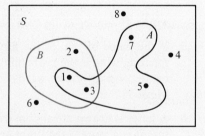

FIGURE 15–3

It also shows two events: *A*, the drawing of an odd number, and *B*, the drawing of a number less than 4. Thus, $A = \{1, 3, 5, 7\}$ and $B = \{1, 2, 3\}$. Notice that $P(A) = \frac{4}{8} = \frac{1}{2}$, and $P(B) = \frac{3}{8}$.

What is the probability that either *A* or *B* (or both) will occur; that is, what is $P(A \cup B)$? Since $A \cup B = \{1, 2, 3, 5, 7\}$, $P(A \cup B) = \frac{5}{8}$. To find a relationship between $P(A)$, $P(B)$, and $P(A \cup B)$, notice first that $A \cap B = \{1, 3\}$, and therefore $P(A \cap B) = \frac{2}{8} = \frac{1}{4}$. Now you can see that $P(A) + P(B) - P(A \cap B) = \frac{1}{2} + \frac{3}{8} - \frac{1}{4} = \frac{5}{8} = P(A \cup B)$.

You can prove that for any two events, $A$ and $B$,

$$P(A \cup B) = P(A) + P(B) - P(A \cap B).$$

**Example 1.** In a certain Swedish city the probability that a person chosen at random is a blue-eyed blond is 0.5, that a person is blond is 0.6, and that a person is blue-eyed or blond or both is 0.9. Find the probability that a person chosen at random has blue eyes.

*Solution:* In the experiment of choosing a person at random, define events $A$ and $B$ as follows:

$A$: The person has blue eyes.
$B$: The person is blond.

Then $P(A \cup B) = 0.9$, $P(B) = 0.6$, and $P(A \cap B) = 0.5$. Substitute into

$$P(A \cup B) = P(A) + P(B) - P(A \cap B).$$
$$0.9 = P(A) + 0.6 - 0.5$$
$$P(A) = 0.9 - 0.6 + 0.5 = 0.8$$

∴ the probability that a person chosen at random has blue eyes is 0.8.

Events $A$ and $B$ are **mutually exclusive** if they have no outcome in common, that is, if $A \cap B = \emptyset$. In this case $P(A \cap B) = P(\emptyset) = 0$. Therefore, if $A$ and $B$ are mutually exclusive events,

$$P(A \cup B) = P(A) + P(B).$$

**Example 2.** A committee of four is to be chosen by lot in a club consisting of seven women and five men. What is the probability that the committee will contain more men than women?

*Solution:* The committee will contain more men than women if it contains exactly 3 men and 1 woman (event $A$), *or else* it contains exactly 4 men and no women (event $B$).

Since $A \cap B = \emptyset$, events $A$ and $B$ are mutually exclusive, and $P(A \cup B) = P(A) + P(B)$.

$$P(A) = \frac{{}_5C_3 \cdot {}_7C_1}{{}_{12}C_4} = \frac{10 \cdot 7}{495} = \frac{14}{99}$$

$$P(B) = \frac{{}_5C_4 \cdot {}_7C_0}{{}_{12}C_4} = \frac{5 \cdot 1}{495} = \frac{1}{99}$$

∴ $P(A \cup B) = \frac{14}{99} + \frac{1}{99} = \frac{15}{99} = \frac{5}{33}$.

*WRITTEN EXERCISES*

**A** **1.** Two dice are rolled. Find the probability that: **(a)** the sum of the numbers is 5; **(b)** at least one 5 is thrown; **(c)** at least one 5 is thrown or the sum of the numbers thrown is 5.

**2.** Two dice are rolled. Find the probability that: **(a)** the sum of the numbers is 7; **(b)** the two numbers thrown are equal; **(c)** the two numbers thrown are equal or have sum 7.

**3.** Two dice are rolled. Find the probability that: **(a)** the sum of the numbers is 8; **(b)** the two numbers thrown are equal; **(c)** the numbers thrown are both 4's; **(d)** the two numbers thrown are equal or have sum 8.

**4.** Three coins are tossed. Find the probability that the number of heads is: **(a)** four; **(b)** three; **(c)** two; **(d)** one; **(e)** zero; **(f)** at least two.

**5.** A subcommittee of 4 members is to be chosen by lot from a Congressional committee consisting of 7 Democrats and 5 Republicans. What is the probability that: **(a)** more than half of the subcommittee members will be Republicans? **(b)** one of the parties fails to be represented on the committee?

**6.** Two cards are drawn from a deck of 52 cards. What is the probability that: **(a)** both are red? **(b)** both are aces? **(c)** either both are red or both are aces?

**B** **7.** If a four-volume set of books is placed at random on a shelf, what is the probability that they will be arranged either in proper or reverse order?

**8.** Tomorrow's forecast states that the probability of rain is 50%, of rain and snow or both is 60%, and of both rain and snow is 10%. What is the probability that it will snow?

**9.** The probability that the Dodgers will win the first game of a double header is $\frac{2}{3}$, that they will win the second game is $\frac{1}{2}$, and that they will win both is $\frac{1}{3}$. What are the odds against their losing both?

**10.** Two men and two women are seated at random in a row. Find the probability that **(a)** two men sit together; **(b)** two women sit together; **(c)** either the men sit together or the women do or both.

**C** **11.** Prove the formula $P(A \cup B) = P(A) + P(B) - P(A \cap B)$ using this fundamental counting principle:

If a set $A$ has $r$ members, a set $B$ has $s$ members, and their intersection $A \cap B$ has $t$ members, then their union $A \cup B$ has $r + s - t$ members.

## 15–9 Independent and Dependent Events

A bag contains $5$ balls, $3$ red and $2$ green. Consider the following two experiments:

**Experiment 1**

A ball is drawn, replaced in the bag, and a ball is drawn a second time.

**Experiment 2**

A ball is drawn, *not* replaced in the bag, and a ball is drawn a second time.

In each experiment let us find the probability that both balls drawn are red; that is, let us find $P(A \cap B)$, where

$A$ is the event that the first ball drawn is red;

$B$ is the event that the second ball drawn is red.

**Experiment 1**

There are $5$ ways of drawing the first ball, and because this ball is replaced, there are $5$ ways of drawing the second ball. By the Fundamental Counting Principle, (page 573) there are $5 \times 5$, or $25$, ways of drawing the two balls. There are $3$ ways of drawing a red ball the first time and $3$ ways the second time. Thus two red balls can be drawn in $3 \times 3$, or $9$, ways. Therefore:

$$P(A \cap B) = \tfrac{9}{25} = 0.36$$

Notice that

$$P(A \cap B) = \frac{3 \times 3}{5 \times 5} = \frac{3}{5} \times \frac{3}{5}$$
$$= P(A) \times P(B).$$

**Experiment 2**

There are $5$ ways of drawing the first ball, but, because this ball is *not* replaced, only $4$ ways of drawing the second. Thus there are $5 \times 4$, or $20$, ways of drawing the two balls. There are $3$ ways of drawing a red ball the first time, but there are only $2$ choices for the second red ball. Thus the two red balls can be drawn in $3 \times 2$, or $6$, ways. Therefore:

$$P(A \cap B) = \tfrac{6}{20} = 0.30$$

In this case,

$$P(A \cap B) = \frac{3 \times 2}{5 \times 4} = \frac{3}{5} \times \frac{2}{4}$$
$$= P(A) \times P(B|A).$$

The symbol $P(B|A)$ introduced in Experiment 2 is the probability that the second ball drawn is red, *given that the first ball drawn is red*. The following general law can be proved (Exercise 18, page 598):

Let $P(B|A)$ denote the probability of event $B$ given that event $A$ has occurred (or is certain to occur). Then the probability that both $A$ and $B$ will occur is

$$P(A \cap B) = P(A) \cdot P(B|A).$$

If $P(B|A) = P(B)$, as in Experiment 1 on page 595, we say that events $A$ and $B$ are **independent**. This means that the probability of the second occurrence does not depend on the first one. That is, two events are independent if and only if

$$P(A \cap B) = P(A) \cdot P(B).$$

Two events which are not independent are called **dependent**.

**Example 1.** A coin is tossed three times.

a. What is the probability of three heads?
b. What is the probability of at least one tail?

*Solution:*  a. The probability of a head on a single toss is $\frac{1}{2}$. The following events are independent: $A$: first toss is a head; $B$: second toss is a head; $C$: third toss is a head. Therefore,

$$P[(A \cap B) \cap C] = [P(A) \cdot P(B)] \cdot P(C) = (\tfrac{1}{2} \cdot \tfrac{1}{2}) \cdot \tfrac{1}{2} = \tfrac{1}{8}.$$

b. The events described in (a) and (b) are mutually exclusive, and it is certain that one of them will occur. Therefore, if $P$ is the probability of at least one tail, $\frac{1}{8} + P = 1$, or $P = 1 - \frac{1}{8} = \frac{7}{8}$.

**Example 2.** Three cards are dealt from a deck of 52 cards. What is the probability they are all hearts?

*Solution 1:*  Let the events be $A$: first card is a heart; $B$: second card is a heart; $C$: third card is a heart. Then

$$P[(A \cap B) \cap C] = [P(A) \cdot P(B|A)] \cdot P(C|A \cap B)$$
$$= [\tfrac{13}{52} \cdot \tfrac{12}{51}] \cdot \tfrac{11}{50} = \tfrac{11}{850}.$$

*Solution 2:*  The number of ways 3 cards can be chosen from 52 is $_{52}C_3$. The number of ways of choosing 3 hearts from 13 is $_{13}C_3$. Therefore the probability is

$$\frac{_{13}C_3}{_{52}C_3} = \frac{\dfrac{13 \cdot 12 \cdot 11}{3 \cdot 2 \cdot 1}}{\dfrac{52 \cdot 51 \cdot 50}{3 \cdot 2 \cdot 1}} = \frac{11}{850}.$$

## WRITTEN EXERCISES

In Exercises 1–4, a red die and a green die are thrown, and in each exercise events $A$ and $B$ are described. (a) Show these events in a sample space, and (b) calculate the probabilities $P(A)$, $P(B)$, $P(A \cap B)$, and $P(B|A)$.

[A]

1. $A$: The sum of the numbers thrown is 7.
   $B$: The red die shows a 2.
   Note that these events are independent.

2. $A$: The sum of the numbers thrown is 7.
   $B$: At least one of the dice shows a 2.
   Note that these events are dependent.

3. $A$: The red die shows a 3 or 4.
   $B$: The sum of the numbers thrown is $< 6$.
   Note that these events are dependent.

4. $A$: The red die shows a 3 or a 4.
   $B$: The sum of the numbers thrown is $> 7$.
   Note that these events are independent.

In Exercises 5–8, a coin is tossed three successive times, and in each exercise events $A$ and $B$ are described. (a) Show these events in a sample space, (b) calculate $P(A)$, $P(B)$ and $P(A \cap B)$, and (c) decide whether or not $A$ and $B$ are independent events.

5. $A$: The first two tosses produce heads.
   $B$: The last toss produces a head.

6. $A$: At least two tosses produce heads.
   $B$: At least one toss produces a tail.

7. $A$: At most two tosses produce heads.
   $B$: At least one toss produces a tail.

8. $A$: The first toss produces a head.
   $B$: The last two tosses produce tails.

[B]

9. A die is thrown and a card is drawn from a 52-card deck. Find the probability that (a) the die shows a 1 or 6 and the card is a heart; (b) event (a) does not occur; (c) the die shows a 2, 3, 4, 5, and the card is not a heart.

10. A coin is tossed, a die is thrown, and a card is drawn from a 52-card pack. Find the probability that (a) the coin shows a head, the die shows a 1 or 6, and the card is a spade; (b) event (a) does not occur; (c) the coin shows a tail, the die shows a 2, 3, 4, or 5, and the card is not a spade.

11. A box contains 6 white and 4 black balls. Two balls are drawn in succession, the first being returned to the box before the second is drawn. Find the probability that both balls drawn are (a) black; (b) the same color.

12. A pair of dice is thrown twice. What is the probability that both throws are: (a) 7's? (b) 11's? (c) the same?

13. The odds are 2 to 1 in favor of a football team winning each of its six remaining games. What is the probability that the team will lose more than one of the six games?

14. A single die is thrown three times. Find the probability that (a) all three numbers thrown are 6's; (b) at least one 6 is thrown; (c) exactly one 6 is thrown; (d) exactly two numbers thrown are the same; (e) exactly two numbers thrown are the same and exactly one 6 is thrown.

15. The odds that Smith can solve a certain problem are 3 to 2, that Jones can solve it are 3 to 1, and that Kelly can solve it are 2 to 1. Find the probability that (a) at least one of the three will solve the problem; (b) Smith will solve it and Jones and Kelly will not; (c) Smith and Jones will solve it and Kelly will not; (d) at least two of them will solve it.

16. A group of men is composed of equal numbers of doctors and lawyers. 15% of the doctors and 40% of the lawyers are Democrats. Find the probability that (a) a man selected at random will be a Democrat; (b) a man selected at random will be a Democratic lawyer; (c) a Democrat selected at random will be a lawyer.

17. Two cards are drawn simultaneously from a 52-card deck, and then two more cards are drawn simultaneously. Find the probability that all four cards are hearts: (a) if the first two cards are replaced in the deck before the second pair is drawn; (b) if the first two cards are not replaced in the deck.

C 18. Denote the number of elements of a finite set $S$ by $N(S)$. Thus if $E$ is any event in a sample space $S$ of an experiment with equally likely outcomes, then $P(E) = \dfrac{N(E)}{N(S)}$, by definition of probability. In considering the probability that an event $B$ will occur given that event $A$ has occurred, where $N(A) \neq 0$, we use $A$ as sample space; hence, $P(B|A) = \dfrac{N(B \cap A)}{N(A)}$. Express both $P(A \cap B)$ and $P(A) \cdot P(B|A)$ in terms of the $N$ function, thereby proving the formula $P(A \cap B) = P(A) \cdot P(B|A)$.

# Chapter Summary

**1.** If one action can be performed in $m$ ways, and for each of these ways a second action can be performed in $n$ ways, then the number of ways the two actions can be performed in the order named is $m \times n$.

Using an extension of the fundamental principle stated above, you can derive a formula for the number of **permutations** of $n$ objects, $r$ at a time:

$$_nP_r = n(n-1)(n-2)\cdots(n-[r-1])$$

**2.** The number of **combination** of $n$ objects, $r$ at a time is given by

$$_nC_r = \frac{n(n-1)\cdots(n-[r-1])}{r!} = \frac{n!}{r!(n-r)!}.$$

The coefficients in the expansion of $(a+b)^n$ are numbers of the form $_nC_r$.

**3.** Probabilities may be discussed by regarding an **event** as a subset of a **sample space** of an experiment. If an experiment has $n$ equally likely outcomes and there are $h$ ways in which an event $A$ can occur, then the **probability** of $A$ is $P(A) = \dfrac{h}{n}$, and odds in favor of $A$ are $h$ to $n - h$.

**4.** The probability that at least one of the events $A$ and $B$ will occur is $P(A \cup B) = P(A) + P(B) - P(A \cap B)$. When $A$ and $B$ are **mutually exclusive**, then $A \cap B = \emptyset$, and therefore $P(A \cap B) = 0$.

**5.** The probability that *both* of the events $A$ and $B$ will occur is given by $P(A \cap B) = P(A) \cdot P(B|A)$, where $P(B|A)$ is the probability that $B$ will occur, given that $A$ has occurred.

## Vocabulary and Spelling

Review each term by reference to the page listed.

permutation $(_nP_r)$ (*p. 576*)
0! (*p. 577*)
circular permutation (*p. 579*)
combination $(_nC_r)$ (*p. 582*)
Pascal's triangle (*p. 586*)
sample space (*p. 587*)
event (*p. 587*)

probability (*p. 589*)
complement (*p. 590*)
odds (*p. 590*)
odds against (*p. 591*)
mutually exclusive (events) (*p. 593*)
independent (events) (*p. 596*)
dependent (events) (*p. 596*)

## Chapter Test and Review

**15–1**   **1.** How many license plates consisting of one letter followed by three different digits are there?

**15–2**   **2.** How many signals can be made by displaying three flags at a time on a vertical flagpole if there are 8 different flags from which to choose?

**15–3**   **3.** How many distinguishable arrangements of all the letters in the word BOOKKEEPER are there?

         **4.** In how many ways can three labor and three management delegates be seated at a round table if labor and management delegates are seated alternately?

**15–4**   **5.** In how many ways can a subcommittee of three Democrats and four Republicans be chosen from a Senate committee of eight Democrats and ten Republicans?

**15–5**   **6.** List in order the binomial coefficients in the expansion of $(a + b)^6$.

         **7.** How many different groups of more than one but less than ten people can be chosen from ten people?

**15–6**   **8.** Experiment: Two letters are drawn at random, one from NO and the other from BET. Describe by roster **(a)** a sample space for the experiment; **(b)** the event that both letters are consonants.

**15–7**   **9.** What is the probability of throwing exactly one six when two dice are thrown?

**15–8**  **10.** A committee of three is to be selected by lot from a group consisting of four men and five women. What is the probability that the committee contains at least two women?

**15–9**  **11.** A pair of dice is thrown twice. Find the probabilities of the following events **(a)** a total of 4 is thrown each time; **(b)** a total of 4 is thrown the first time and a total of 7 the second time; **(c)** a 7 or an 11 is thrown each time.

# Explorations with a Computer

(To be used if you have access to an electronic computer that will accept BASIC.)

The following program will simulate the tossing of a coin. It uses the RND function in BASIC, which allows you to call up random numbers between 0 and 1. Since the probability of a "head" when you toss a coin is $\frac{1}{2}$, you separate the random numbers into two sets. For example, if the random number is $<.5$, the toss may be considered to be a "head"; otherwise, it is considered to be a tail. (Line 30 is the way one system insures random access to the list of random numbers; other systems may use different statements.)

```
10    PRINT "HOW MANY TOSSES";
20    INPUT N
30    LET X=RND(-1)
40    FOR I=1 TO N
50    LET A=RND(X)
60    IF A<.5 THEN 100
70    LET T=T+1
80    PRINT "     T"
90    GOTO 120
100   LET H=H+1
110   PRINT "H"
120   NEXT I
130   PRINT "H =";H,"T =";T,"H/N =";H/N
140   END
```

Run the program several times for N = 20, and compare the results.

Before you try very large numbers of tosses, delete print-out lines 80 and 110.

# Matrices
# and
# Determinants

## MATRICES

### 16–1    Terminology

A **matrix** (plural, *matrices*) is a rectangular array of numbers enclosed by brackets, double lines, or parentheses. Examples of matrices are:

$$\begin{bmatrix} 2 & 0 & -3 \\ 1 & 1 & 4 \end{bmatrix}, \quad \begin{Vmatrix} 2 \\ -2 \end{Vmatrix}, \quad \begin{pmatrix} 3 & 0 \\ \frac{1}{2} & -1 \end{pmatrix}, \quad [3, 1, 0]$$

The individual numbers in the array are the **entries** or **elements** of the matrix. The number of rows (horizontal) and columns (vertical) determine the **dimensions** of the matrix. The sample matrices given above have dimensions $2 \times 3$ (read "two by three"), $2 \times 1$, $2 \times 2$, and $1 \times 3$, respectively. Notice that the number of rows is always given first.

A matrix having only one row or only one column is called a **row matrix** or a **column matrix**, respectively. An $n \times n$ matrix is called a **square matrix** of order $n$.

Two matrices are **equal** if and only if they have the same dimensions and the same corresponding (occupying the same relative position) entries. Thus

$$\begin{bmatrix} \frac{1}{2} & \frac{4}{2} & 0 \\ 1 & 2 & 3 \end{bmatrix} = \begin{bmatrix} \frac{2}{4} & 2 & 0 \\ 1 & 2 & 3 \end{bmatrix} \text{ and } \begin{bmatrix} \sqrt{4} & \sqrt{9} \\ 3 & 0 \end{bmatrix} = \begin{bmatrix} 2 & 3 \\ 3 & 0 \end{bmatrix}$$

but

$$\begin{bmatrix} 1 & 0 \\ 3 & 2 \end{bmatrix} \neq \begin{bmatrix} 1 & 3 \\ 0 & 2 \end{bmatrix} \text{ and } \begin{bmatrix} 0 & 0 \\ 0 & 0 \\ 0 & 0 \end{bmatrix} \neq \begin{bmatrix} 0 & 0 & 0 \\ 0 & 0 & 0 \end{bmatrix}.$$

**Example.** Find $x$ and $y$ so that $\begin{bmatrix} x+1 & 1 \\ 2 & 0 \end{bmatrix} = \begin{bmatrix} 3 & 1 \\ 2 & y-1 \end{bmatrix}$

*Solution:* This matrix equation is satisfied if and only if $x + 1 = 3$ and $0 = y - 1$. Hence $x = 2$ and $y = 1$.

Matrices are often denoted by capital letters, and subscripts are sometimes used to indicate the dimensions. Thus if

$$A = \begin{bmatrix} 2 & 0 \\ 1 & 0 \end{bmatrix}, \quad B = \begin{bmatrix} u \\ v \end{bmatrix}, \quad \text{and} \quad 0 = \begin{bmatrix} 0 & 0 & 0 \\ 0 & 0 & 0 \end{bmatrix},$$

we might write $A_{2\times2}$, $B_{2\times1}$, and $0_{2\times3}$, respectively. Because all its entries are 0, $0_{2\times3}$ is called a **zero matrix**.

Associated with each matrix $A$ is the **transpose** of $A$, written $A^\mathsf{T}$, which is obtained by interchanging the rows and columns of $A$. For example,

$$\text{if } A = \begin{bmatrix} 2 & -1 \\ 3 & 4 \end{bmatrix}, \quad \text{then } A^\mathsf{T} = \begin{bmatrix} 2 & 3 \\ -1 & 4 \end{bmatrix}.$$

*ORAL EXERCISES*

State the dimensions of each matrix.

**1.** $\begin{bmatrix} 3 & 1 & 1 \\ 2 & 0 & 1 \end{bmatrix}$  **3.** $[0 \ 1 \ -2 \ 5]$  **5.** $\begin{bmatrix} 1 & 2 \\ 3 & 4 \end{bmatrix}$  **7.** $[3]$

**2.** $\begin{bmatrix} 2 \\ -1 \\ 3 \end{bmatrix}$  **4.** $\begin{bmatrix} 1 \\ 1 \end{bmatrix}$  **6.** $\begin{bmatrix} 0 & 2 \\ 1 & 3 \\ 1 & 1 \end{bmatrix}$  **8.** $\begin{bmatrix} 1 & 2 & 3 & 4 \\ 1 & 2 & 3 & 4 \end{bmatrix}$

**9–14.** State the dimensions of the transpose of each of the matrices 1–8.

Let $A = \begin{bmatrix} 2 & 0 & 5 \\ 1 & 3 & 9 \\ 6 & 7 & 8 \end{bmatrix}$. State each of the following.

**15.** The entries in the second row.    **16.** The entries in the third column.

**17.** The entry in the first row and third column.

**18.** The entry in the second row and first column.

**19.** The entry in the first row and second column.

**20.** The transpose of $A$ (read off the rows in order).

**21.** The zero matrix of the same dimensions.

## WRITTEN EXERCISES

Find each $x$, $y$, and $z$.

**A** 

**1.** $\begin{bmatrix} x & 2 \\ 0 & 1 \end{bmatrix} = \begin{bmatrix} 3 & 2 \\ 0 & 1 \end{bmatrix}$

**2.** $\begin{bmatrix} 1 & 2 \\ y & -2 \end{bmatrix} = \begin{bmatrix} 1 & 2 \\ 0 & -2 \end{bmatrix}$

**3.** $\begin{bmatrix} x & 3 \\ 2 & y \end{bmatrix} = \begin{bmatrix} 1 & 3 \\ 2 & 4 \end{bmatrix}$

**4.** $\begin{bmatrix} x & 3 & z \\ 2 & 1 & 0 \end{bmatrix} = \begin{bmatrix} 3 & 3 & 3 \\ 2 & y & 0 \end{bmatrix}$

**5.** $\begin{bmatrix} 3 & y \\ 0 & 2 \end{bmatrix}^{\mathsf{T}} = \begin{bmatrix} 3 & 0 \\ 1 & 2 \end{bmatrix}$

**6.** $\begin{bmatrix} 2 & x \\ 3 & 4 \end{bmatrix}^{\mathsf{T}} = \begin{bmatrix} z & y \\ 1 & 4 \end{bmatrix}$

**7.** $\begin{bmatrix} x-2 & 3 \\ y+1 & 2 \end{bmatrix} = \begin{bmatrix} 1 & 3 \\ 2 & 2 \end{bmatrix}$

**8.** $\begin{bmatrix} 1 & x-1 \\ 3 & 0 \end{bmatrix} = \begin{bmatrix} 1 & 4 \\ y+1 & 0 \end{bmatrix}$

**B** 

**9.** $\begin{bmatrix} x+y & 3 \\ x-y & 0 \end{bmatrix} = \begin{bmatrix} 5 & 3 \\ 3 & 0 \end{bmatrix}$

**10.** $\begin{bmatrix} x & 3 \\ 2 & x-1 \end{bmatrix} = \begin{bmatrix} y & 3 \\ 2 & 3-y \end{bmatrix}$

## 16–2   Matrix Addition

Just as the sum of two real numbers is a unique real number, the sum of two matrices having the same dimensions is a unique matrix.

> The sum of two matrices of the same dimensions is the matrix whose entries are the sums of the corresponding entries of the matrices being added.

For example,

$$\begin{bmatrix} 2 & -1 \\ 0 & 2 \\ 3 & 3 \end{bmatrix} + \begin{bmatrix} 1 & 1 \\ 3 & -1 \\ 2 & -2 \end{bmatrix} = \begin{bmatrix} 2+1 & -1+1 \\ 0+3 & 2-1 \\ 3+2 & 3-2 \end{bmatrix} = \begin{bmatrix} 3 & 0 \\ 3 & 1 \\ 5 & 1 \end{bmatrix}.$$

Addition of matrices of different dimensions is not defined. You can show that addition of matrices is commutative and associative.

The sum of the zero matrix $0_{m \times n}$ and any matrix $A_{m \times n}$ is $A_{m \times n}$. For example,

$$\begin{bmatrix} a_1 & a_2 & a_3 \\ b_1 & b_2 & b_3 \end{bmatrix} + \begin{bmatrix} 0 & 0 & 0 \\ 0 & 0 & 0 \end{bmatrix} = \begin{bmatrix} a_1 & a_2 & a_3 \\ b_1 & b_2 & b_3 \end{bmatrix}.$$

Thus $0_{m \times n}$ is the additive identity in the set of all $m \times n$ matrices.

The *negative* (additive inverse) of a matrix $A$ is the matrix $-A$, each of whose entries is the negative of the corresponding entry in $A$. For example,

$$\text{if } A = \begin{bmatrix} 2 & 0 \\ -1 & -3 \end{bmatrix}, \quad \text{then } -A = \begin{bmatrix} -2 & 0 \\ 1 & 3 \end{bmatrix}.$$

Notice that the sum of $A$ and $-A$ is a zero matrix. As with real numbers, we define *subtraction* of matrices in terms of addition: $A_{m \times n} - B_{m \times n} = A_{m \times n} + (-B_{m \times n})$. Thus

$$\begin{bmatrix} x & y \\ u & v \end{bmatrix} - \begin{bmatrix} a & b \\ c & d \end{bmatrix} = \begin{bmatrix} x & y \\ u & v \end{bmatrix} + \begin{bmatrix} -a & -b \\ -c & -d \end{bmatrix} = \begin{bmatrix} x - a & y - b \\ u - c & v - d \end{bmatrix}.$$

When working with matrices we call any real number a scalar.

---

The scalar product of a number $r$ and a matrix $A$ is the matrix $rA$, each of whose entries is $r$ times the corresponding entry of $A$.

---

For example,

$$3 \begin{bmatrix} 1 & 0 \\ -1 & 2 \end{bmatrix} = \begin{bmatrix} 3 & 0 \\ -3 & 6 \end{bmatrix} \quad \text{and} \quad \tfrac{1}{2} \begin{bmatrix} 2 & 0 & 4 \\ 1 & 2 & 8 \end{bmatrix} = \begin{bmatrix} 1 & 0 & 2 \\ \tfrac{1}{2} & 1 & 4 \end{bmatrix}.$$

**Example.** If $A = \begin{bmatrix} 0 & 1 \\ 1 & 0 \end{bmatrix}$ and $B = \begin{bmatrix} -2 & 1 \\ 1 & -2 \end{bmatrix}$, find $3A - 2B$.

*Solution:*
$$3A - 2B = 3 \begin{bmatrix} 0 & 1 \\ 1 & 0 \end{bmatrix} - 2 \begin{bmatrix} -2 & 1 \\ 1 & -2 \end{bmatrix}$$

$$= \begin{bmatrix} 0 & 3 \\ 3 & 0 \end{bmatrix} - \begin{bmatrix} -4 & 2 \\ 2 & -4 \end{bmatrix} = \begin{bmatrix} 4 & 1 \\ 1 & 4 \end{bmatrix}$$

*ORAL EXERCISES*

State the indicated sum, difference, or product. First read off the first row of the answer, then the second row.

**1.** $\begin{bmatrix} 1 & 2 \\ 4 & 3 \end{bmatrix} + \begin{bmatrix} 1 & 1 \\ 2 & 2 \end{bmatrix}$

**2.** $\begin{bmatrix} 2 & 3 \\ -2 & 4 \end{bmatrix} + \begin{bmatrix} -1 & 0 \\ 0 & 2 \end{bmatrix}$

**3.** $\begin{bmatrix} 3 \\ 1 \end{bmatrix} + \begin{bmatrix} 2 \\ -1 \end{bmatrix}$

**4.** $\begin{bmatrix} -2 & 1 & 3 \\ 0 & 1 & 2 \end{bmatrix} + \begin{bmatrix} 2 & 1 & 0 \\ 1 & -1 & -1 \end{bmatrix}$

**5.** $\begin{bmatrix} 2 & 4 \\ 5 & 3 \end{bmatrix} - \begin{bmatrix} 1 & 2 \\ 2 & 3 \end{bmatrix}$

**6.** $\begin{bmatrix} 1 & 1 \\ 1 & 1 \end{bmatrix} - \begin{bmatrix} 1 & 0 \\ 2 & -1 \end{bmatrix}$

**7.** $-1 \begin{bmatrix} 1 & -2 \\ -3 & 0 \end{bmatrix}$

**8.** $2 \begin{bmatrix} 3 & 0 \\ 1 & 2 \end{bmatrix}$

**9.** $0 \begin{bmatrix} 3 & 0 \\ 2 & -1 \end{bmatrix}$

**10.** $5 \begin{bmatrix} 0 & 0 \\ 0 & 0 \end{bmatrix}$

## WRITTEN EXERCISES

Express as a single matrix.

A **1.** $\begin{bmatrix} 3 & 0 \\ 1 & 1 \\ -1 & 2 \end{bmatrix} + \begin{bmatrix} -1 & 1 \\ -1 & 1 \\ -1 & -1 \end{bmatrix}$

**2.** $\begin{bmatrix} 5 & 2 \\ 2 & 2 \\ 0 & 3 \end{bmatrix} - \begin{bmatrix} 4 & 0 \\ 1 & -1 \\ -1 & 3 \end{bmatrix}$

**3.** $2\begin{bmatrix} 3 & 0 \\ 1 & 2 \end{bmatrix} + \begin{bmatrix} 4 & 5 \\ 3 & 6 \end{bmatrix}$

**4.** $\begin{bmatrix} 1 & 2 \\ 5 & 2 \end{bmatrix} + 3\begin{bmatrix} 1 & 0 \\ 0 & -1 \end{bmatrix}$

**5.** $2\begin{bmatrix} 2 & -1 \\ 1 & 3 \end{bmatrix} + 5\begin{bmatrix} 1 & 1 \\ 2 & 0 \end{bmatrix}$

**6.** $3\begin{bmatrix} 1 & 2 \\ 4 & 3 \end{bmatrix} - 2\begin{bmatrix} 0 & 1 \\ 2 & 3 \end{bmatrix}$

**7.** $\begin{bmatrix} 2 & 1 \\ 3 & 2 \\ 4 & 3 \end{bmatrix} + (-1)\begin{bmatrix} 1 & 1 \\ 2 & 2 \\ 3 & 3 \end{bmatrix}$

**8.** $2\begin{bmatrix} 0 & -1 \\ 1 & 3 \\ 2 & 0 \end{bmatrix} + 3\begin{bmatrix} 1 & 1 \\ 1 & 0 \\ -1 & 2 \end{bmatrix}$

**9.** $0_{2\times 2} - \begin{bmatrix} 2 & -1 \\ -3 & 0 \end{bmatrix}$

**10.** $0_{2\times 3} - 2\begin{bmatrix} -1 & 0 & -2 \\ 2 & 0 & 1 \end{bmatrix}$

Solve each matrix equation for the $2 \times 2$ matrix $X$.

**Sample.** $2X + 3\begin{bmatrix} 1 & -1 \\ -2 & 0 \end{bmatrix} = 5\begin{bmatrix} 1 & -3 \\ 2 & 1 \end{bmatrix}$

*Solution:* Perform the multiplications: $2X + \begin{bmatrix} 3 & -3 \\ -6 & 0 \end{bmatrix} = \begin{bmatrix} 5 & -15 \\ 10 & 5 \end{bmatrix}$

To each member add $\begin{bmatrix} -3 & 3 \\ 6 & 0 \end{bmatrix}$, the negative of $\begin{bmatrix} 3 & -3 \\ -6 & 0 \end{bmatrix}$.

$$2X + \begin{bmatrix} 3 & -3 \\ -6 & 0 \end{bmatrix} + \begin{bmatrix} -3 & 3 \\ 6 & 0 \end{bmatrix} = \begin{bmatrix} 5 & -15 \\ 10 & 5 \end{bmatrix} + \begin{bmatrix} -3 & 3 \\ 6 & 0 \end{bmatrix}$$

$$2X + 0_{2\times 2} = \begin{bmatrix} 2 & -12 \\ 16 & 5 \end{bmatrix}$$

$$X = \tfrac{1}{2}\begin{bmatrix} 2 & -12 \\ 16 & 5 \end{bmatrix} = \begin{bmatrix} 1 & -6 \\ 8 & \tfrac{5}{2} \end{bmatrix}$$

B **11.** $X + \begin{bmatrix} 2 & -1 \\ 0 & -2 \end{bmatrix} = \begin{bmatrix} 3 & 1 \\ 2 & 4 \end{bmatrix}$

**12.** $X + \begin{bmatrix} 2 & 1 \\ 6 & 1 \end{bmatrix} = \begin{bmatrix} 1 & 0 \\ 0 & 1 \end{bmatrix}$

**13.** $X - \begin{bmatrix} 1 & 3 \\ 2 & 6 \end{bmatrix} = \begin{bmatrix} 2 & 3 \\ 4 & 6 \end{bmatrix}$

**14.** $X - \begin{bmatrix} 1 & -3 \\ 4 & -2 \end{bmatrix} = 2\begin{bmatrix} 1 & 0 \\ 3 & 1 \end{bmatrix}$

**15.** $2X + \begin{bmatrix} 1 & 0 \\ 0 & 1 \end{bmatrix} = \begin{bmatrix} 3 & 2 \\ 4 & 1 \end{bmatrix}$

**16.** $2X + \begin{bmatrix} 1 & 3 \\ 3 & 1 \end{bmatrix} = 3\begin{bmatrix} 2 & 1 \\ 1 & 2 \end{bmatrix}$

Let $r$ and $s$ be any real numbers, and let $A = \begin{bmatrix} a_1 & b_1 \\ c_1 & d_1 \end{bmatrix}$, $B = \begin{bmatrix} a_2 & b_2 \\ c_2 & d_2 \end{bmatrix}$, $C = \begin{bmatrix} a_3 & b_3 \\ c_3 & d_3 \end{bmatrix}$. Prove that each of the following assertions is true.

☐ **17.** $A + B = B + A$

**18.** $r(A + B) = rA + rB$

**19.** $(A^{\mathsf{T}})^{\mathsf{T}} = A$

**20.** $(A + B) + C = A + (B + C)$

**21.** $(A + B)^{\mathsf{T}} = A^{\mathsf{T}} + B^{\mathsf{T}}$

**22.** $-1 \cdot A = -A$

**23.** $(rs)A = r(sA)$

**24.** Explain why the set of $2 \times 2$ matrices is closed with respect to **(a)** addition; **(b)** scalar multiplication.

## 16–3 Matrix Multiplication

Suppose that an agency sells three models of cars, Lions, Tigers, and Wolves, at profits per car of $a$, $b$, and $c$ respectively. This profit structure can be represented by the $1 \times 3$ matrix $A = [a \ b \ c]$. Let the $3 \times 2$ matrix $B = \begin{bmatrix} u & x \\ v & y \\ w & z \end{bmatrix}$ represent the volume of sales in May and June.

That is, in May the numbers of Lions, Tigers, and Wolves sold were $u$, $v$, and $w$, respectively, while in June it was $x$, $y$, and $z$. In May the profit on Lions was $au$, on Tigers, $bv$, and on Wolves, $cw$, the total being $au + bv + cw$. Similarly, in June, the total profit was $ax + by + cz$. The total-profit matrix $[au + bv + cw \quad ax + by + cz]$ is the *product* of the matrices $A$ and $B$:

$$A \times B = [a \ b \ c] \cdot \begin{bmatrix} u & x \\ v & y \\ w & z \end{bmatrix} = [au + bv + cw \quad ax + by + cz].$$

You obtain the entries of the product matrix by adding the products of elements of a row of the first matrix with the corresponding elements of a column of the second matrix. The process is illustrated below for two $2 \times 2$ matrices:

$$\begin{bmatrix} a & b \\ c & d \end{bmatrix} \cdot \begin{bmatrix} r & s \\ t & u \end{bmatrix} = \begin{bmatrix} ar + bt & as + bu \\ cr + dt & cs + du \end{bmatrix}$$

As the red lines show, you obtain the entry in row 2 and column 1 of the product by multiplying each element in row 2 of the *first* matrix by the corresponding element in column 1 of the *second* and adding the products. The general definition of matrix multiplication is:

> The product of an $m \times n$ matrix $A$ and an $n \times p$ matrix $B$ is the $m \times p$ matrix whose entry in row $i$ and column $j$ is the sum of the products of corresponding elements of row $i$ in $A$ and column $j$ in $B$.

**Example.** Find the product:
$$\begin{bmatrix} 1 & 0 & -2 \\ 4 & 1 & 0 \\ -3 & 2 & 1 \end{bmatrix} \begin{bmatrix} 1 & 2 & -1 \\ 0 & 1 & 1 \\ 3 & 1 & -2 \end{bmatrix}$$

*Solution:*
$$\begin{bmatrix} 1 & 0 & -2 \\ 4 & 1 & 0 \\ -3 & 2 & 1 \end{bmatrix} \begin{bmatrix} 1 & 2 & -1 \\ 0 & 1 & 1 \\ 3 & 1 & -2 \end{bmatrix} = \begin{bmatrix} -5 & & \\ & & \\ & & \end{bmatrix}$$

$$\begin{bmatrix} 1 & 0 & -2 \\ 4 & 1 & 0 \\ -3 & 2 & 1 \end{bmatrix} \begin{bmatrix} 1 & 2 & -1 \\ 0 & 1 & 1 \\ 3 & 1 & -2 \end{bmatrix} = \begin{bmatrix} -5 & 0 & \\ & & \\ & & \end{bmatrix}$$

$$\begin{bmatrix} 1 & 0 & -2 \\ 4 & 1 & 0 \\ -3 & 2 & 1 \end{bmatrix} \begin{bmatrix} 1 & 2 & -1 \\ 0 & 1 & 1 \\ 3 & 1 & -2 \end{bmatrix} = \begin{bmatrix} -5 & 0 & 3 \\ & & \\ & & \end{bmatrix}$$

$$\begin{bmatrix} 1 & 0 & -2 \\ 4 & 1 & 0 \\ -3 & 2 & 1 \end{bmatrix} \begin{bmatrix} 1 & 2 & -1 \\ 0 & 1 & 1 \\ 3 & 1 & -2 \end{bmatrix} = \begin{bmatrix} -5 & 0 & 3 \\ 4 & & \\ & & \end{bmatrix}$$

$$\begin{bmatrix} 1 & 0 & -2 \\ 4 & 1 & 0 \\ -3 & 2 & 1 \end{bmatrix} \begin{bmatrix} 1 & 2 & -1 \\ 0 & 1 & 1 \\ 3 & 1 & -2 \end{bmatrix} = \begin{bmatrix} -5 & 0 & 3 \\ 4 & 9 & -3 \\ 0 & -3 & 3 \end{bmatrix}$$

Notice that when two matrices are to be multiplied each row of the first matrix must have the same number of entries as each column of the second. Thus if

$$A_{1 \times 2} = [1 \quad -2] \quad \text{and} \quad B_{2 \times 4} = \begin{bmatrix} 3 & 0 & 1 & 1 \\ 0 & 1 & 1 & 0 \end{bmatrix},$$

then $A_{1 \times 2} \cdot B_{2 \times 4} = [3 \ -2 \ -1 \ 1]$, but $B_{2 \times 4} \cdot A_{1 \times 2}$ is *not defined*.

Matrix multiplication differs from that of real numbers in several respects. For example, if

$$A = \begin{bmatrix} 2 & -1 \\ -4 & 2 \end{bmatrix} \quad \text{and} \quad B = \begin{bmatrix} 2 & 3 \\ 4 & 6 \end{bmatrix},$$

you can check that

$$AB = \begin{bmatrix} 0 & 0 \\ 0 & 0 \end{bmatrix}, \quad \text{while} \quad BA = \begin{bmatrix} -8 & 4 \\ -16 & 8 \end{bmatrix}.$$

Thus, matrix multiplication is not commutative, and it is necessary to find a product in the order in which it is written. *AB* means *left-multiplication* of *B* by *A*, and *BA* means *right-multiplication* of *B* by *A*. Notice also that the product of two nonzero matrices can be a zero matrix; thus $AB = 0$ does not imply that $A = 0$ or $B = 0$. It can be shown, however, that the associative law $(AB)C = A(BC)$ is valid for matrix multiplication, as are the distributive laws $A(B + C) = AB + AC$ and $(B + C)A = BA + CA$.

A square matrix whose *main diagonal* from upper left to lower right has entries 1, while all other entries are 0, is called an **identity matrix** and denoted by *I*. You can show that the identity matrices

$$I_{2\times 2} = \begin{bmatrix} 1 & 0 \\ 0 & 1 \end{bmatrix} \quad \text{and} \quad I_{3\times 3} = \begin{bmatrix} 1 & 0 & 0 \\ 0 & 1 & 0 \\ 0 & 0 & 1 \end{bmatrix}$$

are the identity elements for multiplication in the sets of 2 × 2 and 3 × 3 matrices, respectively. For example,

$$\begin{bmatrix} 1 & 0 \\ 0 & 1 \end{bmatrix}\begin{bmatrix} a & b \\ c & d \end{bmatrix} = \begin{bmatrix} a & b \\ c & d \end{bmatrix} = \begin{bmatrix} a & b \\ c & d \end{bmatrix}\begin{bmatrix} 1 & 0 \\ 0 & 1 \end{bmatrix}.$$

*WRITTEN EXERCISES*

Write each product as a single matrix.

[A]

1. $[2 \quad 1 \quad 0]\begin{bmatrix} 3 \\ 2 \\ 5 \end{bmatrix}$

2. $[-3 \quad 0 \quad 1]\begin{bmatrix} 0 \\ 2 \\ 0 \end{bmatrix}$

3. $\begin{bmatrix} 2 & -1 \\ 3 & 0 \end{bmatrix}\begin{bmatrix} 1 & 2 \\ 3 & 1 \end{bmatrix}$

4. $\begin{bmatrix} 1 & 2 \\ 3 & 1 \end{bmatrix}\begin{bmatrix} 2 & -1 \\ 3 & 0 \end{bmatrix}$

5. $\begin{bmatrix} 2 & 0 \\ -3 & 0 \end{bmatrix}\begin{bmatrix} 0 & 0 \\ 4 & 1 \end{bmatrix}$

6. $\begin{bmatrix} 0 & 0 \\ 4 & 1 \end{bmatrix}\begin{bmatrix} 2 & 0 \\ -3 & 0 \end{bmatrix}$

7. $\begin{bmatrix} 2 & 1 \\ 3 & 2 \end{bmatrix}\begin{bmatrix} 2 \\ -1 \end{bmatrix}$

8. $[2 \quad 3]\begin{bmatrix} 1 & 2 \\ 4 & 3 \end{bmatrix}$

9. $\begin{bmatrix} -1 & 0 & 2 \\ 3 & 1 & -4 \\ 1 & 0 & 0 \end{bmatrix}\begin{bmatrix} 1 & 0 & -1 \\ 2 & 1 & 3 \\ 3 & -1 & 2 \end{bmatrix}$

10. $\begin{bmatrix} 0 & 0 & 1 \\ 0 & 1 & 0 \\ 1 & 0 & 0 \end{bmatrix}\begin{bmatrix} 2 & 1 & -3 \\ 3 & 0 & 1 \\ 2 & 4 & 5 \end{bmatrix}$

11. $\begin{bmatrix} 2 & 1 & 0 \\ 3 & 0 & 2 \\ 1 & -1 & -2 \end{bmatrix}\begin{bmatrix} 2 \\ 1 \\ -1 \end{bmatrix}$

12. $\begin{bmatrix} 2 & 1 \\ -1 & 3 \\ 0 & 2 \end{bmatrix}\begin{bmatrix} 1 & 0 & 2 & 4 \\ 3 & -1 & 1 & 3 \end{bmatrix}$

In Exercises 13–21, let $A = \begin{bmatrix} 2 & 3 \\ 1 & 2 \end{bmatrix}$, $B = \begin{bmatrix} 1 & -2 \\ 0 & 3 \end{bmatrix}$, and $C = \begin{bmatrix} 2 & -3 \\ -1 & 2 \end{bmatrix}$.

**B** **13.** Find $AB$ and $BA$.          **16.** Find $(A + B)C$ and $AC + BC$.

**14.** Find $AC$ and $CA$.          **17.** Find $A(B + C)$ and $AB + AC$.

**15.** Find $(AB)C$ and $A(BC)$.          **18.** Find $B^2$ and $B^3$.

**19.** Find $(A + B)(A - B)$ and $A^2 - B^2$.

**20.** Find $(A + B)^2$ and $A^2 + 2AB + B^2$.

**21.** Find $(A - B)^2$ and $A^2 - 2AB + B^2$.

**C** **22.** Let $A$ and $B$ be any $2 \times 2$ matrices. Show that $(AB)^{\mathsf{T}} = B^{\mathsf{T}}A^{\mathsf{T}}$.

## SOLVING LINEAR SYSTEMS

## 16–4   Determinants

With each *square matrix* $A$ we associate a number called the **determinant** of $A$ and denoted by det $A$.

$$\text{If} \quad A = \begin{bmatrix} a & b \\ c & d \end{bmatrix}, \quad \text{then} \quad \det A = ad - bc.$$

Thus, you find the determinant of a $2 \times 2$ matrix by adding to the product of the entries on the main diagonal, $ad$, the negative of the product of the entries on the other diagonal, $-bc$ (shown at the right).

**Example 1.**   If $A = \begin{bmatrix} 3 & -2 \\ 4 & -1 \end{bmatrix}$ and $B = \begin{bmatrix} 2 & -3 \\ -6 & 9 \end{bmatrix}$, find det $A$ and det $B$.

*Solution:*          det $A = 3(-1) - (-2) \cdot 4 = -3 + 8 = 5$
det $B = 2 \cdot 9 - (-3)(-6) = 18 - 18 = 0$

For the matrix $A = \begin{bmatrix} a_1 & b_1 & c_1 \\ a_2 & b_2 & c_2 \\ a_3 & b_3 & c_3 \end{bmatrix}$ of order 3 we define

$$\det A = a_1 b_2 c_3 + a_2 b_3 c_1 + a_3 b_1 c_2 - a_1 b_3 c_2 - a_2 b_1 c_3 - a_3 b_2 c_1.$$

Notice that the sum in this definition contains all the products you can form by choosing as factors exactly one entry from each row and column. Here is a way to remember this sum:

1. Copy the first two columns of the matrix, in order, to the right of the third column.

$$
\begin{matrix}
a_1 & b_1 & c_1 & a_1 & b_1 \\
a_2 & b_2 & c_2 & a_2 & b_2 \\
a_3 & b_3 & c_3 & a_3 & b_3
\end{matrix}
$$

2. Form the products of the entries on the diagonals which slant downward to the right. These are the first three terms of the sum.

$$
\begin{matrix}
a_1 & b_1 & c_1 & a_1 & b_1 \\
a_2 & b_2 & c_2 & a_2 & b_2 \\
a_3 & b_3 & c_3 & a_3 & b_3 \\
& & + & + & +
\end{matrix}
$$

3. Form the products of the entries on the diagonals which slant downward to the left. The negatives of these products are the last three terms in the sum.

$$
\begin{matrix}
a_1 & b_1 & c_1 & a_1 & b_1 \\
a_2 & b_2 & c_2 & a_2 & b_2 \\
a_3 & b_3 & c_3 & a_3 & b_3 \\
- & - & - & &
\end{matrix}
$$

**Example 2.** If $A = \begin{bmatrix} 2 & 1 & 0 \\ 1 & -1 & 1 \\ 3 & -4 & 2 \end{bmatrix}$, find det $A$.

*Solution:*

$$
\begin{matrix}
2 & 1 & 0 & 2 & 1 \\
1 & -1 & 1 & 1 & -1 \\
3 & -4 & 2 & 3 & -4
\end{matrix}
$$

$$\det A = (-4) + (3) + (0) - (0) - (-8) - (2) = 5$$

Determinants of higher-order square matrices will be defined in Section 16–7. When the order is greater than 3, however, the crisscross method described above will *not* give the value of the determinant.

The determinant of a matrix is often displayed in the same form as the matrix, but with vertical bars, rather than brackets, enclosing the entries. For example, the determinant of the matrix $\begin{bmatrix} 2 & 3 \\ 1 & 5 \end{bmatrix}$ is

$$
\begin{vmatrix} 2 & 3 \\ 1 & 5 \end{vmatrix} = 2 \cdot 5 - 1 \cdot 3 = 7.
$$

The entries are then called **elements** of the determinant, and the number of rows (or columns) is the **order** of the determinant.

*WRITTEN EXERCISES*

Find the determinant of each matrix.

**A**

1. $\begin{bmatrix} 2 & -3 \\ -1 & 2 \end{bmatrix}$

2. $\begin{bmatrix} 0 & 1 \\ 1 & 0 \end{bmatrix}$

3. $\begin{bmatrix} 78 & 79 \\ 1 & 1 \end{bmatrix}$

4. $\begin{bmatrix} 2 & -3 \\ 1 & 2 \end{bmatrix}$

5. $\begin{bmatrix} 2 & -3 \\ -4 & 6 \end{bmatrix}$

6. $\begin{bmatrix} 2 & 0 & 1 \\ 0 & 1 & 1 \\ 2 & -1 & 1 \end{bmatrix}$

7. $\begin{bmatrix} \frac{2}{3} & 1 \\ -4 & -6 \end{bmatrix}$

8. $\begin{bmatrix} 2 & 1 & 0 \\ 1 & -1 & 1 \\ 0 & 3 & 2 \end{bmatrix}$

**9.** $\begin{bmatrix} 0 & 0 & 2 \\ 0 & 2 & 0 \\ 2 & 0 & 0 \end{bmatrix}$  **10.** $\begin{bmatrix} 2 & 1 & 3 \\ 3 & 0 & 3 \\ -1 & 3 & 2 \end{bmatrix}$  **11.** $\begin{bmatrix} 1 & 2 & 3 \\ 2 & 1 & 3 \\ 3 & 1 & 4 \end{bmatrix}$  **12.** $\begin{bmatrix} 1 & 0 & 0 \\ 1 & 1 & 0 \\ 1 & 1 & 1 \end{bmatrix}$

Let $A$ and $B$ be $2 \times 2$ matrices and $r$ be a scalar.

**B** **13.** Show that det $A^{\mathsf{T}} = $ det $A$.

**14.** Show that det $(rA) = r^2$ det $A$.

**15.** Show that det $(AB) = $ det $A \cdot$ det $B$

**16.** Show that det $(A - A^{\mathsf{T}}) = $ det $(A^{\mathsf{T}} - A)$.

## 16–5   Inverses of Matrices

The product $\begin{bmatrix} 2 & -1 \\ -5 & 3 \end{bmatrix}\begin{bmatrix} 3 & 1 \\ 5 & 2 \end{bmatrix} = \begin{bmatrix} 6 - 5 & 2 - 2 \\ -15 + 15 & -5 + 6 \end{bmatrix} = \begin{bmatrix} 1 & 0 \\ 0 & 1 \end{bmatrix}$ is
the identity matrix $I$. Two real numbers whose product is 1 are said
to be multiplicative inverses of each other. Similarly, any two matrices
$A$ and $B$ such that $AB = BA = I$ are called inverses. Usually $B$ is
denoted by $A^{-1}$, which is called *the inverse of $A$*.

You know that every real number $a$ except 0 has a multiplicative
inverse $a^{-1}$. Does every nonzero $2 \times 2$ matrix $A$ have an inverse $A^{-1}$?
To answer this question let $A = \begin{bmatrix} a & b \\ c & d \end{bmatrix}$ and see if you can find a
matrix $A^{-1} = \begin{bmatrix} x & u \\ y & v \end{bmatrix}$ such that $AA^{-1} = I$. You wish to have:

$$\begin{bmatrix} a & b \\ c & d \end{bmatrix}\begin{bmatrix} x & u \\ y & v \end{bmatrix} = \begin{bmatrix} 1 & 0 \\ 0 & 1 \end{bmatrix}$$

$$\begin{bmatrix} ax + by & au + bv \\ cx + dy & cu + dv \end{bmatrix} = \begin{bmatrix} 1 & 0 \\ 0 & 1 \end{bmatrix}$$

This last equation is true if and only if

$$ax + by = 1 \qquad au + bv = 0$$
$$cx + dy = 0 \qquad cu + dv = 1.$$

If $ad - bc \neq 0$, you can solve these two systems and obtain:

$$x = \frac{d}{ad - bc} \qquad u = \frac{-b}{ad - bc}$$

$$y = \frac{-c}{ad - bc} \qquad v = \frac{a}{ad - bc}$$

Since each denominator is det $A$, you have, if det $A \neq 0$,

$$A^{-1} = \frac{1}{\det A} \begin{bmatrix} d & -b \\ -c & a \end{bmatrix}.$$

You can check that $A^{-1}A = I$, as well as $AA^{-1} = I$.

Thus you can find the inverse of a $2 \times 2$ matrix $A$ provided its determinant is not 0. If det $A = 0$, $A$ has no inverse. Notice what the formula says to do to find $A^{-1}$: interchange $a$ and $d$, replace $b$ and $c$ by their respective negatives, and multiply the resulting matrix by $\dfrac{1}{\det A}$.

**Example.** If $A = \begin{bmatrix} 6 & -2 \\ 4 & -1 \end{bmatrix}$, find $A^{-1}$.

*Solution:* Since det $A = -6 - (-8) = 2$, $A^{-1}$ exists. Now interchange 6 and $-1$, replace $-2$ by 2 and 4 by $-4$, and multiply the resulting matrix by $\frac{1}{2}$:

$$A^{-1} = \tfrac{1}{2} \begin{bmatrix} -1 & 2 \\ -4 & 6 \end{bmatrix} = \begin{bmatrix} -\frac{1}{2} & 1 \\ -2 & 3 \end{bmatrix}.$$

*Check:* $\begin{bmatrix} -\frac{1}{2} & 1 \\ -2 & 3 \end{bmatrix} \begin{bmatrix} 6 & -2 \\ 4 & -1 \end{bmatrix} = \begin{bmatrix} 1 & 0 \\ 0 & 1 \end{bmatrix}$

The inverse of a square matrix $A$ of order $n > 2$ is more difficult to find than the inverse of a $2 \times 2$ matrix, but it always exists if det $A \neq 0$. A matrix which has an inverse is said to be **invertible** or **nonsingular**.

## WRITTEN EXERCISES

Find the inverse if the matrix is invertible. If the matrix is not invertible, so state.

**A**  **1.** $\begin{bmatrix} 2 & 3 \\ 1 & 2 \end{bmatrix}$   **4.** $\begin{bmatrix} 0 & 1 \\ 1 & 0 \end{bmatrix}$   **7.** $\begin{bmatrix} -1 & -4 \\ 2 & 2 \end{bmatrix}$   **10.** $\begin{bmatrix} 2 & -3 \\ \frac{1}{3} & \frac{1}{2} \end{bmatrix}$

**2.** $\begin{bmatrix} 1 & 2 \\ 2 & 3 \end{bmatrix}$   **5.** $\begin{bmatrix} 3 & 1 \\ 1 & 1 \end{bmatrix}$   **8.** $\begin{bmatrix} 3 & 1 \\ 11 & 4 \end{bmatrix}$   **11.** $\begin{bmatrix} 2 & 3 \\ \frac{1}{3} & \frac{1}{2} \end{bmatrix}$

**3.** $\begin{bmatrix} 1 & -2 \\ -2 & 4 \end{bmatrix}$   **6.** $\begin{bmatrix} 3 & -1 \\ 6 & -2 \end{bmatrix}$   **9.** $\begin{bmatrix} 1 & 0 \\ 0 & 1 \end{bmatrix}$   **12.** $\begin{bmatrix} 2 & 3 \\ 5 & 6 \end{bmatrix}$

Solve for the 2 $\times$ 2 matrix Z.

**Sample.** $\begin{bmatrix} 2 & 1 \\ 4 & 3 \end{bmatrix} Z = \begin{bmatrix} 1 & 2 \\ 1 & 4 \end{bmatrix}$

*Solution :* Since the determinant of the matrix $\begin{bmatrix} 2 & 1 \\ 4 & 3 \end{bmatrix}$ is 2, we have

$$\begin{bmatrix} 2 & 1 \\ 4 & 3 \end{bmatrix}^{-1} = \tfrac{1}{2} \begin{bmatrix} 3 & -1 \\ -4 & 2 \end{bmatrix}.$$

$$\therefore \begin{bmatrix} 2 & 1 \\ 4 & 3 \end{bmatrix}^{-1} \begin{bmatrix} 2 & 1 \\ 4 & 3 \end{bmatrix} Z = \begin{bmatrix} 2 & 1 \\ 4 & 3 \end{bmatrix}^{-1} \begin{bmatrix} 1 & 2 \\ 1 & 4 \end{bmatrix}$$

$$IZ = \tfrac{1}{2} \begin{bmatrix} 3 & -1 \\ -4 & 2 \end{bmatrix} \begin{bmatrix} 1 & 2 \\ 1 & 4 \end{bmatrix}$$

$$Z = \tfrac{1}{2} \begin{bmatrix} 2 & 2 \\ -2 & 0 \end{bmatrix} = \begin{bmatrix} 1 & 1 \\ -1 & 0 \end{bmatrix}$$

**B** **13.** $\begin{bmatrix} 2 & 1 \\ 1 & 1 \end{bmatrix} Z = \begin{bmatrix} 3 & -1 \\ 2 & 2 \end{bmatrix}$

**16.** $\begin{bmatrix} 4 & 2 \\ 1 & 1 \end{bmatrix} Z = \begin{bmatrix} 2 & 3 \\ 0 & 1 \end{bmatrix}$

**14.** $\begin{bmatrix} 1 & 2 \\ 1 & 3 \end{bmatrix} Z = \begin{bmatrix} 0 & 1 \\ -1 & 2 \end{bmatrix}$

**17.** $\begin{bmatrix} 5 & 2 \\ 2 & 1 \end{bmatrix} Z - \begin{bmatrix} 1 & 0 \\ 0 & 2 \end{bmatrix} = \begin{bmatrix} 1 & 3 \\ 1 & 1 \end{bmatrix}$

(*Hint:* Transform by addition first.)

**15.** $\begin{bmatrix} 3 & 2 \\ 2 & 2 \end{bmatrix} Z = \begin{bmatrix} 5 & 0 \\ -1 & -1 \end{bmatrix}$

**18.** $\begin{bmatrix} 2 & -5 \\ 1 & -2 \end{bmatrix} Z + \begin{bmatrix} 1 & 2 \\ 2 & 4 \end{bmatrix} = \begin{bmatrix} 3 & 0 \\ 1 & 2 \end{bmatrix}$

**19.** Let $A = \begin{bmatrix} 2 & -1 \\ -5 & 3 \end{bmatrix}$ and $B = \begin{bmatrix} 1 & 3 \\ 1 & 2 \end{bmatrix}$. Find each matrix.

    **a.** $AB$    **b.** $(AB)^{-1}$    **c.** $A^{-1}$    **d.** $B^{-1}$    **e.** $A^{-1}B^{-1}$    **f.** $B^{-1}A^{-1}$

**C** **20.** Show that for any invertible 2 $\times$ 2 matrix $A$, $(A^{\mathsf{T}})^{-1} = (A^{-1})^{\mathsf{T}}$.

**21.** Show that for any invertible 2 $\times$ 2 matrix $A$, $\det (A^{-1}) = \dfrac{1}{\det A}$.

**22.** If $A$ is invertible, and $AX + B = C$, express the matrix $X$ in terms of the matrices $A$, $B$, and $C$.

## 16–6 Matrix Equations

A *single* matrix equation can be used to represent a *system* of linear equations. Because

$$\begin{bmatrix} a & b \\ c & d \end{bmatrix} \begin{bmatrix} x \\ y \end{bmatrix} = \begin{bmatrix} ax + by \\ cx + dy \end{bmatrix},$$

you can write the system

$$ax + by = e$$
$$cx + dy = f$$

in the equivalent form

$$\begin{bmatrix} a & b \\ c & d \end{bmatrix}\begin{bmatrix} x \\ y \end{bmatrix} = \begin{bmatrix} e \\ f \end{bmatrix}.$$

This equation is of the form $AX = B$, where $A$ is a $2 \times 2$ matrix called the **coefficient matrix** of the system and $X$ and $B$ are $2 \times 1$ matrices. If $A$ is invertible, you can solve $AX = B$ for $X$ as follows:

$$AX = B$$
$$A^{-1}AX = A^{-1}B$$
$$IX = A^{-1}B$$
$$X = A^{-1}B$$

This method can be used to solve a system of two linear equations in two variables.

**Example 1.** Solve the system: $\quad x - 2y = 2$
$$3x - 4y = 7$$

*Solution:*  **1.** Write the system as a matrix equation:

$$\begin{bmatrix} 1 & -2 \\ 3 & -4 \end{bmatrix}\begin{bmatrix} x \\ y \end{bmatrix} = \begin{bmatrix} 2 \\ 7 \end{bmatrix}$$

**2.** Find the inverse of the coefficient matrix:

$$\begin{bmatrix} 1 & -2 \\ 3 & -4 \end{bmatrix}^{-1} = \tfrac{1}{2}\begin{bmatrix} -4 & 2 \\ -3 & 1 \end{bmatrix}$$

**3.** Left-multiply both members of the matrix equation by this inverse:

$$\tfrac{1}{2}\begin{bmatrix} -4 & 2 \\ -3 & 1 \end{bmatrix}\begin{bmatrix} 1 & -2 \\ 3 & -4 \end{bmatrix}\begin{bmatrix} x \\ y \end{bmatrix} = \tfrac{1}{2}\begin{bmatrix} -4 & 2 \\ -3 & 1 \end{bmatrix}\begin{bmatrix} 2 \\ 7 \end{bmatrix}$$

$$\begin{bmatrix} 1 & 0 \\ 0 & 1 \end{bmatrix}\begin{bmatrix} x \\ y \end{bmatrix} = \tfrac{1}{2}\begin{bmatrix} 6 \\ 1 \end{bmatrix}$$

$$\begin{bmatrix} x \\ y \end{bmatrix} = \begin{bmatrix} 3 \\ \tfrac{1}{2} \end{bmatrix}$$

$\therefore$ the solution set is $\{(3, \tfrac{1}{2})\}$.

If the coefficient matrix is not invertible, the equations are either dependent or inconsistent.

When you apply the method of Example 1 to the system

$$ax + by = e$$
$$cx + dy = f$$

you obtain

$$\begin{bmatrix} x \\ y \end{bmatrix} = \begin{bmatrix} a & b \\ c & a \end{bmatrix}^{-1} \begin{bmatrix} e \\ f \end{bmatrix} = \frac{1}{\det A} \begin{bmatrix} d & -b \\ -c & a \end{bmatrix} \begin{bmatrix} e \\ f \end{bmatrix}$$

$$\begin{bmatrix} x \\ y \end{bmatrix} = \frac{1}{\det A} \begin{bmatrix} de - bf \\ -ce + af \end{bmatrix}$$

Therefore,

$$x = \frac{de - bf}{\det A}, \quad y = \frac{af - ce}{\det A}$$

or

$$x = \frac{\begin{vmatrix} e & b \\ f & d \end{vmatrix}}{\begin{vmatrix} a & b \\ c & d \end{vmatrix}}, \quad y = \frac{\begin{vmatrix} a & e \\ c & f \end{vmatrix}}{\begin{vmatrix} a & b \\ c & d \end{vmatrix}}.$$

These equations are referred to as **Cramer's Rule** for solving the given system. You can write Cramer's Rule in the form

$$x = \frac{D_x}{D}, \quad y = \frac{D_y}{D}$$

where $D$ is the determinant of the coefficient matrix, and $D_x$ and $D_y$ are obtained from $D$ by replacing the coefficients of $x$ and $y$, respectively, by the constants $e$ and $f$.

**Example 2.** Use Cramer's Rule to solve the system: $\quad x - 4y = 8$
$$2x - 5y = 13$$

*Solution:* 
$$x = \frac{\begin{vmatrix} 8 & -4 \\ 13 & -5 \end{vmatrix}}{\begin{vmatrix} 1 & -4 \\ 2 & -5 \end{vmatrix}}, \quad y = \frac{\begin{vmatrix} 1 & 8 \\ 2 & 13 \end{vmatrix}}{\begin{vmatrix} 1 & -4 \\ 2 & -5 \end{vmatrix}}$$

$$x = \frac{-40 + 52}{-5 + 8} = \frac{12}{3}, \quad y = \frac{13 - 16}{3} = \frac{-3}{3}$$

$$x = 4, \qquad\qquad y = -1. \quad \{(4, -1)\}$$

*WRITTEN EXERCISES*

Transform each system into matrix form and, if the coefficient matrix is invertible, solve it by the method of Example 1, page 616. If the system is dependent or inconsistent, so state.

$\boxed{\text{A}}$

**1.** $2x + 3y = 2$
$\quad x + 2y = 3$

**2.** $x + 2y = -1$
$\quad 2x + 3y = 4$

**3.** $x - 2y = -3$
$\quad -2x + 4y = 6$

**4.** $3x + y = 2$
$\quad x + y = 3$

**5.** $3x + 2y = 1$
$\quad 2x + 3y = 4$

**6.** $3x - y = 1$
$\quad 6x - 2y = 2$

**7.** $2x - 3y = 4$
$\quad \frac{1}{3}x - \frac{1}{2}y = \frac{1}{4}$

**8.** $4x - 5y = 2$
$\quad 3x + 2y = 5$

**9.** $2x = 6 + 3y$
$\quad 6y = 2 + 4x$

**10–18.** Use Cramer's Rule to solve each system in Exercises 1–9, provided that the determinant of the coefficient matrix is not 0.

## MORE ON DETERMINANTS

### 16–7 Expansion by Minors

The minor of an element of a determinant is the determinant which results when you cross out the row and column containing the element. For example, in the determinant

$$\begin{vmatrix} 2 & 3 & -2 \\ 1 & 0 & 5 \\ -1 & -3 & 4 \end{vmatrix}$$

the minor of the element 3 in row 1 and column 2 is

$$\begin{vmatrix} 2 & 3 & -2 \\ 1 & 0 & 5 \\ -1 & -3 & 4 \end{vmatrix}, \quad \text{or} \quad \begin{vmatrix} 1 & 5 \\ -1 & 4 \end{vmatrix},$$

and the minor of the element $-1$ in row 3 and column 1 is

$$\begin{vmatrix} 2 & 3 & -2 \\ 1 & 0 & 5 \\ -1 & -3 & 4 \end{vmatrix}, \quad \text{or} \quad \begin{vmatrix} 3 & -2 \\ 0 & 5 \end{vmatrix}.$$

The value of a third-order determinant as defined in Section 16–4 can be rewritten:

$$\begin{vmatrix} a_1 & b_1 & c_1 \\ a_2 & b_2 & c_2 \\ a_3 & b_3 & c_3 \end{vmatrix} = a_1 b_2 c_3 - a_1 b_3 c_2 - b_1 a_2 c_3 + b_1 a_3 c_2 + c_1 a_2 b_3 - c_1 a_3 b_2$$

By factoring, you obtain

$$a_1(b_2 c_3 - b_3 c_2) - b_1(a_2 c_3 - a_3 c_2) + c_1(a_2 b_3 - a_3 b_2).$$

The expressions in parentheses are the minors of $a_1$, $b_1$, and $c_1$, respectively, and if these minors are denoted by $A_1$, $B_1$, and $C_1$, you have:

$$\begin{vmatrix} a_1 & b_1 & c_1 \\ a_2 & b_2 & c_2 \\ a_3 & b_3 & c_3 \end{vmatrix} = a_1 \begin{vmatrix} b_2 & c_2 \\ b_3 & c_3 \end{vmatrix} - b_1 \begin{vmatrix} a_2 & c_2 \\ a_3 & c_3 \end{vmatrix} + c_1 \begin{vmatrix} a_2 & b_2 \\ a_3 & b_3 \end{vmatrix}$$

$$= a_1 A_1 - b_1 B_1 + c_1 C_1$$

This is the expansion by minors of the first row. It can be shown that you can **expand** a determinant **by minors** of any row or column as follows:

---

1. Multiply each element in the chosen row or column by its minor.
2. Multiply the product obtained by 1 if the sum of the number of the row and the number of the column containing the element is an even integer or by $-1$ if the sum is an odd integer.
3. Add the resulting products.

---

**Example 1.** Expand by minors of the third column to evaluate the determinant.

$$\begin{vmatrix} 4 & 0 & 3 \\ 5 & 1 & 2 \\ -2 & 1 & -1 \end{vmatrix}$$

*Solution:* Write the product of each element in the third column with its minor. Use the negative of the product of the element in row 2, column 3 because $2 + 3 = 5$, which is odd.

$$\begin{vmatrix} 4 & 0 & 3 \\ 5 & 1 & 2 \\ -2 & 1 & -1 \end{vmatrix} = 3 \begin{vmatrix} 5 & 1 \\ -2 & 1 \end{vmatrix} - 2 \begin{vmatrix} 4 & 0 \\ -2 & 1 \end{vmatrix} + (-1) \begin{vmatrix} 4 & 0 \\ 5 & 1 \end{vmatrix}$$

$$= 3 \cdot 7 - 2 \cdot 4 + (-1) \cdot 4 = 9$$

Expansion by minors can be used to *define* fourth- and higher-order determinants. Thus

$$\begin{vmatrix} a_1 & b_1 & c_1 & d_1 \\ a_2 & b_2 & c_2 & d_2 \\ a_3 & b_3 & c_3 & d_3 \\ a_4 & b_4 & c_4 & d_4 \end{vmatrix} = a_1A_1 - b_1B_1 + c_1C_1 - d_1D_1,$$

where

$$A_1 = \begin{vmatrix} b_2 & c_2 & d_2 \\ b_3 & c_3 & d_3 \\ b_4 & c_4 & d_4 \end{vmatrix}, \qquad B_1 = \begin{vmatrix} a_2 & c_2 & d_2 \\ a_3 & c_3 & d_3 \\ a_4 & c_4 & d_4 \end{vmatrix},$$

$$C_1 = \begin{vmatrix} a_2 & b_2 & d_2 \\ a_3 & b_3 & d_3 \\ a_4 & b_4 & d_4 \end{vmatrix}, \qquad D_1 = \begin{vmatrix} a_2 & b_2 & c_2 \\ a_3 & b_3 & c_3 \\ a_4 & b_4 & c_4 \end{vmatrix}.$$

This *inductive definition* process can be continued. For example, the value of a fifth-order determinant can be defined in terms of fourth-order determinants, and these, in turn, can be expressed in terms of third-order determinants, which you know how to evaluate.

**Example 2.** Evaluate:

$$\begin{vmatrix} 0 & 0 & 2 & 3 \\ 3 & 2 & 5 & 4 \\ 0 & 0 & 2 & 0 \\ 1 & 1 & -6 & 1 \end{vmatrix}$$

*Solution:*    Expand by minors of the third row because most of these elements are zero:

$$0 \begin{vmatrix} 0 & 2 & 3 \\ 2 & 5 & 4 \\ 1 & -6 & 1 \end{vmatrix} - 0 \begin{vmatrix} 0 & 2 & 3 \\ 3 & 5 & 4 \\ 1 & -6 & 1 \end{vmatrix} + 2 \begin{vmatrix} 0 & 0 & 3 \\ 3 & 2 & 4 \\ 1 & 1 & 1 \end{vmatrix} - 0 \begin{vmatrix} 0 & 0 & 2 \\ 3 & 2 & 5 \\ 1 & 1 & -6 \end{vmatrix}$$

$$= 2 \begin{vmatrix} 0 & 0 & 3 \\ 3 & 2 & 4 \\ 1 & 1 & 1 \end{vmatrix}$$

$$= 2 \left( 0 \begin{vmatrix} 2 & 4 \\ 1 & 1 \end{vmatrix} - 0 \begin{vmatrix} 3 & 4 \\ 1 & 1 \end{vmatrix} + 3 \begin{vmatrix} 3 & 2 \\ 1 & 1 \end{vmatrix} \right)$$

$$= 6 \begin{vmatrix} 3 & 2 \\ 1 & 1 \end{vmatrix} = 6(3 \cdot 1 - 1 \cdot 2) = 6$$

## WRITTEN EXERCISES

Expand the given determinant by minors of the specified row or column.

A 1. $\begin{vmatrix} 1 & 1 & 2 \\ 2 & 1 & 3 \\ -2 & 2 & 1 \end{vmatrix}$ ; row 1

4. $\begin{vmatrix} 1 & 2 & -1 \\ 1 & -3 & 4 \\ 11 & 2 & 1 \end{vmatrix}$ ; column 1

2. $\begin{vmatrix} 2 & 1 & 2 \\ 1 & 0 & 2 \\ 3 & 2 & 1 \end{vmatrix}$ ; row 2

5. $\begin{vmatrix} 4 & 6 & 1 \\ 1 & 1 & 0 \\ 1 & 3 & 1 \end{vmatrix}$ ; column 3

3. $\begin{vmatrix} 1 & -4 & 1 \\ 3 & 0 & 6 \\ 0 & 2 & 0 \end{vmatrix}$ ; row 3

6. $\begin{vmatrix} 2 & 0 & 1 \\ 3 & 2 & -2 \\ -1 & 1 & 4 \end{vmatrix}$ ; column 2

Expand the given determinant by minors of any row or column.

7. $\begin{vmatrix} 3 & 1 & 2 \\ 2 & 0 & -5 \\ 4 & 1 & 6 \end{vmatrix}$  8. $\begin{vmatrix} 3 & 4 & 2 \\ -2 & 6 & 1 \\ 0 & 1 & 2 \end{vmatrix}$  9. $\begin{vmatrix} 4 & 7 & 0 \\ 3 & 6 & 2 \\ 4 & 2 & 0 \end{vmatrix}$  10. $\begin{vmatrix} 4 & 1 & -2 \\ 0 & 1 & 0 \\ 6 & 1 & 2 \end{vmatrix}$

B 11. $\begin{vmatrix} 1 & 0 & 1 & 0 \\ 3 & 1 & -1 & 2 \\ 3 & -1 & 0 & 2 \\ 1 & 4 & 1 & 1 \end{vmatrix}$

13. $\begin{vmatrix} 1 & 0 & 0 & 0 & 0 \\ 2 & 0 & 0 & 2 & 0 \\ -5 & 1 & 3 & 6 & 1 \\ 4 & 0 & 0 & -1 & 1 \\ 2 & 1 & 2 & 3 & 4 \end{vmatrix}$

12. $\begin{vmatrix} 1 & 0 & 2 & 1 \\ 2 & 0 & 3 & 0 \\ 0 & -2 & 1 & 0 \\ 0 & 2 & 3 & 1 \end{vmatrix}$

14. $\begin{vmatrix} 0 & 0 & 0 & 0 & 1 \\ 0 & 0 & 0 & 1 & 0 \\ 0 & 0 & 1 & 0 & 0 \\ 2 & 0 & 0 & 0 & 0 \\ 0 & 2 & 0 & 0 & 0 \end{vmatrix}$

Solve each equation for x.

15. $\begin{vmatrix} x & x & 0 \\ 2 & 1 & 2 \\ -2 & 1 & 3 \end{vmatrix} = 18$

16. $\begin{vmatrix} x & 0 & 0 \\ 1 & x & 1 \\ 1 & -2 & 1 \end{vmatrix} = 3$

## 16–8 Properties of Determinants

Determinants have some properties which may help you evaluate them. Although these properties are presented here without proof and illustrated with third-order determinants, they apply to determinants of any order. You should verify the examples given below by evaluating the determinants.

**Property 1.** If all the rows and columns of a determinant are interchanged in order (so that the first row becomes the first column; the second row, the second column; and so on), then the resulting determinant equals the original one.

$$\begin{vmatrix} 1 & 0 & 2 \\ 3 & -1 & 4 \\ 5 & -3 & -2 \end{vmatrix} = 6 = \begin{vmatrix} 1 & 3 & 5 \\ 0 & -1 & -3 \\ 2 & 4 & -2 \end{vmatrix}$$

Property 1 enables you to replace "row" by "column" in the statements of the remaining properties.

**Property 2.** If any two rows (columns) of a determinant are interchanged, the resulting determinant is the negative of the original one.

$$\begin{vmatrix} 1 & 0 & 2 \\ 3 & -1 & 4 \\ 5 & -3 & -2 \end{vmatrix} = 6; \qquad \begin{vmatrix} 5 & -3 & -2 \\ 3 & -1 & 4 \\ 1 & 0 & 2 \end{vmatrix} = -6$$

**Property 3.** If two rows (columns) are identical, the determinant is 0.

$$\begin{vmatrix} 2 & 1 & 5 \\ 3 & 4 & 1 \\ 2 & 1 & 5 \end{vmatrix} = 0; \qquad \begin{vmatrix} 5 & 2 & 2 \\ 1 & 6 & 6 \\ 0 & -1 & -1 \end{vmatrix} = 0$$

**Property 4.** If the elements of one row (column) are multiplied by the real number $k$, the resulting determinant is $k$ times the original one.

$$\begin{vmatrix} 1 & 0 & 2 \\ 3 & -1 & 4 \\ 5 & -3 & -2 \end{vmatrix} = 6; \qquad \begin{vmatrix} 1 & 0 & 2 \\ 3 & -1 & 4 \\ -10 & 6 & 4 \end{vmatrix} = -2 \cdot 6 = -12$$

(*Warning:* Do not confuse Property 4 with multiplication of a *matrix* by a scalar.)

**Property 5.** If one row (column) has 0 for every element, the determinant is 0.

$$\begin{vmatrix} 3 & 1 & 2 \\ 0 & 0 & 0 \\ 5 & -4 & 3 \end{vmatrix} = 0; \qquad \begin{vmatrix} 0 & 2 & -1 \\ 0 & 1 & 6 \\ 0 & 7 & 2 \end{vmatrix} = 0$$

**Property 6.** If each element of one row is multiplied by $k$ and each resulting product is added to the corresponding element of a different row, the resulting determinant is equal to the original one. (The corresponding statement with "row" replaced by "column" throughout is also true.)

$$\begin{vmatrix} -3 & 0 & 1 \\ -2 & 5 & 4 \\ 1 & 4 & 2 \end{vmatrix} = \begin{vmatrix} -3+3\cdot1 & 0 & 1 \\ -2+3\cdot4 & 5 & 4 \\ 1+3\cdot2 & 4 & 2 \end{vmatrix} = \begin{vmatrix} 0 & 0 & 1 \\ 10 & 5 & 4 \\ 7 & 4 & 2 \end{vmatrix}$$

$$= 0\begin{vmatrix} 5 & 4 \\ 4 & 2 \end{vmatrix} - 0\begin{vmatrix} 10 & 4 \\ 7 & 2 \end{vmatrix} + 1\begin{vmatrix} 10 & 5 \\ 7 & 4 \end{vmatrix} = \begin{vmatrix} 10 & 5 \\ 7 & 4 \end{vmatrix} = 5$$

You can use Property 6 to obtain an array having at most one non-zero element in a selected row or column. In the Example below, we choose to create zeros in the first column.

**Example.** Evaluate:

$$\begin{vmatrix} 1 & 2 & 4 \\ 3 & 5 & 6 \\ -2 & -1 & -5 \end{vmatrix}$$

*Solution:* **1.** Multiply row 1 by $-3$ and add to row 2.

$$\begin{vmatrix} 1 & 2 & 4 \\ 3+(-3)1 & 5+(-3)2 & 6+(-3)4 \\ -2 & -1 & -5 \end{vmatrix} = \begin{vmatrix} 1 & 2 & 4 \\ 0 & -1 & -6 \\ -2 & -1 & -5 \end{vmatrix}$$

**2.** Multiply row 1 by 2 and add to row 3.

$$\begin{vmatrix} 1 & 2 & 4 \\ 0 & -1 & -6 \\ -2+2\cdot1 & -1+2\cdot2 & -5+2\cdot4 \end{vmatrix} = \begin{vmatrix} 1 & 2 & 4 \\ 0 & -1 & -6 \\ 0 & 3 & 3 \end{vmatrix}$$

**3.** Expand by minors of column 1.

$$1\begin{vmatrix} -1 & -6 \\ 3 & 3 \end{vmatrix} - 0\begin{vmatrix} 2 & 4 \\ 3 & 3 \end{vmatrix} + 0\begin{vmatrix} 2 & 4 \\ -1 & -6 \end{vmatrix} = 15$$

*WRITTEN EXERCISES*

Evaluate the given determinants. Use the properties of determinants stated in this section to simplify the work.

A

**1.** $\begin{vmatrix} 2 & -1 & 3 \\ 1 & 0 & 1 \\ 2 & -1 & 3 \end{vmatrix}$

**3.** $\begin{vmatrix} 1 & 2 & 3 \\ 2 & 4 & 6 \\ -3 & 2 & 5 \end{vmatrix}$

**5.** $\begin{vmatrix} 1 & 2 & 2 \\ -1 & 0 & 1 \\ -1 & 3 & 2 \end{vmatrix}$

**7.** $\begin{vmatrix} 2 & 1 & -2 \\ 1 & 2 & -3 \\ 4 & -2 & 1 \end{vmatrix}$

**2.** $\begin{vmatrix} 3 & 0 & -3 \\ -2 & 1 & 2 \\ -1 & 4 & 1 \end{vmatrix}$

**4.** $\begin{vmatrix} 1 & 3 & -2 \\ -1 & 1 & 2 \\ -2 & 0 & 4 \end{vmatrix}$

**6.** $\begin{vmatrix} 1 & -1 & 0 \\ 1 & 2 & 3 \\ 2 & -1 & -5 \end{vmatrix}$

**8.** $\begin{vmatrix} 2 & 5 & -3 \\ 1 & -2 & 2 \\ 3 & 2 & -4 \end{vmatrix}$

**9.** $\begin{vmatrix} 1 & 3 & -2 & 1 \\ 4 & 6 & -1 & 0 \\ 1 & 3 & -2 & 1 \\ 4 & 6 & 0 & 1 \end{vmatrix}$

**13.** $\begin{vmatrix} 40 & 25 & 15 \\ 6 & -3 & 21 \\ 12 & 8 & 20 \end{vmatrix}$

**14.** $\begin{vmatrix} 25 & 30 & 27 \\ 26 & 29 & 29 \\ 25 & 28 & 26 \end{vmatrix}$

**10.** $\begin{vmatrix} 3 & 1 & -3 & -1 \\ 1 & 6 & -1 & -1 \\ -2 & 2 & 2 & 0 \\ 4 & 3 & -4 & 2 \end{vmatrix}$

**15.** $\begin{vmatrix} 1 & 1 & 1 & 1 \\ 1 & 3 & 6 & 10 \\ 2 & 3 & 4 & 5 \\ 1 & 4 & 10 & 20 \end{vmatrix}$

**11.** $\begin{vmatrix} 1 & 1 & 2 & 0 \\ 2 & 1 & 3 & 5 \\ 3 & 1 & 4 & 3 \\ 4 & 1 & 5 & -2 \end{vmatrix}$

**16.** $\begin{vmatrix} 2 & 1 & 1 & 1 \\ 1 & 2 & 1 & 1 \\ 1 & 1 & 2 & 1 \\ 1 & 1 & 1 & 2 \end{vmatrix}$

**12.** $\begin{vmatrix} 1 & 0 & -1 & 2 \\ -1 & 0 & 1 & -2 \\ 3 & 4 & 0 & 1 \\ 1 & 2 & 1 & 3 \end{vmatrix}$

In each case prove the equality by expanding the determinants involved. (This method of proof of Properties 1–6 becomes much too laborious in the case of higher-order determinants.)

B

**17.** $\begin{vmatrix} a_1 & a_2 \\ b_1 & b_2 \end{vmatrix} = \begin{vmatrix} a_1 & b_1 \\ a_2 & b_2 \end{vmatrix}$  (Property 1)

**18. a.** $\begin{vmatrix} a_2 & b_2 \\ a_1 & b_1 \end{vmatrix} = - \begin{vmatrix} a_1 & b_1 \\ a_2 & b_2 \end{vmatrix}$

**b.** $\begin{vmatrix} b_1 & a_1 \\ b_2 & a_2 \end{vmatrix} = - \begin{vmatrix} a_1 & b_1 \\ a_2 & b_2 \end{vmatrix}$  (Property 2)

**19. a.** $\begin{vmatrix} a_1 & b_1 \\ a_1 & b_1 \end{vmatrix} = 0$   **b.** $\begin{vmatrix} a_1 & a_1 \\ a_2 & a_2 \end{vmatrix} = 0$  (Property 3)

**20. a.** $\begin{vmatrix} ka_1 & kb_1 \\ a_2 & b_2 \end{vmatrix} = k \begin{vmatrix} a_1 & b_1 \\ a_2 & b_2 \end{vmatrix}$

**b.** $\begin{vmatrix} a_1 & kb_1 \\ a_2 & kb_2 \end{vmatrix} = k \begin{vmatrix} a_1 & b_1 \\ a_2 & b_2 \end{vmatrix}$ (Property 4)

**21. a.** $\begin{vmatrix} a_1 & b_1 \\ 0 & 0 \end{vmatrix} = 0$ **b.** $\begin{vmatrix} 0 & b_1 \\ 0 & b_2 \end{vmatrix} = 0$ (Property 5)

**22. a.** $\begin{vmatrix} a_1 & b_1 \\ a_2 + ka_1 & b_2 + kb_1 \end{vmatrix} = \begin{vmatrix} a_1 & b_1 \\ a_2 & b_2 \end{vmatrix}$

**b.** $\begin{vmatrix} a_1 + kb_1 & b_1 \\ a_2 + kb_2 & b_2 \end{vmatrix} = \begin{vmatrix} a_1 & b_1 \\ a_2 & b_2 \end{vmatrix}$ (Property 6)

## 16–9 Cramer's Rule

In Section 16–6 we derived Cramer's Rule for solving systems of two linear equations in two variables. You can use a form of Cramer's Rule to solve a system of $n$ equations in $n$ variables, provided the determinant of the coefficient matrix is not 0. For example, to solve the system

$$a_1x + b_1y + c_1z = d_1$$
$$a_2x + b_2y + c_2z = d_2$$
$$a_3x + b_3y + c_3x = d_3$$

you compute the determinants

$$D = \begin{vmatrix} a_1 & b_2 & c_2 \\ a_2 & b_2 & c_2 \\ a_3 & b_3 & c_3 \end{vmatrix} ; \qquad D_x = \begin{vmatrix} d_1 & b_1 & c_1 \\ d_2 & b_2 & c_2 \\ d_3 & b_3 & c_3 \end{vmatrix} ;$$

$$D_y = \begin{vmatrix} a_1 & d_1 & c_1 \\ a_2 & d_2 & c_2 \\ a_3 & d_3 & c_3 \end{vmatrix} ; \qquad D_z = \begin{vmatrix} a_1 & b_1 & d_1 \\ a_2 & b_2 & d_2 \\ a_3 & b_3 & d_3 \end{vmatrix} .$$

Then if $D \neq 0$, you have

$$x = \frac{D_x}{D} \qquad y = \frac{D_y}{D} \qquad z = \frac{D_z}{D} .$$

(The proof is given on page 626.) Notice that you can obtain each determinant in a numerator from the determinant $D$ by replacing the coefficients of the variable for which you are solving by the constant terms $d_1$, $d_2$, and $d_3$.

**Example.** Use Cramer's Rule to solve the system:

$$x + 2y + 2z = 2$$
$$2x - 2y + z = 1$$
$$2x - 4y - 3z = 6$$

*Solution:*

$$D = \begin{vmatrix} 1 & 2 & 2 \\ 2 & -2 & 1 \\ 2 & -4 & -3 \end{vmatrix} = 6 + 4 - 16 + 8 + 4 + 12 = 18$$

$$D_x = \begin{vmatrix} 2 & 2 & 2 \\ 1 & -2 & 1 \\ 6 & -4 & -3 \end{vmatrix} = 12 + 12 - 8 + 24 + 8 + 6 = 54$$

$$D_y = \begin{vmatrix} 1 & 2 & 2 \\ 2 & 1 & 1 \\ 2 & 6 & -3 \end{vmatrix} = -3 + 4 + 24 - 4 - 6 + 12 = 27$$

$$D_z = \begin{vmatrix} 1 & 2 & 2 \\ 2 & -2 & 1 \\ 2 & -4 & 6 \end{vmatrix} = -12 + 4 - 16 + 8 + 4 - 24 = -36$$

$$x = \frac{D_x}{D} = \frac{54}{18} = 3, \quad y = \frac{D_y}{D} = \frac{27}{18} = \frac{3}{2}, \quad z = \frac{D_z}{D} = \frac{-36}{18} = -2$$

$\therefore$ the solution set is $\{(3, \frac{3}{2}, -2)\}$.

To derive the form of Cramer's Rule used above, assume that $(x, y, z)$ is a solution of this system.

$$a_1x + b_1y + c_1z = d_1$$
$$a_2x + b_2y + c_2z = d_2$$
$$a_3x + b_3y + c_3z = d_3$$

By Property 4, page 622,

$$xD = x \begin{vmatrix} a_1 & b_1 & c_1 \\ a_2 & b_2 & c_2 \\ a_3 & b_3 & c_3 \end{vmatrix} = \begin{vmatrix} a_1x & b_1 & c_1 \\ a_2x & b_2 & c_2 \\ a_3x & b_3 & c_3 \end{vmatrix}.$$

Using Property 6, page 623, you have

$$xD = \begin{vmatrix} a_1x + b_1y + c_1z & b_1 & c_1 \\ a_2x + b_2y + c_2z & b_2 & c_2 \\ a_3x + b_3y + c_3z & b_3 & c_3 \end{vmatrix} = \begin{vmatrix} d_1 & b_1 & c_1 \\ d_2 & b_2 & c_2 \\ d_3 & b_3 & c_3 \end{vmatrix} = D_x.$$

Thus, if $D \neq 0$, $x = \dfrac{D_x}{D}$. You can show in a similar way that $y = \dfrac{D_y}{D}$

and $z = \dfrac{D_z}{D}$. It is possible to verify by substitution that $(x, y, z)$ found above does satisfy the given system.

If $D = 0$, the equations are dependent or inconsistent.

## WRITTEN EXERCISES

Find the solution set of each of the following systems by using Cramer's Rule. If the system contains dependent or inconsistent equations, so state.

A

**1.** $x + y + z = 4$
$2x - y + 2z = 5$
$x - 2y - z = -3$

**2.** $2x - y - z = -6$
$x + 3y - z = 0$
$2x + y + z = -2$

**3.** $x + y = 5$
$y - 2z = 1$
$x + 3z = 5$

**4.** $2x + y = 4$
$2y - z = -5$
$x + 3z = 6$

**5.** $2x - 3y + z = 2$
$x - 2y - 2z = 3$
$3x - 5y - z = 5$

**6.** $4x - y + 2z = 5$
$2x + y - 3z = 7$
$10x - y + z = -2$

B

**7.** $x - 2y \qquad + 2w = 3$
$2x - 4y - 3z \qquad = 1$
$\qquad y + z - w = -1$
$3x - 2y + z - w = 2$

**8.** $x + y + z \qquad = 3$
$3x - y \qquad + 2w = 1$
$3x \qquad - z + 2w = -1$
$x + 5y + z + w = 2$

---

## Chapter Summary

---

**1.** The **sum of two matrices** of the same dimensions is the matrix of the same dimensions whose entries are the sums of the corresponding entries in the two matrices.

**2.** For addition in the set of $m \times n$ matrices, the **identity element** is the **zero matrix** $O_{m \times n}$, all of whose entries are 0, and the **additive inverse** of $A_{m \times n}$ is $-A_{m \times n}$, whose entries are the negatives of the corresponding entries of $A_{m \times n}$.

**3.** The **product of a scalar** $c$ **and a matrix** $A$ is the matrix $cA$ each of whose entries is $c$ times the corresponding entry of $A$.

4. The **product of matrices** $A_{m \times n}$ and $B_{n \times p}$ is the matrix $C_{m \times p}$ whose entry in the $i$th row and $j$th column is the sum of the products of the corresponding entries in the $i$th row of $A$ and the $j$th column of $B$.

5. You may find the determinant, det $A$, of a square matrix $A$ either from the definition or by expansion by minors. Certain properties of determinants are helpful in evaluating them.

6. In the set of $n \times n$ matrices, the matrix $I$ with 1's on main diagonal and 0's elsewhere is the *multiplicative* identity: $AI = IA = A$. If det $A \neq 0$, then $A$ has a multiplicative inverse $A^{-1}$ such that $AA^{-1} = A^{-1}A = I$.

7. You can use Cramer's Rule to solve linear systems of equations.

**Vocabulary and Spelling**

Review each term by reference to the page listed.

matrix (*p. 603*)
entry (of a matrix) (*p. 603*)
dimensions (of a matrix) (*p. 603*)
row matrix (*p. 603*)
column matrix (*p. 603*)
square matrix (*p. 603*)
equal matrices (*p. 603*)
corresponding entries (*p. 603*)
zero matrix (*p. 604*)
transpose ($A^T$) of $A$ (*p. 604*)
scalar (*p. 606*)
product matrix (*p. 609*)

left-multiplication (*p. 610*)
right-multiplication (*p. 610*)
identity matrix ($I$) (*p. 610*)
determinant (det) (*p. 611*)
elements of a determinant (*p. 612*)
order of a determinant (*p. 612*)
inverse matrix ($A^{-1}$) (*p. 613*)
invertible (nonsingular) (*p. 614*)
coefficient matrix (*p. 616*)
Cramer's Rule (*pp. 617, 625*)
minor of an element (*p. 618*)
expansion by minors (*p. 619*)

## *Chapter Test and Review*

**16–1** **1.** Find $x$ and $y$ if $\begin{bmatrix} x & 3 \\ 2 & x+y \end{bmatrix} = \begin{bmatrix} 1 & 3 \\ 2 & 4 \end{bmatrix}$

**16–2** **2.** Write as a single matrix: $\begin{bmatrix} 2 & 0 & -1 \\ 3 & 2 & 1 \end{bmatrix} - 2 \begin{bmatrix} 1 & -1 & 1 \\ 2 & 3 & 1 \end{bmatrix}$

**16–3** Find the indicated products.

**3.** $\begin{bmatrix} 2 & 0 & 1 \\ 1 & -1 & 0 \end{bmatrix} \begin{bmatrix} 3 \\ -2 \\ 1 \end{bmatrix}$

**4.** $\begin{bmatrix} -1 & 2 \\ 2 & 0 \end{bmatrix} \begin{bmatrix} 3 & 3 \\ 3 & 0 \end{bmatrix}$

**16–4**  **5.** If $A = \begin{bmatrix} 1 & -1 \\ -1 & 2 \end{bmatrix}$, find det $(3A)$.

**16–5**  **6.** If $A = \begin{bmatrix} 6 & -4 \\ -4 & 3 \end{bmatrix}$, find $A^{-1}$.

**16–6**  **7.** Solve the system $\begin{aligned} 6x - 4y &= 1 \\ -4x + 3y &= -2 \end{aligned}$

**a.** by writing it in matrix form and using the result of Exercise 6.
**b.** by Cramer's Rule.

**16–7**  **8.** Expand $\begin{vmatrix} 2 & 1 & -3 \\ -1 & 2 & 1 \\ 4 & 0 & 2 \end{vmatrix}$ by minors of the second column.

**16–8**  **9.** Evaluate: $\begin{vmatrix} 1 & 2 & -2 & 3 \\ 0 & 0 & 3 & 0 \\ 3 & 6 & 1 & 4 \\ -2 & -4 & 0 & 1 \end{vmatrix}$

**16–9**  **10.** Use Cramer's Rule to find the $z$-component of the solution of:

$$\begin{aligned} x + 2y \phantom{+ 4z} &= 2 \\ 2x - y + 4z &= 1 \\ -3x + y - 2z &= 0 \end{aligned}$$

## Cumulative Review: Chapters 14–16

**1.** What is the 40th term in the AP 38, 35, 32, ... ?

**2.** Find three arithmetic means between $-9$ and 11.

**3.** Find the sum of 24 terms of the AP 135, 126, 117, ....

**4.** Find the 9th term of the GP 2, $\sqrt{2}$, 1, ....

**5.** Find the geometric mean of 256 and 4.

**6.** Evaluate $\sum\limits_{i=2}^{7} 18(-\tfrac{2}{3})^{i-2}$

**7.** Find the sum of the infinite geometric series $\tfrac{16}{3} + \tfrac{8}{3} + \tfrac{4}{3} + \cdots$.

**8.** Expand $(t^2 - 2r)^5$ and simplify each term.

**9.** How many even natural numbers less than 123 are there?

10. In how many ways can the letters of the word TUCSON be arranged if each letter is used only once in each arrangement?

11. A school has five sections of first-year history. In how many ways can a brother and a sister be assigned to first-year history classes if their parents have requested that they be placed in different classes?

12. Find the number of distinguishable permutations of all the letters in the word SEQUENCE.

13. How many different committees of 4 persons can be appointed from a club with 10 members?

14. From a collection of eight blue marbles and five white marbles, how many ways can you select a set of four blue marbles and two white marbles?

15. What is the probability that two cards drawn at random from a standard deck of 52 cards will both be face cards?

16. A die is rolled and a card drawn from a standard deck of 52 cards. What is the probability that the die will show a 1 or a 2 and that the card will be a 10?

17. Solve for $x$: $\begin{bmatrix} 3 & -1 \\ 2 & 5 \end{bmatrix}^T = \begin{bmatrix} 3 & x \\ -1 & 5 \end{bmatrix}$.

18. Find (a) the sum and (b) the product of $\begin{bmatrix} 3 & -1 \\ 2 & 5 \end{bmatrix}$ and $\begin{bmatrix} -4 & 1 \\ 6 & -2 \end{bmatrix}$.

19. If $A = \begin{bmatrix} 7 & -1 & 2 \\ -3 & 0 & -3 \\ 0 & 4 & 1 \end{bmatrix}$, find det $A$.

20. Solve for $A$: $\begin{bmatrix} 2 & -5 \\ 3 & -7 \end{bmatrix} A = \begin{bmatrix} 4 & -2 \\ 3 & 1 \end{bmatrix}$

21. Solve using matrices: $3x - 5y = 11$
$$2x + y = 3$$

22. Solve for $x$: $\begin{vmatrix} 2 & 1 & x \\ 5 & -1 & 4 \\ 1 & -2 & 3 \end{vmatrix} = -28$

23. Evaluate: $\begin{vmatrix} 1 & 0 & -1 & 2 \\ 2 & 1 & 0 & -1 \\ 1 & 1 & 0 & 0 \\ 2 & 0 & 0 & -1 \end{vmatrix}$

24. Solve using Cramer's Rule: $x - 2y + z = 5$
$$x + 2y - z = -3$$
$$2x + y + 2z = 5$$

# APPENDIX

## TABLE 1

### FORMULAS

| | | | |
|---|---|---|---|
| Circle | $A = \pi r^2, C = 2\pi r$ | Cube | $V = s^3$ |
| Parallelogram | $A = bh$ | Rectangular Box | $V = lwh$ |
| Right Triangle | $A = \frac{1}{2}bh, c^2 = a^2 + b^2$ | Right Circular | |
| Square | $A = s^2$ | Cylinder | $V = \pi r^2 h$ |
| Trapezoid | $A = \frac{1}{2}h(b + b')$ | Pyramid | $V = \frac{1}{3}Bh$ |
| Triangle | $A = \frac{1}{2}bh$ | Right Circular | |
| Sphere | $A = 4\pi r^2$ | Cone | $V = \frac{1}{3}\pi r^2 h$ |
| | | Sphere | $V = \frac{4}{3}\pi r^3$ |

## TABLE 2

### AMERICAN SYSTEM OF WEIGHTS AND MEASURES

| LENGTH | WEIGHT |
|---|---|
| 12 inches = 1 foot | 16 ounces = 1 pound |
| 3 feet = 1 yard | 2000 pounds = 1 ton |
| $5\frac{1}{2}$ yards = 1 rod | 2240 pounds = 1 long ton |
| 5280 feet = 1 land mile | |
| 6076 feet = 1 nautical mile | **CAPACITY** |
| | *Dry Measure* |
| **AREA** | 2 pints = 1 quart |
| 144 square inches = 1 square foot | 8 quarts = 1 peck |
| 9 square feet = 1 square yard | 4 pecks = 1 bushel |
| 160 square rods = 1 acre | |
| 640 acres = 1 square mile | *Liquid Measure* |
| | 16 fluid ounces = 1 pint |
| **VOLUME** | 2 pints = 1 quart |
| 1728 cubic inches = 1 cubic foot | 4 quarts = 1 gallon |
| 27 cubic feet = 1 cubic yard | 231 cubic inches = 1 gallon |

### METRIC SYSTEM OF WEIGHTS AND MEASURES

| | | | | |
|---|---|---|---|---|
| **LENGTH** | 10 millimeters (mm) | = 1 centimeter (cm) | = | 0.3937 inch |
| | 100 centimeters | = 1 meter (m) | = | 39.37 inches |
| | 1000 meters | = 1 kilometer (km) | = | 0.6214 mile |
| **CAPACITY** | 1000 milliliters (ml) | = 1 liter (l) | = | 1.057 quart |
| | 1000 liters (l) | = 1 kiloliter (kl) | = 264.2 | gallons |
| **WEIGHT** | 1000 milligrams (mg) | = 1 gram (g) | = | 0.0353 ounce |
| | 1000 grams | = 1 kilogram (kg) | = | 2.205 pounds |

# TABLE 3 SQUARES AND SQUARE ROOTS

| N | $N^2$ | $\sqrt{N}$ | $\sqrt{10N}$ | N | $N^2$ | $\sqrt{N}$ | $\sqrt{10N}$ |
|---|-------|------------|--------------|---|-------|------------|--------------|
| 1.0 | 1.00 | 1.000 | 3.162 | 5.5 | 30.25 | 2.345 | 7.416 |
| 1.1 | 1.21 | 1.049 | 3.317 | 5.6 | 31.36 | 2.366 | 7.483 |
| 1.2 | 1.44 | 1.095 | 3.464 | 5.7 | 32.49 | 2.387 | 7.550 |
| 1.3 | 1.69 | 1.140 | 3.606 | 5.8 | 33.64 | 2.408 | 7.616 |
| 1.4 | 1.96 | 1.183 | 3.742 | 5.9 | 34.81 | 2.429 | 7.681 |
| 1.5 | 2.25 | 1.225 | 3.873 | 6.0 | 36.00 | 2.449 | 7.746 |
| 1.6 | 2.56 | 1.265 | 4.000 | 6.1 | 37.21 | 2.470 | 7.810 |
| 1.7 | 2.89 | 1.304 | 4.123 | 6.2 | 38.44 | 2.490 | 7.874 |
| 1.8 | 3.24 | 1.342 | 4.243 | 6.3 | 39.69 | 2.510 | 7.937 |
| 1.9 | 3.61 | 1.378 | 4.359 | 6.4 | 40.96 | 2.530 | 8.000 |
| 2.0 | 4.00 | 1.414 | 4.472 | 6.5 | 42.25 | 2.550 | 8.062 |
| 2.1 | 4.41 | 1.449 | 4.583 | 6.6 | 43.56 | 2.569 | 8.124 |
| 2.2 | 4.84 | 1.483 | 4.690 | 6.7 | 44.89 | 2.588 | 8.185 |
| 2.3 | 5.29 | 1.517 | 4.796 | 6.8 | 46.24 | 2.608 | 8.246 |
| 2.4 | 5.76 | 1.549 | 4.899 | 6.9 | 47.61 | 2.627 | 8.307 |
| 2.5 | 6.25 | 1.581 | 5.000 | 7.0 | 49.00 | 2.646 | 8.367 |
| 2.6 | 6.76 | 1.612 | 5.099 | 7.1 | 50.41 | 2.665 | 8.426 |
| 2.7 | 7.29 | 1.643 | 5.196 | 7.2 | 51.84 | 2.683 | 8.485 |
| 2.8 | 7.84 | 1.673 | 5.292 | 7.3 | 53.29 | 2.702 | 8.544 |
| 2.9 | 8.41 | 1.703 | 5.385 | 7.4 | 54.76 | 2.720 | 8.602 |
| 3.0 | 9.00 | 1.732 | 5.477 | 7.5 | 56.25 | 2.739 | 8.660 |
| 3.1 | 9.61 | 1.761 | 5.568 | 7.6 | 57.76 | 2.757 | 8.718 |
| 3.2 | 10.24 | 1.789 | 5.657 | 7.7 | 59.29 | 2.775 | 8.775 |
| 3.3 | 10.89 | 1.817 | 5.745 | 7.8 | 60.84 | 2.793 | 8.832 |
| 3.4 | 11.56 | 1.844 | 5.831 | 7.9 | 62.41 | 2.811 | 8.888 |
| 3.5 | 12.25 | 1.871 | 5.916 | 8.0 | 64.00 | 2.828 | 8.944 |
| 3.6 | 12.96 | 1.897 | 6.000 | 8.1 | 65.61 | 2.846 | 9.000 |
| 3.7 | 13.69 | 1.924 | 6.083 | 8.2 | 67.24 | 2.864 | 9.055 |
| 3.8 | 14.44 | 1.949 | 6.164 | 8.3 | 68.89 | 2.881 | 9.110 |
| 3.9 | 15.21 | 1.975 | 6.245 | 8.4 | 70.56 | 2.898 | 9.165 |
| 4.0 | 16.00 | 2.000 | 6.325 | 8.5 | 72.25 | 2.915 | 9.220 |
| 4.1 | 16.81 | 2.025 | 6.403 | 8.6 | 73.96 | 2.933 | 9.274 |
| 4.2 | 17.64 | 2.049 | 6.481 | 8.7 | 75.69 | 2.950 | 9.327 |
| 4.3 | 18.49 | 2.074 | 6.557 | 8.8 | 77.44 | 2.966 | 9.381 |
| 4.4 | 19.36 | 2.098 | 6.633 | 8.9 | 79.21 | 2.983 | 9.434 |
| 4.5 | 20.25 | 2.121 | 6.708 | 9.0 | 81.00 | 3.000 | 9.487 |
| 4.6 | 21.16 | 2.145 | 6.782 | 9.1 | 82.81 | 3.017 | 9.539 |
| 4.7 | 22.09 | 2.168 | 6.856 | 9.2 | 84.64 | 3.033 | 9.592 |
| 4.8 | 23.04 | 2.191 | 6.928 | 9.3 | 86.49 | 3.050 | 9.644 |
| 4.9 | 24.01 | 2.214 | 7.000 | 9.4 | 88.36 | 3.066 | 9.695 |
| 5.0 | 25.00 | 2.236 | 7.071 | 9.5 | 90.25 | 3.082 | 9.747 |
| 5.1 | 26.01 | 2.258 | 7.141 | 9.6 | 92.16 | 3.098 | 9.798 |
| 5.2 | 27.04 | 2.280 | 7.211 | 9.7 | 94.09 | 3.114 | 9.849 |
| 5.3 | 28.09 | 2.302 | 7.280 | 9.8 | 96.04 | 3.130 | 9.899 |
| 5.4 | 29.16 | 2.324 | 7.348 | 9.9 | 98.01 | 3.146 | 9.950 |
| 5.5 | 30.25 | 2.345 | 7.416 | 10 | 100.00 | 3.162 | 10.000 |

# TABLE 4    CUBES AND CUBE ROOTS

| N | $N^3$ | $\sqrt[3]{N}$ | $\sqrt[3]{10N}$ | $\sqrt[3]{100N}$ | N | $N^3$ | $\sqrt[3]{N}$ | $\sqrt[3]{10N}$ | $\sqrt[3]{100N}$ |
|---|-------|------|-------|--------|---|-------|------|-------|--------|
| 1.0 | 1.000 | 1.000 | 2.154 | 4.642 | 5.5 | 166.375 | 1.765 | 3.803 | 8.193 |
| 1.1 | 1.331 | 1.032 | 2.224 | 4.791 | 5.6 | 175.616 | 1.776 | 3.826 | 8.243 |
| 1.2 | 1.728 | 1.063 | 2.289 | 4.932 | 5.7 | 185.193 | 1.786 | 3.849 | 8.291 |
| 1.3 | 2.197 | 1.091 | 2.351 | 5.066 | 5.8 | 195.112 | 1.797 | 3.871 | 8.340 |
| 1.4 | 2.744 | 1.119 | 2.410 | 5.192 | 5.9 | 205.379 | 1.807 | 3.893 | 8.387 |
| 1.5 | 3.375 | 1.145 | 2.466 | 5.313 | 6.0 | 216.000 | 1.817 | 3.915 | 8.434 |
| 1.6 | 4.096 | 1.170 | 2.520 | 5.429 | 6.1 | 226.981 | 1.827 | 3.936 | 8.481 |
| 1.7 | 4.913 | 1.193 | 2.571 | 5.540 | 6.2 | 238.328 | 1.837 | 3.958 | 8.527 |
| 1.8 | 5.832 | 1.216 | 2.621 | 5.646 | 6.3 | 250.047 | 1.847 | 3.979 | 8.573 |
| 1.9 | 6.859 | 1.239 | 2.668 | 5.749 | 6.4 | 262.144 | 1.857 | 4.000 | 8.618 |
| 2.0 | 8.000 | 1.260 | 2.714 | 5.848 | 6.5 | 274.625 | 1.866 | 4.021 | 8.662 |
| 2.1 | 9.261 | 1.281 | 2.759 | 5.944 | 6.6 | 287.496 | 1.876 | 4.041 | 8.707 |
| 2.2 | 10.648 | 1.301 | 2.802 | 6.037 | 6.7 | 300.763 | 1.885 | 4.062 | 8.750 |
| 2.3 | 12.167 | 1.320 | 2.844 | 6.127 | 6.8 | 314.432 | 1.895 | 4.082 | 8.794 |
| 2.4 | 13.824 | 1.339 | 2.884 | 6.214 | 6.9 | 328.509 | 1.904 | 4.102 | 8.837 |
| 2.5 | 15.625 | 1.357 | 2.924 | 6.300 | 7.0 | 343.000 | 1.913 | 4.121 | 8.879 |
| 2.6 | 17.576 | 1.375 | 2.962 | 6.383 | 7.1 | 357.911 | 1.922 | 4.141 | 8.921 |
| 2.7 | 19.683 | 1.392 | 3.000 | 6.463 | 7.2 | 373.248 | 1.931 | 4.160 | 8.963 |
| 2.8 | 21.952 | 1.409 | 3.037 | 6.542 | 7.3 | 389.017 | 1.940 | 4.179 | 9.004 |
| 2.9 | 24.389 | 1.426 | 3.072 | 6.619 | 7.4 | 405.224 | 1.949 | 4.198 | 9.045 |
| 3.0 | 27.000 | 1.442 | 3.107 | 6.694 | 7.5 | 421.875 | 1.957 | 4.217 | 9.086 |
| 3.1 | 29.791 | 1.458 | 3.141 | 6.768 | 7.6 | 438.976 | 1.966 | 4.236 | 9.126 |
| 3.2 | 32.768 | 1.474 | 3.175 | 6.840 | 7.7 | 456.533 | 1.975 | 4.254 | 9.166 |
| 3.3 | 35.937 | 1.489 | 3.208 | 6.910 | 7.8 | 474.552 | 1.983 | 4.273 | 9.205 |
| 3.4 | 39.304 | 1.504 | 3.240 | 6.980 | 7.9 | 493.039 | 1.992 | 4.291 | 9.244 |
| 3.5 | 42.875 | 1.518 | 3.271 | 7.047 | 8.0 | 512.000 | 2.000 | 4.309 | 9.283 |
| 3.6 | 46.656 | 1.533 | 3.302 | 7.114 | 8.1 | 531.441 | 2.008 | 4.327 | 9.322 |
| 3.7 | 50.653 | 1.547 | 3.332 | 7.179 | 8.2 | 551.368 | 2.017 | 4.344 | 9.360 |
| 3.8 | 54.872 | 1.560 | 3.362 | 7.243 | 8.3 | 571.787 | 2.025 | 4.362 | 9.398 |
| 3.9 | 59.319 | 1.574 | 3.391 | 7.306 | 8.4 | 592.704 | 2.033 | 4.380 | 9.435 |
| 4.0 | 64.000 | 1.587 | 3.420 | 7.368 | 8.5 | 614.125 | 2.041 | 4.397 | 9.473 |
| 4.1 | 68.921 | 1.601 | 3.448 | 7.429 | 8.6 | 636.056 | 2.049 | 4.414 | 9.510 |
| 4.2 | 74.088 | 1.613 | 3.476 | 7.489 | 8.7 | 658.503 | 2.057 | 4.431 | 9.546 |
| 4.3 | 79.507 | 1.626 | 3.503 | 7.548 | 8.8 | 681.472 | 2.065 | 4.448 | 9.583 |
| 4.4 | 85.184 | 1.639 | 3.530 | 7.606 | 8.9 | 704.969 | 2.072 | 4.465 | 9.619 |
| 4.5 | 91.125 | 1.651 | 3.557 | 7.663 | 9.0 | 729.000 | 2.080 | 4.481 | 9.655 |
| 4.6 | 97.336 | 1.663 | 3.583 | 7.719 | 9.1 | 753.571 | 2.088 | 4.498 | 9.691 |
| 4.7 | 103.823 | 1.675 | 3.609 | 7.775 | 9.2 | 778.688 | 2.095 | 4.514 | 9.726 |
| 4.8 | 110.592 | 1.687 | 3.634 | 7.830 | 9.3 | 804.357 | 2.103 | 4.531 | 9.761 |
| 4.9 | 117.649 | 1.698 | 3.659 | 7.884 | 9.4 | 830.584 | 2.110 | 4.547 | 9.796 |
| 5.0 | 125.000 | 1.710 | 3.684 | 7.937 | 9.5 | 857.375 | 2.118 | 4.563 | 9.830 |
| 5.1 | 132.651 | 1.721 | 3.708 | 7.990 | 9.6 | 884.736 | 2.125 | 4.579 | 9.865 |
| 5.2 | 140.608 | 1.732 | 3.733 | 8.041 | 9.7 | 912.673 | 2.133 | 4.595 | 9.899 |
| 5.3 | 148.877 | 1.744 | 3.756 | 8.093 | 9.8 | 941.192 | 2.140 | 4.610 | 9.933 |
| 5.4 | 157.464 | 1.754 | 3.780 | 8.143 | 9.9 | 970.299 | 2.147 | 4.626 | 9.967 |
| 5.5 | 166.375 | 1.765 | 3.803 | 8.193 | 10 | 1000.000 | 2.154 | 4.642 | 10.000 |

# TABLE 5 COMMON LOGARITHMS OF NUMBERS*

| N | 0 | 1 | 2 | 3 | 4 | 5 | 6 | 7 | 8 | 9 |
|----|------|------|------|------|------|------|------|------|------|------|
| 10 | .0000 | .0043 | 0086 | 0128 | 0170 | 0212 | 0253 | 0294 | 0334 | 0374 |
| 11 | 0414 | 0453 | 0492 | 0531 | 0569 | 0607 | 0645 | 0682 | 0719 | 0755 |
| 12 | 0792 | 0828 | 0864 | 0899 | 0934 | 0969 | 1004 | 1038 | 1072 | 1106 |
| 13 | 1139 | 1173 | 1206 | 1239 | 1271 | 1303 | 1335 | 1367 | 1399 | 1430 |
| 14 | 1461 | 1492 | 1523 | 1553 | 1584 | 1614 | 1644 | 1673 | 1703 | 1732 |
| 15 | 1761 | 1790 | 1818 | 1847 | 1875 | 1903 | 1931 | 1959 | 1987 | 2014 |
| 16 | 2041 | 2068 | 2095 | 2122 | 2148 | 2175 | 2201 | 2227 | 2253 | 2279 |
| 17 | 2304 | 2330 | 2355 | 2380 | 2405 | 2430 | 2455 | 2480 | 2504 | 2529 |
| 18 | 2553 | 2577 | 2601 | 2625 | 2648 | 2672 | 2695 | 2718 | 2742 | 2765 |
| 19 | 2788 | 2810 | 2833 | 2856 | 2878 | 2900 | 2923 | 2945 | 2967 | 2989 |
| 20 | 3010 | 3032 | 3054 | 3075 | 3096 | 3118 | 3139 | 3160 | 3181 | 3201 |
| 21 | 3222 | 3243 | 3263 | 3284 | 3304 | 3324 | 3345 | 3365 | 3385 | 3404 |
| 22 | 3424 | 3444 | 3464 | 3483 | 3502 | 3522 | 3541 | 3560 | 3579 | 3598 |
| 23 | 3617 | 3636 | 3655 | 3674 | 3692 | 3711 | 3729 | 3747 | 3766 | 3784 |
| 24 | 3802 | 3820 | 3838 | 3856 | 3874 | 3892 | 3909 | 3927 | 3945 | 3962 |
| 25 | 3979 | 3997 | 4014 | 4031 | 4048 | 4065 | 4082 | 4099 | 4116 | 4133 |
| 26 | 4150 | 4166 | 4183 | 4200 | 4216 | 4232 | 4249 | 4265 | 4281 | 4298 |
| 27 | 4314 | 4330 | 4346 | 4362 | 4378 | 4393 | 4409 | 4425 | 4440 | 4456 |
| 28 | 4472 | 4487 | 4502 | 4518 | 4533 | 4548 | 4564 | 4579 | 4594 | 4609 |
| 29 | 4624 | 4639 | 4654 | 4669 | 4683 | 4698 | 4713 | 4728 | 4742 | 4757 |
| 30 | 4771 | 4786 | 4800 | 4814 | 4829 | 4843 | 4857 | 4871 | 4886 | 4900 |
| 31 | 4914 | 4928 | 4942 | 4955 | 4969 | 4983 | 4997 | 5011 | 5024 | 5038 |
| 32 | 5051 | 5065 | 5079 | 5092 | 5105 | 5119 | 5132 | 5145 | 5159 | 5172 |
| 33 | 5185 | 5198 | 5211 | 5224 | 5237 | 5250 | 5263 | 5276 | 5289 | 5302 |
| 34 | 5315 | 5328 | 5340 | 5353 | 5366 | 5378 | 5391 | 5403 | 5416 | 5428 |
| 35 | 5441 | 5453 | 5465 | 5478 | 5490 | 5502 | 5514 | 5527 | 5539 | 5551 |
| 36 | 5563 | 5575 | 5587 | 5599 | 5611 | 5623 | 5635 | 5647 | 5658 | 5670 |
| 37 | 5682 | 5694 | 5705 | 5717 | 5729 | 5740 | 5752 | 5763 | 5775 | 5786 |
| 38 | 5798 | 5809 | 5821 | 5832 | 5843 | 5855 | 5866 | 5877 | 5888 | 5899 |
| 39 | 5911 | 5922 | 5933 | 5944 | 5955 | 5966 | 5977 | 5988 | 5999 | 6010 |
| 40 | 6021 | 6031 | 6042 | 6053 | 6064 | 6075 | 6085 | 6096 | 6107 | 6117 |
| 41 | 6128 | 6138 | 6149 | 6160 | 6170 | 6180 | 6191 | 6201 | 6212 | 6222 |
| 42 | 6232 | 6243 | 6253 | 6263 | 6274 | 6284 | 6294 | 6304 | 6314 | 6325 |
| 43 | 6335 | 6345 | 6355 | 6365 | 6375 | 6385 | 6395 | 6405 | 6415 | 6425 |
| 44 | 6435 | 6444 | 6454 | 6464 | 6474 | 6484 | 6493 | 6503 | 6513 | 6522 |
| 45 | 6532 | 6542 | 6551 | 6561 | 6571 | 6580 | 6590 | 6599 | 6609 | 6618 |
| 46 | 6628 | 6637 | 6646 | 6656 | 6665 | 6675 | 6684 | 6693 | 6702 | 6712 |
| 47 | 6721 | 6730 | 6739 | 6749 | 6758 | 6767 | 6776 | 6785 | 6794 | 6803 |
| 48 | 6812 | 6821 | 6830 | 6839 | 6848 | 6857 | 6866 | 6875 | 6884 | 6893 |
| 49 | 6902 | 6911 | 6920 | 6928 | 6937 | 6946 | 6955 | 6964 | 6972 | 6981 |
| 50 | 6990 | 6998 | 7007 | 7016 | 7024 | 7033 | 7042 | 7050 | 7059 | 7067 |
| 51 | 7076 | 7084 | 7093 | 7101 | 7110 | 7118 | 7126 | 7135 | 7143 | 7152 |
| 52 | 7160 | 7168 | 7177 | 7185 | 7193 | 7202 | 7210 | 7218 | 7226 | 7235 |
| 53 | 7243 | 7251 | 7259 | 7267 | 7275 | 7284 | 7292 | 7300 | 7308 | 7316 |
| 54 | 7324 | 7332 | 7340 | 7348 | 7356 | 7364 | 7372 | 7380 | 7388 | 7396 |

*Mantissas, decimal points omitted. Characteristics are found by inspection.

# TABLE 5 COMMON LOGARITHMS OF NUMBERS

| N | 0 | 1 | 2 | 3 | 4 | 5 | 6 | 7 | 8 | 9 |
|---|---|---|---|---|---|---|---|---|---|---|
| 55 | 7404 | 7412 | 7419 | 7427 | 7435 | 7443 | 7451 | 7459 | 7466 | 7474 |
| 56 | 7482 | 7490 | 7497 | 7505 | 7513 | 7520 | 7528 | 7536 | 7543 | 7551 |
| 57 | 7559 | 7566 | 7574 | 7582 | 7589 | 7597 | 7604 | 7612 | 7619 | 7627 |
| 58 | 7634 | 7642 | 7649 | 7657 | 7664 | 7672 | 7679 | 7686 | 7694 | 7701 |
| 59 | 7709 | 7716 | 7723 | 7731 | 7738 | 7745 | 7752 | 7760 | 7767 | 7774 |
| 60 | 7782 | 7789 | 7796 | 7803 | 7810 | 7818 | 7825 | 7832 | 7839 | 7846 |
| 61 | 7853 | 7860 | 7868 | 7875 | 7882 | 7889 | 7896 | 7903 | 7910 | 7917 |
| 62 | 7924 | 7931 | 7938 | 7945 | 7952 | 7959 | 7966 | 7973 | 7980 | 7987 |
| 63 | 7993 | 8000 | 8007 | 8014 | 8021 | 8028 | 8035 | 8041 | 8048 | 8055 |
| 64 | 8062 | 8069 | 8075 | 8082 | 8089 | 8096 | 8102 | 8109 | 8116 | 8122 |
| 65 | 8129 | 8136 | 8142 | 8149 | 8156 | 8162 | 8169 | 8176 | 8182 | 8189 |
| 66 | 8195 | 8202 | 8209 | 8215 | 8222 | 8228 | 8235 | 8241 | 8248 | 8254 |
| 67 | 8261 | 8267 | 8274 | 8280 | 8287 | 8293 | 8299 | 8306 | 8312 | 8319 |
| 68 | 8325 | 8331 | 8338 | 8344 | 8351 | 8357 | 8363 | 8370 | 8376 | 8382 |
| 69 | 8388 | 8395 | 8401 | 8407 | 8414 | 8420 | 8426 | 8432 | 8439 | 8445 |
| 70 | 8451 | 8457 | 8463 | 8470 | 8476 | 8482 | 8488 | 8494 | 8500 | 8506 |
| 71 | 8513 | 8519 | 8525 | 8531 | 8537 | 8543 | 8549 | 8555 | 8561 | 8567 |
| 72 | 8573 | 8579 | 8585 | 8591 | 8597 | 8603 | 8609 | 8615 | 8621 | 8627 |
| 73 | 8633 | 8639 | 8645 | 8651 | 8657 | 8663 | 8669 | 8675 | 8681 | 8686 |
| 74 | 8692 | 8698 | 8704 | 8710 | 8716 | 8722 | 8727 | 8733 | 8739 | 8745 |
| 75 | 8751 | 8756 | 8762 | 8768 | 8774 | 8779 | 8785 | 8791 | 8797 | 8802 |
| 76 | 8808 | 8814 | 8820 | 8825 | 8831 | 8837 | 8842 | 8848 | 8854 | 8859 |
| 77 | 8865 | 8871 | 8876 | 8882 | 8887 | 8893 | 8899 | 8904 | 8910 | 8915 |
| 78 | 8921 | 8927 | 8932 | 8938 | 8943 | 8949 | 8954 | 8960 | 8965 | 8971 |
| 79 | 8976 | 8982 | 8987 | 8993 | 8998 | 9004 | 9009 | 9015 | 9020 | 9025 |
| 80 | 9031 | 9036 | 9042 | 9047 | 9053 | 9058 | 9063 | 9069 | 9074 | 9079 |
| 81 | 9085 | 9090 | 9096 | 9101 | 9106 | 9112 | 9117 | 9122 | 9128 | 9133 |
| 82 | 9138 | 9143 | 9149 | 9154 | 9159 | 9165 | 9170 | 9175 | 9180 | 9186 |
| 83 | 9191 | 9196 | 9201 | 9206 | 9212 | 9217 | 9222 | 9227 | 9232 | 9238 |
| 84 | 9243 | 9248 | 9253 | 9258 | 9263 | 9269 | 9274 | 9279 | 9284 | 9289 |
| 85 | 9294 | 9299 | 9304 | 9309 | 9315 | 9320 | 9325 | 9330 | 9335 | 9340 |
| 86 | 9345 | 9350 | 9355 | 9360 | 9365 | 9370 | 9375 | 9380 | 9385 | 9390 |
| 87 | 9395 | 9400 | 9405 | 9410 | 9415 | 9420 | 9425 | 9430 | 9435 | 9440 |
| 88 | 9445 | 9450 | 9455 | 9460 | 9465 | 9469 | 9474 | 9479 | 9484 | 9489 |
| 89 | 9494 | 9499 | 9504 | 9509 | 9513 | 9518 | 9523 | 9528 | 9533 | 9538 |
| 90 | 9542 | 9547 | 9552 | 9557 | 9562 | 9566 | 9571 | 9576 | 9581 | 9586 |
| 91 | 9590 | 9595 | 9600 | 9605 | 9609 | 9614 | 9619 | 9624 | 9628 | 9633 |
| 92 | 9638 | 9643 | 9647 | 9652 | 9657 | 9661 | 9666 | 9671 | 9675 | 9680 |
| 93 | 9685 | 9689 | 9694 | 9699 | 9703 | 9708 | 9713 | 9717 | 9722 | 9727 |
| 94 | 9731 | 9736 | 9741 | 9745 | 9750 | 9754 | 9759 | 9763 | 9768 | 9773 |
| 95 | 9777 | 9782 | 9786 | 9791 | 9795 | 9800 | 9805 | 9809 | 9814 | 9818 |
| 96 | 9823 | 9827 | 9832 | 9836 | 9841 | 9845 | 9850 | 9854 | 9859 | 9863 |
| 97 | 9868 | 9872 | 9877 | 9881 | 9886 | 9890 | 9894 | 9899 | 9903 | 9908 |
| 98 | 9912 | 9917 | 9921 | 9926 | 9930 | 9934 | 9939 | 9943 | 9948 | 9952 |
| 99 | 9956 | 9961 | 9965 | 9969 | 9974 | 9978 | 9983 | 9987 | 9991 | 9996 |

# TABLE 6  VALUES OF TRIGONOMETRIC FUNCTIONS

| Angle | Sin | Cos | Tan | Cot | Sec | Csc | |
|---|---|---|---|---|---|---|---|
| 0° 00′ | .0000 | 1.0000 | .0000 | - - - - | 1.000 | - - - - | 90° 00′ |
| 10′ | .0029 | 1.0000 | .0029 | 343.8 | 1.000 | 343.8 | 50′ |
| 20′ | .0058 | 1.0000 | .0058 | 171.9 | 1.000 | 171.9 | 40′ |
| 30′ | .0087 | 1.0000 | .0087 | 114.6 | 1.000 | 114.6 | 30′ |
| 40′ | .0116 | .9999 | .0116 | 85.94 | 1.000 | 85.95 | 20′ |
| 50′ | .0145 | .9999 | .0145 | 68.75 | 1.000 | 68.76 | 10′ |
| 1° 00′ | .0175 | .9998 | .0175 | 57.29 | 1.000 | 57.30 | 89° 00′ |
| 10′ | .0204 | .9998 | .0204 | 49.10 | 1.000 | 49.11 | 50′ |
| 20′ | .0233 | .9997 | .0233 | 42.96 | 1.000 | 42.98 | 40′ |
| 30′ | .0262 | .9997 | .0262 | 38.19 | 1.000 | 38.20 | 30′ |
| 40′ | .0291 | .9996 | .0291 | 34.37 | 1.000 | 34.38 | 20′ |
| 50′ | .0320 | .9995 | .0320 | 31.24 | 1.001 | 31.26 | 10′ |
| 2° 00′ | .0349 | .9994 | .0349 | 28.64 | 1.001 | 28.65 | 88° 00′ |
| 10′ | .0378 | .9993 | .0378 | 26.43 | 1.001 | 26.45 | 50′ |
| 20′ | .0407 | .9992 | .0407 | 24.54 | 1.001 | 24.56 | 40′ |
| 30′ | .0436 | .9990 | .0437 | 22.90 | 1.001 | 22.93 | 30′ |
| 40′ | .0465 | .9989 | .0466 | 21.47 | 1.001 | 21.49 | 20′ |
| 50′ | .0494 | .9988 | .0495 | 20.21 | 1.001 | 20.23 | 10′ |
| 3° 00′ | .0523 | .9986 | .0524 | 19.08 | 1.001 | 19.11 | 87° 00′ |
| 10′ | .0552 | .9985 | .0553 | 18.07 | 1.002 | 18.10 | 50′ |
| 20′ | .0581 | .9983 | .0582 | 17.17 | 1.002 | 17.20 | 40′ |
| 30′ | .0610 | .9981 | .0612 | 16.35 | 1.002 | 16.38 | 30′ |
| 40′ | .0640 | .9980 | .0641 | 15.60 | 1.002 | 15.64 | 20′ |
| 50′ | .0669 | .9978 | .0670 | 14.92 | 1.002 | 14.96 | 10′ |
| 4° 00′ | .0698 | .9976 | .0699 | 14.30 | 1.002 | 14.34 | 86° 00′ |
| 10′ | .0727 | .9974 | .0729 | 13.73 | 1.003 | 13.76 | 50′ |
| 20′ | .0756 | .9971 | .0758 | 13.20 | 1.003 | 13.23 | 40′ |
| 30′ | .0785 | .9969 | .0787 | 12.71 | 1.003 | 12.75 | 30′ |
| 40′ | .0814 | .9967 | .0816 | 12.25 | 1.003 | 12.29 | 20′ |
| 50′ | .0843 | .9964 | .0846 | 11.83 | 1.004 | 11.87 | 10′ |
| 5° 00′ | .0872 | .9962 | .0875 | 11.43 | 1.004 | 11.47 | 85° 00′ |
| 10′ | .0901 | .9959 | .0904 | 11.06 | 1.004 | 11.10 | 50′ |
| 20′ | .0929 | .9957 | .0934 | 10.71 | 1.004 | 10.76 | 40′ |
| 30′ | .0958 | .9954 | .0963 | 10.39 | 1.005 | 10.43 | 30′ |
| 40′ | .0987 | .9951 | .0992 | 10.08 | 1.005 | 10.13 | 20′ |
| 50′ | .1016 | .9948 | .1022 | 9.788 | 1.005 | 9.839 | 10′ |
| 6° 00′ | .1045 | .9945 | .1051 | 9.514 | 1.006 | 9.567 | 84° 00′ |
| 10′ | .1074 | .9942 | .1080 | 9.255 | 1.006 | 9.309 | 50′ |
| 20′ | .1103 | .9939 | .1110 | 9.010 | 1.006 | 9.065 | 40′ |
| 30′ | .1132 | .9936 | .1139 | 8.777 | 1.006 | 8.834 | 30′ |
| 40′ | .1161 | .9932 | .1169 | 8.556 | 1.007 | 8.614 | 20′ |
| 50′ | .1190 | .9929 | .1198 | 8.345 | 1.007 | 8.405 | 10′ |
| 7° 00′ | .1219 | .9925 | .1228 | 8.144 | 1.008 | 8.206 | 83° 00′ |
| 10′ | .1248 | .9922 | .1257 | 7.953 | 1.008 | 8.016 | 50′ |
| 20′ | .1276 | .9918 | .1287 | 7.770 | 1.008 | 7.834 | 40′ |
| 30′ | .1305 | .9914 | .1317 | 7.596 | 1.009 | 7.661 | 30′ |
| 40′ | .1334 | .9911 | .1346 | 7.429 | 1.009 | 7.496 | 20′ |
| 50′ | .1363 | .9907 | .1376 | 7.269 | 1.009 | 7.337 | 10′ |
| 8° 00′ | .1392 | .9903 | .1405 | 7.115 | 1.010 | 7.185 | 82° 00′ |
| 10′ | .1421 | .9899 | .1435 | 6.968 | 1.010 | 7.040 | 50′ |
| 20′ | .1449 | .9894 | .1465 | 6.827 | 1.011 | 6.900 | 40′ |
| 30′ | .1478 | .9890 | .1495 | 6.691 | 1.011 | 6.765 | 30′ |
| 40′ | .1507 | .9886 | .1524 | 6.561 | 1.012 | 6.636 | 20′ |
| 50′ | .1536 | .9881 | .1554 | 6.435 | 1.012 | 6.512 | 10′ |
| 9° 00′ | .1564 | .9877 | .1584 | 6.314 | 1.012 | 6.392 | 81° 00′ |
| | Cos | Sin | Cot | Tan | Csc | Sec | Angle |

# TABLE 6 VALUES OF TRIGONOMETRIC FUNCTIONS

| Angle | Sin | Cos | Tan | Cot | Sec | Csc | |
|---|---|---|---|---|---|---|---|
| 9° 00′ | .1564 | .9877 | .1584 | 6.314 | 1.012 | 6.392 | 81° 00′ |
| 10′ | .1593 | .9872 | .1614 | 6.197 | 1.013 | 6.277 | 50′ |
| 20′ | .1622 | .9868 | .1644 | 6.084 | 1.013 | 6.166 | 40′ |
| 30′ | .1650 | .9863 | .1673 | 5.976 | 1.014 | 6.059 | 30′ |
| 40′ | .1679 | .9858 | .1703 | 5.871 | 1.014 | 5.955 | 20′ |
| 50′ | .1708 | .9853 | .1733 | 5.769 | 1.015 | 5.855 | 10′ |
| 10° 00′ | .1736 | .9848 | .1763 | 5.671 | 1.015 | 5.759 | 80° 00′ |
| 10′ | .1765 | .9843 | .1793 | 5.576 | 1.016 | 5.665 | 50′ |
| 20′ | .1794 | .9838 | .1823 | 5.485 | 1.016 | 5.575 | 40′ |
| 30′ | .1822 | .9833 | .1853 | 5.396 | 1.017 | 5.487 | 30′ |
| 40′ | .1851 | .9827 | .1883 | 5.309 | 1.018 | 5.403 | 20′ |
| 50′ | .1880 | .9822 | .1914 | 5.226 | 1.018 | 5.320 | 10′ |
| 11° 00′ | .1908 | .9816 | .1944 | 5.145 | 1.019 | 5.241 | 79° 00′ |
| 10′ | .1937 | .9811 | .1974 | 5.066 | 1.019 | 5.164 | 50′ |
| 20′ | .1965 | .9805 | .2004 | 4.989 | 1.020 | 5.089 | 40′ |
| 30′ | .1994 | .9799 | .2035 | 4.915 | 1.020 | 5.016 | 30′ |
| 40′ | .2022 | .9793 | .2065 | 4.843 | 1.021 | 4.945 | 20′ |
| 50′ | .2051 | .9787 | .2095 | 4.773 | 1.022 | 4.876 | 10′ |
| 12° 00′ | .2079 | .9781 | .2126 | 4.705 | 1.022 | 4.810 | 78° 00′ |
| 10′ | .2108 | .9775 | .2156 | 4.638 | 1.023 | 4.745 | 50′ |
| 20′ | .2136 | .9769 | .2186 | 4.574 | 1.024 | 4.682 | 40′ |
| 30′ | .2164 | .9763 | .2217 | 4.511 | 1.024 | 4.620 | 30′ |
| 40′ | .2193 | .9757 | .2247 | 4.449 | 1.025 | 4.560 | 20′ |
| 50′ | .2221 | .9750 | .2278 | 4.390 | 1.026 | 4.502 | 10′ |
| 13° 00′ | .2250 | .9744 | .2309 | 4.331 | 1.026 | 4.445 | 77° 00′ |
| 10′ | .2278 | .9737 | .2339 | 4.275 | 1.027 | 4.390 | 50′ |
| 20′ | .2306 | .9730 | .2370 | 4.219 | 1.028 | 4.336 | 40′ |
| 30′ | .2334 | .9724 | .2401 | 4.165 | 1.028 | 4.284 | 30′ |
| 40′ | .2363 | .9717 | .2432 | 4.113 | 1.029 | 4.232 | 20′ |
| 50′ | .2391 | .9710 | .2462 | 4.061 | 1.030 | 4.182 | 10′ |
| 14° 00′ | .2419 | .9703 | .2493 | 4.011 | 1.031 | 4.134 | 76° 00′ |
| 10′ | .2447 | .9696 | .2524 | 3.962 | 1.031 | 4.086 | 50′ |
| 20′ | .2476 | .9689 | .2555 | 3.914 | 1.032 | 4.039 | 40′ |
| 30′ | .2504 | .9681 | .2586 | 3.867 | 1.033 | 3.994 | 30′ |
| 40′ | .2532 | .9674 | .2617 | 3.821 | 1.034 | 3.950 | 20′ |
| 50′ | .2560 | .9667 | .2648 | 3.776 | 1.034 | 3.906 | 10′ |
| 15° 00′ | .2588 | .9659 | .2679 | 3.732 | 1.035 | 3.864 | 75° 00′ |
| 10′ | .2616 | .9652 | .2711 | 3.689 | 1.036 | 3.822 | 50′ |
| 20′ | .2644 | .9644 | .2742 | 3.647 | 1.037 | 3.782 | 40′ |
| 30′ | .2672 | .9636 | .2773 | 3.606 | 1.038 | 3.742 | 30′ |
| 40′ | .2700 | .9628 | .2805 | 3.566 | 1.039 | 3.703 | 20′ |
| 50′ | .2728 | .9621 | .2836 | 3.526 | 1.039 | 3.665 | 10′ |
| 16° 00′ | .2756 | .9613 | .2867 | 3.487 | 1.040 | 3.628 | 74° 00′ |
| 10′ | .2784 | .9605 | .2899 | 3.450 | 1.041 | 3.592 | 50′ |
| 20′ | .2812 | .9596 | .2931 | 3.412 | 1.042 | 3.556 | 40′ |
| 30′ | .2840 | .9588 | .2962 | 3.376 | 1.043 | 3.521 | 30′ |
| 40′ | .2868 | .9580 | .2994 | 3.340 | 1.044 | 3.487 | 20′ |
| 50′ | .2896 | .9572 | .3026 | 3.305 | 1.045 | 3.453 | 10′ |
| 17° 00′ | .2924 | .9563 | .3057 | 3.271 | 1.046 | 3.420 | 73° 00′ |
| 10′ | .2952 | .9555 | .3089 | 3.237 | 1.047 | 3.388 | 50′ |
| 20′ | .2979 | .9546 | .3121 | 3.204 | 1.048 | 3.356 | 40′ |
| 30′ | .3007 | .9537 | .3153 | 3.172 | 1.049 | 3.326 | 30′ |
| 40′ | .3035 | .9528 | .3185 | 3.140 | 1.049 | 3.295 | 20′ |
| 50′ | .3062 | .9520 | .3217 | 3.108 | 1.050 | 3.265 | 10′ |
| 18° 00′ | .3090 | .9511 | .3249 | 3.078 | 1.051 | 3.236 | 72° 00′ |
| | Cos | Sin | Cot | Tan | Csc | Sec | Angle |

## TABLE 6 VALUES OF TRIGONOMETRIC FUNCTIONS

| Angle | Sin | Cos | Tan | Cot | Sec | Csc | |
|---|---|---|---|---|---|---|---|
| 18° 00' | .3090 | .9511 | .3249 | 3.078 | 1.051 | 3.236 | 72° 00' |
| 10' | .3118 | .9502 | .3281 | 3.047 | 1.052 | 3.207 | 50' |
| 20' | .3145 | .9492 | .3314 | 3.018 | 1.053 | 3.179 | 40' |
| 30' | .3173 | .9483 | .3346 | 2.989 | 1.054 | 3.152 | 30' |
| 40' | .3201 | .9474 | .3378 | 2.960 | 1.056 | 3.124 | 20' |
| 50' | .3228 | .9465 | .3411 | 2.932 | 1.057 | 3.098 | 10' |
| 19° 00' | .3256 | .9455 | .3443 | 2.904 | 1.058 | 3.072 | 71° 00' |
| 10' | .3283 | .9446 | .3476 | 2.877 | 1.059 | 3.046 | 50' |
| 20' | .3311 | .9436 | .3508 | 2.850 | 1.060 | 3.021 | 40' |
| 30' | .3338 | .9426 | .3541 | 2.824 | 1.061 | 2.996 | 30' |
| 40' | .3365 | .9417 | .3574 | 2.798 | 1.062 | 2.971 | 20' |
| 50' | .3393 | .9407 | .3607 | 2.773 | 1.063 | 2.947 | 10' |
| 20° 00' | .3420 | .9397 | .3640 | 2.747 | 1.064 | 2.924 | 70° 00' |
| 10' | .3448 | .9387 | .3673 | 2.723 | 1.065 | 2.901 | 50' |
| 20' | .3475 | .9377 | .3706 | 2.699 | 1.066 | 2.878 | 40' |
| 30' | .3502 | .9367 | .3739 | 2.675 | 1.068 | 2.855 | 30' |
| 40' | .3529 | .9356 | .3772 | 2.651 | 1.069 | 2.833 | 20' |
| 50' | .3557 | .9346 | .3805 | 2.628 | 1.070 | 2.812 | 10' |
| 21° 00' | .3584 | .9336 | .3839 | 2.605 | 1.071 | 2.790 | 69° 00' |
| 10' | .3611 | .9325 | .3872 | 2.583 | 1.072 | 2.769 | 50' |
| 20' | .3638 | .9315 | .3906 | 2.560 | 1.074 | 2.749 | 40' |
| 30' | .3665 | .9304 | .3939 | 2.539 | 1.075 | 2.729 | 30' |
| 40' | .3692 | .9293 | .3973 | 2.517 | 1.076 | 2.709 | 20' |
| 50' | .3719 | .9283 | .4006 | 2.496 | 1.077 | 2.689 | 10' |
| 22° 00' | .3746 | .9272 | .4040 | 2.475 | 1.079 | 2.669 | 68° 00' |
| 10' | .3773 | .9261 | .4074 | 2.455 | 1.080 | 2.650 | 50' |
| 20' | .3800 | .9250 | .4108 | 2.434 | 1.081 | 2.632 | 40' |
| 30' | .3827 | .9239 | .4142 | 2.414 | 1.082 | 2.613 | 30' |
| 40' | .3854 | .9228 | .4176 | 2.394 | 1.084 | 2.595 | 20' |
| 50' | .3881 | .9216 | .4210 | 2.375 | 1.085 | 2.577 | 10' |
| 23° 00' | .3907 | .9205 | .4245 | 2.356 | 1.086 | 2.559 | 67° 00' |
| 10' | .3934 | .9194 | .4279 | 2.337 | 1.088 | 2.542 | 50' |
| 20' | .3961 | .9182 | .4314 | 2.318 | 1.089 | 2.525 | 40' |
| 30' | .3987 | .9171 | .4348 | 2.300 | 1.090 | 2.508 | 30' |
| 40' | .4014 | .9159 | .4383 | 2.282 | 1.092 | 2.491 | 20' |
| 50' | .4041 | .9147 | .4417 | 2.264 | 1.093 | 2.475 | 10' |
| 24° 00' | .4067 | .9135 | .4452 | 2.246 | 1.095 | 2.459 | 66° 00' |
| 10' | .4094 | .9124 | .4487 | 2.229 | 1.096 | 2.443 | 50' |
| 20' | .4120 | .9112 | .4522 | 2.211 | 1.097 | 2.427 | 40' |
| 30' | .4147 | .9100 | .4557 | 2.194 | 1.099 | 2.411 | 30' |
| 40' | .4173 | .9088 | .4592 | 2.177 | 1.100 | 2.396 | 20' |
| 50' | .4200 | .9075 | .4628 | 2.161 | 1.102 | 2.381 | 10' |
| 25° 00' | .4226 | .9063 | .4663 | 2.145 | 1.103 | 2.366 | 65° 00' |
| 10' | .4253 | .9051 | .4699 | 2.128 | 1.105 | 2.352 | 50' |
| 20' | .4279 | .9038 | .4734 | 2.112 | 1.106 | 2.337 | 40' |
| 30' | .4305 | .9026 | .4770 | 2.097 | 1.108 | 2.323 | 30' |
| 40' | .4331 | .9013 | .4806 | 2.081 | 1.109 | 2.309 | 20' |
| 50' | .4358 | .9001 | .4841 | 2.066 | 1.111 | 2.295 | 10' |
| 26° 00' | .4384 | .8988 | .4877 | 2.050 | 1.113 | 2.281 | 64° 00' |
| 10' | .4410 | .8975 | .4913 | 2.035 | 1.114 | 2.268 | 50' |
| 20' | .4436 | .8962 | .4950 | 2.020 | 1.116 | 2.254 | 40' |
| 30' | .4462 | .8949 | .4986 | 2.006 | 1.117 | 2.241 | 30' |
| 40' | .4488 | .8936 | .5022 | 1.991 | 1.119 | 2.228 | 20' |
| 50' | .4514 | .8923 | .5059 | 1.977 | 1.121 | 2.215 | 10' |
| 27° 00' | .4540 | .8910 | .5095 | 1.963 | 1.122 | 2.203 | 63° 00' |
| | Cos | Sin | Cot | Tan | Csc | Sec | Angle |

# TABLE 6 VALUES OF TRIGONOMETRIC FUNCTIONS

| Angle | Sin | Cos | Tan | Cot | Sec | Csc | |
|---|---|---|---|---|---|---|---|
| **27° 00'** | .4540 | .8910 | .5095 | 1.963 | 1.122 | 2.203 | **63° 00'** |
| 10' | .4566 | .8897 | .5132 | 1.949 | 1.124 | 2.190 | 50' |
| 20' | .4592 | .8884 | .5169 | 1.935 | 1.126 | 2.178 | 40' |
| 30' | .4617 | .8870 | .5206 | 1.921 | 1.127 | 2.166 | 30' |
| 40' | .4643 | .8857 | .5243 | 1.907 | 1.129 | 2.154 | 20' |
| 50' | .4669 | .8843 | .5280 | 1.894 | 1.131 | 2.142 | 10' |
| **28° 00'** | .4695 | .8829 | .5317 | 1.881 | 1.133 | 2.130 | **62° 00'** |
| 10' | .4720 | .8816 | .5354 | 1.868 | 1.134 | 2.118 | 50' |
| 20' | .4746 | .8802 | .5392 | 1.855 | 1.136 | 2.107 | 40' |
| 30' | .4772 | .8788 | .5430 | 1.842 | 1.138 | 2.096 | 30' |
| 40' | .4797 | .8774 | .5467 | 1.829 | 1.140 | 2.085 | 20' |
| 50' | .4823 | .8760 | .5505 | 1.816 | 1.142 | 2.074 | 10' |
| **29° 00'** | .4848 | .8746 | .5543 | 1.804 | 1.143 | 2.063 | **61° 00'** |
| 10' | .4874 | .8732 | .5581 | 1.792 | 1.145 | 2.052 | 50' |
| 20' | .4899 | .8718 | .5619 | 1.780 | 1.147 | 2.041 | 40' |
| 30' | .4924 | .8704 | .5658 | 1.767 | 1.149 | 2.031 | 30' |
| 40' | .4950 | .8689 | .5696 | 1.756 | 1.151 | 2.020 | 20' |
| 50' | .4975 | .8675 | .5735 | 1.744 | 1.153 | 2.010 | 10' |
| **30° 00'** | .5000 | .8660 | .5774 | 1.732 | 1.155 | 2.000 | **60° 00'** |
| 10' | .5025 | .8646 | .5812 | 1.720 | 1.157 | 1.990 | 50' |
| 20' | .5050 | .8631 | .5851 | 1.709 | 1.159 | 1.980 | 40' |
| 30' | .5075 | .8616 | .5890 | 1.698 | 1.161 | 1.970 | 30' |
| 40' | .5100 | .8601 | .5930 | 1.686 | 1.163 | 1.961 | 20' |
| 50' | .5125 | .8587 | .5969 | 1.675 | 1.165 | 1.951 | 10' |
| **31° 00'** | .5150 | .8572 | .6009 | 1.664 | 1.167 | 1.942 | **59° 00'** |
| 10' | .5175 | .8557 | .6048 | 1.653 | 1.169 | 1.932 | 50' |
| 20' | .5200 | .8542 | .6088 | 1.643 | 1.171 | 1.923 | 40' |
| 30' | .5225 | .8526 | .6128 | 1.632 | 1.173 | 1.914 | 30' |
| 40' | .5250 | .8511 | .6168 | 1.621 | 1.175 | 1.905 | 20' |
| 50' | .5275 | .8496 | .6208 | 1.611 | 1.177 | 1.896 | 10' |
| **32° 00'** | .5299 | .8480 | .6249 | 1.600 | 1.179 | 1.887 | **58° 00'** |
| 10' | .5324 | .8465 | .6289 | 1.590 | 1.181 | 1.878 | 50' |
| 20' | .5348 | .8450 | .6330 | 1.580 | 1.184 | 1.870 | 40' |
| 30' | .5373 | .8434 | .6371 | 1.570 | 1.186 | 1.861 | 30' |
| 40' | .5398 | .8418 | .6412 | 1.560 | 1.188 | 1.853 | 20' |
| 50' | .5422 | .8403 | .6453 | 1.550 | 1.190 | 1.844 | 10' |
| **33° 00'** | .5446 | .8387 | .6494 | 1.540 | 1.192 | 1.836 | **57° 00'** |
| 10' | .5471 | .8371 | .6536 | 1.530 | 1.195 | 1.828 | 50' |
| 20' | .5495 | .8355 | .6577 | 1.520 | 1.197 | 1.820 | 40' |
| 30' | .5519 | .8339 | .6619 | 1.511 | 1.199 | 1.812 | 30' |
| 40' | .5544 | .8323 | .6661 | 1.501 | 1.202 | 1.804 | 20' |
| 50' | .5568 | .8307 | .6703 | 1.492 | 1.204 | 1.796 | 10' |
| **34° 00'** | .5592 | .8290 | .6745 | 1.483 | 1.206 | 1.788 | **56° 00'** |
| 10' | .5616 | .8274 | .6787 | 1.473 | 1.209 | 1.781 | 50' |
| 20' | .5640 | .8258 | .6830 | 1.464 | 1.211 | 1.773 | 40' |
| 30' | .5664 | .8241 | .6873 | 1.455 | 1.213 | 1.766 | 30' |
| 40' | .5688 | .8225 | .6916 | 1.446 | 1.216 | 1.758 | 20' |
| 50' | .5712 | .8208 | .6959 | 1.437 | 1.218 | 1.751 | 10' |
| **35° 00'** | .5736 | .8192 | .7002 | 1.428 | 1.221 | 1.743 | **55° 00'** |
| 10' | .5760 | .8175 | .7046 | 1.419 | 1.223 | 1.736 | 50' |
| 20' | .5783 | .8158 | .7089 | 1.411 | 1.226 | 1.729 | 40' |
| 30' | .5807 | .8141 | .7133 | 1.402 | 1.228 | 1.722 | 30' |
| 40' | .5831 | .8124 | .7177 | 1.393 | 1.231 | 1.715 | 20' |
| 50' | .5854 | .8107 | .7221 | 1.385 | 1.233 | 1.708 | 10' |
| **36° 00'** | .5878 | .8090 | .7265 | 1.376 | 1.236 | 1.701 | **54° 00'** |
| | Cos | Sin | Cot | Tan | Csc | Sec | Angle |

# TABLE 6 VALUES OF TRIGONOMETRIC FUNCTIONS

| Angle | Sin | Cos | Tan | Cot | Sec | Csc | |
|---|---|---|---|---|---|---|---|
| 36° 00' | .5878 | .8090 | .7265 | 1.376 | 1.236 | 1.701 | 54° 00' |
| 10' | .5901 | .8073 | .7310 | 1.368 | 1.239 | 1.695 | 50' |
| 20' | .5925 | .8056 | .7355 | 1.360 | 1.241 | 1.688 | 40' |
| 30' | .5948 | .8039 | .7400 | 1.351 | 1.244 | 1.681 | 30' |
| 40' | .5972 | .8021 | .7445 | 1.343 | 1.247 | 1.675 | 20' |
| 50' | .5995 | .8004 | .7490 | 1.335 | 1.249 | 1.668 | 10' |
| 37° 00' | .6018 | .7986 | .7536 | 1.327 | 1.252 | 1.662 | 53° 00' |
| 10' | .6041 | .7969 | .7581 | 1.319 | 1.255 | 1.655 | 50' |
| 20' | .6065 | .7951 | .7627 | 1.311 | 1.258 | 1.649 | 40' |
| 30' | .6088 | .7934 | .7673 | 1.303 | 1.260 | 1.643 | 30' |
| 40' | .6111 | .7916 | .7720 | 1.295 | 1.263 | 1.636 | 20' |
| 50' | .6134 | .7898 | .7766 | 1.288 | 1.266 | 1.630 | 10' |
| 38° 00' | .6157 | .7880 | .7813 | 1.280 | 1.269 | 1.624 | 52° 00' |
| 10' | .6180 | .7862 | .7860 | 1.272 | 1.272 | 1.618 | 50' |
| 20' | .6202 | .7844 | .7907 | 1.265 | 1.275 | 1.612 | 40' |
| 30' | .6225 | .7826 | .7954 | 1.257 | 1.278 | 1.606 | 30' |
| 40' | .6248 | .7808 | .8002 | 1.250 | 1.281 | 1.601 | 20' |
| 50' | .6271 | .7790 | .8050 | 1.242 | 1.284 | 1.595 | 10' |
| 39° 00' | .6293 | .7771 | .8098 | 1.235 | 1.287 | 1.589 | 51° 00' |
| 10' | .6316 | .7753 | .8146 | 1.228 | 1.290 | 1.583 | 50' |
| 20' | .6338 | .7735 | .8195 | 1.220 | 1.293 | 1.578 | 40' |
| 30' | .6361 | .7716 | .8243 | 1.213 | 1.296 | 1.572 | 30' |
| 40' | .6383 | .7698 | .8292 | 1.206 | 1.299 | 1.567 | 20' |
| 50' | .6406 | .7679 | .8342 | 1.199 | 1.302 | 1.561 | 10' |
| 40° 00' | .6428 | .7660 | .8391 | 1.192 | 1.305 | 1.556 | 50° 00' |
| 10' | .6450 | .7642 | .8441 | 1.185 | 1.309 | 1.550 | 50' |
| 20' | .6472 | .7623 | .8491 | 1.178 | 1.312 | 1.545 | 40' |
| 30' | .6494 | .7604 | .8541 | 1.171 | 1.315 | 1.540 | 30' |
| 40' | .6517 | .7585 | .8591 | 1.164 | 1.318 | 1.535 | 20' |
| 50' | .6539 | .7566 | .8642 | 1.157 | 1.322 | 1.529 | 10' |
| 41° 00' | .6561 | .7547 | .8693 | 1.150 | 1.325 | 1.524 | 49° 00' |
| 10' | .6583 | .7528 | .8744 | 1.144 | 1.328 | 1.519 | 50' |
| 20' | .6604 | .7509 | .8796 | 1.137 | 1.332 | 1.514 | 40' |
| 30' | .6626 | .7490 | .8847 | 1.130 | 1.335 | 1.509 | 30' |
| 40' | .6648 | .7470 | .8899 | 1.124 | 1.339 | 1.504 | 20' |
| 50' | .6670 | .7451 | .8952 | 1.117 | 1.342 | 1.499 | 10' |
| 42° 00' | .6691 | .7431 | .9004 | 1.111 | 1.346 | 1.494 | 48° 00' |
| 10' | .6713 | .7412 | .9057 | 1.104 | 1.349 | 1.490 | 50' |
| 20' | .6734 | .7392 | .9110 | 1.098 | 1.353 | 1.485 | 40' |
| 30' | .6756 | .7373 | .9163 | 1.091 | 1.356 | 1.480 | 30' |
| 40' | .6777 | .7353 | .9217 | 1.085 | 1.360 | 1.476 | 20' |
| 50' | .6799 | .7333 | .9271 | 1.079 | 1.364 | 1.471 | 10' |
| 43° 00' | .6820 | .7314 | .9325 | 1.072 | 1.367 | 1.466 | 47° 00' |
| 10' | .6841 | .7294 | .9380 | 1.066 | 1.371 | 1.462 | 50' |
| 20' | .6862 | .7274 | .9435 | 1.060 | 1.375 | 1.457 | 40' |
| 30' | .6884 | .7254 | .9490 | 1.054 | 1.379 | 1.453 | 30' |
| 40' | .6905 | .7234 | .9545 | 1.048 | 1.382 | 1.448 | 20' |
| 50' | .6926 | .7214 | .9601 | 1.042 | 1.386 | 1.444 | 10' |
| 44° 00' | .6947 | .7193 | .9657 | 1.036 | 1.390 | 1.440 | 46° 00' |
| 10' | .6967 | .7173 | .9713 | 1.030 | 1.394 | 1.435 | 50' |
| 20' | .6988 | .7153 | .9770 | 1.024 | 1.398 | 1.431 | 40' |
| 30' | .7009 | .7133 | .9827 | 1.018 | 1.402 | 1.427 | 30' |
| 40' | .7030 | .7112 | .9884 | 1.012 | 1.406 | 1.423 | 20' |
| 50' | .7050 | .7092 | .9942 | 1.006 | 1.410 | 1.418 | 10' |
| 45° 00' | .7071 | .7071 | 1.000 | 1.000 | 1.414 | 1.414 | 45° 00' |
| | Cos | Sin | Cot | Tan | Csc | Sec | Angle |

# TABLE 7
## FOUR-PLACE LOGARITHMS OF VALUES OF TRIGONOMETRIC FUNCTIONS*

| Angle | L Sin | d | L Tan | cd | L Cot | d | L Cos | Angle |
|---|---|---|---|---|---|---|---|---|
| 0° 0' | —— | | —— | | —— | 0 | 10.0000 | 90° 0' |
| 10' | 7.4637 | 3011 | 7.4637 | 3011 | 12.5363 | 0 | 10.0000 | 50' |
| 20' | 7.7648 | 1760 | 7.7648 | 1761 | 12.2352 | 0 | 10.0000 | 40' |
| 30' | 7.9408 | 1250 | 7.9409 | 1249 | 12.0591 | 0 | 10.0000 | 30' |
| 40' | 8.0658 | 969 | 8.0658 | 969 | 11.9342 | 0 | 10.0000 | 20' |
| 50' | 8.1627 | 792 | 8.1627 | 792 | 11.8373 | 0 | 10.0000 | 10' |
| 1° 0' | 8.2419 | 669 | 8.2419 | 670 | 11.7581 | 1 | 9.9999 | 89° 0' |
| 10' | 8.3088 | 580 | 8.3089 | 580 | 11.6911 | 0 | 9.9999 | 50' |
| 20' | 8.3668 | 511 | 8.3669 | 512 | 11.6331 | 0 | 9.9999 | 40' |
| 30' | 8.4179 | 458 | 8.4181 | 457 | 11.5819 | 0 | 9.9999 | 30' |
| 40' | 8.4637 | 413 | 8.4638 | 415 | 11.5362 | 1 | 9.9998 | 20' |
| 50' | 8.5050 | 378 | 8.5053 | 378 | 11.4947 | 0 | 9.9998 | 10' |
| 2° 0' | 8.5428 | 348 | 8.5431 | 348 | 11.4569 | 1 | 9.9997 | 88° 0' |
| 10' | 8.5776 | 321 | 8.5779 | 322 | 11.4221 | 0 | 9.9997 | 50' |
| 20' | 8.6097 | 300 | 8.6101 | 300 | 11.3899 | 1 | 9.9996 | 40' |
| 30' | 8.6397 | 280 | 8.6401 | 281 | 11.3599 | 0 | 9.9996 | 30' |
| 40' | 8.6677 | 263 | 8.6682 | 263 | 11.3318 | 1 | 9.9995 | 20' |
| 50' | 8.6940 | 248 | 8.6945 | 249 | 11.3055 | 0 | 9.9995 | 10' |
| 3° 0' | 8.7188 | 235 | 8.7194 | 235 | 11.2806 | 1 | 9.9994 | 87° 0' |
| 10' | 8.7423 | 222 | 8.7429 | 223 | 11.2571 | 1 | 9.9993 | 50' |
| 20' | 8.7645 | 212 | 8.7652 | 213 | 11.2348 | 0 | 9.9993 | 40' |
| 30' | 8.7857 | 202 | 8.7865 | 202 | 11.2135 | 1 | 9.9992 | 30' |
| 40' | 8.8059 | 192 | 8.8067 | 194 | 11.1933 | 1 | 9.9991 | 20' |
| 50' | 8.8251 | 185 | 8.8261 | 185 | 11.1739 | 1 | 9.9990 | 10' |
| 4° 0' | 8.8436 | 177 | 8.8446 | 178 | 11.1554 | 1 | 9.9989 | 86° 0' |
| 10' | 8.8613 | 170 | 8.8624 | 171 | 11.1376 | 0 | 9.9989 | 50' |
| 20' | 8.8783 | 163 | 8.8795 | 165 | 11.1205 | 1 | 9.9988 | 40' |
| 30' | 8.8946 | 158 | 8.8960 | 158 | 11.1040 | 1 | 9.9987 | 30' |
| 40' | 8.9104 | 152 | 8.9118 | 154 | 11.0882 | 1 | 9.9986 | 20' |
| 50' | 8.9256 | 147 | 8.9272 | 148 | 11.0728 | 1 | 9.9985 | 10' |
| 5° 0' | 8.9403 | 142 | 8.9420 | 143 | 11.0580 | 2 | 9.9983 | 85° 0' |
| 10' | 8.9545 | 137 | 8.9563 | 138 | 11.0437 | 1 | 9.9982 | 50' |
| 20' | 8.9682 | 134 | 8.9701 | 135 | 11.0299 | 1 | 9.9981 | 40' |
| 30' | 8.9816 | 129 | 8.9836 | 130 | 11.0164 | 1 | 9.9980 | 30' |
| 40' | 8.9945 | 125 | 8.9966 | 127 | 11.0034 | 1 | 9.9979 | 20' |
| 50' | 9.0070 | 122 | 9.0093 | 123 | 10.9907 | 2 | 9.9977 | 10' |
| 6° 0' | 9.0192 | 119 | 9.0216 | 120 | 10.9784 | 1 | 9.9976 | 84° 0' |
| 10' | 9.0311 | 115 | 9.0336 | 117 | 10.9664 | 1 | 9.9975 | 50' |
| 20' | 9.0426 | 113 | 9.0453 | 114 | 10.9547 | 2 | 9.9973 | 40' |
| 30' | 9.0539 | 109 | 9.0567 | 111 | 10.9433 | 1 | 9.9972 | 30' |
| 40' | 9.0648 | 107 | 9.0678 | 108 | 10.9322 | 1 | 9.9971 | 20' |
| 50' | 9.0755 | 104 | 9.0786 | 105 | 10.9214 | 2 | 9.9969 | 10' |
| 7° 0' | 9.0859 | 102 | 9.0891 | 104 | 10.9109 | 1 | 9.9968 | 83° 0' |
| 10' | 9.0961 | 99 | 9.0995 | 101 | 10.9005 | 2 | 9.9966 | 50' |
| 20' | 9.1060 | 97 | 9.1096 | 98 | 10.8904 | 2 | 9.9964 | 40' |
| 30' | 9.1157 | 95 | 9.1194 | 97 | 10.8806 | 1 | 9.9963 | 30' |
| 40' | 9.1252 | 93 | 9.1291 | 94 | 10.8709 | 2 | 9.9961 | 20' |
| 50' | 9.1345 | 91 | 9.1385 | 93 | 10.8615 | 2 | 9.9959 | 10' |
| 8° 0' | 9.1436 | 89 | 9.1478 | 91 | 10.8522 | 1 | 9.9958 | 82° 0' |
| 10' | 9.1525 | 87 | 9.1569 | 89 | 10.8431 | 2 | 9.9956 | 50' |
| 20' | 9.1612 | 85 | 9.1658 | 87 | 10.8342 | 2 | 9.9954 | 40' |
| 30' | 9.1697 | 84 | 9.1745 | 86 | 10.8255 | 2 | 9.9952 | 30' |
| 40' | 9.1781 | 82 | 9.1831 | 84 | 10.8169 | 2 | 9.9950 | 20' |
| 50' | 9.1863 | 80 | 9.1915 | 82 | 10.8085 | 2 | 9.9948 | 10' |
| 9° 0' | 9.1943 | | 9.1997 | | 10.8003 | 2 | 9.9946 | 81° 0' |
| | L Cos | d | L Cot | cd | L Tan | d | L Sin | Angle |

* These tables give the logarithms increased by 10. Hence in each case 10 should be subtracted.

# TABLE 7

## FOUR-PLACE LOGARITHMS OF VALUES OF TRIGONOMETRIC FUNCTIONS

| Angle | L Sin | d | L Tan | cd | L Cot | d | L Cos | |
|---|---|---|---|---|---|---|---|---|
| 9° 0′ | 9.1943 | | 9.1997 | | 10.8003 | | 9.9946 | 81° 0′ |
| 10′ | 9.2022 | 79 | 9.2078 | 81 | 10.7922 | 2 | 9.9944 | 50′ |
| 20′ | 9.2100 | 78 | 9.2158 | 80 | 10.7842 | 2 | 9.9942 | 40′ |
| 30′ | 9.2176 | 76 | 9.2236 | 78 | 10.7764 | 2 | 9.9940 | 30′ |
| 40′ | 9.2251 | 75 | 9.2313 | 77 | 10.7687 | 2 | 9.9938 | 20′ |
| 50′ | 9.2324 | 73 | 9.2389 | 76 | 10.7611 | 2 | 9.9936 | 10′ |
| 10° 0′ | 9.2397 | 73 | 9.2463 | 74 | 10.7537 | 2 | 9.9934 | 80° 0′ |
| 10′ | 9.2468 | 71 | 9.2536 | 73 | 10.7464 | 3 | 9.9931 | 50′ |
| 20′ | 9.2538 | 70 | 9.2609 | 73 | 10.7391 | 2 | 9.9929 | 40′ |
| 30′ | 9.2606 | 68 | 9.2680 | 71 | 10.7320 | 2 | 9.9927 | 30′ |
| 40′ | 9.2674 | 68 | 9.2750 | 70 | 10.7250 | 3 | 9.9924 | 20′ |
| 50′ | 9.2740 | 66 | 9.2819 | 69 | 10.7181 | 2 | 9.9922 | 10′ |
| 11° 0′ | 9.2806 | 66 | 9.2887 | 68 | 10.7113 | 3 | 9.9919 | 79° 0′ |
| 10′ | 9.2870 | 64 | 9.2953 | 66 | 10.7047 | 2 | 9.9917 | 50′ |
| 20′ | 9.2934 | 64 | 9.3020 | 67 | 10.6980 | 3 | 9.9914 | 40′ |
| 30′ | 9.2997 | 63 | 9.3085 | 65 | 10.6915 | 2 | 9.9912 | 30′ |
| 40′ | 9.3058 | 61 | 9.3149 | 64 | 10.6851 | 3 | 9.9909 | 20′ |
| 50′ | 9.3119 | 61 | 9.3212 | 63 | 10.6788 | 2 | 9.9907 | 10′ |
| 12° 0′ | 9.3179 | 60 | 9.3275 | 63 | 10.6725 | 3 | 9.9904 | 78° 0′ |
| 10′ | 9.3238 | 59 | 9.3336 | 61 | 10.6664 | 3 | 9.9901 | 50′ |
| 20′ | 9.3296 | 58 | 9.3397 | 61 | 10.6603 | 2 | 9.9899 | 40′ |
| 30′ | 9.3353 | 57 | 9.3458 | 61 | 10.6542 | 3 | 9.9896 | 30′ |
| 40′ | 9.3410 | 57 | 9.3517 | 59 | 10.6483 | 3 | 9.9893 | 20′ |
| 50′ | 9.3466 | 56 | 9.3576 | 59 | 10.6424 | 3 | 9.9890 | 10′ |
| 13° 0′ | 9.3521 | 55 | 9.3634 | 58 | 10.6366 | 3 | 9.9887 | 77° 0′ |
| 10′ | 9.3575 | 54 | 9.3691 | 57 | 10.6309 | 3 | 9.9884 | 50′ |
| 20′ | 9.3629 | 54 | 9.3748 | 57 | 10.6252 | 3 | 9.9881 | 40′ |
| 30′ | 9.3682 | 53 | 9.3804 | 56 | 10.6196 | 3 | 9.9878 | 30′ |
| 40′ | 9.3734 | 52 | 9.3859 | 55 | 10.6141 | 3 | 9.9875 | 20′ |
| 50′ | 9.3786 | 52 | 9.3914 | 55 | 10.6086 | 3 | 9.9872 | 10′ |
| 14° 0′ | 9.3837 | 51 | 9.3968 | 54 | 10.6032 | 3 | 9.9869 | 76° 0′ |
| 10′ | 9.3887 | 50 | 9.4021 | 53 | 10.5979 | 3 | 9.9866 | 50′ |
| 20′ | 9.3937 | 50 | 9.4074 | 53 | 10.5926 | 3 | 9.9863 | 40′ |
| 30′ | 9.3986 | 49 | 9.4127 | 53 | 10.5873 | 4 | 9.9859 | 30′ |
| 40′ | 9.4035 | 49 | 9.4178 | 51 | 10.5822 | 3 | 9.9856 | 20′ |
| 50′ | 9.4083 | 48 | 9.4230 | 52 | 10.5770 | 3 | 9.9853 | 10′ |
| 15° 0′ | 9.4130 | 47 | 9.4281 | 51 | 10.5719 | 4 | 9.9849 | 75° 0′ |
| 10′ | 9.4177 | 47 | 9.4331 | 50 | 10.5669 | 3 | 9.9846 | 50′ |
| 20′ | 9.4223 | 46 | 9.4381 | 50 | 10.5619 | 3 | 9.9843 | 40′ |
| 30′ | 9.4269 | 46 | 9.4430 | 49 | 10.5570 | 4 | 9.9839 | 30′ |
| 40′ | 9.4314 | 45 | 9.4479 | 49 | 10.5521 | 3 | 9.9836 | 20′ |
| 50′ | 9.4359 | 45 | 9.4527 | 48 | 10.5473 | 4 | 9.9832 | 10′ |
| 16° 0′ | 9.4403 | 44 | 9.4575 | 48 | 10.5425 | 4 | 9.9828 | 74° 0′ |
| 10′ | 9.4447 | 44 | 9.4622 | 47 | 10.5378 | 3 | 9.9825 | 50′ |
| 20′ | 9.4491 | 44 | 9.4669 | 47 | 10.5331 | 4 | 9.9821 | 40′ |
| 30′ | 9.4533 | 42 | 9.4716 | 47 | 10.5284 | 4 | 9.9817 | 30′ |
| 40′ | 9.4576 | 43 | 9.4762 | 46 | 10.5238 | 3 | 9.9814 | 20′ |
| 50′ | 9.4618 | 42 | 9.4808 | 46 | 10.5192 | 4 | 9.9810 | 10′ |
| 17° 0′ | 9.4659 | 41 | 9.4853 | 45 | 10.5147 | 4 | 9.9806 | 73° 0′ |
| 10′ | 9.4700 | 41 | 9.4898 | 45 | 10.5102 | 4 | 9.9802 | 50′ |
| 20′ | 9.4741 | 41 | 9.4943 | 45 | 10.5057 | 4 | 9.9798 | 40′ |
| 30′ | 9.4781 | 40 | 9.4987 | 44 | 10.5013 | 4 | 9.9794 | 30′ |
| 40′ | 9.4821 | 40 | 9.5031 | 44 | 10.4969 | 4 | 9.9790 | 20′ |
| 50′ | 9.4861 | 40 | 9.5075 | 44 | 10.4925 | 4 | 9.9786 | 10′ |
| 18° 0′ | 9.4900 | 39 | 9.5118 | 43 | 10.4882 | 4 | 9.9782 | 72° 0′ |
| | L Cos | d | L Cot | cd | L Tan | d | L Sin | Angle |

# TABLE 7

## FOUR-PLACE LOGARITHMS OF VALUES OF TRIGONOMETRIC FUNCTIONS

| Angle | L Sin | d | L Tan | cd | L Cot | d | L Cos | |
|---|---|---|---|---|---|---|---|---|
| 18° 0' | 9.4900 | | 9.5118 | | 10.4882 | | 9.9782 | 72° 0' |
| 10' | 9.4939 | 39 | 9.5161 | 43 | 10.4839 | 4 | 9.9778 | 50' |
| 20' | 9.4977 | 38 | 9.5203 | 42 | 10.4797 | 4 | 9.9774 | 40' |
| 30' | 9.5015 | 38 | 9.5245 | 42 | 10.4755 | 4 | 9.9770 | 30' |
| 40' | 9.5052 | 37 | 9.5287 | 42 | 10.4713 | 5 | 9.9765 | 20' |
| 50' | 9.5090 | 38 | 9.5329 | 42 | 10.4671 | 4 | 9.9761 | 10' |
| | | 36 | | 41 | | 4 | | |
| 19° 0' | 9.5126 | | 9.5370 | | 10.4630 | | 9.9757 | 71° 0' |
| 10' | 9.5163 | 37 | 9.5411 | 41 | 10.4589 | 5 | 9.9752 | 50' |
| 20' | 9.5199 | 36 | 9.5451 | 40 | 10.4549 | 4 | 9.9748 | 40' |
| 30' | 9.5235 | 36 | 9.5491 | 40 | 10.4509 | 5 | 9.9743 | 30' |
| 40' | 9.5270 | 35 | 9.5531 | 40 | 10.4469 | 4 | 9.9739 | 20' |
| 50' | 9.5306 | 36 | 9.5571 | 40 | 10.4429 | 5 | 9.9734 | 10' |
| | | 35 | | 40 | | 4 | | |
| 20° 0' | 9.5341 | | 9.5611 | | 10.4389 | | 9.9730 | 70° 0' |
| 10' | 9.5375 | 34 | 9.5650 | 39 | 10.4350 | 5 | 9.9725 | 50' |
| 20' | 9.5409 | 34 | 9.5689 | 39 | 10.4311 | 4 | 9.9721 | 40' |
| 30' | 9.5443 | 34 | 9.5727 | 38 | 10.4273 | 5 | 9.9716 | 30' |
| 40' | 9.5477 | 34 | 9.5766 | 39 | 10.4234 | 5 | 9.9711 | 20' |
| 50' | 9.5510 | 33 | 9.5804 | 38 | 10.4196 | 5 | 9.9706 | 10' |
| | | 33 | | 38 | | 4 | | |
| 21° 0' | 9.5543 | | 9.5842 | | 10.4158 | | 9.9702 | 69° 0' |
| 10' | 9.5576 | 33 | 9.5879 | 37 | 10.4121 | 5 | 9.9697 | 50' |
| 20' | 9.5609 | 33 | 9.5917 | 38 | 10.4083 | 5 | 9.9692 | 40' |
| 30' | 9.5641 | 32 | 9.5954 | 37 | 10.4046 | 5 | 9.9687 | 30' |
| 40' | 9.5673 | 32 | 9.5991 | 37 | 10.4009 | 5 | 9.9682 | 20' |
| 50' | 9.5704 | 31 | 9.6028 | 37 | 10.3972 | 5 | 9.9677 | 10' |
| | | 32 | | 36 | | 5 | | |
| 22° 0' | 9.5736 | | 9.6064 | | 10.3936 | | 9.9672 | 68° 0' |
| 10' | 9.5767 | 31 | 9.6100 | 36 | 10.3900 | 5 | 9.9667 | 50' |
| 20' | 9.5798 | 31 | 9.6136 | 36 | 10.3864 | 6 | 9.9661 | 40' |
| 30' | 9.5828 | 30 | 9.6172 | 36 | 10.3828 | 5 | 9.9656 | 30' |
| 40' | 9.5859 | 31 | 9.6208 | 36 | 10.3792 | 5 | 9.9651 | 20' |
| 50' | 9.5889 | 30 | 9.6243 | 35 | 10.3757 | 5 | 9.9646 | 10' |
| | | 30 | | 36 | | 6 | | |
| 23° 0' | 9.5919 | | 9.6279 | | 10.3721 | | 9.9640 | 67° 0' |
| 10' | 9.5948 | 29 | 9.6314 | 35 | 10.3686 | 5 | 9.9635 | 50' |
| 20' | 9.5978 | 30 | 9.6348 | 34 | 10.3652 | 6 | 9.9629 | 40' |
| 30' | 9.6007 | 29 | 9.6383 | 35 | 10.3617 | 5 | 9.9624 | 30' |
| 40' | 9.6036 | 29 | 9.6417 | 34 | 10.3583 | 6 | 9.9618 | 20' |
| 50' | 9.6065 | 29 | 9.6452 | 35 | 10.3548 | 5 | 9.9613 | 10' |
| | | 28 | | 34 | | 6 | | |
| 24° 0' | 9.6093 | | 9.6486 | | 10.3514 | | 9.9607 | 66° 0' |
| 10' | 9.6121 | 28 | 9.6520 | 34 | 10.3480 | 5 | 9.9602 | 50' |
| 20' | 9.6149 | 28 | 9.6553 | 33 | 10.3447 | 6 | 9.9596 | 40' |
| 30' | 9.6177 | 28 | 9.6587 | 34 | 10.3413 | 6 | 9.9590 | 30' |
| 40' | 9.6205 | 28 | 9.6620 | 33 | 10.3380 | 6 | 9.9584 | 20' |
| 50' | 9.6232 | 27 | 9.6654 | 34 | 10.3346 | 5 | 9.9579 | 10' |
| | | 27 | | 33 | | 6 | | |
| 25° 0' | 9.6259 | | 9.6687 | | 10.3313 | | 9.9573 | 65° 0' |
| 10' | 9.6286 | 27 | 9.6720 | 33 | 10.3280 | 6 | 9.9567 | 50' |
| 20' | 9.6313 | 27 | 9.6752 | 32 | 10.3248 | 6 | 9.9561 | 40' |
| 30' | 9.6340 | 27 | 9.6785 | 33 | 10.3215 | 6 | 9.9555 | 30' |
| 40' | 9.6366 | 26 | 9.6817 | 32 | 10.3183 | 6 | 9.9549 | 20' |
| 50' | 9.6392 | 26 | 9.6850 | 33 | 10.3150 | 6 | 9.9543 | 10' |
| | | 26 | | 32 | | 6 | | |
| 26° 0' | 9.6418 | | 9.6882 | | 10.3118 | | 9.9537 | 64° 0' |
| 10' | 9.6444 | 26 | 9.6914 | 32 | 10.3086 | 7 | 9.9530 | 50' |
| 20' | 9.6470 | 26 | 9.6946 | 32 | 10.3054 | 6 | 9.9524 | 40' |
| 30' | 9.6495 | 25 | 9.6977 | 31 | 10.3023 | 6 | 9.9518 | 30' |
| 40' | 9.6521 | 26 | 9.7009 | 32 | 10.2991 | 6 | 9.9512 | 20' |
| 50' | 9.6546 | 25 | 9.7040 | 31 | 10.2960 | 7 | 9.9505 | 10' |
| | | 24 | | 32 | | 6 | | |
| 27° 0' | 9.6570 | | 9.7072 | | 10.2928 | | 9.9499 | 63° 0' |
| | L Cos | d | L Cot | cd | L Tan | d | L Sin | Angle |

# TABLE 7

## FOUR-PLACE LOGARITHMS OF VALUES OF TRIGONOMETRIC FUNCTIONS

| Angle | L Sin | d | L Tan | cd | L Cot | d | L Cos | |
|---|---|---|---|---|---|---|---|---|
| 27° 0′ | 9.6570 | | 9.7072 | | 10.2928 | | 9.9499 | 63° 0′ |
| 10′ | 9.6595 | 25 | 9.7103 | 31 | 10.2897 | 7 | 9.9492 | 50′ |
| 20′ | 9.6620 | 25 | 9.7134 | 31 | 10.2866 | 6 | 9.9486 | 40′ |
| 30′ | 9.6644 | 24 | 9.7165 | 31 | 10.2835 | 7 | 9.9479 | 30′ |
| 40′ | 9.6668 | 24 | 9.7196 | 31 | 10.2804 | 6 | 9.9473 | 20′ |
| 50′ | 9.6692 | 24 | 9.7226 | 30 | 10.2774 | 7 | 9.9466 | 10′ |
| 28° 0′ | 9.6716 | 24 | 9.7257 | 31 | 10.2743 | 7 | 9.9459 | 62° 0′ |
| 10′ | 9.6740 | 24 | 9.7287 | 30 | 10.2713 | 6 | 9.9453 | 50′ |
| 20′ | 9.6763 | 23 | 9.7317 | 30 | 10.2683 | 7 | 9.9446 | 40′ |
| 30′ | 9.6787 | 24 | 9.7348 | 31 | 10.2652 | 7 | 9.9439 | 30′ |
| 40′ | 9.6810 | 23 | 9.7378 | 30 | 10.2622 | 7 | 9.9432 | 20′ |
| 50′ | 9.6833 | 23 | 9.7408 | 30 | 10.2592 | 7 | 9.9425 | 10′ |
| 29° 0′ | 9.6856 | 23 | 9.7438 | 30 | 10.2562 | 7 | 9.9418 | 61° 0′ |
| 10′ | 9.6878 | 22 | 9.7467 | 29 | 10.2533 | 7 | 9.9411 | 50′ |
| 20′ | 9.6901 | 23 | 9.7497 | 30 | 10.2503 | 7 | 9.9404 | 40′ |
| 30′ | 9.6923 | 22 | 9.7526 | 29 | 10.2474 | 7 | 9.9397 | 30′ |
| 40′ | 9.6946 | 23 | 9.7556 | 30 | 10.2444 | 7 | 9.9390 | 20′ |
| 50′ | 9.6968 | 22 | 9.7585 | 29 | 10.2415 | 7 | 9.9383 | 10′ |
| 30° 0′ | 9.6990 | 22 | 9.7614 | 29 | 10.2386 | 8 | 9.9375 | 60° 0′ |
| 10′ | 9.7012 | 22 | 9.7644 | 30 | 10.2356 | 7 | 9.9368 | 50′ |
| 20′ | 9.7033 | 21 | 9.7673 | 29 | 10.2327 | 8 | 9.9361 | 40′ |
| 30′ | 9.7055 | 22 | 9.7701 | 28 | 10.2299 | 7 | 9.9353 | 30′ |
| 40′ | 9.7076 | 21 | 9.7730 | 29 | 10.2270 | 8 | 9.9346 | 20′ |
| 50′ | 9.7097 | 21 | 9.7759 | 29 | 10.2241 | 7 | 9.9338 | 10′ |
| 31° 0′ | 9.7118 | 21 | 9.7788 | 29 | 10.2212 | 8 | 9.9331 | 59° 0′ |
| 10′ | 9.7139 | 21 | 9.7816 | 28 | 10.2184 | 8 | 9.9323 | 50′ |
| 20′ | 9.7160 | 21 | 9.7845 | 29 | 10.2155 | 8 | 9.9315 | 40′ |
| 30′ | 9.7181 | 21 | 9.7873 | 28 | 10.2127 | 7 | 9.9308 | 30′ |
| 40′ | 9.7201 | 20 | 9.7902 | 29 | 10.2098 | 8 | 9.9300 | 20′ |
| 50′ | 9.7222 | 21 | 9.7930 | 28 | 10.2070 | 8 | 9.9292 | 10′ |
| 32° 0′ | 9.7242 | 20 | 9.7958 | 28 | 10.2042 | 8 | 9.9284 | 58° 0′ |
| 10′ | 9.7262 | 20 | 9.7986 | 28 | 10.2014 | 8 | 9.9276 | 50′ |
| 20′ | 9.7282 | 20 | 9.8014 | 28 | 10.1986 | 8 | 9.9268 | 40′ |
| 30′ | 9.7302 | 20 | 9.8042 | 28 | 10.1958 | 8 | 9.9260 | 30′ |
| 40′ | 9.7322 | 20 | 9.8070 | 28 | 10.1930 | 8 | 9.9252 | 20′ |
| 50′ | 9.7342 | 20 | 9.8097 | 27 | 10.1903 | 8 | 9.9244 | 10′ |
| 33° 0′ | 9.7361 | 19 | 9.8125 | 28 | 10.1875 | 8 | 9.9236 | 57° 0′ |
| 10′ | 9.7380 | 19 | 9.8153 | 28 | 10.1847 | 8 | 9.9228 | 50′ |
| 20′ | 9.7400 | 20 | 9.8180 | 27 | 10.1820 | 9 | 9.9219 | 40′ |
| 30′ | 9.7419 | 19 | 9.8208 | 28 | 10.1792 | 8 | 9.9211 | 30′ |
| 40′ | 9.7438 | 19 | 9.8235 | 27 | 10.1765 | 8 | 9.9203 | 20′ |
| 50′ | 9.7457 | 19 | 9.8263 | 28 | 10.1737 | 8 | 9.9194 | 10′ |
| 34° 0′ | 9.7476 | 19 | 9.8290 | 27 | 10.1710 | 8 | 9.9186 | 56° 0′ |
| 10′ | 9.7494 | 18 | 9.8317 | 27 | 10.1683 | 9 | 9.9177 | 50′ |
| 20′ | 9.7513 | 19 | 9.8344 | 27 | 10.1656 | 8 | 9.9169 | 40′ |
| 30′ | 9.7531 | 18 | 9.8371 | 27 | 10.1629 | 9 | 9.9160 | 30′ |
| 40′ | 9.7550 | 19 | 9.8398 | 27 | 10.1602 | 9 | 9.9151 | 20′ |
| 50′ | 9.7568 | 18 | 9.8425 | 27 | 10.1575 | 9 | 9.9142 | 10′ |
| 35° 0′ | 9.7586 | 18 | 9.8452 | 27 | 10.1548 | 8 | 9.9134 | 55° 0′ |
| 10′ | 9.7604 | 18 | 9.8479 | 27 | 10.1521 | 9 | 9.9125 | 50′ |
| 20′ | 9.7622 | 18 | 9.8506 | 27 | 10.1494 | 9 | 9.9116 | 40′ |
| 30′ | 9.7640 | 18 | 9.8533 | 27 | 10.1467 | 9 | 9.9107 | 30′ |
| 40′ | 9.7657 | 17 | 9.8559 | 26 | 10.1441 | 9 | 9.9098 | 20′ |
| 50′ | 9.7675 | 18 | 9.8586 | 27 | 10.1414 | 9 | 9.9089 | 10′ |
| 36° 0′ | 9.7692 | 17 | 9.8613 | 27 | 10.1387 | 9 | 9.9080 | 54° 0′ |
| | L Cos | d | L Cot | cd | L Tan | d | L Sin | Angle |

# TABLE 7

## FOUR-PLACE LOGARITHMS OF VALUES OF TRIGONOMETRIC FUNCTIONS

| Angle | L Sin | d | L Tan | cd | L Cot | d | L Cos | |
|---|---|---|---|---|---|---|---|---|
| 36° 0′ | 9.7692 | | 9.8613 | | 10.1387 | | 9.9080 | 54° 0′ |
| 10′ | 9.7710 | 18 | 9.8639 | 26 | 10.1361 | 10 | 9.9070 | 50′ |
| 20′ | 9.7727 | 17 | 9.8666 | 27 | 10.1334 | 9 | 9.9061 | 40′ |
| 30′ | 9.7744 | 17 | 9.8692 | 26 | 10.1308 | 9 | 9.9052 | 30′ |
| 40′ | 9.7761 | 17 | 9.8718 | 26 | 10.1282 | 10 | 9.9042 | 20′ |
| 50′ | 9.7778 | 17 | 9.8745 | 27 | 10.1255 | 9 | 9.9033 | 10′ |
| 37° 0′ | 9.7795 | 17 | 9.8771 | 26 | 10.1229 | 10 | 9.9023 | 53° 0′ |
| 10′ | 9.7811 | 16 | 9.8797 | 26 | 10.1203 | 9 | 9.9014 | 50′ |
| 20′ | 9.7828 | 17 | 9.8824 | 27 | 10.1176 | 10 | 9.9004 | 40′ |
| 30′ | 9.7844 | 16 | 9.8850 | 26 | 10.1150 | 9 | 9.8995 | 30′ |
| 40′ | 9.7861 | 17 | 9.8876 | 26 | 10.1124 | 10 | 9.8985 | 20′ |
| 50′ | 9.7877 | 16 | 9.8902 | 26 | 10.1098 | 10 | 9.8975 | 10′ |
| 38° 0′ | 9.7893 | 16 | 9.8928 | 26 | 10.1072 | 10 | 9.8965 | 52° 0′ |
| 10′ | 9.7910 | 17 | 9.8954 | 26 | 10.1046 | 10 | 9.8955 | 50′ |
| 20′ | 9.7926 | 16 | 9.8980 | 26 | 10.1020 | 10 | 9.8945 | 40′ |
| 30′ | 9.7941 | 15 | 9.9006 | 26 | 10.0994 | 10 | 9.8935 | 30′ |
| 40′ | 9.7957 | 16 | 9.9032 | 26 | 10.0968 | 10 | 9.8925 | 20′ |
| 50′ | 9.7973 | 16 | 9.9058 | 26 | 10.0942 | 10 | 9.8915 | 10′ |
| 39° 0′ | 9.7989 | 16 | 9.9084 | 26 | 10.0916 | 10 | 9.8905 | 51° 0′ |
| 10′ | 9.8004 | 15 | 9.9110 | 26 | 10.0890 | 10 | 9.8895 | 50′ |
| 20′ | 9.8020 | 16 | 9.9135 | 25 | 10.0865 | 11 | 9.8884 | 40′ |
| 30′ | 9.8035 | 15 | 9.9161 | 26 | 10.0839 | 10 | 9.8874 | 30′ |
| 40′ | 9.8050 | 15 | 9.9187 | 26 | 10.0813 | 10 | 9.8864 | 20′ |
| 50′ | 9.8066 | 16 | 9.9212 | 25 | 10.0788 | 11 | 9.8853 | 10′ |
| 40° 0′ | 9.8081 | 15 | 9.9238 | 26 | 10.0762 | 10 | 9.8843 | 50° 0′ |
| 10′ | 9.8096 | 15 | 9.9264 | 26 | 10.0736 | 11 | 9.8832 | 50′ |
| 20′ | 9.8111 | 15 | 9.9289 | 25 | 10.0711 | 11 | 9.8821 | 40′ |
| 30′ | 9.8125 | 14 | 9.9315 | 26 | 10.0685 | 11 | 9.8810 | 30′ |
| 40′ | 9.8140 | 15 | 9.9341 | 26 | 10.0659 | 11 | 9.8800 | 20′ |
| 50′ | 9.8155 | 15 | 9.9366 | 25 | 10.0634 | 10 | 9.8789 | 10′ |
| 41° 0′ | 9.8169 | 14 | 9.9392 | 26 | 10.0608 | 11 | 9.8778 | 49° 0′ |
| 10′ | 9.8184 | 15 | 9.9417 | 25 | 10.0583 | 11 | 9.8767 | 50′ |
| 20′ | 9.8198 | 14 | 9.9443 | 26 | 10.0557 | 11 | 9.8756 | 40′ |
| 30′ | 9.8213 | 15 | 9.9468 | 25 | 10.0532 | 11 | 9.8745 | 30′ |
| 40′ | 9.8227 | 14 | 9.9494 | 26 | 10.0506 | 12 | 9.8733 | 20′ |
| 50′ | 9.8241 | 14 | 9.9519 | 25 | 10.0481 | 11 | 9.8722 | 10′ |
| 42° 0′ | 9.8255 | 14 | 9.9544 | 25 | 10.0456 | 11 | 9.8711 | 48° 0′ |
| 10′ | 9.8269 | 14 | 9.9570 | 26 | 10.0430 | 12 | 9.8699 | 50′ |
| 20′ | 9.8283 | 14 | 9.9595 | 25 | 10.0405 | 11 | 9.8688 | 40′ |
| 30′ | 9.8297 | 14 | 9.9621 | 26 | 10.0379 | 12 | 9.8676 | 30′ |
| 40′ | 9.8311 | 14 | 9.9646 | 25 | 10.0354 | 11 | 9.8665 | 20′ |
| 50′ | 9.8324 | 13 | 9.9671 | 25 | 10.0329 | 12 | 9.8653 | 10′ |
| 43° 0′ | 9.8338 | 14 | 9.9697 | 26 | 10.0303 | 12 | 9.8641 | 47° 0′ |
| 10′ | 9.8351 | 13 | 9.9722 | 25 | 10.0278 | 12 | 9.8629 | 50′ |
| 20′ | 9.8365 | 14 | 9.9747 | 25 | 10.0253 | 11 | 9.8618 | 40′ |
| 30′ | 9.8378 | 13 | 9.9772 | 25 | 10.0228 | 12 | 9.8606 | 30′ |
| 40′ | 9.8391 | 13 | 9.9798 | 26 | 10.0202 | 12 | 9.8594 | 20′ |
| 50′ | 9.8405 | 14 | 9.9823 | 25 | 10.0177 | 12 | 9.8582 | 10′ |
| 44° 0′ | 9.8418 | 13 | 9.9848 | 25 | 10.0152 | 13 | 9.8569 | 46° 0′ |
| 10′ | 9.8431 | 13 | 9.9874 | 26 | 10.0126 | 12 | 9.8557 | 50′ |
| 20′ | 9.8444 | 13 | 9.9899 | 25 | 10.0101 | 12 | 9.8545 | 40′ |
| 30′ | 9.8457 | 13 | 9.9924 | 25 | 10.0076 | 13 | 9.8532 | 30′ |
| 40′ | 9.8469 | 12 | 9.9949 | 25 | 10.0051 | 12 | 9.8520 | 20′ |
| 50′ | 9.8482 | 13 | 9.9975 | 26 | 10.0025 | 13 | 9.8507 | 10′ |
| 45° 0′ | 9.8495 | 13 | 10.0000 | 25 | 10.0000 | 12 | 9.8495 | 45° 0′ |
| | L Cos | d | L Cot | cd | L Tan | d | L Sin | Angle |

**Abscissa:** The first coordinate in an ordered pair of numbers which is associated with a point in the coordinate plane. (p. 89)

**Absolute value:** For every nonzero real number $a$, its absolute value is defined to be the greater number of the pair $a$ and $-a$. (p. 39) See also **Modulus of a complex number.**

**Acute angle:** An angle whose measure is between $0°$ and $90°$ (called, in particular, a positive acute angle). (p. 440)

**Additive inverse:** The additive inverse of a real number $a$ is the real number whose sum with $a$ is 0. (p. 20)

**Algebraic expressions:** Numerical expressions, variables, and indicated sums, products, differences, and quotients containing variables. (p. 54)

**Amplitude of a complex number:** A position angle $\theta$ of a point $P$ which represents a complex number in the complex plane. (p. 466)

**Amplitude of a periodic function:** When a periodic function has maximum value $M$ and minimum value $m$, its amplitude is $\dfrac{M-m}{2}$. (p. 522)

**Angle:** The set of points composing two rays with a common vertex, together with a rotation that sends one ray into the other. (pp. 427–428)

**Angle of depression:** The angle between the horizontal ray through an observer and the line of sight from the observer through an object below the observer. (p. 445)

**Angle of elevation:** The angle between the horizontal ray through an observer and the line of sight from the observer through an object above the observer. (p. 445)

**Antilogarithm:** If $\log x = a$, then $x$ is called the antilogarithm of $a$. (p. 405)

**Arithmetic means:** The terms between two given terms in an arithmetic progression. (p. 547) A single arithmetic mean inserted between two numbers is the *average* or *arithmetic mean* of the two numbers. (p. 547)

**Arithmetic progression:** A sequence in which the difference between any two successive terms is a constant, $d$. (p. 544)

**Asymptotes (of a hyperbola):** In the hyperbola $\dfrac{x^2}{a^2} - \dfrac{y^2}{b^2} = 1$, the asymptotes are the diagonals of the rectangle formed by the lines $x = a$, $x = -a$, $y = b$ and $y = -b$. (p. 371)

**Axiom (postulate):** A basic statement that is assumed to be true. (p. 16) See p. 29 for a listing of the axioms for the real numbers.

**Axis:** A reference line, such as the horizontal and vertical axes in a coordinate plane. (p. 88)

**Axis of symmetry (of a parabola):** The line through the focus perpendicular to the directrix. (p. 363)

**Base:** In the expression $a^n$, $a$ is called the base. (p. 148) *See also* Logarithm.

**Binomial:** A polynomial of two terms. (p. 149)

**Characteristic:** The integral part of a logarithm. (p. 405)

**Circle:** A set of points at a given distance from a given point, called the center. (p. 359)

**Circular function:** A function, such as the *sine function over the set of real numbers*, for which values in the domain are real numbers which are measures of arcs on a unit circle. (p. 518) Compare with the *trigonometric sine function* for which elements in the domain are angles.

**Cofunctions:** The following pairs of trigonometric functions are cofunctions: sine and cosine, tangent and cotangent, secant and cosecant. (p. 441)

**Combination:** A subset containing $r$ elements of a set with $n$ elements is often called a combination of $n$ elements taken $r$ at a time. (p. 582)

**Combined variation:** Any combination of direct and inverse variation. (p. 267)

**Common logarithms:** Logarithms to the base 10. (p. 404)

**Complementary angles:** Two angles are complementary if the sum of their degree measures is 90. (p. 72)

**Completing the square:** Transforming a quadratic expression into the square of a binomial. (p. 310)

**Complex fraction:** A fraction whose numerator or denominator (or both) contains one or more fractions or powers involving negative exponents. (p. 227)

**Complex number:** A number of the form $a + bi$ where $a$ and $b$ are real numbers and $i$ is the *imaginary unit*, with the property $i^2 = -1$. (p. 291)

**Complex plane:** When vectors in standard position are considered to represent complex numbers, the $xy$-plane is called the complex plane. The $x$-axis is called the *real axis* and the $y$-axis is called the *imaginary axis*. (p. 465)

**Components of a vector:** Two vectors whose sum is another vector, $\vec{V}$, are called components of $\vec{V}$. (p. 461)

**Conjugates:** Two complex numbers in the form $a + bi$ and $a - bi$ are called conjugates of each other. (p. 297)

**Consistent equations:** Equations which have at least one common solution. (p. 105)

**Constant function:** A function which has the same second coordinate in all its ordered pairs. (p. 139)

**Constant of proportionality (variation):** In an equation of the form $y = mx$, $m \neq 0$, which specifies a direct variation, $m$ is the constant of proportionality. (p. 144)

**Constant term:** The numerical term, or term of degree zero, in a polynomial in simple form. (p. 180)

**Converse statements:** Each of two statements is the converse of the other if the hypothesis of each statement is the conclusion of the other. (p. 55)

**Coordinate:** The number paired with a point on the number line. (p. 12)

**Corollary:** A theorem which follows easily from another theorem. (p. 24)

**Cosecant function:** The set of ordered pairs $\{(\theta, \csc \theta); \csc \theta = \dfrac{1}{\sin \theta}$, provided $\sin \theta \neq 0\}$. (p. 435)

**Cosine function:** If $\theta$ is any angle in standard position, and if $(a, b)$ denotes the coordinates of the point one unit from the origin on the terminal side of $\theta$, then the set of all ordered pairs $(\theta, a)$ is called the cosine function. (p. 431)

**Cotangent function:** The set of ordered pairs $\{(\theta, \cot \theta): \cot \theta = \dfrac{\cos \theta}{\sin \theta}$, provided $\sin \theta \neq 0\}$. (p. 435)

**Coterminal angles:** Angles having the same initial and terminal sides. (p. 429)

**Decimal numeral:** The decimal numerals for rational numbers are either *terminating*, such as 3.5625, or *repeating*, such as $.3\overline{18}$. (p. 242)

**Degree:** A unit of angle measure. One degree is $\dfrac{1}{360}$ of a complete counterclockwise rotation. (p. 428)

**Degree of a monomial:** The sum of the exponents of the variables in the monomial. (p. 149)

**Degree of a polynomial:** The greatest of the degrees of the terms of the polynomial. (p. 149)

**Dependent equations:** Two equations in a system which are equivalent. The graphs of such equations coincide. (p. 105)

**Dependent events:** In probability, two events which are not independent. (p. 596)

**Determinant of a matrix:** A particular real number associated with each square matrix. (p. 611)

**Direct variation:** Any linear function specified by an equation of the form $y = mx$, where $m$ is a nonzero constant. (p. 144)

**Discriminant:** The expression $b^2 - 4ac$ is called the discriminant of the quadratic equation $ax^2 + bx + c = 0$. (p. 323)

**Domain (of a relation):** The set of all the first coordinates of the ordered pairs in the relation. (p. 131)

**Domain (of a variable):** The set whose elements may serve as replacements for a variable. Also called the *replacement set* or *universe*. (p. 9)

**Ellipse:** In a plane, a set of points for each of which the sum of the distances from two fixed points (the foci) is a given constant. (p. 366)

**Empty set:** The set which has no members. Also called the *null* set. (p. 6)

**Equation:** A statement that two numerical expressions designate the same number. (p. 2)

**Equivalent expressions:** Expressions which represent the same number for every value of the variable(s) they contain. (p. 55)

**Equivalent inequalities:** Inequalities which have the same solution set. (p. 61)

**Equivalent open sentences:** Two open sentences which have the same solution set. (p. 55)

**Equivalent systems:** Two systems of equations which have the same solution set. (p. 105)

**Equivalent vectors:** Vectors which have the same magnitude and direction. (p. 458)

**Event:** See **Sample space.**

**Exponent:** In a power, the number of times the base occurs as a factor. In the power $2^3$, 3 is the exponent. (p. 148)

**Exponential form:** A form in which radical expressions are written as powers, or as products of powers. For example, the radical expression $\sqrt{a^5bc^{-6}}$ in exponential form is $a^{\frac{5}{2}}b^{\frac{1}{2}}c^{-3}$. (p. 390)

**Exponential function with base $b$:** The function of the form $\{(x, y): y = b^x, b > 0, b \neq 1\}$. (p. 393)

**Extremes of a proportion:** In the proportion $\dfrac{y_1}{x_1} = \dfrac{y_2}{x_2}$, $y_1$ and $x_2$ are called the extremes. (p. 144)

**Extremum:** The *minimum* or *maximum* value of a function. (pp. 339–340)

**Factorial notation:** A notation for the product of certain successive natural numbers. For example, 5! (five factorial) is the product $5 \cdot 4 \cdot 3 \cdot 2 \cdot 1$. (p. 565)

**Fractional equation:** An equation in which a variable appears in the denominator of a fraction. (p. 236)

**Function:** A relation which assigns to each element in the domain a *single* element in the range. (p. 135)

**Geometric means:** The terms between two given terms in a geometric progression are called the geometric means between the given terms. A *single* geometric mean inserted between two numbers is called *the geometric mean*, or the *mean proportional*, of the two numbers. (pp. 553–554)

**Geometric progression:** A sequence in which the ratio $r$ of every pair of two successive terms is a constant. (p. 553)

**Graph:** On a number line, the point associated with a number. (p. 12) On a coordinate plane, the point associated with an ordered pair of numbers. (p. 88)

**Graph of an open sentence:** The set of all points whose coordinates satisfy the open sentence. (p. 89)

**Greatest common factor:** The greatest integer that is a factor of two or more given integers. (p. 174) The monomial with the greatest constant coefficient and the greatest degree which is a factor of each of several given monomials. (p. 174)

**Half-plane:** Part of a plane, bounded by a line in the plane. (p. 114)

**Hyperbola:** A set of points in a plane such that for each point of the set the absolute value of the difference of its distances from two fixed points (called foci) is a constant. (p. 369)

**Hypotenuse:** The side opposite the right angle in a right triangle. (p. 352)

**Identity:** An equation which is true for every numerical replacement of the variable(s) in the equation. (p. 55)

**Identity matrix:** A square matrix whose main diagonal from upper left to lower right consists of entries of 1, while all other entries are 0. (p. 610)

**Inconsistent equations:** Equations which have no common solution. (p. 105)

**Independent equations:** Two equations which are not equivalent. (p. 105)

**Independent events:** Two events, $A$ and $B$, are independent if the probability of the second occurence does not depend on the probability of the first. In symbols, $P(A \cap B) = P(A) \cdot P(B)$. (p. 596)

**Index:** In the expression $\sqrt[n]{a}$, $n$ is the root index. (p. 272)

**Inequality:** A statement that two numerical expressions represent different numbers. (p. 12)

**Infinite set:** A set for which the process of counting elements continues without end. (p. 6)

**Intersection (of sets):** The set of all elements belonging to both of two given sets. (p. 65)

**Inverse functions:** Two functions, $f$ and $g$, are inverses of each other if they are related in the following manner. $f(u) = v$ if and only if $u = g(v)$. (p. 397)

**Inverse matrices:** Any two matrices $A$ and $B$ such that $AB = BA = I$ (the identity matrix) are called inverse matrices. (p. 613)

**Inverse variation:** Any function $\{(x, y): xy = k\}$ where $k$ is a nonzero constant. (p. 267)

**Invertible (non-singular) matrix:** A matrix that has an inverse. (p. 614)

**Irrational number:** A real number which is not rational. (p. 276)

**Irreducible polynomial:** A polynomial which cannot be expressed as a product of polynomials of lower degree belonging to a given set. (p. 181)

**Joint variation:** A variable varies jointly as two or more other variables if it varies directly as the product of those variables. (p. 267)

**Least common multiple:** The least positive integer that has two or more given integers as a factor. (p. 174) The monomial with the least constant coefficient and the least degree which has each of several given monomials as a factor. (p. 174)

**Lemma:** A theorem introduced mainly to help in proving another theorem. (p. 24)

**Linear equation:** An equation in which each term is either a constant or a monomial of degree 1. (p. 89)

**Linear function:** A function $F$ is a linear function provided there exist real numbers $b$ and $m$ such that for every $x$ in the domain of $F$, $F(x) = mx + b$. (p. 139)

**Linear interpolation:** A process in which we assume that a small part of the graph of a function is a straight line in order to approximate the value of the function at a given point. It is used to find approximate values of logarithms. (p. 407)

**Linear programming:** A branch of mathematics concerned with solving practical problems involving linear inequalities. (p. 127)

**Linear term:** The term of first degree in a polynomial in simple form. (p. 180)

**Logarithm of $u$ to the base $v$:** If $u$ denotes any positive real number and $v$ any positive real number except 1, there is a unique real number called the logarithm of $u$ to the base $v$ ($\log_v u$) which is the exponent in the power of $v$ that equals $u$; that is, $\log_v u = n$ if and only if $u = v^n$. (p. 395)

**Logarithmic function with base $b$:** A function of the form $\{(x, y): y = \log_b x, b > 0, b \neq 1, x > 0\}$. (p. 395)

**Mantissa:** The decimal, or fractional, part of a logarithm. (p. 405)

**Matrix:** A rectangular array of numbers exhibited between brackets or double lines. Individual numbers that make up a matrix are called *entries* or *elements*. (p. 603)

**Means of a proportion:** In the proportion $\dfrac{y_1}{x_1} = \dfrac{y_2}{x_2}$, $x_1$ and $y_2$ are called the means. (p. 144)

**Modulus of a complex number:** The *modulus* or *absolute value* of $a + bi$ is $\sqrt{a^2 + b^2}$; in the complex plane, it is the distance from the origin to the point representing $a + bi$. (p. 466)

**Monomial:** A term which is either a constant, a variable, or a product of a constant and one or more variables. (p. 149)

**Multiplicative inverse:** Two numbers whose product is 1 are called *multiplicative inverses*, or *reciprocals*, of each other. (p. 21)

**Mutually exclusive events:** Two events which have no outcome in common. (p. 593)

**Natural numbers:** Members of the set $\{1, 2, 3, 4, \ldots\}$. (p. 1)

**$n$th root:** For every positive integer $n$, any solution of $x^n = b$ is called an $n$th root of $b$. (p. 271)

**Numeral (Numerical expression):** A symbol used to designate a number. (p. 2)

**One-to-one function:** A function $f$ is one-to-one if $x_1 \neq x_2$ implies $f(x_1) \neq f(x_2)$. (p. 400)

**Open sentence:** A sentence which contains a variable. (p. 9)

**Ordered pair of numbers:** A pair of numbers of the form $(x, y)$ in which the order is important; $x$ is called the *first coordinate* and $y$ the *second* coordinate of the ordered pair. (p. 84)

**Ordinate:** The second coordinate in an ordered pair of numbers which is associated with a point in the coordinate plane. (p. 89)

**Origin:** On the number line, the point $O$, corresponding to zero. (p. 12) In the Cartesian plane, the point $O$, corresponding to the ordered pair $(0, 0)$. (p. 88)

**Parabola:** Any curve consisting of the set of points equidistant from a fixed line (the directrix) and a fixed point (the focus) not on the line. (p. 362)

**Parallel lines:** In a plane, lines having the same slope, or no slope. (p. 101)

**Per cent:** A per cent denotes a fraction whose denominator is 100. (p. 231)

**Periodic function:** A function $f$ such that for a nonzero constant $p$, $f(x + p) = f(x)$ for each $x$ in the domain of $f$. $p$ is called the *period* of $f$. (p. 521)

**Permutation:** Any arrangement of the elements of a set in a definite order. (p. 576) A *circular permutation* is a special type of arrangement about a circular framework. (p. 579)

**Polynomial:** A monomial or a sum of monomials. (p. 149)

**Power:** The number named by an expression in the form of $a^n$, where $n$ denotes the number of times $a$ is used as a factor. (p. 148)

**Precision of a measurement:** The precision of a measurement is given by the unit used in making it. The *accuracy* of a measurement is the ratio of the *maximum possible error* to the measurement itself. (p. 245)

**Prime number:** An integer greater than 1 that has no positive integral factors other than itself and 1. (p. 173)

**Prime polynomial:** An irreducible polynomial whose greatest monomial factor is 1. (p. 181)

**Probability of an event:** In an experiment with $n$ equally likely outcomes, the probability of an event E with $h$ outcomes is $\dfrac{h}{n}$. (p. 589)

**Proportion:** An equality of ratios. (p. 144)

**Quadrant:** One of the four regions into which the plane is separated by the coordinate axes. (p. 88)

**Quadrantal angle:** An angle in standard position whose terminal side lies on a coordinate axis. (p. 463)

**Quadratic formula:** $x = \dfrac{-b \pm \sqrt{b^2 - 4ac}}{2a}$ ; this equation is equivalent to the general quadratic equation, $ax^2 + bx + c = 0$. (p. 313)

**Quadratic function:** A function whose values are given by a quadratic polynomial, $ax^2 + bx + c$, where $a$, $b$, and $c$ are real numbers, $a \neq 0$. (p. 335)

**Quadratic term:** The term of second degree in a polynomial in simple form. (p. 180)

**Radian:** An angular measure. The measure of 1 radian is assigned to an angle subtended by an arc of measure 1 on a circle of radius 1. (p. 514)

**Radical:** The symbol $\sqrt[n]{b}$ is called a radical and is read "the principal $n$th root of $b$," or "the $n$th root of $b$." (p. 271)

**Radical equation:** An equation having a variable in a radicand. (p. 325)

**Radicand:** The expression under a radical sign. (p. 359)

**Radius of a circle:** The distance between each of its points and the center. (p. 359)

**Range (of a relation):** The set of all the second coordinates of the ordered pairs in the relation. (p. 131)

**Rational algebraic expression:** Any expression that is the quotient of two polynomials. (p. 217) Any rational algebraic expression defines a *rational function*. (p. 218)

**Rational number:** Any number which can be represented by a fraction whose numerator is an integer and whose denominator is a nonzero integer. (pp. 1, 217)

**Rationalizing the denominator:** Transforming a term involving radicals and fractions into an equivalent fraction with denominator free of radicals. (p. 288)

**Real number:** Any positive or negative number or 0. (p. 1)

**Reciprocal:** See **Multiplicative inverse.**

**Rectangular (Cartesian) coordinate system:** A plane with two mutually perpendicular number lines, called *axes*. Every point in the plane is associated with an ordered pair of real numbers. (p. 88)

**Reference angle:** With each angle $\theta$ in standard position a reference angle $\alpha$ in the first quadrant is associated such that the absolute values of the trigonometric functions of $\alpha$ and $\theta$ are equal. (p. 453)

**Relation:** Any set of ordered pairs. (p. 131)

**Relatively prime:** Two integers are relatively prime if their greatest common factor is 1. (p. 275)

**Resultant:** The sum of two vectors. (p. 458)

**Sample space:** In probability, a set $S$ of elements that correspond one-to-one with all possible outcomes of an experiment. An *event* is any subset of a sample space. (p. 587)

**Scalar:** In working with matrices, a term used for a real number. (p. 606)

**Secant function:** The set of ordered pairs $\{(\theta, \sec \theta): \sec \theta = \dfrac{1}{\cos \theta}, \text{provided } \cos \theta \neq 0\}$. (p. 435)

**Sequence:** The values of any function whose domain is a set of consecutive natural numbers are said to form a sequence. Each value is called a *term* of the sequence. (p. 543)

**Series:** The indicated sum of the terms of a sequence. (p. 549)

**Set:** A collection of objects. The objects in a set are called the *members* or *elements* of the set. (p. 5)

**Sine function:** If $\theta$ is any angle in standard position, and if $(a, b)$ denotes the coordinates of the point one unit from the origin on the terminal side of $\theta$, then the set of all ordered pairs $(\theta, b)$ is called the sine function. (p. 431)

**Slope:** If $(x_1, y_1)$ are the coordinates of a point $P$ in the plane, and $(x_2, y_2)$ are the coordinates of a different point $Q$, then the slope of line $PQ$ is $\dfrac{y_2 - y_1}{x_2 - x_1}$. (p. 94)

**Solution set (truth set):** For an open sentence in one variable, the subset of the domain for which the open sentence is true. (p. 9) For an open sentence in two variables, the set of ordered pairs of numbers which belong to the replacement sets of the variables for which the sentence is true. (p. 84)

**Standard position (of an angle):** The placement of an angle on a coordinate plane such that the vertex of the angle is at the origin and the initial side of the angle is on the positive side of the $x$-axis. (p. 429)

**Subset:** If each element of set $A$ is also an element of set $B$, we say that $A$ is a subset of $B$. One of the subsets of $A$ is $A$ itself; every other subset of $A$ is a *proper subset*. (pp. 5–6)

**System of simultaneous equations:** Two or more equations which represent two or more conditions imposed at the same time on the same variables. (p. 105)

**Tangent function:** The set of ordered pairs $\{(\theta, \tan \theta): \tan \theta = \dfrac{\sin \theta}{\cos \theta}$ , provided $\cos \theta \neq 0\}$. (p. 435)

**Theorem:** An assertion to be (or that has been) proved. (p. 24)

**Trigonometric functions:** See **Sine, Cosine, Tangent, Cotangent, Secant,** and **Cosecant.**

**Trinomial:** A polynomial of three terms. (p. 149)

**Union (of sets):** The set consisting of all elements belonging to at least one of two or more given sets. (p. 66)

**Value of $f$ at $x$:** The number which the function $f$ assigns to $x$. (p. 136)

**Variable:** A symbol which may represent any member of a specified set. The members of the set are called the values of the variable. A variable with just one value is called a **constant**. (p. 9)

**Vector:** A directed line segment or arrow. Vectors are used to represent *vector quantities*, whose designation requires both magnitude and direction. (p. 456)

**Vertex (of an angle):** The common endpoint of the two rays of the angle. (p. 428)

**Vertex (of a parabola):** The point in which a parabola intersects its axis. (pp. 331, 363)

**$x$-intercept:** The abscissa of the point in which a line or curve intersects the $x$-axis. (pp. 101, 336)

**$y$-intercept:** The ordinate of the point in which a line intercepts the $y$-axis. (p. 101)

**Zero of a function:** In the domain of a function $f$, any value of $x$ which satisfies the equation $f(x) = 0$. (p. 162)

**Zero matrix:** A matrix all of whose entries are 0. (p. 604)

## Chapter 1. Real Numbers

**Page 4 Written Exercises A 1.** 15 **3.** 2 **5.** 2 **7.** 80 **9.** 10 **11.** 16 **13.** 0 **15.** 6 **17.** 19 **19.** 11 **21.** 6 **23.** 0 **25.** 8 **27.** 1 **29.** $\neq$ **31.** $\neq$ **33.** $\neq$ **B 35.** 36 **37.** 9

**Pages 7–8 Written Exercises A 1.** {a, l, g, e, b, r} **3.** {George Washington, John Adams, Thomas Jefferson} **5.** {United States of America} **7.** Ø **9.** {Positive even integers less than 8} **11.** {Natural numbers} **13.** Answers will vary. **15.** {Capitals of Canada and U.S.A.} **17.** = **19.** $\neq$ **B 21.** 0 **23.** $\neq$ **25.** $\neq$ **27.** {x}, {y}, {z} **29.** Ø **C 31.** $8 = 2^3$; $16 = 2^4$

**Pages 11–12 Written Exercises A 1.** {3} **3.** {$\frac{2}{2}, \frac{4}{2}$} **5.** {0} **7.** {5} **9.** {x: x is an integer between 1 and 5} **11.** {x: x is an odd integer between 0 and 6} **13.** {x: x is a natural number greater than 9} **15.** {x: x is an odd integer greater than 10} **17.** {2} **19.** {$\frac{5}{2}$} **21.** {5, 10, 15, . . .} **23.** {2, 4, 6, . . .} **B 25.** {5} **27.** {$\frac{29}{7}$} **29.** {2} **31.** {2}

**Page 15 Written Exercises A 11.** {5, 6, 7, 8} **13.** {7} **15.** Ø **17.** $n \leq 5$ **19.** $8 < z < 10$ **21.** $-5 < s < 0$ **23.** $4 \leq x < 10$

**Pages 19–20 Written Exercises A 1.** Closure under Mult. **3.** Assoc. Axiom of Add. **5.** Comm. Axiom of Add. **7.** Assoc. Axiom of Add. **9.** Comm. Axiom of Add. **11.** 55 **13.** $12zk$ **15.** $8 + 6y$ **17.** 1 **19. a.** Closed, **b.** Closed **21. a.** Closed, **b.** Closed **B 23. a.** Not closed, **b.** Closed **25. a.** Not closed, **b.** Closed **C 27. a.** 6, **b.** Closed, **c.** (1) Yes (2) Yes **29. a.** 8, **b.** Closed, **c.** (1) No (2) No

**Pages 22–23 Written Exercises A 1.** Axiom of Negatives **3.** Axiom of Reciprocals **5.** Axiom of Reciprocals **7.** Axiom of Reciprocals **9.** Comm. Axiom of Add. **11.** {4} **13.** {−8} **15.** {−1} **17.** {0, 1} **19.** {−1, −2, −3, . . .} **B. 21.** {4} **23.** {0} **25.** {4} **27.** {−3} **29.** {x: $x \neq 0$, $x \in \Re$}

**Pages 26–28 Written Exercises B 1.** (1) Assoc. Axiom of Add., (2) Axiom of Negatives, (3) Axiom of Zero **3.** (3) Mult. Property of Equality, (4) Assoc. Axiom of Mult., (5) Axiom of Reciprocals, (6) Axiom of One **5.** (1) Hypothesis, (3) Assoc. Axiom of Mult., (4) Axiom of Reciprocals, (5) Axiom of One **7.** (1) Comm. Axiom of Mult., (2) Assoc. Axiom of Mult., (3) Assoc. Axiom of Mult., (4) Axiom of Reciprocals, (5) Axiom of One, (6) Axiom of Reciprocals, (8) Symmetric Property of Equality

**Pages 31–32 Chapter Test 1.** $-3, 0, \sqrt{1}, \sqrt{4}$ **3.** 63 **5.** {x: x is an integer between 5 and 9} **7.** {4} **11.** Comm. Axiom of Add. **13.** Assoc. Axiom of Mult. **15.** Distributive Axiom **17.** {4} **19.** (1) Hypothesis, (2) Mult. Property of Equality, (3) Assoc. Axiom of Mult., (4) Axiom of Reciprocals, (5) Axiom of One, (6) Zero-factor Property

**Pages 32–34 Chapter Review 1.** Natural **3.** Numerals, numerical **5.** Grouping **7.** 48 **9.** Members, elements **11.** Subset **13.** {e, r, o} **15.** Ø **17.** Open sentence **19.** Domain or replacement set **21.** {3} **23.** Graph, coordinate **25.** 2, x, 4 **29.** Addition, multiplication **31.** Closure under Mult. **33.** Assoc. Axiom of Add. **35.** Identity element, identity element **37.** $-7, \frac{1}{7}$ **39.** Distributive Axiom **41.** Axiom of One **43.** {3} **45.** {−2} **47.** Theorem **49.** 0, 0 **51.** Corollary

## Chapter 2. Applications of Real-Number Properties

**Pages 42–43 Written Exercises A 1.** $-4 < a < 4$ **3.** $n > \frac{3}{2}$ or $n < -\frac{3}{2}$ **5.** $|x| = \frac{5}{6}$ **7.** $|v| = 5$ **9.** $|y| > 3$ **11.** {7, −7} **13.** {1, −1} **15.** Ø **17.** {1, −1} **19.** {0, −6} **21.** {2, −2} **C 43.** Ø

**Pages 46–49 Written Exercises A 1.** 91 **3.** −30 **5.** 12.9 **7.** −13.1 **9.** 26.7 **11.** 7.3 **13.** 14 **15.** −16 **17.** −6 **19.** 8 **21.** −35 **23.** 44.3 **25.** 325 ft. **27.** 7:31:23 A.M. **29.** Closed **31.** Closed **B 33.** (1) Def. of Sub., (2) Property of the Negative of a Sum, (3) Addition fact **35.** (1) Property of the Negative of a Sum, (2) Def. of Sub. **37.** (1) Def. of Sub., (2) Assoc. Axiom of Add., (3) Property of the Negative of a Sum, (4) Def. of Sub.

**Pages 52–54 Written Exercises A 1.** 210 **3.** 30 **5.** 180 **7.** −48 **9.** 9 **11.** 0 **13.** 5 **15.** $\frac{12}{7}$ **17.** 0 **19.** 0 **21.** Not closed **23.** Closed **25. a.** $6 \div 2 \neq 2 \div 6$ **b.** $(12 \div 6) \div 2 \neq 12 \div (6 \div 2)$ **B 27.** (1) Def. of Div. (2) Axiom of Reciprocals **29. a.** (1) Even number of negative factors, (2) Axiom of One, (3) Uniqueness of Reciprocal **b.** (1) Mult. Property of −1, (2) Part a. of Proof, (3) Mult. Property of −1. **37.** $\frac{1}{c}$ does not exist if $c = 0$.

**Pages 58–59 Written Exercises 1.** {7} **3.** {5} **5.** {−5} **7.** {16} **9.** {−47} **11.** {−2} **13.** {18} **15.** {−8} **17.** {−35} **19.** {1} **21.** {2} **23.** {43} **25.** {212} **27.** {$\frac{1}{4}$} **29.** Ø **B 31.** {−2} **33.** {2} **35.** {−10} **C 37.** $z = -4s$ **39.** $h = \dfrac{-vg - r}{v}$ **41.** $b = \dfrac{2A - ha}{h}$

**Pages 67–68  Written Exercises  B 13.** $\{0, 1, 2, 3\}$  **15.** $\{-2, -1, 0, 1, 2, 3, 4\}$  **17.** $\{1, 2\}$

**Pages 72–74  Problems  A 1.** 17, 19, 21  **3.** 8 ft. by 19 ft.  **5.** Triangle: $10''$ by $10''$ by $8''$; Square: $8''$ on a side.  **7.** 18  **9.** 73, 42, 65  **11.** 8 5-minute and 5 6-minutes  **13.** 44  **15.** $3000  **17.** 7 tons and 11 tons  **19.** 64 mph and 32 mph  **21.** Between 48 and 98  **B 23.** More than 50 gal. and less than $133\frac{1}{3}$ gal.  **25.** 2 minutes

**Page 77  Chapter Test  1. a.** True **b.** Not true **c.** True  **5.** $-20$  **7.** $-15$  **9.** $\{-\frac{3}{2}\}$  **11.** $-\frac{13}{3} < x < 3$  **13.** 6 ft. by 8 ft.

**Pages 78–79  Chapter Review  1.** Greater, $a$, $-a$, 0  **3.** $|a|$, right, left  **5.** True  **9.** $\emptyset$  **11.** Additive inverse  **13.** 7  **15.** $-18$  **17.** (1) Def. of Sub., (2) Property of the Negative of a Sum, (3) Additive Inverse of $-b$ is $b$  **19.** Product  **21.** $-4$  **23.** $-1$  **25.** If $ab$ is positive, then $a$ and $b$ are positive. No.  **27.** $\frac{21}{11}$  **29.** $t > 1$  **31.** $-3 < x \le 5$  **33.** 10 girls, 16 boys

**Pages 85–86  Written Exercises  A 1.** $y = -2x + 3$, **a.** 5, **b.** 3, **c.** 1, **d.** $-1$  **3.** $y = \frac{3}{5}x$, **a.** $-\frac{3}{5}$, **b.** 0, **c.** $\frac{3}{5}$, **d.** $\frac{6}{5}$  **5.** $y = -\frac{2}{3}x + \frac{10}{3}$, **a.** 4, **b.** $\frac{10}{3}$, **c.** $\frac{8}{3}$, **d.** 2  In Ex. 7–26 answers vary.  **7.** (0, 4), (4, 1), (8, $-2$)  **9.** (0, 6), (1, 5), (2, 4)  **11.** (0, 4), (3, 2), (6, 0)  **13.** (1, 1), (5, $-2$), ($-3$, 4)  **15.** (0, $-4$), (1, $-4$), (2, $-4$)  **17.** (0, 0), (1, 5), ($-3$, 12)  **B 19.** (0, 0), (0, 2), (1, $\frac{1}{2}$)  **21.** (0, 1), (0, 2), (1, $-1$)  **23.** ($-2$, $-4$), ($-4$, $-2$), ($-5$, $-1$)  **25.** ($-1$, 2), ($-2$, $-2$), (6, $-4$)  **27.** $r = 4$, $s = 1$  **29.** $r = -1$, $s = 1$  **C 31.** $r = 1$ or $r = -1$, $s = 0$  **33.** $r = -1$, $s = \frac{3}{2}$ or $s = -\frac{3}{2}$

## Chapter 3.  Systems of Linear Open Sentences

**Pages 86–88  Problems  A 1.** Answers vary. **a.** 6 and 4, **b.** $-3$ and $-7$, **c.** 12 and $-2$  **3.** 4 dimes and 3 quarters or 9 dimes and 1 quarter.  **5.** 613, 442, 271  **B 7.** $5'' \times 5'' \times 6''$, $6'' \times 6'' \times 4''$, $7'' \times 7'' \times 2''$  **9.** In the order (nickels, dimes, quarters) the solutions are (9, 3, 7), (8, 6, 6), (7, 9, 5), (6, 12, 4), (5, 15, 3), (4, 18, 2), and (3, 21, 1).  **11.** In the order (newspaper, glass, cans): (2, 4, 2), (5, 2, 1).

**Pages 92–93  Written Exercises  A 1.** $3x + y = 5$  **3.** $3x + 4y = 6$  **5.** $x + 0 \cdot y = 2$  **B 27.** 3  **29.** $-3$

**Pages 97–99  Written Exercises  A 1.** 1  **3.** 2  **5.** 0  **7.** No slope  **9.** $-\frac{2}{3}$  **11.** $-\frac{3}{2}$  **13.** $-\frac{1}{2}$  In Ex. 15–20 answers vary.  **15.** (2, 3), (3, 5), (4, 7)  **17.** (7, $\frac{2}{3}$), (9, 2), (12, 4)  **19.** (0, $-7\frac{1}{4}$), (3, $-8$), (7, $-9$)  **27.** Yes  **29.** No  **31.** $m = 2$, (0, 1)  **33.** $m = -\frac{2}{3}$, (0, $\frac{4}{3}$)  **B 34.** Previously proved.  **37.** Mult. prop. of equality  **39.** Add. prop. of equality  **41.** Substitution  **43.** $\frac{2}{25}$  **C 45.** No solution

**Pages 102–104  Written Exercises  A 1.** $x - y = -1$  **3.** $2x - y = 1$  **5.** $x - 2y = 0$  **7.** $y = 2$  **9.** $x = 6$  **11.** $x - y = -4$  **13.** $x + y = 1$  **15.** $4x + y = 5$  **17.** $y = 1$  **19.** $x = 2$  **B 21.** $x + y = 3$  **23.** $y = -\frac{1}{2}x + 5$  **25.** $y = -3$  **27.** $2x + 5y = 4$  **29.** $y = 2x - 2$  **31.** $2x - y = 1$  **33.** $x + y = 4$ or $3x - y = 0$  **35.** $x + by = 2b^2 + b$  **37.** $x$-intercept is $C/A$ and $y$-intercept is $C/B$.

**Pages 109–110  Written Exercises  1.** inconsistent  **3.** consistent, dependent  **5.** consistent, independent  **7.** $\{(1, 1)\}$  **9.** $\{(-2, -2)\}$  **11.** $\{(3, -1)\}$  **13.** $\{(2, -3)\}$  **15.** $\emptyset$  **17.** $\{(x, y): 3x - 5y = -7\}$  **19.** $\{(\frac{1}{3}, \frac{1}{2})\}$  **21.** $\{(1, \frac{1}{2})\}$  **23.** $\{(k, 0)\}$  **25.** $\{(b - a, 2b)\}$  **27.** $\left\{\left(\dfrac{ak_1 + bk_2}{a^2 + b^2}, \dfrac{bk_1 - ak_2}{a^2 + b^2}\right)\right\}$  **29.** $\left\{\left(\dfrac{s}{r}, \dfrac{r}{s}\right)\right\}$

**Pages 112–113  Problems  A 1.** 12, 6  **3.** 5 nickels, 3 quarters  **5.** $50°$, $40°$  **7.** 40  **9.** 3 ft. by 5 ft.  **11.** Airspeed 595 mph, windspeed 35 mph.  **13.** Earth 7918 mi., Mars 4200 mi.  **B 15.** 25  **17.** Rate in still water is 7 mph, rate of current is 1 mph.  **19.** $a = 2$, $b = 1$  **21.** $m = 2$, $b = -1$  **23.** $a = 5$, $b = 1$

**Pages 120–121  Written Exercises  A 1.** $\{(1, 0, -1)\}$  **3.** $\{(1, 1, -1)\}$  **5.** $\{(-2, 1, -3)\}$  **7.** $\{(-2, 1, -3)\}$  **9.** $\{(\frac{1}{2}, 1, \frac{3}{2})\}$  **B 11.** $\{(\frac{1}{2}, 1, 2)\}$  **13.** $a = 100$, $b = 60$, $c = -16$  **15.** $a = -1$, $b = 1$, $c = 3$  **C 17.** $\{(2, -1, 0)\}$  **19.** $\{(\frac{1}{18}, -\frac{3}{2}, \frac{11}{18})\}$

**Page 121  Problems  A 1.** $\frac{1}{2}$ cup tomato juice, $1\frac{1}{2}$ cups vinegar, 6 cups olive oil.  **3.** 2, 3, $-4$  **5.** 4 nickels, 2 dimes, 5 quarters.

**Pages 123–124  Chapter Test  1.** Answers vary, (0, $-4$), (3, 0)  **7.** $x - y = -4$  **9.** $\{(x, y): x - 4y = -5\}$

**Pages 124–125  Chapter Review  1.** $a = 2$, $b = -4$  **3.** Answers vary, ($-1$, $-1$)  **5.** No.  **7.** 2, 5  **11.** $2x - 3y = 1$  **13.** $x - 2y = 2$  **15.** $\{(1, -1)\}$  **17.** 5 nickels, 3 dimes  **19.** Above  **23.** 1, $-2$, 0

**Page 129  Extra for Experts  1.** 175 Leopards, 125 Tigers  **3.** Maximum value 15, minimum value 2  **5.** 300 Brand 1, 100 Brand 2

## Chapter 4. Introduction to Functions

**Page 134** **Written Exercises** **1.** $\{-4, -2, 0, 2, 4, 6, 8\}$ **3.** $\{y: y \geq -2\}$ **5.** $\{-4\}$ **7.** $\{0, 2, 4, 6\}$
**9.** $\{y: y \leq 10\}$ **11.** $\{1, 2, 3, 4\}$

**Pages 137–138** **Written Exercises** **A 1.** 0 **3.** 5 **5.** 24 **7.** $(a+1)(3a+1)$ **9.** 1 **11.** $-5$ **13.** 0
**15.** 20 **B 17.** 3 **19.** 1 **21.** 5 **23.** $2|a+2|+1$ **25.** 0, 0 **27.** $a, a$

**Pages 142–143** **Written Exercises** **A 1.** $f(x) = 3x - 5$ **3.** $h(x) = -3x + 4$ **5.** $f(x) = -\frac{2}{3}x - \frac{7}{3}$
**7.** $f(x) = 1$ **9.** 8 **11.** 6 **13.** $\frac{7}{4}$ **15.** $-\frac{8}{3}$ **B 17.** \$15 **19.** \$22,000 **21.** \$19,600

**Pages 145–147** **Written Exercises** **A 1.** 39 **3.** $\frac{20}{3}$ **5.** 32 **7.** 10 **9.** 3 **C 19.** 3

**Pages 147–148** **Problems** **A 1.** \$37,400 **3.** 65 mph **5.** 50,000 **7.** \$30 **9.** 2 hrs. 24 min. **B 11.** 2.64
**13.** 258 in eastern plant, 210 in northern plant **15.** $6\frac{2}{3}$

**Pages 150–151** **Written Exercises** **A 1.** $6x - 2$ **3.** $-2x^2y^3 + 6xy^2 + 3x^2y - 2$ **5.** $4m^2n - 4mn - 5$
**7.** $5x^5 + x^4 - 5x^3 - x$ **9.** $3x^4y - 2xy^4 - 3$ **11.** $-7z^3 + z$ **13.** $3d^2 - 6d - 3$ **15.** $-6t^3 + 2t^2 + 6$
**B 17.** $-\frac{1}{2}x^2 + \frac{3}{10}x + \frac{51}{8}$ **19.** $-1.62x^2 + 9.3x - 1.3$ **21.** $-14.575y^2 - 1.22y + 2.1$

**Pages 154–155** **Written Exercises** **A 1.** $6z^2 + z - 3$ **3.** $5t^2 + 4t$ **5.** $4y^2 + 6$ **7.** $-5z^2 - 4z - 22$
**9.** $r^3 - 10r^2 + 2r - 3$ **11.** $11x^2 - 20x + 32$ **13.** $5x^2 + 3x + 2$ **15.** $-5x^2 - 22x + 22$ **17.** $-11x^2 + 6x - 2$ **19.** $12x^2 - 6x + 12$ **B 21.** $-2x^3 - 3x^2 - 2x + 1$ **23.** $-3x^3 - 8x^2 - 5x + 3$ **25.** $x^3 - 8x^2 - x + 3$

**Page 157** **Written Exercises** **A 1.** $-45x^5 - 180x^4 + 150x^3$ **3.** $192b^5 - 72b^4 + 48b^3$ **5.** $24a^2 + 19a - 35$ **7.** $48 - 22c - 15c^2$ **9.** $15x^4 + x^2 - 28$ **11.** $2x^3 - 5x^2 + x + 2$ **13.** $6 + 13t - 31t^2 + 4t^3$
**15.** $12y^3 + 20y^2 - 33y - 20$ **17.** $2x^4 - 13x^3 + 20x^2 - 11x + 2$ **B 19.** $a^{4n+1} + 4a^{5n} - 3a^{3n}$ **21.** $x^{2n} + x^n - 2$ **23.** $x^{3n} + 5x^{2n} + 2x^n - 8$ **25.** $2z^{3n} - z^{2n} - 7z^n + 6$ **27.** $x^3 - 7x^2 + 13x - 4$ **29.** $x^3 - 4x^2 + 2x + 3$ **31.** $16x^3 - 56x^2 + 49x - 20$

**Pages 160–161** **Written Exercises** **A 1.** $a^8x^4$ **3.** $9a^7n^8$ **5.** $27x^{22}y^7$ **7.** $9x^{4n}$ **9.** $a^{2n^2}x^{n^2}$ **11.** $a^{3n^2}$
**13.** $9x^4 - 12x^2y + 4y^2$ **15.** $49x^4 - 4$ **17.** $64a^3 + 8$ **19.** $x^6 - 8$ **21.** $16x^4 - 8x^2a^2 + a^4$ **23.** $-2xy$
**B 25.** $8z^4 + 6z^2 + 1$ **27.** $r^3 + 6r^2 + 12r + 8 + t^3$ **29.** $a^2 + b^2 + c^2 + 2ab + 2ac + 2bc$ **31.** $a^2 + b^2 + c^2 + d^2 + 2ab + 2ac + 2ad + 2bc + 2bd + 2cd$ **33.** $a^3 + b^3 + c^3 + 3a^2b + 3a^2c + 3ab^2 + 3ac^2 + 6abc + 3b^2c + 3bc^2$ **C 35.** $a^3 + 3a^2b + 3ab^2 + b^3$ **37.** $a^4 + 4a^3b + 6a^2b^2 + 4ab^3 + b^4$

**Page 163** **Written Exercises** **A 1.** $\{-2, 3\}$ **3.** $\{-5\frac{1}{3}, -\frac{2}{3}\}$ **5.** $\{0, \frac{7}{3}\}$ **7.** $\{0\}$ **9.** $\{-\frac{3}{2}, \frac{7}{5}\}$ **11.** $\{1\}$
**13.** $\{\frac{8}{9}\}$ **15.** $\{5\}$ **B 17.** $\{-\frac{1}{3}\}$ **19.** $\{-\frac{5}{9}\}$

**Page 165** **Chapter Test** **1.** $\{y: y \geq -7\}$ **3.** 2 **5.** $-1$ **7.** $1\frac{1}{8}''$ **9.** $-8x + 14$ **11.** $2x^3 - x^2 - 13x + 15$
**13.** $z^6 - 1$ **15.** $\{0, \frac{3}{2}\}$

**Pages 166–167** **Chapter Review** **1.** Ordered pairs **3.** $-2, 2$ **5.** Unique **7.** Range, 2 **9.** Linear **11.** Range
**13.** Proportion **15.** 60 **17.** Monomial **19.** $3, 2x^3 - 4x^2 + 3x + 5$ or $5 + 3x - 4x^2 + 2x^3$ **21.** $f(x) + g(x)$,
$f(x) - g(x)$ **23.** $a^{m+n}$ **25.** $x^4 - 2x^3 + x^2 + 4x - 6$ **27.** $x^2 - 4x + 4, 4x^2 - 9$ **29.** $9x^{10}y^{14}$ **31.** Zero, root

## Chapter 5. Factoring Polynomials

**Page 176** **Written Exercises** **A 1.** $3 \cdot 5^2$ **3.** $2^3 \cdot 3 \cdot 5$ **5.** $7^3$ **7.** $2^2 \cdot 3^2 \cdot 5 \cdot 13$ **9. a.** 9 **b.** 135 **11. a.** 80
**b.** 240 **13. a.** $3ab^2$ **b.** $21a^2b^3$ **15. a.** $5m^2n$ **b.** $60m^4n^3$ **17. a.** 1 **b.** $6x^2$ **19. a.** $19ac$ **b.** $114a^2b^2c$ **B 21.** 1, 2, 3
**23.** $1, 3, p, 3p$ **25.** 132

**Page 179** **Written Exercises** **A 1.** $2(x+2)(x-2)$ **3.** $a(a+3)(a-3)$ **5.** $(x^2+9)(x+3)(x-3)$
**7.** $3(a-2)^2$ **9.** $-(x-7)^2$ **11.** $(3a+5d)(b+c)$ **13.** $5(a+4)^2$ **15.** $2(s-25)^2$ **17.** $(x-y)(x^2+xy+y^2)$
**19.** $(3a+1)(9a^2-3a+1)$ **21.** $(x-4)(x^2+4x+16)$ **23.** $(2a+b)(x-3y)$ **25.** $(10-x)(100+10x+x^2)$
**27.** $(7-2a)(49+14a+4a^2)$ **B 29.** $(x^2+2)(x^4-2x^2+4)$ **31.** $(a+bc)(a^2-abc+b^2c^2)$
**33.** $(3m^2+1)(9m^4-3m^2+1)$ **35.** $(x-2)(x+1)(x-1)$ **37.** $(s-2-t)(s-2+t)$ **39.** $(5+x)^2$
**41.** $(x+2+y)(x+2-y)$ **43.** $(x-y)(x+y+1)$ **C 45.** $(a+x)(a^2-ax+x^2)(a-x)(a^2+ax+x^2)$
**47.** $(x^n+1)^2$ **49.** $x^2(x^2+1)$ **51.** $(x-3)^2(4-x)(7-5x+x^2)$

**Pages 182–183 Written Exercises A 1.** $(x + 7)(x + 5)$ **3.** $(x - 3)(x - 1)$ **5.** $(x - 4)(x - 5)$
**7.** $(w + 2)(w - 1)$ **9.** $(k + 9)(k - 6)$ **11.** $(9 + m)(2 - m)$ **13.** $(2x + 1)(x + 2)$ **15.** $(2y + 3)^2$
**17.** $(-4z - 1)(z - 1)$ **19.** $(3q + 4)(2q - 1)$ **21.** $7(2x + 3)(x - 1)$ **23.** $4(2w + 1)(w + 5)$
**B 25.** $(k - 5l)(k + 2l)$ **27.** $x^4(1 - x + x^2)$ **29.** $2y(2x + 1)(x - 2)$ **31.** $\frac{1}{6}(3y - 1)(2y + 1)$
**33.** $(x^2 - 8)(x^2 + 1)$ **35.** $x^3(x + 5)(x - 5)$ **C 37.** $(2y^n + 1)(y^n + 1)$ **39.** $(a + b - 10)(a + b - 4)$
**41.** $(2x + 9)(x - 2)^2$

**Pages 184–185 Written Exercises A 1. a.** $y^2$, **b.** $y^5$ **3. a.** 1, **b.** $(2a + b)(a + 2b)$ **5. a.** $3x^2 - 2$,
**b.** $15(3x^2 - 2)$ **7. a.** $x + 6$, **b.** $5(x + 6)(x - 6)$ **9. a.** $16x^3(y + z)^2$, **b.** $192x^5(y + z)^3$ **11. a.** $x + 3$,
**b.** $(x + 3)(x - 2)(2x - 5)$ **B 13. a.** $x + 3$, **b.** $(x + 3)(x - 3)(x^2 - 3x + 9)$ **15. a.** 1, **b.** $(w + 1)(w - 1)^2$
**C 17. a.** 1, **b.** $x^2y^3(x - y)(x^2 + xy + y^2)(2x^4 - 4x^2y^2 + y^4)$ **19. a.** 1, **b.** $(3r + s)(r - 3s)(3r - s)(r + 3s)$

**Page 187 Written Exercises A 1.** $\{-1, -3\}$ **3.** $\{1, 5\}$ **5.** $\{-5, 3\}$ **7.** $\{-3, 16\}$ **9.** $\{-2, 2\}$ **11.** $\{-6, 6\}$
**13.** $\{-\frac{3}{2}, 1\}$ **15.** $\{\frac{3}{2}, 2\}$ **17.** $\{-\frac{1}{5}, -2\}$ **19.** $\{-\frac{1}{2}, -2\}$ **21.** $\{-5, 3\}$ **23.** $\{-9, 6\}$ **25.** $\{2, 10\}$ **27.** $\{\frac{19}{5}\}$
**B 29.** $\{-5, 0, 8\}$ **31.** $\{-2, 0, 2\}$ **C 33.** $\left\{\dfrac{1}{a + b}\right\}$

**Page 188 Problems A 1.** 21, 22 or $-22$, $-21$ **3.** 3, 4, 5, 6 or 1, 2, 3, 4 **5.** 6 yds. by 10 yds. **7.** 11 yds.
by 10 yds. **9.** 3 seconds and 5 seconds after thrown. **B 11.** 2 seconds, and 3 minutes and 18 seconds after
fired. **13.** 9' and 12' **15.** 20 yds.

**Page 191 Written Exercises A 1.** $\{x: x < 0 \text{ or } x > 7\}$ **3.** $\{x: x < -3 \text{ or } x > 5\}$ **5.** $\{y: 1 \le y \le 4\}$
**7.** $\{a: a \le \frac{2}{5} \text{ or } a \ge 2\}$ **9.** $\{x: x < 2 \text{ or } x > \frac{5}{2}\}$ **B 11.** $\{y: y < -4 \text{ or } 0 < y < 4\}$ **13.** $\{x: x \le -\frac{4}{3} \text{ or }$
$0 \le x \le \frac{1}{2}\}$ **15.** $\{x: x < 0 \text{ or } 0 < x < 1\}$ **C 17.** First $2\frac{1}{2}$ seconds **19.** First 5 seconds

**Pages 193–194 Written Exercises A 1.** $12x^2 - 8x + 1$ **3.** $4x^6 - 3x^4 + \frac{1}{2}x^3$ **5.** $6x^3 - 4x^2 + 5$
**7.** $-x^{11} + 2x^5 + 5x^2 - \frac{1}{5}$ **9.** $w - 7$ **11.** $n + 5$ **13.** $2x + 8 + \dfrac{49}{2x - 5}$ **15.** $x - 7 - \dfrac{10}{5x + 2}$
**17.** $3x^2 + 8x + 6$ **19.** $2k^2 + 3k - 1$ **B 21.** $7d + 3c$ **23.** $5y - x$ **25.** $x^2 + ax + a^2$ **27.** $x^2 - 3xy - y^2$
**C 29.** $a^4 - a^3b + a^2b^2 - ab^3 + b^4$ **31.** 1 **33.** $a = 15$; $b = -5$

**Page 197 Written Exercises A 1.** quotient $= x^2 - 4x - 1$; remainder $= -2$ **3.** quot. $= x^3 - 4x^2 +$
$12x - 13$; rem. $= 7$ **5.** quot. $= 2x^3 - 9x^2 + 31x - 92$; rem. $= 275$ **7.** quot. $= x^2 - 4x + 15$; rem. $= -55$
**B 9.** quot. $= x^2 + \frac{5}{2}$; rem. $= \frac{11}{2}$ **11.** quot. $= 3x^2 - 5x + 10$; rem. $= -23$
**13.** quot. $= 2x^3 + \frac{1}{2}x^2 - \frac{9}{8}x - \frac{33}{32}$; rem. $= -\frac{97}{32}$ **C 15.** 3 **17.** 102

**Page 199 Written Exercises A 1.** $(x - 3)(x + 3)(x + 1)$ **3.** $(x + 1)(x - 7)(x + 1)$ **5.** $(x - 4) \times$
$(x - 4)(x + 1)$ **7.** $(x - 5)(x + 3)(x + 2)$ **9.** $(x - 3)(x - 3)(x - 2)$ **11.** $(x - 6)(x^2 + 1)$ **B 13.** $(x - 2) \times$
$(x - 4)(x + 2)(x + 1)$ **15.** $(x - 7)(x + 1)(x + 2)(x + 1)$ **C 17.** $(x - 2)(x + 2)(x^2 + 6)$ **19.** $(x - 1) \times$
$(x + 2)(x^3 - x^2 + 2x - 4)$

**Page 201 Chapter Test 1.** GCF $= 6xz$; LCM $= 252x^4y^3z^2$ **3.** $(y + 4)(y^2 - 4y + 16)$ **5.** $(2x + 5)^2$
**7.** GCF $= x + 2$; LCM $= 7x(2x - 1)(x + 2)(3x + 5)$ **9.** $\{0, -3, 2\}$ **11.** $\{x: x < -\frac{5}{3} \text{ or } x > 1\}$
**13.** quot. $= 2x^2 - 5x + 13$; rem. $= -51$

**Pages 201–203 Chapter Review 1.** 2, 3, and 5 **3. a.** $4xy^2$, **b.** $144x^3y^4$ **5.** $(y - 9)^2$ **7.** $(3z + 2) \times$
$(9z^2 - 6z + 4)$ **9.** $2(3x + 4)(x - 3)$ **11.** $2(5x - 2y)(x + y)$ **13. a.** $x + 3$, **b.** $x^2(2x - 1)(x + 3)(x - 5)$
**15.** 0; 0 **17.** $\{-\frac{1}{2}, -2\}$ **19.** (b) and (c) **21.** $\{x: x < \frac{1}{2} \text{ or } x > 2\}$ **23.** $x^2 - 3x - 10$ **25.** quot. $=$
$5x^3 + 12x^2 + 36x + 112$; rem. $= 335$ **27.** quot. $= \frac{5}{2}x^3 - 9x^2 + 27x - 79$; rem. $= 473$ **29.** $(x - 2) \times$
$(x + 2)(x + 3)$ **31.** $(x - 3)(x + 2)(x + 3)(x^2 - 2x + 4)$

**Page 205 Extra for Experts 1.** $\{1, 3, -2\}$ **3.** $\{-2, -3, 3\}$ **5.** 6 and $-\frac{1}{2}$ **7.** No other real roots.
**9.** $-3$ **11.** 2 and $-2$

**Pages 206–207 Cumulative Review 1.** 24 **3.** $x \le 3$ **5.** Axiom of zero **7.** Commutative axiom for mult.
**11.** $-9$ **13.** 16 **15.** Answers vary; for example, $(0, -12)$, $(8, 0)$, $(10, 3)$. **17.** Slope $= \frac{1}{2}$; $y$-intercept $= -\frac{5}{4}$
**19.** $\{(-2, 3)\}$ **21. a.** $-5$, **b.** 5, **c.** $12a^2 + 2a - 5$ **23.** $-2x^2 + 3x - 7$ **25.** $t^5 + t^3 + 4t^2 - 2t + 8$
**27.** $3(2x - 3y)(3x + y)$ **29.** $3t(1 - 3t)(1 + 3t + 9t^2)$ **31.** 8 **33.** 40 pounds **35.** Rate in still water $=$
$7\frac{1}{2}$ mph; current $= 1\frac{1}{2}$ mph. **37.** $14''$ **39.** 12 benches; 6 tables

## Chapter 6. Rational Numbers, Expressions, and Functions

**Page 212  Written Exercises  A 1.** $3t^3$  **3.** $-\dfrac{3n}{4m^2}$  **5.** $\dfrac{s^3}{3r^2}$  **7.** $\dfrac{6n^2p}{5tq^2}$  **9.** $\dfrac{x^8}{4y^2}$  **11.** $x$  **13.** $\dfrac{4a^6}{9}$  **15.** $\dfrac{a^7x^{10}}{z^4}$

**B 17.** $\dfrac{a^7b}{125c^2}$  **19.** $-\dfrac{3087t^7}{200r^6}$  **21.** $\dfrac{27a^6}{8c^3}$

**Pages 215–217  Written Exercises  A 1.** $\frac{125}{27}$  **3.** $\frac{1}{81}$  **5.** $-64$  **7.** $\frac{1}{576}$  **9.** $36$  **11.** $0.0783$  **13.** $\frac{9}{4}$  **15.** $17$

**17.** $\dfrac{b}{a^4c^3}$  **19.** $\dfrac{6u^4w^2}{7v^3}$  **21.** $\dfrac{y^7}{x^5y^8}$  **23.** $11s^2$  **25.** $\frac{1}{2}\neq 2$  **27.** $\frac{3}{16}\neq\frac{1}{1296}$  **B 29.** $3+5t-3t^2$

**31.** $2x^5-x^3+3x-1$  **33.** $5+7x-2x^2+3x^3-x^4$  **C 35.** $-3$  **37.** $5+11x-4x^2$

**Pages 219–220  Written Exercises  A 1.** $\dfrac{x+3}{2(x+2)}$, $x\neq-2$, $x\neq 3$  **3.** $\dfrac{1}{t-1}$, $t\neq 1$  **5.** $\dfrac{(2x+1)(2x-1)}{(2x-3)(x-2)}$,

$x\neq\frac{3}{2}, x\neq 2$  **7.** $\dfrac{x+2}{x-1}$, $x\neq 1$, $x\neq-5$  **9.** $\dfrac{y(3y+1)}{y+3}$, $y\neq-3$, $y\neq\frac{1}{3}$  **11.** $\dfrac{2}{a+b}$, $a\neq-b$, $b\neq 2a$

**13.** $\dfrac{x(x-2)}{5(x-6)}$, $x\neq 6$  **15.** $\dfrac{t^2+4t+16}{s(t+4)}$, $s\neq 0$, $t\neq-4$, $t\neq 4$  **17.** $\dfrac{(-1)(y+4)}{6+y}$, $y\neq-6$, $y\neq 4$

**19.** $\dfrac{1}{(2t+1)(2t-1)}$, $t\neq-\frac{1}{2}$, $t\neq\frac{1}{2}$  **21.** $\{x: x\neq\frac{1}{2}$ and $x\neq 5\}$, zeros: $-5, 0$.  **23.** $\{x: x\neq 2\}$, zeros: $3$.

**25.** $\{x: x\neq-1$ and $x\neq 1\}$, no real zeros.  **27.** $\{x: x$ is real$\}$, zeros: $-2, 2$.  **B 29.** $\dfrac{3(a+b)(a-b)}{a^2+b^2}$

**31.** $\dfrac{x^2-xy+y^2+y-x}{3(x+y)}$  **33.** $-\dfrac{a-b+1}{b+a}$  **C 35.** $\dfrac{4(2x-3)(-3x^2+6x-1)}{(x^2-1)^5}$  **37.** $\dfrac{16+15z^2-2z^3}{(2z+5)(2z^2+5z)^2}$

**Pages 222–223  Written Exercises  A 1.** $-\frac{5}{9}$  **3.** $\dfrac{x}{a^2}$  **5.** $-\dfrac{y^2}{x}$  **7.** $1$  **9.** $\dfrac{2y}{y+5}$  **11.** $\dfrac{4(x+y)}{x-y}$

**13.** $\dfrac{(x-7)^2(x+2)(x+4)}{(x-3)^2(x+8)(x-5)}$  **15.** $\dfrac{3x(x-1)(2x-3)^2}{(1-2x)(2x+3)(2x+1)^2}$  **17.** $20x^2y^2-xy-12$  **19.** $\dfrac{(b-3)(a^2+4)}{a-2}$

**B 21.** $\dfrac{1}{(a+2b)^2}$  **23.** $\dfrac{3a+1}{a+1}$  **25.** $\dfrac{(a+3)^2(a^2-3a+9)^2}{3(a+4)^2(a-3)^2}$

**Pages 225–227  Written Exercises  A 1.** $\frac{10}{11}$  **3.** $\frac{17}{27}$  **5.** $\dfrac{24x-1}{24}$  **7.** $\dfrac{17}{x-2}$  **9.** $\dfrac{17}{12x}$  **11.** $\dfrac{8x^2+10z^2-15y^2}{20xyz}$

**13.** $\dfrac{7-4x}{6x}$  **15.** $\dfrac{4}{x+1}$  **17.** $\dfrac{30x^2-2x}{(3x-4)(5x+6)}$  **19.** $\dfrac{7}{3x}$  **21.** $\dfrac{2x^2+8x+7}{(x+2)(x+3)}$  **23.** $\dfrac{3x^2-4x+4}{(x-2)(x+2)}$

**25.** $\dfrac{-3a^2-3b^2}{(2a-b)(a-2b)}$  **27.** $\dfrac{y^2-2y+3}{y(y+1)(y-1)}$  **B 29.** $\dfrac{4x}{(x+y)(x-y)}$  **31.** $\dfrac{-3y+20}{(y-3)(y-2)(y+5)}$

**33.** $\dfrac{x^3-3}{(x+1)(x-1)}$  **35.** $\dfrac{10x^2+x+3}{(1-2x)(1+2x)}$  **37.** $-\dfrac{12}{(x+1)(x-1)}$  **C 39.** $\dfrac{x^2+2x-2}{(x-3)(x+2)(x-1)}$

**41.** $\dfrac{2y^2+18y+3}{(3y+1)(2y-1)(y+2)}$

**Pages 228–229  Written Exercises  A 1.** $\frac{52}{105}$  **3.** $\dfrac{x}{2a}$  **5.** $x$  **7.** $\dfrac{r^2+r+1}{r}$  **9.** $\dfrac{z+5}{z+2}$  **11.** $\dfrac{x-4}{3x+4}$  **13.** $1$

**B 15.** $\dfrac{2xy}{y+x}$  **17.** $\dfrac{1+uv-u+v}{1+u^2}$  **19.** $\dfrac{4a^2-3a}{4a+1}$  **21.** $\dfrac{x^3+xy^2}{x^2y+2xy^2-y^3}$  **C 23.** $\dfrac{(x+y)(y^2+x^2)}{x^2y}$

**Pages 232–233  Written Exercises  A 1.** $(2x-5)(x+1)=0$; $\{\frac{5}{2}, -1\}$  **3.** $(5a-7)(a+1)=0$; $\{\frac{7}{5}, -1\}$

**5.** $(9x-5)(2x+3)=0$; $\{\frac{5}{9}, -\frac{3}{2}\}$  **7.** $(5z+3)(3z+2)=0$; $\{-\frac{3}{5}, -\frac{2}{3}\}$  **9.** $(7z-2)(z+4)=0$; $\{\frac{2}{7}, -4\}$

**11.** $(y+3)(y-1)=0$; $\{-3, 1\}$

**Pages 233–235  Problems  A 1.** 12 kilometers  **3.** 42,000 standard, 8,000 deluxe  **5.** 35 and 21  **7.** 2 hours  **9.** 16 miles  **11.** 20,000 addition operations; 12,000 multiplication operations  **13.** 6 ounces of water  **15.** 240 gallons  **B 17.** 9 hours  **19.** $600  **21.** $32\frac{1}{22}$ miles  **C 23.** $26,455

**Pages 238–239  Written Exercises  A 1.** $\{3\}$  **3.** $\{6\}$  **5.** $\{\frac{1}{3}\}$  **7.** $\{5\}$  **9.** $\emptyset$  **11.** $\{2\}$  **13.** $\emptyset$  **15.** $\{13\}$  **B 17.** $\emptyset$  **19.** $\{\frac{3}{2}, -7\}$  **21.** $\{\frac{5}{2}, 4\}$  **C 23.** $\{(-3, 2)\}$  **25.** $\{(-162, 42)\}$  **27.** $\{(\frac{1}{3}, \frac{1}{2})\}$  **29.** $\{(-3, 5)\}$

**Pages 239–241  Problems  A 1.** 12 mph  **3.** 29 students  **5.** $\frac{6}{18}$  **7.** $-\frac{3}{5}$ or $\frac{5}{6}$  **9.** 15 and 36  **11.** $1\frac{1}{2}$ mph.  **13.** 5%, $11,000  **B 15.** 50 books  **17.** 2 hours  **19.** $\frac{15}{6}$  **C 21.** 14.4 mph.  **23.** $18.75\leq x\leq 40$, where $x=$ no. of lbs. of pure tin.

**Page 244   Written Exercises   A 1.** 0.875   **3.** $-0.\overline{7}$   **5.** $-1.3125$   **7.** $2.9\overline{4}$   **9.** $0.3\overline{571428}$   **11.** $\frac{793}{250}$
**13.** $-\frac{13}{1250}$   **15.** 1   **17.** $\frac{26}{99}$   **B 19.** $-\frac{3709}{999}$   **21.** $\frac{21059}{9900}$   **23.** $-\frac{6121997}{99900}$   **25.** $\frac{7}{10000}$   **C 27.** $x = \frac{3}{11}$

**Page 248   Written Exercises   A 1.** $6.18 \times 10^1$   **3.** $2.1 \times 10^{-1}$   **5.** $6.92 \times 10^0$   **7.** $8.52 \times 10^{-3}$
**9.** $4.289 \times 10^5$   **11.** $1 \times 10^3$   **13.** $2.13 \times 10^{-6}$   **15.** $4.213 \times 10^8$   **17.** 1000   **19.** 0.0001   **21.** 32,100
**23.** 0.0021   **25.** 1   **27.** 0.0000000004   **29.** 720   **31.** 500,000   **33.** 30,000   **35.** 3000   **37. a.** $1 \times 10^{-3}$ m.,
**b.** $5 \times 10^{-4}$ m., **c.** 10%   **39. a.** $1 \times 10^1$ m., **b.** $5 \times 10^0$ m., **c.** 1%   **41. a.** $1 \times 10^{-3}$ m., **b.** $5 \times 10^{-4}$ m.,
**c.** 0.6%   **43. a.** $1 \times 10^4$ m., **b.** $5 \times 10^3$ m., **c.** 0.1%   **45. a.** $1 \times 10^{-4}$ m., **b.** $5 \times 10^{-5}$ m., **c.** 0.1%
**47. a.** $1 \times 10^5$ m., **b.** $5 \times 10^4$ m., **c.** 0.01%

**Page 249   Problems   A 1.** $\frac{3.412}{4.378} = 77.9\%$   **3.** $5.87 \times 10^{12}$ mi.   **5.** Wave length, $2.19 \times 10^{-5}$ in.; precision,
$1 \times 10^{-12}$ in.; accuracy, 0.0001.   **7.** $2.26 \times 10^{22}$; precision, $1 \times 10^{16}$; accuracy, 0.0008.   **B 9.** 12.25 sq. in. $\le$
area $\le$ 20.25 sq. in.

**Pages 251–252   Chapter Test   1.** $\frac{-3s}{t}$   **3.** $-\frac{6bc^3}{x^2}$   **5.** $\frac{t^2 - 4}{(t - 3)(t + 1)}$; $t \ne 3, t \ne -1$   **7.** 1   **9.** $\frac{1}{12}$
**11.** $\frac{3(x - 3)}{x}$   **13.** $9\frac{9}{13}$ min.   **15.** 75 mph   **17.** $\frac{32}{99}$   **19.** 4000

**Pages 252–254   Chapter Review   1.** $\frac{p}{q} \cdot \frac{r}{s}$   **3.** $2y^5$   **5.** Division   **7.** $\frac{1}{x} + \frac{1}{y}$ or $\frac{y + x}{xy}$; $\frac{1}{x + y}$   **9.** $\frac{2}{x + 3}$
**11.** $6x^2a^3$   **13.** $\frac{x + 3}{x + y}$   **15.** $\frac{15x - 25}{x(x + 5)(x - 5)}$   **17.** $\frac{3(x - 3)}{x}$   **19.** 1   **21.** $\{-\frac{33}{7}\}$   **23.** $\{1\}$   **25.** $\frac{8}{3}, \frac{3}{2}$
**27.** Repeating; 2; 5   **29.** $\frac{1}{1000}$   **31.** 0.007%

**Page 257   Exercises   1.** $P(3) = 10$   **3.** $P(2) = 13$   **5.** $ax^2 + (ar + b)x + (ar^2 + br + c) +$
$\dfrac{ar^3 + br^2 + cr + d}{x - r}$

## Chapter 7.  Irrational and Complex Numbers

**Pages 263–264   Written Exercises   A 9.** $a = 2$   **11.** $a = \frac{1}{2}$   **13.** $a = -\frac{1}{32}$   **15.** $y = 200$   **17.** $r = \frac{32}{3}$
**19.** $A = 25\pi$   **B 21.** $x = \pm\frac{2}{27}$   **23.** $s$ is four times greater   **25.** $k$ is divided by 8.   **C 27.** $a = \pm\frac{1}{2}$

**Pages 264–266   Problems   A 1.** 270 lbs.   **3.** $6.75 \times 10^5$ ft.-lbs.   **5.** 3.125 lbs.   **7.** $3\frac{1}{8}$ tons   **9.** Safe load of
the 4″ shaft is 8 times that of the 2″ shaft.   **B 11.** Moment of inertia of 3″ disk is $\frac{1}{81}$ that of 9″ disk.   **13.** $\frac{343}{125}$
**15. a.** 540 in.², **b.** $\frac{8}{27}$

**Pages 268–269   Written Exercises   A 1.** $C = 4$   **3.** $C = 10$   **5.** $C = \frac{2}{3}$   **7.** $C = 4.65$   **9.** $t = 4$   **11.** $y = 36$
**13.** $r = \frac{3}{2}$   **B 15.** $z$ is 4 times greater.   **17.** $r$ is doubled.   **19.** $r$ is halved.

**Pages 269–270   Problems   A 1.** $1000   **3.** 160 c.c.   **5.** 968 rpm   **7.** 167 lbs.   **B 9.** 1350 lbs.   **11.** $2\frac{2}{3}$ ft.

**Pages 274–275   Written Exercises   A 1.** 2.6   **3.** 0.5   **5.** 3.7   **7.** 0.3   **9.** 0.9   **11.** 4.3   **13.** $\{-\frac{3}{8}, \frac{3}{8}\}$
**15.** $\{\frac{3}{2}\}$   **17.** $\{-\frac{1}{2}, \frac{1}{2}\}$   **19.** $\{\frac{5}{6}, -\frac{5}{6}\}$   **B 21.** $\{-9\}$   **23.** $\emptyset$   **25.** 1   **27.** 3

**Pages 277–278   Written Exercises   B 17.** $\{1, -2, 3\}$   **19.** $\{-1, -\frac{1}{2}, \frac{1}{2}\}$   **21.** $\{0, -\frac{3}{2}\}$

**Page 281   Written Exercises   A 1.** 2.64   **3.** 1.44   **5.–13.** Answers will vary.   **C 15.** Yes

**Pages 284–286   Written Exercises   A 1.** 36.7   **3.** 3.54   **5.** 0.837   **7.** 2.88   **9.** 1.17   **11.** 127   **13.** 4.90
**15.** 103   **17.** $6\sqrt{6}$   **19.** $3\sqrt[3]{5}$   **21.** $2xy\sqrt[3]{xy^2}$   **23.** $\frac{1}{x^2}\sqrt[6]{x^5}$   **25.** $\frac{2y^2}{x^2}\sqrt[6]{x^3y}$   **27.** $4ab\sqrt{b}$   **29.** $2ab^2\sqrt[3]{9b}$
**31.** $5b^2|a|\sqrt{b|a|}$   **33.** $6a^2\sqrt{2}$   **35.** $\frac{3}{2}\sqrt[3]{2xy^2}$   **37.** $\sqrt[3]{9}$   **39.** $2\sqrt{|x|}$   **41.** $\sqrt[4]{5|y|}$   **B 43.** $|x - 3|$
**45.** $(x + 2)\sqrt{x - 2}$   **47.** $243x^4\sqrt[3]{3}$   **49.** $\frac{1}{2}\sqrt[3]{2x^3}$   **51.** $\frac{1}{y^2(y - 1)}\sqrt{y - 1}$   **53.** $\frac{1}{|xy|}\sqrt{y^2 - x^2}$   **55.** $\frac{5x}{\sqrt{35x}}$
**57.** $\frac{2(y - 2)^2}{\sqrt[3]{4(y - 2)}}$   **59.** $\frac{2(x^2 + 1)}{\sqrt[4]{8(x^2 + 1)^3}}$

**Pages 286–287   Problems   A 1.** 3 in.   **3.** 4 sec.   **5.** 2 mi.   **B 7.** 20 mps   **9.** $\frac{15}{21}$

**Pages 289–291  Written Exercises  A  1.** $7\sqrt{2}$  **3.** 0  **5.** $5\sqrt[3]{5}$  **7.** $-2\sqrt{x}$  **9.** $\frac{27}{2}\sqrt[4]{5}$  **11.** $3\sqrt{2}+\sqrt{6}$
**13.** $1+\sqrt{5}$  **15.** $28-\sqrt{5}$  **17.** $42\sqrt[3]{3}+12\sqrt[3]{9}-2$  **19.** $9-6\sqrt{2}$  **21.** $-1+\sqrt{3}$  **23.** $\frac{2+\sqrt{y}}{4-y}$
**25.** $\frac{a+\sqrt{ab}}{a-b}$  **27.** $\frac{\sqrt{6}}{2}$  **B  29.** $-2\sqrt[3]{3x}$  **31.** $\frac{3}{y^2}\sqrt{y}$  **33.** $b$  **35.** $\frac{\sqrt{x-a}+x-a}{1-x+a}$  **37.** $2\sqrt{a+1}+2\sqrt{a}$
**39.** $\frac{1}{2-\sqrt{2}}$  **41.** $\frac{2-x}{2\sqrt{x-1}+x}$  **43.** $\{\sqrt{3}\}$  **45.** $\{1\}$  **47.** $\{2+2\sqrt{6},2-2\sqrt{6}\}$  **C  49.** $\frac{1}{\sqrt{y+1}}$
**51.** $\frac{3s^2-r^2}{\sqrt{r^2+s^2}}$  **53.** $(t+4\sqrt{2})(t-4\sqrt{2})$  **55.** $(n+\sqrt{3})(n+\sqrt{3})$  **57.** $(z+5\sqrt{2})(z-3\sqrt{2})$

**Page 293  Written Exercises  A  1.** $2i\sqrt{6}$  **3.** $7i\sqrt{2}$  **5.** $6i\sqrt{2}$  **7.** $18i\sqrt{5}$  **9.** 0  **11.** $-2$  **13.** 24  **15.** $-216i$
**17.** $-18i$  **19.** $-8\sqrt{3}$  **B  21.** $-i$  **23.** $-2$  **25.** $6i$  **27.** $-i\sqrt{2}$  **29.** $-\frac{3}{4}i\sqrt{2}$  **31.** $-i\sqrt{2}$  **33.** $4\sqrt{2}$  **35.** $-3$
**37.** $-\frac{2}{3}$  **39.** $i\sqrt{5}$  **C  41.** $14z^2i$  **43.** $-2|z|i\sqrt{2}$

**Pages 296–297  Written Exercises  A  1.** $14+4i$  **3.** $12+3i$  **5.** $46+6i$  **7.** $-10-11i$  **9.** $17-7i$
**11.** $34-12i\sqrt{2}$  **13.** $2i$  **15.** $-10$  **17.** 3  **19.** 3  **21.** Sum: $1+i$; diff: $-1+5i$; prod. $6+3i$  **23.** Sum:
4; diff: $14i$; prod: 53  **25.** Sum: 2; diff: $2i$; prod: 2  **B  27.** $x=1,y=3$  **29.** $x=-1,y=2$  **31.** $x=\frac{4}{25}$,
$y=\frac{3}{25}$

**Page 299  Written Exercises  A  1.** $\frac{1}{2}-\frac{1}{2}i$  **3.** $\frac{9}{13}-\frac{6}{13}i$  **5.** $-\frac{6}{5}+\frac{3}{5}i$  **7.** $\frac{2}{5}+\frac{1}{5}i$  **9.** $-\frac{5}{37}-\frac{30}{37}i$
**11.** $0+i$  **13.** $\frac{7}{5}+\frac{4}{5}i$  **15.** $-3-i$  **B  17.** $\frac{9}{74}+\frac{17}{74}i$  **19.** $-1-i$  **21.** $\frac{1}{2}+\frac{1}{2}i$  **23.** $\frac{1}{2}+0i$  **25.** $\frac{1}{5}+\frac{11}{10}i$
**27.** $4+0i$  **C  29.** $0+0i$

**Pages 301–302  Chapter Test  1.** $\frac{9}{4}$  **3.** $y=\pm\frac{8}{11}$  **5.** Answers vary.  **7.** $2\sqrt{2xy}$  **9.** $-\frac{16-7\sqrt{5}}{11}$  **11.** 3
**13.** $17+i$  **15.** $\frac{5}{13}-\frac{1}{13}i$

**Pages 302–304  Chapter Review  1.** Power  **3.** Proportionality  **5.** $y=32$  **7.** 2; no  **9.** Radicand; index
**11.** $2x\sqrt[3]{2xy}$  **13.** $1,-1,2,-2$  **15.** 2.23, 2.24  **17.** Answers vary.  **19.** $2b\sqrt[3]{3a^3};\frac{1}{2}\sqrt{2}$  **21.** $2x\sqrt[3]{2x}$
**23.** $6\sqrt{2}-4\sqrt{3}$  **25.** $2i\sqrt{6}$  **27.** Complex  **29.** Imaginary; pure imaginary  **31.** $ac-bd;(ad+bc)i$
**33.** $-\frac{\sqrt{2}}{2}+3i$

**Pages 305–306  Exercises  1.** $2i$  **3.** $2-i$; $x^3-3x^2+x+5=0$  **5.** $-i\sqrt{2},3-i$  **7.** $2+i$; 3
**9.** $-4-i$; $-2$; $-1$  **11. a.** $(x-1)(x^2+1)$  **b.** $(x-1)(x+i)(x-i)$  **13. a.** $(x+2)(x-2)(x^2+1)$
**b.** $(x+2)(x-2)(x+i)(x-i)$  **15.** False  **17.** True by the Fundamental Theorem.

# Chapter 8.  Quadratic Equations and Functions

**Pages 312–313  Written Exercises  A  1.** $\{-\sqrt{15},\sqrt{15}\}$  **3.** $\{-7i,7i\}$  **5.** $\left\{-\frac{\sqrt{6}}{2},\frac{\sqrt{6}}{2}\right\}$
**7.** $\left\{-\frac{3i}{2}\sqrt{2},\frac{3i}{2}\sqrt{2}\right\}$  **9.** $\{-1,7\}$  **11.** $\{-4,-1\}$  **13.** $\{-\frac{2}{3}+\frac{2}{3}\sqrt{3},-\frac{2}{3}-\frac{2}{3}\sqrt{3}\}$  **15.** $\{-2+i,-2-i\}$
**17.** $\{-6,2\}$  **19.** $\left\{-\frac{3}{2}+\frac{\sqrt{13}}{2},-\frac{3}{2}-\frac{\sqrt{13}}{2}\right\}$  **21.** $\left\{\frac{3}{2}+\frac{3\sqrt{3}}{2},\frac{3}{2}-\frac{3\sqrt{3}}{2}\right\}$  **23.** $\left\{-\frac{5}{2}+\frac{3\sqrt{5}}{2},-\frac{5}{2}-\frac{3\sqrt{5}}{2}\right\}$
**25.** $\left\{-\frac{3}{4}+\frac{\sqrt{41}}{4},-\frac{3}{4}-\frac{\sqrt{41}}{4}\right\}$  **B  27.** $\{-\frac{5}{2},\frac{3}{2}\}$  **29.** $\{-3,4\}$  **31.** $\left\{\frac{6}{11}+\frac{\sqrt{3}}{11},\frac{6}{11}-\frac{\sqrt{3}}{11}\right\}$
**33.** $\left\{\frac{3}{2}-\frac{\sqrt{5}}{2},-\frac{3}{2}-\frac{\sqrt{5}}{2}\right\}$  **35.** $\{-\sqrt{3}+\sqrt{11},-\sqrt{3}-\sqrt{11}\}$  **C  37.** $\left\{-\frac{1}{4}+\frac{\sqrt{8c+1}}{4},-\frac{1}{4}-\frac{\sqrt{8c+1}}{4}\right\}$
**39.** $\left\{-\frac{1}{2a}+\frac{\sqrt{1-4a}}{2a},-\frac{1}{2a}-\frac{\sqrt{1-4a}}{2a}\right\}$  **41.** $(x+3)^2+(y+2)^2=19$  **43.** $(x+\frac{5}{2})^2+(y-\frac{3}{2})^2=\frac{13}{2}$

**Pages 315–316  Written Exercises  A  1.** $\{1,2\}$  **3.** $\left\{-\frac{1}{3}+\frac{\sqrt{13}}{3},-\frac{1}{3}-\frac{\sqrt{13}}{3}\right\}$  **5.** $\{\frac{1}{2}+\sqrt{2},\frac{1}{2}-\sqrt{2}\}$
**7.** $\{-\frac{2}{3}+\frac{1}{3}\sqrt{10},-\frac{2}{3}-\frac{1}{3}\sqrt{10}\}$  **9.** $\{4+\sqrt{17},4-\sqrt{17}\}$  **11.** $\{1+i\sqrt{2},1-i\sqrt{2}\}$  **13.** $\left\{-\frac{3}{8}+\frac{\sqrt{393}}{8},\right.$
$\left.-\frac{3}{8}-\frac{\sqrt{393}}{8}\right\}$  **15.** $\{\frac{4}{5},1\}$  **B  17.** $\{2\}$  **19.** $\{-4,4\}$  **21.** $\left\{1+\frac{\sqrt{57}}{3},1-\frac{\sqrt{57}}{3}\right\}$  **23.** $\{-1+3\sqrt{5},-1-3\sqrt{5}\}$

*(Cont. on next page)*

**25.** $\{2\sqrt{3} + 2\sqrt{2}, 2\sqrt{3} - 2\sqrt{2}\}$ **27.** $\left\{\dfrac{\sqrt{6}}{4} + \dfrac{\sqrt{2}}{4}, \dfrac{\sqrt{6}}{4} - \dfrac{\sqrt{2}}{4}\right\}$ **29.** $\{3 + \sqrt{9 + 2\sqrt{7}}, 3 - \sqrt{9 + 2\sqrt{7}}$

**C 31.** $\left\{\dfrac{3\sqrt{10}}{2} + \dfrac{9\sqrt{2}}{2}, \dfrac{3\sqrt{10}}{2} - \dfrac{9\sqrt{2}}{2}\right\}$ **33.** $\left\{\dfrac{3 + 2i}{2} + \dfrac{\sqrt{-11 + 16i}}{2}, \dfrac{3 + 2i}{2} - \dfrac{\sqrt{-11 + 16i}}{2}\right\}$

**35.** $\{2, -\sqrt{3} - 1\}$ **37.** $\{-i, 7i\}$ **39.** $\{2y + 4, -2y\}$ **41.** $\{1, 2a - 1\}$

**Pages 317–318 Problems A 1.** $2 + \sqrt{10} \doteq 5.2$ ft. **3.** Length $= \dfrac{5 + 3\sqrt{5}}{2}$ cm. $\doteq 5.9$ cm.; width $=$ $\dfrac{-5 + 3\sqrt{5}}{2}$ cm. $\doteq 0.9$ cm. **5.** 9 ft. by 12 ft. **B 7.** $12''$ **9.** 45 mph going, 30 mph returning **11.** $n = 1$

**Pages 320–321 Written Exercises A 1.** $x^2 - 5x - 6 = 0$ **3.** $3x^2 + 7x - 6 = 0$ **5.** $9x^2 - 9x - 10 = 0$ **7.** $x^2 - 4x = 0$ **9.** $x^2 - 2x - 2 = 0$ **11.** $x^2 - 14x + 49 = 0$ **13.** $x^2 - 2x + 2 = 0$ **15.** $4x^2 - 16x + 19 = 0$ **17.** $(x - 2)(9x - 5)$ **19.** $(7n - 1)(n + 3)$ **21.** $(2n - 3)(4n + 5)$ **B 23.** Root, $-\frac{4}{5}$; $b = -6$ **25.** Root, $-3 - \sqrt{7}$; $c = 2$ **27.** roots, $-5, -2$; $c = 10$

**Page 324 Written Exercises A 1.** Real, unequal, irrational. **3.** Real, unequal, rational **5.** Conjugate, imaginary **7.** Real, unequal, irrational **9.** Real, unequal **11.** Real, equal **B 13.** $k = \pm\sqrt{7}$ **15.** $k = 0$, $k = -5, k = -12, k = -21, \ldots$ **17.** $k = -\frac{9}{4}$

**Pages 326–327 Written Exercises A 1.** $\{28\}$ **3.** $\{7\}$ **5.** $\{32\}$ **7.** $\{3\}$ **9.** $\{4\}$ **11.** $\{1, 5\}$ **13.** $\emptyset$ **15.** $\{-\frac{8}{9}\}$ **17.** $\{2\}$ **19.** $\{1\}$ **21.** $\emptyset$ **23.** $\{1, 3\}$ **B 25.** $\{-\frac{17}{2}, 5\}$ **27.** $\{4\}$ **29.** $\{-\frac{13}{3}\}$ **31.** $\{-2\}$ **33.** $A = \pi r^2$ **35.** $l = \dfrac{P^2 g}{\pi^2}$ **37.** $\{a\}$ **C 39.** $x = a + b$, if $a \geq 0, b \geq 0$ **41.** $x^2 - 6x - 11 = 0$

**Page 329 Written Exercises A 1.** $\{25\}$ **3.** $\{2, -2, i, -i\}$ **5.** $\left\{\dfrac{\sqrt{2}}{2}, -\dfrac{\sqrt{2}}{2}, 3i, -3i\right\}$ **7.** $\{-2, -1, 1, 2\}$ **9.** $\{-1, 27\}$ **11.** $\{-\frac{1}{2}, 2\}$ **13.** $\{-1, \frac{3}{2}\}$ **B 15.** $\{626\}$ **17.** $\{-2, -1, \frac{1}{2}, \frac{3}{2}\}$ **19.** $\{-2, -1, 1, 2\}$ **C 21.** $\{-2, 3\}$ **23.** $\{-2, -1\}$ **25.** $\{-\frac{9}{5}, -3\}$

**Pages 333–335 Written Exercises A 11.** $y = \frac{1}{2}(x - 2)^2$ **13.** $y = 3(x - 2)^2 - 7$ **15.** $y = 9(x - 3)^2 + 1$ **17.** $k = -3$ **19.** $h = 1, h = 3$ **21.** $a = \frac{1}{2}$ **23.** $k = 4$ **25.** $h = -5; h = 3$ **27.** $x = -6, x = -2$, $y = 24$ **29.** $x = -5, x = 3, y = -\frac{15}{2}$ **31.** $x = \frac{5}{2}, x = \frac{7}{2}, y = -70$ **B 33.** $a = -1, k = 2$, $y = -(x - 2)^2 + 2$ **35.** $a$ is any real number, $k = 5 - 4a, y = a(x - 1)^2 + 5 - 4a$ **37.** $a = 3, k = -5$, $y = 3x^2 - 5$

**Pages 338–339 Written Exercises A 1.** $V(-1, -1)$ **3.** $V(2, -16)$ **5.** $V(3, 9)$ **7.** $V(2, 0)$ **9.** $V(-1, -1)$ **11.** $V(\frac{7}{6}, \frac{25}{12})$ **13. a.** 2 points, **b.** Vertex below $x$-axis. **15. a.** 2 points, **b.** Vertex above $x$-axis. **17. a.** 2 points **b.** Vertex above $x$-axis. **B 19.** $f(x) = x^2 + 4x - 5$ **21.** $f(x) = -x^2 + 7x - 6$ **23.** $f(x) = -x^2 + 5$

**Page 342 Written Exercises A 1.** 4, maximum, 2 $x$-intercepts **3.** $1\frac{3}{4}$, minimum, no $x$-intercepts **5.** 3, minimum, no $x$-intercepts **7.** 0, minimum, 1 $x$-intercept **9.** $\frac{13}{8}$, minimum, no $x$-intercepts **11.** $-\frac{49}{4}$, minimum, 2 $x$-intercepts **13.** $\frac{9}{4}$, maximum, 2 $x$-intercepts **15.** $-\frac{49}{8}$, minimum, 2 $x$-intercepts **17.** 8, maximum, 2 $x$-intercepts

**Pages 342–343 Problems B 1.** 9 and 9 **3.** 14,400 ft. **5.** 225 $w$ **C 7. a.** $z = 16 - 4x + x^2$ **b.** $z = 12$ **9.** 200 by 300 ft.

**Page 344 Chapter Test 1.** $\left\{\dfrac{3}{2} \pm \dfrac{\sqrt{13}}{2}\right\}$ or $\{-0.3, 3.3\}$ **3.** $\frac{4}{5}, \frac{5}{4}$ **5.** Real, unequal, irrational **7.** $\{4\}$ **11.** 4; maximum

**Pages 345–346 Chapter Review 1.** $\sqrt{d}, -\sqrt{d}$ **3.** $2, -1$ **5.** $\dfrac{-b \pm \sqrt{b^2 - 4ac}}{2a}$ **7.** $\dfrac{-2 \pm \sqrt{4 + 72}}{6}$ **9.** $-\dfrac{b}{a}, \dfrac{c}{a}$ **11.** $x^2 - 4x + 13 = 0$ **13.** $b^2 - 4ac$ **15.** $-24$ **17.** $\sqrt{x - 4} = 10 - x$ **19.** 13, 8 **21.** $x^{-1}$ **23.** $\frac{1}{3}, \frac{1}{2}$ **25.** Parabola, $(h, k)$, $x = h$ **27.** $(-3, -4)$, $x = -3$, downward **29.** Quadratic **31.** 2 **33.** 156.25

**Pages 348–349 Exercises 1. a.** 3 changes in sign for $P(x)$; no changes for $P(-x)$, **b.** upper bound 6; lower bound $-1$ **3. a.** 1 change in sign for $P(x)$; no changes for $P(-x)$, **b.** Upper bound 4; lower bound $-1$ **5. a.** 2 changes in sign for $P(x)$; no changes for $P(-x)$, **b.** Upper bound 4; lower bound $-1$ **7. a.** 1 change in sign for $P(y)$; 2 changes for $P(-y)$, **b.** Upper bound 2; lower bound $-2$ **9.** At most 3 real roots, at least 1 real root. **11.** No changes in sign for either $P(x)$ or $P(-x)$.

## Chapter 9. Quadratic Relations and Systems

**Pages 353–354** **Written Exercises** **A 1. a.** $(\frac{15}{2}, -4)$, **b.** 5 **3. a.** $(4, \frac{7}{6})$, **b.** $\sqrt{5}$ **5. a.** $(\frac{5}{2}, -\frac{3}{2})$, **b.** 5
**7. a.** $\left(\frac{3\sqrt{3}}{2}, \frac{1}{2}\right)$, **b.** $2\sqrt{3}$ **9. a.** $\left(\frac{3\sqrt{3}}{2}, \frac{3\sqrt{2}}{2}\right)$, **b.** $\sqrt{5}$ **11. a.** $\left(\frac{3r}{2}, 0\right)$, **b.** $\sqrt{r^2 + 4s^2}$ **13.** $A(-1, 4)$
**15.** $A(4, -16)$ **B 17. a.** $18\sqrt{2}$, **b.** Isosceles, **c.** Not a right triangle **19. a.** $15 + 9\sqrt{5}$, **b.** Not isosceles, **c.** Right triangle, area 45 **21. a.** 48, **b.** Not isosceles, **c.** Not a right triangle **23. a.** $15 + \sqrt{41} + 4\sqrt{5}$, **b.** Not isosceles, **c.** Not a right triangle

**Pages 354–356** **Problems** **A 1.** 50 ft. **3.** $4\sqrt{10} \doteq 12.6$ ft **5.** $\frac{\sqrt{3}}{4}s^2$ **7.** 200 sq. cm. **B 9.** $x - y = 0$
**11.** $3x^2 - 32x + 3y^2 = -64$ **13.** $x^2 - 6x + y^2 - 6y - 18 = 0$

**Pages 358–359** **Written Exercises** **A 1.** 1 **3.** $-\frac{3}{4}$ **5.** 5 **7.** $\frac{1}{4}$ **9.** $2x + 3y = 23$ **11.** $x + 3y = 0$
**B 13.** $3x - y = 3$

**Pages 361–362** **Written Exercises** **A 1.** $x^2 + y^2 - 8x - 2y + 1 = 0$ **3.** $x^2 + y^2 + 4x + 3 = 0$
**5.** $x^2 + y^2 + 6x + 14y + 54 = 0$ **7.** $x^2 + y^2 + 4x - 10y + \frac{439}{16} = 0$ **9.** $x^2 + y^2 - x - 3y + \frac{3}{2} = 0$
**11.** $x^2 + y^2 - \frac{4}{3}x - \frac{6}{5}y - \frac{494}{225} = 0$ **B 25.** $r = \sqrt{58}$; $x^2 + y^2 = 58$ **27.** $r = \sqrt{18}$; $(x - 2)^2 + (y - 1)^2 = 18$

**Pages 364–365** **Written Exercises** **B 15.** $y = \frac{1}{4}(x - 2)^2 - 1$ **17.** $x = \frac{1}{4}y^2 - 1$ **C 19.** $2x - y = -5$
**21.** $y = -\frac{1}{2}(x - 2)^2 - \frac{1}{2}$

**Pages 368–369** **Written Exercises** **B 19.** $\frac{x^2}{9} + \frac{y^2}{5} = 1$ **21.** $\frac{x^2}{25} + \frac{y^2}{9} = 1$ **C 23.** $\frac{x^2}{a^2} + \frac{y^2}{b^2} = 1, b^2 = a^2 - c^2$

**Pages 372–373** **Written Exercises** **C 23.** $\frac{x^2}{9} - \frac{y^2}{7} = 1$ **25.** $\frac{x^2}{a^2} - \frac{y^2}{b^2} = 1, b^2 = c^2 - a^2$

**Pages 374–375** **A 1.** $\{(-1, -4), (5, 20)\}$ **3.** $\{(-3.5, -2.5), (2.5, 3.5)\}$ **5.** $\{(2, 3), (3, 2)\}$ **7.** $\{(1, 2),$
$(2.5, -1)\}$ **9.** $\emptyset$ **11.** $\{(-2.5, 4.5), (2.5, 4.5)\}$ **B 13.** $\{(-2, 1), (-2, -1), (2, 1), (2, -1)\}$ **15.** $\{(-1, 0.5),$
$(-1, -0.5), (1, 0.5), (1, -0.5)\}$ **17.** $\{(-2, -2), (2, 2)\}$ **C 19.** $\{(1.5, 10)\}$

**Pages 376–377** **Written Exercises** **A 1.** $\{(2, 7), (1, 4)\}$ **3.** $\{(-3, 4), (4, -3)\}$ **5.** $\{(-\frac{46}{13}, -\frac{9}{12}), (2, 3)\}$
**7.** $\{(\frac{7}{3}, \frac{2}{3}), (3, 2)\}$ **9.** $\{(0, 6), (-3, 0)\}$ **11.** $\{(\frac{1}{20}, -\frac{6}{5}), (\frac{3}{4}, \frac{2}{3})\}$ **13.** $\{(-3, -4), (1, 4)\}$ **15.** $\{(-\frac{4}{5}, \frac{13}{25}), (3, -1)\}$
**17.** $\{(-\frac{6}{5}, \frac{13}{5}), (2, 1)\}$ **B 19.** $\{(\frac{5}{7}, \frac{2}{7}), (5, 6)\}$ **21.** $\{(\frac{1}{8}, -\frac{3}{4}), (2, 3)\}$ **23.** $\{(3, 9)\}$ **25.** $\{(-2, 1), (6, 9)\}$
**C 27.** $x = \pm b$

**Page 377** **Problems** **A 1.** 2 ft. by 3 ft. **3.** 24 in. and 7 in. **5.** 5 and 2 or $-5$ and $-2$ **7.** 2 and 6;
12 and $-4$ **B 9.** 5 and 7 **11.** $k = -1$

**Pages 379–380** **Written Exercises** **A 1.** $\{(1, 3), (1, -3), (-1, 3), (-1, -3)\}$ **3.** $\{(2, 1), (2, -1), (-2, 1),$
$(-2, -1)\}$ **5.** $\{(4, 3), (3, 4), (-4, -3), (-3, -4)\}$ **7.** $\{(4, \frac{1}{2}), (1, 2), (-4, -\frac{1}{2}), (-1, -2)\}$ **9.** $\{(3, 9)\}$
**11.** $\{(7, 1), (-7, 1)\}$ **13.** $\left\{(\frac{1}{2}, 3), (\frac{1}{2}, -3), \left(-\frac{5}{4}, \frac{\sqrt{29}}{2}\right), \left(-\frac{5}{4}, -\frac{\sqrt{29}}{2}\right)\right\}$ **15.** $\{(\sqrt{3}, 1), (\sqrt{3}, -1),$
$(-\sqrt{3}, 1), (-\sqrt{3}, -1)\}$ **B 17.** $\{(4, 20), (-20, -4)\}$ **19.** $\{(-8, -1), (\frac{2}{3}, 12)\}$ **21.** $\{(2, 7)\}$
**C 23.** $\{(1, -1), (-1, 1)\}$

**Pages 381–382** **Problems** **A 1.** 8, 12 **3.** 5″ by 12″ **5.** 18 cm. by 27 cm. **7.** 40 mph going, 50 mph
returning **9.** 5%, $3000 **B 11.** $x^2 + (y - 3)^2 = 25$ **13.** $\frac{110}{21}$

**Page 383** **Chapter Test** **1.** $2\sqrt{29}$; $(3, 8)$ **3.** $(1, -2)$; $r = 3$ **7.** $\{4, -1\}$ **9.** $\{(2, 3), (2, -3), (-2, 3),$
$(-2, -3)\}$

**Pages 384–385** **Chapter Review** **1.** Legs, square, hypotenuse **3.** $(1, 4)$ **5.** $4x - 3y = -17$ **7.** Circle;
interior **9.** Symmetric; $x$ **11.** $y = 2$ **13.** $5, -5; 4, -4$ **15.** $5, -5$ **17.** $\{(-2, 8), (1, 5)\}$
**19.** $\{(\sqrt{14}, 2\sqrt{14} - 5), (-\sqrt{14}, -2\sqrt{14} - 5)\}$ **21.** $\{(4, 2), (4, -2), (-4, 2), (-4, -2)\}$

**Page 386** **Exercises** **1.** $y = \frac{1}{8}x^2 - 1$ **3.** $\frac{x^2}{4} - \frac{y^2}{12} = 1$

## Chapter 10. Exponential Functions and Logarithms

**Page 391  Written Exercises  A 1.** $x^{\frac{2}{3}}y^{\frac{1}{3}}$  **3.** $-5a^{\frac{3}{4}}b^{\frac{1}{4}}$  **5.** $3x^{-2}y^{\frac{2}{3}}z^{-\frac{2}{3}}$  **7.** $2\sqrt[6]{32}$  **9.** $2\sqrt[4]{2}\sqrt[4]{9}$  **11.** $\sqrt[36]{2^{11}}$
**13.** $\sqrt[4]{3}$  **15.** $\frac{1}{8}$  **17.** Not defined  **19.** $9\frac{1}{9}$  **21.** $-\frac{7}{4}$  **23.** $10^{4.5}$  **25.** $10^{0.59}$  **B 27.** $5^{\frac{3}{8}}$  **29.** $a^{\frac{7}{4}}$  **31.** $a^{\frac{1}{6}}$
**33.** $x^{\frac{1}{2}} + y^{\frac{1}{2}}$  **35.** $\{\frac{1}{8}\}$  **37.** $\{-1\}$  **39.** $\{16\}$

**Pages 393–394  Written Exercises  A 1.** 16  **3.** 1  **5.** $2^{6-2\sqrt{5}}$  **7.** $3\sqrt{3}-\sqrt{2}$  **9.** $3^{2\sqrt{5}}$  **11.** $\{8\}$  **13.** $\{-22\}$
**15.** $\{\frac{6}{7}\}$  **17.** $\{\frac{5}{3}\}$  **C 27.** $\{(\frac{1}{2}, \frac{3}{2})\}$  **29.** $\{(\frac{3}{2}, 5)\}$

**Pages 396–397  Written Exercises  A 1.** 6  **3.** $\frac{1}{2}$  **5.** $-2$  **7.** 4  **9.** 3  **11.** $-2$  **13.** 5  **15.** 4  **17.** 5
**19.** 64  **21.** $-3$  **23.** $\frac{1}{5}$  **B 25.** 27  **27.** 16  **29.** 6  **31.** $-10, 1$

**Pages 399–400  Written Exercises  A 1.** $f^{-1}(x) = 3x + 6$  **3.** No inverse function  **5.** No inverse function
**7.** $f^{-1}(x) = \dfrac{1}{x}$  **9.** No inverse function  **B 11.** No inverse function  **13.** $f^{-1}(x) = \log_{10} x$

**Pages 402–403  Written Exercises  A 1.** $\{343\}$  **3.** $\{128\}$  **5.** $\{6\}$  **7.** $\{1\}$  **9.** $\{\sqrt[4]{72}\}$  **B 11.** 0.9542
**13.** $-0.5229$  **15.** 1.1761  **17.** $-0.2219$  **19.** 6  **21.** 64  **23.** $\frac{5}{4}$  **C 25.** $\{5\}$  **27.** $\{6\}$  **29.** $\{3\}$

**Pages 405–406  A 1.** 0.5328  **3.** 1.3345  **5.** $9.9096 - 10$  **7.** 2.5340  **9.** $7.0792 - 10$  **11.** 5.8451
**13.** 3.4786  **15.** $8.8254 - 10$  **17.** 7.46  **19.** 5090  **21.** 0.00916  **23.** 0.00512  **25.** 0.000981  **27.** 9970
**29.** 0.000235  **31.** 8,000,000  **B 33.** 3.07  **35.** 9350  **37.** 0.00396  **C 39.** $\{x: x > 2 \text{ or } 0 < x < 1\}$

**Page 408  Written Exercises  A 1.** 0.7646  **3.** 1.1045  **5.** $9.7885 - 10$  **7.** 4.5066  **9.** 0.9130
**11.** $7.2954 - 10$  **13.** 247.3  **15.** 4952  **17.** 60,530  **19.** 3.022  **21.** 0.0004983  **23.** 0.1191

**Page 412  Written Exercises  A 1.** 175  **3.** 5.09  **5.** 7.84  **7.** 0.2718  **9.** $-121$  **11.** 0.373  **13.** 14360
**15.** $-27.40$  **17.** 0.4126  **B 19.** 1.84  **21.** 0.844  **23.** 43.5  **25.** 1180

**Page 415  Written Exercises  A 1.** 232  **3.** 5.64  **5.** 0.382  **7.** 3.55  **9.** 134  **B 11.** 2.50  **13.** 0.200
**15.** 31.7  **17.** 3.49  **19.** 13.7

**Pages 415–416  Problems  A 1.** \$7397  **3.** \$16,230  **5.** \$5000  **7.** 133.1  **9.** $1.07 \times 10^{10}$  **B 11.** 8.07 g.
**13.** 19.8 in.  **15.** $4.93 \times 10^6$

**Page 418  Written Exercises  A 1.** 2.387  **3.** 0.5589  **5.** 341  **7.** 10  **9.** 4.31  **11.** $-1.75$  **13.** 1.47
**15.** 0.783  **17.** 2.63  **19.** 1.97  **B 21.** $-2.209$  **23.** 4.8363  **25.** 8.0087  **C 27. 1.** $a^x = n$ if and only if
$x = \log_a n$  **2.** Log function is one-to-one  **3.** $\log_b x^n = n \log_b x$  **4.** Division property of equality  **5.** Substitution of $\log_a n$ for $x$

**Page 419  Problems  A 1.** 4.5 years  **3.** 12.5 years  **5.** 14 years  **B 7.** $9.69 \times 10^{-5}$

**Page 420  Chapter Test and Review  1.** $3x^{\frac{3}{4}}y^{\frac{5}{4}}$  **3.** $-3$  **5.** $\left\{(x, y): y = \dfrac{x + 4}{6}\right\}$  **7.** 0.6191  **9.** $8.3644 - 10$
**11.** 15.5  **13.** 1.5113

**Pages 421–422  Cumulative Review: Chapters 1–10  1.** $\{13.5\}$  **3.** $\{2, 3\}$  **5.** $\{\frac{1}{2}, -3\}$  **7.** $\dfrac{x - 2}{x - 1}$  **9.** $\frac{1}{6}$
**11.** 24  **13.** $-8$  **15.** $-\frac{1}{5} + \frac{8}{5}i$  **17.** $5x^2 - 14x - 3 = 0$  **23.** 1  **25.** $\{-\frac{2}{3}\}$  **27.** $-8, -6$  **29.** 12 lbs.
**31.** 21 hrs.  **33.** 8 yd., 15 yd.

**Pages 424–425  Exercises  1.** $-1$ and 0; 2 and 3  **3.** 0 and 1  **5.** $-1$ and 0; 1 and 2  **7.** 0 and 1
**9.** $-3$ and $-2$  **11.** $\{0.58\}$  **13.** $\{1.20\}$  **15.** $\{-2.47\}$  **17.** $\{0.68\}$

## Chapter 11. Trigonometric Functions and Vectors

**Pages 429–430  Written Exercises  A 11.** $0, 480°; 0, -240°$  **B 19.** $\{(x, y): y = \frac{2}{5}x \text{ and } x \geq 0\}$
**21.** $\{(x, y): y = -\frac{1}{4}x \text{ and } x \geq 0\}$  **23.** $\{(x, y): y = 3x \text{ and } x \leq 0\}$  **25.** $\{(x, y): y = 0 \text{ and } x \leq 0\}$

**Pages 433–434  Written Exercises**

| | $\theta$ | $\cos\theta$ | $\sin\theta$ |
|---|---|---|---|
| **A 1.** | 45° | 0.7 | 0.7 |
| 3. | 110° | −0.3 | 0.9 |
| 5. | 150° | −0.9 | 0.5 |
| 7. | 240° | −0.5 | −0.9 |
| 9. | −60° | 0.5 | −0.9 |
| 11. | −135° | −0.7 | −0.7 |
| 13. | 520° | −0.9 | 0.3 |
| 15. | −810° | 0 | −1 |

**17.** $\cos\theta = -\frac{8}{17}$; $\sin\theta = \frac{15}{17}$ **19.** $\cos\theta = -\frac{15}{17}$; $\sin\theta = -\frac{8}{17}$

**21.** $\cos\theta = 0$; $\sin\theta = -1$ **23.** $\cos\theta = -\frac{\sqrt{5}}{5}$; $\sin\theta = -\frac{2\sqrt{5}}{5}$

**B 25.** $\cos\theta = \frac{5\sqrt{29}}{29}$; $\sin\theta = -\frac{2\sqrt{29}}{29}$ **27.** $\cos\theta = -\frac{\sqrt{2}}{2}$; $\sin\theta = \frac{\sqrt{2}}{2}$

**29.** $\cos\theta = -1$; $\sin\theta = 0$ **31.** $\cos\theta = \frac{2}{5}\sqrt{6}$ or $-\frac{2}{5}\sqrt{6}$

**33.** $\sin\theta = \frac{1}{2}\sqrt{2}$ or $-\frac{1}{2}\sqrt{2}$

**Pages 438–439  Written Exercises A  1.** $\sin\theta = \frac{4}{5}$  $\csc\theta = \frac{5}{4}$ **3.** $\sin\theta = \frac{\sqrt{2}}{2}$  $\csc\theta = \sqrt{2}$

$\cos\theta = \frac{3}{5}$  $\sec\theta = \frac{5}{3}$    $\cos\theta = \frac{\sqrt{2}}{2}$  $\sec\theta = \sqrt{2}$

$\tan\theta = \frac{4}{3}$  $\cot\theta = \frac{3}{4}$    $\tan\theta = 1$  $\cot\theta = 1$

**5.** $\sin\theta = \frac{5\sqrt{34}}{34}$  $\csc\theta = \frac{\sqrt{34}}{5}$ **7.** $\sin\theta = \frac{2\sqrt{29}}{29}$  $\csc\theta = \frac{\sqrt{29}}{2}$

$\cos\theta = -\frac{3\sqrt{34}}{34}$  $\sec\theta = -\frac{\sqrt{34}}{3}$    $\cos\theta = -\frac{5\sqrt{29}}{29}$  $\sec\theta = -\frac{\sqrt{29}}{5}$

$\tan\theta = -\frac{5}{3}$  $\cot\theta = -\frac{3}{5}$    $\tan\theta = -\frac{2}{5}$  $\cot\theta = -\frac{5}{2}$

**9.** $\sin\theta = -\frac{3}{5}$  $\csc\theta = -\frac{5}{3}$ **11.** $\sin\theta = \frac{\sqrt{3}}{2}$  $\csc\theta = \frac{2\sqrt{3}}{3}$ **B 13.** $\sin\theta = -\frac{1}{5}$  $\csc\theta = -5$

$\cos\theta = -\frac{4}{5}$  $\sec\theta = -\frac{5}{4}$    $\cos\theta = -\frac{1}{2}$  $\sec\theta = -2$    $\cos\theta = \frac{2\sqrt{6}}{5}$  $\sec\theta = \frac{5\sqrt{6}}{12}$

$\tan\theta = \frac{3}{4}$  $\cot\theta = \frac{4}{3}$    $\tan\theta = -\sqrt{3}$  $\cot\theta = -\frac{\sqrt{3}}{3}$    $\tan\theta = -\frac{\sqrt{6}}{12}$  $\cot\theta = -2\sqrt{6}$

**15.** $\sin\theta = \frac{\sqrt{3}}{2}$  $\csc\theta = \frac{2\sqrt{3}}{3}$ **17.** $\sin\theta = -\frac{1}{2}$  $\csc\theta = -2$

$\cos\theta = \frac{1}{2}$  $\sec\theta = 2$    $\cos\theta = -\frac{\sqrt{3}}{2}$  $\sec\theta = -\frac{2\sqrt{3}}{3}$

$\tan\theta = \sqrt{3}$  $\cot\theta = \frac{\sqrt{3}}{3}$    $\tan\theta = \frac{\sqrt{3}}{3}$  $\cot\theta = \sqrt{3}$

**19.** $\sin\theta = -\frac{\sqrt{51}}{10}$  $\csc\theta = -\frac{10\sqrt{51}}{51}$ **21.** $\sin\theta = -\frac{\sqrt{55}}{8}$  $\csc\theta = -\frac{8\sqrt{55}}{55}$

$\cos\theta = \frac{7}{10}$  $\sec\theta = \frac{10}{7}$    $\cos\theta = \frac{3}{8}$  $\sec\theta = \frac{8}{3}$

$\tan\theta = -\frac{\sqrt{51}}{7}$  $\cot\theta = -\frac{7\sqrt{51}}{51}$    $\tan\theta = -\frac{\sqrt{55}}{3}$  $\cot\theta = -\frac{3\sqrt{55}}{55}$

**23.** $\sin\theta = \frac{\sqrt{2}}{4}$  $\csc\theta = 2\sqrt{2}$ **25.** $(-6, 8)$ **27.** $(2\sqrt{2}, -2\sqrt{2})$ **C 29.** $\frac{\sqrt{2}}{2}, -\frac{\sqrt{2}}{2}$ **31.** $-\frac{\sqrt{3}}{2}$

$\cos\theta = \frac{\sqrt{14}}{4}$  $\sec\theta = \frac{2\sqrt{14}}{7}$

$\tan\theta = \frac{\sqrt{7}}{7}$  $\cot\theta = \sqrt{7}$

**Pages 442–443  Written Exercises  A 1.** $\frac{3}{2}$ **3.** 2 **5.** 2 **7.** 1 **17.** $a = \frac{4\sqrt{3}}{3}$ **19.** $\frac{5\sqrt{106}}{106}$

**Pages 446–447  Written Exercises  A 1.** 0.9186 **3.** 0.2144 **5.** 0.7501 **7.** 1.119 **9.** 1.002 **11.** 1.512 **13.** 21°6′ **15.** 74°14′ **17.** 54°53′ **19.** 44°58′ **B 21.** $\angle A = 58°$; $a = 20$; $b = 13$ **23.** $\angle B = 42°$; $a = 39$; $c = 52$ **25.** $\angle A = 37°$; $\angle B = 53°$; $c = 10$

**Pages 447–449  Problems  A 1. a.** 81,990 ft., **b.** 82,000 ft. **3. a.** 1.082 mi., **b.** 1.1 mi. **5. a.** 1526 yds., **b.** 1530 yds. **7. a.** 14°29′, **b.** 14° **B 9. a.** Point $A$ is 5.08 feet higher than point $C$., **b.** South bank higher by 5 feet **11.** Case 1: **a.** 16,740 ft., **b.** 16,740 ft. Case 2: **a.** 82,860 ft., **b.** 82,860 ft. **13. a.** 5.36 in., **b.** 5 in. **15. a.** 399.2 ft., **b.** 399 ft.

**Pages 450–451  Written Exercises  A 1.** 186.6 **3.** 0.7273 **5.** 1.396 **7.** 11.13 **9.** 0.001773 **11.** 2.311 **13.** 55°43′ **15.** 26°28′ **17.** 48°39′ **19.** 3°8′

**Pages 451–452  Problems  A 1.** 47.23 ft. **3.** 47°53′ **5.** 24.00 in. **7.** 50°40′ **B 9.** 2.553 in. **11.** 13,720 ft.

**Pages 455–456  Written Exercises**

**A 13.**

| $\theta$ | 30° | 150° | 210° | 330° |
|---|---|---|---|---|
| $\sin \theta$ | $\frac{1}{2}$ | $\frac{1}{2}$ | $-\frac{1}{2}$ | $-\frac{1}{2}$ |
| $\cos \theta$ | $\frac{\sqrt{3}}{2}$ | $-\frac{\sqrt{3}}{2}$ | $-\frac{\sqrt{3}}{2}$ | $\frac{\sqrt{3}}{2}$ |
| $\tan \theta$ | $\frac{\sqrt{3}}{3}$ | $-\frac{\sqrt{3}}{3}$ | $\frac{\sqrt{3}}{3}$ | $-\frac{\sqrt{3}}{3}$ |

**15.**

| $\theta$ | 45° | 135° | 225° | 315° |
|---|---|---|---|---|
| $\sin \theta$ | $\frac{\sqrt{2}}{2}$ | $\frac{\sqrt{2}}{2}$ | $-\frac{\sqrt{2}}{2}$ | $-\frac{\sqrt{2}}{2}$ |
| $\cos \theta$ | $\frac{\sqrt{2}}{2}$ | $-\frac{\sqrt{2}}{2}$ | $-\frac{\sqrt{2}}{2}$ | $\frac{\sqrt{2}}{2}$ |
| $\tan \theta$ | $1$ | $-1$ | $1$ | $-1$ |

**17.** $\sin 30°$  **19.** $\tan 30°$  **21.** $-\sec 70°$  **23.** $\cos 70°$  **25.** $\cot 35°$  **27.** $-\csc 15°12'$  **29.** $-0.4695$  **31.** $-0.6157$  **33.** $9.7811 - 10$  **35.** $341°10'$  **37.** $261°20'$  **39.** $217°40'$  **B 41.** $(-2.9, 4.1)$  **43.** $(-0.4, -8.3)$  **45.** $(9.7, -2.6)$  **47.** $5.0; 37°$  **49.** $7.3; 196°$  **51.** $7.1; 356°$

**Pages 459–460  Written Exercises  A 1.** 38 mph at 302°  **3.** 57 mi. at 322°  **5.** 155 lbs. at 345°  **B 7.** 458 knots at 228°  **9.** 44 feet at 105°  **11.** 81 kg. at 52°

**Page 460  Problems  A 1.** 1332 mi., 1339 mi. from beacon  **3.** 148 nautical miles, 27°25'  **B 5.** 353°23', 347 mph  **7.** 73°18', yes

**Pages 463–464  Written Exercises  A 1.** Horiz. 10.61; vert. $-10.61$  **3.** Horiz. 27.71; vert. $-16$  **5.** Horiz. 1000; vert. 0  **7.** Horiz. 10.6; vert. 14.6  **9.** Horiz. 61.1; vert. 38.2  **11.** Horiz. 160.7; vert. 191.5  **B 13.** [7, 0]  **15.** $[-2, 5]$  **17.** $[-4, 6]$  **19.** [0, 0]

**Pages 464–465  Problems  A 1.** 117.35 lbs.  **3.** 34.66 mi. north; 34.41 mi. west  **B 5.** 5 lb. to 15 lb.  **7. a.** 3.008 mi., **b.** 3.000 mph

**Page 467  Written Exercises  A 7.** $3\sqrt{2}(\cos 45° + i \sin 45°)$  **9.** $2(\cos 120° + i \sin 120°)$  **11.** $5(\cos 36°50' + i \sin 36°50')$  **13.** $13(\cos 247°20' + i \sin 247°20')$  **15.** $\frac{3}{2} + \frac{\sqrt{3}}{2} i$  **17.** $0 + 2i$  **19.** $-2 - 2i$

**Page 469  Chapter Test and Review  5.** $\{20°\}$  **7.** $82°20'$  **9.** 18.87 meters at 212°  **11.** $3\sqrt{2}(\cos 135° + i \sin 135°)$

## Chapter 12. Trigonometric Identities and Formulas

**Pages 474–475  Written Exercises  A 1.** $\sec A \sin A = \tan A$  **3.** $\frac{\sin \alpha}{\tan \alpha} = \cos \alpha$  **5.** $\cot^2 \alpha \sin^2 \alpha = 1 - \sin^2 \alpha$  **7.** $\tan \theta = \pm \frac{\sqrt{1 - \cos^2 \theta}}{\cos \theta}$  **9.** $\sec \alpha = \pm\sqrt{1 + \tan^2 \alpha}$  **11.** $\cos^2 \beta = 1 - \frac{1}{\csc^2 \beta}$  **13.** $\cot^2 \theta(\sec^2 \theta - 1) + \cot \theta \sin \theta = 1 + \cos \theta$  **15.** $\frac{1 + \tan^2 \alpha}{\csc^2 \alpha} = \frac{1 - \cos^2 \alpha}{\cos^2 \alpha}$  **B 17.** 0  **19.** 1  **21.** $\frac{14{,}112}{7225}$  **23.** $\frac{97}{85}$

**Page 480  Written Exercises  A 1.** 132.2  **3.** 32.7  **5.** 69.8  **7.** 32.6  **9.** 108.8  **11.** 257.4

**Pages 483–484  Written Exercises  A 1.** $-\cos 80° = -0.1736$  **3.** $-\cos 80° = -0.1736$  **5.** $\cos 35° = 0.8192$  **7.** $\cos (45° - 30°) = \frac{\sqrt{6} + \sqrt{2}}{4}$  **9.** $\cos (225° - 30°) = -\frac{\sqrt{6} + \sqrt{2}}{4}$  **B 17.** $\cos (\alpha - \beta)$  **19.** $-\sec \beta + \csc \alpha$

**Pages 486–487  Written Exercises  A 1.** $\frac{\sqrt{3}}{2}$  **3.** $-\frac{\sqrt{2}}{2}$  **5.** $\frac{\sqrt{3}}{2}$  **7.** $-\sqrt{3}$  **9.** $\tan 15° = \tan (45° - 30°) = \frac{\tan 45° - \tan 30°}{1 + \tan 45° \tan 30°} = 2 - \sqrt{3}$  **11.** $\tan 75° = \tan (45° + 30°) = \frac{\tan 45° + \tan 30°}{1 - \tan 45° \tan 30°} = 2 + \sqrt{3}$  **13.** $\cos 195° = \cos (150° + 45°) = -\frac{\sqrt{6} + \sqrt{2}}{4}$  **15.** $\cos 165° = \cos (120° + 45°) = -\frac{\sqrt{2} + \sqrt{6}}{4}$  **17.** $\tan 285° = \tan (225° + 60°) = \frac{1 + \sqrt{3}}{1 - \sqrt{3}}$  **19.** $\sin 345° = \sin (300° + 45°) = \frac{\sqrt{2} - \sqrt{6}}{4}$  **21. a.** $\frac{216}{745}$, **b.** $-\frac{713}{745}$, **c.** $-\frac{624}{745}$, **d.** $\frac{407}{745}$, **e.** $-\frac{216}{713}$, **f.** $-\frac{624}{407}$  **23. a.** $\frac{84}{13}$, **b.** $\frac{13}{84}$, **c.** $-\frac{13}{85}$, **d.** $-\frac{36}{85}$

**Pages 490–492   Written Exercises   A   1.** $\frac{1}{2}$   **3.** $-\frac{\sqrt{2}}{2}$   **5.** $-\frac{\sqrt{3}}{2}$   **7.** $-\sqrt{3 + 2\sqrt{2}}$ or $-1 - \sqrt{2}$
**9.** $-\frac{1}{2}\sqrt{2 - \sqrt{3}}$   **11.** $\frac{7}{25}$   **13.** $\frac{24}{25}$   **15.** $\frac{336}{625}$   **17.** $\frac{2}{5}\sqrt{5}$   **19.** 2   **21.** $\frac{1}{10}\sqrt{50 + 10\sqrt{5}}$
**B   31. b.** $\cos 110° \cos 10° = \frac{1}{2} \cos 120° + \frac{1}{2} \cos 100°$   **C   33. b.** $\sin 36° + \sin 24° = 2 \sin 30° \cos 6°$
**35. b.** $\cos 130° - \cos 40° = -2 \sin 85° \sin 45°$

**Pages 495–497   Written Exercises   A   1.** 2   **3.** 26   **5.** 8   **7.** 25   **9.** 30°   **11.** 93°   **13.** 123 mi.   **15.** 154 mi.,
7°   **17.** 44 lbs.   **B   19.** 79°   **21.** 85°   **27.** $\angle A \doteq 52°6'$, $\angle B \doteq 82°4'$, $\angle C \doteq 45°48'$   **29.** $\angle A = 27°00'$,
$\angle B = 41°50'$, $\angle C = 111°10'$

**Pages 499–500   Written Exercises   A   1.** 1.2   **3.** 12.2   **5.** 41°50'   **7.** 32°50'   **9.** 1.1   **11.** 98 in.   **13.** 46 mi.
**B   15.** 48 min.   **17.** $BD = 6.3$ in., $DE = 5.4$ in., $EC = 6.3$ in.   **C   23.** 102°49', 29°57

**Page 504   Written Exercises   A   1.** $A = 23°35'$, $C = 126°25'$, $c = 8.048$   **3.** $B = 72°33'$, $C = 49°27'$,
$c = 14.33$   **5.** $b = 17.89$, $A = 36°22'$, $C = 98°38'$   **7.** $B = 68°21'$, $C = 53°59'$, $c = 19.15$ or $B = 111°39'$,
$C = 10°41'$, $c = 4.388$   **9.** $B = 47°16'$, $C = 97°54'$, $c = 24.28$ or $B = 132°44'$, $C = 12°26'$, $c = 5.277$
**11.** $A = 69°24'$, $C = 60°11'$, $a = 164$ or $A = 9°46'$, $C = 119°49'$, $a = 29.71$   **13.** No triangle with the given
parts   **B   15. a.** $b < 14.14$, **b.** $b = 14.14$ or $b > 20$, **c.** $14.14 < b < 20$

**Pages 504–506   Problems   A   1.** 47°, 56°, 77°   **3.** $40\frac{1}{4}$ ft.   **5.** Bearing 132°, 393 mph   **7.** 220°33'   **9.** 148 ft.
**B   11.** 22.7 mi. on course of 283°   **13.** 26°55', 428.5 mph   **15.** 1086 ft.   **C   17.** N 64° E, 1.4 hr.

**Page 507   Written Exercises   A   1.** 8.048   **3.** 109.4   **5.** 132.6   **7.** Acute triangle, 178; obtuse triangle, 40.79
**9.** Acute triangle, 124.8; obtuse triangle, 27.13   **11.** Acute triangle, 9604; obtuse triangle, 1740   **13.** No
triangle   **B   15.** 6°51'   **17.** 3678 sq. cm.

**Pages 508–509   Chapter Test and Review   1.** $\sec^2 \theta = \frac{1}{1 - \sin^2 \theta}$   **3.** 61   **5.** $\frac{12 - 3\sqrt{7}}{20}$   **7.** 8.185   **9.** 2

**Page 510   Exercises   1.** product, $0 - 4i$; quotient, $2 + 0i$   **3.** product, $-8 - 8i$; quotient, $\frac{1}{4} - \frac{1}{4}i$
**5.** $-8 - 8\sqrt{3}\,i$   **7.** $0.0601 + 0.0858i$

## Chapter 13.  Circular Functions and Their Inverses

**Pages 515–517   Written Exercises   A   1.** $\frac{\pi}{6}^R$   **3.** $\frac{\pi}{2}^R$   **5.** $-\pi^R$   **7.** $\frac{7\pi}{6}^R$   **9.** $-\frac{8\pi}{3}^R$   **11.** $5\pi^R$   **13.** $1.40^R$
**15.** $-3.80^R$   **17.** 36°   **19.** $-75°$   **21.** 648°   **23.** $-105°$   **25.** 115°   **27.** 29°   **29.** $\frac{1}{2}$   **31.** $\frac{1}{2}$   **33.** 0   **35.** $-\frac{2\sqrt{3}}{3}$
**37.** 37.7 cm.   **39.** 20.9 in.   **41.** 62.8 yd.   **43.** 4 in.   **45.** 24 ft.   **47.** 17.2 ft.   **B   49.** 94.25 ft./min.
**51.** 400 mi./min.

**Pages 519–520   Written Exercises   A   1.** $\sin \frac{\pi}{6} = \frac{1}{2}$, $\cos \frac{\pi}{6} = \frac{\sqrt{3}}{2}$, $\tan \frac{\pi}{6} = \frac{\sqrt{3}}{3}$, $\csc \frac{\pi}{6} = 2$, $\sec \frac{\pi}{6} = \frac{2\sqrt{3}}{3}$,
$\cot \frac{\pi}{6} = \sqrt{3}$   **3.** $\sin \frac{\pi}{3} = \frac{\sqrt{3}}{2}$, $\cos \frac{\pi}{3} = \frac{1}{2}$, $\tan \frac{\pi}{3} = \sqrt{3}$, $\csc \frac{\pi}{3} = \frac{2\sqrt{3}}{3}$, $\sec \frac{\pi}{3} = 2$, $\cot \frac{\pi}{3} = \frac{\sqrt{3}}{3}$
**5.** $\sin\left(-\frac{2\pi}{3}\right) = -\frac{\sqrt{3}}{2}$, $\cos\left(-\frac{2\pi}{3}\right) = -\frac{1}{2}$, $\tan\left(-\frac{2\pi}{3}\right) = \sqrt{3}$, $\csc\left(-\frac{2\pi}{3}\right) = -\frac{2\sqrt{3}}{3}$, $\sec\left(-\frac{2\pi}{3}\right) = -2$,
$\cot\left(-\frac{2\pi}{3}\right) = \frac{\sqrt{3}}{3}$   **7.** $\sin\left(-\frac{3\pi}{4}\right) = -\frac{\sqrt{2}}{2}$, $\cos\left(-\frac{3\pi}{4}\right) = -\frac{\sqrt{2}}{2}$, $\tan\left(-\frac{3\pi}{4}\right) = 1$, $\csc\left(-\frac{3\pi}{4}\right) = -\sqrt{2}$,
$\sec\left(-\frac{3\pi}{4}\right) = -\sqrt{2}$, $\cot\left(-\frac{3\pi}{4}\right) = 1$   **9.** $\sin \frac{5\pi}{4} = -\frac{\sqrt{2}}{2}$, $\cos \frac{5\pi}{4} = -\frac{\sqrt{2}}{2}$, $\tan \frac{5\pi}{4} = 1$, $\csc \frac{5\pi}{4} = -\sqrt{2}$,
$\sec \frac{5\pi}{4} = -\sqrt{2}$, $\cot \frac{5\pi}{4} = 1$   **11.** 0.17   **13.** $-1.12$   **15.** 14.22   **17.** $\frac{1}{3}$ ft.; 0 ft./sec.; $-3$ ft./sec.$^2$   **19.** 0 ft.;
1 ft./sec.; 0 ft./sec.$^2$   **21.** 0 volts; $-6.43$ amp   **23.** 96.36 volts; 21.56 amp

**Page 524   Written Exercises   A   1.** $a = 3$; $p = 2\pi$   **3.** $a = \frac{1}{2}$; $p = 2\pi$   **5.** $a = 1$; $p = \frac{2\pi}{3}$   **7.** $a = 2$;
$p = 6\pi$   **9.** $a = 2$; $p = 2\pi$   **11.** $a = 1$; $p = \pi$   **B   13.** $a = 1$; $p = 2$   **15.** $a = 1$; $p = 4$   **17.** $a = 1$; $p = 2\pi$
**C   19.** $a = 1$; $p = 2\pi$   **21.** $x \doteq 1.1$ and $x \doteq 4.2$   **23.** $x = 0.4$   **25.** $x = 0.7$

**Page 527   Written Exercises   A   1.** $p = \pi$   **3.** $p = \pi$   **5.** $p = \pi$   **7.** $p = 2\pi$   **9.** $p = 2\pi$   **11.** $p = 4\pi$
**B   19.** $x \doteq -0.1$, $x \doteq 0.1$

**Pages 530–531   Written Exercises   A 1.** $-\dfrac{\pi}{2}$   **3.** $\dfrac{\pi}{2}$   **5.** $\dfrac{\pi}{6}$   **7.** $\dfrac{\pi}{4}$   **9.** $\dfrac{3}{5}$   **11.** $\dfrac{3}{5}$   **13.** 0   **15.** $-1$
**B 17. a.** $y = 1 - x^2; y \geq 0$   **19. a.** $y = x; -1 \leq y \leq 1$   **21. a.** $y = x; -\dfrac{\pi}{2} \leq y \leq \dfrac{\pi}{2}$   **23. a.** $y = \dfrac{\pi}{2} - x;$
$-\dfrac{\pi}{2} \leq y \leq \dfrac{\pi}{2}$   **C 25. a.** $y = 2\pi - x; 0 \leq y \leq \pi$

**Pages 534–535   Written Exercises   A 1.** $\dfrac{3}{2}$   **3.** $\dfrac{3\sqrt{10}}{10}$   **5.** $\dfrac{10}{13}$   **7.** $\dfrac{120}{169}$   **9.** $4\sqrt{5}$   **11.** $\sqrt{2}$   **13.** 1   **15.** $\dfrac{63}{16}$
**B 17.** $\sqrt{1 - 4x^2}$   **19.** $1 - 2x^2$   **21.** $\dfrac{x + y}{1 - xy}$, for $xy \neq 1$   **23.** $y\sqrt{1 - x^2} - x\sqrt{1 - y^2}, |x| \leq 1, |y| \leq 1$
**35.** $\{1, -1\}$   **C 37.** $\{\frac{1}{3}\}$

**Pages 536–537   Written Exercises   A 1.** $\{0, 3.1\}$   **3.** $\{1.0, 4.2\}$   **5.** $\{240°, 300°\}$   **7.** $\{0°, 180°\}$
**9.** $\{4.2, 5.2\}$   **11.** $\left\{\dfrac{7\pi}{6}, \dfrac{3\pi}{2}, \dfrac{11\pi}{6}\right\}$   **13.** $\{30°, 150°, 210°, 330°\}$   **15.** $\{0, 2.0, 3.1, 5.2\}$   **17.** $\{90°, 120°, 240°, 270°\}$
**19.** $\{0° < \theta < 45°, 135° < \theta < 225°, 315° < \theta < 360°\}$   **B 21. a.** $\left\{x: x = \dfrac{\pi}{4} + n\pi\right\}$,   **b.** $\{\theta: \theta = 45° + n \cdot 180°\}$
**23. a.** $\left\{x: x = \dfrac{\pi}{4} + \dfrac{n\pi}{2}\right\}$,   **b.** $\{\theta: \theta = 45° + n \cdot 90°\}$   **C 25.** $\left\{x: -\pi < x \leq -\dfrac{3\pi}{4} \text{ or } 0 < x \leq \dfrac{\pi}{4}\right\}$
**27.** $\left\{x: -\dfrac{3\pi}{4} \leq x \leq -\dfrac{\pi}{2} \text{ or } -\dfrac{\pi}{2} < x \leq \dfrac{\pi}{4}\right\}$   **29.** $\left\{x: -\pi < x \leq -\dfrac{3\pi}{4} \text{ or } -\dfrac{\pi}{4} \leq x \leq \dfrac{\pi}{4} \text{ or } \dfrac{3\pi}{4} \leq x \leq \pi\right\}$

**Pages 538–539   Written Exercises   A 1.** $\{22\frac{1}{2}°, 157\frac{1}{2}°, 202\frac{1}{2}°, 337\frac{1}{2}°\}$   **3.** $\{0°, 60°, 120°, 180°, 240°, 300°\}$
**5.** $\left\{\dfrac{\pi}{2}, \dfrac{3\pi}{2}\right\}$   **7.** $\left\{0, \dfrac{\pi}{6}, \dfrac{\pi}{2}, \dfrac{5\pi}{6}, \pi, \dfrac{7\pi}{6}, \dfrac{3\pi}{2}, \dfrac{11\pi}{6}\right\}$   **9.** $\left\{0, \dfrac{2\pi}{3}, \dfrac{4\pi}{3}\right\}$   **11.** $\left\{\dfrac{\pi}{18}, \dfrac{5\pi}{18}, \dfrac{13\pi}{18}, \dfrac{17\pi}{18}, \dfrac{25\pi}{18}, \dfrac{29\pi}{18}\right\}$
**B 13.** $\{0, \pi\}$   **15.** $\left\{0, \dfrac{\pi}{6}, \dfrac{5\pi}{6}\right\}$   **17.** $\left\{x: x = \dfrac{5\pi}{8} + n\pi \text{ or } x = \dfrac{7\pi}{8} + n\pi\right\}$   **19.** Identity   **C 21.** Identity
**23.** $\{30°, 150°, 210°, 330°\}$   **25.** $\{20°, 100°, 140°, 220°, 260°, 340°\}$

**Page 540   Chapter Test and Review   1.** $-\dfrac{3\pi^R}{4}$   **3. a.** $-\frac{1}{2}$, **b.** $\sqrt{3}$   **7.** $\dfrac{\sqrt{2}}{2}$   **9.** $\{\theta: \theta = 120° + n \cdot 360°,$
$240° + n \cdot 360°, n \cdot 360°\}$

**Pages 540–541   Cumulative Review: Chapters 11–13   3.** $-\dfrac{2\sqrt{6}}{7}$   **5.** 46°13′   **7.** Horizontal, 14.85; vertical,
2.09   **9.** $\dfrac{1 - \sin^2 x}{\sin^2 x}$   **11. a.** $\dfrac{12\sqrt{7} + 15}{52}$, **b.** $\dfrac{960 - 509\sqrt{7}}{1121}$   **13.** $A = 33°41′, B = 56°19′, c = 10.8$
**15.** $4.25\pi^R$   **17.** 700 miles   **21.** $\dfrac{\sqrt{2}}{10}$   **23.** $\left\{\dfrac{7\pi}{6}, \dfrac{11\pi}{6}\right\}$   **25.** $\left\{\dfrac{\pi}{6}, \dfrac{5\pi}{6}, \dfrac{3\pi}{2}\right\}$

## Chapter 14.   Sequences, Series and Binomial Expansions

**Page 545   Written Exercises   A 1.** 3, 5, 7, 9   **3.** 4, $-1, -6, -11$   **5.** $a, a + d, a + 2d, a + 3d$   **7.** 31
**9.** $\frac{15}{2}$ or $7\frac{1}{2}$   **11.** $-92$   **13.** $-22i$   **15.** 9   **17.** 27   **19.** $x + 51a$   **21.** 21st term

**Pages 545–546   Problems   A 1.** $9700   **3.** 21 machines   **5.** 22 rows   **B 7.** $(3700 - 200n)$; after $18\frac{1}{2}$ yrs.
**9.** $-7, -3, 1$

**Pages 547–548   Written Exercises   A 1.** $32\frac{1}{2}$   **3.** $1\frac{1}{4}, 1\frac{1}{2}, 1\frac{3}{4}, 2, 2\frac{1}{4}, 2\frac{1}{2}, 2\frac{3}{4}$   **5.** $-5\frac{1}{2}, -5, -4\frac{1}{2}$   **7.** $-2\frac{8}{9}$,
$\frac{2}{9}, 3\frac{1}{3}, 6\frac{4}{9}, 9\frac{5}{9}, 12\frac{2}{3}, 15\frac{7}{9}, 18\frac{8}{9}$   **9.** 57   **11.** $-15$   **B 13.** 3   **15.** $16\frac{1}{2}$   **17.** $-3x$

**Page 548   Problems   A 1.** $2\frac{1}{2}$ ft. apart   **3.** 6, 10, 14, 18, 22, and 26 grams   **B 5.** 8, 14   **7.** $\frac{2}{5}$   **9.** $b = 3$,
$c = -7$

**Pages 551–552   Written Exercises   A 1.** $\displaystyle\sum_{k=1}^{4} (3k - 1)$   **3.** $\displaystyle\sum_{k=1}^{5} (k - \frac{1}{2})$   **5.** $\displaystyle\sum_{k=1}^{4} (9 - 3k)$   **7.** $\displaystyle\sum_{k=1}^{4} (0.88 -$
$0.06k)$   **9.** 77   **11.** 154   **13.** 297   **15.** 5035   **17.** $-324$   **19.** 194,218   **B 21.** 4   **23.** $\frac{7}{3}$   **C 25.** $a = 4, b = -3$

**Page 552   Problems   A 1.** $399   **3.** 225   **5.** 5 test flights   **B 7.** 23 days   **9.** $5.50

**Pages 555–556   Written Exercises   A 1.** 54,162,486; $t_n = 2(3)^{n-1}$   **3.** $\frac{16}{3}, \frac{32}{3}, \frac{64}{3}$; $t_n = \frac{2}{3}(2)^{n-1}$   **5.** 4, 2, 1;
$t_n = 32(\frac{1}{2})^{n-1}$   **7.** $a^3x, a^4x, a^5x$; $t_n = x(a)^{n-1}$   **9.** 3, 6, 12, 24, 48 or 3, $-6, 12, -24, 48$   **11.** $\frac{1}{4}, \frac{1}{2}, 1, 2, 4, 8$
**13.** 3, 15, 75   **15.** $a^2, -a^5, a^8$   **17.** $\pm 2$   **19.** 5   **21.** 5th   **23.** 3   **25.** 54   **C 29.** $\dfrac{1 + \sqrt{5}}{2}$

**Pages 556–557 Problems A 1.** 120% **3.** 2048 **5.** 32,200 **B 7.** 1.875 in. **C 9.** 23

**Pages 558–559 Written Exercises A 1.** 39 **3.** $31\frac{7}{8}$ **5.** $\frac{63}{128}$ **7.** $\frac{85}{64}$ **9.** $47\frac{1}{4}$ **11.** $\frac{111,111}{5000} = 22.2222$
**13.** 3 **B 15.** $r = -2, n = 7$ **17.** $r = \frac{1}{6}, s_6 = \frac{9331}{32}$ **19.** $r = 4, l = 16$ or $r = -5, l = 25$ **21.** $a = 5$, $l = 135$ **C 23.** 301 **25.** 42

**Pages 559–560 Problems A 1.** 73.8 ft. **3.** 124.992 cu. in. **5.** Approx. 215 cm. **7.** $70 + 30\sqrt{2}$ in.
**9.** $\frac{1}{32}$ **B 11.** $1245

**Page 562 Written Exercises A 1.** 24 **3.** $\frac{18}{5}$ **5.** No sum **7.** No sum **9.** $\frac{10}{3}$ **11.** $\frac{4}{9}$ **13.** 1, $\frac{1}{3}$, $\frac{1}{9}$
**15.** 11, $-1$, $\frac{1}{11}$ **17.** $\frac{1}{3}$ **19.** $\frac{31}{99}$ **21.** $\frac{11}{90}$ **23.** $\frac{1}{37}$

**Pages 562–563 Problems A 1.** 200 cm. **3.** $96 + 48\sqrt{2}$ in. **5.** 288 sq. in. **B 7.** $\frac{2\sqrt{3}}{5}$ sq. units
**C 9.** $\frac{1 + x}{(1 - x^2)}$

**Page 565 Written Exercises A 1.** $r^4 + 4r^3s + 6r^2s^2 + 4rs^3 + s^4$ **3.** $1 - 6t + 15t^2 - 20t^3 + 15t^4 - 6t^5 + t^6$ **5.** $x^4 - 8x^3 + 24x^2 - 32x + 16$ **7.** $\frac{x^5}{243} + \frac{5x^4}{27} + \frac{10x^3}{3} + 30x^2 + 135x + 243$ **9.** $8z^3 - 12a^2z^2 + 6a^4z - a^6$ **11.** $8x^3 - 6x^2y + \frac{3y^2}{2} - \frac{y^3}{8}$ **13.** $a^{12} - 24a^{11}b + 264a^{10}b^2$ **15.** $\frac{z^8}{256} + \frac{z^7}{2} + 28z^6$
**17.** $x^{18} + 18x^{17}y + 153x^{16}y^2$ **B 19.** 1.22 **21.** 0.92

**Page 567 Written Exercises A 1.** 9 **3.** 6 **5.** $\frac{1}{6}$ **7.** $n$ **9.** $15a^2t^4$ **11.** $-12x^5$ **13.** $792t^7$ **15.** $-960x^{14}y^3$
**17.** $\frac{35}{16}r^6x^8$

**Pages 568–569 Chapter Test and Review 1.** $-68$ **3.** 16 years **5.** 66 times **7.** 756 **9.** $z^4 + z^3 + \frac{3}{8}z^2 + \frac{1}{16}z + \frac{1}{256}$

## Chapter 15. Permutations, Combinations, and Probability

**Pages 574–575 Written Exercises A 1.** 36 **3.** 243 **5.** 120 **7.** 125 **9.** 45 **B 11.** 36 **13.** 599,040
**15.** 155 **17.** 145 **19.** 9576

**Pages 577–578 Written Exercises A 1.** 720 **3.** 11,880 **5.** 12,144 **7.** 48 **9.** 840 **11.** 1099 **13.** 36
**15.** 576 **17.** 8640 **19.** 3456 **C 27.** $n = 3$ **29.** $n = 7$

**Page 581 Written Exercises A 1.** 840 **3.** 180 **5.** 3360 **7.** 210 **9.** 6 **11.** 10 **13.** 15 **15.** 60
**B 17.** 144 **19.** 48

**Pages 583–584 Written Exercises A 1.** 10; CO, CU, CN, CT, OU, ON, OT, UN, UT, TN **3.** 230,300
**5.** 66 **7.** 161,700 **9.** 15 **11.** 10 **B 13.** 840 **15.** 105 **17.** $1.05 \times 10^7$ **19.** $n = 90$

**Page 586 Written Exercises A 1.** $x^4 + 4x^2 + 6 + \frac{4}{x^2} + \frac{1}{x^4}$ **3.** $x^5 - 5x^3 + 10x - \frac{10}{x} + \frac{5}{x^3} - \frac{1}{x^5}$
**B 5.** 10; 15; $\frac{n}{2}[1 + n]$ **7.** 1002

**Pages 588–589 Written Exercises A 1.** $S = \{1, 2, 3, 4, 5, 6, 7, 8, 9\}$; $E_1 = \{1, 3, 5, 7, 9\}$; $E_2 = \{3, 6, 9\}$
**3.** $S = \{S, Q, U, A, R, E\}$; $E = \{S, R, Q\}$ **5.** (order in each outcome may be reversed) $S = \{(-2, -1),$
$(-2, 0), (-2, 1), (-2, 2), (-1, 0), (-1, 1), (-1, 2), (0, 1), (0, 2), (1, 2)\}$; $E = \{(-2, 2), (-1, 1)\}$
**B 9.** $S = \{(M, E), (M, A), (M, R), (Y, E), (Y, A), (Y, R)\}$; $E_1 = \{(Y, E), (Y, A)\}$; $E_2 = \{(M, R)\}$
**C 11.** $S = \{(M, T), (M, O_1), (M, O_2), (E, T), (E, O_1), (E, O_2)\}$; $E_1 = \{(E, O_1), (E, O_2)\}$; $E_2 = \{(M, O_1),$
$(M, O_2), (E, T)\}$ **13.** $S = \{(b_1, b_1), (b_1, b_2), (b_1, w_1), (b_1, w_2), (b_2, b_1), (b_2, b_2), (b_2, w_1), (b_2, w_2), (w_1, b_1),$
$(w_1, b_2), (w_1, w_1), (w_1, w_2), (w_2, b_1), (w_2, b_2), (w_2, w_1), (w_2, w_2)\}$; $E_1 = \{(b_1, b_1), (b_1, b_2), (w_1, w_1),$
$(w_1, w_2), (b_2, b_1), (b_2, b_2), (w_2, w_1), (w_2, w_2)\}$; $E_2 = \{(b_1, w_1), (b_1, w_2), (b_2, w_1), (b_2, w_2)\}$

**Pages 591–592 Written Exercises A 1.** a. $\frac{1}{13}$, b. 12, c. $\frac{1}{4}$, d. 7 **3.** a. $\frac{1}{2}$, b. $\frac{2}{3}$, c. $\frac{2}{3}$ **5.** a. $\frac{2}{25}$, b. $\frac{1}{5}$,
c. $\frac{2}{25}$ **7.** a. $\frac{1}{3}$, b. 2 to 1 **9.** $\frac{1}{8}$ **11.** a. $\frac{1}{45}$, b. $\frac{1}{15}$, c. $\frac{2}{15}$, d. $\frac{1}{3}$ **13.** $\frac{1}{5}$ **15.** 3 to 2

**Page 594** **Written Exercises** **A 1. a.** $\frac{1}{9}$, **b.** $\frac{11}{36}$, **c.** $\frac{5}{12}$ **3. a.** $\frac{5}{36}$, **b.** $\frac{1}{6}$, **c.** $\frac{1}{36}$, **d.** $\frac{5}{18}$, **5. a.** $\frac{5}{33}$, **b.** $\frac{8}{99}$ **B 7.** $\frac{1}{12}$ **9.** 5 to 1

**Pages 597–598** **Written Exercises** **A 1. b.** $P(A) = \frac{1}{6}$; $P(B) = \frac{1}{6}$; $P(A \cap B) = \frac{1}{36}$; $P(B|A) = \frac{1}{6}$ **3. b.** $P(A) = \frac{1}{3}$, $P(B) = \frac{5}{12}$, $P(A \cap B) = \frac{1}{12}$, $P(B|A) = \frac{1}{4}$ **5. a.** $A = \{HHH, HHT\}$; $B = \{HHH, THH\}$; **b.** $P(A) = \frac{1}{4}$, $P(B) = \frac{1}{2}$, $P(A \cap B) = \frac{1}{8}$; **c.** Independent **7. a.** $A = \{HHT, HTH, HTT, THH, THT, TTH, TTT\}$; $B = \{HHT, HTH, HTT, THH, THT, TTH, TTT\}$; **b.** $P(A) = \frac{7}{8}$, $P(B) = \frac{7}{8}$; $P(A \cap B) = \frac{7}{8}$; **c.** Dependent **9. a.** $\frac{1}{12}$, **b.** $\frac{11}{12}$, **c.** $\frac{1}{2}$ **11. a.** $\frac{4}{25}$, **b.** $\frac{13}{25}$ **13.** $\frac{473}{729}$ **15. a.** $\frac{29}{30}$, **b.** $\frac{1}{20}$, **c.** $\frac{3}{20}$, **d.** $\frac{3}{4}$ **17. a.** $\frac{1}{289}$, **b.** $\frac{11}{4165}$

**Page 600** **Chapter Test and Review** **1.** 18,720 **3.** 75,600 **5.** 11,760 **7.** 1012 **9.** $\frac{5}{18}$ **11. a.** $\frac{1}{144}$, **b.** $\frac{1}{72}$, **c.** $\frac{4}{81}$

## Chapter 16. Matrices and Determinants

**Page 605** **Written Exercises** **A 1.** $x = 3$ **3.** $x = 1, y = 4$ **5.** $y = 1$ **7.** $x = 3, y = 1$ **B 9.** $x = 4, y = 1$

**Pages 607–608** **Written Exercises** **A 1.** $\begin{bmatrix} 2 & 1 \\ 0 & 2 \\ -2 & 1 \end{bmatrix}$ **3.** $\begin{bmatrix} 10 & 5 \\ 5 & 10 \end{bmatrix}$ **5.** $\begin{bmatrix} 9 & 3 \\ 12 & 6 \end{bmatrix}$ **7.** $\begin{bmatrix} 1 & 0 \\ 1 & 0 \\ 1 & 0 \end{bmatrix}$ **9.** $\begin{bmatrix} -2 & 1 \\ 3 & 0 \end{bmatrix}$ **B 11.** $X = \begin{bmatrix} 1 & 2 \\ 2 & 6 \end{bmatrix}$ **13.** $X = \begin{bmatrix} 3 & 6 \\ 6 & 12 \end{bmatrix}$ **15.** $X = \begin{bmatrix} 1 & 1 \\ 2 & 0 \end{bmatrix}$

**Pages 610–611** **Written Exercises** **A 1.** [8] **3.** $\begin{bmatrix} -1 & 3 \\ 3 & 6 \end{bmatrix}$ **5.** $\begin{bmatrix} 0 & 0 \\ 0 & 0 \end{bmatrix}$ **7.** $\begin{bmatrix} 3 \\ 4 \end{bmatrix}$ **9.** $\begin{bmatrix} 5 & -2 & 5 \\ -7 & 5 & -8 \\ 1 & 0 & -1 \end{bmatrix}$ **11.** $\begin{bmatrix} 5 \\ 4 \\ 3 \end{bmatrix}$ **B 13.** $AB = \begin{bmatrix} 2 & 5 \\ 1 & 4 \end{bmatrix}$; $BA = \begin{bmatrix} 0 & -1 \\ 3 & 6 \end{bmatrix}$ **15.** $(AB)C = \begin{bmatrix} -1 & 4 \\ -2 & 5 \end{bmatrix}$; $A(BC) = \begin{bmatrix} -1 & 4 \\ -2 & 5 \end{bmatrix}$ **17.** $A(B + C) = \begin{bmatrix} 3 & 5 \\ 1 & 5 \end{bmatrix}$; $AB + AC = \begin{bmatrix} 3 & 5 \\ 1 & 5 \end{bmatrix}$ **19.** $(A + B)(A - B) = \begin{bmatrix} 4 & 14 \\ 6 & 0 \end{bmatrix}$; $A^2 - B^2 = \begin{bmatrix} 6 & 20 \\ 4 & -2 \end{bmatrix}$ **21.** $(A - B)^2 = \begin{bmatrix} 6 & 0 \\ 0 & 6 \end{bmatrix}$; $A^2 - 2AB + B^2 = \begin{bmatrix} 4 & -6 \\ 2 & 8 \end{bmatrix}$

**Pages 612–613** **Written Exercises** **A 1.** 1 **3.** $-1$ **5.** 0 **7.** 0 **9.** $-8$ **11.** 0

**Pages 614–615** **Written Exercises** **A 1.** $\begin{bmatrix} 2 & -3 \\ -1 & 2 \end{bmatrix}$ **3.** Not invertible **5.** $\begin{bmatrix} \frac{1}{2} & -\frac{1}{2} \\ -\frac{1}{2} & \frac{3}{2} \end{bmatrix}$ **7.** $\begin{bmatrix} \frac{1}{3} & \frac{2}{3} \\ -\frac{1}{3} & -\frac{1}{6} \end{bmatrix}$ **9.** $\begin{bmatrix} 1 & 0 \\ 0 & 1 \end{bmatrix}$ **11.** Not invertible **B 13.** $Z = \begin{bmatrix} 1 & -3 \\ 1 & 5 \end{bmatrix}$ **15.** $Z = \begin{bmatrix} 6 & 1 \\ -\frac{13}{2} & -\frac{3}{2} \end{bmatrix}$ **17.** $Z = \begin{bmatrix} 0 & -3 \\ 1 & 9 \end{bmatrix}$ **19. a.** $AB = \begin{bmatrix} 1 & 4 \\ -2 & -9 \end{bmatrix}$, **b.** $(AB)^{-1} = \begin{bmatrix} 9 & 4 \\ -2 & -1 \end{bmatrix}$, **c.** $A^{-1} = \begin{bmatrix} 3 & 1 \\ 5 & 2 \end{bmatrix}$, **d.** $B^{-1} = \begin{bmatrix} -2 & 3 \\ 1 & -1 \end{bmatrix}$, **e.** $A^{-1}B^{-1} = \begin{bmatrix} -5 & 8 \\ -8 & 13 \end{bmatrix}$, **f.** $B^{-1}A^{-1} = \begin{bmatrix} 9 & 4 \\ -2 & -1 \end{bmatrix}$

**Page 618** **Written Exercises** **A 1.** $\{(-5, 4)\}$ **3.** Dependent or inconsistent system **5.** $\{(-1, 2)\}$ **7.** Dependent or inconsistent system **9.** Dependent or inconsistent system **11.** $\{(11, -6)\}$ **13.** $\{(-\frac{1}{2}, \frac{7}{2})\}$ **15.** Dependent or inconsistent system **17.** $\{(\frac{29}{23}, \frac{14}{23})\}$

**Page 621** **Written Exercises** **A 1.** $-1$ **3.** $-6$ **5.** 0 **7.** $-13$ **9.** 40 **B 11.** $-15$ **13.** $-2$ **15.** $-2$

**Pages 624–625** **Written Exercises** **A 1.** 0 **3.** 0 **5.** $-7$ **7.** $-1$ **9.** 0 **11.** 0 **13.** $-4560$ **15.** $-1$

**Page 627** **Written Exercises** **A 1.** $\{(1, 1, 2)\}$ **3.** $\{(2, 3, 1)\}$ **5.** Dependent or inconsistent system **7.** $\{(3, 2, 2, -1)\}$

**Pages 628–629** **Chapter Test and Review** **1.** $x = 1, y = 3$ **3.** $\begin{bmatrix} 7 \\ 5 \end{bmatrix}$ **5.** 9 **7. a.** $\{(-\frac{5}{2}, -4)\}$, **b.** $\{(-\frac{5}{2}, -4)\}$ **9.** 0

**Pages 629–630** **Cumulative Review: Chapters 14–16** **1.** $-79$ **3.** 756 **5.** 32 **7.** $\frac{32}{3}$ **9.** 61 **11.** 20 **13.** 210 **15.** $\frac{11}{221}$ **17.** $x = 2$ **19.** 57 **21.** $\{(2, -1)\}$ **23.** $-1$

# Law of Cosines

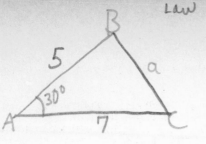

$$a^2 = b^2 + c^2 - 2bc \cos A$$
$$a^2 = 49 + 25 - 70 (.8660)$$
$$a^2 = 74 - 60.9$$
$$a^2 = 13.1$$
$$\boxed{a \approx 3.6}$$

## Right Triangle Ratio

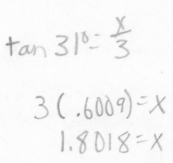

$$\tan 31° = \frac{x}{3}$$
$$3(.6009) = x$$
$$1.8018 = x$$

## Law of Sines

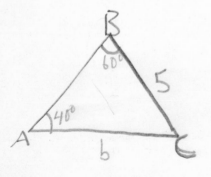

$$\frac{\sin A}{a} = \frac{\sin B}{b}$$

$$\frac{\sin 40°}{5} \times \frac{\sin 60°}{b}$$

$$b(\sin 40°) = 5(\sin 60°)$$

$$b(.6428) = 5(.8660)$$
$$b = \frac{4.33}{.6428}$$

$$.6428 \overline{)4.3300} \quad \text{use calculator}$$

8-5

5